Theory of the Earth

Theory of the Earth

Don L. Anderson

Seismological Laboratory
California Institute of Technology
Pasadena, California

BLACKWELL SCIENTIFIC PUBLICATIONS

Boston Oxford London Edinburgh Melbourne

Editorial Offices

Osney Mead, Oxford, OX2, 0EL, UK
8 John Street, London WC1N 2ES, UK
23 Ainslie Place, Edinburgh, EH3 6AJ, UK
3 Cambridge Center, Suite 208, Cambridge, MA 02142
107 Barry Street, Carlton, Victoria 3053, Australia

Distributors

USA and Canada
Blackwell Scientific Publications
c/o P.B.S.
P.O. Box 447
Brookline Village, MA 02147
(617) 524-7678

Australia
Blackwell Scientific Publications (Australia) Pty Ltd
107 Barry Street, Carlton
Victoria 3053

UK
Blackwell Scientific Publications
Osney Mead
Oxford OX2 0EL
011 44 865-240201

Sponsoring Editor: John Staples
Manuscript Editor: Andrew Alden
Production Coordinator: Partners in Publishing/Pat Waldo and Susan Swanson
Interior and Cover Design: Gary Head Design
Compositor: G&S Typesetters, Inc.
Printer: Arcata Graphics/Kingsport

90 91 92 9 8 7 6 5 4 3 2

Library of Congress Cataloging-in-Publication Data

Anderson, Don L.
Theory of the earth.
Includes bibliographies and index.
1. Earth 2. Geophysics I. Title.
QC806.A515 1989 551 88-7903
ISBN 0-86542-335-0 (hardcover)
0-86542-123-4 (paperback)

Contents

10
Isotopes 197

11
Evolution of the Mantle 217

12
The Shape of the Earth, Heat Flow and Convection 239

13
Heterogeneity of the Mantle 259

14
Anelasticity 279

15
Anisotropy 303

It was a long time before man came to understand that any true theory of the earth must rest upon evidence furnished by the globe itself, and that no such theory could properly be framed until a large body of evidence had been gathered together.

Sir Archibald Geikie, 1905

Preface

The maturing of the Earth sciences has led to a fragmentation into subdisciplines which speak imperfectly to one another. Some of these subdisciplines are field geology, petrology, mineralogy, geochemistry, geodesy and seismology, and these in turn are split into even finer units. The science has also expanded to include the planets and even the cosmos. The practitioners in each of these fields tend to view the Earth in a completely different way. Discoveries in one field diffuse only slowly into the consciousness of a specialist in another. In spite of the fact that there is only one Earth, there are probably more Theories of the Earth than there are of astronomy, particle physics or cell biology where there are uncountable samples of each object. Even where there is cross-talk among disciplines, it is usually as noisy as static. Too often, one discipline's unproven assumptions or dogmas are treated as firm boundary conditions for a theoretician in a slightly overlapping area. The data of each subdiscipline are usually consistent with a range of hypotheses. The possibilities can be narrowed considerably as more and more diverse data are brought to bear on a particular problem. The questions of origin, composition and evolution of the Earth require input from astronomy, cosmochemistry, meteoritics, planetology, geology, petrology, mineralogy, crystallography, materials science and seismology, at a minimum. To a student of the Earth, these are artificial divisions, however necessary they are to make progress on a given front.

Examples abound. A seismologist struggling with the meaning of velocity anomalies beneath various tectonic provinces, or in the vicinity of a deeply subducting slab, is apt to interpret seismic results in terms of temperature variations in a homogeneous, isotropic half-space or series of layers. However, the petrological aspects—variations in mineralogy, crystal orientation or partial melt content—are much more important. These, in turn, require knowledge of phase equilibria and material properties.

An isotope geochemist, upon finding evidence for several ancient isolated reservoirs in the rocks and being generally aware of the geophysical evidence for a crust and a 650-km discontinuity, will tend to interpret the chemical data in terms of crustal contamination or recycling, a "normal" mantle source and a lower mantle source. This "standard" petrological model is a homogeneous peridotite mantle which contains about 20 percent basalt, available as needed, to fuel the midocean ridges with uniform magmas. Exotic basalts are assumed to be from the core-mantle boundary or to have interacted with the crust. The crust and shallow mantle may be inhomogeneous, but the rest of the mantle is viewed as well homogenized by convection.

The convection theoretician treats the mantle as a homogeneous fluid or as a two-layered system, with constant physical properties, driven by temperature-induced bouyancy, ignoring melting and phase changes and even pressure.

In *Theory of the Earth* I attempt to assemble the bits and pieces from a variety of disciplines which are relevant to an understanding of the Earth. Rocks and magmas are our most direct source of information about the interior, but they are biased toward the properties of the crust and shallow mantle. Seismology is our best source of information about the deep interior; however, the interpretation of seismic data for purposes other than purely structural requires input from solid-state physics and experimental petrology. Although this is not a book about seismology, it uses seismology in a variety of ways.

The "Theory of the Earth" developed here differs in

many respects from conventional views. Petrologist's models for the Earth's interior usually focus on the composition of mantle samples contained in basalts and kimberlites. The "simplest" hypothesis based on these samples is that the observed basalts and peridotites bear a complementary relation to one another, that peridotites are the source of basalts or the residue after their removal, and that the whole mantle is identical in composition to the inferred chemistry of the upper mantle and the basalt source region. The mantle is therefore homogeneous in composition, and thus all parts of the mantle eventually rise to the surface to provide basalts. Subducted slabs experience no barrier in falling through the mantle to the core-mantle boundary.

Geochemists recognize a variety of distinct reservoirs, or source regions, usually taken as the upper mantle for midocean-ridge basalts and the lower mantle for hotspot, plume or ocean-island basalts. In some models the mantle is still grossly homogeneous but contains blobs of isotopically distinct materials so that it resembles a marble cake.

Seismologists recognize large lateral heterogeneity in the upper mantle and several major seismic discontinuities. The discontinuities, in the homogeneous-mantle scenario, represent equilibrium phase changes. They could also be, in part, due to changes in chemistry. The ocean and continental lithospheres, or high-velocity layers, may represent material different from the underlying mantle.

It is difficult to take data which refer to the present Earth and to extrapolate conditions very far back into the past. The "present is the key to the past" training of Earth scientists leads them naturally to the view that the Earth was always pretty much the same and, if it assumed that the mantle is homogeneous now, it always was. A modification of this view is that the crust formed, in early Earth history, from the upper mantle, and the lower mantle is still "primordial" or undifferentiated, and always has been.

If one considers the circumstances of the Earth's birth, however, one comes up with a different "simplest" hypothesis. The energy of accretion is so great, and the melting temperatures and densities of the products are so different, that chemical stratification is the logical outcome. Basalts and the incompatible elements are expected to be concentrated toward the surface, and dense refractory crystals are expected to settle toward the interior.

These various hypotheses can be tested with modern data. It is only by considering a wide-variety of data that we can narrow the possibilities.

The title of this book was not picked casually. This year is the two-hundredth anniversary of the publication of "Theory of the Earth; or an Investigation of the Laws Observable in the Composition, Dissolution, and Restoration of Land Upon the Globe" by James Hutton, the founder of modern geology. It was not until much progress had been made in all the physical and natural sciences that geology could possess any solid foundations or real scientific status. Hutton's knowledge of chemistry and mineralogy was considerable, and his powers of observation and generalization were remarkable, but the infancy of the other basic sciences made his "Theory of the Earth" understandably incomplete. In the present century the incorporation of physics, chemistry and biology into geology and the application of new tools of geophysics and geochemistry has made geology a science that would be unrecognizable to the Founder, although the goals are the same. Hutton's uniformitarian principle demanded an enormous time period for the processes he described to shape the surface of the Earth, and Hutton could see that the differend kinds of rocks had been formed by diverse processes. These are still valid concepts, although we now recognize catastrophic events as well. Hutton's views prevailed over the then current precipitation theory which held that all rocks were formed by mineral deposits from the oceans. Ironically, a currently emerging view is that crystallization from a gigantic magma ocean was an important process in times that predate the visible geological record. Uniformitarianism apparently cannot be carried too far.

The word "theory" is used in two ways. A theory is the collection of facts, principles and assumptions which guide the workers in a given field. Well-established theories from physics, chemistry, biology and astrophysics, as well as from geology, are woven into the Earth sciences. Students of the Earth must understand solid-state physics, crystallography, thermodynamics, Hooke's Law, optics and so on. Yet these collections of theories do not provide a "Theory of the Earth." They provide the tools for unraveling the secrets of the Earth and for providing the basic facts which in turn are only clues to how the Earth operates. By assembling these clues we hope to gain a better understanding of the origin, structure, composition and evolution of our planet. This better understanding is all that we can hope for in developing a "Theory of the Earth."

ACKNOWLEDGMENTS

I wish to express my sincere appreciation to many colleagues, including students, who have collaborated with me over the years, particularly David Harkrider, Nafi Toksöz, Charles Archambeau, Hiroo Kanamori, Toshiro Tanimoto, Hartmuth Spetzler, Robert Hart, Tom Jordan, Jeff Given, Adam Dziewonski, Charles Sammis, Ichiro Nakanishi, Henri-Claude Nataf, Janice Regan, Jay Bass and Bernard Minster.

The chapters on anisotropy and heterogeneity include material based on joint work with Adam Dziewonski, Janice Regan, Toshiro Tanimoto and Bernard Minster. I thank them and the American Geophysical Union, Physics of the Earth and Planetary Interiors and the Geophysical Journal for permission to use figures and other material.

The chapter on anelasticity is based partly on joint studies with Hiroo Kanamori and Jeff Given. I thank them and the AGU and Nature for permission to use material from these publications.

I also thank Earth and Planetary Science Letters and Science for permission to use material from articles I have published in these journals in Chapter 12.

Thanks are also due to Roberta Eager and Ann Freeman for working through numerous drafts of the manuscript.

Finally I acknowledge my debts to Nancy. While I was doing this she was doing everything else.

Theory of the Earth

1

The Terrestrial Planets

I want to know how God created this world. I am not interested in this or that phenomenon, in the spectrum of this or that element. I want to know his thoughts, the rest are details.

—EINSTEIN

Earth is part of the solar system. Although it is the most studied planet, it cannot be completely understood in isolation. The chemistry of meteorites and the Sun provide constraints on the composition of the bulk of the Earth. The properties of other planets provide ideas for and tests of theories of planetary formation and evolution. In trying to understand the origin and structure of the Earth, one can take the geocentric approach or the *ab initio* approach. In the former, one describes the Earth and attempts to work backward in time. For the latter, one attempts to track the evolution of the solar nebula through collapse, cooling, condensation and accretion, hoping that one ends up with something resembling the Earth and other planets. In Chapter 1 I develop the external evidence that might be useful in understanding the Earth. In Chapter 2 I describe the Earth and Moon.

THEORIES OF PLANETARY FORMATION

The nature and evolution of the solar nebula and the formation of the planets are complex and difficult subjects. The fact that terrestrial planets did in fact form is a sufficient motivation to keep a few widely dispersed scientists working on these problems. There are several possible mechanisms of planetary growth. Either the planets were assembled from smaller bodies (planetesimals), a piece at a time, or diffuse collections of these bodies, clouds, became gravitationally unstable and collapsed to form planetary sized objects. The planets, or protoplanetary nuclei, could have formed in a gas-free environment or in the presence of a large amount of gas that was subsequently dissipated.

The planets are now generally thought to have originated in a slowly rotating disk-shaped "solar nebula" of gas and dust with solar composition. The temperature and pressure in the hydrogen-rich disk decrease both radially from its center and outward from its plane. The disk cools by radiation, mostly in the direction normal to the plane, and part of the incandescent gas condenses to solid "dust" particles. As the particles grow, they settle to the median plane by processes involving collisions with particles in other orbits, by viscous gas drag and gravitational attraction by the disk. The total pressure in the vicinity of Earth's orbit may have been of the order of 10^{-3} to 10^{-4} bar. The particles in the plane probably formed rings and gaps. The sedimentation time is fairly rapid, but the processes and time scales involved in the collection of small objects into planetary sized objects are not clear. The common thread of all cosmogonic theories is that the planets formed from dispersed material, that is, from a protoplanetary nebula. Comets, some meteorites and some small satellites may be left over from these early stages of accretion.

The following observations are the main constraints on theories of planetary origin:

1. Planetary orbits are nearly circular, lie virtually in a single plane, and orbit in the same sense as the Sun's rotation. The Sun's equatorial plane is close to the orbital plane. The planets exhibit a preferred sense of rotation.
2. The distribution of planetary distances is regular (Bode's Law).
3. The planets group into compositional classes related to distance from the Sun. The inner, or terrestrial planets (Mercury, Venus, Earth and Mars), are small, have high density, slow rotation rates and few satellites. The Moon is often classified as a terrestrial planet. The giant planets (Jupiter, Saturn, Uranus and Neptune) are large, have low density, rotate rapidly and have numerous satellites. Although the Sun contains more than 99 percent of the

mass of the solar system, the planets contain more than 98 percent of the angular momentum.

Apart from the mechanisms of accretion and separation of planetary from solar material, there are several important unresolved questions.

How dense was the protoplanetary nebula? A lower limit is found by taking the present mass of the planets and adding the amount of light elements necessary to achieve solar composition. This gives about 10^{-2} solar mass. Young stars in the initial stages of gravitational contraction expel large quantities of matter, possibly accounting for several tens of percent of the star's mass. Some theories therefore assume a massive early nebula that may equal twice the mass of the Sun including the Sun's mass. T-Tauri stars, for example, expel about 10^{-6} solar mass per year for 10^5 to 10^6 years.

What are the time scales of cooling, separation of dust from gas, growth of asteroidal size bodies, and growth of planets from meter- to kilometer-size objects? If cooling is slow compared to the other processes, then planets may grow during cooling and will form inhomogeneously. If cooling is fast, then the planets may form from cold material and grow from more homogeneous material.

The accretion-during-condensation, or inhomogeneous accretion, hypothesis would lead to radially zoned planets with refractory and iron-rich cores, and a compositional zoning away from the Sun; the outer planets would be more volatile-rich. Superimposed on this effect is a size effect; the larger planets, having a larger gravitational cross section, collect more of the later condensing (volatile) material.

The Safronov (1972) cosmogonical theory is currently the most popular. It is assumed that the Sun initially possessed a uniform gas-dust nebula. The nebula evolves into a torus and then into a disk. Particles with different eccentricities and inclinations collide and settle to the median plane within a few orbits. As the disk gets denser, it goes unstable and breaks up into many dense accumulations where the self-gravitation exceeds the disrupting tidal force of the Sun. As dust is removed from the bulk of the nebula, the transparency of the nebula increases, and a large temperature gradient is established in the nebula.

The mechanism for bringing particles together and keeping them together to form large planets is obscure. A large body, with an appreciable gravity field, can attract and retain planetesimals. Small particles colliding at high speed disintegrate and have such small gravitational cross section that they can attract only nearby particles. Large collections of co-rotating particles, with minimum relative velocities, seems to be a prerequisite condition. Self-gravitation of the aggregate can then bring the particles together. Small bodies might also act as condensation nuclei and therefore add material directly from the gaseous phase. In the Safronov theory, accumulation of 97–98 percent of the Earth occurred in about 10^8 years. In other theories the accretion time is much shorter, 10^5–10^6 years.

If the relative velocity between planetesimals is too high, fragmentation rather than accumulation will dominate and planets will not grow. If relative velocities are too low, the planetesimals will be in nearly concentric orbits and the collisions required for growth will not take place. Safronov (1972) showed that for plausible assumptions regarding dissipation of energy in collisions and size distribution of the bodies, mutual gravitation causes the mean relative velocities to be only somewhat less than the escape velocities of the larger bodies. Thus, throughout the entire course of planetary growth, the system regenerates itself such that the larger bodies would always grow.

The initial stage in the formation of a planet is the condensation in the cooling nebula. The first solids appear in the range 1750–1600 K and are oxides, silicates and titanates of calcium and aluminum (such as Al_2O_3, $CaTiO_3$, $Ca_2Al_2Si_2O_7$) and refractory metals such as the platinum group. These minerals (such as corundum, perovskite, melilite) and elements are found in white inclusions (chondrules) of certain meteorites, most notably in Type III carbonaceous chondrites. Metallic iron condenses at relatively high temperature followed shortly by the bulk of the silicate material as forsterite and enstatite. FeS and hydrous minerals appear only at very low temperature, less than 700 K. Volatile-rich carbonaceous chondrites have formation temperatures in the range 300–400 K, and at least part of the Earth must have accreted from material that condensed at these low temperatures. The presence of CO_2 and H_2O on the Earth has led some to propose that the Earth was made up entirely of cold carbonaceous chondritic material—the cold accretion hypothesis. Turekian and Clark (1969) assume that volatile-rich material came in as a late veneer—the inhomogeneous accretion hypothesis. Even if the Earth accreted slowly, compared to cooling and condensation times, the later stages of accretion could involve material that condensed further out in the nebula and was later perturbed into the inner solar system. The Earth and the Moon are deficient in not only the very volatile elements that make up the bulk of the Sun and the outer planets, but also the moderately volatile elements such as sodium, potassium, rubidium and lead.

A large amount of gravitational energy is released as the particles fall onto a growing Earth, enough to raise the temperature by tens of thousands of degrees and to evaporate the Earth back into space as fast as it forms. There are mechanisms for buffering the temperature rise and to retain material even if it vaporizes, but melting and vaporization are likely once the proto-Earth has achieved a given size, say lunar size. The mechanism of accretion and its time scale determine the fraction of the heat that is retained, and therefore the temperature and heat content of the growing Earth. The "initial" temperature of the Earth is likely to have been high even if it formed from cold planetesimals. A rapidly growing Earth retains more of the gravitational energy of accretion, particularly if there are large impacts that can bury a large fraction of their gravitational energy. Evidence for early and widespread melting on such small objects as the Moon and various meteorite parent bodies attests to the importance of high initial temperatures, and the energy of accretion of the Earth is more than 15 times greater than that for the Moon. The intensely cratered surfaces of the solid planets provide abundant testimony of the importance of high-energy impacts in the later stages of accretion.

The initial temperature distribution in a planet can be estimated by using the equation of conservation of energy acquired during accretion:

$$\rho \frac{GM(r)}{r} \mathrm{d}r = \varepsilon\sigma[T^4(r) - T_b^4]\mathrm{d}t + \rho C_p[T(r) - T_b]\mathrm{d}t$$

where t is the time, ρ is the density of the accreting particles, G is the gravitational constant, $M(r)$ is the mass of a growing planet of radius r, σ is the Stefan-Boltzmann constant, ε is the emissivity, C_p is the specific heat, T is the temperature at radius r, and T_b is the blackbody radiation temperature. The equation gives the balance between the gravitational energy of accretion, the energy radiated into space and the thermal energy produced by heating of the body. Latent heats associated with melting and vaporization are also involved when the surface temperature gets high enough. The ability of the growing body to radiate away part of the heat of accretion depends on how much of the incoming material remains near the surface and how rapidly it is covered or buried. An impacting body cannot bury all of its heat since heat is transferred to the planetary material, and both the projectile and target fragments are thrown large distances through the atmosphere, cooling during transit and after spreading over the surface. Parts of the projectile and ejecta are buried and must, of course, conduct their heat to the surface before the heat can be radiated back to space. Gardening by later impacts can also bring buried hot material to the surface. Devolatization and heating associated with impact can be expected to generate a hot, dense atmosphere that serves to keep the surface temperature hot and to trap solar radiation.

Intuitively, one would expect the early stages of accretion to be slow, because of the small gravitational cross section and absence of atmosphere, and the terminal stages to be slow, because the particles are being used up. A convenient expression for the rate of accretion that has these characteristics is (Hanks and Anderson, 1969)

$$\frac{\mathrm{d}r}{\mathrm{d}t} = k_1 t^2 \sin k_2 t$$

where k_1 and k_2 can be picked to give a specified final radius and accretion time. The temperature profile resulting from this growth law gives a planet with a cold interior, a temperature peak at intermediate depth, and a cold outer layer. Superimposed on this is the temperature increase with depth due to self-compression and possibly higher temperatures of the early accreting particles. However, large late impacts, even though infrequent, can heat up and melt the upper mantle.

One limiting case is that at every stage of accretion the surface temperature of the Earth is such that it radiates energy back into the dust cloud at precisely the rate at which gravitational energy is released by dust particles free-falling onto its surface. By assuming homogeneous accretion spread out over 10^6 years, the maximum temperature is 1000 K. For this type of model a short accretion time is required to generate high temperatures. The Earth, however, is unlikely to grow in radiative equilibrium. Higher internal temperatures can be achieved if the Earth accumulated partly by the continuing capture of planetesimal swarms of meteoritic bodies. These bodies hit the Earth at velocities considerably higher than free fall and, by shock waves, generate heat at depth in the impacted body.

Modern accretional calculations, taking into account the energy partitioning during impact, have upper-mantle temperatures in excess of the melting temperature during most of the accretion time (Figure 1-1). If melting gets too extensive, the melt moves toward the surface, and some fraction reaches the surface and radiates away its heat. A hot atmosphere, a thermal boundary layer and the presence of chemically buoyant material at the Earth's surface, however, insulates most of the interior, and cooling is slow. Extensive cooling of the upper mantle can only occur if cold surface material is subducted into the mantle. This requires an unstable surface layer, that is, a very cold, thick thermal boundary layer that is denser than the underlying mantle. An extensive accumulation of basalt or olivine near the Earth's surface during accretion forms a buoyant layer that resists subduction. An extensively molten upper mantle is

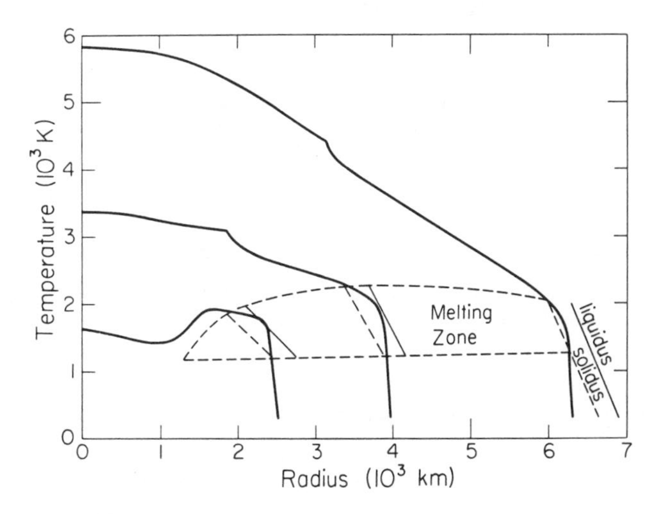

FIGURE 1-1
Schematic temperatures as a function of radius at three stages in the accretion of a planet (heavy lines). Temperatures in the interior are initially low because of the low energy of accretion. The solidi and liquidi and the melting zone in the upper mantle are also shown. Upper-mantle melting and melt-solid separation is likely during most of the accretion process. Silicate melts, enriched in incompatible elements, will be concentrated toward the surface throughout accretion. Temperature estimates provided by D. Stevenson (personal communication).

therefore likely during accretion. As a thick basalt crust cools, the lower portions eventually convert to dense eclogite, and at this point portions of the upper mantle can be rapidly cooled. A dense primitive atmosphere is also an effective insulating agent and serves to keep the crust and upper mantle from cooling and crystallizing rapidly.

All things considered, it is likely that impact melting and gravitational separation combined with internal radioactive heating resulted in terrestrial planets that were being differentiated while they were accreting. Extensive upper-mantle melting gives an upward concentration of melts and the incompatible and volatile elements and burial of dense refractory crystals and melts. Dense melts include Fe, FeS and FeO-MgO-rich melts that form at high pressure. The enrichment of volatiles and incompatible elements in the Earth's atmosphere and crust and the depletion of siderophile elements in the upper mantle point toward a very effective chemical separation of this type, as does the presence of a ligh crust and a dense core. There is evidence from both the Earth and the Moon that these bodies were covered by deep magma oceans early in their history. The lifetime of such an ocean depends on the temperature of the atmosphere, the thickness of an insulating crust and the rate of energy delivery, from outside, by impacts and, from inside, through the thermal boundary layer at the base of the ocean. Removal of crystals from a crystallizing magma ocean and drainage of melt from a cooling crystal mush (also, technically a magma) are very much faster processes than cooling and crystallization times. Therefore, an expected result of early planetary differentiation is a stratified composition.

The following is a plausible variant of the inhomogeneous accretion hypothesis. Planets accrete as the nebula cools, and the accreting material has the composition of the solids that are in equilibrium with the nebula at that temperature plus the more refractory material that has condensed earlier and escaped accretion. The mean composition of a planet therefore becomes less refractory with time and with radius. After dissipation of the nebula, the terrestrial planets continue to slowly accrete material that has condensed in their vicinity and the more volatile material that condensed farther out in the solar system. It is not necessary that all of the refractories and iron be accreted before the silicates. There is always unaccreted material available for interaction with the gas. Iron, for example, is accreted as metal at the early stages but reacts with the silicates to form the ferromagnesian silicates that are accreted later. Likewise, calcium, aluminum, uranium and thorium are available for incorporation into the later condensates, but they are enriched in the early condensates. FeS will condense and accrete at low temperatures unless all the Fe metal has been removed by earlier processes.

The viability of the inhomogeneous accretion hypothesis depends on the relative time scales of nebula cooling and accretion in the early stages of condensation. If cooling is slow relative to accretion rates, then the iron and refractories will form the initial nuclei of the planets. Alternatively, cooling can be rapid if temperatures in the vicinity of the Earth do not drop far into the olivine-pyroxene stability field before dissipation of the nebula. In this case, the majority of the mantle would be added by material perturbed into Earth orbit from cooler parts of the nebula. The earliest condensates also have more time for accretion and possibly experience more viscous drag in the dense, early nebula. The presence of water at the Earth's surface and the siderophile content of the upper mantle indicate that the later stages of accretion did not involve substantial amounts of metallic iron (Lange and Ahrens, 1984). Thus, there are several arguments supporting the view that the proto-Earth was refractory and became more volatile-rich with time. The atmospheres of the terrestrial planets are apparently secondary, formed by outgassing of the interior and devolatilization of late impacts. Primitive atmospheres are generally thought to have been blown away either by a strong solar wind or by giant impacts. Giant impacts in early Earth history may have blasted material into orbit to form the Moon (see Chapter 2) and, in later Earth history, been responsible for the various extinctions that punctuate the paleontological record. Some meteorites found on Earth are thought to have been derived from the surface of the Moon and Mars by large impacts on these bodies.

There is no particular reason to believe that there were originally only four or five terrestrial planets of Mars or Moon size or greater. The sweeping up of multiple small planets by the remaining objects is, in effect, a mechanism of rapid accretion. The early history of the surviving terrestrial planets is therefore violent and characterized by melting and remelting events.

METEORITES

Using terrestrial samples, we cannot see very far back in time or very deep into a planet's interior. Meteorites offer us the opportunity to extend both of these dimensions. Some meteorites, the chondrites, are chemically *primitive,* having compositions—volatile elements excluded—very similar to that of the sun. The volatile-rich carbonaceous chondrites are samples of slightly altered, ancient planetesimal material that condensed at moderate to low temperatures in the solar nebula. The nonchondritic meteorites are *differentiated* materials of nonsolar composition that have undergone chemical processing like that which has affected all known terrestrial and lunar rocks.

Meteorites are assigned to three main categories. Irons (or siderites) consist primarily of metal; stones (or aerolites) consist of silicates with little metal; stony irons (or siderolites) contain abundant metal and silicates. These are further subdivided in various classification schemes, as listed in Table 1-1.

TABLE 1-1
Classification and Characteristics of Stony Meteorites and Iron-rich Meteorites
(Elements, Weight Percent; Ratios Based on Atomic Percent)

Meteorite class	Minerals	S	O	Fe^T	Fe^0	$\frac{Fe}{Fe+Mg}$	Al/Si	Ca/Si	Mg/Si	Remarks
Stones (95 percent of meteorite population)										
Chondrites (86 percent)										
Carbonaceous (5 percent)	Layer-silicates	5.9	46.0	18.4	0		0.086	0.072–	1.05–	No
C1 = CI	("clays")							0.088	1.06	chondrules
e.g. Ivuna; Orgueil	18–22 pct. H_2O									
C2 = CM = CII	6–16 pct. H_2O	3.4	41.5	21.9	0–1		0.084–	0.072	1.02–	
e.g. Mighei; Murchison							0.094		1.04	
C3 = CIII	Olivine,	2.2	36.5	25.2	0–8	6–23	0.092,	0.074,	1.05	
e.g. CO = Ornans	refractory						0.12–	0.09	1.12	
CV = Vigarano	minerals						0.14			
Allende	<4 pct. H_2O		35.9							
Ordinary (81 percent)										
E = Enstatite	Enstatite, Fe-Ni	3.3–	29.3	25–	19–25	.04–	0.048	0.036	0.73–	Fs_0, SiO_2,
e.g. Abee; Khairpur		5.9		33		1.4			0.81	FeS, CaS
H = High iron = Bronzite	Olivine,	2.1	35.1	27.6	15–19	16–19	0.06–	0.05	0.95–	Fa_{15-19},
	bronzite, Fe-Ni						0.07		0.97	Fs_{15-17}
L = Low iron =	Olivine,	2.1	38.1	21.8	4–9	21–25	0.061	0.048	0.92–	Fa_{22-25},
"Hypersthene"	bronzite, Fe-Ni								0.94	Fs_{19-22}
LL = Very low iron =	Olivine	2.3	38.8	20.0	0.3–3	27–32	0.062	0.046	0.92–	Fa_{26-32},
Amphoterite									0.93	Fs_{22-25}
= Soko-Banja										
Achondrites (9 percent)										
Calcium-poor	Olivine,	0.5	45.4							
Ureilites = Olivine-	pigeonite,			14.5	0.3–6	10	0.016	0.023	1.37	
pigeonite	Fe-Ni			20.6	1	25	0.042	0.015	1.28	
†Chassignites = Basaltic										
Aubrites = Enstatite	Enstatite			1.0	<1	<0.03	0.02	0.021	1.0	
*Diogenites =	Orthopyroxene	0.4		13.5		25–27	0.034	0.029	0.74	Fs_{23-27},
"Hypersthene,"										An_{85-90}
"Cumulates"										
Calcium-rich			42.4							
*Howardites and	Orthopyroxene,	0.27		13.9	<0.3	25–40	0.191	0.137	0.47	
*Eucrites = Basaltic	pigeonite,	0.20		14.4	<0.1		0.290	0.234	0.22	
e.g. Juvinas,	plagioclase					50–67				
Pasamonte, Stannern,										
Moore Co.										
Angrites = Augite	Augite	0.45		7.5			0.233	0.595	0.34	
†Nakhlites = Basaltic	cpx, ol	0.06		16.6			0.042	0.332	0.37	
†Shergottites = Basaltic	cpx, plag									
Stony Irons (1 percent)										
*Mesosiderites	Orthopyroxene, plagioclase, Fe-Ni	1.1			44–52					
Pallasites	Olivine, Fe-Ni	0.19			69					
Irons (4 percent)										
Octahedrites	Fe-Ni	0.02–0.09			90.5–92					
Hexahedrites	Fe-Ni	0.06			93.6					
Nickel-rich Ataxites	Fe-Ni	0.08			79.6					

*The eucritic association.
†The SNC association.

TABLE 1-2
Major Minerals of Calcium-Aluminum-Rich Inclusions (CAI)

Mineral	Volume Percent	Condensation Temperature*
Spinel ($MgAl_2O_4$)	15–30	1513–1362 K
Melilite ($Ca_2Al_2SiO_7$)	0–85	1625–1450
Perovskite ($CaTiO_3$)	0–2	1647–1393
Anorthite ($CaAl_2Si_2O_8$)	0–50	1362
Pyroxenes	0–60	1450

Grossman (1972).

*Lower temperature is temperature at which phase reacts with nebular gas to form new phase.

Carbonaceous Chondrites

Carbonaceous chondrites contain unusually high abundances of volatile components such as water and organic compounds, have low densities, and contain the heavier elements in nearly solar proportions. They also contain carbon and magnetite. These characteristics show that they have not been strongly heated, compressed or altered since their formation; that is, they have not been buried deep inside planetary objects.

The C1 or CI meteorites are the most extreme in their primordial characteristics and are used to supplement solar values in the estimation of cosmic composition. The other categories of carbonaceous chondrites, CII (CM) and CIII (CO and CV), are less volatile-rich.

Some carbonaceous chondrites contain calcium-aluminum-rich inclusions (CAI), which appear to be high-temperature condensates from the solar nebula. The minerals (Table 1-2) include anorthite ($CaAlSi_2O_8$), spinel, diopside, melilite, perovskite ($CaTiO_3$), hibonite ($CaAl_{12}O_{19}$) and the Al-Ti pyroxene, fassaite. These inclusions are found in CV and CO chondrites, most notably (because of the total volume of recovered material) the Allende meteorite. Theoretical calculations show that compounds rich in Ca, Al and Ti, including the above minerals, are among the first to condense in a cooling solar nebula. Highly refractory elements are strongly enriched in the CAI compared to C1 meteorites, but they occur in C1, or cosmic, ratios.

Chondrites are named after the rounded fragments, or chondrules, that they contain. Some of these chondrules appear to be frozen drops of silicate liquid and others resemble hailstones in their internal structure. Whatever their origin, the presence of chondrules indicates the composite nature of meteorites and the melting or remelting episodes that characterized the history of at least some of their components. C1 "chondrites" are fine grained and do not contain chondrules. They are chemically similar, however, to the true chondrites (see Table 1-3).

Ordinary Chondrites

As the name suggests, ordinary chondrites are more abundant, at least in Earth-crossing orbits, than all other types of meteorites. They are chemically similar but differ in their contents of iron and other siderophiles, and in the ratio of

TABLE 1-3
Compositions of Chondrites (Weight Percent)

		Ordinary		Carbonaceous			
	Enstatite	H	L	CI	CM	CO	CV
Si	16.47–20.48	17.08	18.67	10.40	12.96	15.75	15.46
Ti	0.03–0.04	0.06	0.07	0.04	0.06	0.10	0.09
Al	0.77–1.06	1.22	1.27	0.84	1.17	1.41	1.44
Cr	0.24–0.23	0.29	0.31	0.23	0.29	0.36	0.35
Fe	33.15–22.17	27.81	21.64	18.67	21.56	25.82	24.28
Mn	0.19–0.12	0.26	0.27	0.17	0.16	0.16	0.16
Mg	10.40–13.84	14.10	15.01	9.60	11.72	14.52	14.13
Ca	1.19–0.96	1.26	1.36	1.01	1.32	1.57	1.57
Na	0.75–0.67	0.64	0.70	0.55	0.42	0.46	0.38
K	0.09–0.05	0.08	0.09	0.05	0.06	0.10	0.03
P	0.30–0.15	0.15	0.15	0.14	0.13	0.11	0.13
Ni	1.83–1.29	1.64	1.10	1.03	1.25	1.41	1.33
Co	0.08–0.09	0.09	0.06	0.05	0.06	0.08	0.08
S	5.78–3.19	1.91	2.19	5.92	3.38	2.01	2.14
H	0.13	—	—	2.08	1.42	0.09	0.38
C	0.43–0.84	—	—	3.61	2.30	0.31	1.08
Fe^0/Fe_{tot}	0.70–0.75	0.60	0.29	0.00	0.00	0.09	0.11

Mason (1962).

TABLE 1-4
Normative Mineralogy of Ordinary Chondrites

Species	High Iron	Low Iron
Olivine	36.2	47.0
Hypersthene	24.5	22.7
Diopside	4.0	4.6
Feldspar	10.0	10.7
Apatite	0.6	0.6
Chromite	0.6	0.6
Ilmenite	0.2	0.2
Troilite	5.3	6.1
Ni-Fe	18.6	7.5

Mason (1962).

oxidized to metallic iron. As the amount of oxidized iron decreases, the amount of reduced iron increases. Olivine is the most abundant mineral in chondrites, followed by hypersthene, feldspar, nickel-iron, troilite and diopside with minor apatite, chromite and ilmenite (Table 1-4). The composition of the olivine varies widely, from 0 to 30 mole percent Fe_2SiO_4 (Fa). Enstatite chondrites are distinguished from ordinary chondrites by lower Mg/Si ratios (Table 1-5), giving rise to a mineralogy dominated by $MgSiO_3$ and having little or no olivine. They formed in a uniquely reducing environment and contain silicon-bearing metal and very low FeO silicates. They contain several minerals not found elsewhere (CaS, TiN, Si_2N_2O). In spite of these unusual properties, enstatite chondrites are within 20 percent of solar composition for most elements. They are extremely old and have not been involved in major planetary processing.

Achondrites

The achondrites are meteorites of igneous origin that are thought to have been dislodged by impact from small bodies in the solar system. Some of these may have come from the asteroid belt, others are almost certainly from the Moon, and one subclass (the SNC group) may have come from Mars. Many of the achondrites crystallized between 4.4 and 4.6 billion years ago. They are extremely diverse and are chemically dissimilar to chondrites. They range from almost monomineralic olivine and pyroxene rocks to objects that resemble lunar and terrestrial basalts. Two important subgroups, classified as basaltic achondrites, are the *eucrites* and the *shergottites*. Two groups of meteoritic breccias, the *howardites* and the *mesosiderites,* also contain basaltic material. The eucrites, howardites, mesosiderites and diogenites appear to be related and may come from different depths of a common parent body. They comprise the *eucritic association*. The shergottites, nakhlites and chassignites form another association and are collectively called the SNC meteorites.

TABLE 1-5
Element Ratios (by Weight) in Four Subtypes of Chondritic Meteorites

Ratio	C1	H	L	E6
Al/Si	0.080	0.063	0.063	0.044
Mg/Si	0.91	0.80	0.79	0.71
Ca/Al	1.10	1.11	1.08	1.06
Cr/Mg	0.025	0.025	0.026	0.024

Eucrites are primarily composed of anorthite, $CaAl_2Si_2O_8$, and pigeonite, $MgFeSi_2O_6$, and are therefore plagioclase-pyroxene rocks similar to basalts. They also have textures similar to basalts. However, terrestrial basalts have higher abundances of sodium, potassium, rubidium and other volatile elements and have more calcium-rich pyroxenes. Eucrite plagioclase is richer in calcium and poorer in sodium than terrestrial basaltic feldspar. The presence of free iron in eucrites demonstrates that they are more reduced than terrestrial basalts.

Eucrites have reflectance spectra similar to Vesta, the second-largest asteroid, and this or a similar asteroid may be the parent body. Trace-element and petrological studies suggest that eucrites could form by about 5 to 15 percent melting, at low pressure, of material similar to chondritic meteorites, leaving behind olivine, pigeonite and plagioclase. Some eucrites exhibit high vesicularity, suggesting crystallization at low pressure. Others have characteristics of cumulates in layered igneous intrusions. Some are similar to lunar and ancient terrestrial calcic anorthosites.

Diogenites are achondrites that consist almost entirely of magnesian orthopyroxene (Fs_{23-27}), bronzite, and minor plagioclase (An_{85-90}). They appear to be metamorphosed cumulates related to eucrites, howardites, and mesosiderites, which have similar isotopic characteristics.

Howardites are mineralogically similar to eucrites, but are breccias that, from a variety of evidence (solar rare-gas content, particle tracks, micrometeorite craters), appear to be from the surface regolith of their parent body. They appear to be essentially mixtures of diogenites and eucrites.

Mesosiderites are stony irons (17–80 wt. percent metal) with silicate fractions similar to diogenites and eucrites. They appear to be a mixture of mantle and core-forming material from their parent body.

Studies of basalts from the Moon and the eucrite parent body have several important implications for the early history of the Earth and the other terrestrial planets. They show that even very small bodies can melt and differentiate. The energy source must be due to impact, rapid accretion, short-lived radioactive isotopes or formation in a hot nebula. The widespread occurrence of chondrules in chondritic meteorites also is evidence for high temperatures and melting in the early solar system. The depletion of volatiles in eucrites and lunar material suggests that small planets, and the early planetesimal stage of planet formation, may be characterized by volatile loss. These extraterrestrial basalts also contain evidence that free iron was removed from their source region. The process of core formation must start very early and is probably contemporaneous with accretion.

Shergottites are remarkably similar to terrestrial basalts. They are unusual, among meteorites, for having very low crystallization ages, about 10^9 years, and, among basalts, for having abundant maskelynite or glassy shocked plagioclase. Shergottites also contain augite, pigeonite and magnetite. In contrast to eucrites, the plagioclase-maskelynite contains terrestrial-type abundances of sodium and calcium. The shergottites are so similar to terrestrial basalts that their source regions must be similar to the upper mantle of the Earth. The similarities extend to the trace elements, be they refractory, volatile or siderophile, suggesting a similar evolution for both bodies. The young crystallization ages imply that the shergottites are from a large body, one that could maintain igneous processes for 3 billion years. Cosmic-ray exposure ages show that they were in space for several million years after ejection from their parent body. Shergottites are slightly richer in iron and manganese than terrestrial basalts, and, in this respect, they are similar to the eucrites. They contain no water and have different oxygen isotopic compositions than terrestrial basalts. The major-element chemistry is similar to that inferred for the martian soil. The rare-gas contents of shergottites are also similar to the martian atmosphere, giving strong circumstantial support to the idea that these meteorites may have come from the surface of Mars. In any case, these meteorites provide evidence that other objects in the solar system have similar chemistries and undergo similar processes as the Earth's upper mantle. The growing Earth probably always had basalt at the surface and, consequently, was continuously zone-refining the incompatible elements toward the surface. The corollary is that the deep interior of a planet is refractory and depleted in volatile and incompatible elements. The main difference between the Earth and the other terrestrial planets, including any meteorite parent body, is that the Earth can recycle material back into the interior. Present-day basalts on Earth may be recycled basaltic material that formed during accretion and in early Earth history rather than initial melts from a previously unprocessed peridotitic parent. Indeed, no terrestrial basalt shows evidence, if all the isotopic and geochemical properties are taken into account, of being from a primitive, undifferentiated reservoir.

PLANETARY ATMOSPHERES

The very volatile elements are concentrated near the surface of a planet and provide important clues as to the average composition of the interior, the mode of formation and the outgassing history. In this section I use the term "volatile" to refer to the very volatile (atmospheric) elements. It is not yet known whether the present planetary atmospheres formed while the planets were accreting, or in early catastrophic events, or in a continuous fashion over geological time. It also is not known if the volatiles were uniformly distributed in the accreting material or if most of the volatiles were brought in as a thin veneer in the terminal stages of accretion. In most theories of atmospheric origin, the early atmospheres are thought to have been dissipated, either by solar activity or by violent impacts, and the present atmospheres are secondary, having been formed by outgassing of the interior. This outgassing must be relatively efficient for the Earth since an appreciable fraction of the argon-40 produced by the decay of potassium-40 in the interior resides in the atmosphere. On the other hand, primordial gases such as helium-3 are still being expelled from the Earth's mantle. Estimates of the extent of outgassing of the Earth and the efficiency of crustal formation are in the 30–70 percent range. This refers to secondary processes, not the accretional outgassing and melting. Accretional devolatilization and melting concentrate the volatiles into the atmosphere and such LIL elements as uranium and potassium into the near-surface layers.

The present atmospheres contain only part of the volatile inventory. Earth has a large amount of CO_2 tied up in organic limestones, and this must be counted as part of the prebiotic atmosphere. Most of the water, of course, is presently in the oceans, ice caps and porous near-surface rocks. Mars has appreciable CO_2 and some H_2O in its polar caps. Venus apparently had an appreciable water content as evidenced from its high deuterium-hydrogen abundance ratio (Donahue and others, 1982). When these factors are taken into account, the early atmospheres of Earth and Venus, similar-sized planets, may have been more similar than they are now.

Differences in the atmospheric abundances and compositions between the terrestrial planets may be due to (1) chemistry of the accreting planetesimals, (2) incomplete outgassing of the interior, (3) trapping in surface regions, (4) catastrophic loss of an early atmosphere and (5) gradual escape of the lighter constituents from the top of the atmosphere.

Among the terrestrial planets there is little correlation of volatile content, as inferred from atmospheric composition, and distance from the sun, but there is a tendency for the larger bodies to be better endowed with volatiles than the smaller planets. Gravitational escape may, in part, be responsible for this trend but cannot explain the persistence of the trend to higher molecular weight species. The larger planets may be more efficiently outgassed or may have been able to accrete more of the volatile-rich late condensates. The study of isotopes, particularly of the noble gases, provides important clues.

In the currently available data on the absolute abundances and ratios of constituents of the volatile inventory of Earth, Venus, Mars and the Sun (Table 1-6) are three important trends: (1) The absolute abundances of N_2 and CO_2 are essentially the same for Venus and Earth and much lower for Mars. (2) The absolute abundance of argon-36 and the $^{36}Ar/^{14}N$ ratio decrease by several orders of magnitude from Venus to Earth to Mars, and a similar increase occurs for $^{40}Ar/^{36}Ar$. (3) Ratios such as $^{20}Ne/^{36}Ar$, $^{38}Ar/^{36}Ar$, $^{18}O/^{16}O$ and $^{13}C/^{12}C$ are similar for the three planets. The ratios of primordial rare-gas species are similar to those of chondritic meteorites but differ from solar values.

The similarity of the planetary rare-gas ratios and their differences from solar values, and the argon, carbon and nitrogen trends dictate against the trapping of solar nebula or solar wind by the accreting planetesimals, the "primary atmosphere model." The volatile content of the terrestrial planets is best explained if the volatiles are brought in as a late veneer by carbonaceous chondritic material (Anders and Owen, 1977). This implies that some carbon and nitrogen has entered the cores of the terrestrial planets. These elements are important in carbonaceous chondrites but are less so in the atmosphere-crust-mantle system.

The high $^{40}Ar/^{36}Ar$ of the Earth suggests that late outgassing may have been more efficient than on Venus. This is probably a consequence of plate tectonics and continuous overturning of the mantle. The high surface temperature on Venus (about 740 K) may preclude deep subduction and the consequent displacement of a large part of the upper mantle to the near surface where it can be outgassed.

Mars has an $^{40}Ar/^{36}Ar$ ratio 10 times greater than the terrestrial values but on the basis of absolute abundances appears to be less outgassed than the Earth or endowed with much lower primitive abundances of the volatiles. The most likely explanations are that Mars is depleted in the very volatile elements such as argon-36 or that it lost its early atmosphere. In addition, it is also likely that a small body such as Mars, with little evidence of plate tectonics, is less outgassed than the Earth.

The high $^{15}N/^{14}N$ ratio for Mars indicates that material has been lost from the atmosphere, presumably by thermal escape. Thermal escape is most likely for low atomic weight elements such as hydrogen, helium and neon. There are also various nonthermal mechanisms, such as impact erosion, for the escape of atmospheric material. Reactions such as

$$2FeO + H_2O \rightarrow Fe_2O_3 + H_2$$

$$CaCO_3 + SiO_3 \rightarrow CaSiO_3 + CO_2$$

between the atmosphere and the surface can also alter the composition and amount of the atmosphere.

The similarity in nitrogen, argon and CO_2 abundances between Venus and Earth suggest that both planets experienced a similar degree of outgassing, but the higher $^{40}Ar/^{36}Ar$ of Earth suggests that later outgassing was more efficient for this body. The "missing water" on Venus apparently escaped (Donahue and others, 1982). The lifetime for helium-4 in Venus's atmosphere is about 10^9 years, about 300 times longer than for the Earth. The longer lifetime reflects the higher abundance and the lower escape efficiency for Venus. The present ^4He abundance in the atmosphere of Venus is about that expected to have been produced over the past 10^9 years if the uranium and thorium content of Venus is similar to the Earth. The relative average abundances of the argon-40 in the terrestrial and venusian atmospheres suggest that the rate of argon-40 outgassing from Venus is about a factor of four less than from the Earth (Donahue and others, 1982).

TABLE 1-6
Volatile Abundances

	Earth	Venus	Mars	Chondrites	Sun
N_2	3×10^{-6}	4×10^{-6}	10^{-7}	—	1.3×10^{-3}
Ar	10^{-8}	7×10^{-9}	5×10^{-10}	—	10^{-4}
^{40}Ar	10^{-8}	3×10^{-9}	5×10^{-10}	—	—
Ne	10^{-11}	2×10^{-10}	4×10^{-14}	—	1.7×10^{-3}
Kr	3×10^{-12}	10^{-10}	2×10^{-14}	—	9.7×10^{-8}
CO_2	2×10^{-4}	0.95×10^{-4}	$>3.5 \times 10^{-8}$	—	10^{-2}
H_2O	2.8×10^{-4}	$>10^{-6}$	$>5 \times 10^{-6}$	—	10^{-2}
$^{40}Ar/^{36}Ar$	292	1.2	3000	—	—
$^{38}Ar/^{36}Ar$	0.2	0.2	0.2	0.2	—
$^{20}Ne/^{36}Ar$	0.5	0.5	0.5	0.2	31
$^{84}Kr/^{36}Ar$	0.036	0.01	0.03	0.022	0.00027
$^{18}O/^{16}O$	2×10^{-3}	2×10^{-3}	2×10^{-3}	—	—
$^{13}C/^{12}C$	.011	0.012	0.011	—	—
$^{14}N/^{15}N$	277	277	165	—	—
D/H	1.6×10^{-4}	1.6×10^{-2}	—	—	—

Donahue and others (1982), Anders and Owen (1977), Pollack and Black (1979).

Nonreactive components of the atmosphere such as the rare gases are presumably incorporated into a planet by being dissolved in crystal interiors and absorbed in grain boundaries and surfaces, or by rapid accretion of a gas-dust mixture. Reactive components such as H_2O and CO_2 are contained in silicates, such as $Mg(OH)_2$, $Mg_3Si_2O_5(OH)_4$, and in carbonate minerals. H_2O and CO_2 are released from these minerals by high-velocity impacts. The high temperatures associated with impact, however, can cause such reactions as

$$MgSiO_3 + Fe + H_2O \rightarrow 1/2Mg_2SiO_4 + 1/2Fe_2SiO_4 + H_2 \quad (1)$$

that, for chrondritic abundances, uses up the available water and allows FeO to be incorporated into the mantle. Free water will only be available at the end of accretion if there is a deficiency of free iron in the late-stage accreted materials (Lange and Ahrens, 1984). Reaction 1 leads to a high FeO content for the mantle unless the core contains appreciable oxygen. Free H_2O in the atmosphere and hydrosphere therefore requires slightly inhomogeneous accretion (more free iron in the early stages of accretion) and a core containing iron and FeO.

Estimates of the accretion time of the Earth range from about 10^7 to 2.5×10^8 years, but the bulk of the Earth was accreted on a much shorter time scale. For most of the accretion time, the gravitational energy of accretion was sufficient to melt and vaporize infalling material. Much outgassing was therefore contemporaneous with accretion. Likewise, the process of core formation was probably synchronous with accretion rather than a later event occurring after the Earth had assembled. Rapid infall of material may have allowed some volatiles to be trapped and buried by subsequent debris-forming events. Rapid accretion and accretion of large, deeply penetrating bodies also causes the early Earth to be a hot body. Since accretional energy increases as the planet grows, it is possible that the early temperature gradient of the mantle was negative.

There are, therefore, several processes that are responsible for the incorporation of volatiles in the interior of a planet: (1) the initial low-energy accretion that allows infalling particles to retain their volatiles, (2) subsequent high-energy accretion that allows a fraction of the incoming volatile inventory to be deeply buried or covered, and (3) late-stage accretion of volatile-rich bodies. Correspondingly, there are several time scales of degassing: (1) the devolatilization of incoming material that is contemporaneous with accretion, (2) the slow and inefficient process of outgassing that involves convection and the cycling of material to the near surface.

The steady decrease in the atmospheric argon-36 abundance, per gram of planet, from Venus to Earth to Mars may represent a chemical gradient in the solar nebula, different degrees of outgassing or atmospheric erosion. The abundance of argon-40, which represents late outgassing from the decay of potassium-40 in the interior, decreases from Earth to Venus to Mars. The factor of 16 difference between

TABLE 1-7
Cosmic Abundances of the Elements (Atoms/10^6Si)

1	H	2.72×10^{10}	24	Cr	1.34×10^4	48	Cd	1.69	72	Hf	0.176
2	He	2.18×10^9	25	Mn	9510	49	In	0.184	73	Ta	0.0226
3	Li	59.7	26	Fe	9.00×10^5	50	Sn	3.82	74	W	0.137
4	Be	0.78	27	Co	2250	51	Sb	0.352	75	Re	0.0507
5	B	24	28	Ni	4.93×10^4	52	Te	4.91	76	Os	0.717
6	C	1.21×10^7	29	Cu	514	53	I	0.90	77	Ir	0.660
7	N	2.48×10^6	30	Zn	1260	54	Xe	4.35	78	Pt	1.37
8	O	2.01×10^7	31	Ga	37.8	55	Cs	0.372	79	Au	0.186
9	F	843	32	Ge	118	56	Ba	4.36	80	HG	0.52
10	Ne	3.76×10^6	33	As	6.79	57	La	0.448	81	Tl	0.184
11	Na	5.70×10^4	34	Se	62.1	58	Ce	1.16	82	Pb	3.15
12	Mg	1.075×10^6	35	Br	11.8	59	Pr	0.174	83	Bi	0.144
13	Al	8.49×10^4	36	Kr	45.3	60	Nd	0.836	90	Th	0.0335
14	Si	1.00×10^6	37	Rb	7.09	62	Sm	0.261	92	U	0.0090
15	P	1.04×10^4	38	Sr	23.8	63	Eu	0.0972			
16	S	5.15×10^5	39	Y	4.64	64	Gd	0.331			
17	Cl	5240	40	Zr	10.7	65	Tb	0.0589			
18	Ar	1.04×10^5	41	Nb	0.71	66	Dy	0.398			
19	K	3770	42	Mo	2.52	67	Ho	0.0875			
20	Ca	6.11×10^4	44	Ru	1.86	68	Er	0.253			
21	Sc	33.8	45	Rh	0.344	69	Tm	0.0386			
22	Ti	2400	46	Pd	1.39	70	Yb	0.243			
23	V	295	47	AG	0.529	71	Lu	0.0369			

Anders and Ebihara (1982).

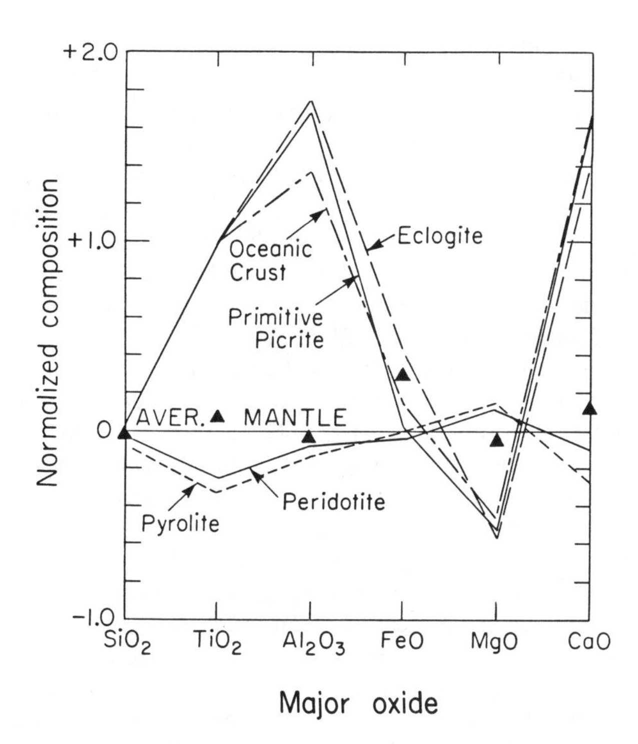

FIGURE 1-2
Bulk chemistry of ultramafic rocks (peridotite) and basic, or basaltic, rocks (oceanic crust, picrite, eclogite) normalized to average mantle composition based on cosmochemical considerations and an assumption about the FeO content of the mantle. Pyrolite is a hypothetical rock proposed by some to be representative of the whole mantle. A composition equivalent to 80 percent peridotite and 20 percent eclogite (or basalt), shown by triangles, is a mix that reconciles petrological and cosmochemical major-element data. Allowance for trace-element data and a possible $MgSiO_3$-rich lower mantle reduces the allowable basaltic component to 15 weight percent or less.

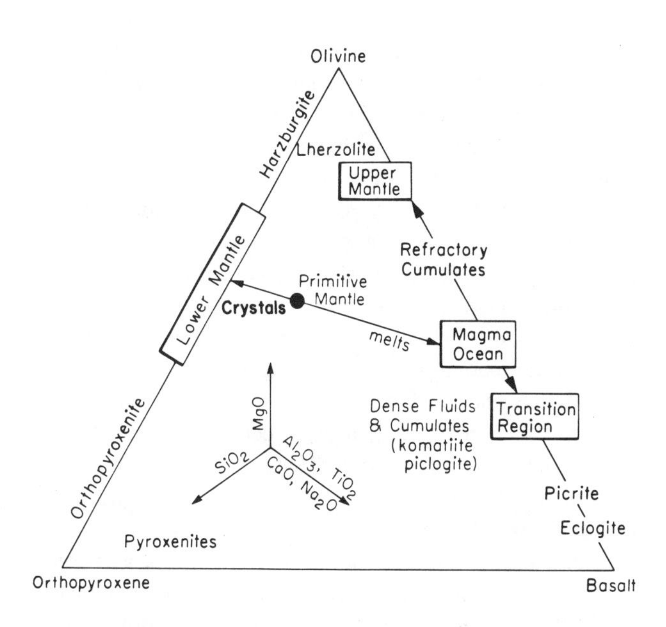

FIGURE 1-3
Representation of mantle components in terms of olivine and orthopyroxene (the high melting point minerals) and basalt (the most easily fusible component). Primitive mantle is based on cosmic abundances. Melting of chondritic material at high temperature gives an MgO-rich melt (basalt + olivine) and a dense refractory residual (olivine + orthopyroxene). Crystallization of a magma ocean separates clinopyroxene and garnet from olivine. Melting during accretion tends to separate components according to density and melting temperature, giving a chemically zoned planet. If melting and melt-crystal separation occur primarily at low pressure, the upper mantle will be enriched in basalt, olivine and the incompatible elements relative to the lower mantle and relative to the chondritic starting material.

Earth and Mars probably represents a lower volatile content, including potassium, for Mars and the lesser outgassing expected for a small, relatively cold planet.

COMPOSITION OF THE TERRESTRIAL PLANETS

It is now generally believed that with the exception of a few elements such as Li, Be and B, the composition of the solar atmosphere is essentially equal to the composition of the material out of which the solar system formed. (This ignores the possibility that the Sun is a chemically zoned object.) The planets are assumed to accrete from material that condensed from a cooling primitive solar nebula. Various attempts have been made to compile tables of "cosmic" abundances. The Sun contains most of the mass of the solar system; therefore, when we speak of the elemental abundances in the solar system, we really refer to those in the Sun, assuming that the abundances in the Sun, its surface, and in the primitive solar nebula are the same. The spectroscopic analyses of elemental abundances in the solar photosphere do not have as great an accuracy as chemical analyses of solid materials. C1 carbonaceous chondrite meteorites, which appear to be the most representative samples of the relatively nonvolatile constituents of the solar system, are used for compilations of the abundances of most of the elements (Table 1-7). For the very abundant volatile elements, solar abundance values are used.

The very light and volatile elements (such as H, He, C and N) are extremely depleted in the Earth relative to the Sun or carbonaceous chondrites. Moderately volatile elements (such as K, Na, Rb, Cs and S) are moderately depleted in the Earth. Refractory elements (such as Ca, Al, Sr, Ti, Ba, U and Th) are generally assumed to be retained by the planets in their cosmic ratios. It is also likely that magnesium and silicon occur in a planet in chondritic or cosmic ratios with the more refractory elements. The Mg/Si ratio, however, varies somewhat among meteorite classes. Sometimes it is assumed that magnesium, iron and silicon may be fractionated by accretional or pre-accretional processes, but these effects, if they exist, are slight. The upper mantle is olivine-rich and has a high Mg/Si ratio compared to the cosmic ratio (Figures 1-2 and 1-3). If the Earth is chondritic in major-element chemistry, then the deeper mantle must be rich in pyroxene.

It is not clear that the planets accreted homogeneously from material of uniform composition. The mean uncompressed densities of the terrestrial planets decreases in the order Mercury, Earth, Venus, Mars, Moon (Figure 1-4 and Table 1-8). The high density of Mercury probably means a high iron content, and the reverse is implied for the Moon. The other planets may have a variable iron content or may differ in the oxidation state of the iron, that is, the FeO/Fe ratio. Venus and Earth are so close in mass and mean density that they may have nearly identical major-element chemistries. The high surface temperature of Venus decreases the density of near-surface materials, and this along with the low pressures depress the depths of phase changes such as basalt-eclogite, making Venus overall a less dense body. The uncompressed density of Mars is less than that of Earth and Venus, and it therefore differs in composition. The outer planets and satellites are much more volatile-rich than the inner planets. Meteorites also vary substantially in composition. The above considerations suggest that there may be an element of inhomogeneity in the accretion of the planets, perhaps caused by temperature and pressure gradients in the early solar nebula. Early forming planetesimals would have been refractory- and iron-rich and the later forming planetesimals more volatile-rich. If planetary accretion was occurring simultaneously with cooling and condensation, then the planets would have formed inhomogeneously. As a planet grows, the gravitational energy of accretion increases, and impact vaporization becomes more

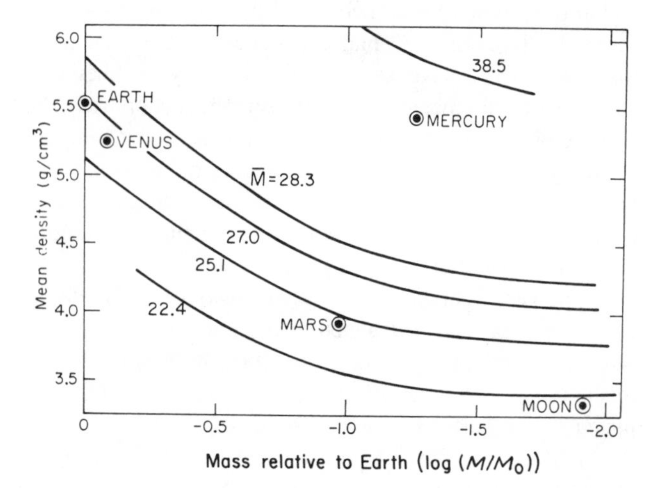

FIGURE 1-4
Mean density versus mass, relative to Earth, of planets having the same structure as the Earth and various metal/silicate ratios, expressed as $\bar{M}$, mean atomic weight. Earth and Venus have similar bulk chemistries while Mars and Moon are clearly deficient in iron. Mercury is enriched in iron.

important for the larger planets and for the later stages of accretion. The assumption that Earth has cosmic abundances of the elements is therefore only a first approximation but is likely to be fairly accurate for the involatile elements. There is little dispersion of the refractory elements among the various stony meteorite classes, suggesting that these elements are not appreciably fractionated by pre-accretional processes. Fortunately, the bulk of a terrestrial planet is iron, magnesium, silicon, calcium, and aluminum and their oxides. The bulk composition of a terrestrial planet can therefore be discussed with some confidence.

During large solar flares a sample of the corona can be accelerated to high energies. Solar energetic particles can be used to infer the composition of the corona and photosphere (Breneman and Stone, 1985). Some recent results are given in Table 1-9. The iron abundance is about 40 percent higher than previous estimates of cosmic or solar abundances. This agrees with recent spectroscopic values for the photosphere (Grevasse, 1984). The refractory elements calcium and titanium are also more abundant than chondritic values. A terrestrial planet formed from the solar nebula can therefore have much more free iron, FeO and diopside ($CaMgSi_2O_6$) than previously supposed. With these new values Mercury is still enriched in iron, and Mars and Moon are depleted in iron compared to the Sun.

It is usually assumed that the Sun and planets formed more or less contemporaneously from a common mass of interstellar dust and gas. There is a close similarity in the relative abundances of the condensable elements in the atmosphere of the Sun, in chondritic meteorites and in the Earth. To a first approximation one can assume that the planets incorporated the condensable elements in the proportions observed in the Sun and the chondrites. On the other hand, the differences in the mean densities of the planets, corrected for differences in pressure, show that they cannot all be composed of materials having exactly the same composition. Variations in iron content and oxidation state of iron can cause large density variations among the terrestrial planets. The giant, or Jovian planets, must contain much larger proportions of low-atomic-weight elements than Mercury, Venus, Earth, Moon and Mars. The condensation behavior of the elements is given in Figure 1-5.

The equilibrium assemblage of solid compounds that exists in a system of solar composition depends on temperature and pressure and, therefore, with location and time. At a nominal nebular pressure of 10^{-3} atm, the material would be a vapor at temperatures greater than about 1900 K (Larimer, 1967; Grossman, 1972). The first solids to condense at lower temperature or higher pressure are the refractory metals (such as W, Re, Ir and Os). Below about 1750 K refractory oxides of aluminum, calcium, magnesium and titanium condense, and metallic iron condenses near 1470 K (Table 1-10 and Figure 1-6). Below about 1000 K, sodium and potassium condense as feldspars, and a portion of the iron is stable as fayalite and ferrosilite with the proportion increasing with a further decrease in temperature. FeS

TABLE 1-8
Properties of the Terrestrial Planets

	GM $10^{18}cm^3/s^2$	R km	ρ g/cm^3	I/MR^2	D* km
Earth	398.60	6371	5.514	0.3308	14
Moon	4.903	1737	3.344	0.393	75
Mars	42.83	3390	3.934	0.365	>28
Venus	324.86	6051	5.24	?	?
Mercury	22.0	2440	5.435	?	?

*Estimated crustal thickness.

condenses below about 750 K. Hydrated silicates condense below about 300 K. Differences in planetary composition may depend on the location of the planet, the location and width of its feeding zone and the effects of other planets in sweeping up material or perturbing the orbits of planetesimals. In general, one would expect planets closer to the Sun and the median plane of the nebula to be more refractory-rich than the outer planets. On the other hand, if the final stages of accretion involve coalescence of large objects of different eccentricities, then there may be little correspondence between bulk chemistry and the present position of the terrestrial planets.

There are several ways in which interactions between the gaseous nebula and solid condensate particles might have controlled the composition of the planets. In one extreme, all or most of the material joining a planet may have equilibrated in a relatively narrow range of temperatures peculiar to that planet. In another extreme, one mineral af-

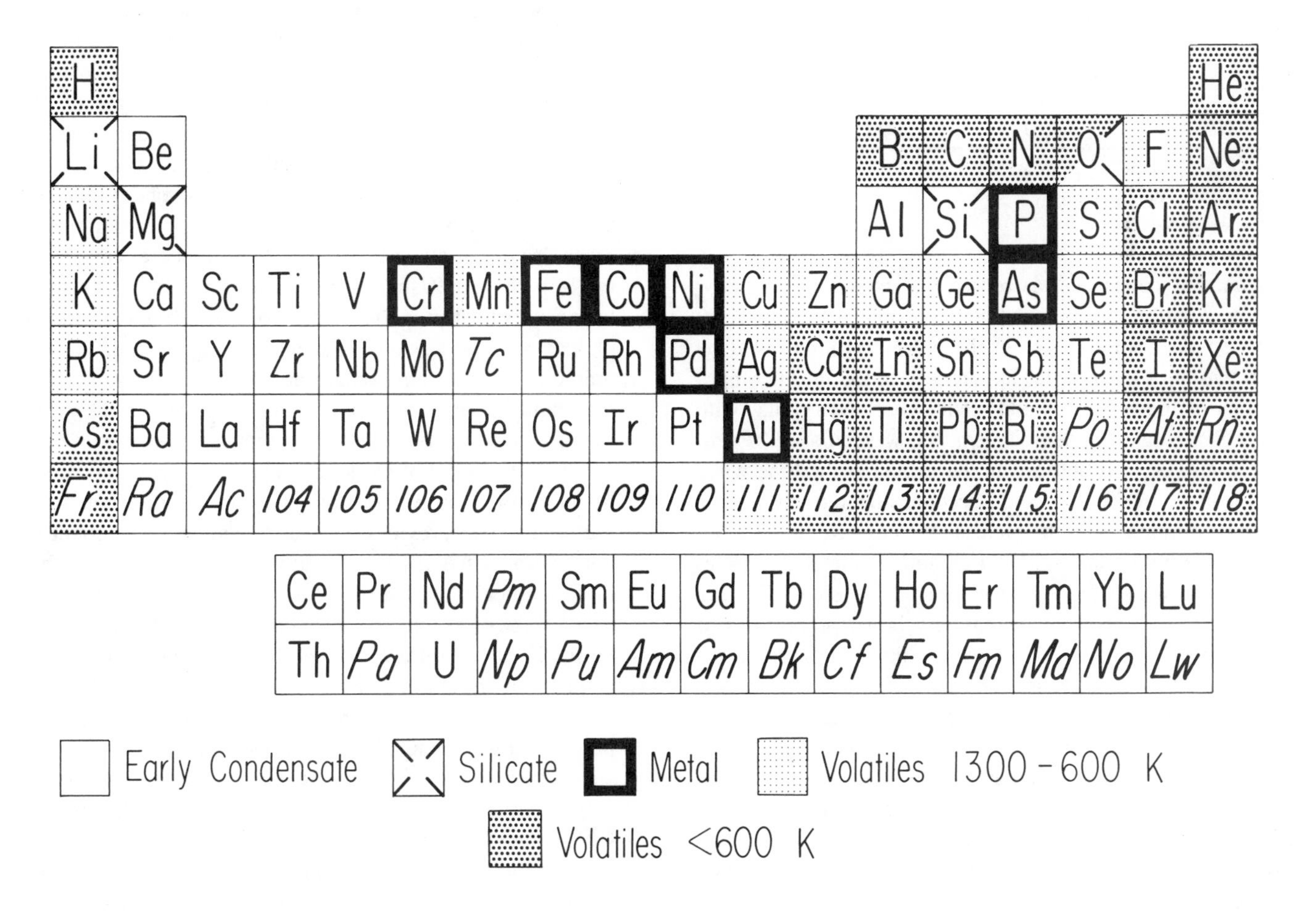

FIGURE 1-5
Condensation behavior of the elements. Short-lived radioactive elements are shown in italics (after Morgan and Anders, 1980).

TABLE 1-9
Solar and "Cosmic" Abundances in Atoms/1000 Si Atoms

Z	Element	Corona (1)	Photosphere (1)	"Cosmic" (2)
6	C	2350	6490	12,100
7	N	700	2775	2480
8	O	5680	22,900	20,100
9	F	0.28	1.1	0.843
10	Ne	783	3140	3760
11	Na	67.0	67.0	57.0
12	Mg	1089	1089	1075
13	Al	83.7	83.7	84.9
14	Si	1000	1000	1000
15	P	4.89	9.24	10.4
16	S	242	460	515
17	Cl	2.38	9.6	5.24
18	Ar	24.1	102	104
19	K	3.9	3.9	3.77
20	Ca	82	82	61.1
21	Sc	0.31	0.31	0.034
22	Ti	4.9	4.9	2.4
23	V	0.48	0.48	0.295
24	Cr	18.3	18.3	13.4
25	Mn	6.8	6.8	9.51
26	Fe	1270	1270	900
27	Co	<18.1	<18.1	2.25
28	Ni	46.5	46.5	49.3
29	Cu	0.57	0.57	0.514
30	Zn	1.61	1.61	1.26

(1) Breneman and Stone (1985).
(2) Anders and Ebihara (1982).

ter another condenses, as required to maintain thermodynamic equilibrium in the nebula, and immediately accretes into a planet. At some point the process is interrupted by dissipation of the nebula. Differences in the mean compositions of the planets would result when the nebula, with any remaining uncondensed elements, was removed. If temperatures declined outward in the nebula at the time when condensation ended, it could account qualitatively for the density differences in the planets. Mercury would have accreted mainly the calcium and aluminum silicates and metallic iron; planets farther out in the solar system would have condensed and accreted increasingly larger proportions of low-density silicates and volatiles. This process, described earlier in this chapter, is termed heterogeneous accretion (Turekian and Clark, 1969; Anderson, 1972a,b; Clark and others, 1972). The planets formed would be layered from the outset, having the highest temperature condensates at their centers and successively lower temperature condensates closer to their surfaces. Thus, planets would contain substantial metallic cores mainly as a result of accretion rather than subsequent interior melting and differentiation. At the early stages of accretion, it might be easier to accrete and retain material on a ductile iron nucleus than on a brittle silicate nucleus.

The chondritic meteorites may be samples of relatively primitive planetary material. Many of them consist of several components that have experienced very different temperature histories (Wood, 1962), which are mingled on the scale of a millimeter. Observed differences between the bulk compositions of the several chondrite subclasses can be explained as resulting from (1) processes that physically fractionated the various components from one another before they accreted into the parent chondrite planets and (2) differing nebula temperatures of accretion for the chondrite subclasses, which had the effect of excluding elements that required temperatures lower than the accretion temperature to condense. For example, siderophiles, which behave coherently during magmatic processes in a planet, might be fractionated in the nebula because they span a large range of condensation temperatures. Nebular fractionation processes include (1) gain or loss of early condensate by settling to the median plane of the nebula and (2) gain or loss of metal relative to silicate, by ferromagnetic or gravitational effects, and remelting of the primary condensate. Some typical compositions of all these components are listed in Table 1-11.

Various authors have suggested that the same physical components and fractionating processes that occurred in the chondrite subclasses are responsible for differences in the compositions and mean densities of the terrestrial planets (Mason, 1962; Morgan and Anders, 1980). Some of the differences among meteorite classes may be due to magmatic processes that occurred inside the parent body.

There are some constraints on the amounts or ratios of a number of key elements in a planet. For example, the mean density of a planet, or the size of the core, constrains the iron content. Using cosmic ratios of elements of similar geochemical properties (say Co, Ni, refractory siderophiles), a whole group of elements can be constrained. The uranium and thorium content are constrained by the heat flow and thermal history calculations. The K/U ratio, roughly constant in terrestrial magmas, is a common constraint in this kind of modeling. The Pb/U ratio can be estimated from lead isotope data. The amount of argon-40 in the atmosphere provides a lower bound on the amount of potassium in the crust and mantle. (This and some other rare-gas species are listed in Table 1-12.) Most of these are very weak constraints, but they do allow rough estimates to be made of the refractory, siderophile, volatile and other contents of the Earth and terrestrial planets. The elements that are correlated in magmatic processes have very similar patterns of geochemical behavior, even though they may be strongly fractionted during nebular condensation. Thus, some abundance patterns established during condensation tend not to be disturbed by subsequent planetary melting and igneous fractionation. On the other hand, some elements are so strongly fractionated from one another by magmatic and core formation processes that discovering a "cosmic" or "chondritic" pattern can constrain the nature of these processes.

TABLE 1-10
Approximate Sequence of Condensation of Phases and Elements from a Gas of Solar Composition at 10^{-3} atm Total Pressure

Phase	Formula	Temperature
Hibonite	$CaAl_{12}O_{19}$	1770 K
Corundum	Al_2O_3	1758 K
Platinum metals	Pt, W, Mo, Ta	
	Zr, REE, U, Th	
	Sc, Ir	
Perovskite	$CaTiO_3$	1647 K
Melilite	$Ca_2Al_2SiO_7$-	
	$Ca_2Mg_2Si_2O_7$	1625 K
	Co	
Spinel	$MgAl_2O_4$	1513 K
	Al_2SiO_5	
Metallic iron	Fe, Ni	1473 K
Diopside	$CaMgSi_2O_6$	1450 K
Forsterite	Mg_2SiO_4	1444 K
Anorthite	$CaAl_2Si_2O_8$	1362 K
	Ca_2SiO_4	
	$CaSiO_3$	
Enstatite	$MgSiO_3$	1349 K
	Cr_2O_3	
	P, Au, Li	
	$MnSiO_3$	
	MnS, Ag	
	As, Cu, Ge	
Feldspar	$(Na,K)AlSi_3O_8$	
	Ag, Sb, F, Ge	
	Sn, Zn, Se, Te, Cd	
Reaction products	$(Mg,Fe)_2SiO_4$	1000 K
	$(Mg,Fe)SiO_3$	
Troilite, pentlandite	FeS, (Fe, Ni)S	700 K
	Pb, Bi, In, Tl	
Magnetite	Fe_3O_4	405 K
Hydrous minerals	$Mg_3Si_2O_72H_2O$, etc.	
Calcite	$CaCO_3$	<400 K
Ices	H_2O, NH_3, CH_4	<200 K

Anders (1968), Grossman (1972), Fuchs and others (1973), Grossman and Larimer (1974).

There is strong evidence that the most refractory elements condensed from the solar nebula as a group, unfractionated from one another, at temperatures above the condensation temperature of the Mg-silicates (Morgan and Anders, 1980). Hence, the lithophile refractory elements (Al, Ca, Ti, Be, Sc, V, Sr, Y, Zr, Nb, Ba, rare-earth elements, Hf, Ta, Th, and U and, to some extent, W and Mo) can be treated as one component. From the observed abundance ratios of correlated elements in samples from the Moon, Earth and achondrites, there is excellent proof that these elements are present in the same ratios as in C1 chondrites. The abundance of the refractory elements in a given planet can be calculated from the inferred abundance of their heat-producing members uranium and thorium, provided reliable data and interpretations of the global heat flux are available (Ganapathy and Anders, 1974). This assumes that surface heat flow represents the current rate of heat production. A large fraction of the present heat flow, however, may be due to cooling of the Earth, which means that only an upper bound can be placed on the uranium and thorium content. Nevertheless, this is a useful constraint particularly when combined with the lower bound on potassium provided by argon-40 and estimates of K/U and Th/U provided by magmas and the crust.

There is little justification for assuming that the volatile elements joined the planets in constant proportions. In this context the volatiles include the alkali metals, sulfur and so forth in addition to the gaseous species.

All of these complexities make planetary models somewhat variable. Table 1-13 lists typical model compositions for the terrestrial planets as well as one for the eucrite parent body. These and other models will be considered below in more detail.

Venus

Venus is 320 km smaller in radius than the Earth and is about 4.9 percent less dense. Most of the difference in density is due to the lower pressure, giving a smaller amount of self-compression and deeper phase changes. Venus is a much smoother planet than the Earth but has a measurable triaxiality of figure and a 0.34 km offset of the center of the figure from the center of mass. This offset is much smaller than those of the Moon (2 km), Mars (2.5 km) and Earth (2.1 km). The moment of inertia of Venus is not known.

In contrast to the bimodal distribution of Earth's topography, representing continent-ocean differences, Venus has a narrow unimodal height distribution with 60 percent of the surface lying within 500 m of the mean elevation. This difference is probably related to erosion and isostatic differences caused by the presence of an ocean on Earth. For both Earth and Venus the topography is dominated by long-wavelength features. Most of the surface of Venus is gently rolling terrain. The gravity and topography are positively correlated at all wavelengths. On Earth most of the long-wavelength geoid is uncorrelated with surface topography and is due to deep mantle dynamics.

The other respects in which Venus differs markedly from the Earth are its slow rotation rate, the absence of a satellite, the virtual absence of a magnetic field, the low abundance of water in the atmosphere and at the surface, the abundance of primordial argon, the high surface temperature and the lack of obvious signs of subduction (Phillips and others, 1980; Anderson, 1980, 1981; Head and Solomon, 1981). There is some evidence for spreading, ridge-like features and compressional features.

The moderate differences of the mass, mean density and solar distance of Venus from those of the Earth make it appropriate to discuss its bulk constitution (Table 1-14) in terms of differences from the Earth.

If Venus had an identical bulk composition and structure to the Earth (see Figure 1-4), then its mean density

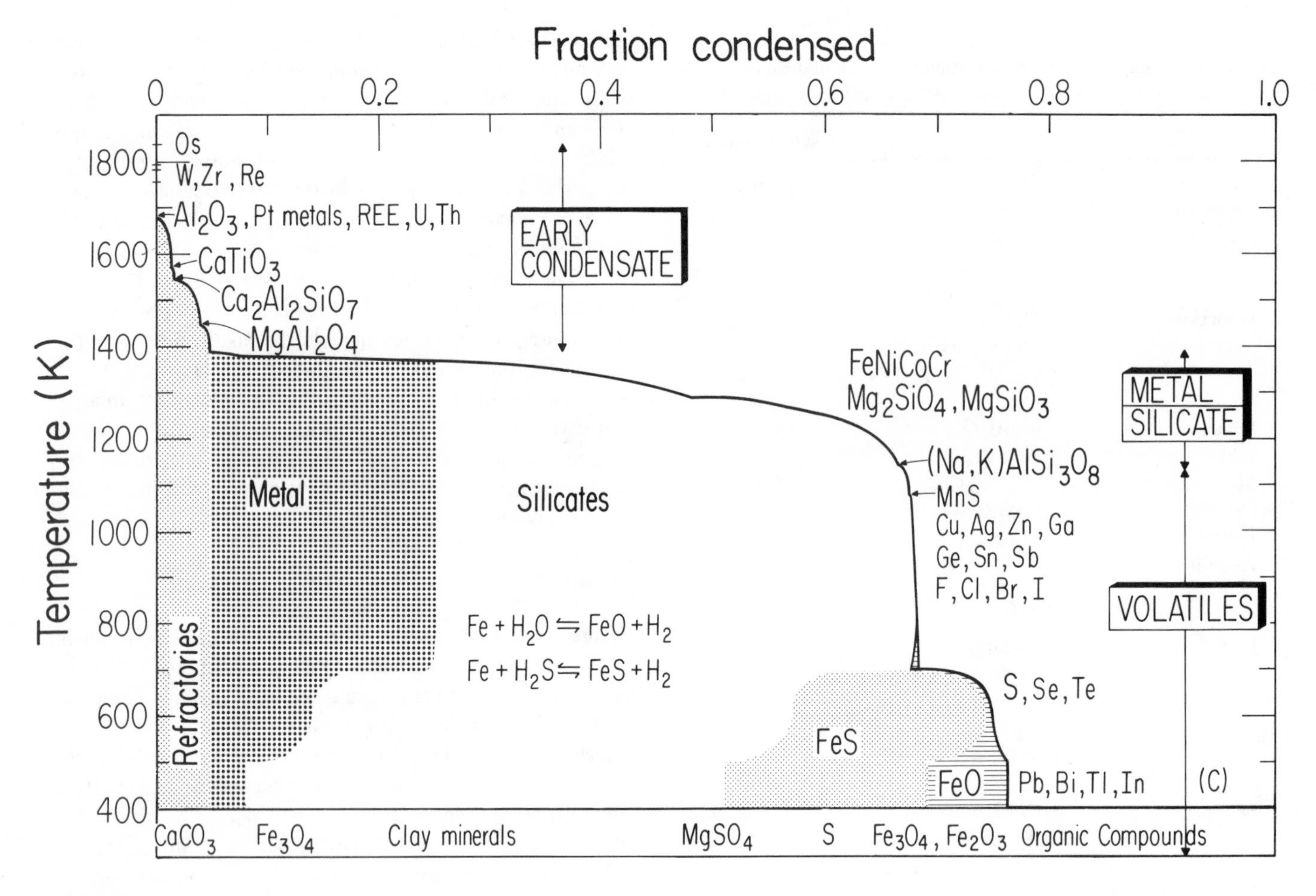

FIGURE 1-6
Condensation of a solar gas at 10^{-4} atm (after Morgan and Anders, 1980).

would be about 5.34 g/cm³. By "identical structure" I mean that (1) most of the iron is in the core, (2) the crust is about 0.4 percent of the total mass and (3) the deep temperature gradient is maintained by convection. The high surface temperature of Venus, about 740 K, would have several effects; it would reduce the depth at which the convectively controlled gradient is attained, it would deepen temperature-sensitive phase changes and it may prevent mantle cooling by subduction.

The density of Venus is 1.2 to 1.9 percent less than that of the Earth after correcting for the difference in pressure. This has been attributed to differences in iron content, sulfur content, oxidization state of the mantle and deepening of the basalt-eclogite phase change. Most of the original basaltic crust of the Earth subducted when the upper mantle temperatures cooled into the eclogite stability field. The density difference between basalt and eclogite is about 15 percent. Because of the high surface temperature on Venus, the upper-mantle temperatures are likely to be 200–400 K hotter in the outer 300 km or so than at equivalent depths on Earth, or melting is more extensive.

Since the deep interior temperatures of the two planets are likely to be similar, the near-surface thermal gradient is controlled by the thickness, δ, of the conductive thermal boundary layer. This is related to the Rayleigh number, *Ra*, and the thickness, *D*, of the underlying convecting layer by

$$\delta \approx Ra^{-1/3}D$$

where

$$Ra = \frac{\alpha g \Delta T D^3}{\kappa \nu},$$

which is written in terms of the gravitational acceleration g, temperature rise across the layer ΔT, the thermal expansion and diffusivity α and κ, and the kinematic viscosity ν. The largest difference between Earth and Venus is likely to be the viscosity since this depends exponentially on temperature and pressure. The higher temperatures in the outer layer of Venus and the lower pressure throughout serve to decrease the average viscosity of the Venus mantle by at least several orders of magnitude. The other parameters in the Rayleigh number are only weakly dependent on temperature or on the size of the planet. Because of the thinner boundary layer, the near-surface thermal gradient is steeper in Venus than the Earth. The net result is that the deep mantle adiabat is brought closer to the surface. This has inter-

TABLE 1-11
Compositions of Possible Components of the Terrestrial Planets (Percent or ppm)

Species	C1	EC	HTC
SiO_2	30.9	39.1	20.2
TiO_2	0.11	0.06	1.9
Al_2O_3	2.4	1.9	36.5
Cr_2O_3	0.38	0.35	—
MgO	20.8	21.3	7.1
FeO	32.5	1.7	—
MnO	0.25	0.14	—
CaO	2.0	1.6	34.1
Na_2O	1.0	1.0	—
K (ppm)	800	920	—
U (ppm)	0.013	0.009	0.19
Th (ppm)	0.059	0.034	0.90
Fe	0	26.7	—
Ni	1.3	1.7	—
S	8.3	4.5	—

C1: Average C1 carbonaceous chondrite, on a C-, H_2O-free basis (Wood, 1962).
EC: Average enstatite chondrite (Wood, 1962).
HTC: High-temperature condensate (Grossman, 1972).

esting implications for the phase relations in the upper mantle and the evolution of the planet. In particular, partial melting in the upper mantle of Venus is probably much more extensive than is the case for the Earth. Crust can be much thicker because of the deepening of the basalt-eclogite phase boundary.

Schematic geotherms are shown in Figure 1-7 for surface temperatures appropriate for Earth and Venus. With the phase diagram shown, the high-temperature geotherm crosses the solidus at about 85 km. With other phase relations the eclogite field is entered at a depth of about 138 km. For Venus, the lower gravity and outer layer densities increase these depths by about 20 percent; thus, we expect a surface layer of 100 to 170 km thickness on Venus composed of basalt and partial melt. On the present Earth, the eclogite stability field is entered at a depth of 40–60 km.

A large amount of basalt has been produced by the Earth's mantle, but only a thin veneer is at the surface at any given time. There must therefore be a substantial amount of eclogite in the mantle, the equivalent of about 200 km in thickness. If this were still at the surface as basalt, the Earth would be several percent less dense. Correcting for the difference in temperature, surface gravity and mass and assuming that Venus is as well differentiated as Earth, only a fraction of the basalt in Venus would have converted to eclogite. This would make the uncompressed density of Venus about 1.5 percent less than Earth's without invoking any differences in composition or oxidation state. Thus, Venus may be close to Earth in composition. It is possible that the present tectonic style on Venus is similar to that of Earth in the Archean, when temperatures and temperature gradients were higher.

The high degree of correlation between gravity and elevation on Venus might suggest that surface loads are supported by a thick, strong lithosphere. Because of the high surface temperature, this is unlikely. In the other extreme, in a purely viscous, convecting planet, such a correlation is expected if hot upwellings deform the surface upward to an extent that more than compensates the lower density. This, in general, is the expected situation. A thick buoyant crust would also give this effect if isostasy prevails. The relationship between elevation and geoid, or gravity, gives the depth of compensation. High topography may also be maintained dynamically by horizontal compression, but this is expected to be a transient situation.

The depth of compensation on Venus has been estimated to be 115 ± 30 km (Phillips and others, 1980). A thick buoyant crust on Venus is therefore a distinct possibility. If the depth of compensation corresponds to the crustal thickness, then it would be six times the average crustal thickness on Earth and comparable in thickness to the seismic lithosphere under continental shields, which has been interpreted as olivine-rich residual material that is lighter than "normal" fertile (high garnet content) mantle. The amount of implied crust for Venus is not unreasonable, considering the amount of CaO, Al_2O_3, Na_2O and so on that is

TABLE 1-12
Rare-gas Isotopes in Planetary Atmospheres and Chondrites

	^{36}Ar (10^{-10}cm^3/g)	^{40}Ar (10^{-8}cm^3/g)	^{40}Ar/^{36}Ar	^{129}Xe (10^{-12}cm^3/g)	^{129}Xe/^{132}Xe
Mars	1.6	48	3000	1.12	1.49
Earth	210	612	291	1.04	0.067
Venus	21,000	225	1.07	—	—
Chondrites					
C1	7700	476	6.2	380	0.05
C3V	3440	1500	44	780	0.26
E4	3310	8606	260	3890	4.05

Mazor and others (1970), Anders and Owen (1977), Morgan and Anders (1980), Wacker and Marti (1983), von Zahn and others (1983).

TABLE 1-13
Representative Model Compositions of Terrestrial Planets and Other Bodies Based Mainly on Cosmochemical Considerations (Weight Percent and ppm)

Species	Mercury	Venus	Earth	Mars	Moon	EPB
			Mantle + Crust			
SiO_2	43.8	53.9	48.0	40.0	45.6	46.0
TiO_2	—	0.20	0.27	0.1	0.2	0.15
Al_2O_3	4.4	3.9	5.2	3.1	4.6	3.1
Cr_2O_3	—	—	1.1	0.6	0.4	0.8
MgO	47.7	38.3	34.3	27.4	32.4	32.5
FeO	0	2.1	7.9	24.3	13.0	14.4
MnO	—	—	0.12	0.2	0.18	0.4
CaO	4.1	3.6	4.2	2.5	3.8	2.5
Na_2O	—	1.5	0.33	0.8	0.06	0.07
H_2O	—	—	0.11	0.9	—	—
K (ppm)	—	1318	262	573	60	62
U (ppm)	0.024	0.021	0.028	0.017	0.027	0.018
Th (ppm)	0.11	0.096	0.100	0.077	—	0.07
			Core			
Fe	94.5	84.7	89.2	72.0	—	—
Ni	5.5	5.2	5.7	9.3	—	—
S	0	10.0*	5.1*	18.6*	—	—
			Relative Masses			
Mantle + Crust	33.0	69.1	63.9	88.1	—	—
Core	67.0	30.9	36.1	11.9	—	—

Mercury: Cosmic Al:Mg ratio, sufficient SiO_2 for anorthite, diopside, and forsterite phases in mantle, core/mantle mass ratio satisfying mean density (BVP, 1980).

Venus: Equilibrium condensation modified by use of feeding zones (Weidenschilling, 1976; BVP, 1980). Venus can be identical in composition to the Earth (Anderson, 1980, 1981).

Earth: Ganapathy and Anders (1974).

Mars: Anderson (1972a).

Moon: Wänke and others (1977).

Eucrite Parent Body: Dreibus and others (1977), Dreibus and Wänke (1979).

*A fraction of the cosmic complement of sulfur is assumed to be in the core.

likely to be incorporated into planetary interiors and considering the relative thickness of the martian and lunar crusts. It is Earth that is anomalous in total crustal thickness, and this can be explained by crustal recycling and the shallowness of the basalt-eclogite boundary in the Earth. Most of the Earth's "crust" probably resides in the transition region of the mantle. Estimates of bulk Earth chemistry can yield a basaltic layer of about 10 percent of the mass of the mantle.

There is little evidence for Earth-style plate tectonics on Venus, but this does not mean that Venus is tectonically dead. If the surface is choked with thick buoyant crust, the manifestation of mantle convection would be quite different than on Earth, where oceanic lithosphere can subduct to make room for the new lithosphere formed at midocean ridges. The rifts and highlands on Venus may be recent transient features, and the surface may be constantly reorganizing itself more on the style of pack ice in the polar oceans. If the crust and lithosphere on Venus is buoyant, the cooling effect of subduction is precluded, and the venerian mantle may be hotter than the Earth's mantle.

The very slow retrograde spin of Venus and its great abundance of primordial argon might be explained by the impact of a major body (Cameron, 1982). The slow rotation rate may also be at least partly responsible for the small magnetic field, which is smaller than either Earth's or Mars's. The magnetic dipole moment of Venus is at least four orders of magnitude less than Earth's. Other factors that might be involved in the small magnetic field of Venus include the roughness of the core-mantle boundary, the temperature of the core and the absence of chemical sedimentation in the core. It is also remotely possible that the field is temporarily low, as occurs on Earth when the field is reversing polarity.

The Soviet gamma-ray data from the surface of Venus (Table 1-15) are consistent with those expected for basaltic rocks.

Mars

Mars is about one-tenth of the mass of Earth. The uncompressed density is substantially lower than that of Earth or

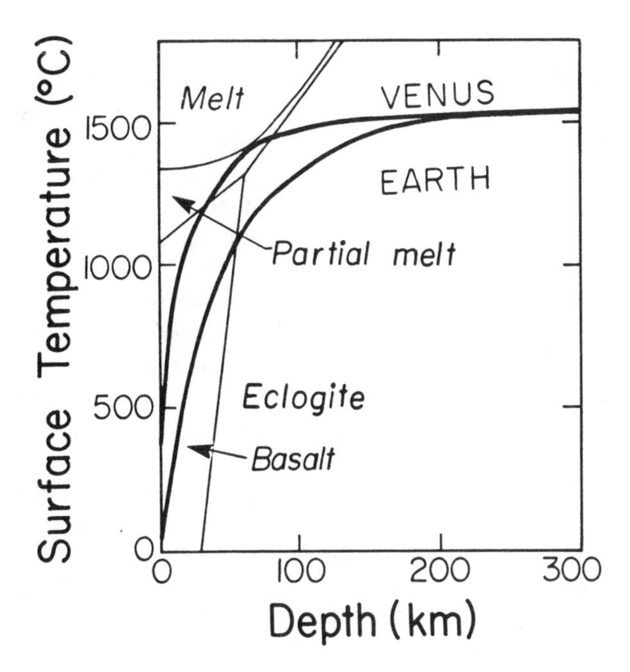

FIGURE 1-7
Schematic geotherms for the Earth with different surface temperatures. Note that the eclogite stability field is deeper for the higher geotherms and that a partial melt field intervenes between the basaltic crust and the rest of the upper mantle. Basaltic material in the eclogite field will probably sink through the upper mantle and be replaced by peridotite. Shallow subduction of basaltic crust leads to remelting in the case of Venus and the early Earth but conversion to eclogite and deep subduction for the present Earth. The depth scale is for an Earth-size planet with the colder geotherm and present crust and upper-mantle densities. For Venus, with smaller *g*, higher temperatures and low-density crust replacing part of the upper mantle, the depths are increased by about 20 percent.

Venus and is very similar to the inferred density of a fully oxidized (less C and H_2O) chondritic meteorite. The moment of inertia, however, requires an increase in density with depth over and above that due to self-compression and phase changes, indicating the presence of a small core. This in turn indicates that Mars is a differentiated planet.

Mars has long been known as the red planet. Its soils are apparently rich in iron oxides, possibly formed by weathering of an FeO-rich basalt. Models of the mantle of Mars are also rich in FeO.

The tenuous atmosphere of Mars suggests that it either is more depleted in volatiles or has experienced less outgassing than Earth or Venus. It could also have lost much of its early atmosphere by large impacts. Geological evidence for running water on the surface of Mars suggests that a large amount of water is tied up in permafrost and ground water as well as in the polar caps. The northern polar cap of Mars is mainly solid CO_2, while water ice is an important component of the "permanent" southern polar cap.

The high $^{40}Ar/^{36}Ar$ ratio on Mars, 10 times the terrestrial value, suggests either a high potassium-40 content plus efficient outgassing, or a net depletion of argon-36 and, possibly, other volatiles. Early outgassed argon-36 could also have been removed from the planet.

The SNC meteorites, described earlier, have trapped rare-gas and nitrogen contents that differ from other meteorites but closely match those in the martian atmosphere (Bogard and Johnson, 1983; Becker and Pepin, 1984). The discovery of meteorites in Antarctica that appear to have come from the Moon increases the possibility that impacts on other planets can launch fragments that eventually land on the Earth. If SNC meteorites do come from Mars (Dreibus and Wänke, 1985), then a relatively volatile-rich planet is implied, and the atmospheric evidence for a low volatile content for Mars would have to be rationalized by the loss of the early accretional atmosphere. Mars, of course, is more susceptible to atmospheric escape than Venus or Earth owing to its low gravity. A model for Mars based on SNC composition is compared with other models in Table 1-16.

The surface of Mars appears to be basaltic, and the large volcanoes on Mars are similar in form to shield volcanoes on Earth. The composition of the soil is consistent with weathering from basaltic parent materials (Table 1-17).

The topography and gravity field of Mars indicate that parts of Mars are grossly out of hydrostatic equilibrium and that the crust is highly variable in thickness. If variations in the gravity field are attributed to variations in crustal thickness, with a constant density ratio between crust and mantle, then reasonable values of the density contrast imply that

TABLE 1-14
Cosmochemical Model Compositions of Venus

Species	V1	V3
Mantle Plus Crust		
SiO_2	52.9	49.8
TiO_2	0.20	0.21
Al_2O_3	3.8	4.1
Cr_2O_3	—	0.87
MgO	37.6	35.5
FeO	0.24	5.4
MnO	—	—
CaO	3.6	3.3
Na_2O	1.6	0.28
H_2O	0	0.22
K (ppm)	1442	221
U (ppm)	0.020	0.022
Th (ppm)	0.094	0.079
Core		
Fe	94.4	88.6
Ni	5.6	5.5
S	0	5.1
Relative Masses		
Mantle + Crust	69.8	68.0
Core	30.2	32.0

V1: Equilibrium condensation (BVP, 1980).
V3: Morgan and Anders (1980).

the average crustal thickness is at least 30 km (Bills and Ferrari, 1978). This minimal bound is based on the assumption of zero crustal thickness in the Hellas basin. An impact large enough to excavate the Hellas basin would easily remove a 30-km-thick crustal layer. This minimal average crustal thickness on Mars gives a crust/planet mass ratio that is more than five times the terrestrial value (0.4 percent), indicating a well-differentiated planet.

The crust of the Earth is enriched in CaO, Al_2O_3, K_2O, and Na_2O in comparison to the mantle, and ionic radii considerations and experimental petrological results (Chapter 16) suggest that the crust of any planet will be enriched in these constituents. A minimal average crustal thickness for a fully differentiated chondritic planet can be obtained by removing all of the CaO possible, with the available Al_2O_3, as anorthite to the surface. This operation gives a crustal thickness of about 100 km for Mars. Incomplete differentiation and retention of CaO and Al_2O_3 in the mantle will reduce this value, which is likely to be the absolute upper bound. (Earth's crust is much thinner due to crustal recycling and the basalt-eclogite phase change.)

TABLE 1-15
Surface Composition of Venus

	Venera 8	Venera 9	Venera 10
K, percent	4.0 ± 1.2	0.47 ± 0.08	0.30 ± 0.16
U, ppm	2.2 ± 0.7	0.60 ± 0.16	0.46 ± 0.26
Th, ppm	6.5 ± 0.2	3.65 ± 0.42	0.70 ± 0.34
K/U	20,000	8000	7000
Th/U	3.0	6.1	1.5

Surkov (1977).

TABLE 1-16
Model Compositions of Silicate Portions of Mars and Earth

Species	Mars		Earth
	(1)	(2)	(3)
SiO_2 (percent)	44.4	40.0	47.9
TiO_2	0.14	0.1	0.2
Al_2O_3	3.02	3.1	3.9
Cr_2O_3	0.76	0.6	0.9
MgO	30.2	27.4	34.1
FeO	17.9	24.3	8.9
MnO	0.46	0.2	0.14
CaO	2.45	2.5	3.2
Na_2O	0.50	0.8	0.20
P_2O_5	0.16	0.34	0.01
K (ppm)	315	573	200
Rb	112	—	0.4
Zn	74	—	46
Ga	6.6	—	4.6
In (ppb)	14	—	10
Cl (ppm)	44	—	8
Th	0.56	0.077	0.060
U	0.016	0.017	0.020

(1) SNC model (Dreibus and Wänke, 1985).
(2) Meteorite mix model plus geophysical constraints (Anderson, 1972a).
(3) Morgan and Anders (1980).

The average thickness of the crust of the Earth is 15 km, which amounts to 0.4 percent of the mass of the Earth. The crustal thickness is 5–10 km under oceans and 30–50 km under older continental shields. The situation on the Earth is complicated, since new crust is constantly being created at the midoceanic ridges and consumed at island arcs. It is probable that some of the crust is being recycled. If the present rate of crustal genesis was constant over the age of the Earth and none of the crust was recycled, then 17 percent of the Earth would be crustal material.

The Moon apparently has a mean crustal thickness greater than that of the Earth. If the average composition of the Moon is similar to chondrites minus the Fe-Ni-S, then the crust could be as thick as an average of 62 km. This is about the thickness of the crust determined by seismic experiments on the nearside of the Moon but less than the average thickness inferred from the gravity field.

Atmospheric analyses suggest that Mars is a less outgassed planet than Earth, but the high ratio of argon-40 relative to the other inert gases suggests that the martian crust is enriched in potassium-40 in relation to the Earth's crust. The only direct evidence concerning the internal structure of Mars is the mean density, moment of inertia, topography, and gravity field. It is possible to calculate models of the martian interior with plausible chemical models and temperature profiles that satisfy these few constraints; however, the process is highly nonunique (Anderson and others, 1971; Anderson, 1972a, 1977; BVP, 1980).

The mean density of Mars, corrected for pressure, is less than that of Earth, Venus and Mercury but greater than that of the Moon. This implies either that Mars has a small total Fe-Ni content, in keeping with its density, or that the FeO/Fe ratio varies. Plausible models for Mars can be constructed that have solar or chondritic values for iron, if most or all of it is taken to be oxidized (Anderson, 1972a).

With such broad chemical constraints, mean density and moment of inertia and under the assumption of a differentiated planet, it is possible to trade off the size and density of the core and density of the mantle. Most models favor an FeO-enrichment of the martian mantle relative to the mantle of the Earth. Anderson (1972a) concluded that Mars has a total iron content of about 25 weight percent, which is significantly less than the iron content of Earth, Mercury or Venus but is close to the total iron content of ordinary and carbonaceous chondrites. The high zero-pressure density of the mantle suggests a relatively high FeO content in the silicates of the martian mantle. The radius of the core can range from as small as one-third the martian radius for an iron core, or a core similar in composition to the Earth's core, to more than half the radius of the planet if it is pure FeS. With chondritic abundances of Fe-FeS, the size of the

TABLE 1-17
Mars Surface Chemistry

Species	Soil	Igneous Component
SiO_2 (percent)	43	53.3
TiO_2	0.63	0.78
Al_2O_3	7.2	8.9
Fe_2O_3	18.0	2.23
MgO	6	7.4
CaO	5.7	7.1
SO_3	7.6	—
Cl	0.75	—
Rb (ppm)	<30	—
Sr	60 ± 30	—
Y	70 ± 30	—
Zr	30 ± 20	—
K	2500	—
U	1.1 ± 0.8	—

Taylor (1982).

core would be about 45 percent of the planet's radius, or about 12 percent by mass. A small dense core would imply a high-temperature origin or early history because of the high melting temperature of nickel-iron, while a larger light core, presumably rich in sulfur, would allow a cooler early history, since sulfur substantially reduces the melting temperature. A satisfactory model for the interior of Mars can be obtained by exposing ordinary chondrites to moderate temperature, allowing the iron to form a core. The mechanisms for placing oxygen in the core and for reducing FeO to metallic iron would be less effective in Mars than Earth.

We know the equations of state (Chapter 5) and the locations of phase changes (Chapter 16) for most of the materials that might be expected to be important in the interiors of the terrestrial planets. With this information, we can completely define the structure of a two-zone planet (for example, one containing a mantle and a core) in terms of the zero-pressure densities of the mantle and the core and the radius of the core. For Mars two of these parameters can be found as a function of the third. If sulfur is the light alloying element in the core, the density of the core will increase as it grows and become richer in iron because of the nature of the Fe-FeS phase diagram.

The mantle of Mars is presumably composed mainly of silicates, which can be expected to undergo one or two major phase changes, each involving a 10 percent increase in density. To a good approximation, these phase changes will occur at one-third and two-thirds of the radius of Mars. The deeper phase change will not occur if the radius of the core exceeds one-third of the radius of the planet. With these parameters we can solve for the radius and density of the core, given the density of the mantle and the observable mass, radius and moment of inertia for Mars. The results are given in Table 1-18.

The curve in Figure 1-8, giving these results in terms of the density of the core and the radius of the core, is the locus of possible Mars models. Clearly, the data can accommodate a small dense core or a large light core. The upper limit to the density of the core is probably close to the density of iron, in which case the core would be 0.36 of Mars's radius, or about 8 percent of its mass. To determine a lower limit to the density, one must consider possible major components of the core. Of the potential core-forming materials, iron, sulfur, oxygen and nickel are by far the most abundant elements.

The assumption of a chondritic composition for Mars leads to values of the relative radius and mass of the core: $R_c/R = 0.50$ and $M_c/M = 0.21$. The density of the mantle is less than the density of the silicate phase of most ordinary chondrites.

Three kinds of chondrites, HL (high iron, low metal or ornasites), LL (low iron, low metal or amphoterites) and L (low iron or hypersthene-olivine) chondrites, all have lower amounts of potentially core-forming material than is required for Mars, although HL and LL have about the right silicate density (3.38 g/cm³). If completely differentiated, H (high iron) chondrites have too much core and too low a silicate density (3.26–3.29 g/cm³). We can match the properties of H chondrites with Mars if we assume that the planet is incompletely differentiated. If the composition of the core-forming material is on the Fe side of the Fe-FeS eutectic, and temperatures in the mantle are above the eutectic composition, but below the liquidus, then the core will be more sulfur-rich and therefore less dense than the potential core-forming material.

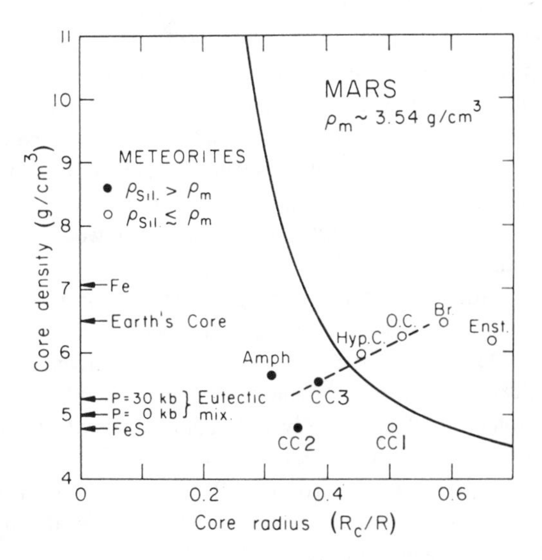

FIGURE 1-8
Radius of the core versus density of core for Mars models. The points are for meteorites with all of the FeS and free iron and nickel differentiated into the core. The dashed line shows how core density is related to core size in the Fe-FeS system (after Anderson, 1972a).

TABLE 1-18
Model Compositions of Mars

Species	Ma1	Ma2	Ma3	Ma4	Ma5
		Mantle Plus Crust			
SiO_2	43.6	43.9	41.6	40.0	41.0
TiO_2	0.16	0.16	0.3	0.1	—
Al_2O_3	3.1	3.2	6.4	3.1	3.2
Cr_2O_3	—	—	0.6	0.6	—
MgO	31.0	31.2	29.8	27.4	34.1
FeO	17.2	16.7	15.8	24.3	16.8
MnO			0.15	0.2	—
CaO	2.9	3.0	5.2	2.5	2.8
Na_2O	1.4	1.4	0.1	0.8	—
H_2O	0.47	0.44	0.001	0.9	—
K (ppm)	1190	1199	69	573	1200
U (ppm)	0.017	0.017	0.033	0.017	0.017
Th (ppm)	0.078	0.078	0.113	0.077	0.079
		Core			
Fe	59.8	60.4	88.1	72.0	—
Ni	5.9	5.8	8.0	9.3	—
S	34.3	33.8	3.5	18.6	—
		Relative Masses			
Mantle + Crust	74.7	74.3	81.0	88.1	85.0
Core	25.3	25.7	19.0	11.9	15.0

Ma1: Equilibrium condensation (BVP, 1980).
Ma2: Equilibrium condensation modified by use of feeding zones (BVP, 1980).
Ma3: Morgan and Anders (1980).
Ma4: Anderson (1972a).
Ma5: McGetchin and Smyth, mantle only (in BVP, 1980).

Carbonaceous chondrites are extremely rich in such low-temperature condensates as H_2O and FeS, as well as carbon and "organic matter." If we ignore the water and carbon components, a fully differentiated planet of this composition would have a core of 15 percent by mass, composed mainly of FeS (13.6 percent FeS, 1.4 percent Ni), and a mantle with a density of about 3.5 g/cm^3. However, these meteorites also contain about 19 percent H_2O, most of which must have escaped if Mars is to be made up primarily of this material. Otherwise, the mantle would not be dense enough.

In ordinary high-iron chondrites, the free iron content averages 17.2 percent by weight. The FeS content is approximately 5.4 percent (3.4 percent Fe, 2.0 percent S) and the nickel content is 1.6 percent. A planet assembled from such material, if completely differentiated, would yield a core of 24 percent of the mass of the planet, with Fe:S:Ni in the approximate proportions of 21:2:2 by weight. Low-iron chondrites would yield a core of 15 percent of the mass of the planet, with proportions of 12:1:2.

Carbonaceous chondrites have little or no free iron but contain 7–25 percent by weight FeS and about 1.5 percent nickel. The average core size for a planet made of carbonaceous chondrites would be 15 percent by mass, Fe:S:Ni being in the proportions 18:10:3. An absolute minimum core density can probably be taken as 4.8 g/cm^3, corresponding to a pure FeS core with a fractional core radius of 0.6 and a fractional mass of 26 percent. On these grounds, the mass of the martian core can be considered to lie between 8 and 24 percent of the mass of the planet.

This range can be narrowed considerably by further consideration of the compositions of meteorites. The points in Figure 1-8 represent most of the major categories of stony meteorites. The size and density of the "core" are computed from the amounts of iron, sulfur, and nickel in the meteorite. No single class of meteorites, fully differentiated into core and mantle, would satisfy the data for Mars, although carbonaceous chondrites and hypersthene (low-iron) chondrites come close. The open circles indicate silicate (mantle) densities less than the inferred density for the mantle of Mars; the closed circles indicate silicate densities that are too high. The meteorites above the curve can be migrated downward and to the left by placing some of the iron of the core in the mantle and thereby increasing the density of the mantle and decreasing the density and radius of the core. Physically, the result would correspond to a meteorite model that has been incompletely differentiated. If chondrites are an appropriate guide to the composition of Mars, possible core sizes would be further restricted to 12–15 percent by mass. Alternatively, the closed circles could be migrated to the locus of possible Mars models by reducing some of the FeO in the silicate phase and allowing the iron to enter the core. This procedure is much more drastic and requires high temperature and, probably, the presence of carbon to effect the reduction. It is interesting that the meteorites in question do contain substantial amounts of carbon. The resulting cores would be about the same size as was previously inferred.

A third possibility would be to assemble Mars from a mixture of meteorites that fall above and below the curve. The meteorites below the curve, however, are relatively rare, although Earth may not be collecting a representative sample.

The size of the core and its density can be traded off. By using the density of pure iron and the density of pure troilite (FeS) as reasonable upper and lower bounds for the density of the core, its radius can be considered to lie between 0.36 and 0.60 of the radius of the planet.

The zero-pressure density of the mantle implies an FeO content of 21–24 weight percent unless some free iron has been retained by the mantle. The presence of CO_2 and H_2O, rather than CO and H_2, in the martian atmosphere suggests that free iron is not present in the mantle.

If chondrites are an appropriate guide to the major-element composition of Mars, the core of Mars is smaller and less dense than the core of Earth, and the mantle of Mars is denser than that of Earth. Mars contains 25–28 percent iron, independent of assumptions about the overall composition or distribution of the iron. Earth is clearly enriched in iron or less oxidized when compared with Mars or most classes of chondritic meteorites.

The lithosphere on Mars is apparently much thicker than on Earth and is capable of supporting large surface loads. The evidence includes the roughness of the gravity field, the heights of the shield volcanoes, the lack of appreciable seismicity and thermal history modeling. There is some evidence that the lithosphere has thickened with time. Olympus Mons is a volcanic construct with a diameter of 700 km and at least 20 km of relief, making it the largest known volcano in the solar system. It is nearly completely encircled by a prominent scarp several kilometers in height and it coincides with the largest gravity anomaly on Mars. The load is apparently primarily supported by thick lithosphere, perhaps greater than 150 km in thickness (Thurber and Toksöz, 1978).

The surface of Mars is much more complex than those of the Moon and Mercury. There is abundant evidence for volcanic modification of large areas after the period of heavy bombardment, subsequent to 3.8 Ga. Mars has a number of gigantic shield volcanoes and major fault structures. In contrast to Mercury there are no large thrust or reverse faults indicative of global contradiction; all of the large tectonic features are extensional. The absence of terrestrial-style plate tectonics is probably the result of a thick cold lithosphere. The youngest large basins on Mars have gravity anomalies suggesting incomplete isostatic compensation and therefore a lithosphere of finite strength (Head and Solomon, 1981).

The data regarding an intrinsic magnetic field for Mars are inconclusive but allow, at most, only a small permanent field in spite of the rapid rotation rate and the probable presence of a dense core. There is no information yet on the physical state of the core.

Mercury

The planet Mercury has a mass of 3.30×10^{26} g and a mean radius of 2444 km, giving a density of 5.43 g/cm^3. Although Mercury is only 5.5 percent of the mass of the Earth, it has a very similar density. Any plausible bulk composition satisfying this density (Table 1-19) is about 60 percent iron. This iron is largely differentiated into a core, because (1) Mercury has a perceptible magnetic field, appreciably more than either Venus or Mars; and (2) Mercury's surface has the appearance of being predominantly silicate. A further inference is that the iron core existed early in its history; a late core-formation event would have resulted in a significant expansion of Mercury. The presence of an internally generated magnetic field implies that the iron core is at least partially fluid.

Mercury's shape may have significantly changed over the history of the planet. Tidal despinning results in a less oblate planet and compressional tectonics in the equatorial regions. Cooling and formation of a core both cause a change in the mean density and radius. A widespread system of arcuate scarps on Mercury, which appear to be thrust

TABLE 1-19
Cosmochemical Model Compositions of Mercury
(Weight Percent or ppm)

Species	Me4	Me3
	Mantle Plus Crust	
SiO_2	43.5	47.1
TiO_2	—	0.33
Al_2O_3	4.7	6.4
Cr_2O_3	—	3.3
MgO	47.7	33.7
FeO	0	3.7
MnO	—	0.06
CaO	4.1	5.2
Na_2O	0	0.08
H_2O	—	0.016
K (ppm)	0	69
U (ppm)	0.026	0.034
Th (ppm)	0.12	0.122
	Core	
Fe	94.5	93.5
Ni	5.5	5.4
S	0	0.35
	Relative Masses	
Mantle + Crust	35.2	32.0
Core	64.8	68.1

Me3: Morgan and Anders (1980).
Me4: Refractory condensate mixing model (Wood, 1962).

faults, provides evidence for compressional stresses in the crust (Solomon, 1977). The absence of normal faults, the result of extension, suggest that Mercury has contracted, perhaps by as much as 2 km in radius. This perhaps is evidence for cooling of the interior.

The primary factor affecting the bulk composition of Mercury is the probably high temperature in its zone of the solar nebula, so that it is formed of predominantly high-temperature condensates. If the temperature was held around 1300 K until most of the uncondensed material was blown away, then a composition satisfying Mercury's mean density can be obtained, since most of the iron will be condensed, but only a minor part of the magnesian silicates. Since the band of temperatures at which this condition prevails is quite narrow, other factors must be considered. Two of these are (1) dynamical interaction among the material in the terrestrial planet zones, leading to compositional mixing, and (2) collisional differentiation.

If the composition of Mercury is controlled by high-temperature condensation, it is unlikely to have significant SiO_2 beyond that necessary to combine with the MgO, Al_2O_3, and CaO to make forsterite, diopside, and anorthite. Any such model of Mercury must have a ratio of Fe + Ni to Al_2O_3 + CaO + MgO + SiO_2 higher than calculated from cosmic abundances to satisfy its mean density (BVP).

Mercury probably has an Al:Mg ratio appreciably higher than cosmic but an Al:Fe ratio somewhat lower, leading to an anorthite content on the order of 5–10 percent of its mass. If Mercury had a differentiation efficiency similar to that of the Moon (the nearest terrestrial planet in terms of size), it would have a crust approximately 75 to 150 km thick.

The areal density and size distribution of craters on the surface of Mercury are similar to lunar highland values. Both Moon and Mercury, and by inference the other terrestrial bodies, were subjected to a high flux of impacting objects in early planetary history. The end of the high-flux period can be dated from lunar studies at about 3.8 billion years ago. The large basins on the surface of Mercury probably formed during the period of high bombardment. Later cooling and contraction apparently were responsible for global compression of the outer surface and may have shut off volcanism. Tidal despinning may have affected the distribution of fault scarps over the surface.

COMPARATIVE PLANETOLOGY

Before the advent of space exploration, Earth scientists had a handicap almost unique in science: They had only one object to study. Compare this with the number of objects available to astronomers, particle physicists, biologists and sociologists. Earth theories had to be based almost entirely on evidence from Earth itself. Although each object in the solar system is unique, we have learned some lessons that can now be applied to Earth.

Study of the Moon, Mars and the basaltic achondrites demonstrated that basaltic volcanism is ubiquitous, even on very small bodies. Planets apparently form, or become, hot and begin to differentiate at a very early stage in their evolution, probably during accretion. Although primitive objects, such as the carbonaceous chondrites, have survived for the age of the solar system, there is no evidence for the survival of primitive material once it has been in a planet. One would hardly expect large portions of the Earth to have escaped this planetary differentiation.

The magma ocean concept was developed to explain the petrology and geochemistry of the Moon. It will prove fruitful to apply this to the Earth, taking into account the differences required by the higher pressures on the Earth.

We now know that the total crustal volume on the Earth is anomalously small, but it nevertheless contains a large fraction of the terrestrial inventory of incompatible elements.

The difference in composition of the atmospheres of the terrestrial planets shows that the original volatile compositions, the extent of outgassing or the subsequent processes of atmospheric escape have been quite different. The importance of great impacts in the early history of the planets is now clear; it has implications ranging from early, dense, insulating atmospheres, atmospheric escape and siderophile and volatile composition to formation of the Moon and angular momentum of the planets.

We now know that plate tectonics, at least the recycling kind, is unique to Earth. The thickness and average temperature of the lithosphere and the role of phase changes in basalt seem to be important. Any theory of plate tectonics must explain why the other terrestrial planets do not behave like Earth.

The compositions of lunar KREEP and terrestrial kimberlite are so similar that a similar explanation is almost demanded.

These lessons, some of which are just hints, should be kept in mind as we turn our attention to Earth.

References

Anders, E. (1968) Chemical processes in the early solar system, as inferred from meteorites, *Acct. Chem. Res.*, *1*, 289–298.

Anders, E. and M. Ebihara (1982) Solar system abundances of the elements, *Geochim. Cosmochim. Acta*, *46*, 2363–2380.

Anders, E. and T. Owen (1977) Mars and Earth: Origin and abundance of volatiles, *Science*, *198*, 453–465.

Anderson, D. L. (1972a) The internal composition of Mars, *J. Geophys. Res.*, *77*, 789–795.

Anderson, D. L. (1972b) Implications of the inhomogeneous planetary accretion hypothesis, *Comments on Earth Sciences: Geophysics*, *2*, 93–98.

Anderson, D. L. (1977) Composition of the mantle and core, *Ann. Rev. Earth Planet. Sci., 5,* 179–202.

Anderson, D. L. (1980) Tectonics and composition of Venus, *Geophys. Res. Lett., 7,* 101–102.

Anderson, D. L. (1981) Plate tectonics on Venus, *Geophys. Res. Lett., 8,* 309–311.

Anderson, D. L., C. G. Sammis and T. H. Jordan (1971) Composition and evolution of the mantle and core, *Science, 171,* 1103–1112.

BVP, Basaltic Volcanism Study Project (1980) *Basaltic Volcanism on the Terrestrial Planets,* Pergamon, New York, 1286 pp.

Becker, R. and R. Pepin (1984) The case for a martian origin of the shergottites, *Earth Planet. Sci. Lett., 69,* 225–242.

Bills, B. G. and A. J. Ferrari (1978) Mars topography harmonics and geophysical implications, *J. Geophys. Res., 83,* 3497–3507.

Bogard, D. D. and P. Johnson (1983) Martian gases in an Antarctic meteorite? *Science, 221,* 651–654.

Breneman, H. H. and E. C. Stone (1985) Solar coronal and photospheric abundances from solar energetic particle measurements, *Astrophys. J. (Letters),* December 1.

Cameron, A. G. W. (1982) Elementary and nuclidic abundances in the solar system. In *Essays in Nuclear Astrophysics* (C. A. Barnes, et al., eds.), Cambridge University Press, Cambridge.

Clark, S. P., K. K. Turekian and L. Grossman (1972) Model for the early history of the Earth. In *The Nature of the Solid Earth* (E. C. Robertson, ed.), 3–18, McGraw-Hill, New York.

Donahue, T., V. Hoffman, R. Hodges, Jr. and A. Watson (1982) Venus was wet: A measurement of the ratio of deuterium to hydrogen, *Science, 216,* 630–633.

Dreibus, G. and H. Wänke (1985) Mars, a volatile-rich planet, *Meteoritics, 20,* 367–381.

Dreibus, G., H. Kruse, B. Spettel and H. Wänke (1977) The bulk composition of the Moon and the eucrite parent body. *Proc. Lunar Sci. Conf. 8th,* 211–227.

Dreibus, G. and H. Wänke (1979) On the chemical composition of the Moon and the eucrite parent body and a comparison with the composition of the Earth, the case of Mn, Cr, and V (abstract). In *Lunar and Planetary Science X,* 315–317, Lunar and Planetary Institute, Houston.

Fuchs, L. H., E. Olsen and K. J. Jensen (1973) Mineralogy, mineral-chemistry, and composition of the Murchison (C2) meteorite, *Smithsonian Contrib. Earth Sci.* No. 10, 39 pp.

Ganapathy, R. and E. Anders (1974) Bulk compositions of the Moon and Earth estimated from meteorites, *Proc. Lunar Sci. Conf., 5,* 1181–1206.

Grevesse, N. (1984) *Physica Scripta T8, 49.*

Grossman, L. (1972) Condensation in the primitive solar nebula, *Geochim. Cosmochim. Acta, 36,* 597–619.

Grossman, L. and J. W. Larimer (1974) Early chemical history of the solar system, *Rev. Geophys. Space Phys., 12,* 71–101.

Hanks, T. and D. L. Anderson (1969) The early thermal history of the Earth, *Phys. Earth Planet. Interiors, 2,* 19–29.

Head, J. and S. Solomon (1981) Tectonic evolution of the terrestrial planets, *Science, 213,* 62–76.

Lange, M. A. and T. J. Ahrens (1984) FeO and H_2O and the homogeneous accretion of the earth, *Earth and Planetary Science Letters, 71,* 111–119.

Larimer, J. W. (1967) Chemical fractionations in meteorites—I. Condensation of the elements, *Geochim. Cosmochim. Acta, 31,* 1215–1238.

Mason, B. (1962) *Meteorites,* John Wiley & Sons, New York, 274 pp.

Mazor, E., D. Heymann and E. Anders (1970) Noble gases in carbonaceous chondrites, *Geochim. Cosmochim. Acta, 34,* 781–824.

Morgan, J. W. and E. Anders (1980) Chemical composition of the Earth, Venus, and Mercury, *Proc. Natl. Acad. Sci., 77,* 6973.

Phillips, R., W. Kaula, G. McGill and M. Malin (1980) Tectonics and evolution of Venus, *Science, 212,* 879–887.

Pollack, J. and D. Black (1979) Implications of the gas compositional measurements of Pioneer Venus for the origin of planetary atmospheres, *Science, 205,* 56–59.

Ringwood, A. E. (1977) *Composition and Origin of the Earth.* Publication No. 1299, Research School of Earth Sciences, Australian National University, Canberra.

Safronov, V. S. (1972) Accumulation of the planets. In *On the Origin of the Solar System* (H. Reeves, ed.), 89–113, Centre Nationale de Recherche Scientifique, Paris.

Solomon, S. C. (1977) The relationship between crustal tectonics and internal evolution in the Moon and Mercury, *Phys. Earth Planet. Inter., 15,* 135–145.

Surkov, Yu. A. (1977) Geochemical studies of Venus by Venera 9 and 10 automatic interplanetary stations, *Proc. Lunar Sci. Conf., 8,* 2665–2689.

Taylor, S. R. (1982) *Planetary Science, a Lunar Perspective,* Lunar and Planetary Institute, Houston, 482 pp.

Thurber, C. and M. N. Toksöz (1978) Martian lithospheric thickness from elastic flexure theory, *Geophys. Res. Lett., 5,* 977–980.

Turekian, K. K. and S. P. Clark (1969) Inhomogeneous accretion of the Earth from the primitive solar nebula, *Earth Planet. Sci. Lett., 6,* 346–348.

von Zahn, V., S. Kumar, H. Niemann and R. Prinn (1983) Composition of the Venus atmosphere. In *Venus* (D. M. Hunten, L. Colin, T. Donahue and V. Moroz, eds.), 299–430, University of Arizona Press, Tucson.

Wacker, J. and K. Marti (1983) Noble gas components of Albee meteorite, *Earth Planet. Sci. Lett., 62,* 147–158.

Wänke, H., H. Baddenhausen, K. Blum, M. Cendales, G. Dreibus, H. Hofmeister, H. Kruse, E. Jagoutz, C. Palme, B. Spettel, R. Thacker and E. Vilcsek (1977) On chemistry of lunar samples and achondrites; Primary matter in the lunar highlands; A re-evaluation, *Proc. Lunar Sci. Conf. 8th,* 2191–2213.

Weidenschilling, S. J. (1976) Accretion of the terrestrial planets. II, *Icarus, 27,* 161–170.

Wood, J. A. (1962) Chondrules and the origin of the terrestrial planets, *Nature, 194,* 127–130.

2

Earth and Moon

Strange all this difference should be
Twixt tweedle-dum and tweedle-dee.

—JOHN BYROM

BULK COMPOSITION OF THE EARTH

Recent compilations of cosmic abundances (Table 2-1) agree fairly closely for the more significant rock-forming elements. From these data a simple model can be made (Table 2-2) of the Earth's bulk chemical composition.

The mantle is nearly completely oxidized and is therefore composed of compounds primarily of MgO, SiO_2, Al_2O_3, CaO, Na_2O and Fe-oxides. The major minerals in rocks from the upper mantle are

$$MgO + SiO_2 = MgSiO_3 \text{ (enstatite)}$$

and

$$2MgO + SiO_2 = Mg_2SiO_4 \text{ (forsterite)}$$

Iron occurs in various oxidation states, FeO, Fe_2O_3, Fe_3O_4, and typically replaces about 10 percent of the MgO in the above minerals:

$$(Mg_{0.9}Fe_{0.1})\ SiO_3 \text{ (orthopyroxene)}$$

$$(Mg_{0.9}Fe_{0.1})_2\ SiO_4 \text{ (olivine)}$$

These minerals can be considered as solid solutions between $MgSiO_3$ and $FeSiO_3$ (ferrosilite) and Mg_2SiO_4 and Fe_2SiO_4 (fayalite). Another important solid-solution series is between MgO (periclase) and FeO (wüstite) giving (Mg,Fe)O (magnesiowüstite or ferropericlase). These oxides usually combine with SiO_2 to form silicates, but they may occur as separate phases in the lower mantle.

The approximate composition of the mantle and the relative sizes of the mantle and core can be determined from the above consideration. Oxygen is combined with the major rock-forming elements, and then the relative weight fractions can be determined from the atomic weights. The preponderance of silica (SiO_2) and magnesia (MgO) in mantle rocks gave rise to the term "sima," an archaic term for the chemistry of the mantle. In contrast, crustal rocks were called "sial" for silica and alumina.

The composition in the first column in Table 2-2 is based on Cameron's (1982) cosmic abundances. These are converted to weight fractions via the molecular weight and renormalization. The Fe_2O requires some comment. Based on cosmic abundances, it is plausible that the Earth's core is mainly iron; however, from seismic data and from the total mass and moment of inertia of the Earth, there must be a light alloying element in the core. Of the candidates that have been proposed (O, S, Si, N, H, He and C), only oxygen and silicon are likely to be brought into a planet in refractory solid particles—the others are all very volatile elements and will tend to be concentrated near the surface or in the atmosphere. The hypothetical high-pressure phase Fe_2O has about the right density to match core values. There is also the possibility that the oxidation state of iron decreases with pressure from Fe^{3+} to Fe^{2+} to Fe^{+} to Fe^{0}. At high pressure FeO is soluble in molten iron, giving a composition similar to Fe_2O at core pressures (Ohtani and Ringwood, 1984).

If most of the iron is in the core, in Fe_2O proportions, then the mass of the core will be 30 to 34 weight percent of the planet. The actual mass of the core is 33 percent, which is excellent agreement. There may also, of course, be some sulfur, carbon, and so on in the core, but little or none seems necessary. The FeO content of the mantle is less than 10 percent so that does not change the results much. The mass of the core will be increased by about 5 percent if it

TABLE 2-1
Short Table of Cosmic Abundances (Atoms/Si)

Element	Cameron (1982)	Anders and Ebihara (1982)
O	18.4	20.1
Na	0.06	0.057
Mg	1.06	1.07
Al	0.085	0.0849
Si	1.00	1.00
K	0.0035	0.00377
Ca	0.0625	0.0611
Ti	0.0024	0.0024
Fe	0.90	0.90
Ni	0.0478	0.0493

has a cosmic Ni/Fe ratio, and will be larger still if the newer estimates of solar iron abundances are used, unless the iron content of the mantle is greater than is evident from upper-mantle rocks. The new solar values for Fe/Si give 15% weight percent FeO for the mantle if the core is 32% with an Fe_2O composition. This FeO content is similar to estimates of the composition of the lunar and Martian mantles.

Note that the cosmic atomic ratio Mg/Si is 1.06. This means that the mantle (the iron-poor oxides or silicates) is predicted to be mainly pyroxene, $MgSiO_3$, rather than olivine, Mg_2SiO_4, which has a Mg/Si ratio of 2 (atomic). The presence of FeO in the mantle means that slightly more olivine can be present. It is the (Fe + Mg)/Si ratio that is important.

Garnet also has an Mg/Si ratio near 1, although garnet usually has more FeO than do olivine and pyroxene. This suggests that much of the mantle is composed of pyroxene and garnet, rather than olivine, which is the main constituent of the shallow mantle. The CaO and Al_2O_3 are contained in garnet and clinopyroxene and the Na_2O in the jadeite molecule at upper mantle pressures. Seismic data are consistent with a mainly "perovskite" lower mantle. $MgSiO_3$ transforms to a perovskite structure, an analog of $CaTiO_3$, at lower-mantle pressures. $MgSiO_3$ "perovskite" is therefore the dominant mineral of the mantle.

TABLE 2-2
Simple Earth Model Based on Cosmic Abundances

Oxides	Molecules	Molecular Weight	Grams	Weight Fraction
MgO	1.06	40	42.4	0.250
SiO_2	1.00	60	60.0	0.354
Al_2O_3	0.0425	102	4.35	0.026
CaO	0.0625	56	3.5	0.021
Na_2O	0.03	62	1.84	0.011
Fe_2O	0.45	128	57.6	0.339
Total			169.7	1.001

The crust is rich in SiO_2 and Al_2O_3 (thus "sial"), CaO and Na_2O. The Al_2O_3 and CaO and Na_2O content of the "cosmic" mantle totals to 5.8 percent. The crust represents only 0.5 percent of the mantle. Thus, there must be considerable potential crust in the mantle. The thinness of the terrestrial crust, compared to the Moon and Mars and relative to the potential crust, is partially related to the presence of plate tectonics on Earth, which continuously recycles crustal material. It is also due to high pressures in the Earth, which convert crustal materials such as basalt to high-density eclogite.

Estimates of upper-mantle compositions based on peridotites are deficient in SiO_2, TiO_2, Al_2O_3 and CaO relative to cosmic abundances. Basalts, picrites and eclogites are enriched in these oxides. Cosmic ratios of the refractory oxides can be obtained with a mixture of approximately 80 percent peridotite and 20 percent eclogite. If the lower mantle is mainly $MgSiO_3$ ("perovskite"), then a much smaller eclogitic, or basaltic, component is required to balance the Mg/Si ratio of upper-mantle peridotites. A chondritic Earth contains about 6 to 10 percent of a basaltic component.

EVOLUTION OF THE EARTH'S INTERIOR

The evolution of Earth, its outgassing and its differentiation into a crust, mantle and core depend on whether it accreted hot or cold and whether it accreted homogeneously or inhomogeneously. It now appears unavoidable that a large fraction of the gravitational energy of accretion was trapped by the Earth, and it therefore started life as a hot body relative to the melting point. The processes of differentiation, including core formation, were probably occurring while the Earth was accreting, and it is therefore misleading to talk of a later "core formation event." Melting near the surface leads to a zone refining process with light melts rising to the surface and dense melts and residual crystals sinking toward the center. What is not so clear is whether most of the Earth accreted homogeneously or inhomogeneously. A certain amount of inhomogeneity in the chemistry of the accreting material is required, otherwise reactions between H_2O and free iron would oxidize all the iron and no water would exist at the surface. This can be avoided by having more iron accrete in the early stages and more H_2O in the later stages. Iron metal, of course, condenses earlier in a cooling nebula, and because of its density and ductility may have formed the earliest planetesimals and perhaps the initial nuclei of the planets.

In earlier theories of Earth evolution, it was assumed that cold volatile-rich material similar to type I carbonaceous chondrites accreted to form a homogeneous planet, perhaps with some reduction and vaporization at the surface to form reduced iron, which subsequently warmed up and differentiated into an iron-rich core and a crust. In these theories it proved difficult to transport molten iron to the center of the planet because of the effect of pressure on the melting point: Molten iron in the upper mantle would freeze before it reached the lower mantle. A central iron-rich nucleus mixed with or surrounded by refractory-rich material including aluminum-26, uranium and thorium would alleviate this thermal problem as well as the iron-oxidation problem. The presence of an ancient magnetic field, as recorded in the oldest rocks, argues for a sizeable molten iron-rich core early in Earth's history. The abundance of siderophile elements in the upper mantle also suggests that this region was not completely stripped of these elements by molten iron draining to the core.

The early and extensive melting and differentiation of the Moon and some meteorite parent bodies attests to the importance of melting and differentiation of small bodies in the early solar system. This melting may have been caused by the energy of accretion or the presence of extinct, short-lived radioactive nuclides. In some cases, as in present-day Io, tidal pumping is an important energy source. Isotopic studies indicate that distinct geochemical reservoirs formed in the mantle early in its history. Magmas from these reservoirs retain their isotopic identity, and this proves that the mantle has not been homogenized in spite of the fact that it has presumably been convecting throughout its history. A chemically stratified mantle is one way to keep reservoirs distinct and separate until partial melting processes allow separation of magmas, which rise quickly to the surface.

Zone refining during accretion and crystallization of a deep magma ocean are possible ways of establishing a chemically zoned planet (Figure 2-1). At low pressures basaltic melts are less dense than the residual refractory crystals, and they rise to the surface, taking with them many of the trace elements. The refractory crystals themselves are also less dense than undifferentiated mantle and tend to concentrate in the shallow mantle. At higher pressure there is a strong likelihood that melts become denser than the crystals they are in equilibrium with. Such melts, trapped at depth and insulated by the overlying rock, may require considerable time to crystallize.

Most scientists agree that simplicity is a desirable attribute of a theory. Simplicity, however, is in the eye of the beholder. The end results of natural processes can be incredibly complex, even though the underlying principles may be very simple. A uniform, or homogeneous, Earth is probably the "simplest" theory, but it violates the simplest tests. Given the most basic observations about the Earth, the next simplest theory is a three-part Earth: homogeneous crust, homogeneous mantle and homogeneous core. How the Earth might have achieved this simple state, however, involves a complex series of ad hoc mechanisms to separate core and crust from the primitive mantle and then to homogenize the separate products.

On the other hand, one can assume that the processes

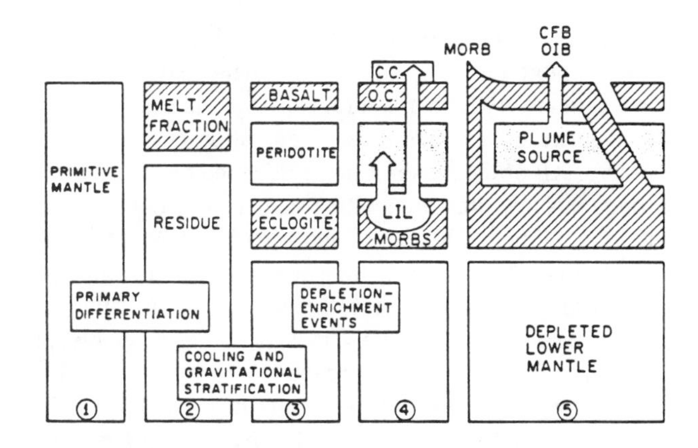

FIGURE 2-1
A model for the evolution of the mantle. Primitive mantle (1) is partially molten either during accretion or by subsequent whole-mantle convection, which brings the entire mantle across the solidus at shallow depths. Large-ion lithophile (LIL) elements are concentrated in the melt. The deep magma ocean (2) fractionates into a thin plagioclase-rich surface layer and deeper olivine-rich and garnet-rich cumulate layers (3). Late-stage melts in the eclogite-rich cumulate are removed (4) to form the continental crust (C.C.), enrich the peridotite layer and deplete MORBS, the source region of oceanic crust (O.C.) and lower oceanic lithosphere. Partial melting of the plume source (5) generates continental flood basalts (CFB), ocean-island basalts (IOB) and other enriched magmas, leaving a depleted residue (harzburgite) that stays in the upper mantle. The MORB source may be the main basalt reservoir in the mantle, and enriched or hot-spot magmas (OIB, CFB) may be MORB contaminated by interaction with shallow mantle.

that formed and differentiated the Earth were uniform and ask what the resulting Earth would look like. For example, the crust is made of the lighter and more fusible materials and was formed as a result of igneous differentiation involving the upward migration of light melts. The core is an iron-rich alloy that also melts at low temperature but drains to the interior because of its high density. This suggests a simple hypothesis: The stratification of the Earth is a result of gravitational separation of materials according to their melting points and density. The materials accreting to form the Earth *may* have been uniform, but the high temperatures associated with accretion, even a violent accretion, would result in a chemically differentiated planet. Thus, a simple, even obvious, process gives a complex result.

The mantle itself is neither expected to be nor observed to be homogeneous. Let us follow this single, simple hypothesis further. As the Earth grows, the crustal elements are continuously concentrated into the melts and rise to the surface. When these melts freeze, they form the crustal minerals that are rich in silicon, calcium, aluminum, potassium and the large-ion lithophile (LIL) elements. Melts generally are also rich in FeO compared to primitive material. This plus the high compressibility of melts means that the densities of melts and residual crystals converge, or even cross, as the pressure increases. Melt separation is therefore more difficult at depth, and melts may even drain downward at very high pressure. However, during accretion the majority of the melt-crystal separation occurs at low pressure. All of the material in the deep interior has passed through this low-pressure melting stage in a sort of continuous zone refining. The magnesium-rich minerals, Mg_2SiO_4 and $MgSiO_3$, have high melting temperatures and are fed through the melting zone into the interior. Even if the accreting material is completely melted during assembly of the Earth, these minerals will be the first to freeze, and they will still separate from the remaining melt. $(Mg,Fe)_2SiO_4$ has a slightly higher melting or freezing point than $(Mg,Fe)SiO_3$ and a slightly higher density, so that olivine may even separate from orthopyroxene. This separation, however, is expected to be much less effective than the separation of olivine and orthopyroxene from the melts rich in SiO_2, Al_2O_3, CaO, Na_2O and K_2O which differ substantially from these minerals in both density and melting point. The downward separation of iron-rich melts, along with nickel, cobalt, sulfur and the trace siderophile elements, strips these elements out of the crust and mantle. The Al_2O_3, CaO and Na_2O content in chondritic and solar material is adequate to form a crust some 200 km thick. The absence of such a massive crust on the Earth might suggest that the Earth has not experienced a very efficient differentiation. On the other hand, the size of the core and the extreme concentration of the large-ion, magmaphile elements into the crust suggest that differentiation has been extremely efficient.

The solution to this apparent paradox does not require special pleading. At pressures corresponding to depths of the order of 50 km, the low-density minerals of the crust convert to a mineral assemblage denser than olivine and orthopyroxene. Most of the original crust therefore is unstable and sinks into the mantle. Any melts below 200–400 km may suffer the same fate. Between 50 and 500 km the Al_2O_3-CaO-Na_2O-rich materials crystallize as clinopyroxene and garnet, a dense eclogite assemblage that is denser than peridotite. Eclogite transforms to a garnet solid solution, which is still denser and which is stable between about 500 and 800 km. Peridotite also undergoes a series of phase changes that prevent eclogite from sinking deeper than about 650 km. However, when the Earth was about Mars' size and smaller, the eclogite could sink to the core-mantle boundary.

The base of the present mantle, a region called D″, is anomalous and may be composed of high-pressure eclogite. The earliest stages of accretion probably involved the most refractory materials under nebular conditions. These include compounds rich in Fe and CaO, Al_2O_3 and TiO_2. The refractory lithophiles would have been excluded from a molten Fe-rich core, and these may also be concentrated in D″. In either case D″ would be enriched in Al_2O_3 and CaO. Equilibration between material in D″ and the core may also result in a high FeO content for this region. If D″ is intrinsically denser than the rest of the lower mantle, it would be gravitationally stable at the base of the mantle. On the other hand, it is embedded in the thermal boundary layer between mantle and core and therefore has a high temperature that may locally permit D″ material to rise into the lower mantle until it becomes neutrally buoyant. As it cools, it will sink back to D″.

The end result for a planet experiencing partial melting, gravitational separation and phase changes is chemical stratification. The possibility of three "basaltic" regions (high CaO and Al_2O_3 and possibly FeO, relative to MgO) has been identified. These regions are the crust, the transition region (between upper and lower mantle) and D″. The latter two may be the result of solid subduction or sinking of high-density melts.

The bulk of the upper mantle and lower mantle may therefore be sandwiched between basalt-rich layers. If melting during accretion extended below some 300 km, the composition of the melt and residual refractory phases changes. Orthopyroxene transforms to majorite, a garnet-like phase that replaces olivine on the liquidus. Melts, therefore, are MgO rich, and we have another mechanism for separating major elements in the mantle and concentrating olivine (Mg_2SiO_4) in the shallow mantle. Giant impacts in early Earth history have the potential for melting mantle to great depth and for concentrating dense refractory residual phases such as orthopyroxene-majorite in the lower mantle.

The subsequent cooling and crystallization of the Earth introduces additional complications. A chemically stratified mantle cools more slowly than a homogeneous Earth. Phase change boundaries are both temperature and pressure dependent, and these migrate as the Earth cools. A thick basalt crust, stable at high temperature, converts to eclogite at its base as it cools through the basalt-eclogite phase boundary. The initial crust of the Earth, or at least its deeper portions, therefore can become unstable and plunge into the mantle. This is an effective way to cool the mantle and to displace lighter and hotter material to the shallow mantle where it can melt by pressure release, providing a continuous mechanism for bringing melts to the surface.

The separation of melts and crystals is a process of differentiation. Convection is often thought of as a homogenization process, tantamount to stirring. Differentiation, however, can be irreversible. Melts that are separated from the mantle when the Earth was smaller or from the present upper mantle crystallize to assemblages that have different phase relations than the residual crystals or original mantle material. If these rocks are returned to the mantle, they will not in general have neutral buoyancy, nor are they necessarily denser than "normal" mantle at all depths. Eclogite, for example, is denser than peridotite when the latter is in the olivine, β-spinel and γ-spinel fields but is less dense than the lower mantle, and it transforms to dense perovskite-bearing assemblages at higher pressure than peridotite.

ORIGIN OF THE CRUST

Although the crust represents less than 0.5 percent of the mass of the Earth, it contains a large fraction of the elements that preferentially enter the melt when a silicate is melted—the large-ion lithophile elements (LIL). For example, it can be estimated that the continental crust contains 58 percent of the rubidium, 53 percent of the cesium, 46 percent of the potassium, 37 percent of the barium and 35 percent of the uranium and thorium in the crust-mantle system. Other highly concentrated elements include bismuth (34 percent), lead (32 percent), tantalum (30 percent), chlorine (including that in seawater, 26 percent), lanthanum (19 percent), and strontium (13 percent). These high concentrations in such a small volume mean that the Earth is an extensively differentiated body. Apparently, most or all of the mantle has been processed by partial melting and upward melt extraction. It has also been estimated that 77 percent of the argon-40 produced by the decay of potassium-40 in the mantle and crust resides in the atmosphere. This also points toward a well-differentiated, and outgassed, Earth. It is possible that most of the H_2O is in the ocean and the crust.

The age of the oldest continental crust is at least 3.8×10^9 years, and perhaps as old as 4.2×10^9 years; isotopic results from mantle rocks are also consistent with ancient differentiation. Yet the mean age of the continental crust is 1.5×10^9 years (Jacobsen and Wasserberg, 1979). The energy of accretion of the Earth is great enough to melt a large fraction of the incoming material, therefore the processes of melting, crust formation and outgassing were probably contemporaneous with accretion. The absence of older crust may reflect extensive bombardment in the later stages of accretion and a high-temperature crust and upper mantle rather than a late onset of the crust-forming process. A stable crust may have been delayed by the freezing of a deep magma ocean.

Most of the continental crust formed between 2.5 and 3 Ga ago. In more recent times a small amount of continental crust has been added by accretion of island arcs, oceanic islands and plateaus and by continental flood basalts. Some crustal material is eroded, subducted and recycled into the mantle, but the net effect is nearly constant crustal volume over the past 1–2 Ga. Estimates of growth rates of the continental crust are 0.1–0.5 km^3/yr at present, 0.3–0.7 km^3/yr over the past 2.5 Ga and 5.7–6.6 km^3/yr at the end of the Archean, or approximately ten times the present rate (McLennan and Taylor, 1980).

Isotopic ages of the crust suggest that crustal growth has been episodic with major additions in the time intervals 3.8–3.5, 2.8–2.5, 1.9–1.6, 1.6–1.2, 1.2–0.9 and 0.5–0 Ga. The composition of the crust has varied with time, most abruptly at the Archean-Proterozoic boundary. In particular there are decreases in chromium and nickel and increases in REE, thorium, uranium, Th/U and $^{87}Sr/^{86}Sr$ (see Weaver and Tarney, 1984, and McLennan and Taylor, 1980, for reviews).

The formation of the original crust is probably linked to the formation and evolution of a magma ocean. However, it is harder to form a crust on a magma ocean than on a watery ocean because a silicate crust is probably denser than the liquid it freezes from. Feldspar crystals of sufficient size might be able to break away from convection and float, but they only form after extensive freezing and grow only slowly. The density contrast is also relatively low. Gas-crystal packets, or foams, may provide sufficient buoyancy even though the individual crystals are small. The interstitial fluids formed in deep olivine-orthopyroxene cumulates may be the source of the protocrust—the crust may have formed by the freezing of light fluids that rose to the surface of the magma ocean. Certainly, the crust is extremely enriched in the incompatible elements, and forming it from a liquid from which the refractory crystals have been removed is one way to account for this enrichment.

The major additions to the continental crust in recent times are due to the lateral accretion of island arcs. The average composition of the continental crust may therefore be similar to the average island-arc andesite (Taylor and McLennan, 1981). The andesite model can be used to estimate the composition of the lower crust if it is assumed that andesite melts during orogenesis to form a granodioritic upper crust and a refractory, residual lower crust. The total average crust, in these models, is average island-arc andesite, and the upper-crust composition is approximated by the average post-Archean sediment. There is some evidence, however, that island arcs are more basic than andesite. Seismic data also suggest that the average crust is closer to quartz diorite and granodiorite than to andesite; that is, it is more siliceous. Heat flow suggests that the lower crust is rather depleted in the radioactive heat-producing elements (K, U, Th). An alternative model for continental crust chemistry was derived by Weaver and Tarney (1984) using a variety of xenolith and geophysical data. Both models are listed in Table 2-3.

ORIGIN OF THE MANTLE AND CORE

Larimer (1967) has outlined the condensation history of a cooling gas of cosmic composition. Compounds such as $CaTiO_3$, $MgAl_2O_4$, Al_2SiO_5, and $CaAl_2Si_2O_3$ condense first at temperatures between 1740 and 1620 K (see Figure 1-6). Iron condenses next at 1620 K. Magnesium-rich pyroxenes and olivines condense between 1740 and 1420 K; FeS condenses at 680 K and H_2O at 210 K. All the above temperatures were calculated on the assumption of a total pressure of 6.6×10^{-3} atmosphere. Larimer and Anders (1970) concluded that the fractionation patterns in meteorites occurred in the solar nebula as it cooled from high temperatures and could not be produced in the meteorite parent bodies. Especially, they inferred the following accretion temperatures from the abundance patterns: carbonaceous chondrites, 400 K or below; enstatite and ordinary chondrites, 400 to 650 K; and the major fraction of iron meteorites, 1100 K or above.

Accretion of the planets presumably involved planetesimals that condensed over the entire temperature range, although the different planets may have incorporated different proportions of the various condensates. As a planet grows, the accretional energy increases and the temperature at the surface of the body is controlled by a balance between the available gravitational or kinetic energy, the heat capacity and thermal conductivity of the surface layer, the heats of reaction involved in chemical reactions occurring at the surface, and the reradiation of energy to the dust-gas cloud. As the planet grows, it will be less and less capable of retaining the volatiles brought in by the accreting particles, and the fractionation, in time, will be roughly the reverse of the condensation procedure. Refractories enter the planet and the volatiles contribute to the atmosphere. When the surface temperature reaches 680 K, the reaction $FeS + H_2 \rightarrow Fe + H_2S$ will occur if the planet is accreting in a H_2-rich environment; the planet will, thereafter, not be able to incorporate much FeS into its interior.

Reactions such as

$$Mg_3Si_4(OH)_2O_{10} \rightleftarrows 3MgSiO_3 + SiO_2 + H_2O$$

$$M(OH)_2 \rightleftarrows MO + H_2O \quad (M = Mg,Ca,Fe)$$

$$CH_4 \rightleftarrows C + 2H_2$$

$$FeS_2 + MgSiO_3 + H_2O \rightleftarrows FeS + FeMgSiO_4 + H_2S$$

between the atmosphere and the surface of the accreting planet buffer the surface temperature and keep it below 500 K for long periods of time during the accretion process. Substantial amounts of heat, buried in the planet by the impacting bodies, are unavailable for reradiation if the rate of accretion is faster than the rate of heat conduction to the surface. The above considerations prolong the period available for trapping FeS in the interior. Eventually, however, dense impact-generated atmospheres serve to keep the surface insulated and hot during most of the accretion process.

Implicit in the above discussion is the assumption that the planets did not start to accrete until the condensation process was complete and that the solar nebula was relatively cold. A growing planet would incorporate the solid portion of available material—that is, the "local cosmic abundances"—in its interior in its initial stages of growth but could retain only the more refractory compounds as it grew. A large-scale redistribution of material must occur later in its history in order to form a core and to transport the retained volatiles and the less refractory compounds from the center of the body to the surface to form the crust and atmosphere. Thus the Earth, for example, must have turned itself inside out to obtain its present configuration.

TABLE 2-3
Estimates of Average Bulk Composition of the Continental Crust

Species	A	B	C
SiO_2 (percent)	58.0	63.7	57.3
TiO_2	0.8	0.5	0.9
Al_2O_3	18.0	15.8	15.9
FeO	7.5	4.7	9.1
MnO	0.14	0.07	—
MgO	3.5	2.7	5.3
CaO	7.5	4.5	7.4
Na_2O	3.5	4.3	3.1
K_2O	1.5	2.0	1.1
P_2O_5	—	0.17	—
Rb (ppm)	42	55	32
Sr	400	498	260
Th	4.8	5.1	3.5
U	1.25	1.3	0.91
Pb	10	15	8

A: Andesite model (Taylor and McLennan, 1985).
B: Amphibolite-granulite lower crustal model (Weaver and Tarney, 1984).
C: Theoretical model (Taylor and McLennan, 1985).

If the planets accreted while condensation was taking place, a chemically zoned planet would result, with compounds that condense above 1620 K, such as $CaTiO_3$, $MgAl_2O_4$, Al_2SiO_5, $CaAl_2Si_2O_8$, and iron, forming the central part of the body. In the next 160 K range of cooling, compounds such as Ca_2SiO_4, $CaSiO_3$, $CaMgSi_2O_6$, $KAlSi_3O_8$, $MgSiO_3$, SiO_2, and nickel condense. Between 1420 and 1200 K the rest of the important mantle minerals, $MgSiO_4$, $NaAlSi_3O_8$, and $(Na,K)_2SiO_3$ condense. Further cooling would bring in various metals (Cu, Ge, Au, Ga, Sn, Ag, Pb), the sulfides of zinc, iron and cadmium, and, finally, Fe_3O_4 (400 K), the hydrated silicates ($\sim$300 K), H_2O (210 K), and the rare gases.

In both of the above models of accretion, the composition of the planet changes as it grows, and there is no compelling reason to assume that the composition of the various planets should be the same or the same as any single class of meteorites. Carbonaceous chondrites probably represent the last material to condense from a cooling nebula rather than representative planetary material. In the cold-accretion model, this material would be mixed with material that condensed at higher temperatures, such as ordinary chondritic and iron meteorites, to form the nucleus of the accreting planet, which would contain most of the volatiles such as H_2O and FeS. As the planet grows, the material that can be retained on the surface and incorporated into the interior becomes progressively less representative of the material available.

The amount of volatiles and low-temperature minerals retained by a planet depends on the fraction of the planet that accretes during the initial and terminal stages. In the accretion-during-condensation model, the full complement of volatiles and sulfides must be brought in during the terminal stage. In both of these models much of the material that forms the core (Fe + FeS in the cold-accretion model and Fe in the condensation model) is already near the center of the Earth. In alternative proposals the core-forming material is either distributed evenly throughout the mantle or at the surface of the accreting planet.

One important boundary condition for the formation of the Earth is that the outer core, now and probably 3 Ga ago, be molten. If we take the temperature in the mantle to be 1880°C at 620-kilometer depth and assume an adiabatic temperature gradient of 0.5°C per kilometer from that depth to the core-mantle boundary, plus 1000° across D″, the temperature at the top of the core is about 4000°C, which is of the order of the melting temperature of iron at core pressures. For the condensation model the temperature of the primitive iron core is 1150° to 1350°C and increases to 2300° to 2500°C, owing to adiabatic compression, as the planet assumes its present size. These temperatures can be compared with estimates of about 7000°C for the melting temperature of pure iron at the boundary of the inner core. These temperatures are all subject to great uncertainty, but it appears difficult with the condensation model to have a molten iron core. It is even more difficult to raise the central part of Earth above the melting temperature of pure iron in the cold-accretion model. It appears that the extra component in the core must also serve to decrease its melting temperature. This is easily accomplished with sulfur and, possibly, with oxygen as well, particularly if the core behaves as a eutectic system. An Fe-O-S core probably melts at a lower temperature than a two-component core. The presence of siderophiles in the upper mantle and water at the surface also provides constraints on the mechanism and timing of core formation.

The eutectic temperature for the system Fe-FeS is 990°C and is remarkably insensitive to pressure up to at least 30 kbar (Usselman, 1975). The eutectic composition is 31 percent by weight of sulfur at 1 atmosphere and 27 percent by weight of sulfur at 30 kbar. In an Earth of meteoritic composition, a sulfur-rich iron liquid would be the first melt to be formed. Core formation could proceed under these conditions at a temperature some 600°C lower than would be required to initiate melting in pure iron. If the Earth accreted from cold particles under conditions of radiative equilibrium, the temperatures would be highest in the upper mantle, which is where melting would commence. In most plausible thermal history calculations, the melting point of iron increases with depth much more rapidly than the actual temperatures, and any sinking molten iron will refreeze. Unless the Earth was very hot during most of the accretional process, it is difficult to get the iron to the center of the planet. This difficulty does not occur for the iron-sulfur model of the core, since gravitational accretion energy, adiabatic compression, and radioactive heating bring the temperatures throughout most of the Earth above 1000°C early in its history and, probably, during most of the accretional process. Since the eutectic temperature increases only slowly with pressure, core formation is self-sustaining. The increase of gravitational energy due to core formation leads to a temperature rise of about 1600° to 2000°C throughout the Earth (Birch, 1965). Such temperatures would be adequate to melt the rest of the iron in the mantle, which would drain into the core, and to cause extensive melting of silicates and differentiation of the crust and upper mantle. The short time scale of accretion (or giant impacts) required by Hanks and Anderson (1969) to form a molten iron core within the first 10^9 years of earth history is not required for a sulfur-rich core.

Even if most of the light element in the core is oxygen, as seems probable, the above suggests that some sulfur will be present. This would serve to decrease the melting point of the core and would help explain the excessive depletion of sulfur in the crust-mantle system. An Fe-O-S core is likely, but there are few data on this system at high pressure.

One of the early criticisms of inhomogeneous accretion for the Earth was that it did not seem possible to melt the initial metallic core after it was accreted and buried. Attempts to do this by the thermal history of accretion were unsatisfactory. It was pointed out that the condensation sequence predicts a refractory calcium-aluminum-rich initial condensate before the condensation of metallic iron (Anderson and Hanks, 1972). Therefore, there would be calcium-aluminum-rich silicates accreting with the metallic iron, and both the theoretical calculations and measurements of the Allende inclusions show this initial condensate to be enriched in uranium and thorium as well as other heavy metals. Therefore, the inhomogeneous accretion model also predicts a long-lived heat source within the initial core. The short-lived nuclide aluminum-26, if present, could dominate the early thermal history and be responsible for melting of the core. Although uranium and thorium are unlikely to be in the present core, they may be concentrated at the core-mantle boundary, if held in sufficiently dense silicates.

The extensive melting of the Earth by giant planet-sized impactors would remove many of the problems addressed above.

How might the core of Mars have formed? The average temperature in Mars below some 600 km is about 1500°C if the planet accreted in 3×10^5 years (Hanks and Anderson, 1969), which is well above the Fe-FeS eutectic temperature. The liquidus for an FeS-rich FeS-Fe system is 1100°C at atmospheric pressure, 1600°C at 60 kbar, and about 3400°C at the center of Mars. Thus, most of the deep interior of Mars is closer to the eutectic temperature than to the liquidus temperature, and only a fraction of the planet's complement of iron and FeS will be molten. For a carbonaceous chondrite I composition, the total FeS plus iron content is 24 percent by weight. The core of a small planet of that composition would therefore be richer in sulfur than the core of the Earth, and much FeS would be left in the mantle. The small size of the martian core and the small size of the planet suppress the importance of gravitational heating due to core formation. If the density of the martian core is bracketed by the density of Earth's core and by the density of the eutectic mix in the system Fe-FeS at 30 kbar, then the core can be 9 to 15 percent, by mass, of the planet. A satisfactory model for Mars can be obtained by melting most of the FeS and some of the iron in ordinary chondrites (Anderson, 1971). Most of the arguments for putting substantial amounts of oxygen into the core do not apply to Mars.

MINERALOGY OF THE MANTLE

The composition of the mantle is conventionally given in terms of oxides (SiO_2, MgO, Al_2O_3 and so forth), and these oxides are the building blocks of mantle minerals. The main minerals in the upper mantle are olivine, orthopyroxene, clinopyroxene and garnet with minor ilmenite and chromite. Clinopyroxene is basically a solid solution of $CaMgSiO_3$ (diopside) and $NaAlSiO_2O_6$ (jadeite). Al_2O_3 is mainly in garnet, clinopyroxene and spinel, $MgAl_2O_4$. These minerals are only stable over a limited pressure and temperature range. The mineralogy of the mantle therefore changes with depth due to solid-solid phase changes. The elastic properties and density of the mantle are primarily controlled by the proportions of the above minerals or their high-pressure equivalents. To a lesser extent the physical properties depend on the compositions of the individual minerals.

Several cosmochemical estimates of mineralogy are given in Table 2-4. These tend to be less rich in olivine than estimates of the composition of the upper mantle (column 4). The mantle is unlikely to be uniform in composition, and therefore the mineralogy will change with depth, not only because of solid-solid phase changes, but also because of intrinsic chemical differences.

Although Al_2O_3, CaO and Na_2O are minor constituents of the average mantle, their presence changes the mineralogy, and this in turn affects the physical properties such as the seismic velocities. The effect on density can result in chemical stratification of the mantle and concentration of these elements into certain layers. They also influence the melting point and tend to be concentrated in melts.

Olivine is an essential component in most groups of meteorites except the irons. Pallasites are composed of nickel-iron and olivine. Olivine is a major constituent in all the chondrites except the enstatite chondrites and some of the carbonaceous chondrites, and it is present in some achondrites. For these reasons olivine is usually considered to be the major constituent of the mantle. The olivines in meteorites, however, are generally much richer in Fe_2SiO_4 than mantle olivines. The olivine compositions in chondrites generally lie in the range 19 to 24 mole percent Fe_2SiO_4 (fayalite or Fa) but range up to Fa_{31}. The achondrites have even higher fayalite contents in the olivines. The seismic velocities in the mantle would be too low and the densities would be too high if the olivine were so iron rich. Mantle olivines are typically Fa_{10}. Removal of iron from meteoritic olivine, either as Fe or FeO, would decrease the olivine content of the mantle:

$$Fe_2SiO_4 \text{ (olivine)} + H_2 \rightarrow H_2O + Fe + FeSiO_3 \text{ (orthopyroxene)}$$

$$Fe_2SiO_4 \text{ (olivine)} + CO \rightarrow CO_2 + Fe + FeSiO_3$$

$$Fe_2SiO_4 \rightarrow FeO + FeSiO_3$$

TABLE 2-4
Mineralogy of Mantle

Species	Whole—Mantle Models (1)	(2)	(3)	Upper Mantle (4)
Olivine	47.2	36.5	37.8	51.4
Orthopyroxene	28.3	33.7	33.2	25.6
Clinopyroxene	12.7	14.6	11.8	11.0
Jadeite	9.8	2.2	1.8	0.65
Ilmenite	0.2	0.5	0.24	0.57
Garnet	1.53	11.6	14.2	9.6
Chromite	0.0	1.6	0.94	0.44

(1) Equilibrium condensation (BVP, 1980).
(2) Cosmochemical model (Ganapathy and Anders 1974).
(3) Cosmochemical model (Morgan and Anders, 1980).
(4) Pyrolite (Ringwood, 1977).

Earth models based on cosmic abundances, with Mg/Si approximately 1 (molar), give relatively low total olivine contents. For example, the model of Morgan and Anders (1980) yields 37.8 mole percent olivine. Since the upper mantle appears to be olivine rich, this results in an even more olivine-poor transition region and lower mantle. It is usually assumed, however, that the basaltic fraction of the Earth is still mostly dispersed throughout the mantle. This is the assumption behind the pyrolite model (Ringwood, 1977) discussed in Chapter 8. Basalts were probably liberated during accretion of the Earth and concentrated in the upper mantle. Efficient remixing or rehomogenization can be ruled out with the isotopic evidence for ancient isolated reservoirs.

A given mineral is stable only over a restricted range of temperature and pressure. As pressure is increased, the atoms rearrange themselves, and ultimately a new configuration of atoms is energetically favorable, usually with a denser packing (see Chapter 16). Some of the phases of common mantle minerals are described below. The formulas and cation coordinations are given in Table 2-5.

TABLE 2-5
Formulas and Coordination of Mantle Minerals

Formula	Species
$^{IV}Mg\ ^{VI}Al_2O_4$	Spinel
$^{VI}Mg\ ^{IV}SiO_3$	Enstatite
$^{VIII}Mg_3\ [^{VI}Al]_2\ ^{IV}Si_3O_{12}$	Pyrope
$^{VIII}Mg_3[^{VI}Mg\ ^{VI}Si]\ ^{IV}Si_3O_{12}$	Majorite
$^{VI}Mg\ ^{VI}SiO_3$	"Ilmenite"
$^{VIII-XII}Mg\ ^{VI}SiO_3$	"Perovskite"
^{VI}MgO	Periclase
$^{VI}Mg_2\ ^{IV}SiO_4$	Olivine
$^{VI}Mg_2\ ^{IV}SiO_4$	β-spinel
$^{VI}Mg_2\ ^{IV}SiO_4$	γ-spinel
$^{VIII}Ca\ ^{VI}Mg\ ^{IV}Si_2O_6$	Diopside
$^{VIII}Na\ ^{VI}Al\ ^{IV}Si_2O_6$	Jadeite
$^{VI}SiO_2$	Stishovite

$(Mg,Fe)_2SiO_4$–Olivine

The olivine structure is based on a nearly hexagonal-closest packing of oxygen ions:

$$^{VI}Mg_2\ ^{IV}SiO_4$$

The MgO_6 octahedra are not all equivalent. Each $Mg(1)O_6$ octahedron shares four edges with adjacent Mg-octahedra and two edges with Si-tetrahedra. Each $Mg(2)O_6$ octahedron shares two edges with adjacent Mg-octahedra and one edge with a Si-tetrahedron. The $Mg(1)O_6$ octahedra are linked to each other by sharing edges, making a chain along the *c* axis. The $Mg(2)O_6$ octahedra are larger and more distorted than the $Mg(1)O_6$ octahedra. Olivine is formed of separate $(SiO_4)^{4-}$ groups. The four oxygen atoms surrounding a silicon atom are not linked to any other silicon atom. Olivine is a very anisotropic mineral and apparently is easily aligned by recrystallization or flow, making peridotites anisotropic as well.

$(Mg,Fe)_2SiO_4$–Spinel

In 1936 Bernal suggested that common olivine (α) might transform under high pressure to a polymorph possessing the spinel ($^{VI}Al_2\ ^{IV}MgO_4$) structure with an increase of density of about nine percent. Jeffreys (1937) adopted this suggestion to explain the rapid increase in seismic velocity at a depth of 400 km—"the 20° discontinuity" so named because the effect is seen in an increase of apparent velocity of seismic waves at an arc distance of 20° from a seismic event. Many A_2BO_4 compounds have now been transformed to the spinel structure at high pressure, and natural olivine does indeed transform at about the right temperature and pressure to explain the 400-km discontinuity. Although Fe_2SiO_4 transforms directly to the spinel structure, olivines having high MgO contents occur in two modifications related to the spinel structure, β-spinel and γ-spinel. Both of these have much higher elastic wave velocities than appropriate for the mantle just below 400 km, so there must be other components in the mantle that dilute the effect of the $\alpha-\beta$ phase change. Although normal spinel is a structural analog to silicate spinel, it should be noted that Mg occurs in 4-coordination in $MgAl_2O_4$ and in 6-coordination in β- and γ-spinel with a consequent change in the Mg–O distance and the elastic properties. ^{IV}Mg is an unusual coordination for Mg, and the elastic properties and their derivatives cannot be assumed to be similar for ^{IV}MgO and ^{VI}MgO compounds. In particular, normal spinel has an unusually low pressure derivative of the rigidity, a property shared by other 4-coordinated compounds but not ^{VI}MgO-bearing compounds.

The spinel structure consists of an approximate cubic close packing of oxygen anions. In the *normal* spinel structure, exhibited by many $A_2^{2+}B^{4+}O_4$ compounds, we have $^{VI}A_2\ ^{IV}BO_4$. In *inverse* spinels one of the A^{2+} ions occupies the tetrahedral site, and the octahedral sites are shared by A^{2+} and B^{4+} ions. Intermediate configurations are possible.

β-Spinel is crystallographically orthorhombic, but the oxygen atoms are in approximate cubic close packing. It is sometimes referred to as the β-phase or the distorted or modified spinel structure. It is approximately 7½ percent denser than α-phase, or olivine. The arrangements of the A and B ions are different from those in spinel. Two SiO_4 tetrahedra, which would be isolated in the spinel structure, share one of their oxygen atoms, resulting in a Si_2O_7 group and an oxygen not bonded to any Si. β-Spinel is an elastically anisotropic mineral. The transformation to γ-spinel results in a density increase of about 3½ percent with no overall change in coordination.

Garnets

Garnets are cubic minerals of composition

$$^{VIII}M_3^{2+}\ ^{VI}Al_2^{3+}\ ^{IV}Si_3O_{12}$$

where M is Mg, Fe^{2+} or Ca for the common garnets, but can be almost any 2+ element. Some natural garnets have Cr^{3+} or Fe^{3+} instead of Al^{3+}. Garnets are stable over an enormous pressure range, reflecting their close packing and stable cubic structure. They are probably present over most of the upper mantle and, perhaps, into the lower mantle. Furthermore, they dissolve pyroxene at high pressure, so their volume fraction expands with pressure. Garnets are the densest common upper-mantle mineral, and therefore eclogites and fertile (undepleted) peridotites are denser than basalt-depleted peridotites or harzburgites. On the other hand, they are less dense than other phases that are stable at the base of the transition region. Therefore, eclogite can become less dense than other rock types at great depth. Garnet has a low melting point and is eliminated from peridotites in the upper mantle at small degrees of partial melting. The large density change associated with partial melting of a garnet-bearing rock is probably one of the most important sources of buoyancy in the mantle. Garnets are elastically isotropic.

$MgSiO_3$–Majorite

Pyroxene enters the garnet structure at high pressure via the substitution

$$^{VI}[MgSi] \rightarrow Al_2$$

The mineral

$$^{VIII}Mg_3\ ^{VI}[MgSi]\ ^{IV}Si_3O_{12}$$

is known as majorite, and it exhibits a wide range of solubility in garnet. Note that one-fourth of the Si atoms are in 6-coordination. However, the elastic properties of MgO plus SiO_2-stishovite are similar to Al_2O_3, so we expect the elastic properties of majorite to be similar to garnet. Majorite also has a density similar to that of garnet.

$MgSiO_3$–Ilmenite

The structural formula of "ilmenite," the hexagonal high-pressure form of enstatite, is

$$^{VI}Mg\ ^{VI}SiO_3$$

making it isostructural with true ilmenite ($FeTiO_3$) and corundum (Al_2O_3). Silicon is also 6-coordinated in stishovite and perovskite, dense high-elasticity phases. Because of similarity in ionic radii, we expect that extensive substitution of $^{VI}Al_2$ for $^{VI}(MgSi)$ is possible. $MgSiO_3$–ilmenite is a platy mineral, suggesting that it may be easily oriented in the mantle. It is also a very anisotropic mineral, rivaling olivine in its elastic anisotropy.

The arrangement of oxygen atoms is based on a distorted hexagonal closest packing having a wide range of O–O distances.

"Ilmenite" is a stable phase at the base of the transition region and the top of the lower mantle. The transformations γ-spinel plus stishovite, or majorite, to "ilmenite" and "ilmenite" to "perovskite" (see below) may be responsible for the high gradients of seismic velocity found in this depth range. "Ilmenite" appears to be a lower temperature phase than majorite and appears to occur mainly in deeply subducted slabs, contributing to the high velocities found in the vicinity of deep-focus earthquakes.

$MgSiO_3$–Perovskite

The structural formula of the high-pressure phase of enstatite, or "perovskite," is

$$^{VIII-XII}Mg\ ^{VI}SiO_3$$

It appears to be stable throughout the lower mantle and is therefore the most abundant mineral in the mantle. Related structures are expected for $CaSiO_3$ and garnet at high pressure. Olivine also transforms to a "perovskite"-bearing assemblage

$$MgSiO_3\text{–perovskite} + MgO$$

$MgSiO_3$–perovskite is 3.1 percent denser than the isochemical mixture stishovite plus periclase. There are a variety of Mg–O and Si–O distances. The mean Si–O distance is similar to that of stishovite.

The structure is orthorhombic and represents a distortion from ideal cubic perovskite. In the ideal cubic perovskite the smaller cation has a coordination number of 6, whereas the larger cation is surrounded by 12 oxygens. In $MgSiO_3$–perovskite the 12 Mg–O distances are divided into 4 short distances, 4 fairly long distances and 4 intermediate distances, giving an average distance appropriate for a mean coordination number of 8.

MOON

The Moon is one of the more obvious of our neighbors in space and is certainly the most accessible. In spite of intensive analysis and probing by virtually every conceivable chemical and physical technique, the maneuvering room for speculation on lunar origin was scarcely diminished as a result of the Apollo and Luna programs. This was not primarily due to lack of information but to the unexpected and confusing nature of the new data, most of which were open to multiple interpretations.

Composition and Structure

The Moon's unique characteristics have become even more unique as a result of lunar exploration. It is like no other body in the solar system that we know about. It has long been known that the Moon is deficient in iron (in comparison with the proportions of iron in the Earth and the other terrestrial planets). It is also deficient in all elements and compounds more volatile than iron.

The density of the Moon is considerably less than that of the other terrestrial planets, even when allowance is made for pressure. Venus, Earth and Mars contain about 30 percent iron, which is consistent with the composition of stony meteorites and the nonvolatile components of the Sun. They therefore fit into any scheme that has them evolve from solar material. Mercury is even better endowed with iron. Because iron is the major dense element occurring in the Sun, and presumably in the preplanetary solar nebula, the Moon is clearly depleted in iron. Many theories of lunar origin have been based on this fact, and numerous attempts have been made to explain how iron can be separated from other elements and compounds. Density, magnetic properties and ductility have all been invoked to rationalize why iron should behave differently than silicates in early solar-system processes.

Once samples were returned from the Moon, however, it became clear that the Moon was deficient not only in iron but also in a number of other elements as well (Figure 2-2). A common characteristic of many of these elements and their compounds is volatility. The returned samples showed that the Moon is depleted in compounds more volatile than iron. Calcium, aluminum and titanium are the major elements involved in high-temperature condensation processes in the solar nebula; minor refractory elements include barium, strontium, uranium, thorium and the rare-earth elements. The Moon is enriched in all these elements, and we are now sure that more than iron-silicate separation must be involved in lunar origin.

The abundance of titanium in the returned lunar samples was one of the first surprises of the Apollo program. Titanium is not exactly rare on Earth, but it is usually considered a minor or trace element. The first samples returned from the Moon contained 10 percent of titanium-rich compounds. The surface samples were also remarkably depleted in such volatile elements as sodium, potassium, rubidium, and other substances that, from terrestrial and laboratory experience, we would expect to find concentrated in the crust, such as water and sulfur. The refractory trace elements—such as barium, uranium and the rare-earth elements—are concentrated in lunar surface material to an extent several orders of magnitude over that expected on the basis of cosmic or terrestrial abundances.

Some of these elements, such as uranium, thorium, strontium and barium, are large-ion elements, and one would expect them to be concentrated in melts that would be intruded or extruded near the surface. However, other volatile large-ion elements such as sodium and rubidium are clearly deficient, in most cases, by at least several orders of magnitude from that expected from cosmic abundances. The enrichment of refractory elements in the surface rocks is so pronounced that several geochemists proposed that refractory compounds were brought to the Moon's surface in great quantity in the later stages of accretion. The reason behind these suggestions was the belief that the Moon, overall, must resemble terrestrial, meteoritic or solar material and that it was unlikely that the whole Moon could be enriched in refractories. In these theories the volatile-rich materials must be concentrated toward the interior. In a

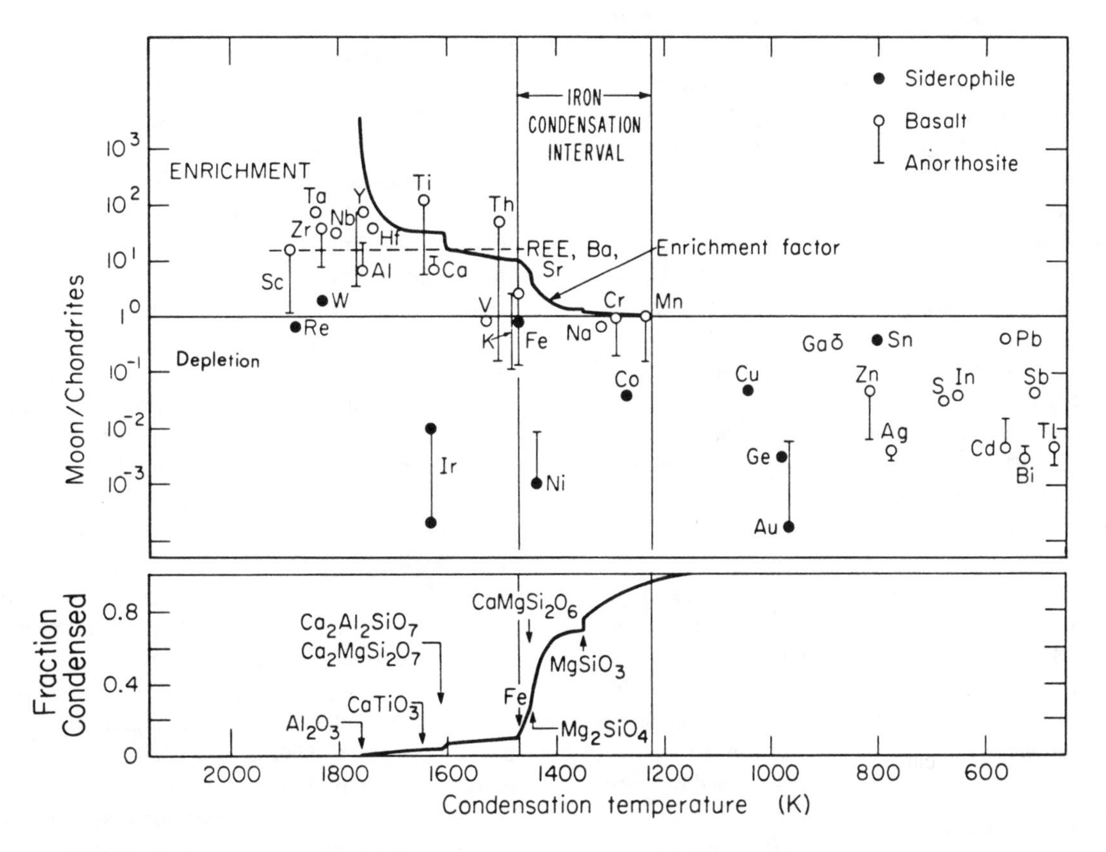

FIGURE 2-2
Chondrite-normalized lunar abundances as a function of condensation temperature of the element or the first condensing compound containing this element at 10^{-3} atm total pressure. Ratios greater than unity indicate enrichment and less than unity indicate depletion (after Anderson, 1973b).

cooling-nebula model of planetary formation, the refractories condense before the volatiles, and it was therefore implied that the Moon was made inside out!

However, it now appears that the depletion of iron and volatiles can be taken at face value and that the whole Moon is deficient in elements and compounds more volatile than iron. Petrological considerations show that not only the surface rocks but also their igneous source regions, deep in the Moon, are also depleted in volatiles. The Moon is probably enriched in calcium, aluminum, titanium and the refractory trace elements throughout. This composition would explain the mean density of the Moon and the high heat flow, and it would help to explain why the Moon melted and differentiated very rapidly. However, if the Moon's origin is linked to a high-energy impact on the Earth, we have a ready explanation both for an initially molten Moon and a silicate and refractory-rich Moon.

In view of the abundant geological evidence that the surface rocks resulted from melting processes in the interior, it was no surprise that the geophysical evidence indicated that the Moon has a low-density crust. Its great age, as measured by geochemical techniques, and great thickness, as required by the physical evidence, were, however, unexpected. These are important boundary conditions on the origin and composition of the Moon and, probably, the Earth as well.

The Moon's principal moments of inertia indicate that crustal thickness varies from about 40 km at the poles to more than 150 km on the lunar farside. Large variations in crustal thickness are also required to satisfy gravity data and the noncoincidence of the centers of mass and figure of the Moon. The present orientation of the Moon and the restriction of basalt-filled maria to the Earth-facing hemisphere are undoubtedly the result of this asymmetry in crustal thickness.

The maria are remarkably smooth and level; slopes of less than one-tenth of a degree persist for hundreds of kilometers, and topographic excursions from the mean are generally less than 150 m. By contrast, elevation differences in the highlands are commonly greater than 3 km. The mean altitude of the terrace, or highlands, above maria is also about 3 km.

The center of mass is displaced toward the Earth and slightly toward the east by about 2 km. This gross asymmetry of the Moon has long been known from consideration of the principal moments of inertia. The differences between the principal moments of inertia are more than an order of magnitude greater than can be accounted for by a simple homogeneous body, rotating and stretched by Earth tides. The simplest interpretation is in terms of a crust of highly variable thickness, an interpretation supported by nearside gravity results.

Asymmetry is not a unique characteristic of the Moon; the asymmetric distribution of continents and oceans on the Earth is well known, and Mars, likewise, is very asymmetric in both its topography and its gravity field. Large-scale convection associated with early gravitational differentiation could lead to the observed asymmetries and may be one common characteristic of all the terrestrial planets.

In the case of the Earth and the Moon, and probably for Mars as well, the physical asymmetry correlates well with, and is probably the result of, chemical asymmetry. The lunar highlands are dominantly plagioclase-feldspar-rich rocks with densities considerably less than the nearside mare basalts and the mean density of the Moon. These feldspars crystallize at higher temperatures than basalt and can therefore be expected to float to the surface of their parent liquid. The residual liquids would likely be the source region of the mare basalts, which erupt to the surface later.

This scenario explains not only the physical measurements but also some subtle details of the chemistry. Large-ion refractory elements are preferentially retained by the liquid, and therefore such elements as barium, strontium, uranium and thorium would be concentrated in the last liquid to crystallize. These elements are concentrated in the lunar-mare basalts by several orders of magnitude over the highland plagioclase-rich material, with the notable exception of europium, which is retained by plagioclase. Compared to the other rare-earth elements, europium is depleted in basalts and enriched in anorthosites. The "europium anomaly" was one of the early mysteries of the lunar sample-return program and implied that plagioclase was abundant somewhere on the Moon. The predicted material was later found in the highlands.

Seismic activity of the Moon is much less than the Earth's, both in numbers of quakes and their size, or magnitude. Their times of occurrence appear to correlate with tidal stresses caused by the varying distance between the Moon and the Earth. Compared with the Earth, they seem to occur at great depth, about half the lunar radius. Both the age-dating evidence and the seismic data indicate that the Moon today is a relatively inactive body. This conclusion is consistent with the absence of obvious tectonic activity and with the low level of stresses in the lunar interior implied by gravity and moment-of-inertia data.

The outer kilometer has extremely low seismic velocities, less than 1 km/s. This value is more appropriate for rubble than consolidated rock. Velocities increase to 4 km/s at 1 km depth and 6 km/s at 20 km; the lower velocity is appropriate for consolidated rubble or extensively fractured igneous rock, such as basalt. The increase of velocity with depth is the result of consolidation and crack closure. The 6-km/s velocity is consistent with laboratory measurements on returned samples of lunar basalt.

At 20 km depth the velocity increases abruptly to about 6.7 km/s, and it remains relatively constant to 60 km depth. The constancy of velocity means that most cracks have been eliminated and also that the effects of temperature and pressure are either small or mutually canceling. In this region the velocities can be matched by anorthositic gabbro, a plagioclase-rich rock type that has a low iron content and relatively low density (about 2.9 g/cm^3). This layer may be similar in composition to the lunar highlands.

At 60 km the velocity jumps, at least locally, to about 9 km/s. When all the seismic data are considered together, we may find that this layer is very thin (less than 40 km) or it occurs only locally, or both. Perhaps it occurs only as pods or lenses under maria basins or only under mascon basins; at the moment we have no way of telling. In any event, such high velocities are unexpected and are unusual by any standards. They may also be fictitious, because seismic waves refracted by dipping interfaces can give apparent velocities slower or faster than real velocities. A velocity of 9 km/s is much greater than the 8 km/s velocity typical of the Earth's upper mantle and of rocks thought to be common in the upper lunar mantle. Only a few minerals, exotic by terrestrial standards, have such high velocities. These include spinel ($MgAl_2O_4$), corundum (Al_2O_3), kyanite (Al_2SiO_5) and Ca-rich garnet ($Ca_3Al_2Si_3O_{12}$). These are all calcium- and/or aluminum-rich minerals and occur as the dense residual crystals when a Ca-Al-rich liquid partially solidifies. The 9-km/s layer may therefore be related petrologically to the overlying crustal layer.

The apparent seismic velocity at greater depth is only 7.7 km/s—intermediate between the velocities we usually associate with the crust and the mantle. This velocity continues to a depth of at least 150 km (the deepest depth of penetration of seismic energy from artificial impacts to date) and is appropriate for pyroxene or for plagioclase-rich (meaning crustal) rocks. In the latter case we would still be monitoring the crust and therefore would have only a lower bound, 150 km, on its thickness. It should be recalled that the crust is thicker on the farside. Even if the crust is only 60 km thick, the conventional interpretation of the seismic results, it is much thicker than the average terrestrial crust, particularly in relative terms; this great thickness indicates that the Moon is extensively differentiated. In combination with the age data this means that the Moon was extensively melted early in its history. The source of this early heat is a matter of some controversy. Forming the Moon from debris resulting from a giant impact on Earth is the current "best idea."

Evidence for the constitution of the deeper interior is very sketchy. Seismic shear waves apparently cannot pass efficiently below some 1000 km depth. This can be taken as tentative evidence for a hot, if not molten or partially molten, deep interior. One of the long-standing controversies regarding the Moon is whether its interior is hot or cold. The widespread occurrence of basalt certainly indicates that it was at least partially molten early in its history. Conduction alone is only efficient in lowering the internal temperatures of the outer 300 km.

The lunar crust represents about 10 percent of the lunar volume. The mare basalts constitute about one percent of the crust. The average seismic velocities are about 7.7 km/s for *P*-waves and 4.45 km/s for *S*-waves down to about 1000 km. There may be a minor discontinuity at a depth of 400–490 km in the Moon, with a drop in velocity below that depth range. Most of the moonquakes occur between 800 and 1000 km depth. There is some evidence for increased attenuation of seismic waves below 1100 km (Goins et al, 1981). The geophysical data place a weak upper bound of about 500 km radius on the size of a dense, conductive core. The depletion of iridium in lunar rocks is consistent with either removal by iron draining to the core or an overall depletion of siderophiles. The moment of inertia, 0.391, is consistent with a density increase in the interior over and above the crust-mantle density increase. This could be due to a large eclogite core or a small iron-rich core.

The anorthositic highlands are probably also the result of igneous and differentiation processes. But their emplacement mechanism would be plagioclase flotation, resulting in a thick insulating blanket for the remaining liquid. The age of maria material indicates that melts still existed at moderate depths for more than 10^9 years after creation of the plagioclase-rich highlands.

It is quite possible that the mare-forming igneous episodes were a result of thermal, tidal and impact stresses, all of which were intense in the earliest history of the Moon. Igneous activity may have ceased when stresses were no longer adequate to breach the thick lithosphere. If this is true, then the Moon below some 300 to 1000 km may still be partially molten. A lithosphere of this thickness could easily support the stresses implied by the nonhydrostatic shape of the Moon and the presence of mass concentrations.

The nonequilibrium shape of the Moon, the offset of the center of mass from the center of figure and the presence of large surface concentrations of mass (mascons) have been used as arguments that the Moon is a cold, strong body. However, when viewed more carefully, all of this evidence suggests just the contrary. The stresses required to support the nonequilibrium shape are only some tens of bars, and a relatively thin, strong, outer layer would suffice to support these stresses. The Earth, by contrast, supports stress differences of hundreds of bars, and stresses at kilobar levels are required to break rocks in the laboratory. Thus, taken at face value, the lunar data suggest that the Moon is a hot weak body. This conclusion is consistent with the lunar heat-flow values, the low level of seismic stresses and the high radioactivity inferred for the interior.

Early History

If a small body such as the Moon formed hot, it is likely that the other planets also formed hot and that they too are differentiated. This is an important boundary condition on the vaguely understood processes of planetary formation. The accretion of planetesimals into planets must have been extremely rapid, at least on the time scale of cooling of the nebula. Otherwise the gravitational energy of accretion would be conducted to the surface and radiated away. A rapid time scale of accretion is consistent with modern theories of planetary accretion in a collapsing gas-dust cloud. High initial temperatures are also consistent with the hypothesis of accretion during condensation, explained earlier in this chapter.

How can we explain the early high-temperature history of the Moon and at the same time its rapid death as an active body? Prior to lunar exploration the preferred theory for the formation of the planets involved the gradual accretion of cold particles with subsequent heating, melting and differentiation resulting from decay of radioactive elements. Because the first 10^9 years of terrestrial history are inaccessible to us, there was no compelling reason to suppose that the Earth was initially hot. Some indirect clues, however, point to a high-temperature origin. The oldest terrestrial rocks are clearly the result of igneous processes, and they contain evidence that the Earth had a magnetic field—hence a molten core—when they crystallized. Many ancient meteorites also have clearly gone through a very early high-temperature event or are the result of magmatic processes, which must have taken place in a body much smaller than the Earth.

Age dating of lunar material indicates that most of the differentiation of the Moon occurred in the first 10^9 years of its existence. For the Earth nearly all of this early record is erased by later activity, but the Moon has been remarkably quiescent for the last 3×10^9 years. This is true not only for internal processes but also for the external bombardment processes that are mainly responsible for the surface features of the Moon. Both the internal structure and exterior morphology were apparently the result of an extensive early history of activity.

Many sources have been proposed as early heating processes. These include short-lived radioactive elements, solar radiation, high temperatures associated with the early nebula, and energy of gravitational accretion. For any of the latter three processes to be effective, the bodies in the solar system must have assembled very quickly. Gravitational collapse of a gas-dust cloud is an attractive way both to localize protoplanetary material and to have initially high temperatures. The main problem is to be able to build planetary nuclei to a size big enough that they can effectively scavenge the remaining material in their vicinity.

Some of the most obvious facts about the Moon may also be among the most relevant for our attempts to understand its origin. It is small, light and is close to a more massive body. These simple facts may all be related, for it is unlikely that preplanetary processes resulted in a single nucleus for each of the present planets. The nebula probably evolved from a disk to a series of rings; each ring in turn collapsed to a series of local gas-dust concentrations that collapsed further to form protoplanets, the building blocks of the planets.

Variations in the eccentricities and inclinations of the orbits of protoplanets at this stage of development ensure that they periodically approach each other; encounter velocities between bodies in a ring are low, and concentration rather than dispersal is the natural result. It is not difficult to believe that eventually one body will predominate; the remaining bodies will impact, orbit temporarily before impact, or orbit permanently. The scenario repeats for those bodies in orbit about the primary body. The largest nucleus, the Earth in this case, grows at the expense of the smaller particles, and if all this is happening in a cooling nebula it will inherit most of the later, lower-temperature condensates.

In a cooling nebula of solar composition, the first compounds to condense are calcium-, aluminum- and titanium-rich oxides, silicates and titanates. These compounds comprise approximately 6 percent of the nonvolatile composition of the nebula, which is roughly everything but hydrogen, helium, carbon and that oxygen not tied up in the refractory compounds. Carbonaceous chondrites are usually taken as an approximation to the "nonvolatile" content of the nebula. Type I and II carbonaceous chondrites contain about 10 percent of unique white inclusions, composed primarily of exotic Ca-Al-rich minerals such as gehlenite, spinel and anorthosite. These inclusions are rich in barium, strontium, uranium and the rare-earth elements, and they have most of the properties that have been inferred for the Moon. They are also very ancient. This material may represent the most primitive material in the solar system and be the initial building blocks of the planets. At the very least they are analogs of the refractory material in the solar system.

Planetesimals formed at this point will be deficient in iron, magnesium, silicon, sulfur, sodium and potassium, all of which are still held in the gaseous phase. Solid particles will rapidly concentrate toward the median plane to accrete into refractory-rich planetary nuclei. While cooling of the nebula continues, iron and magnesium silicates condense; these are the most abundant constituents of meteorites and of the terrestrial planets. The largest body at any distance from the Sun will obtain the major share of these later condensates. In this scenario the Moon is one of the original smaller bodies that avoided impact with the Earth. It may also represent the coagulation of many smaller bodies that were trapped into Earth orbit from Earth-crossing solar orbits. In this hypothesis many of the satellites in the solar system may be more refractory than their primary bodies. The angular momentum of the Earth-Moon system is difficult to understand with this scenario. The Moon may also have formed by condensation of material vaporized from the Earth by a large impact.

The Lunar Crust

The Moon is such a small body that both its major-element chemistry and its incompatible trace-element chemistry are affected by the crustal composition. The thick lunar highland crust (Taylor, 1982) is estimated to have 13 percent aluminum, 11 percent calcium and 3400 ppm titanium, which, combined with concentrations of trace elements and estimates of crustal thickness, immediately indicate that the Moon is both a refractory-rich body and an extremely well-differentiated body. The amount of aluminum in the highland crust may represent about 40 percent of the total lunar budget. This is in marked contrast to the Earth, where the amount of such major elements as aluminum and calcium in the crust is a trivial fraction of the total in the planet. On the other hand the amount of the very incompatible elements such as rubidium, uranium and thorium in the Earth's crust is a large fraction of the terrestrial inventory. This dichotomy between the behavior of major elements and incompatible trace elements can be understood by considering the effect of pressure on the crystallization behavior of calcium- and aluminum-rich phases. At low pressures these elements enter low-density phases such as plagioclase, which are then concentrated toward the surface. At higher pressures these elements enter denser phases such as clinopyroxene and garnet. At still higher pressures, equivalent to depths greater than about 300 km in the Earth, these phases react to form a dense garnet-like solid solution that is denser than such upper-mantle phases as olivine and pyroxene. Therefore, in the case of the Earth, much of the calcium and aluminum is buried at depth. The very incompatible elements, however, do not readily enter any of these phases, and they are concentrated in light melts. The higher pressures in the Earth's magma ocean and the slower cooling rates of the larger body account for the differences in the early histories of the Earth and Moon.

In the case of the Moon, the anorthositic component is due to the flotation of plagioclase during crystallization of the ocean. Later basalts are derived from cumulates or cumulus liquids trapped at depth, and the KREEP (K, REE, P rich) component represents the final residual melt. The isotopic data (Pb, Nd, Sr) require large-scale early differentiation and uniformity of the KREEP component. About 50 percent of the europium and potassium contents of the bulk Moon now reside in the highland crust, which is less than 9 percent of the mass of the Moon. The energy associated with the production of the highland crust is much greater than needed for the mare basalts, which are an order of magnitude less abundant. Estimates of the required thickness of the magma ocean are generally in excess of 200 km, and mass-balance calculations require that most or all of the Moon has experienced partial melting and melt extraction. A 200-km-thick magma ocean will freeze in about 10^7–10^8 years. Other evidence in support of the magma ocean concept, or at least widespread and extensive melting, include: (1) the complementary highland and mare basalt trace-element patterns, particularly the europium anomaly; (2) the enrichment of incompatible elements in the crust and KREEP; (3) the isotopic uniformity of KREEP; and (4) the isotopic evidence for early differentiation of the mare basalt source region, which was complete by about 4.4 Ga.

There is thus a considerable body of evidence in support of large-scale and early, prior to 4.4 Ga, lunar differentiation (Taylor, 1982). The separation of crustal material and fractionation of trace elements is so extreme that the concept of a deep magma ocean plays a central role in theories of lunar evolution. The cooling and crystallization of such an ocean permits efficient separation of various density crystals and magmas and the trace elements that accompany these products of cooling. This concept does not require a continuous globally connected ocean that extends to the surface nor one that is even completely molten. The definition of magma extends to melt-crystal slurries, and oceans can be covered by a thin crust of light crystals. Such semantic niceties have resulted in unproductive quibbling about whether it is a magma ocean or a series of magma marshes, unconnected magma lakes, buried magma chambers or widespread partial melt zones in the mantle. I refer to all of these possibilities under the collective name "magma ocean," which is descriptive and serves to focus attention on the geochemical requirements for widespread and extensive partial melting of the Moon that leads to at least local accumulations of magma and cumulative crystal layers and efficient separation of certain elements. Part of the evidence for a magma ocean on the Moon is the thick anorthositic highland crust and the widespread occurrence of KREEP, an incompatible-element-rich material best interpreted as the final liquid dregs of a Moon-wide melt zone. The absence of an extensive early terrestrial anorthositic crust and the presumed absence of a counterpart to KREEP have kept the magma ocean concept from being adopted as a central principle in theories of the early evolution of the Earth. I show in Chapter 11, however, that a magma ocean is also quite likely for the Earth and probably the other terrestrial planets as well.

The gross physical properties of the Moon such as its mass and moment of inertia do not place significant constraints on the major-element abundances of the deep lunar interior, except that the Moon is clearly deficient in iron in relation to cosmic, chondritic, or terrestrial abundances. In particular, this kind of data cannot be used to place upper bounds on the CaO and Al_2O_3 content. However, the surface evidence suggests that the Moon is also deficient in elements and compounds more volatile than iron. This led to the suggestion (Anderson, 1973a, b) that the interior of the Moon was also enriched in CaO and Al_2O_3. Such a body would be highly enriched in the refractory trace elements relative to chondritic abundances and upon melting or fractional crystallization would yield a plagioclase-pyroxene outer shell of the order of 250 km in thickness. A mixture of about 14 percent basalt and 86 percent anorthosite satisfies the bulk chemistry of this layer and eliminates the europium anomaly. These proportions are broadly consistent with the areal extent of the maria and the volumetric relationships implied by elevation differences, offset of center of mass, and the moments of inertia. High-CaO-and-Al_2O_3 peridotites have broad intermediate density fields, and

TABLE 2-6
Crustal Compositions in the Moon and Earth

	Lunar Highland Crust	Lunar Mare Basalt	Terrestrial Continental Crust	Terrestrial Oceanic Crust
	Major Elements (percent)			
Mg	4.10	3.91	2.11	4.64
Al	13.0	5.7	9.5	8.47
Si	21.0	21.6	27.1	23.1
Ca	11.3	8.4	5.36	8.08
Na	0.33	0.21	2.60	2.08
Fe	5.1	15.5	5.83	8.16
Ti	0.34	2.39	0.48	0.90
	Refractory Elements (ppm)			
Sr	120	135	400	130
Y	13.4	41	22	32
Zr	63	115	100	80
Nb	4.5	7	11	2.2
Ba	66	70	350	25
La	5.3	6.8	19	3.7
Yb	1.4	4.6	2.2	5.1
Hf	1.4	3.9	3.0	2.5
Th	0.9	0.8	4.8	0.22
U	0.24	0.22	1.25	0.10
	Volatile Elements (ppm)			
K	600	580	12,500	1250
Rb	1.7	1.1	42	2.2
Cs	0.07	0.04	1.7	0.03

Taylor (1982), Taylor and McLennan (1985).

the gabbro-eclogite transformation pressure increases with Al_2O_3 content. Therefore the Moon can have a thick plagioclase-rich outer shell without violating its density. The seismic velocities in the lunar mantle can be compared with the range 7.5–7.6 km/s for terrestrial anorthosites with more than 95 volume percent plagioclase. A material composed of about 80 percent gabbroic anorthosite and about 20 percent Ca-rich pyroxene yields a velocity of about 7.7 km/s.

Table 2-6 gives comparisons between the crusts of the Moon and the Earth. In spite of the differences in size, the bulk composition and magmatic history of these two bodies, the products of differentiation, are remarkably similar. The lunar crust is less silicon rich and poorer in volatiles, probably reflecting the overall depletion of the Moon in volatiles. The lunar highland crust and the mare basalts are both more similar to the terrestrial oceanic crust than to the continental crust. Depletion of the Moon in siderophiles and similarity of the Earth and Moon in oxygen isotopes are consistent with the Moon forming from the Earth's mantle, after separation of the core. A giant impact on Earth is the most recent variant of the "Moon came from the Earth" scenario. Earlier variants included tidal disruption and rotational instability.

Origin of the Moon

The Moon has several unique attributes, in addition to its large size, which place constraints on its origin. The substantial prograde angular momentum of the Earth-Moon system seems to rule out co-accretion from a gas-dust cloud or from planetesimals. On the other hand oxygen-isotope data favor formation of both bodies from a similar cosmic pool of material. There are similarities in the refractory-element abundances between the mantles of Earth and Moon, but the Moon is extremely depleted in the volatile elements and iron compared to the Earth and most meteorites. The Moon is a highly differentiated body, so much so that extensive melting of all or most of the Moon and the presence of a deep magma ocean appear likely.

The small mass of the Moon indicates that the event or process that formed it was an energetic one; it was probably responsible for the extensive and early melting and depletion in volatiles. The angular difference between Earth's equator and the lunar orbital plane (about 5°) suggests a role for large impacts in early Earth-Moon history.

The low density of the Moon restricts not only the size of an iron-rich core but also the total iron content. The maximum allowable size of the lunar core is one-fifteenth, by mass fraction, of the terrestrial core. The lunar mantle silicates must also be less iron rich than those in chondritic meteorites. The depletion of refractory siderophiles in the Moon suggests that the Moon was formed from material poor in iron, not just in volatiles.

Much attention has recently been focused on the role of giant impacts in early planetary evolution. Giant impactors, say 10 to 30 percent of the mass of a planet, can extensively melt the parent body, remove primitive atmospheres, and even melt, vaporize and remove large portions of the crust and upper mantle. They would also affect, and possibly control, the angular momentum of the impacted body. If even a modest fraction of the mass of a terrestrial planet, say 30 percent, is brought in by giant impactors, then the initial condition will be hot, just as if the whole planet accreted rapidly. A giant impact is, of course, an example of very rapid accretion.

A sufficiently energetic impactor, greater than the mass of Mars, would melt the crust and much of the mantle and throw enough material into orbit to form the Moon. The objection that such an event would be extremely rare is actually a point in its favor, since the Moon is unique. A single such event could explain the geochemical similarities and differences between the Earth and the Moon, the absence of an early terrestrial geological record, the angular momentum problem, the lunar depletion of volatiles and iron and an early magma ocean on both bodies.

The main conventional theories of lunar origin include fission from a rapidly rotating Earth, capture of a fully formed moon into Earth orbit and co-formation with the

Earth from the solar nebula. There are apparently insurmountable problems associated with each of these mechanisms, mainly involving the distribution of angular momentum and geochemical similarities and differences between the two bodies. One oft-stated difficulty with the various theories of lunar origin is their improbability. This, by itself, is not a serious problem since the uniqueness of the Moon in the solar system shows that its formation was a rare event. If formation of moons was a high-probability event, we would have to explain why the Earth has only one moon and why Mercury, Venus and Mars do not have massive moons. On the other hand, no theory of lunar origin can violate the laws of physics or the observables regarding the properties of the Earth and Moon and their orbits.

The most significant clues pertinent to this problem are not the detailed studies resulting from the Apollo program but more mundane observations such as the large number of impacts on the surfaces of the Moon and other bodies and the small number of terrestrial planets. The planets must have grown from something, and at present there is relatively little material to accrete and impacts are very infrequent. We must conclude that there were large numbers of bodies in the early solar system and that the present planets grew by intercepting the orbits of other bodies of all sizes. If the initial orbits of planetesimals had a variety of ellipticities and inclinations, encounters and close encounters would be frequent. Encounters would be both destructive and constructive, but the end result would be to regularize the orbits and to place more objects in a single plane with more nearly circular orbits. Intersection of these orbits would be more likely to result in accretion than in destruction, and eventually the smaller bodies would be perturbed into the orbits of the larger bodies and swept up. The present bimodal distribution of objects—a few large objects and many small objects—is the result of these competing processes. A plausible initial distribution, the result of comminution, would have an increase of about an order of magnitude in the numbers of objects for each order of magnitude decrease in the mass. For example, for every Earth-sized body there would be about 10 Mars-sized bodies and 100 Moon-sized bodies. The impacts of equal-sized bodies would probably destroy both, if they were in dissimilar orbits, because of the high impact velocities. These bodies would be more likely to coalesce if they were in similar orbits. The net result, however, is a few larger bodies and many smaller bodies.

An Earth-sized body could probably survive an impact by a Mars-sized body, but it would be traumatic to both. The Earth would be partially melted and vaporized, and considerable amounts of both the Earth and the impactor would be thrown out of the Earth. If orbital mechanics alone were responsible for the subsequent fate of this material, it would either go into hyperbolic orbit and escape or reimpact the Earth. However, the massive atmosphere caused by such an impact, the mutual interactions of the many small solid fragments, and the freezing and condensation of liquids and gases ejected from the surface before they escape from the Earth's gravitational attraction results in a small fraction of the material being trapped in Earth orbit, where it accretes into a small version of the Moon. The presence of a protomoon in Earth orbit provides a nucleus for sweeping up debris from later Earth impacts. In this scenario, the material forming the Moon is a mixture of material melted and vaporized from the Earth and the impactor and recondensed in Earth orbit. The Moon would presumably be richer in the early condensing and freezing material and would therefore be refractory rich, as in the condensation scenario, but the material would have undergone a prior stage of differentiation, including siderophile removal, on a large planet.

References

Anderson, D. L. (1971) Internal constitution of Mars, *Jour. Geophys. Res.*, *77*, 789–795.

Anderson, D. L. (1973a) Removal of a constraint on the composition of the lunar interior, *J. Geophys. Res.*, *78*, 3222-3225.

Anderson, D. L. (1973b) The composition and origin of the Moon, *Earth Planetary Sci. Lett.*, 18, 301–316.

Anderson, D. L. and T. C. Hanks (1972) Formation of the Earth's core, *Nature*, *237*, 387–388.

Birch, F. (1965) Energetics of core formation, *J. Geophys. Res.*, *70*, 6217–6221.

BVP, Basaltic Volcanism Study Project (1980) *Basaltic Volcanism on the Terrestrial Planets*, Pergamon, New York, 1286 pp.

Cameron, A. G. W. (1982) Elementary and nuclidic abundances in the solar system. In *Essays in Nuclear Astrophysics* (C. A. Barnes, et al., eds.), Cambridge University Press, Cambridge.

Ganapathy, R. and E. Anders (1974) Bulk compositions of the Moon and Earth estimated from meteorites, *Proc. Lunar Sci. Conf.*, *5*, 1181–1206.

Goins, N., A. Dainty and M. Toksöz (1981) Lunar Seismology, *J. Geophys. Res.*, *86*, 5061–5074.

Hanks, T. C. and D. L. Anderson (1969) The early thermal history of the Earth, *Phys. Earth Planet. Interiors*, *2*, 19–29.

Hartman, W., R. Phillips and G. Taylor (1986) Origin of the Moon, Lunar and Planetary Institute, Houston, 781 pp.

Jacobsen, S. B. and G. J. Wasserburg (1979), The mean age of mantle and crustal reservoirs, *J. Geophys. Res.*, *84*, 7411–7427.

Larimer, J. W. (1967) Chemical fractionation in meteorites—I. Condensation of the elements, *Geochim. Cosmochim. Acta*, *31*, 1215–1238.

McLennan, S. M. and S. R. Taylor (1980) Th and U in sedimentary rocks; crustal evolution and sedimentary recycling, *Nature*, *285*, 621–624.

Morgan, J. W. and E. Anders (1980) Chemical composition of the Earth, Venus, and Mercury, *Proc. Natl. Acad. Sci.*, *77*, 6973.

Ohtani, E. and A. E. Ringwood (1984), Composition of the core, *Earth Planetary Sci. Lett., 71,* 85–93.

Taylor, S. R. (1982) *Planetary Science, A Lunar Perspective,* Lunar and Planetary Institute, Houston, 482 pp.

Taylor, S. R. and S. M. McLennan (1981) The composition and evolution of the Earth's crust; rare earth element evidence from sedimentary rocks, *Phil. Trans. Roy. Soc. Lond. A, 301,* 381–399.

Taylor, S. R. and S. M. McLennan (1985) *The Continental Crust: Its Composition and Evolution,* Blackwell, London.

Usselman, T. M. (1975) Experimental approach to the state of the core; Part 1. The liquidus relations of the Fe-rich portion of the Fe-Ni-S system from 30 to 100 kb, *Am. J. Sci., 275,* 278–290.

Weaver, B. L. and J. Tarney (1984) Major and trace element composition of the continental lithosphere, *Phys. Chem Earth, 15,* 39–68.

3

The Crust and Upper Mantle

ZOE: Come and I'll peel off.
BLOOM: (*feeling his occiput dubiously with the unparalleled embarrassment of a harassed pedlar gauging the symmetry of her peeled pears*) Somebody would be dreadfully jealous if she knew.

—JAMES JOYCE, *ULYSSES*

The structure of the Earth's interior is fairly well known from seismology, and knowledge of the fine structure is improving continuously. Seismology not only provides the structure, it also provides information about the composition, crystal structure or mineralogy and physical state. In subsequent chapters I will discuss how to combine seismic with other kinds of data to constrain these properties. A recent seismological model of the Earth is shown in Figure 3-1. Earth is conventionally divided into crust, mantle and core, but each of these has subdivisions that are almost as fundamental (Table 3-1). The lower mantle is the largest subdivision, and therefore it dominates any attempt to perform major-element mass balance calculations. The crust is the smallest solid subdivision, but it has an importance far in excess of its relative size because we live on it and extract our resources from it, and, as we shall see, it contains a large fraction of the terrestrial inventory of many elements. In this and the next chapter I discuss each of the major subdivisions, starting with the crust and ending with the inner core.

THE CRUST

The major divisions of the Earth's interior—crust, mantle and core—have been known from seismology for about 70 years. These are based on the reflection and refraction of P- and S-waves. The boundary between the crust and mantle is called the Mohorovičić discontinuity (M-discontinuity or Moho for short) after the Croatian seismologist who discovered it in 1909. It separates rocks having P-wave velocities of 6–7 km/s from those having velocities of about 8 km/s. The term "crust" has been used in several ways. It initially referred to the brittle outer shell of the Earth that extended down to the asthenosphere ("weak layer"); this is now called the lithosphere ("rocky layer"). Later it was used to refer to the rocks occurring at or near the surface and acquired a petrological connotation. Crustal rocks have distinctive physical properties that allow the crust to be mapped by a variety of geophysical techniques.

The term "crust" is now used to refer to that region of the Earth above the Moho. It represents 0.4 percent of the Earth's mass. In a strict sense, knowledge of the existence of the crust is based solely on seismological data. The Moho is a sharp seismological boundary and in some regions appears to be laminated. There are three major crustal types—continental, transitional and oceanic. Oceanic crust generally ranges from 5 to 15 km in thickness and comprises 60 percent of the total crust by area and more than 20 percent by volume. In some areas, most notably near oceanic fracture zones, the oceanic crust is as thin as 3 km. Oceanic plateaus and aseismic ridges may have crustal thicknesses greater than 30 km. Some of these appear to represent large volumes of material generated at oceanic spreading centers or hotspots, and a few seem to be continental fragments. Although these anomalously thick crust regions constitute only about 10 percent of the area of the oceans, they may represent up to 50 percent of the total volume of the oceanic crust. Islands, island arcs and continental margins are collectively referred to as transitional crust and generally range from 15 to 30 km in thickness. Continental crust generally ranges from 30 to 50 km thick,

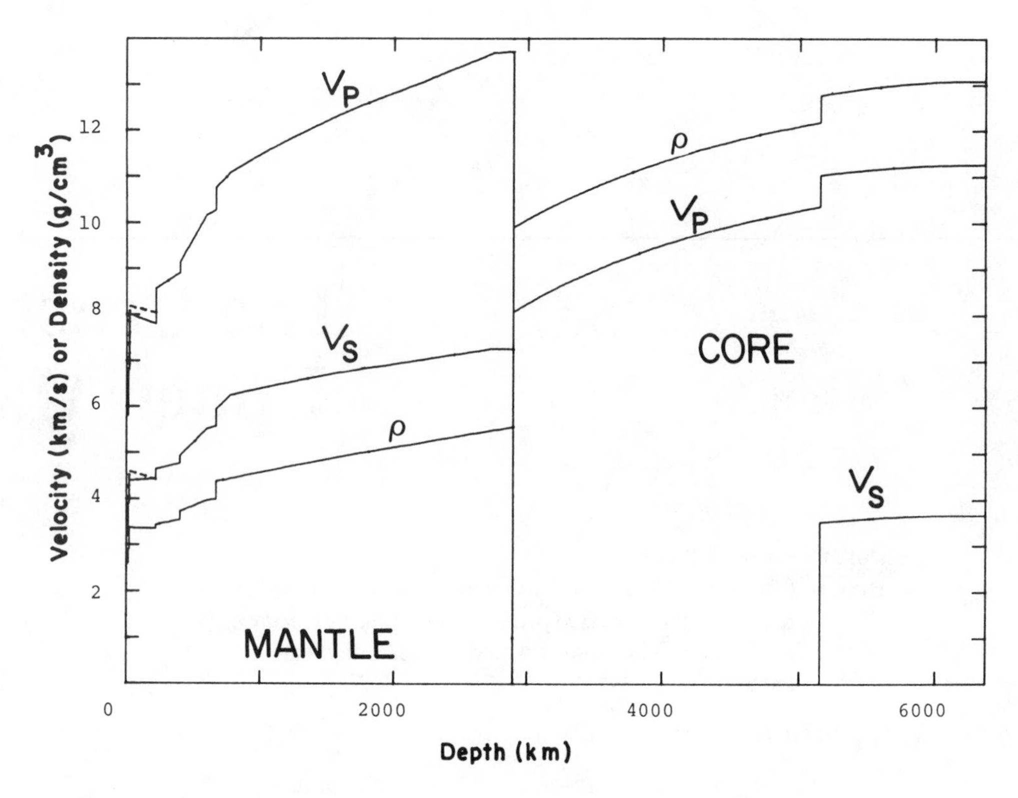

FIGURE 3-1
The Preliminary Reference Earth Model (PREM). The model is anisotropic in the upper 220 km, as shown in Figure 3-3. Dashed lines are the horizontal components of the seismic velocity (after Dziewonski and Anderson, 1981).

but thicknesses up to 80 km are reported in some convergence regions. Based on geological and seismic data, the main rock type in the upper continental crust is granodiorite or tonalite in composition. The lower crust is probably diorite, garnet granulite and amphibolite. The average composition of the continental crust is thought to be similar to andesite or diorite. The upper part of the continental crust is enriched in such "incompatible" elements as potassium, rubidium, barium, uranium and thorium and has a marked negative europium anomaly relative to the mantle. Niobium and tantalum are apparently depleted relative to other normally incompatible trace elements. It has recently been recognized that the terrestrial crust is unusually thin compared to the Moon and Mars and compared to the amount of potential crust in the mantle. This is related to the fact that crustal material converts to dense garnet-rich assemblages at relatively shallow depth. The maximum theoretical thickness of material with crust-like physical properties is about 50–60 km, although the crust may temporarily achieve somewhat greater thickness because of the sluggishness of phase changes at low temperature.

TABLE 3-1
Summary of Earth Structure

Region	Depth (km)	Fraction of Total Earth Mass	Fraction of Mantle and Crust
Continental crust	0–50	0.00374	0.00554
Oceanic crust	0–10	0.00099	0.00147
Upper mantle	10–400	0.103	0.153
Transition region	400–650	0.075	0.111
Lower mantle	650–2890	0.492	0.729
Outer core	2890–5150	0.308	—
Inner core	5150–6370	0.017	—

Composition

Mineralogically, feldspar (K-feldspar, plagioclase) is the most abundant mineral in the crust, followed by quartz and hydrous minerals (such as the micas and amphiboles) (Table 3-2). The minerals of the crust and some of their physical properties are given in Table 3-3. A crust composed of these minerals will have an average density of about 2.7 g/cm³. There is enough difference in the velocities and V_p/V_s ratios of the more abundant minerals that seismic velocities provide a good mineralogical discriminant. One uncertainty is the amount of serpentinized ultramafic rocks in the lower crust since serpentinization decreases the velocity of olivine

TABLE 3-2
Crustal Minerals

Mineral	Composition	Range of Crustal Abundances (vol. pct.)
Plagioclase		31–41
Anorthite	$Ca(Al_2Si_2)O_8$	
Albite	$Na(Al,Si_3)O_8$	
Orthoclase		9–21
K-feldspar	$K(Al,Si_3)O_8$	
Quartz	SiO_2	12–24
Amphibole	$NaCa_2(Mg,Fe,Al)_5[(Al,Si)_4O_{11}]_2(OH)_2$	0–6
Biotite	$K(Mg,Fe^{2+})_3(Al,Si_3)O_{10}(OH,F)_2$	4–11
Muscovite	$K\ Al_2(Al,Si_3)O_{10}(OH)$	0–8
Chlorite	$(Mg,Fe^{2+})_5Al(Al,Si_3)O_{10}(OH)_8$	0–3
Pyroxene		
Hypersthene	$(Mg,Fe^{2+})SiO_3$	0–11
Augite	$Ca(Mg,Fe^{2+})(SiO_3)_2$	
Olivine	$(Mg,Fe^{2+})_2SiO_4$	0–3
Oxides		~2
Sphene	$CaTiSiO_5$	
Allanite	$(Ce,Ca,Y)(Al,Fe)_3(SiO_4)_3(OH)$	
Apatite	$Ca_5(PO_4,CO_3)_3(F,OH,Cl)$	
Magnetite	$FeFe_2O_4$	
Ilmenite	$FeTiO_3$	

to crustal values. In some regions the seismic Moho may not be at the base of the basaltic section but at the base of the serpentinized zone in the mantle.

Estimates of the composition of the oceanic and continental crust are given in Table 3-4; another that covers the trace elements is given in Table 3-5. Note that the continental crust is richer in SiO_2, TiO_2, Al_2O_3, Na_2O and K_2O than the oceanic crust. This means that the continental crust is richer in quartz and feldspar and is therefore intrinsically less dense than the oceanic crust. The mantle under stable continental-shield crust has seismic properties that suggest that it is intrinsically less dense than mantle elsewhere. The elevation of continents is controlled primarily by the density and thickness of the crust and the intrinsic density and temperature of the underlying mantle. It is commonly assumed that the seismic Moho is also the petrological Moho, the boundary between sialic or mafic crustal rocks and ultramafic mantle rocks. However, partial melting, high pore pressure and serpentinization can reduce the velocity of mantle rocks, and increased abundances of olivine and pyroxene can increase the velocity of crustal rocks. High pressure also increases the velocity of mafic rocks, by the gabbro-eclogite phase change, to mantle-like values. The increase in velocity from "crustal" to "mantle" values in regions of thick continental crust may be due, at least in part, to the appearance of garnet as a stable phase. The situation is complicated further by kinetic considerations. Garnet is a common metastable phase in near-surface intrusions such as pegmatites and metamorphic teranes. On the other hand, feldspar-rich rocks may exist at depths greater than the gabbro-eclogite equilibrium boundary if temperatures are so low that the reaction is sluggish.

The common assumption that the Moho is a chemical boundary is in contrast to the position commonly taken with regard to other mantle discontinuities. It is almost univer-

TABLE 3-3
Average Crustal Abundance, Density and Seismic Velocities of Major Crustal Minerals

Mineral	Volume percent	ρ (g/cm^3)	V_p (km/s)	V_s (km/s)
Quartz	12	2.65	6.05	4.09
K-feldspar	12	2.57	5.88	3.05
Plagioclase	39	2.64	6.30	3.44
Micas	5	2.8	5.6	2.9
Amphiboles	5	3.2	7.0	3.8
Pyroxene	11	3.3	7.8	4.6
Olivine	3	3.3	8.4	4.9

TABLE 3-4
Estimates of the Chemical Composition of the Crust (Weight Percent)

Oxide	Oceanic Crust	Continental Crust	
	(1)	(2)	(3)
SiO_2	47.8	63.3	58.0
TiO_2	0.59	0.6	0.8
Al_2O_3	12.1	16.0	18.0
Fe_2O_3	—	1.5	—
FeO	9.0	3.5	7.5
MgO	17.8	2.2	3.5
CaO	11.2	4.1	7.5
Na_2O	1.31	3.7	3.5
K_2O	0.03	2.9	1.5
H_2O	1.0	0.9	—

(1) Elthon (1979).
(2) Condie (1982).
(3) Tayor and McLennan (1985).

sally assumed that the major mantle discontinuities represent equilibrium solid-solid phase changes in a homogeneous material. It should be kept in mind that chemical changes may also occur in the mantle. It is hard to imagine how the Earth could have gone through a high-temperature accretion and differentiation process and maintained a homogeneous composition throughout. It is probably not a coincidence that the maximum crustal thicknesses are close to the basalt-eclogite boundary. Eclogite is denser than peridotite, at least in the shallow mantle, and will tend to fall into normal mantle, thereby turning a phase boundary (basalt-eclogite) into a chemical boundary (basalt-peridotite).

Seismic Velocities in the Crust and Upper Mantle

Seismic velocities in the crust and upper mantle are typically determined by measuring the transit time between an earthquake or explosion and an array of seismometers. Crustal compressional wave velocities in continents, beneath the sedimentary layers, vary from about 5 km/s at shallow depth to about 7 km/s at a depth of 30 to 50 km. The lower velocities reflect the presence of pores and cracks more than the intrinsic velocities of the rocks. At greater depths the pressure closes cracks and the remaining pores are fluid-saturated. These effects cause a considerable increase in velocity. A typical crustal velocity range at depths greater than 1 km is 6–7 km/s. The corresponding range in shear velocity is about 3.5 to 4.0 km/s. Shear velocities can be determined from both body waves and the dispersion of short-period surface waves. The top of the mantle under

TABLE 3-5
Composition of the Bulk Continental Crust, by Weight

SiO_2	57.3 pct.	Co	29 ppm	Ce	33 ppm
TiO_2	0.9 pct.	Ni	105 ppm	Pr	3.9 ppm
Al_2O_3	15.9 pct.	Cu	75 ppm	Nd	16 ppm
FeO	9.1 pct.	Zn	80 ppm	Sm	3.5 ppm
MgO	5.3 pct.	Ga	18 ppm	Eu	1.1 ppm
CaO	7.4 pct.	Ge	1.6 ppm	Gd	3.3 ppm
Na_2O	3.1 pct.	As	1.0 ppm	Tb	0.6 ppm
K_2O	1.1 pct.	Se	0.05 ppm	Dy	3.7 ppm
Li	13 ppm	Rb	32 ppm	Ho	0.78 ppm
Be	1.5 ppm	Sr	260 ppm	Er	2.2 ppm
B	10 ppm	Y	20 ppm	Tm	0.32 ppm
Na	2.3 pct.	Zr	100 ppm	Yb	2.2 ppm
Mg	3.2 pct.	Nb	11 ppm	Lu	0.30 ppm
Al	8.41 pct.	Mo	1 ppm	Hf	3.0 ppm
Si	26.77 pct.	Pd	1 ppb	Ta	1 ppm
K	0.91 pct.	Ag	80 ppb	W	1 ppm
Ca	5.29 pct.	Cd	98 ppb	Re	0.5 ppb
Sc	30 ppm	In	50 ppb	Ir	0.1 ppb
Ti	5400 ppm	Sn	2.5 ppm	Au	3 ppb
V	230 ppm	Sb	0.2 ppm	Tl	360 ppb
Cr	185 ppm	Cs	1 ppm	Pb	8 ppb
Mn	1400 ppm	Ba	250 ppm	Bi	60 ppb
Fe	7.07 pct.	La	16 ppm	Th	3.5 ppm
				U	0.91 ppm

Taylor and McLennan (1985).

continents usually has velocities in the range 8.0 to 8.2 km/s for compressional waves and 4.3 and 4.7 km/s for shear waves.

The compressional velocity near the base of the oceanic crust usually falls in the range 6.5–6.9 km/s. In some areas a thin layer at the base of the crust with velocities as high as 7.5 km/s has been identified. The oceanic upper mantle has a *P*-velocity (P_n) that varies from about 7.9 to 8.6 km/s. The velocity increases with oceanic age, because of cooling, and varies with azimuth, presumably due to crystal orientation. The fast direction is generally close to the inferred spreading direction. The average velocity is close to 8.2 km/s, but young ocean has velocities as low as 7.6 km/s. Tectonic regions also have low velocities.

Since water does not transmit shear waves and since most velocity measurements use explosive sources, it is difficult to measure the shear velocity in the oceanic crust and upper mantle. There are therefore fewer measurements of shear velocity, and these have higher uncertainty and lower resolution than those for *P*-waves. The shear velocity increases from about 3.6–3.9 to 4.4–4.7 km/s from the base of the crust to the top of the mantle.

Ophiolite sections found at some continental margins are thought to represent upthrust or obducted slices of the oceanic crust and upper mantle. These sections grade downward from pillow lavas to sheeted dike swarms, intrusives, pyroxene and olivine gabbro, layered gabbro and peridotite and, finally, harzburgite and dunite. Laboratory velocities in these rocks are given in Table 3-6. There is good agreement between these velocities and those actually observed in the oceanic crust and upper mantle. The sequence of extrusives, intrusives and cumulates is consistent with what is expected at a midocean-ridge magma chamber. Many ophiolites apparently represent oceanic crust formed near island arcs in marginal basins. They might not be typical of crustal sections formed at mature midoceanic spreading centers. Marginal basin basalts, however, are very similar to midocean-ridge basalts, at least in major-element chemistry, and the bathymetry, heat flow and seismic crustal structure in marginal basins are similar to values for the major ocean basins.

TABLE 3-6
Density, Compressional Velocity and Shear Velocity in Rock Types Found in Ophiolite Sections

Rock Type	ρ (g/cm^3)	V_p (km/s)	V_s (km/s)	Poisson's Ratio
Metabasalt	2.87	6.20	3.28	0.31
Metadolerite	2.93	6.73	3.78	0.27
Metagabbro	2.95	6.56	3.64	0.28
Gabbro	2.86	6.94	3.69	0.30
Pyroxenite	3.23	7.64	4.43	0.25
Olivine gabbro	3.30	7.30	3.85	0.32
Harzburgite	3.30	8.40	4.90	0.24
Dunite	3.30	8.45	4.90	0.25

Salisbury and Christensen (1978), Christensen and Smewing (1981).

The velocity contrast between the lower crust and upper mantle is commonly smaller beneath young orogenic areas (0.5 to 1.5 km/s) than beneath cratons and shields (1 to 2 km/s). Continental rift systems have thin crust (less than 30 km) and low P_n velocities (less than 7.8 km/s). Thinning of the crust in these regions appears to take place by thinning of the lower crust. In island arcs the crustal thickness ranges from about 5 km to 35 km. In areas of very thick crust such as in the Andes (70 km) and the Himalayas (80 km), the thickening occurs primarily in the lower crustal layers. Paleozoic orogenic areas have about the same range of crustal thicknesses and velocities as platform areas.

THE SEISMIC LITHOSPHERE OR LID

Uppermost mantle velocities are typically 8.0 to 8.2 km/s, and the spread is about 7.9–8.6 km/s. Some long refraction profiles give evidence for a deeper layer in the lithosphere having a velocity of 8.6 km/s. The seismic lithosphere, or LID, appears to contain at least two layers. Long refraction profiles on continents have been interpreted in terms of a laminated model of the upper 100 km with high-velocity layers, 8.6–8.7 km/s or higher, embedded in "normal" material (Fuchs, 1977). Corrected to normal conditions these velocities would be about 8.9–9.0 km/s. The *P*-wave gradients are often much steeper than can be explained by self-compression. These high velocities require oriented olivine or large amounts of garnet. The detection of 7–8 percent azimuthal anisotropy for both continents and oceans suggests that the shallow mantle at least contains oriented olivine. Substantial anisotropy is inferred to depths of at least 50 km depth in Germany (Bamford, 1977).

The average values of V_p and V_s at 40 km when corrected to standard conditions are 8.72 km/s and 4.99 km/s, respectively. The corrections for temperature amount to 0.5 and 0.3 km/s, respectively, for V_p and V_s. The pressure corrections are much smaller. Short-period surface wave data have better resolving power for V_s in the LID. Applying the same corrections to surface wave data (Morris and others, 1969), we obtain 4.48–4.55 km/s and 4.51–4.64 km/s for 5-Ma-old and 25-Ma-old oceanic lithosphere. Presumably, velocities can be expected to increase further for older regions. A value for V_p of 8.6 km/s is commonly observed near 40 km depth in the oceans. This corresponds to about 8.87 km/s at standard conditions. These values can be compared with 8.48 and 4.93 km/s for olivine-rich aggregates. Eclogites have V_p and V_s as high as 8.8 and 4.9 km/s in certain directions and as high as 8.61 and 4.86 km/s as average values.

All of the above suggests that corrected velocities of at least 8.6 and 4.8 km/s, for V_p and V_s, respectively, occur in the lower lithosphere, and this requires substantial amounts of garnet at relatively shallow depths. At least 26 percent garnet is required to satisfy the compressional velocities. The density of such an assemblage is about 3.4 g/cm^3. If one is to honor the higher seismic velocities, even greater proportions of garnet are required. The lower lithosphere may therefore be gravitationally unstable with respect to the underlying mantle, particularly in oceanic regions. The upper mantle under shield regions is consistent with a very olivine-rich peridotite which is buoyant and therefore stable relative to "normal" mantle.

Most refraction profiles, particularly at sea, sample only the uppermost lithosphere. P_n velocities of 8.0–8.2 km/s are consistent with peridotite or harzburgite, thought by some to be the refractory residual after basalt removal. Anisotropies are also appropriate for olivine-pyroxene assemblages. The sequence of layers, at least in oceanic regions, seems to be basalt, peridotite, eclogite.

Anisotropy of the upper mantle is a potentially useful petrological constraint, although it can also be caused by organized heterogeneity, such as laminations or parallel dikes and sills or aligned partial melt zones, and stress fields. Under oceans the uppermost mantle, the P_n region, exhibits an anisotropy of 7 percent (Morris and others, 1969). The fast direction is in the direction of spreading, and the magnitude of the anisotropy and the high velocities of P_n arrivals suggest that oriented olivine crystals control the elastic properties. Pyroxene exhibits a similar anisotropy, whereas garnet is more isotropic. The preferred orientation is presumably due to the emplacement or freezing mechanism, the temperature gradient or to nonhydrostatic stresses. A peridotite layer at the top of the oceanic mantle is consistent with the observations.

The average anisotropy of the upper mantle is much less than the values given above. Forsyth (1975) studied the dispersion of surface waves and found shear-wave anisotropies, averaged over the upper mantle, of 2 perent. Shear velocities in the LID vary from 4.26 to 4.46 km/s, increasing with age; the higher values correspond to a lithosphere 10–50 Ma old. This can be compared with shear-wave velocities of 4.30–4.86 km/s and anisotropies of 1–4.7 percent found in relatively unaltered eclogites (Manghnani, et al, 1974). The compressional velocity range in the same samples is 7.90–8.61 km/s, reflecting the large amounts of garnet. Surface waves exhibit both azimuthal and polarization anisotropy for at least the upper 200 km of the mantle.

On the basis of data available at the time, Ringwood suggested in 1975 that the V_p/V_s ratio of the lithosphere was smaller than measured on eclogites. He therefore ruled out eclogite as an important constituent of the lithosphere. The eclogite-peridotite controversy regarding the composition of the suboceanic mantle is long standing and still unresolved. Newer and more complete data on the V_p/V_s ratio in eclogites show that the high-V_p/V_s eclogites are generally of low density and contain plagioclase or olivine. The higher-density eclogites are consistent with the properties of the lower lithosphere. Garnet and clinopyroxene may be important components of the lithosphere. A lithosphere composed primarily of olivine and (Mg,Fe)SiO_3, that is, pyrolite or lherzolite, does not satisfy the seismic data for the bulk of the lithosphere. The lithosphere, therefore, is not just cold mantle, or a thermal boundary layer alone.

Oceanic crustal basalts represent only part of the basaltic fraction of the upper mantle. The peridotite layer represents depleted mantle or refractory cumulates but may be of any age. Basaltic material may also be intruded at depth. It is likely that the upper mantle is layered, with the volatiles and melt products concentrated toward its top. As the lithosphere cools, this basaltic material is incorporated onto the base of the plate, and as the plate thickens it eventually transforms to eclogite, yielding high velocities and increasing the thickness and mean density of the oceanic plate. Eventually the plate becomes denser than the underlying asthenosphere, and conditions become appropriate for subduction or delamination.

O'Hara (1968) argued that erupted lavas are not the original liquids produced by partial melting of the upper mantle, but are residual liquids from processes that have left behind complementary eclogite accumulates in the upper mantle. Such a model is consistent with the seismic obser-

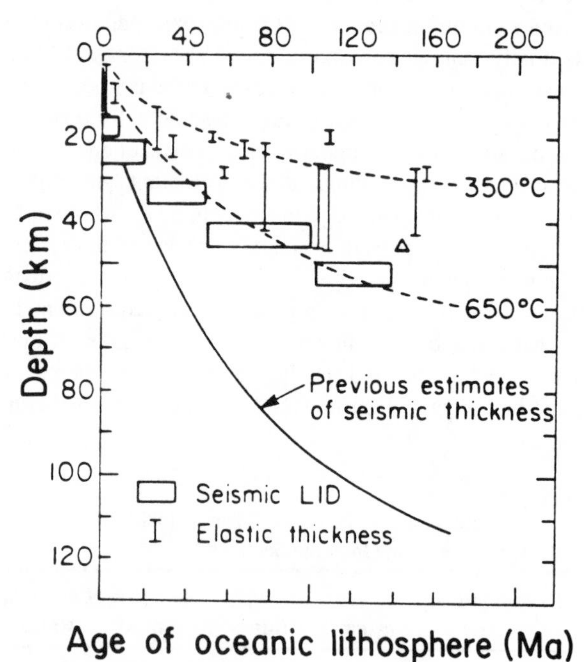

FIGURE 3-2
The thickness of the lithosphere as determined from flexural loading studies and surface waves. The upper edges of the open boxes gives the thickness of the seismic LID (high-velocity layer, or seismic lithosphere). The lower edge gives the thickness of the mantle LID plus the oceanic crust (Regan and Anderson, 1984). The triangle is a refraction measurement of oceanic seismic lithosphere thickness (Shimamura and others, 1977). The LID under continental shields is about 150 km thick (see Figure 3-4).
thick (see Figure 3-4).

vations of high velocities for P_n at midlithospheric depths and with the propensity of oceanic lithosphere to plunge into the asthenosphere. The latter observation suggests that the average density of the lithosphere is greater than that of the asthenosphere.

The thickness of the seismic lithosphere, or high-velocity LID, is about 150 km under continental shields. Some surface-wave results give a much greater thickness. A thin low-velocity zone (LVZ) at depth, as found from body-wave studies, however, cannot be well resolved with long-period surface waves. The velocity reversal between about 150 and 200 km in shield areas is about the depth inferred for kimberlite genesis, and the two phenomena may be related.

There is very little information about the deep oceanic lithosphere from body-wave data. Surface waves have been used to infer a thickening with age of the oceanic lithosphere to depths greater than 100 km (Figure 3-2). However, when anisotropy is taken into account, the thickness may be only about 50 km for old oceanic lithosphere (Regan and Anderson, 1984). This is about the thickness inferred for the "elastic" lithosphere from flexural bending studies around oceanic islands and at trenches.

The seismic velocities of some upper-mantle minerals and rocks are given in Tables 3-7 and 3-8, respectively. Garnet and jadeite have the highest velocities, clinopyroxene and orthopyroxene the lowest. Mixtures of olivine and orthopyroxene (the peridotite assemblage) can have velocities similar to mixtures of garnet-diopside-jadeite (the eclogite assemblage). Garnet-rich assemblages, however, have velocities higher than orthopyroxene-rich assemblages. The V_p/V_s ratio is greater for the eclogite minerals than for the peridotite minerals. This ratio plus the anisotropy are useful diagnostics of mantle mineralogy. High velocities alone do not necessarily discriminate between garnet-rich and olivine-rich assemblages. Olivine is very anisotropic, having compressional velocities of 9.89, 8.43 and 7.72 km/s along the principal crystallographic axes. Orthopyroxene, likewise, has velocities ranging from 6.92 to 8.25 km/s, depending on direction. In natural olivine-rich aggregates (Table 3-9), the maximum velocities are about 8.7 and 5.0 km/s for *P*-waves and *S*-waves, respectively. With 50 percent orthopyroxene the velocities are reduced to 8.2 and 4.85 km/s, and the composite is nearly isotropic. Eclogites are also nearly isotropic.

The "standard model" for the oceanic lithosphere assumes 24 km of depleted peridotite, complementary to and contemporaneous with the basaltic crust, between the crust and the presumed fertile peridotite upper mantle. There is no direct evidence for this hypothetical model. The lower oceanic lithosphere may be much more basaltic or eclogitic than in this simple model.

TABLE 3-7
Densities and Elastic-wave Velocities in Upper-mantle Minerals

Mineral	ρ (g/cm^3)	V_p	V_s (km/s)	V_p/V_s
Olivine				
Fo	3.214	8.57	5.02	1.71
Fo_{93}	3.311	8.42	4.89	1.72
Fa	4.393	6.64	3.49	1.90
Pyroxene				
En	3.21	8.08	4.87	1.66
En_{80}	3.354	7.80	4.73	1.65
Fs	3.99	6.90	3.72	1.85
Di	3.29	7.84	4.51	1.74
Jd	3.32	8.76	5.03	1.74
Garnet				
Py	3.559	8.96	5.05	1.77
Al	4.32	8.42	4.68	1.80
Gr	3.595	9.31	5.43	1.71
Kn	3.85	8.50	4.79	1.77
An	3.836	8.51	4.85	1.75
Uv	3.85	8.60	4.89	1.76

Sumino and Anderson (1984).

TABLE 3-8
Densities and Elastic-wave Velocities of Upper-mantle Rocks

Rock	ρ	V_p	V_s	V_p/V_s
Garnet	3.53	8.29	4.83	1.72
lherzolite	3.47	8.19	4.72	1.74
	3.46	8.34	4.81	1.73
	3.31	8.30	4.87	1.70
Dunite	3.26	8.00	4.54	1.76
	3.31	8.38	4.84	1.73
Bronzitite	3.29	7.89	4.59	1.72
	3.29	7.83	4.66	1.68
Eclogite	3.46	8.61	4.77	1.81
	3.61	8.43	4.69	1.80
	3.60	8.42	4.86	1.73
	3.55	8.22	4.75	1.73
	3.52	8.29	4.49	1.85
	3.47	8.22	4.63	1.78
Jadeite	3.20	8.28	4.82	1.72

Clark (1966), Babuska (1972), Manghnani and others (1974), Jordan (1979).

THE LOW-VELOCITY ZONE OR LVZ

A region of diminished velocity or negative velocity gradient in the upper mantle was proposed by Beno Gutenberg in 1959. Earlier, just after isostasy had been established, it had been concluded that a weak region underlay the relatively strong lithosphere. This has been called the asthenosphere. The discovery of a low-velocity zone strengthened

TABLE 3-9
Anisotropy of Upper-mantle Rocks

Mineralogy	Direction	V_p	V_{s_1}	V_{s_2}	V_p/V_s	
			Peridotites			
100 pct. ol	1	8.7	5.0	4.85	1.74	1.79
	2	8.4	4.95	4.70	1.70	1.79
	3	8.2	4.95	4.72	1.66	1.74
70 pct. ol,	1	8.4	4.9	4.77	1.71	1.76
30 pct. opx	2	8.2	4.9	4.70	1.67	1.74
	3	8.1	4.9	4.72	1.65	1.72
100 pct. opx	1	7.8	4.75	4.65	1.64	1.68
	2	7.75	4.75	4.65	1.63	1.67
	3	7.78	4.75	4.65	1.67	1.67
			Eclogites			
51 pct. ga,	1	8.476		4.70		1.80
23 pct. cpx,	2	8.429		4.65		1.81
24 pct. opx	3	8.375		4.71		1.78
47 pct. ga,	1	8.582		4.91		1.75
45 pct. cpx	2	8.379		4.87		1.72
	3	8.30		4.79		1.73
46 pct. ga,	1	8.31		4.77		1.74
37 pct. cpx	2	8.27		4.77		1.73
	3	8.11		4.72		1.72

Manghnani and others (1974), Christensen and Lundquist (1982).

the concept of an asthenosphere, even though a weak layer is not necessarily a low-velocity layer.

Gutenberg based his conclusions primarily on amplitudes and apparent velocities of waves from earthquakes in the vicinity of the low-velocity zone. He found that at distances from about 1° to 15° the amplitudes of longitudinal waves decrease about exponentially with distance. At 15° they increase suddenly by a factor of more than 10 and then decrease at greater distances. These results can be explained in terms of a low-velocity region, which defocuses seismic energy, underlain by a higher gradient that serves to focus the rays.

Most recent models of the velocity distribution in the upper mantle include a region of high gradient between 250 and 350 km depth. Lehmann (1961) interpreted her results for several regions in terms of a discontinuity at 220 km (sometimes called the Lehmann discontinuity), and many subsequent studies give high-velocity gradients near this depth.

It is difficult to study details of the velocity distribution in and just below a low-velocity zone, and it is still not clear if the base of the low-velocity zone is gradual or abrupt. Reflections have been reported from depths between 190 and 250 km by a number of authors (Anderson, 1979). This situation is further complicated by the extreme lateral heterogeneity of the upper 200 km of the mantle. This region is also low Q (high attenuation) and anisotropic. Some recent results are shown in Figures 3-3 and 3-4.

Various interpretations have been offered for the low-velocity zone. This is undoubtedly a region of high thermal gradient, the boundary layer between the near surface where heat is transported by conduction and the deep interior where heat is transported by convection. If the temperature gradient is high enough, the effects of pressure can be overcome and velocity can decrease with depth. It can be shown, however, that a high temperature gradient alone is not an adequate explanation. Partial melting and dislocation

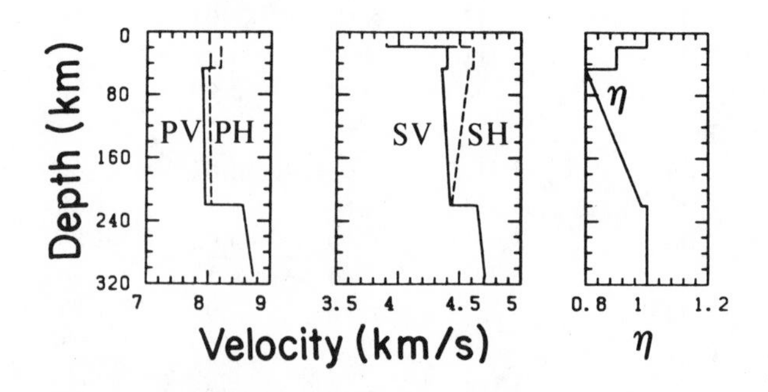

FIGURE 3-3
Velocity-depth profiles for the average Earth, as determined from surface waves (Regan and Anderson, 1984). From left to right, the graphs show P-wave velocities (vertical and horizontal), S-wave velocities (vertical and horizontal), and the anisotropy parameter η (see Chapter 15), where 1 represents isotropy.

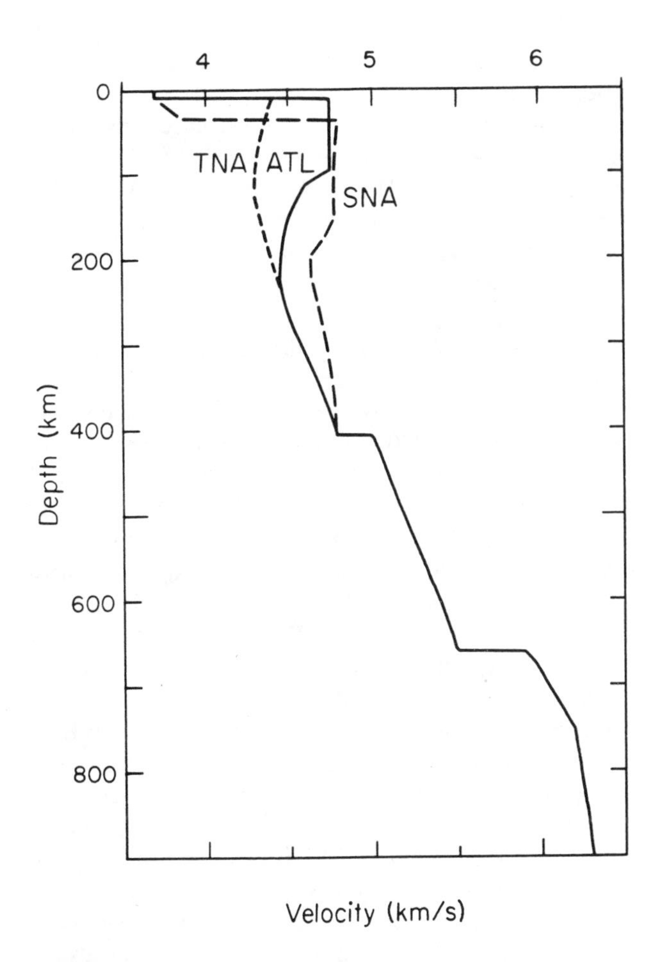

FIGURE 3-4
High-resolution shear-wave velocity profiles for various tectonic provinces; TNA is tectonic North America, SNA is shield North America, ATL is north Atlantic (after Grand and Helmberger, 1984).

relaxation both cause a large decrease in velocity. For partial melting to be effective the melt must occur, microscopically, as thin grain boundary films or, macroscopically, as narrow dikes or sills. Melting experiments suggest that melting occurs at grain corners and is more likely to occur in interconnected tubes. This also seems to be required by electrical conductivity data. However, numerous thin dikes and sills act macroscopically as thin films for long-wavelength seismic waves. High attenuation is associated with relaxation processes such as grain boundary relaxation, including partial melting, and dislocation relaxation. Allowance for anelastic dispersion increases the velocities in the low-velocity zone determined by free-oscillation and surface-wave techniques (Hart and others, 1976), but partial melting is still required to explain the regions of very low velocity. Allowance for anisotropy results in a further upward revision for the velocities in this region (Dziewonski and Anderson, 1981), as discussed below. This plus the recognition that subsolidus effects, such as dislocation relaxation, can cause a substantial decrease in velocity has complicated the interpretation of seismic velocities in the shallow mantle. Velocities in tectonic regions and under some oceanic regions, however, are so low that partial melting is implied. In most other regions a subsolidus mantle composed of oriented olivine-rich aggregates can explain the velocities and anisotropies to depths of about 200 km. There is, as yet, no detailed information on anisotropy below 220 km in any single geographic region. The global inversions of Nataf and others (1986) involve a laterally heterogeneous velocity and anisotropy structure to depths as great as 400 km, but anisotropy resolution below 200 km is poor.

The rapid increase in velocity below 220 km may be due to chemical or compositional changes or to transition from relaxed to unrelaxed moduli. The latter explanation will involve an increase in Q, and some Q models exhibit this characteristic. However, the resolving power for Q is low, and most of the seismic Q data can be satisfied with a constant-Q upper mantle, at least down to 400 km.

The low-velocity zone is very thin under shields, extending from about 150 to 200 km. Under the East Pacific Rise low velocities persist to 400 km, and the Lehmann discontinuity appears to be absent. One possible interpretation is that material from below 200 km rises to the near surface under ridges, thereby breaking through any chemical discontinuity. The fact that earthquakes associated with subduction of young oceanic lithosphere do not extend below 200 km suggests that this may be a buoyancy, or chemical, discontinuity. Old dense lithosphere, however, penetrates deeper.

The velocities between 200 and 400 km can be satisfied by either an olivine-rich aggregate, such as peridotite, or a garnet-clinopyroxene aggregate such as eclogite. Deep seated kimberlites bring xenoliths of both types to the surface, although the eclogite nodules are much rarer. Oceanic ridges have low velocities throughout this depth range, suggesting that the source region for midocean ridge basalts is in the transition region (middle mantle) rather than in the low-velocity zone. Many petrologists assume that the low-velocity zone is the source region for most basalts. This is based on early seismological interpretations of a global layer of partial melt at that depth. Basalts from a deeper layer must, of course, traverse the low-velocity zone, and this is probably where melt-crystal separation occurs and where increased melting due to adiabatic ascent occurs. Under the East Pacific Rise the maximum V_p/V_s ratio occurs at about 100 km, and that is where melting caused by adiabatic ascent from deeper levels is most pronounced in this region and probably other ridges as well.

Recent surface-wave tomographic results show that the lateral variations of velocity in the upper mantle are as pronounced as the velocity variations that occur with depth (Nataf and others, 1986). Thus, it is misleading to think of the mantle as a simple layered system. Below 400 km there is little evidence from body waves for large lateral variations. Detailed body-wave modeling for regions as diverse as the Canadian and Baltic shields, western North America–East Pacific Rise, northwestern North America and the western Atlantic, while exhibiting large changes above 200 km, converge from 200 to 400 km. Surface waves have detected small lateral changes between depths of 400 and 650 km. Lateral changes are, of course, expected in a convecting mantle because of variations in temperature and anisotropy due to crystal orientation.

The geophysical data (seismic velocities, attenuation, heat flow) can be explained if the low-velocity regions in the shallow mantle are permeated by partial melt (Anderson and Sammis, 1970; Anderson and Bass, 1984). This explanation, in turn, suggests the presence of volatiles in order to depress the solidus of mantle materials, or a high-temperature mantle. The top of the low-velocity zone may mark the crossing of the geotherm with the wet solidus of peridotite. Its termination would be due to (1) a crossing in the opposite sense of the geotherm and the solidus, (2) the absence of water or other volatiles, or (3) the removal of water into high-pressure hydrous or hydroxylated phases. In all of these cases the boundaries of the low-velocity zone would be expected to be sharp. Small amounts of melt (about 1 percent) can explain the velocity reduction if the melt occurs as thin grain-boundary films. Considering the wavelength of seismic waves, magma-filled dikes and sills, rather than intergranular melt films, would also serve to decrease the seismic velocity by the appropriate amount.

The melting that is inferred for the lower velocity regions of the upper mantle may be initiated by adiabatic ascent from deeper levels. The high compressibility and high iron content of melts means that the density difference between melts and residual crystals decreases with depth. High temperatures and partial melting tend to decrease the garnet content and thus to lower the density of the mantle. Buoyant diapirs from depths greater than 200 km will extensively melt on their way to the shallow mantle. Therefore, partially molten material as well as melts can be delivered to the shallow mantle. The ultimate source of basaltic melts may be below 300–400 km even if melt-solid separation does not occur until shallower depths.

The Base of the LVZ

The major seismic discontinuities in the mantle are near 400 and 670 km, bracketing the transition region. There is another important region of high velocity gradient at a depth near 220 km, the base of the low-velocity zone. A discontinuity at 232 km depth was proposed in 1917 by Galitzin. The most detailed early studies, by Inge Lehmann (1961), indicated the presence of a discontinuity under North America and Europe near 215–220 km, and this is sometimes referred to as the Lehmann discontinuity. This is confusing since the outer core–inner core boundary is also sometimes given this name.

Many recent studies have found evidence for a discontinuity or high-gradient region between 190 and 230 km from body-wave data (Drummond and others, 1982). The increase in velocity is on the order of 3.5–4.5 percent. Niazi (1969) demonstrated that the Lehmann discontinuity in California and Nevada is a strong reflector and found a depth of 227 ± 22 km. Reflectors at 140–160 km depth may represent an upwarping of the Lehmann discontinuity caused by hotter than normal mantle.

Converted phases have been reported from a discontinuity at a depth of 200–250 km under the Canadian and Baltic shields (Jordan and Frazer, 1975; Sacks and others, 1977). Reflections from a similar depth have been reported from P′P′ precursors for Siberia, western Europe, North Atlantic, Atlantic-Indian Rise, Antarctica, and the Ninetyeast Ridge (Whitcomb and Anderson, 1970). Evidence now exists for the Lehmann discontinuity in the eastern and western United States, Canadian Shield, Baltic Shield, oceanic ridges, normal ocean, Australia, the Hindu Kush, the Alps, and the African rift. The V_p/V_s ratio of recent Earth models reverses trend near 220 km. This is indicative of a change in composition, phase, or temperature gradient.

There is a variety of indirect evidence in support of an important boundary near 220 km. This boundary affects seismicity and may be a density or mechanical impediment to slab penetration. It marks the depth above which there are large differences between continental shields and oceans. Few earthquakes occur below this depth in continental collision zones and in regions where the subducting lithosphere is less than about 50 Ma old. In most seismic regions, earthquakes do not occur deeper than about 250 km. This applies to oceanic, continental, and mixed domains. The maximum depths are 200 km in the South Sandwich arc, Burma, Rumania, the Hellenic arc, and the Aleutian arc; 250 km in the west Indian arc; and 300 km in the Ryukyu arc and the Hindu Kush. There are large gaps in seismicity between 250 km and 500–650 km in New Zealand, New Britain, Mindanao, Sunda, New Hebrides, Kuriles, North Chile, Peru, South Tonga, and the Marianas. In the New Hebrides there is a concentration of seismic activity between 190 and 280 km that moves up to 110 and 150 km in the region where a buoyant ridge is attempting to subduct. In the Bonin-Mariana region there is an increase in activity at 280–340 km to the south and a general decrease in activity with depth down to about 230 km. In the Tonga-Kermadec region, seismic activity decreases rapidly down to 230 km and, in the Tonga region, picks up again at 400 km. In Peru most of the seismicity occurs above 190–230 km, and there is a pronounced gap between this depth and 500 km. In Chile the activity is confined to above 230 km and below 500 km. Cross sections of seismicity in these regions suggest impediments to slab penetration at depths of about 230 and 600 km. Oceanic lithosphere with buoyant ridges appears to penetrate only to 150 km.

Compressional stresses parallel to the dip of the seismic zone are prevalent everywhere that the zone exists below about 300 km, indicating resistance to downward motion below about this depth (Isacks and Molnar, 1971). Actually, between 200 and 300 km about half the focal mechanisms indicate downdip compression, and most of the mechanisms below 215 km are compressional, suggesting that the slabs encounter stronger or denser material that resists their sinking. Resistance at a much deeper level, however, may also explain the seismicity.

In the regions of continent-continent collision, the distribution of earthquakes should define the shape and depth of the collision zone. The Hindu Kush is characterized by a seismicity pattern terminating in an active zone at 215 km. A pronounced minimum in seismic activity occurs at 160 km.

The most conclusive body-wave evidence for the Lehmann discontinuity comes from the study of earthquakes near this depth (Hales and others, 1980). These studies determine a velocity contrast of 0.2 to 0.3 km/s. If the contrast is truly sharp, a chemical discontinuity may occur, at least locally. Compressional velocities in garnet lherzolite, a rock type thought to make up the shallow mantle, are typically 8.2 to 8.3 km/s under normal conditions. These rocks are generally anisotropic with directional velocities ranging from about 8.1 to 8.4 km/s. Natural eclogites have V_p velocities in the range 8.22 to 8.61 km/s. Natural garnets have velocities of the order of 8.8 to 9.0 km/s; jadeite, a component of eclogite clinopyroxenes, has V_p of 8.7 km/s. A garnet- and clinopyroxene-rich eclogite can therefore have extremely high velocities. Therefore, a transition from garnet lherzolite to eclogite would give a seismic discontinuity. Because of the low melting point of garnet and clinopyroxene and the rapid decrease in density of garnet-clinopyroxene aggregates as the pressure is decreased or the temperature increased, a deep eclogite layer is potentially unstable. At low temperature eclogite is denser than lherzolite, but as the temperature rises it can become less dense and an instability will develop. In high-temperature regions of the mantle, the discontinuity may be destroyed.

The V_p/V_s ratios for lherzolites are generally lower (1.73) than for eclogites (~ 1.8). This is another possible diagnostic and suggests that the shear-wave velocity jump associated with such a chemical discontinuity will be less than the jump in V_p.

In Australia the Lehmann discontinuity is underlain by a low-velocity zone that has a gradual onset about 30 km below the discontinuity. Leven and others (1981) suggested that the velocity knee represents a zone of decoupling of the continental lithosphere from the deeper mantle and that the high velocities result from anisotropy due to alignment of olivine crystals along the zone of movement. They argued that a lherzolite-eclogite chemical change could not explain the high-velocity contrast, but they used theoretical estimates of velocity in lherzolite and measured values on natural eclogites and discarded the higher eclogite values. Natural rocks have lower velocities than theoretical aggregates made out of the same minerals because of the effects of pores and cracks.

The Lehmann discontinuity is enigmatic—it is difficult to observe with surface-focus events and is apparently not present in some areas. In contrast to most other seismic discontinuities, the waves refracted from the Lehmann discontinuity generally do not form first arrivals. Reflections and arrivals from intermediate-focus earthquakes are therefore the best sources of data. If there is a layer of eclogite in the upper mantle, it should show up as a discontinuity in *P*-wave velocities but may not appear in *S*-wave profiles because of the change in V_p/V_s. The discontinuity also may not show up at all azimuths because, in certain directions, olivine-rich aggregates can be faster than eclogite. I will show, later in this chapter, that the 400-km discontinuity may, on average, separate olivine-rich peridotite from a more eclogite-rich transition region. In the hotter parts of the mantle, the eclogite-rich layer may rise into the shallow mantle, generating midocean-ridge basalts.

Effect of Anisotropy on the LVZ

The pronounced minimum in the group velocity of long-period mantle Rayleigh waves is one of the classic arguments for the presence of an upper-mantle low-velocity zone. This argument, however, is invalid if the upper mantle is anisotropic (Anderson, 1966; Dziewonski and Anderson, 1981; Anderson and Dziewonski, 1982). The most recent dispersion data can be satisfied with anisotropic models that have only modest gradients in seismic velocities in the upper 200 km of the mantle (Anderson and Dziewonski, 1982). These models differ considerably from those that assume the mantle to be isotropic. In particular, they do not have a pronounced LID of high velocity and have appreciably higher velocities in the vicinity of the low-velocity zone than the isotropic models. Evidence for a high-velocity layer at the top of the mantle must come from shorter period waves (Regan and Anderson, 1984).

It has long been known that isotropic models cannot simultaneously satisfy mantle Love wave and Rayleigh wave data. The Love wave–Rayleigh wave discrepancy, in fact, is the best evidence for widespread anisotropy of the upper 200 km or so of the mantle. It has been common practice in recent years to fit the Love wave and Rayleigh wave data separately and to take the difference in the resulting isotropic models as a measure of the anisotropy. This procedure is not valid since the equations of motion do not decouple in that way. In even the simplest departure from isotropy, transverse isotropy, five elastic constants must be determined to specify the velocities of propagation of the quasi-longitudinal and two quasi-shear waves in all directions (see Chapter 15). This requires simultaneous inversion of Love wave and Rayleigh wave data including, if possible, higher modes.

Absorption and the LVZ

It is well known that elastic-wave velocities are independent of frequency only for a non-dissipative medium. In a real solid dispersion must accompany absorption. The effect is small when the seismic quality factor Q is large or unimportant if only a small range of frequencies is being considered. Even in these cases, however, the measured velocities or inferred elastic constants are not the true elastic properties but lie between the high-frequency and low-frequency limits or the so-called "unrelaxed" and "relaxed" moduli.

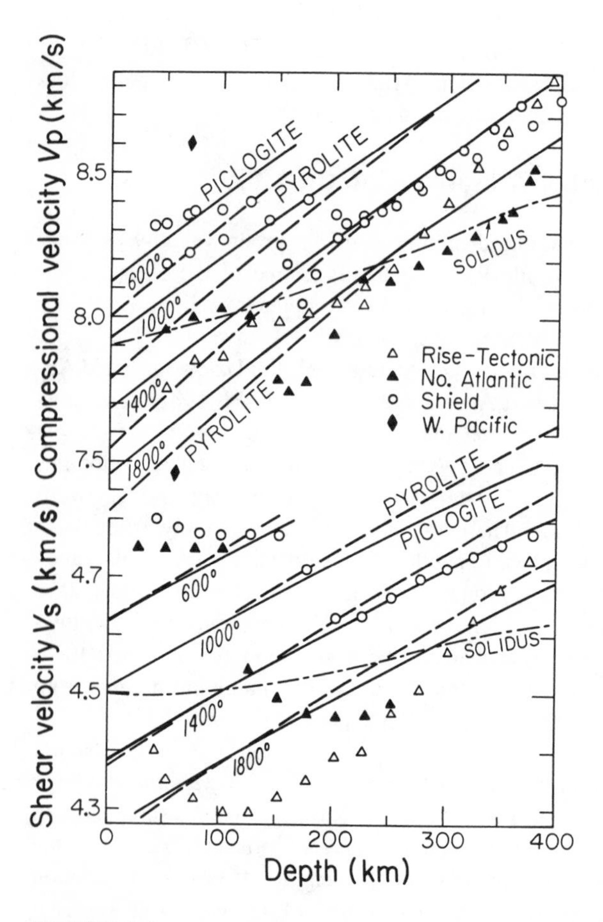

FIGURE 3-5
Compressional and shear velocities for two petrological models, pyrolite and piclogite, along various adiabats. The temperature (°C) are for zero pressure. The portions of the adiabats below the solidus curves are in the partial melt field. The seismic profiles are for two shields (Given and Helmberger, 1981; Walck, 1984), a tectonic-rise area (Grand and Helmberger, 1984a; Walck, 1984), and the North Atlantic (Grand and Helmberger, 1984b) region; two isolated points are Pacific Ocean data (Shimamura and others, 1977; after Anderson and Bass, 1984).

The magnitude of the effect depends on the nature of the absorption band and the value of Q. When comparing data taken over a wide frequency band, the effect of absorption can be considerable, especially considering the accuracy of present body-wave and free-oscillation data. Liu and others (1976) and Anderson and others (1977) showed that dispersion depends to first order on absorption in the seismic frequency band and derived a linear superposition model that gives a Q that is independent of frequency. They showed how to correct surface-wave and free-oscillation data for physical dispersion. Much of the early support for the existence of an upper-mantle low-velocity zone came from the inversion of normal-mode data uncorrected for physical dispersion due to absorption.

Anelasticity alone does not remove the necessity for a low-velocity zone, or a negative velocity gradient in the upper mantle. Allowance for anelastic dispersion (that is, frequency-dependent seismic velocities), however, makes it possible to reconcile normal-mode and body-wave models. The low upper-mantle velocities found by surface-wave and free-oscillation techniques were partially a result of low Q in the shallow mantle. Upper-mantle velocities are greater at short periods. The mechanism for low Q may involve dislocation relaxation or other subsolidus mechanisms.

The presence of physical dispersion complicates the problem of inferring chemistry and mineralogy by comparing seismic data with high-frequency ultrasonic data. This is less a problem if only the bulk modulus or seismic parameter Φ is used.

Although seismic data alone are ambiguous regarding the presence or absence of partial melting, there are other constraints that can be brought to bear on the problem. Electrical conductivity, heat flow and the presence of volcanism often suggest the presence of partial melting in regions of the upper mantle where the seismic velocities are particularly low.

MINERALOGICAL MODELS OF 50–400 km DEPTH

Because of the intervention of partial melting and other relaxation phenomena in parts of the upper mantle, it is difficult to determine the mineralogy in this region. Figure 3-5 shows calculations for the seismic velocities for two different mineral assemblages. *Pyrolite* is a garnet peridotite composed mainly of olivine and orthopyroxene. *Piclogite* is a clinopyroxene- and garnet-rich aggregate with some olivine. Note the similarity in the calculated velocities. Below 200 km the seismic velocities under shields lie near the 1400° adiabat. Above 150 km the shield lithosphere is most consistent with cool olivine-rich material. The lower velocity regions have velocities so low that partial melting or some other high-temperature relaxation mechanism is implied. The adiabats falling below the solidus curves are predicted to fall in the partial melt field.

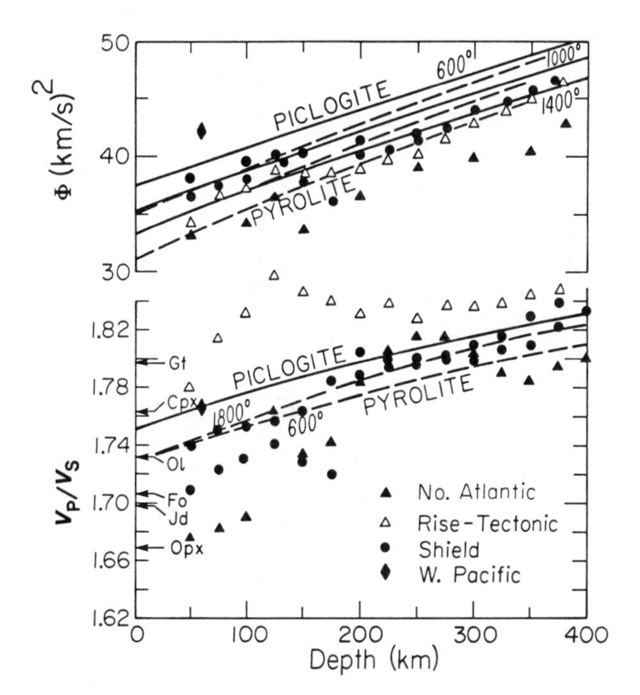

FIGURE 3-6
Seismic parameter Φ and V_p/V_s for two petrological models and various seismic models. Symbols and sources are the same as in Figure 3-5. V_p/V_s ratios for various minerals are shown in the lower panel (after Anderson and Bass, 1984).

Figure 3-6 shows the calculated and observed bulk modulus Φ and V_p/V_s. The high V_p/V_s ratio for the rise-tectonic mantle is consistent with partial melting in the upper mantle under these regions.

The upper 200 km or so of the mantle is anisotropic. Deeper levels may be as well, but it is more difficult to detect anisotropy at depth. The anisotropy of the shallow mantle and the low density of olivine and orthopyroxene, combined with their refractory nature, compared to garnet-rich aggregates, are indirect arguments in favor of a peridotite shallow mantle. Kimberlite pipes contain fragments that appear to have come from below the continental lithosphere. Peridotites are the most common xenolith, but some pipes contain abundant eclogite. The eclogite could be samples of oceanic crust that have been subducted under the continental lithosphere, or trapped melts which froze before they made their way to the surface.

THE TRANSITION REGION

The transition region of the upper mantle, Bullen's region C, is generally defined as that part of the mantle between the 400-km and 650-km discontinuities. Sometimes the mantle below the bottom of the low-velocity zone (~190–250 km) is included. The 400-km discontinuity is often equated with the olivine-spinel phase change, considered as an equilibrium phase boundary in a homogeneous mantle, but there are serious problems with this interpretation. The seismic velocity jump is much smaller than predicted for this phase change (Duffy and Anderson, 1988). The orthopyroxene-garnet reaction leading to a garnet solid solution is also complete near this depth, possibly contributing to the rapid increase of velocity and density at the top of the transition region. For these reasons the 400-km discontinuity should not be referred to as the olivine-spinel phase change. If the discontinuity is as small as in recent seismic models, then a change in chemistry near 400 km is implied, or the olivine content of this part of the mantle is low.

In the classical mantle models of Harold Jeffreys and Beno Gutenberg, the velocity gradients between 400 and

TABLE 3-10
Measured and Estimated Properties of Mantle Minerals

Mineral	ρ (g/cm^3)	V_p (km/s)	V_s (km/s)	V_p/V_s
Olivine ($Fa_{.12}$)	3.37	8.31	4.80	1.73
β–Mg_2SiO_4	3.63	9.41	5.48	1.72
γ–Mg_2SiO_4	3.72	9.53	5.54	1.72
Orthopyroxene ($Fs_{.12}$)	3.31	7.87	4.70	1.67
Clinopyroxene ($Hd_{.12}$)	3.32	7.71	4.37	1.76
Jadeite	3.32	8.76	5.03	1.74
Garnet	3.68	9.02	5.00	1.80
Majorite	3.59	9.05*	5.06*	1.79*
Perovskite	4.15	10.13*	5.69*	1.78*
($Mg_{.19}Fe_{.21}$)O	4.10	8.61	5.01	1.72
Stishovite	4.29	11.92	7.16	1.66
Corundum	3.99	10.86	6.40	1.70

*Estimated.
Duffy and Anderson (1988), Weidner (1986).

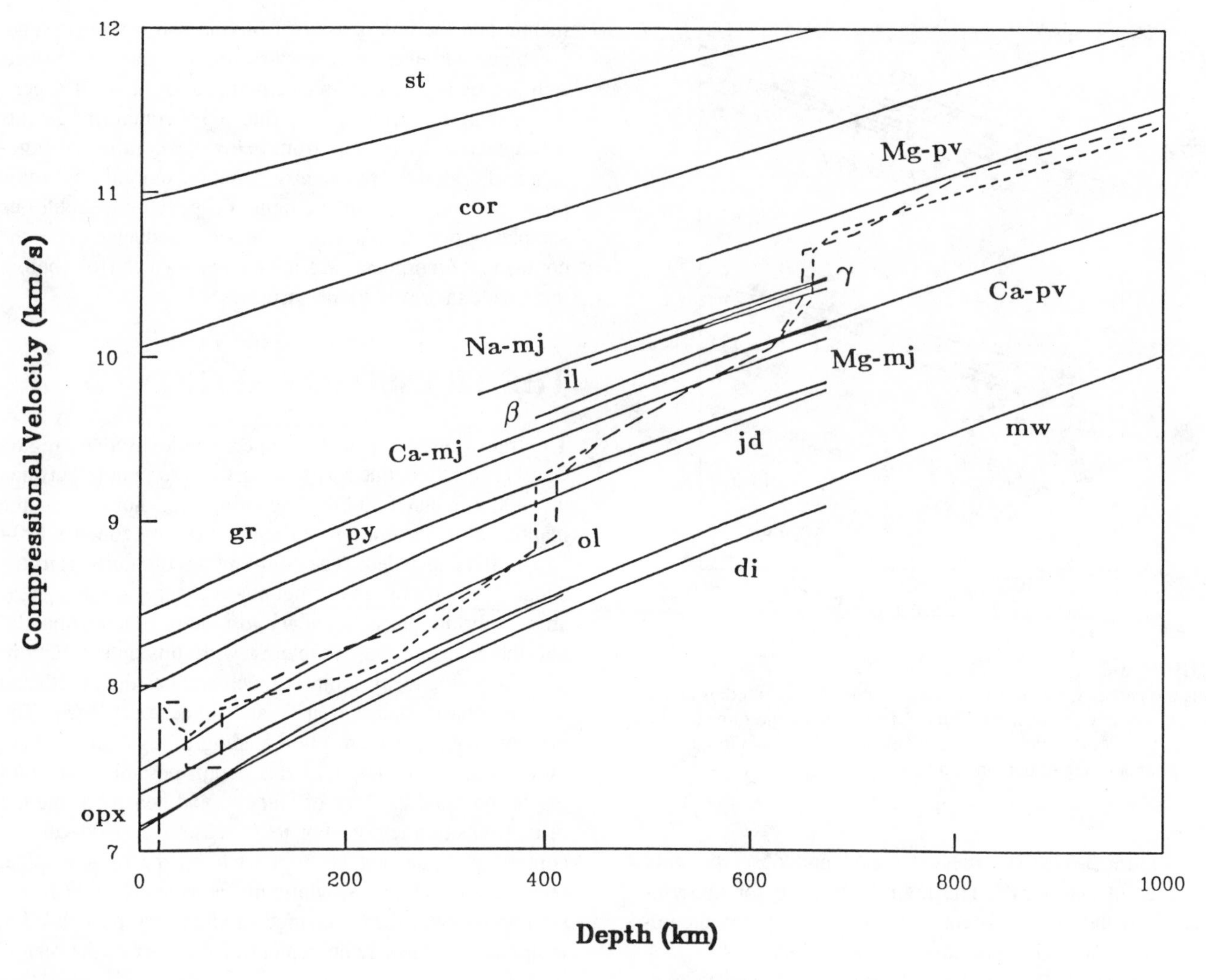

FIGURE 3-7
Calculated compressional velocity versus depth for various mantle minerals. "Majorite" (mj), "perovskite" (pv) and "ilmenite" (il) are structural, not mineralogical terms. The dashed lines are two recent representative seismic profiles (after Duffy and Anderson, 1988).

800 km were too high to be the result of self-compression; hence, it was called the transition region and was interpreted by Francis Birch as a region of phase changes. This region was later found to contain two major seismic discontinuities (Anderson and Toksöz, 1963; Niazi and Anderson, 1965; Johnson, 1967, 1969), one near 400 km and one near 650 km, which were initially attributed to the olivine-spinel and spinel–post-spinel phase changes, respectively, in an olivine-rich mantle (Anderson, 1967). (Properties of these and other deep mantle phases are listed in Table 3-10.) These phase changes are probably spread out over depth intervals of about 20 km and therefore result in diffuse seismic boundaries rather than sharp discontinuities. It was subsequently found that the 650-km discontinuity is a good reflector of seismic energy (Whitcomb and Anderson, 1970), requiring that its width be less than 4 km and that the large increase in elastic properties was not consistent with any phase change in olivine. There is also a high-gradient region below the discontinuity. The spinel–post-spinel transformation therefore is not an adequate explanation for the 650-km discontinuity. There appears to be no phase change in a chemically homogeneous mantle that has the requisite properties.

The velocity gradients between 400 and 650 km are higher than expected for a homogeneous self-compressed region. This region may represent the gradual conversion of diopside and jadeite to an Al_2O_3-poor garnet structure. In the presence of Al_2O_3-rich garnet, diopside is stable to much higher pressures than are calcium-poor pyroxenes.

In the transition zone the stable phases are garnet solid solution, β- and γ-spinel and, possibly, jadeite. Garnet solid solution is composed of ordinary Al_2O_3-rich garnet and SiO_2-rich garnet (majorite). The extrapolated elastic properties of the spinel forms of olivine are higher than

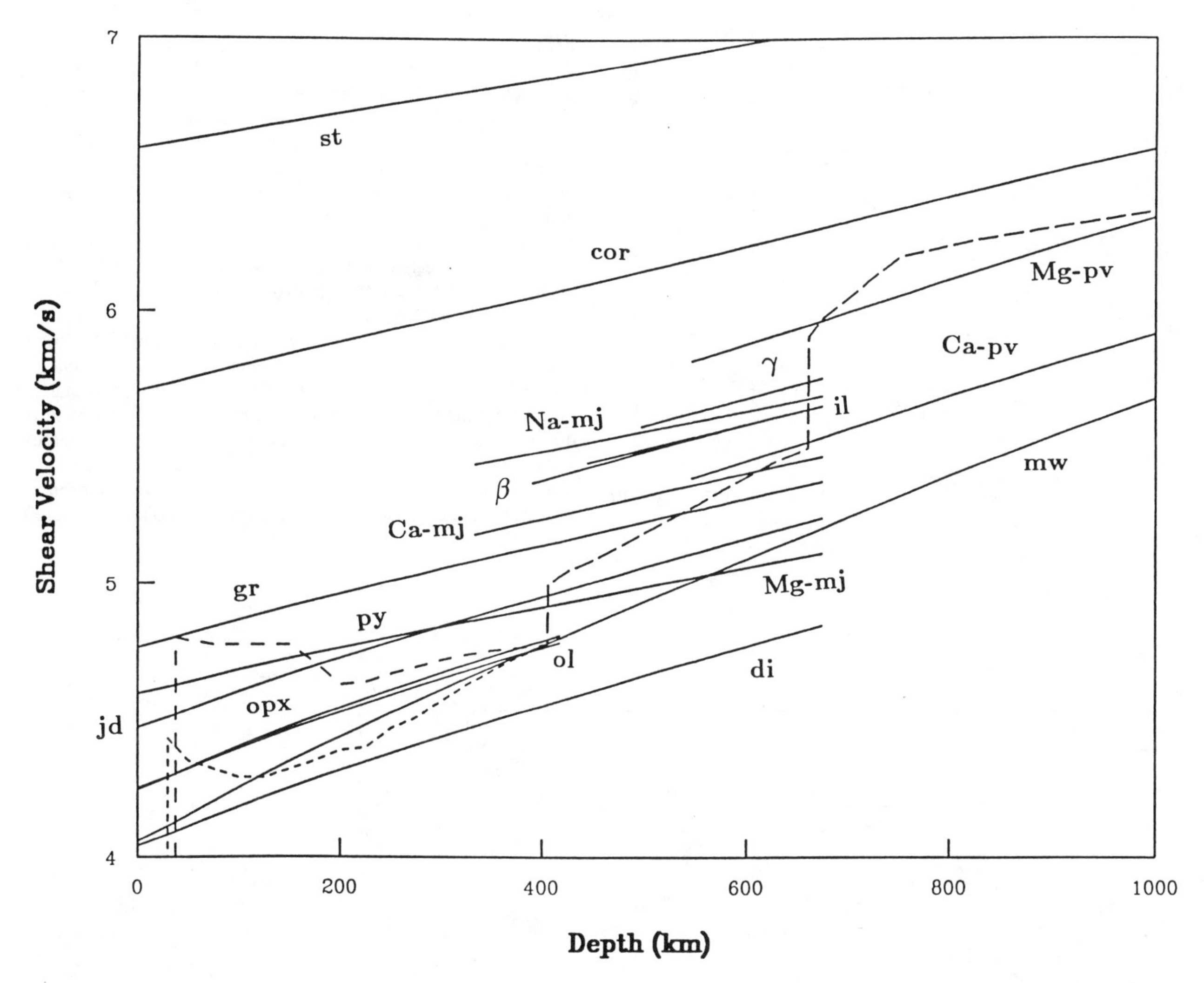

FIGURE 3-8
Same as Figure 3-7 but for the shear velocities (after Duffy and Anderson, 1988).

those observed (Figures 3-7 and 3-8). Pyroxenes in the garnet structure probably have elastic properties similar to ordinary garnet. (Mg,Fe)SiO_3 in the garnet structure is called "majorite"; I shall sometimes use this term to refer to any Al_2O_3-poor, SiO_2-rich garnet. The high velocity gradients throughout the transition zone imply a continuous change in chemistry or phase. Appreciable Al_2O_3-rich garnet is implied in order to match the velocities. A spread-out phase change involving clinopyroxene (diopside plus jadeite) transforming to Ca-rich majorite can explain the high velocity gradients. Detailed modeling (Bass and Anderson, 1984, Duffy and Anderson, 1988) suggests that olivine (in the β- and γ-structures) is not the major constituent of the transition zone. The assemblage appears to be more eclogitic than pyrolite. The best fitting mineralogy contains less than 50 percent olivine.

The changes in elastic properties at the α–β phase boundary are large, but those at the β–γ phase change appear to be minor (Weidner and others, 1984). A small amount of olivine and orthopyroxene are adequate to explain the magnitude of the 400-km discontinuity by changing phase at this depth (Figures 3-9 and 3-10). Phase changes, however, in general, are smeared out over a considerable depth range and do not result in sharp discontinuities. Recent work suggests that the olivine–β-spinel transition may occur over a rather narrow pressure interval, but the predicted increase in seismic velocities is much greater than observed (Bina and Wood, 1986). Other calculations favor a spread-out α-to-β transition.

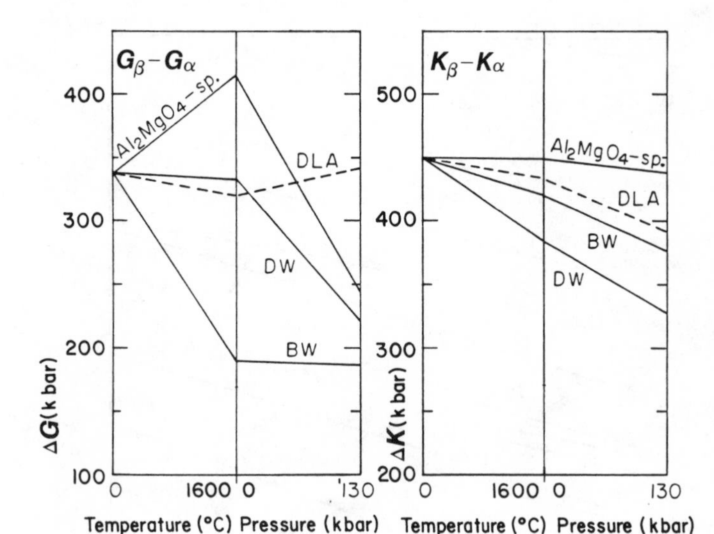

FIGURE 3-9
The 400-km seismic discontinuity is partially due to the α–β phase change in olivine. The amount of olivine implied by the size of the discontinuity depends on the as yet unmeasured pressure and temperature derivatives of the β-phase. The changes in rigidity (ΔG) and in bulk modulus (ΔK) associated with the phase change at zero pressure and low temperature are shown along the left axes. The changes with temperature and pressure are shown for various assumptions about the β-derivatives. If β-spinel has olivine-like derivatives, the ΔG and ΔK follow the dashed lines (DLA). The derivatives of Al_2MgO_4-spinel give the upper curves. Extreme values of the β-derivatives give the lines labeled DW and BW; in these two cases the ΔG and ΔK always decrease rightward and an olivine content of over 60 percent is allowed. For more normal values of the derivatives, less than 50 percent olivine is allowed. The conventional interpretation of the 400-km discontinuity is in terms of the olivine-spinel (β-phase) phase change in a peridotite mantle (olivine over 60 percent). These results suggest either a lower olivine content or a chemical transition to less olivine-rich material (as in an olivine eclogite).

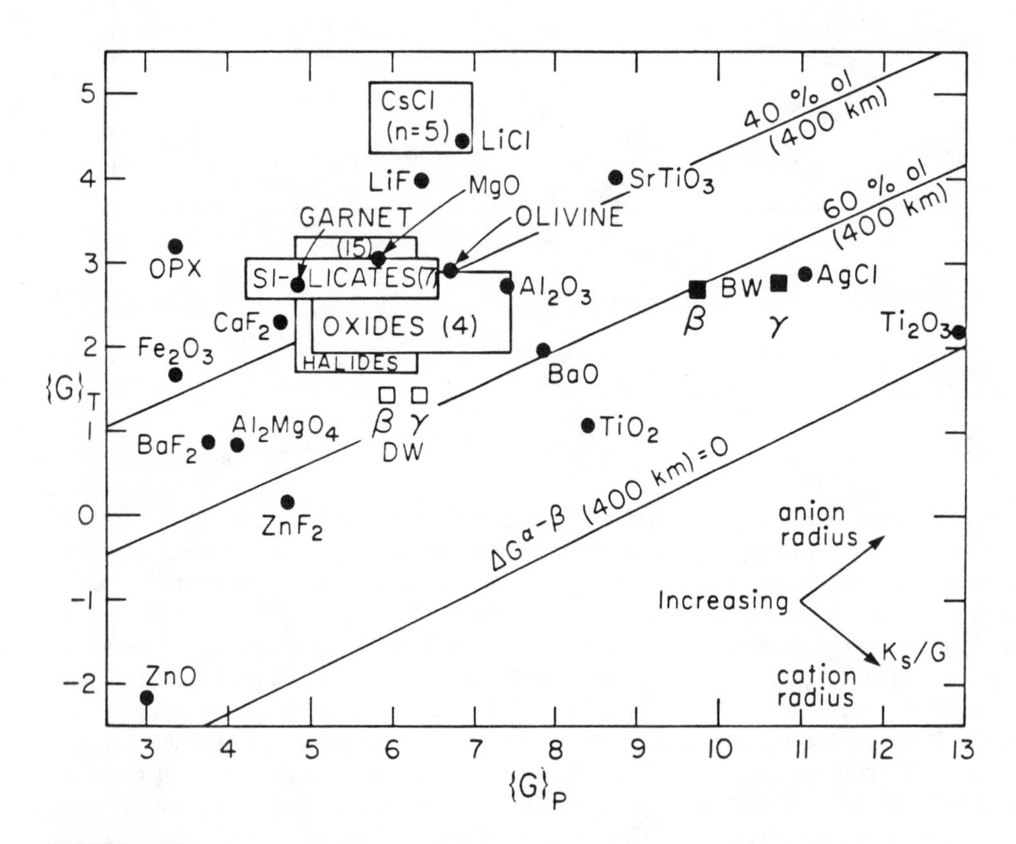

FIGURE 3-10
Normalized rigidity derivatives $\{G\}_T = (\partial \ln G/\partial \ln \rho)_T$ and $\{G\}_P = (\partial \ln G/\partial \ln \rho)_P$, for various minerals. The upper curve shows the $\{G\}_T$:$\{G\}_P$ relationship for silicates and oxides pertinent to the mantle. These values are consistent with an olivine content of 40 percent at 400 km based on the size of the seismic velocity jump at the 400-km discontinuity. The middle curve shows the parameter combinations required if the olivine content is 60 percent at 400 km. The lower curve assumes a zero rigidity increase at 400 km associated with the α–β phase change. BW and DW are parameters used by Bina and Wood (1986) and Weidner (1986).

General References

Anderson, D. L. (1982) Chemical composition and evolution of the mantle. In *High-Pressure Research in Geophysics* (S. Akimoto and M. Manghnani, eds.), 301–318, D. Reidel, Dordrecht, Neth.

Condie, K. C. (1982) *Plate Tectonics and Crustal Evolution,* 2nd ed., Pergamon, New York, 310 pp.

Dziewonski, A. M. and D. L. Anderson (1981) Preliminary reference Earth model, *Phys. Earth Planet. Inter., 25,* 297–356.

Jacobs, J. A. (1975) *The Earth's Core,* Academic Press, London, 253 pp.

Levy, E. H. (1976) Kinematic reversal schemes for the geomagnetic dipole, *Astrophys. J., 171,* 635–642.

Melchior, P. (1986) *The Physics of the Earth's Core,* Pergamon, New York, 256 pp.

Parker, E. N. (1983) Magnetic fields in the cosmos, *Sci. Am., 249,* 44–54.

Ringwood, A. E. (1979) *Origin of the Earth and Moon,* Springer-Verlag, New York, 295 pp.

Taylor, S. R. (1982) *Planetary Science, a Lunar Perspective,* Lunar and Planetary Institute, Houston, 482 pp.

Taylor, S. R. and S. M. McLennan (1985) *The Continental Crust: Its Composition and Evolution,* Blackwell, London.

References

Anderson, D. L. (1966) Recent evidence concerning the structure and composition of the Earth's mantle. In *Physics and Chemistry of the Earth, 6,* 1–131, Pergamon, Oxford.

Anderson, D. L. (1967) Phase changes in the upper mantle, *Science, 157,* 1165–1173.

Anderson, D. L. (1979) The deep structure of continents, *J. Geophys. Res., 84,* 7555–7560.

Anderson, D. L. and J. D. Bass (1984) Mineralogy and composition of the upper mantle, *Geophys. Res. Lett., 11,* 637–640.

Anderson, D. L. and A. M. Dziewonski (1982) Upper mantle anisotropy: Evidence from free oscillations, *Geophys. Jour. Roy. Astr. Soc., 69,* 383–404.

Anderson, D. L., H. Kanamori, R. S. Hart and H. P. Liu (1977) The Earth as a seismic absorption band, *Science, 196,* 1104–1106.

Anderson, D. L. and C. G. Sammis (1970) Partial melting in the upper mantle, *Phys. Earth Planet. Inter., 3,* 41–50.

Anderson, D. L. and M. N. Toksöz (1963) Upper mantle structure from Love waves, *Jour. Geophys. Res., 68,* 3483–3500.

Babuska, V. (1972) Elasticity and anisotropy of dunite and bronzitite, *J. Geophys. Res., 77,* 6955–6965.

Bamford, D. (1977) P_n velocity anisotropy in a continental upper mantle, *Geophys. J.R. Astron. Soc., 49,* 29–48.

Bass, J. D. and D. L. Anderson (1984) *Geophys. Res. Lett., 11,* 237–240.

Bina, C. and B. Wood (1986) *Nature, 324,* 449.

Christensen, N. I. and J. N. Lundquist (1982) Pyroxene orientation within the upper mantle, *Geol. Soc. Amer. Bull., 93,* 279–288.

Christensen, N. I. and J. D. Smewing (1981) Geology and seismic structure of the northern section of the Oman ophiolite, *J. Geophys. Res., 86,* 2545–2555.

Clark, S. P., Jr. (1966) *Handbook of Physical Constants.* Geol. Soc. Amer. Mem. 97, 587 pp.

Condie, K. L. (1982) *Plate Tectonics and Crustal Evolution,* 2nd ed., Pergamon, New York, 310 pp.

Drummond, B., K. Muirhead and A. L. Hales (1982) *Geophys. J. R. Astr. Soc., 70,* 67–77.

Duffy, T. and D. L. Anderson (1988) in press, *J. Geophys. Res.*

Dziewonski, A. M. and D. L. Anderson (1981) Preliminary reference Earth model, Phys. Earth Planet. Inter., 25, 297–356.

Forsyth, D. W. (1975) The early structural evolution and anisotropy of the oceanic upper mantle, *Geophys. J.R. Astron. Soc., 43,* 103–162.

Fuchs, K. (1977) Seismic anisotropy of the subcrustal lithosphere as evidence for dynamical processes in the upper mantle, *Geophys. J.R. Astron. Soc., 49,* 167–179.

Given, J. and D. Helmberger (1981) *J. Geophys. Res., 85,* 7183–7194.

Grand, S. and D. Helmberger (1984a) *Geophys. J.R.A.S., 76,* 399–438.

Grand, S. and D. Helmberger (1984b) *J. Geophys. Res., 88,* 801–804.

Gutenberg, B. (1959) *Physics of the Earth's Interior,* Academic Press, New York, 240 pp.

Hales, A. L., K. Muirhead and J. Rynn (1980) *Technophysics, 63,* 309–348.

Hart, R., D. L. Anderson and H. Kanamori (1976) The effect of attenuation on gross Earth models, *Earth and Planet. Sci. Lett., 32,* 25–34.

Isacks, B. and P. Molnar (1971) Distribution of stresses in the descending lithosphere from a global survey of focal-mechanism solutions of mantle earthquakes, *Rev. Geophys. Space Phys., 9,* 103–174.

Johnson, L. R. (1969) *Bull. Seis. Soc. Amer., 59,* 973–1008.

Johnson, L. R. (1967) Array measurements of P velocities in the upper mantle, *J. Geophys. Res., 72,* 6309–6325.

Jordan, T. H. (1979) In *The Mantle Sample* (F. R. Boyd and H. O. A. Meyer, eds.), American Geophysical Union, Washington, D.C.

Jordan, T. H. and L. N. Frazer (1975) Crustal and upper mantle structure from Sp phases, *J. Geophys. Res., 80,* 1504–1518.

Lehmann, I. (1961) S and the structure of the upper mantle, *Geophys. J. Roy. Astron. Soc., 4,* 124–138.

Leven, J. H., I. Jackson and A. Ringwood (1981) Upper mantle seismic anisotropy and lithospheric decoupling, *Nature, 289,* 234–239.

Liu, H. P., D. L. Anderson and H. Kanamori (1976) Velocity dispersion due to anelasticity; implications for seismology and mantle composition, *Geophys. J.R. Astron. Soc., 47,* 41–58.

Manghnani, M. et al. (1974) *J. Geophys. Res., 79,* 5427.

Morris, G. B., R. W. Raitt and G. G. Shor (1969) Velocity anisotropy and delay-time maps of the mantle near Hawaii, *J. Geophys. Res., 74,* 4300–4316.

Morse, S. A. (1986) *Earth Planet. Sci. Lett., 81,* 118–126.

Nataf, H.-C., I. Nakanishi and D. L. Anderson (1986) Measurements of mantle wave velocities and inversion for lateral heterogeneities and anisotropy, Part III. Inversion, *J. Geophys. Res., 91,* 7261–7307.

Niazi, M. (1969) Use of source arrays in studies of regional structure, *Bull. Seismol. Soc. Amer., 59,* 1631–1643.

Niazi, M. and D. L. Anderson (1965) Upper mantle structure of western North America from apparent velocities of P waves, *Jour. Geophys. Res., 70,* 4633–4640.

O'Hara, M. J. (1968) The bearing of phase equilibria studies in synthetic and natural systems on the origin and evolution of basic and ultrabasic rocks, *Earth Sci. Rev., 4,* 69–133.

Regan, J. and D. L. Anderson (1984) Anisotropic models of the upper mantle, *Phys. Earth and Planet. Int., 35,* 227–263.

Ringwood, A. E. (1975) *Composition and Petrology of the Earth's Mantle,* McGraw-Hill, New York, 618 pp.

Sacks, I. S., J. A. Snoke and E. S. Husebye (1977) Lithospheric thickness beneath the Baltic Shield, *Carnegie Inst. Yearb., 76,* 805–813.

Salisbury, M. and N. I. Christensen (1978) The seismic velocity structure of a traverse through the Bay of Islands ophiolite complex, Newfoundland, an exposure of oceanic crust and upper mantle, *J. Geophys. Res., 83,* 805–817.

Shimamura, H., T. Asada and M. Kumazawa (1977) *Nature, 269,* 680–682.

Sumino, Y. and O. L. Anderson (1984) Elastic constants of minerals. In *Handbook of Physical Properties of Rocks,* v. 3 (R. S. Carmichael, ed.), 39–138, CRC Press, Boca Raton, Florida.

Taylor, S. R. and S. M. McLennan (1985) *The Continental Crust: Its Composition and Evolution,* Blackwell, London.

Walck, M. (1984) *Geophys. J.R.A.S. 76,* 697–723.

Weidner, D. J. (1986) in *Chemistry and Physics of Terrestrial Planets,* Ed. S. K Saxena, Springer-Verlag, New York, 405 pp.

Weidner, D. J., H. Sawamoto, S. Sasaki and M. Kumazawa (1984) Single-crystal elastic properties of the spinel phase of Mg_2SiO_4, *J. Geophys. Res., 89,* 7852–7860.

Whitcomb, J. H. and D. L. Anderson (1970) Reflection of P′P′ seismic waves from discontinuities in the mantle, *J. Geophys. Res., 75,* 5713–5728.

4

The Lower Mantle and Core

I must be getting somewhere near the centre of the earth. Let me see: that would be four thousand miles down. I think—

—ALICE

The lower mantle starts just below the major mantle discontinuity near 650 km. The depth of this discontinuity varies, perhaps by as much as 100 km and is variously referred to as the "650-km discontinuity" or "670-km discontinuity." In recent Earth models there is a region of high velocity gradient for another 50 to 100 km below the discontinuity. This is probably due to phase changes, but it could represent a chemical gradient. The "lower mantle proper" therefore does not start until a depth of about 750 or 800 km. Below this depth the lower mantle is relatively homogeneous until about 300 km above the core-mantle boundary. If there is a chemical difference between the upper and lower mantle then, in a convecting dynamic mantle, the boundary will not be at a fixed depth. This clarification is needed because of the controversy about whether slabs penetrate into the lower mantle or whether they just push down the discontinuity.

COMPOSITION OF THE LOWER MANTLE

Several methods can be used to estimate the composition of the lower mantle from seismic data; perhaps the most direct is to compare shock-wave densities at high pressure of various silicates and oxides with seismically determined densities. The shock-wave Hugoniot data must be corrected to adiabats. There is a trade-off between temperature and composition, so this exercise is nonunique. Materials of quite different compositions, say (Mg,Fe)SiO_3 (perovskite) and (Mg,Fe)O, can have identical densities, and mixtures involving different proportions of MgO, FeO and SiO_2 can satisfy the density constraints. In addition, the density in the Earth is not as well determined as such parameters as the compressional and shear velocities. Nevertheless, many authors have used density alone to argue for specific compositional models for the lower mantle or to argue that the mantle is chemically homogeneous. The density of the lower mantle and the density jump at 650 km are very weak constraints on the chemistry of the lower mantle or the change in chemistry between the upper and lower mantle. Arguments based on viscosity and mean atomic weight are even weaker. The mineralogy of the lower mantle is even harder to determine since oxide mixtures, such as MgO + SiO_2 (stishovite), have densities, at high pressure, similar to compounds such as perovskite having the same stoichiometry.

The bulk modulus K_s can be determined by differentiating shock-wave data, $(\rho \partial P/\partial \rho)_s$, but this, of course, is subject to uncertainties. Nevertheless, using both ρ and K_s in comparisons with seismic data reduces the ambiguities. A more direct comparison uses the seismic parameter Φ, which can be determined from both seismology and shock-wave data:

$$\Phi = \left(V_p^2 - \frac{4}{3} V_s^2\right) = (\partial P/\partial \rho)_s$$

It has been shown that a chondritic composition for the lower mantle gives satisfactory agreement between shock-wave and seismic data (Anderson, 1977). Pyrolite, with 46 percent SiO_2, can not simultaneously satisfy both Φ and ρ. Watt and Ahrens (1982) also concluded that the SiO_2 content of the lower mantle is closer to chondritic than pyrolitic.

Another approach is to extrapolate seismic data to zero pressure with the assumption that the lower mantle is homogeneous and adiabatic (Anderson and Jordan, 1970; Anderson and others, 1971; Butler and Anderson, 1978). A variety of equations of state are available (discussed in Chapter 5) that can be used to fit ρ, G, K_s, V_p and V_s in the lower mantle, and the zero-pressure parameters can be compared with values inferred or measured for various candidate minerals and compositions. Although a large extrapolation is required, there is a large range of compressions available over the extent of the lower mantle, and the parameters of the equations of state can be estimated more accurately than they can over the available range of compressions in most static experiments. The temperature corrections to be applied to the extrapolated lower-mantle values are, of course, uncertain. Butler and Anderson (1978) concluded that pure "perovskite," $MgSiO_3$, was consistent with the seismic data. A range of $(Mg,Fe)SiO_3$ compositions is also permitted because of the uncertainty in the moduli of "perovskite" and lower mantle temperature.

The next approach is to use measured or inferred values of physical properties of various candidate lower-pressure phases (such as perovskite or magnesiowüstite) and to extrapolate to lower-mantle conditions. This method suffers from an extensive reliance on systematics involving analog compounds. Gaffney and Anderson (1973) and Burdick and Anderson (1975) concluded that the lower mantle was richer in SiO_2 than the upper mantle or olivine-rich assemblages. Bass and Anderson (1984) found that pyrolite and $(Mg,Fe)SiO_3$ (perovskite) gave similar results at the top of the lower mantle. The relatively homogeneous part of the lower mantle, however, does not set in until about 800 km depth.

In all of these approaches there is a trade-off between temperature and composition. If the lower mantle falls on or above the 1400° adiabat, then chondritic or pyroxenitic compositions are preferred. If temperatures are below the 1200°C adiabat, then more olivine ("perovskite" plus (MgFe)O) can be accommodated. A variety of evidence suggests that the higher temperatures are more appropriate.

Velocity-density systematics can also be applied to the problem (Anderson, 1970a). There are systematic variations between density, velocity and mean atomic weight $\overline{M}$. The lower mantle has higher Φ and ρ than inferred for the high-pressure forms of olivine and peridotite. This has been used to argue for iron enrichment in the lower mantle. It was later recognized that these systematics could not be applied through a phase change that involves an increase in coordination. An increase in coordination involves a large increase in density but only a small increase, or even a decrease, in seismic velocity (Anderson, 1970b). This weakens the arguments for FeO enrichment in the lower mantle but strengthens the arguments for SiO_2 enrichment.

Attempts to compute velocity throughout the mantle, assuming chemical homogeneity but allowing for phase changes, have not satisfied the seismic data, at least for an olivine-rich composition (Lees and others, 1983). A difference in composition between the upper and lower mantle is implied. The sharpness of the 650-km discontinuity implies either a univariant phase change, for which there is no laboratory evidence, or a compositional boundary. The absence of earthquakes below 690 km is indirect evidence, although inconclusive, for a chemical boundary that prevents penetrative convection.

The seismic velocities in the upper 150 km of the lower mantle exhibit a high gradient. This is probably due to the continuous transformation of garnet solid solution (garnet plus majorite) to "perovskite" and γ-spinel to "perovskite" plus (Mg,Fe)O. Reactions involving the ilmenite structure may also be involved.

The mantle between about 800 and 2600 km depth appears to be relatively homogeneous, although a slight increase with depth of FeO may be permitted (Gaffney and Anderson, 1973; Anderson, 1977).

Region D″, just above the core-mantle boundary, has a different gradient than the overlying mantle and may contain one or more discontinuities. It is also laterally inhomogeneous, causing scatter in the travel times and amplitudes of seismic waves that traverse it. It may be a region of high thermal gradient and small-scale convection, but its properties cannot be entirely explained by thermal boundary theory. Phase or compositional changes or both are probably involved. There is also some evidence that Q in this region is lower than in the overlying mantle (Anderson and Given, 1982).

D″ is a logical site for a chemically distinct layer. Light material from the core can be plated to the base of the mantle, and if denser than the mantle, there it will remain. Chemically dense blobs from the mantle would also settle on the core-mantle boundary. As the Earth was accreting, the denser silicates, as well as iron, would probably sink through the mantle. A basaltic layer at the surface would transform to eclogite at high pressure and could sink to the protocore-mantle boundary, unimpeded by the spinel–post-spinel or garnet-perovskite phase changes until the Earth was Mars-size or larger. This assumes that the perovskite phase change in eclogite occurs at a higher pressure than in Al_2O_3-poor material and the high-pressure form of eclogite is less dense than Al_2O_3-poor assemblages. Subduction today probably cannot provide material to the lower mantle. D″ may therefore be the site of ancient subducted lithosphere.

In the inhomogeneous accretion model the deep interior of the Earth would be initially rich in Fe and CaO-Al_2O_3-rich silicates. D″ may therefore be more calcium- and aluminum-rich than the bulk of the mantle. At D″ pressures this may be denser than "normal" mantle (Ruff and Anderson, 1980).

The seismic parameter Φ_o for the lower mantle ranges from about 61 to 63 km²/s², depending on the temperature assumed. For comparison $MgSiO_3$ (perovskite), Al_2O_3 and SiO_2 (stishovite) are 63, 63.2 and 73.7 km²/s², respectively.

(Mg,Fe)O ranges from 40.7 to 47.4 km^2/s^2 for reasonable ranges in iron content. Increasing FeO decreases Φ unless Fe is in the low-spin state (see discussion below). Therefore, it appears that MgO and SiO_2 in approximately equal molar proportions are implied for the lower mantle.

There is a slight drop in Poisson's ratio across the 650-km discontinuity. Temperature and pressure both increase Poisson's ratio in a homogeneous material, so this drop is an indication of a change in chemistry or mineralogy. Increasing the packing efficiency of atoms in a lattice and increasing coordination both serve to decrease the Poisson's ratio (Anderson and Julian, 1969). Spinels and garnets, the major minerals of the transition region, have zero-pressure Poisson's ratios of about 0.24 to 0.27. Stishovite and most perovskites have values in the range 0.22–0.23. The difference is about that observed across the discontinuity. MgO has a very low value, 0.18, and is estimated to be about 0.25 at 670 km. The observed value at the top of the lower mantle is about 0.27.

Two measures of homogeneity are dK/dP and the Bullen parameter (B.P.). These are tabulated in the Appendix for the Preliminary Reference Earth Model (PREM) of Dziewonski and Anderson. In homogeneous self-compressed regions we expect dK/dP to be a smoothly decreasing function of depth and B.P. to be close to unity. These conditions are satisfied approximately between 770 km and 2500 km depth. Velocity gradients are very low in the lower 200 km of the mantle. The region at the top of the lower mantle has high gradients, possibly due to the garnet-perovskite or garnet-ilmenite transitions.

The partitioning of trace elements into a (Mg,Fe)SiO_3-perovskite should be evident in upper-mantle chemistry if a deep (greater than 700 km) magma ocean existed or if material from the deep mantle is brought into the upper mantle. The trace-element patterns of the refractory elements can, however, be explained by partitioning between melts and the common upper-mantle minerals. This suggests a chemically zoned planet, formed by a low-pressure zone refining process, and a chemically isolated lower mantle.

CaO and Al_2O_3

According to arguments based on cosmic abundance, the major components of the lower mantle are MgO and SiO_2. CaO and Al_2O_3 are likely to be the next most abundant components, but their concentrations are expected to be low, particularly if the material in the lower mantle has experienced low-pressure melting and removal of the basaltic components. There may be regions, however, such as D″, that are enriched in refractories such as CaO and Al_2O_3. CaO and Al_2O_3 have densities and elastic properties similar to those inferred for the lower mantle, and therefore appreciable amounts may be accommodated without affecting the seismic properties. Thus they are essentially invisible and arbitrary amounts can be accommodated. Ca-rich perovskites, however, may have lower Φ than (MgFe)SiO_3-perovskite (Ruff and Anderson, 1980) and this may contribute to the low seismic gradients observed in D″. The low density of $CaSiO_3$-perovskite, compared to (Mg,Fe)SiO_3-perovskite, may prevent Ca-rich material such as eclogite from sinking into the lower mantle. The CaO content of planets constructed from new solar abundances of the refractories (see Chapter 1) is higher than chondritic and the CaO/Al_2O_3 ratio is higher.

$CaSiO_3$ transforms to a perovskite structure with a density about the same or slightly greater than $MgSiO_3$-perovskite (Ringwood, 1975, 1982). At lower pressures $CaSiO_3$ combines with $MgSiO_3$ to form diopside and Ca-garnets. CaO also combines with Al_2O_3 to form compounds at low pressure, but Ringwood (1975) argued that Al_2O_3 will not be accommodated in $CaSiO_3$-perovskite related compounds. $CaSiO_3 \cdot xAl_2O_3$ (garnet) will therefore disproportionate to $CaSiO_3$(perovskite) + xAl_2O_3 at high pressure. This transformation occurs at a much higher pressure than the $CaSiO_3$-perovskite transformation. In spite of the above comments, Liu (1977) apparently synthesized a phase $Ca_2Al_2SiO_7$ related to perovskite with a density of 4.43 g/cm^3. Weng and others (1983), on the basis of high-pressure measurements on the system $MgSiO_3 \cdot CaSiO_3 \cdot Al_2O_3$, concluded that $CaSiO_3$ forms a separate phase. Shock-wave measurements (Svendsen, 1987) on $CaSiO_3$ and $CaMgSi_2O_6$ give high-pressure phases with densities consistent with either mixed oxides or perovskite that have zero-pressure densities of 4.0 to 4.13 g/cm^3. At high pressure the measured densities are considerably less than the lower mantle. There was no evidence for a superdense phase. In fact, CaO-rich material approaches the density of the lower mantle only for high iron contents (about 18 mole percent $FeSiO_3$) (Svendsen, 1987). The disproportionation products of $CaSiO_3 \cdot xAl_2O_3$ will be even less dense because of the low density of Al_2O_3. The low inferred zero-pressure density for $CaSiO_3$ from high-pressure shock-wave experiments suggests that $CaSiO_3$ has not completely transformed even at pressures as high as 1.8 megabars.

On balance, it appears that $CaSiO_3$-perovskite exists as a separate phase in the lower mantle. At lower pressure $CaSiO_3$ forms garnet solid solutions with (Mg,Fe)$SiO_3 \cdot xAl_2O_3$ and therefore disproportionation reactions are involved in the formation of $CaSiO_3$-perovskite. Because of the broad stability interval of garnet, the transformation pressure is much higher than for pure $CaSiO_3$. $CaSiO_3$-perovskite, because of its low density relative to (Mg,Fe)SiO_3-perovskite and apparently high transformation pressure, does not contribute much to the negative buoyancy of eclogite in subducted slabs. Stishovite has a ρ_o of 4.29 g/cm^3 and will serve to substantially increase the density of subducted quartz-bearing eclogite, but not SiO_2-poor basalt/eclogite.

Ringwood assumed that the grossularite portion of garnet disproportionates to $CaSiO_3$ (perovskite) plus Al_2O_3 at depths above 670 km. However, this assemblage is less dense than the lower mantle or (Mg,Fe)SiO_3(perovskite). This suggests that the basaltic and pyroxenitic portions of subducted lithosphere, and the eclogite cumulates formed in early Earth history are trapped in the upper mantle. Depleted peridotite is also trapped in the upper mantle because of its low density. Subducted slabs will tend to depress a chemical interface at 650 km, and convection will also deform this boundary. In fact, the "650-km" discontinuity may vary in depth by more than 80 km. Depths reported in seismological studies range from 640 to 720 km. The sharpness of the discontinuity is consistent with a chemical discontinuity.

Garnet has an extensive stability field in a silicate of eclogite composition; the transformation from garnet to perovskite is probably not complete until about 750 km. This is much deeper than the transformation in olivine and Al_2O_3-poor pyroxene. Grossularite plus $CaSiO_3$ forms a garnet solid solution that is probably stable to 900 km (Liu, 1979) and is considerably less dense than (Mg,Fe)SiO_3 (perovskite), which is stable at shallower depths. The bulk density of eclogite at 650 km is therefore less than the density of the lower mantle. An eclogite layer is gravitationally stable at midmantle depths. Transformations in CaO and Al_2O_3-rich silicates probably contribute to the high velocity gradient found between the 400- and 650-km discontinuities.

It is often assumed that overridden oceanic lithosphere disappears out of the bottom of the Wadati-Benioff zone. Some aspects of continental geology, however, invoke the presence of subducted lithosphere 1000–3000 km into the continental interior (Dickinson and Snyder, 1978). The thermal lifetime of overridden oceanic lithosphere is very long. It heats up with time but cools the adjacent mantle, so that if the slab remains in the upper mantle it should show up as a high-velocity anomaly. If subducted material is trapped in the upper mantle, the western Atlantic and the Brazilian and Canadian shields will be underlain by oceanic lithosphere that represents Jurassic Pacific Ocean. In fact, these parts of the world are in geoid lows and have high upper-mantle velocities. The fate of subducted oceanic lithosphere is intimately related to the problems of whole-mantle versus layered mantle convection and chemical inhomogeneity of the mantle. There is, as yet, no convincing evidence that slabs sink into the lower mantle.

Low-Spin Fe^{2+}

Two alternate electronic configurations, high-spin and low-spin, are possible for Fe^{2+}. The high-spin (H.S.) state is usually stable in silicates and oxides at normal pressures. The ionic radius of the low-spin (L.S.) state is much smaller than the high-spin state, and a spin-pairing transition is induced by increased pressure. A large increase in density accompanies this phase transformation. For example, the volume change accompanying a phase change in Fe_2O_3 at 500 kbar, attributed to the high-spin–low-spin transition, is 11–15 percent.

Gaffney and Anderson (1973) proposed that spin-pairing is likely in the mantle at depths below 1700 km and perhaps at higher levels as well. The small ionic radius of Fe^{2+}(L.S.) probably means that Fe^{2+} will not readily substitute for Mg^{2+} under lower-mantle conditions. Additional Fe^{2+} (L.S.)O-bearing phases will form with high densities and bulk modulus. Assuming that Fe^{2+} spin-pairing occurs below 670 km, Gaffney and Anderson (1973) showed that the lower mantle could be enriched in FeO and SiO_2 relative to the upper mantle. The magnesium-rich phases of the lower mantle may be relatively iron free:

$$MgFeSiO_4 \rightarrow MgSiO_3 \text{ (perovskite)} + FeO(L.S.)$$

which would facilitate the entry of FeO into molten iron and removal to the core.

The possible presence of low-spin Fe^{2+} in the lower mantle complicates the interpretation of seismic data in terms of chemistry and mineralogy. The lower mantle may be chondritic or "solar" in major elements or it may be residual refractory material remaining after extraction of the basaltic elements, calcium, aluminum and sodium. In the latter case it would be expected to be depleted in the radioactive elements, uranium, thorium and potassium. At very high pressure FeO may become metallic and, therefore, readily enter the core.

REGION D″

The lowermost 200 km of the mantle, region D″, has long been known to be a region of generally low seismic gradient and increased scatter in travel times and amplitudes. Lay and Helmberger (1983) found a shear-velocity jump of 2.8 percent in this region that may vary in depth by up to 40 km. They concluded that a large shear-velocity discontinuity exists about 280 km above the core, in a region of otherwise low velocity gradient. The basic feature of a 2.75 ± 0.25 percent velocity discontinuity is present for each of several distinct paths. There appears to be a lateral variation in the velocity increase and sharpness of the structure, but the basic character of the discontinuity seems to be well established. Wright and Lyons (1981) found a rapid increase in compressional wave velocity of 2.5 to 3.0 percent about 200 km above the core-mantle boundary.

D″ may represent a chemically distinct region of the mantle. If so it may vary laterally, and the discontinuity in D″ would vary considerably in radius, the hot regions being elevated with respect to the cold regions. A chemically distinct layer at the base of the mantle that is only marginally denser than the overlying mantle would be able to rise into the lower mantle when it is hot and sink back when it cools off. The mantle-core boundary, being a chemical interface, is a region of high thermal gradient, at least in the colder parts of the lower mantle.

I argued earlier that neither the peridotitic nor the eclogitic portions of subducted oceanic lithosphere can sink into the lower mantle. However, while the Earth was accreting, conditions would have been more favorable for deep subduction of eclogite. D″ may therefore be the repository for ancient subducted lithosphere. Likewise, light material from the core may have underplated the mantle. In either case D″ would be more refractory (Ca-, Al-, Ti-rich) than the average mantle.

Because the core is a good conductor and has low viscosity, it is nearly isothermal. Lateral temperature variations can be maintained in the mantle, but they converge at the base of D″. This means that temperature gradients are variable in D″. In some places, in hotter mantle, the gradient may even be negative in D″. Regions of negative shear velocity gradient in D″ are probably regions of high temperature gradient and high heat loss from the core.

THE CORE

The core is approximately half the radius of the Earth and is about twice as dense as the mantle. It represents 32 percent of the mass of the Earth. A large dense core can be inferred from the mean density and moment of inertia of the Earth, and this calculation was performed by Emil Wiechert in 1891. The existence of stony meteorites and iron meteorites had earlier led to the suggestion that the Earth may have an iron core surrounded by a silicate mantle. The first seismic evidence for the existence of a core was presented in 1906 by Oldham, although it was some time before it was realized that the core does not transmit shear waves and is therefore probably a fluid. It was recognized that the velocity of compressional waves dropped considerably at the core-mantle boundary. Beno Gutenberg made the first accurate determination of the depth of the core, 2900 km, in 1912, and this is remarkably close to current values. The mantle-core boundary is sometimes referred to as the Gutenberg discontinuity and sometimes as the CMB.

Although the idea that the westward drift of the magnetic field might be due to a liquid core goes back 300 years, the fluidity of the core was not established until 1926 when Jeffreys pointed out that tidal yielding required a smaller rigidity for the Earth as a whole than indicated by seismic waves for the mantle. It was soon agreed by most that the transition from mantle to core involves both a change in composition and a change in state. Subsequent work has shown that the boundary is extremely sharp. There is some evidence for variability in depth, in addition to hydrostatic ellipticity. Variations in lower-mantle density and convection in the lower mantle can cause at least several kilometers of relief on the core-mantle boundary. The outer core has extremely high Q and transmits P-waves with very low attenuation. Evidence that the outer core is mainly an iron-rich fluid also comes from the magnetohydrodynamic requirement that the core be a good electrical conductor.

Although the outer core behaves as a fluid, it does not necessarily follow that temperatures are above the liquidus. It would behave as a fluid even if it contained 30 percent or more of suspended particles. All we know for sure is that at least part of the outer core is above the solidus or eutectic temperature and that the outer core, on average, has a very low rigidity and low viscosity. Because of the effect of pressure on the liquidus temperature, a homogeneous core can only be adiabatic if it is above the liquidus throughout. An initially homogeneous core with an adiabatic temperature profile that lies between the solidus and liquidus will contain suspended particles that will tend to rise or sink, depending on their density. The resulting core will be on the liquidus throughout and will have a radial gradient in iron content. The core will be stably stratified if the iron content increases with depth.

Inge Lehmann (1936) used seismic data from the "core shadow" to infer the presence of a higher velocity inner core. Although no waves have yet been identified that have traversed the inner core unambiguously as shear waves, indirect evidence indicates that the inner core is solid (Birch, 1952). Julian and others (1972) reported evidence for PKJKP, a compressional wave in the mantle and outer core that traverses the inner core as a shear wave, but this has yet to be confirmed. Early free-oscillation models (Jordan and Anderson, 1973) gave very low shear velocities for the inner core, 2 to 3 km/s, and some models (Backus and Gilbert, 1970) had entirely fluid cores. More recent models give shear velocities in the inner core ranging from 3.46 to 3.7 km/s (Anderson and Hart, 1976; Dziewonski and Anderson, 1981).

Gutenberg (1957) suggested that the boundary of the inner core is frequency dependent and, therefore, that the inner core might be a highly viscous fluid rather than a crystalline solid. The boundary of the inner core is also extremely sharp (Engdahl and others, 1970). The Q of the inner core is relatively low, and appears to increase with depth.

The high Poisson's ratio of the inner core, 0.44, has been used to argue that it is not a crystalline solid, or that it is near the melting point or partially molten or that it involves an electronic phase change. However, Poisson's ratio increases with both temperature and pressure and is expected to be high at inner core pressures, particularly if it is metallic (Anderson, 1977; Brown and McQueen, 1982). Some metals have Poisson's ratios of 0.43 to 0.46 even under laboratory conditions.

Table 4-1 presents numerous properties of the core.

Composition of the Core

Butler and Anderson (1978) fit a variety of equations of state to the seismic data for the outer core. Third-order finite strain theory was shown to be inadequate, and the best fits

TABLE 4-1
Properties of Core

Symbol	Property	Outer Core	Inner Core	Uncertainty
R	Radius (km)	3480	1221	
P	Pressure (Mbar)	1.36	3.29–3.64	
ρ	Density (g/cm^3)	9.90–12.17	12.76–13.09	
ρ_o		6.6–6.73	7.6	± 0.2
K_s	Bulk modulus (Mbar)	6.4–13.0	13.4–14.3	
K_o		1.2–1.4		
G	Shear modulus (Mbar)	<0.02	1.57–1.76	± 0.2
K_o'		4.3–4.8	1.76	± 0.2
V_p	Compressional velocity (km/s)	8.06–10.36	11.03–11.26	
V_s	Shear velocity (km/s)	~0	3.50–3.67	
V_ϕ	Bulk velocity (km/s)	8.06–10.36	10.26–10.44	
V_{ϕ_o}		4.3–4.6		± 0.35
γ	Grüneisen ratio	1.7	1.6	20 pct.
c_p	Specific heat (erg/g.K)	5×10^6		10 pct.
α	Expansivity (K^{-1})	10^{-5}		30 pct.
k	Thermal conductivity (erg/cm.K.s)	4×10^6		$\times 2$
σ	Electrical resistivity ($\mu\Omega$cm)	100–160		$\times 2$
ν	Shear viscosity (cm^2/s)	8×10^{-3}		$\times 10^2$
T_m	Melting temperature (K)	2600–5000	6150–7000	
R_m	Magnetic Reynolds number	200–600		$\times 10^2$
	Decay time (years)	15,000		
	Ohmic dissipation (W)			
	Poloidal	10^8		
	Toroidal	10^{10}–10^{12}	10^{11}	
	Heat loss (W)	10^{12}–10^{13}		
	Rotation rate (rad S^{-1})	7.29×10^{-5}		
	Westward drift	0.2°/yr		
	Dipole in core (Wb m^{-2})	3.8×10^4		
H_r	Poloidal field (gauss)	6		
H_ϕ	Toroidal field (gauss)	50–2400	$<10^6$	
μ	Permeativity	1		
	Heat of fusion (erg/g)	4×10^9		
	Ekman number	10^{-15}		
	Reynolds number	3×10^8		
	Rossby number	4×10^{-7}		
	Magnetic Rossby number	2×10^{-9}		

Verhoogen (1973), Ruff and Anderson (1980), Stevenson (1981), Jacobs (1975), Dziewonski and Anderson (1981), Melchior (1986), Gubbins (1977).

were obtained for fourth-order finite strain, Bardeen's equation of state and an equation of state involving an exponential repulsive potential. Their best fits for the region between 2200 and 3200 km radius gave the following values for zero-pressure quantities:

$$\rho_o = 6.60\text{–}6.71 \text{ g/cm}^3$$

$$K_o = 1.22\text{–}1.40 \text{ Mbar}$$

$$V_{p_o} = 4.30\text{–}4.57 \text{ km/s}$$

$$\Phi_o = 18.5\text{–}20.9 \text{ km}^2/\text{s}^2$$

$$K_o' = 4.5\text{–}4.8$$

These are uncorrected for temperature and therefore represent high-temperature values. Butler and Anderson concluded that a pure iron-nickel core has too high a density and too low a bulk sound velocity to be compatible with the seismic data. A lighter alloying element that increases the bulk sound speed seems to be required. The pressure derivative at K_o at $P = 0$ is K_o' and this appears to have normal values.

If the ratios of nonvolatile elements in the Earth are similar to those in the Sun and chondritic meteorites, then an iron-rich core is required. Some early workers proposed that silicates may undergo metallic phase changes and that material of high density, high electrical conductivity and

TABLE 4-2
Properties of Iron

Property	Units	Value
ρ_0	g/cm^3	7.02 (liq. at 1810 K)
		8.35 (ε)
α	K^{-1}	11.9×10^{-5} (liq.)
K_0	Mbar	1.40
		0.85 (liq.)
		1.95 (ε)
V_{ϕ_0}	km/s	3.80
γ	—	2.2–2.4
Electrical resistivity	$\mu\Omega$cm	140
Thermal conductivity	erg/cm K s	3.22×10^6
Shear viscosity	poises	3×10^{-3} (liq. at MP)

Ahrens (1979), Jeanloz and Knittle (1986), Stevenson (1981).

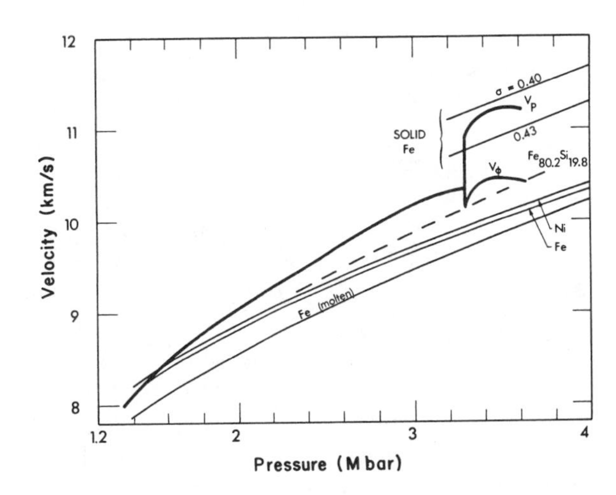

FIGURE 4-2
Compressional velocities (V_Φ) in the outer core and compressional (V_p) and bulk sound speeds (V_Φ) in the inner core (heavy lines) compared to estimates for iron and nickel. Values are shown for two Poisson's ratios σ in the inner core (after Anderson, 1977).

low melting point might be formed from silicates at high pressure. However, material of sufficiently high density has not been observed in any shock-wave or static-compression experiment on silicates or oxides, and the iron hypothesis is the most reasonable one. Properties of pure iron are listed in Table 4-2.

Figures 4-1 and 4-2 show that the properties of the core closely parallel the properties of iron but that a light alloying element is required that also serves to increase the compressional wave velocity. This alloying element should also serve to decrease the melting point, since the melting point of pure iron is probably higher than temperatures in the outer core. Elements such as nickel and cobalt are likely to be in the core, but if they occur in cosmic ratios with iron they will not affect the seismic properties and melting temperature very much. Candidate elements should dissolve in iron in order to affect the melting point and to avoid separating out of the core. Material held in suspension could reduce the velocity, but unless the core is turbulent, or the particles are very small, such material would rapidly settle out because of the presumed low viscosity of the core. This mechanism cannot be ruled out completely, because new suspended material may be constantly replenished by convection across the liquidus or by erosion of the lower mantle and inner core.

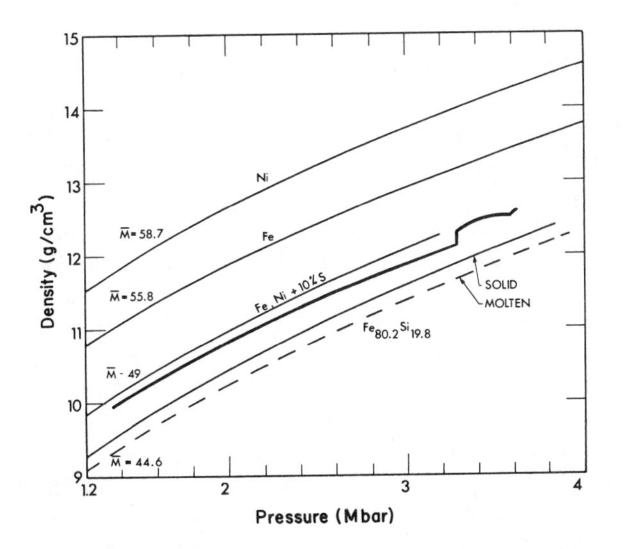

FIGURE 4-1
Estimated densities of iron, nickel and some iron-rich alloys, compared with core densities (heavy line). The estimated reduction in density due to melting is shown (dashed line) for one of the alloys (after Anderson, 1977).

Candidate materials, based on cosmic abundances alone, are hydrogen, helium, carbon, nitrogen, silicon, magnesium, oxygen and sulfur. The volatiles hydrogen, helium and possibly carbon, nitrogen and sulfur, which form volatile compounds under appropriate conditions, are depleted in the Earth relative even to the amount in the infalling planetesimals because of devolatilization during the accretional process. Silicon and magnesium are likely to partition strongly into the silicate phase, in preference to iron, at core pressure just as they do at low pressure. Some carbon, nitrogen, silicon and sulfur may enter the core since they form iron alloys. Sulfur and oxygen (perhaps as FeO or some other oxide) appear to be the strongest candidates for large concentrations in the core.

Sulfur depresses the melting point substantially ($\sim$1000°C) at low pressure. Shock-wave results indicate that 6 to 12 percent of sulfur can explain the density in the core (Anderson, 1977; Ahrens, 1979). This range has been confirmed by more recent data (Brown and McQueen, 1982). The density of α-iron (7.87 g/cm^3) is much greater than the sulfides of iron; compare, for instance, FeS (troilite), 4.83 g/cm^3; FeS (sphalerite structure), 3.60 g/cm^3; FeS (würtzite structure), 3.54 g/cm^3; FeS_2 (pyrite), 5.02 g/cm^3; and FeS_2 (marcasite), 4.89 g/cm^3. The seismic velocities of molten iron-sulfur alloys are unknown; velocities in solid sulfides are greater than in the corresponding metals, but it is not clear if this carries over to the molten state. Pyrite, for example, has a V_p of about 8 km/s, compared to 6 km/s for pure iron. Pyrrhotite has a bulk sound velocity $(\partial P/\partial \rho)_s$, or c, about 20 percent greater than ε-iron at high pressure (Brown and others, 1984). Zero-pressure ultrasonic data on pyrite (FeS_2) give a c_o of 5.23 km/s, which is much higher than the shock-wave speed of 3.45 km/s for pure iron at zero pressure. The bulk sound speeds in such sulfides as CdS and ZnS are 40 to 45 percent greater than in the metal. The approximate zero-pressure bulk sound speed of an FeS-Fe core is 3.9 km/s. For Ringwood's FeSi core the corresponding value is about 4.2 km/s. Butler and Anderson (1978) and Anderson and others (1971) estimated that c_o in the outer core is 4.35 to 5.2 km/s. Anderson and others (1971) estimated 3.1 to 3.7 km/s for shocked iron-nickel alloys with a possible further decrease of 7 to 15 percent to allow for melting. Values estimated for pyrrhotite are 4.4–4.9 km/s (Brown and others, 1984).

Thus, sulfur appears to have the appropriate characteristic to be the light alloying element in the core. Sulfur, however, is a volatile element and will tend to be lost upon accretion. Other volatiles that are unlikely to be sequestered in the core are also depleted in the crust-mantle system relative to carbonaceous chondrites, and it is difficult to argue that sulfur is immune to this depletion process. The mantle is not particularly depleted in chalcophiles relative to other volatiles.

The depletion of sulfur in the crust-mantle system, relative to carbonaceous chondrites, is quite remarkable, roughly 10^{-3}. This is about an order of magnitude more depletion than other volatiles such as thallium, lead, bismuth and indium. The depletion is comparable to that of siderophile refractories such as rhenium, osmium and iridium. It is not clear at this point whether it is primarily the volatile nature of sulfur that prevented it from being accreted by the Earth, or its siderophile nature that allowed it to be removed efficiently to the core as FeS. Some sulfur could have been incorporated into the early Earth as the refractory CaS. In order to explain the density of the core, about 20 percent to 50 percent of the cosmic complement of sulfur must have been retained by the Earth, and this seems excessive considering the depletion of other volatiles.

An Earth composed of cosmic or chondritic abundances gives the proper mantle/core mass ratio if the core composition is about Fe_2O or 50 mole percent FeO. This also gives about the right density for the core. It would be of interest to know if the hypothetical intermetallic compound Fe_2O is stable at high pressure.

Goto and others (1982) estimated values of ρ_o and c_o for a high-pressure phase of Fe_2O_3 of 6.22 g/cm^3 and 6.7 km/s, respectively. This can be compared with the zero-pressure values of 6.6 ± 0.15 g/cm^3 and 4.35 ± 0.35 km/s estimated by Butler and Anderson for the outer core. Earlier estimates for ρ_o gave 6.4 to 7.2 g/cm^3. Brown and McQueen (1982) obtained ρ_o = 8.28 g/cm^3 and c_o = 4.64 km/s for ε-iron. The high-pressure form of FeO has an estimated density of 6.7–8.4 g/cm^3 (McCammon and others, 1983). At the core-mantle boundary ρ and c are approximately 9.9 g/cm^3 and 8.1 km/s. Values estimated for liquid iron at comparable pressures are 10.8 g/cm^3 and 7.5 km/s (Brown and McQueen, 1982). Thus, it appears that iron alloyed with oxygen will have lower density and higher velocity than pure iron.

Reactions such as

$$FeO \rightarrow \frac{1}{3}(Fe_2O_3 + \varepsilon\text{-Fe})$$

$$FeO \rightarrow \frac{1}{4}(Fe_3O_4 + \varepsilon\text{-Fe})$$

may be energetically favorable at high pressure and could permit the FeO component of mantle silicates to disproportionate and remove ε-iron to the core. The low-spin transition in Fe^{2+} would favor the creation of separate iron-rich phases (Gaffney and Anderson, 1973), which might then be involved directly in the above reactions. Ahrens (1979) concluded that the density of the core permitted 7–8 percent oxygen, slightly less than the allowable range for sulfur.

If iron-bearing silicates can disproportionate to separate iron-rich phases at high pressure, then it may be possible to form a core without invoking reduction of iron oxides at the surface or having free iron drain through the upper mantle. The presence of siderophiles in the upper mantle and water at the surface both argue against free iron near the surface, at least in the terminal stages of accretion. If fully oxidized material, such as carbonaceous chondrites, accreted to form the Earth, then there must be a mechanism for reducing the high fayalite content of meteoritic olivine to values appropriate for the mantle and, at the same time, preventing the complete stripping of siderophiles from the upper mantle.

One apparent problem with the oxygen-rich core hypothesis is the very limited solubility of oxygen in molten iron at low temperatures and pressures. However, at high temperature and pressure molten iron can dissolve a considerable amount of oxygen. At 2400°C, for example, molten iron can contain 40 mole percent FeO (Ohtani and Ringwood, 1984). At 2800°C molten iron in equilibrium with $(Mg_{0.8}Fe_{0.2})O$ is predicted to contain about 40 mole percent of FeO. Solubility of FeO in molten iron also increases sharply with pressure. The Fe-FeO phase diagram should resemble a simple eutectic system above about 20 GPa (McCammon and others, 1983). The solubility of FeO in molten iron in equilibrium with $(Mg_{0.8}Fe_{0.2})O$ at 2500°C increases from 14 mole percent at P = 0 to 25 mole percent at 20 GPa. Since the core is presumably in equilibrium with the silicates and oxides at the base of the mantle, it is likely that the core contains considerable oxygen. It appears that the core can dissolve enough FeO to explain its low density and to considerably lower its melting point.

Ringwood (1966) rejected hydrogen, helium, carbon, oxygen and nitrogen as important elements in the core because they form interstitial solid solutions with iron and would therefore not decrease the density. The applicability of this argument to molten iron at core pressures and temperatures is obscure, but it led Ringwood to favor silicon as the light element in the core. Ringwood also argued strongly against sulfur. By putting some silicon in the core and vaporizing more silicon in the terminal stages of accretion, he managed to generate an olivine-rich mantle from cosmic abundances. If the core of the Earth is formed *in situ* by reduction, the reaction products, H_2 and CO, plus silicon would form a massive atmosphere totaling more than half the mass of the core, and an efficient dissipation mechanism must be postulated.

Balchan and Cowan (1966) determined the density of shocked iron-silicon alloys at conditions comparable to those in the core and concluded that their results were consistent with a core containing 14 to 20 percent silicon in iron by weight. The zero-pressure, room-temperature densities of these compositions are 7.02 and 7.25 g/cm^3. The zero-pressure bulk sound speed, c_o, of the iron-silicon alloys lies between 4.1 ± 0.4 (4 percent silicon) and 5.4 ± 0.1 km/s (20 percent silicon). These values bracket estimates for the core.

There are other possible meteorite-based models for the core. Mixing of 40 percent carbonaceous I, 46 percent ordinary, and 14 percent iron meteorites, for example, yields the proper core-mantle ratio. The mean atomic weight and the zero-pressure density of the resulting core are 50.5 and 6.34 g/cm^3, respectively; the sulfur content is 14 percent by weight. The density will be reduced by up to 5 percent upon melting. The mantle, for this mix, contains 18.4 percent by weight of FeO. Another approach is to reduce some of the FeO and SiO_2 of a carbon-, sulfur- and H_2O-free type I carbonaceous chondrite in order to obtain a mantle composition similar to pyrolite and to obtain the proper silicate/metal or mantle/core ratio. The resulting core has 11 percent silicon by weight, a mean atomic weight of 50.4, and a zero-pressure density of 6.24 g/cm^3; the last two values are very close to those estimated above for the iron-sulfur core. There is no particular reason, however, for postulating a mantle that is deficient in silicon, as in the pyrolite model.

Thus it appears that silicon, oxygen and sulfur all serve to decrease the density and increase the velocity of iron. These estimates are very crude and do not completely take into account phase changes or melting. The point is that the various alloys all have similar physical properties. More shock-wave and static-compression data on mixtures may be able to resolve the possibilities, particularly if accurate values for c_o can be determined. The "chondritic coincidence," the fact that the core is in contact with the mantle, the depletion of the Earth in volatiles and the high solubility of FeO in molten iron at high temperature and pressure all favor oxygen as the major light element in the core. A possible implication is that the lower mantle, in particular region D″, may be deficient in FeO. If FeO has been preferentially stripped out of the lowermost mantle, then the parts so affected would be rich in MgO and pyroxene relative to primitive mantle; for example,

$$(Mg_{0.5}Fe_{0.5})_2\,SiO_4 \rightarrow FeO + MgSiO_3$$

$$\text{Iron-rich olivine} \rightarrow \text{Wüstite} + \text{Enstatite}$$

The motivation for placing silicon in the core is that the upper mantle is deficient in silicon relative to cosmic abundances. However, there are magmatic processes for concentrating olivine in the shallow mantle and seismic evidence in favor of a chondritic Mg/Si ratio for the mantle as a whole. The melting point of Fe + S + O, at high-pressure, has not yet been determined. This is likely to be much lower than Fe + O or Fe + S and the core may have a much lower temperature than generally assumed.

The Inner Core

The inner core has a radius of 1222 km and a density about 13 g/cm^3. It represents about 1.7 percent of the mass of the Earth. The density and velocity jumps at the inner core–outer core boundary are large enough, and the boundary is sharp enough, so that the inner core boundary is a good reflector of short-period seismic energy.

There is a jump in V_p at the boundary, but the bulk sound speed $\sqrt{K/\rho}$ is nearly continuous. The increase in V_p may therefore be almost entirely due to the presence of a rigidity term, that is, $V_p = \sqrt{(K + 4/3G)/\rho}$, with no change in composition (Figure 4-2).

Because of the small size of the core, it is difficult to determine an accurate value for density. The main constraint on composition is therefore the compressional velocity. Within the uncertainties the inner core may be simply a frozen version of the outer core, Fe_2O or FeNiO, pure iron or an iron-nickel alloy. If the inner core froze out of the outer core, then the light alloying element may have been excluded from the inner core during the freezing or sedimentation process. An inner core growing over time could therefore cause convection in the outer core and may be an important energy source for maintaining the dynamo.

The possibility that the outer core is below the liquidus, with iron in suspension, presents an interesting dynamic problem. The iron particles will tend to settle out unless held in suspension by turbulent convection. If the composition of the core is such that it is always on the iron-rich side of the eutectic composition, the iron will settle to the inner core–outer core boundary and increase the size of the solid inner core. Otherwise it will melt at a certain depth in the core. The end result may be an outer core that is chemically inhomogeneous and on the liquidus throughout. The effect of pressure on the liquidus and the eutectic composition may, however, be such that solid iron particles can form in the upper part of the core and melt as they sink. In such a situation the core may oscillate from a nearly chemically homogeneous adiabatic state to a nearly chemically stratified unstable state. Such complex behavior is well known in other nonlinear systems. The apparently erratic behavior of the Earth's magnetic field may be an example of chaos in the core, oscillations controlled by nonlinear chemistry and dynamics.

Since the outer core is a good thermal conductor and is convecting, the lateral temperature gradients are expected to be quite small. The mantle, however, with which the outer core is in contact, is a poor conductor and is convecting much less rapidly. Seismic data for the lowermost mantle indicate large lateral changes in velocity and, possibly, a chemically distinct layer of variable thickness. Heat can only flow across the core-mantle boundary by conduction. A thermal boundary layer, a layer of high temperature gradient, is therefore established at the base of the colder parts of the mantle. That in turn can cause small-scale convection in this layer if the thermal gradient and viscosity combine to give an adequately high Rayleigh number. It is even possible for material to break out of the thermal boundary layer, even if it is also a chemical boundary, and ascend into the lower mantle above D″. The lateral temperature gradient near the base of the mantle also affects convection in the core. This may result in an asymmetric growth of the inner core. Hot upwellings in the outer core will deform and possibly erode or dissolve the inner core. Iron precipitation in cold downwellings could serve to increase inner-core growth rates in these areas. These considerations suggest that the inner-core boundary might not be a simple surface in rotational equilibrium.

The orientation of the Earth's spin axis is controlled by the mass distribution in the mantle. The most favorable orientation of the mantle places the warmest regions around the equator and the coldest regions at the poles. Insofar as temperatures in the mantle control the temperatures in the core, the polar regions of the core will also be the coldest regions. Precipitation of solid iron is therefore most likely in the axial cylinder containing the inner core.

There are two processes that could create a solid inner core: (1) Core material was never completely molten and the solid material coalesced into the solid inner core, and (2) the inner core solidified due to gradual cooling, increase of pressure as the Earth grew, and the increase of melting temperature with pressure. It is possible that both of these processes have occurred; that is, there was an initial inner core due to inhomogeneous accretion, incomplete melting or pressure freezing and, over geologic time, there has been some addition of solid precipitate. The details are obviously dependent on the early thermal history, the abundance of aluminum-26 and the redistribution of potential energy. The second process is controlled by the thermal gradient and the melting gradient. The inner core is presently 5 percent of the mass of the core, and it could either have grown or eroded with time, depending on the balance between heating and cooling. Whether or not the core is thermally stable depends on the distribution of heat sources and the state of the mantle. If all the uranium and thorium is removed with the refractories to the lowermost mantle, then the only energy sources in the core are cooling, a growing inner core and further gravitational separation in the outer core.

In the inhomogeneous accretion model the early condensates, calcium-aluminum-rich silicates, heavy refractory metals, and iron accreted to form the protocore (Ruff and Anderson, 1980). The early thermal history is likely to be dominated by aluminum-26, which could have produced enough heat to raise the core temperatures by 1000 K and melt it even if the Earth accreted 35 Ma after the Allende meteorite, the prototype refractory body. Melting of the protocore results in unmixing and the emplacement of refractory material (including uranium, thorium and possibly ^{26}Al) into the lowermost mantle. Calculations of the physical properties of the refractory material and normal mantle suggest that the refractories would be gravitationally stable in the lowermost mantle but would have a seismic velocity difference of a few percent.

Depending upon the available heat energy, the iron core could have been either completely or partially molten at the time of unmixing. Therefore, the present solid inner core could be remnant solid iron (or iron-nickel) from the segregation event, or it may have grown through geologic time from the precipitation of the solid phase from the fluid core.

Ruff and Anderson (1980) proposed that aluminum-26 dominated the early thermal history and that long-lived radioactive heat sources are distributed irregularly in the lowermost mantle and drive the fluid motions in the core that are responsible for the geodynamo. The anomalous lower-mantle velocity gradient suggests chemical inhomogeneity and/or a high thermal gradient. The seismic evidence for lateral variation at the base of the mantle is evidence for either variable temperature or varying composition. A new driving mechanism, differential cooling from above, was proposed to sustain the dynamo.

The lowest seismic velocity regions of the lowermost mantle are preferentially located in the equatorial regions. If these are due to high temperature, then downwellings in the outer core will be preferentially located in high latitudes where the lowermost mantle appears to be coldest. Lateral variations in D″ temperature, temperature gradient and radioactivity probably control the pattern of convection in the core, even if the dynamo is not driven from above.

Eventually, one would hope to see similarities in lower-mantle tomographic maps and maps of the magnetic field. Temperature differences in D″ and at the top of the core may also generate contributions to the magnetic field by the thermoelectric effect.

MANTLE-CORE EQUILIBRATION

Upper mantle rocks are extremely depleted in the siderophile elements such as cobalt, nickel, osmium, iridium and platinum, and it can be assumed that these elements have mostly entered the core. This implies that material in the core had at one time been in contact with material currently in the mantle, or at least the upper mantle. Alternatively, the siderophiles could have experienced preaccretional separation, with the iron, from the silicate material that formed the mantle. In spite of their low concentrations, these elements are orders of magnitude more abundant than expected if they had been partitioned into core material under low-pressure equilibrium conditions. The presence of iron in the mantle would serve to strip the siderophile elements out of the silicates. The magnitude of the partitioning depends on the oxidation state of the mantle. The "overabundance" of siderophiles in the the upper mantle is based primarily on observed partitioning between iron and silicates in meteorites. The conclusion that has been drawn is that the entire upper mantle could never have equilibrated with metallic iron, which subsequently settled into the core. Various scenarios have been invented to explain the siderophile abundances in the mantle; these include rapid settling of large iron blobs so that equilibration is not possible or a late veneer of chondritic material that brings in siderophiles after the core is formed. The trouble with the latter explanation is that the siderophiles do not occur in the mantle in chondritic ratios, although they are not fractionated as strongly as one would expect if they had been exposed to molten iron. Some groups of siderophiles do have chondritic ratios.

Brett (1971) took another look at this problem. He argued that the iron-rich liquid involves the system Fe-S-O and looked at the partitioning of several metals (Co, Cu, Ni, Ga and Au) between this liquid and olivine and basaltic melts (Table 4-3). The calculated abundances for the silicate phase were remarkably close to upper-mantle abundances, and thus it appears that protocore material could have been in equilibrium with the upper mantle. Further, a protocore containing sulfur and oxygen seems likely.

Since the upper-mantle siderophile abundances fit a local equilibration model, the implication is that the upper mantle has not been mixed with the rest of the mantle since core formation. The partition coefficients depend on temperature, pressure and oxidation state, and it is unlikely that they are constant throughout the mantle. This is relevant to the question of whole-mantle versus layered-mantle convection and the chemical isolation of the lower mantle from the upper mantle. Since core formation was an early process, the implication is that subsequent convection did not homogenize the mantle. When a larger number of siderophile elements is considered, the original problem reemerges.

The highly siderophile elements (Os, Re, Ir, Ru, Pt, Rh, Au, Pd) have high metal-silicate partition coefficients and therefore strongly partition into any metal that is in contact with a silicate. These elements are depleted in the crust-mantle system by almost three orders of magnitude compared to cosmic abundances but occur in roughly chondritic proportions. If the mantle had been in equilibrium with an iron-rich melt, which was then completely removed to form the core, they would be even more depleted and would not occur in chondritic ratios. Either part of the melt remained in the mantle or part of the mantle, the part we sample, was not involved in core formation and has never been in contact with the core. Many of the moderately siderophile elements (including Co, Ni, W, Mo and Cu) also occur in nearly chondritic ratios, but they are depleted by about an order of magnitude less than the highly siderophile elements. They are depleted in the crust-mantle system to about the extent that iron is depleted. These elements have a large range of metal-silicate partition coefficients, and their relatively constant depletion factors suggest, again, that the upper mantle has not been exposed to the core or that some core-forming material has been trapped in the upper mantle.

It is not clear why the siderophiles should divide so clearly into two groups with chondritic ratios occurring among the elements within, but not between, groups. The least depleted siderophiles are of intermediate volatility, and very refractory elements occur in both groups.

TABLE 4-3
Partitioning Between Sulfide Melt and Silicates

M	$M_{sulfide}/M_{silicate}$ (1)	$M_{Fe}/M_{silicate}$ (2)
Ni	150–560	1700
Cu	50–330	330–50
Co	7–80	200
Ga	4	—
Ge	—	1000
Re	2×10^3	$\sim 10^5$
Au	10^4	$\sim 10^5$
W	100	—
Ir	400	—
Mo	10^5	—
P	200	—
Ag	250	—
Pb	16	—

(1) Brett (1984), Jones and Drake (1985).
(2) Ringwood (1979).

THE MAGNETIC FIELD

The magnetic fields of planets and stars are generally attributed to the dynamo action of a convecting, conducting core. The study of the interaction of a moving electrically conducting fluid and a magnetic field is called *magnetohydrodynamics*. Magnetic fields entrained in a conducting fluid are stretched and folded by the fluid motion, gaining energy in the process, and thus acting as a *dynamo*, a device that converts mechanical energy into the energy of an electric current and a magnetic field. A moving conductive fluid can amplify a magnetic field. The dynamo mechanism does not explain how the magnetic field originated, only how it is amplified and maintained in spite of the losses caused by the dissipation of the associated current. Fluid motions of the conducting liquid in the presence of the magnetic field induce currents that themselves generate the field. The fluid motions may be due to a variety of causes including precession, thermal convection and chemical convection.

The magnetic field can be visualized as lines of force, the closed loops along which a compass needle aligns itself. The strength of the field in any given volume can be represented by the density of lines in the volume. One may regard the field lines as being "frozen" into the conducting fluid or attached to the particles of which the fluid is composed. The field moves with the fluid, and the stretching of the field lines corresponds to a gain in strength of the field. The energy of the motion of the particles is converted into the energy of a magnetic field, and induced electromotive forces drive the current associated with the field.

The first requirements for a magnetohydrodynamic dynamo are the presence of a magnetic field and an electrically conducting fluid capable of supporting the currents associated with the field. The second requirement is a pattern of fluid motion that amplifies the magnetic field. The naturally occurring combination of *nonuniform rotation* and *cyclonic convection* seems to be particularly effective since these occur in planets, stars and galaxies, all of which can exhibit magnetic fields. A rotating body containing a convecting fluid exhibits differential rotation and cyclonic convection.

The dipolar magnetic field of the Earth is associated with circular electric currents of about 2×10^9 amperes flowing from east to west in the molten iron core. Local anomalies in the field, having dimensions of several thousand kilometers and amplitudes of about 10 percent of the main field, change slowly with time, drifting westward at about 20 km per year. This surface drift rate corresponds to a fluid velocity at the surface of the core of about a meter per hour. The nondipole field is generally attributed to a dozen or so cells in the core. The most obvious explanation for the slow rotation of the core is the action of the Coriolis force on the rising and sinking fluid in the convective cells. The conservation of angular momentum requires that the angular velocity of the rising fluid decrease as it moves further from the spin axis. Therefore, the surface of the core rotates faster at high latitudes than at low latitudes and the inner part of the core rotates faster than the surface.

The primary magnetic field in the core is an east-west field, at right angles to the mean component of the field at the surface. It is called the *azimuthal* or *toroidal field;* the part lying in the planes through the axis is called a *meridional field*. The azimuthal field is created by the stretching of the north-south lines of force of the dipole field as they are carried around in the rotating fluid of the core. The part of a field lying near the axis is carried around further than the parts lying away from the axis; this nonuniform rotation stretches the north-south lines in an east-west direction. As the field lines are carried around, the azimuthal field gets stronger. The amplification continues until it is balanced by the tension of the magnetic lines of force or the resistive decay of the associated electric current. The azimuthal field in the core may be hundreds of times stronger than the dipole field observed at the surface, perhaps 100 gauss or more. The dipole field observed at the surface of the Earth is therefore a secondary effect of the azimuthal field, which is shielded from view by the insulating mantle.

In the 1930s T. G. Cowling proved that fluid motions cannot generate a perfect dipole field or any field with rotational symmetry about an axis. However, cyclonic convection can generate a dipole field. Cyclonic motion raises and rotates the lines of force of the azimuthal field, deforming them into helixes. Intermittent cyclonic convection generates a net dipole field. The essential ingredient for the generation of a field is that the motion of the fluid be helical with the field rotating about its direction of motion as it streams along.

A constraint on the terrestrial dynamo is that it must amplify the dipole field at a rate high enough to balance the decay of the field by ohmic dissipation. The magnetic field in a current-carrying body decays in a characteristic time that is proportional to the conductivity times the cross-sectional area. The strength of the magnetic field is determined by the number of times the field lines can be wrapped around the Earth in their lifetime.

Another property of the Earth's magnetic field is its ability to reverse its polarity abruptly in 1000 years or less, at apparently random intervals of about 10^5 to 10^7 years. Reversals of the magnetic field might be caused by sudden increases in the velocity of convection in the core. This in turn might be triggered by convection in the mantle, through instabilities in the thermal boundary layer at the base of the mantle or changes in core-mantle coupling caused by convection-induced irregularities in the shape of the boundary. There is some evidence that magnetic field variations are correlated with plate tectonic and magmatic events. Reversals might also be the result of the intrinsic nonlinear behavior of the core: nonperiodic chaotic behavior.

Due to the mathematical difficulties in treating the

complete dynamical system, dynamo models derived for the terrestrial and astrophysical magnetic fields are generally kinematic models. The kinematic approach neglects the equations of fluid motion and heat transfer and considers just the hydromagnetic equation,

$$\frac{\partial \mathbf{B}}{\partial t} = \nabla \times (\mathbf{V} \times \mathbf{B}) + \eta^{-2}\,\mathbf{B}$$

where **B** is the magnetic field, **V** is the fluid velocity, and η is the magnetic diffusivity. A particular velocity field is prescribed along with an initial magnetic field, and a regenerative solution to the above equation is then sought. This approach has yielded several successful models (see Levy, 1976 for a review). The successful velocity fields found vary from large-scale nearly axisymmetric motions to small-scale turbulence with a particular statistical nature. Cyclonic fluid motions with a radial component of velocity have appeared in several dynamo models (Parker, 1983, Levy, 1972) and in the limit of small length scale can be likened to the turbulent dynamo model. The cyclonic model also has the capability of producing a self-reversing dynamo (Levy 1976), an important observed feature of the Earth's magnetic field. Although other velocity fields can produce a dynamo, the cyclonic model is particularly pertinent to models driven by differential heating from above.

Kinematic models can thus describe velocity fields necessary for a dynamo, but they do not indicate the source of fluid motions. These are usually assumed to result from thermal or chemical convection within the core. Efforts toward a dynamic treatment including thermal convection have produced a few results, notably the "convective rolls" dynamo of Busse (1964) and the Rossby wave dynamo of Gilman (1969) as extended by Braginsky and Roberts (1973).

One strong constraint on the geodynamo is that adequate energy be supplied to maintain the magnetic field. Due to ohmic losses, energy must be supplied to the magnetic field through the velocity field. Since the magnetic field has existed at nearly the same intensity for at least 2.7 billion years (McElhinny and others, 1968) and the decay time for the fundamental mode of the magnetic field has been estimated at about 10,000 years (Cox, 1972), there has seemingly been a near-constant energy supply over geologic time.

Gubbins reviewed the energy requirements of the magnetic field and provided lower and upper bounds on the energy supply. The upper bound is of order 10^{20} erg/s, which is the observed surface heat flux, and the lower bound is 2×10^{17} erg/s by consideration of conduction and electric currents. This requires an energy source acting over geologic time of considerable size. The precessional dynamo has been eliminated on the basis of energy constraints (Rochester and others, 1975). Latent heat released from the supposed growth of the inner core is marginal as an energy supply, but this mechanism produces motions restricted to near the inner core (Verhoogen, 1973) unless the precipitation mechanism discussed in previous sections is operative. The secular variations of the magnetic field require substantial fluid motions in the outermost core (Elsasser, 1946).

The only other potential energy sources are radiogenic heating and, possibly, gravitational mechanical stirring. It was in the context of searching for an energy source that potassium-40 was suggested to be in the outer core (Lewis, 1971). This suggestion is rather arbitrary and is not consistent with any known differentiation process. The observational evidence argues against significant potassium in the metallic phase at low pressures. This issue is still controversial, but aside from whether or not potassium would partition into the metallic phase is the problem of the amount required. Murthy and Hall (1972) required three-fourths of the potassium within the Earth to be segregated into the metallic core. To partition that amount of potassium into the core is inconsistent with any accretion and evolution model for the Earth.

The idea of a mechanically stirred core has been suggested. The basic idea is that the inner core has grown continuously over the age of the Earth by precipitating Fe and Ni, excluding the lighter element from the inner core. This process releases a lighter fraction near the inner core boundary, which then causes fluid motions. If the inner core has grown with time and if there is a compositional difference between the inner and outer core, this process may well occur. However, it is not clear that it would be important for the magnetic field, particularly if the core is stratified. The quantitative calculation of the potential energy release (Loper and Roberts, 1977) assumed an adiabatic temperature gradient throughout the core over geologic time, and this assumption conflicts with many recent results, including those of Gubbins. Any stability within the core, even if only in the outermost part, seriously affects the gravitational energy available for fluid motions. An alternative to this model for inner core growth is continual freezing out of metallic iron from a sub-liquidus outer core, which then sinks to the inner core (Figure 4-3). This only works if core compositions are on the Fe side of the eutectic composition or if an intermetallic compound such as Fe_2O or FeNiO is stable at high pressure near the liquidus temperature.

At present there is no consensus on the energy source, or on details of the fluid motions. A considerable advance would be made if the topographies of the outer- and inner-core boundaries could be mapped and if lateral seismic velocity variations in D″ and the outer core could be mapped. Seismic tomography is relevant to these questions.

Of the terrestrial planets, Earth and probably Mercury possess substantial intrinsic magnetic fields generated by core dynamos, while Venus and Mars apparently lack such fields. Thermal history calculations suggest that sulfur must be present in the core of Mercury if it is to be molten and capable of sustaining a dynamo.

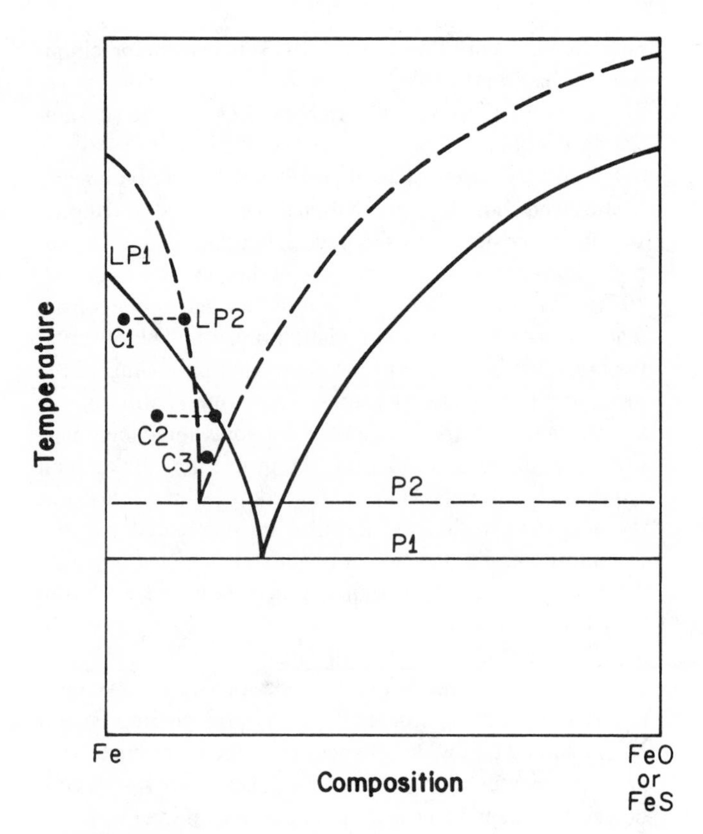

FIGURE 4-3
Possible eutectic phase relations for the core at two pressures. Three possible core compositions are shown (C1, C2 and C3). P1 and P2 are solidus curves for low and high pressures; LP1 and LP2 are the corresponding liquidus curves. For C1 the core is closer to the liquidus at low pressure; the reverse is true for C2. For C3 the core is above the liquidus at high pressure. Depending on the composition and the effect of pressure on the phase relations, one can have the solid content of the core increase or decrease with depth. If the solid particles become large enough they will settle out, giving a compositionally stratified core that may be gravitationally stable or unstable. An adiabatic temperature gradient may alternate with chemical homogeneity.

General References

Anderson, D. L. (1982) Chemical composition and evolution of the mantle. In *High-Pressure Research in Geophysics* (S. Akimoto and M. Manghnani, eds.), 301–318, D. Reidel, Dordrecht, Neth.

Condie, K. C. (1982) *Plate Tectonics and Crustal Evolution,* 2nd ed., Pergamon, New York, 310 pp.

Dziewonski, A. M. and D. L. Anderson (1981) Preliminary Reference Earth Model, *Phys. Earth Planet. Inter., 25,* 297–356.

Gubbins, D. (1974) Theories of the geomagnetic and solar dynamos, *Rev. Geophys. Space Phys., 12,* 137.

Jacobs, J. A. (1975) *The Earth's Core,* Academic Press, London, 253 pp.

Levy, E. H. (1976) Kinematic reversal schemes for the geomagnetic dipole, *Astrophys. J., 171,* 635–642.

Melchior, P. (1986) *The Physics of the Earth's Core,* Pergamon, New York, 256 pp.

Parker, E. N. (1983) Magnetic fields in the cosmos, *Sci. Am., 249,* 44–54.

Ringwood, A. E. (1979) *Origin of the Earth and Moon,* Springer-Verlag, New York, 295 pp.

Taylor, S. R. (1982) *Planetary Science, a Lunar Perspective,* Lunar and Planetary Institute, Houston, 482 pp.

Taylor, S. R. and S. M. McLennan (1985) *The Continental Crust: Its Composition and Evolution,* Blackwell, London.

References

Ahrens, T. J. (1979) Equations of state of iron sulfide and constraints on the sulfur content of the Earth, *J. Geophys. Res., 84,* 985–998.

Anderson, D. L. (1966) Recent evidence concerning the structure and composition of the Earth's mantle. In *Physics and Chemistry of the Earth, 6,* 1–131, Pergamon, Oxford.

Anderson, D. L. (1970a) *Mineralog. Soc. America Spec. Paper, 3,* 85–93.

Anderson, D. L. (1970b) Velocity-density relations, *Jour. Geophys. Res., 75,* 1623–1624.

Anderson, D. L. (1977) Composition of the mantle and core, *Ann. Rev. Earth Planet. Sci., 5,* 179–202.

Anderson, D. L. and A. M. Dziewonski (1982) Upper mantle anisotropy: Evidence from free oscillations, *Geophys. J. Roy. Astron. Soc., 69,* 383–404.

Anderson, D. L. and J. W. Given (1982) Absorption band *Q* model for the Earth, *Jour. Geophys. Res., 87,* 3893–3904.

Anderson, D. L. and R. S. Hart (1976) An Earth model based on free oscillations and body waves, *Jour. Geophys. Res., 81,* 1461–1475.

Anderson, D. L. and T. H. Jordan (1970) The composition of the lower mantle, *Phys. Earth Planet. Inter., 3,* 23–35.

Anderson, D. L. and B. R. Julian (1969) Shear velocities and elastic parameters of the mantle, *Jour. Geophys. Res., 74,* 3281–3286.

Anderson, D. L., C. G. Sammis and T. H. Jordan (1971) Composition and evolution of the mantle and core, *Science, 171,* 1103–1112.

Backus, G. and F. Gilbert (1970) *Phil. Trans. Roy. Soc. London, A 266,* 123–192.

Balchan, A. S. and G. R. Cowan (1966) Shock compression of two iron-silicon alloys to 2.7 megabars, *J. Geophys. Res., 71,* 3577–3588.

Bass, J. D. and D. L. Anderson (1984) *Geophys. Res. Lett., 11,* 237–240.

Birch, F. (1952) Elasticity and constitution of the Earth's interior, *J. Geophys. Res., 57,* 227–286.

Braginskii, S. (1964) *Geomag. Aeron., IV,* 572.

Brett, R. (1971) The Earth's core: Speculations on its chemical equilibration with the mantle, *Geochim. Cosmochim. Acta, 35,* 203–221.

Brett, R. (1984) Chemical equilibration of the Earth's core and upper mantle, *Geochim. Cosmochim. Acta, 48,* 1183–1188.

Brown, J. M., T. J. Ahrens and D. L. Shampine (1984) Hugoniot data for pyrrhotite and the Earth's core, *J. Geophys. Res., 89,* 6041–6048.

Brown, J. M. and R. G. McQueen (1982) The equation of state of iron and the Earth's core. In *High Pressure Research in Geophysics* (S. Akimoto and M. H. Manghnani, eds.), 611–624, Center for Academic Publications.

Burdick, L. J. and D. L. Anderson (1975) Interpretation of velocity profiles of the mantle, *Jour. Geophys. Res., 80,* 1070–1074.

Busse, F. (1973) *J. Fluid Mech., 57,* 529.

Butler, R. and D. L. Anderson (1978) Equation of state fits to the lower mantle and outer core, *Phys. Earth Planet. Inter., 17,* 147–162.

Cox, A. (1972) Geomagnetic reversals: Characteristic time constants and stochastic processes, *Eos, 53,* 613 (abstract).

Dickinson, W. R. and W. Snyder (1978) Plate tectonics of the Larimide orogeny, *Geol. Soc. Am. Mem, 151,* 355–366.

Dziewonski, A. M. and D. L. Anderson (1981) Preliminary reference Earth model, *Phys. Earth Planet. Inter., 25,* 297–356.

Elsasser, W. (1946) *Phys. Rev., 70,* 202.

Engdahl, E. R., E. A. Flinn and C. Romney (1970) Seismic waves reflected from the Earth's inner core, *Nature, 228,* 852.

Gaffney, E. S. and D. L. Anderson (1973) Effect of low-spin Fe^{2+} on the composition of the lower mantle, *Jour. Geophys. Res., 78,* 7005–7014.

Goto, T., J. Sato and Y. Syono, in High-pressure research in geophysics, ed. S. Akimoto, M. Manghnani, Reidel Publishing Co., Dordrecht, 595-610.

Gilman, P. (1969) *Solar Phys., 8,* 316–330.

Gubbins, D. (1977) Energetics of the Earth's Core, *J. Geophys., 43,* 453.

Gutenberg, B. (1957) The "boundary" of the Earth's inner core, *Trans. Am. Geophys. Jn., 38,* 750–753.

Jacobs, J. A. (1975) The Earth's Core, Academic Press, N.Y., 253 pp.

Jeanloz, R. and E. Knittle (1986) Reduction of mantle and core properties to a standard state by adiabatic decompression. In *Chemistry and Physics of Terrestrial Planets* (S. K. Saxena, ed.), 275–305, Springer-Verlag, Berlin.

Jones, J. H. and M. J. Drake (1986) Geochemical constraints on core formation in the Earth, *Nature, 322,* 221–228.

Jordan, T. H. and D. L. Anderson (1973) Earth structure from free oscillations and travel times, *Geophys. Jour. Roy. Astr. Soc., 36,* 411–459.

Julian, B. R., D. Davies, and R. Sheppard (1972) PKJKP, *Nature, 235,* 317–318.

Lay, T. and D. V. Helmberger (1983) *Geophys. J. R. Astron. Soc., 75,* 799–837.

Lees, A. C., M. S. Bukowinski and R. Jeanloz (1983) Reflection properties of phase transition and compositional change models of the 670-km discontinuity, *J. Geophys. Res., 88,* 8145–8159.

Lehmann, I. (1936) *Publ. Bur. Cent. Seism. Int. Ser. A, 14,* 3.

Levy, E. (1972) *Astrophys. J., 171,* 621.

Levy, E. (1972) *Astrophys. J., 171,* 635.

Levy, E. H. (1976) Kinematic reversal schemes for the geomagnetic dipole, *Astrophys. J., 171,* 635–642.

Lewis, J. S. (1971) Consequences of the presence of sulfur in the core of the Earth, *Earth Planet. Sci. Lett., 11,* 130–134.

Liu, L.-G. (1977) The system enstatite-pyrope at high pressures and temperatures and mineralogy of the Earth's mantle, *Earth Planet. Sci. Lett., 36,* 237–245.

Liu, L. G. (1979) In *The Earth, Its Origin, Structure and Evolution* (M. W. McElhinny, ed.), 117–202, Academic Press, New York.

Loper, D. E. and P. H. Roberts (1977) Possible and plausible thermal states of the Earth's core, *Eos Trans. Am. Geophys. U., 58,* 1129.

McCammon, C. A., A. E. Ringwood and I. Jackson (1983) Thermodynamics of the system Fe-FeO-MgO at high pressure and temperature and a model for formation of the Earth's core, *Geophys. J. Roy. Astron. Soc., 72,* 577–595.

McElhinny, M. W., J. C. Briden, D. L. Jones and A. Brock (1968) Geological and geophysical implications of paleomagnetic results from Africa, *Rev. Geophys., 6,* 201.

Murthy, V. R. and H. T. Hall (1972) *Phys. Earth Planet. Inter., 6,* 125–130.

Ohtani, E. and A. E. Ringwood (1984) Composition of the core, I, Solubility of oxygen in molten iron at high temperatures; II, Effect of high pressure on solubility of FeO in molten iron, *Earth Planet. Sci. Lett., 71,* 85–103.

Parker, E. (1955) *Astrophys. J., 122,* 293.

Parker, E. (1969) *Astrophys. J., 158,* 815.

Parker, E. (1971) *Astrophys. J., 164,* 491.

Ringwood, A. E. (1966) The chemical composition and origin of the Earth, In *Advances in Earth Sciences* (P. M. Hurley, ed.), 287–356, MIT Press, Cambridge, Mass.

Ringwood, A. E. (1975) *Composition and Petrology of the Earth's Mantle,* McGraw-Hill, New York, 618 pp.

Ringwood, A. E. (1979) *Origin of the Earth and Moon,* Springer-Verlag, New York, 295 pp.

Ringwood, A. E. (1982) Phase transformations and differentiation in subducting lithosphere, *J. Geology, 90,* 611–643.

Rochester, M., J. Jacobs, D. Smylie and K. Chong (1975) Can precession power the geomagnetic dynamo? *Geophys. J. Roy. Astron. Soc., 43,* 661–678.

Ruff, L. J. and D. L. Anderson (1980) Core formation, evolution, and convection; a geophysical model, *Phys. Earth and Planet. Inter., 21,* 181–201.

Stevenson, D. J. (1981) *Science, 214,* 611–618.

Svendsen, B. (1987) Thesis, Caltech, Pasadena, California, 250 pp.

Verhoogen, J. (1973) Thermal regime of the Earth's core, *Phys. Earth Planet. Inter.*, *7*, 47–58.

Watt, J. P. and T. J. Ahrens (1982) The role of iron partitioning in mantle composition, evolution, and scale of convection, *J. Geophys. Res.*, *87*, 5631–5644.

Weng, K., J. Xu, H.-K. Mao and P. M. Bell (1983) Preliminary Fourier-transform infrared spectral data on the SiO_6 octahedral group in silicate-perovskites, *Carnegie Inst. Wash. Yearb. 82*, 355–356.

Wright, C. and J. A. Lyons (1981) *Pageoph, 119*, 137–162.

5

Thermodynamics and Equations of State

The art of narrative consists in concealing from your audience everything it wants to know until after you expose your favorite opinions on topics foreign to the subject. I will now begin, if you please, with a horoscope located in the Cherokee Nation; and end with a moral tune on the phonograph.

—O. HENRY, "CABBAGES AND KINGS"

THERMODYNAMICS

The total energy contained in a system, such as a mineral, is called the *internal energy, U,* and includes the kinematic and potential energy of all the atoms. It depends on temperature, pressure and position in the field of gravity. For an infinitesimal change of the system, the law of conservation of energy, or the *first law of thermodynamics,* is

$$dU = dQ - d\mathcal{W}$$

where Q is the heat flow and $\mathcal{W}$ is the mechanical work, for example the change of volume acting against a hydrostatic pressure

$$d\mathcal{W} = P\, dV$$

The *enthalpy* or heat content of a system is

$$H = U + PV$$

$$dH = dU + P\, dV + V\, dP$$

The energy contents cannot be determined in absolute terms; they are only known as differences. The usual, but arbitrary, zero point is known as the *standard state* and is denoted $\Delta H°$.

The *heat capacity* or *specific heat* is the heat required to raise a unit mass of the material by one degree. This can be done at constant volume or at constant pressure and the corresponding symbols are C_V and C_P,

$$dU = C_V\, dT$$

$$dH = C_P\, dT$$

For minerals,

$$C_V \approx C_P \approx 0.3 \text{ cal/°C g}$$

A certain fraction of the heat entering a system, dQ/T, is not available for mechanical work. The integral of this is the entropy, S, defined from

$$dS = dQ/T$$

giving the *second law of thermodynamics,*

$$T\, dS = dU + d\mathcal{W}$$

which applies to reversible processes, processes that do not lose energy to the environment. In irreversible processes,

$$T\, dS > dU + d\mathcal{W}$$

Entropy is a measure of the energy associated with the random arrangement and thermal motion of the atoms and that is therefore unavailable for external work. At absolute zero temperature a perfectly ordered crystal has zero entropy; with increasing temperature a certain disorder or randomness is introduced. The entropy at temperature T is

$$S = \int_0^T (C/T)\, dT$$

At high temperature,

$$S \approx C \ln T$$

When a mineral undergoes a change of phase at temperature T involving a change in enthalpy or latent heat of transformation ΔH, there is a discontinuous change of entropy:

$$\Delta S = \Delta H/T$$

The mechanical part of the free energy U is the Helmholtz free energy F:

$$F = U - TS$$

$$dF = dU - T\,dS - S\,dT$$

$$= -P\,dV - S\,dT$$

giving

$$P = -(\partial F/\partial V)_T$$

$$S = -(\partial F/\partial T)_V$$

When using P and T as independent variables, instead of V and T, it is convenient to use the *Gibbs free energy, G:*

$$G = H - TS = U + PV - TS = F + PV$$

For a reversible process,

$$dG = V\,dP - S\,dT$$

$$V = (\partial G/\partial P)_T$$

$$S = (\partial G/\partial T)_P$$

If W is any thermodynamic function, the volume and pressure derivatives at constant temperature may be related by writing

$$(\partial W/\partial V)_T = (\partial W/\partial P)_T\,(\partial P/\partial V)_T$$

or

$$(\partial W/\partial V)_T = -(K_T/V)\,(\partial W/\partial P)_T$$

We can also write

$$(\partial W/\partial T)_V = (\partial W/\partial T)_P + \alpha K_T(\partial W/\partial P)_T$$

where α is the volume expansion coefficient.

Thermodynamic Identities

There are a variety of relations between the partial differentials of the standard thermodynamic parameters. Some of the standard forms are:

TABLE 5-1
Differentials of Thermodynamic Parameters

Differential element	Constant			
	T	**P**	**V**	**S**
∂T	—	1	1	γT
∂P	$-K_T/V$	—	$\alpha K_T=\gamma\rho C_v$	K_s
∂V	1	αV	—	$-V$
∂S	$\alpha K_T=\gamma\rho C_v$	mC_p/T	mC_v/T	—
∂U	$\alpha K_T T-P=\gamma\rho C_v T-P$	$mC_p-\alpha VP$	mC_v	PV
∂H	$-K_T(1-\alpha T)$	mC_p	$mC_v(1+\gamma)$	$K_s V$
∂F	$-P$	$-S-\alpha VP$	$-S$	$PV-\gamma TS$
∂G	$-K_T$	$-S$	$-S+\alpha K_T V=-S+\gamma mC_v$	$K_s V-\gamma TS$
	U	**H**	**F**	**G**
∂T	$P-\alpha TK_T=P-\gamma\rho C_v T$	$1-\alpha T$	P	1
∂P	$-\rho C_v(K_s-\gamma P)$	$-\rho C_p$	$K_T(S/V+\alpha P)$	S/V
∂V	mC_v	$\alpha V(1+1/\gamma)$	$-S$	$\alpha V-S/K_T=(1/K_T)(\gamma mC_v-S)$
∂S	$mC_v P/T$	mC_p/T	$mC_v(P/T-\gamma S/V)$	$mC_p/T-\alpha S$
∂U	—	$mC_p-PV\alpha(1+1/\gamma)$ $=\alpha V[K_s/\gamma-P(1+1/\gamma)]$	$mC_v P-S\gamma\rho C_v T+SP$	$mC_p-\alpha TS-P\alpha V+PS/K_T$
∂H	$mC_v[P(1+\gamma)-K_s]$	—	$SK_T(1-\alpha T)+mC_v P(1+\gamma)$	$mC_p+S(1-\alpha T)$
∂F	$\rho C_v(\gamma TS-PV)-PS$	$-S(1-\alpha T)-PV\alpha(1+1/\gamma)$	—	$-S(1-P/K_T)-P\alpha V$
∂G	$mC_v[\gamma TS/V+\alpha P-K_s]-PS$	$-S(1-\alpha T)-mC_p$	$S(K_T-P)+PV\alpha K_T$	—

Stacey (1977).

U	Internal energy*	V	Volume*
H	Enthalpy*	γ	Gruneisen parameter
F	Helmholtz free energy*	α	Volume expansion coefficient
G	Gibbs free energy*	ρ	Density
S	Entropy*	m	Mass of material*
T	Absolute temperature	K	Bulk modulus = incompressibility
P	Pressure	C	Specific heat

Subscripts signify parameters held constant.
*Parameters proportional to mass.

$$dU = (\partial U/\partial S)_V\, dS + (\partial U/\partial V)_S\, dV = T\, dS - P\, dV$$

$$dH = T\, dS + V\, dP$$

$$dF = -S\, dT - P\, dV$$

$$dG = -S\, dT + V\, dP$$

The Maxwell relations are:

$$(\partial T/\partial V)_S = -(\partial P/\partial S)_V = -\gamma T/V$$

$$(\partial S/\partial V)_T = (\partial P/\partial T)_V = \gamma \rho C_V = \alpha K_T$$

$$(\partial T/\partial P)_S = (\partial V/\partial S)_P = \gamma T/K_S$$

$$-(\partial S/\partial P)_T = (\partial V/\partial T)_P = \alpha V$$

Table 5-1 represents all possible partial differentials of the standard parameters. The individual entries are to be taken in pairs. Thus $(\partial T/\partial P)_S$ is ∂T at constant S (that is, γT) divided by ∂P at constant S (that is, K_S) giving

$$(\partial T/\partial P)_S = \gamma T/K_S$$

The following partial differentials are of particular interest:

$$(\partial P/\partial T)_V = \gamma \rho C_V$$

is the differential form of the Mie-Grüneisen equation and gives the variation in pressure in heating at constant volume.

$$(\partial \alpha/\partial P)_T = (1/K_T^2)(\partial K_T/\partial T)_P$$

connects the pressure dependence of the coefficient of thermal expansion with the temperature dependence of the bulk modulus. The relation

$$T(\partial \alpha/\partial T)_V = -\rho(\partial C_V/\partial P)_T$$

is useful in the high-temperature limit where $C_V \approx 3R$ (R being the gas constant) and α is independent of T at constant V and nearly independent of T at constant P.

Table 5-2 gives thermodynamic data for a few minerals.

The combination αK_T occurs in many thermodynamic relationships. The following second derivative thermodynamic identities are therefore useful:

$$\left[\frac{\partial(\alpha K_T)}{\partial V}\right]_T = -\frac{1}{V}\left(\frac{\partial K_T}{\partial T}\right)_V$$

$$\left[\frac{\partial(\alpha K_T)}{\partial P}\right]_T = \frac{1}{K_T}\left(\frac{\partial K_T}{\partial T}\right)_V$$

$$\left[\frac{\partial(\alpha K_T)}{\partial T}\right]_V = \frac{1}{T}\left(\frac{\partial C_V}{\partial V}\right)_T = -\frac{\rho K_T}{T}\left(\frac{\partial C_V}{\partial P}\right)_T$$

$$\left[\frac{\partial(\alpha K_T)}{\partial T}\right]_P = K_T\left(\frac{\partial \alpha}{\partial T}\right)_V$$

Chemical Equilibria

The fact that a mineral assemblage changes into a different assemblage means that the new association has a lower free energy than the old. At equilibrium both assemblages have the same free energy. The stable phase has the lowest free energy, at the given pressure and temperature and mineral association, of all alternative phases. In general, the denser phases are favored at high pressure and low temperature.

The *partial molal free energy* or *chemical potential* per mole of species i is F_i,

$$F_i = RT \ln a_i + F_i^\circ$$

where a_i is the *activity* of a chemical species, and F_i° is the free energy in a standard state. The total energy is

$$G = \sum n_i F_i$$

where G is Gibbs free energy and n_i is the number of moles of species i. At constant temperature and pressure,

$$dG = \sum F_i\, dn_i$$

$$F_i = M_i \mu_i$$

TABLE 5-2
Thermodynamic Properties of Minerals

Mineral	C_P erg/g/K (10^6)	C_V erg/g/K (10^6)	α ($K^{-1} \times 10^{-6}$)	K_s (kbar)	γ	θ (K)
MgO	9.25	9.11	31.5	942	1.55	940
CaO	7.50	7.42	29.0	675	1.31	680
Al_2O_3	7.79	7.74	16.4	1035	1.33	1040
Mg_2SiO_4	8.38	8.31	24.7	760	1.18	760
$MgSiO_3$	7.8	7.60	47.7	706	1.89	710
$MgAl_2O_4$	8.15	8.07	20.8	863	1.41	860
SiO_2	7.41	7.35	36.6	572	0.7	570
Garnet	7.61	7.55	21.6	745	1.21	750
Garnet	7.0	6.96	18.3	739	1.09	740

and M_i is the molecular weight and μ_i is the *chemical potential per gram*. The change of activity with pressure is

$$\left(\frac{\partial \ln a}{\partial P}\right)_T = \frac{V}{RT}$$

The total Gibbs free energy of a system of C components and p phases is

$$G = \sum_{i=1}^{C} \sum_{j=1}^{p} \mu_i^j n_i^j$$

where n_i^j is the number of moles of the component i in phase j and μ_i^j is its chemical potential in phase j. The equilibrium assemblage, at a given pressure and temperature, is found by minimizing G. Taking the standard state of i to be pure i in phase j at the pressure and temperature of interest,

$$\mu_i^j = \mu^0 + RT \ln a_i^j$$

$$\mu^0 = H_{0,T}^j - TS_{0,T}^j + \int_0^P V_T^j \, dP$$

where $H_{0,T}^j$, $S_{0,T}^j$ and V_T^j are the enthalpy and entropy of pure i in phase j at $P = 0$ and T and volume at T. With the chosen standard state for the activity, a_i^j in phase j containing pure i is one. The activity of pure liquids or pure solids is unity. In an ideal solution a_i is equal to the mole fraction of component i.

At equilibrium the standard-state free-energy change at the pressure and temperature of interest is

$$\Delta G_{P,T}^0 = -RT \ln K$$

where K is the equilibrium constant. Consider the hypothetical reaction

$$2A \rightleftarrows C + D$$

where two molecules of A react to form one molecule each of C and D. The rate of reaction is proportional to the collisional probability between any two molecules, which is related to the product of the concentrations. For equilibrium the rates of the two reversible reactions are the same, and

$$K = \frac{(C) \times (D)}{(A)^2}$$

where (A), (C) and (D) are the concentrations or activities. For the general reaction,

$$nA + mB = qC + rD$$

then

$$K = \frac{(C)^q \times (D)^r}{(A)^n \times (B)^m}$$

The equation for equilibrium

$$d(\Delta G) = \Delta V \, dP - \Delta S \, dT = 0$$

yields

$$\frac{dP}{dT} = \frac{\Delta S}{\Delta V} = \frac{\Delta H}{T\Delta V}$$

which is the Clausius-Clapeyron equation.

For a system in equilibrium the following relation holds between the number of coexisting phases p, components c, and degrees of freedom f:

$$p = c + 2 - f$$

This is the phase rule of J. Willard Gibbs. The *phases* are the parts of the system that can be mechanically separated, for example, the minerals and any coexisting liquid and gas. The *components* are the smallest number of chemical species necessary to make up all the phases. The *degrees of freedom* are generally the temperature, pressure and composition.

The components distribute themselves over all the phases of the system. No phase can be without some contribution from all components since the chemical potential or activity of each component must be the same in all phases of the system. The phase rule places a limitation on the number of minerals that can occur in equilibrium in a given rock. The maximum number of phases can be attained only in an invariant system, one with P and T fixed. If both P and T vary during the process of formation of a rock, then

$$p \leq c$$

which is the *mineralogical phase rule* of Goldschmidt. Because of the phenomenon of solid solution, the number of different minerals in a rock is less than the number of components.

Table 5-3 is a compilation of the terms and relations introduced in this section.

THEORETICAL EQUATIONS OF STATE

The equation of state of a substance gives the pressure P as a function of volume V and temperature T:

$$P = P(V,T)$$

The general expression for the free energy of a crystal can be written in terms of three functions

$$F(X,T) = U(X) + f_1(\theta/T) + f_2(X,T)$$

where $X = V_o/V = \rho/\rho_o$ is the dimensionless volume relative to the volume at normal conditions and θ is a characteristic temperature, such as the Debye or Einstein temperature. $U(X)$ is the potential part of the free energy, which depends only on the volume. The second term is the phonon term and is usually calculated from the Debye or Einstein theory. The third term represents high-temperature corrections to the equation of state. This term, which is generally

TABLE 5-3
Notation and Basic Relationships

V = Specific volume

$V_o = V$ at $P = 0$

ρ = Density = $\bar{M}/V$

$\rho_o = \rho$ at $P = 0$

T = Absolute temperature

P = Pressure

S = Entropy

$$\alpha = \frac{1}{V}\left(\frac{\partial V}{\partial T}\right)_P = -\frac{1}{\rho}\left(\frac{\partial \rho}{\partial T}\right)_P = \text{Volume thermal expansion}$$

$$K_T = -V\left(\frac{\partial P}{\partial V}\right)_T = \rho\left(\frac{\partial P}{\partial \rho}\right)_T = \text{Isothermal bulk modulus}$$

$$K_S = -V\left(\frac{\partial P}{\partial V}\right)_S = \rho\left(\frac{\partial P}{\partial \rho}\right)_S = \text{Adiabatic bulk modulus}$$

$$\Phi_S = K_S/\rho = V_p^2 - \frac{4}{3}V_s^2 = \left(\frac{\partial P}{\partial \rho}\right)_S = \text{elastic ratio}$$

$\gamma = \alpha K_T/\rho C_V = \alpha K_S/\rho C_P$ = Grüneisen ratio

C_V = Specific heat at constant volume

C_P = Specific heat at constant pressure

V_p, V_s = Velocity of compressional and shear waves

$K_S = K_T(1+\alpha\gamma T)$

$$\left(\frac{\partial P}{\partial T}\right)_V = K_T\alpha$$

$$\left(\frac{1}{K_T}\right)\left(\frac{\partial K_T}{\partial T}\right)_P = -V\left(\frac{\partial \alpha}{\partial V}\right)_T$$

$$\left(\frac{1}{K_T}\right)\left(\frac{\partial K_T}{\partial T}\right)_P = K_T\left(\frac{\partial \alpha}{\partial P}\right)_T$$

θ = Characteristic temperature

$\bar{M}$ = Mean atomic weight

Bulk modulus = Incompressibility = 1/Compressibility.

$\delta = [\partial \ln K/\partial \ln \rho]_p$ = Second Grüneisen ratio

small, is due to anharmonic lattice oscillations, formation of point defects and thermal excitation of conduction electrons in metals. For most geophysical problems $U(X)$ is the dominant term.

The potential energy of a crystal can be written as the sum of an attractive potential, which holds the atoms together, and a repulsive potential, which keeps the crystal from collapsing:

$$U = -\frac{A}{r^m} + \frac{B}{r^n} = -\frac{A}{V^{m/3}} + \frac{B}{V^{n/3}}$$

where r is the interatomic spacing and A, B, m and n are constants, different from those in the last section. The functional form of the repulsive potential is uncertain, and an exponential form is also often used.

The pressure is obtained by differentiation:

$$P = \left(\frac{\partial U}{\partial V}\right)_T$$

The isothermal bulk modulus, K_T, is

$$K_T = -V(\partial P/\partial V)_T$$

The bulk modulus is also called the incompressibility. At $P = 0$, $V = V_o$ and $K_T = K_T(0)$. The PV equation of state can therefore be written as

$$P = \frac{3K_T(0)}{(m-n)}\left[\left(\frac{V_o}{V}\right)^{(m+3)/3} - \left(\frac{V_o}{V}\right)^{(n+3)/3}\right]$$

and the bulk modulus as

$$K_T = \frac{K_T(0)}{(m-n)} \times \left[(m+3)\left(\frac{V_o}{V}\right)^{(m+3)/3} - (n+3)\left(\frac{V_o}{V}\right)^{(n+3)/3}\right]$$

The pressure derivative of K_T at $P = 0$ is

$$K_T'(0) = (m + n + 6)/3$$

$K_T'(0)$ is approximately 4 for many substances. Since the repulsive potential is a stronger function of r than the attractive potential, $n > m$ and $3 < n < 6$ for $K_T'(0) = 4$.

THE GRÜNEISEN RELATIONS

Grüneisen (1912) introduced the concept of a "thermal pressure" derived from the pressure of a collection of atoms vibrating under the excitation of the energy associated with nonzero temperature. A crystalline solid composed of N atoms has $3N$ degrees of freedom, and the solid can be viewed as a collection of harmonic oscillators. The energy levels of a harmonic oscillator are $nh\nu$, where n are successive integers and h is Planck's constant. In thermal equilibrium a given energy level is populated with the probability $\exp(-nh\nu/kT)$, where k is Boltzmann's constant. The individual oscillators have a frequency ν_i, and these are considered to be independent of temperature but dependent on the volume, V.

The quantity

$$\gamma_i = -\frac{\text{d}\log \nu_i}{\text{d}\log V}$$

is involved in calculations of the thermal pressure and is known as the Grüneisen ratio. If it is assumed that all the γ_i are the same, then

$$\nu_i \approx V^{-\gamma}$$

The Grüneisen equation of state is

$$P = P_o + \frac{\gamma U_D}{V}$$

where P_o is the pressure at absolute zero and U_D is the internal energy of the oscillators in a volume V due to the elevated temperature:

$$U_D = \frac{h\nu_i}{\exp(h\nu_i/kT) - 1}$$

Differentiating P with respect to temperature gives

$$\left(\frac{dP}{dT}\right)_V = \frac{\gamma C_V}{V}$$

where C_V is the specific heat at constant V. From the thermodynamic relations

$$\alpha = \frac{1}{V}\left(\frac{\partial V}{\partial T}\right)_P = \frac{1}{K_T}\left(\frac{\partial P}{\partial T}\right)_V$$

where α is the volume coefficient of thermal expansion, the following relation can be derived:

$$\gamma = \frac{VK_T\alpha}{C_V}$$

which is called the Grüneisen relation.

The thermal energy of a crystal is equal to the sum over all oscillators and, therefore, over all pertinent frequencies. In the Debye theory the sum is replaced by an integral, and it is assumed that all frequencies of vibration are bounded by some maximum value $\nu_i < \nu_m$.

In an elastic solid three modes of wave motion are permitted, one compressional mode and two shear modes having orthogonal particle motions. The total thermal energy is therefore

$$U_D = \frac{9N_o}{\nu_m^3}\int_0^{\nu_m} \frac{h\nu^3\, d\nu}{\exp(h\nu/kT) - 1}$$

where N_o is the number of atoms per unit volume.

The maximum oscillation frequency is related to the volume available to the oscillator and the velocity of elastic waves. In the Debye theory a mean sound velocity is implied, and thus

$$\nu_m^3 = \frac{9N_A\rho}{4\pi\overline{M}}\left(\frac{1}{V_p^3} + \frac{2}{V_s^3}\right)^{-1}$$

where N_A is Avogadro's number, ρ is the density, and $\overline{M}$ is the mean atomic weight (molecular weight divided by the number of atoms in the molecule). In the Debye theory it is assumed that velocity is isotropic and nondispersive, that is, independent of direction and frequency.

The Debye temperature is defined as

$$\theta = h\nu_m/k$$

and therefore

$$\theta = \frac{h}{k}\left(\frac{3N_A\rho}{4\pi\overline{M}}\right)^{1/3} V_m$$

where V_m is mean velocity:

$$V_m = \left[\frac{1}{3}\left(\frac{2}{V_s^3} + \frac{1}{V_p^3}\right)\right]^{-1/3}$$

The Debye temperature can be estimated from the velocities of elastic waves and, therefore, can be estimated for the mantle from seismic data. In principle, the velocities should be measured at frequencies near ν_m ($\sim 10^{13}$ Hz) since there is some dispersion. Ignoring dispersion, however, is consistent with Debye's original assumption. There are also optical modes, as well as acoustic modes, and these are ignored in the simple theories. By differentiation of U we obtain for the thermal pressure

$$P^* = \frac{U_D}{V}\frac{d\log\theta}{d\log V}$$

and, therefore,

$$\gamma = \frac{-d\log\theta}{d\log V} = \frac{-d\log\nu_m}{d\log V}$$

At high temperature, $\theta/T \ll 1$,

$$P^* = \frac{3NkT\gamma\rho}{\overline{M}} = P^*(\mathrm{HT})$$

At very low temperature,

$$P^* = \frac{3}{5}\pi^4\left(\frac{T}{\theta}\right)^3 \frac{NkT\gamma\rho}{\overline{M}}$$

At intermediate temperature,

$$P^* = 3P^*(\mathrm{HT})\left(\frac{T}{\theta}\right)^3 \int_0^{\theta/T} \frac{\xi^3\, d\xi}{e^{\xi} - 1}$$

where $\xi = h\nu/kT$.

The thermal pressure in the mantle is estimated to be between 10 and 200 kilobars, increasing with depth. The Debye temperature increases by about a factor of 2 through the mantle, and the Grüneisen parameter probably remains close to 1.

The specific heat can be written

$$C_V = \left(\frac{\partial U}{\partial T}\right)_V = 9N_o k\left(\frac{T}{\theta}\right)^3 \int_0^{\theta/T} \frac{\xi^4 e^{\xi}\, d\xi}{(e^{\xi} - 1)^2}$$

At $T \gg \theta$ we have the classic high-temperature limit,

$$C_V = 3kN_o$$

Silicates show a close approach to the "classical" values at temperatures greater than about 1000°C. Under these conditions C_V approaches 6 cal/°C g atom for each particle of the chemical formula. The mean atomic weight for most rock-forming minerals is close to 20, so the specific heat at high temperatures is close to 0.3 cal/°C g. The variation of specific heat with pressure is

$$\frac{1}{C_P}\left(\frac{\partial C_P}{\partial P}\right)_T = -\frac{T\alpha\gamma}{K_S}\left[1 + \frac{1}{\alpha^2}\left(\frac{\partial\alpha}{\partial T}\right)_P\right]$$

The specific heat probably only decreases about 10 percent at the highest mantle pressures, and its variation is therefore small relative to the changes expected for bulk modulus and thermal expansion.

Most of the interior of the Earth is hot, well above the Debye temperature. This means that the Earth's interior probably can be treated with classical solid-state physics concepts. I say "probably" because the interior of the Earth is at simultaneous high temperature and high pressure and these are competing effects. The quantization of lattice vibrations and the departures from classical behavior that are of interest to quantum and low-temperature physicists are not relevant except, in some cases, when extrapolating from laboratory measurements to the high temperatures in the interior. The close relationship between γ and the elastic constants and their pressure derivatives means that γ can be estimated from seismology.

The thermal pressure, P^*, can be viewed as the radiation pressure exerted on the solid by completely diffuse elastic waves, that is,

$$P^* = \frac{U_P}{V}\left(\frac{1}{3} - \frac{V}{V_P}\frac{\partial V_P}{\partial V}\right) + 2\frac{U_S}{V}\left(\frac{1}{3} - \frac{V}{V_P}\frac{\partial V_S}{\partial V}\right)$$

$$= \frac{-U_P}{\theta_P}\frac{\partial \theta_P}{\partial V} - 2\frac{U_S}{\theta_S}\frac{\partial \theta_S}{\partial V}$$

where the U_m, V_m and θ_m are the thermal energies, elastic wave velocities and characteristic temperatures associated with the longitudinal (P) and transverse (S) waves. At high temperature we have

$$P^* = \frac{RT}{V}\gamma_P + 2\frac{RT}{V}\gamma_S$$

or, for $\gamma_P = \gamma_S$,

$$P^* = \frac{3RT\gamma\rho}{M}$$

The thermal pressure can be written in a form analogous to the perfect gas equation:

$$P^* = \frac{Q}{V}RT, \quad Q = \gamma_P + 2\gamma_S$$

where Q is of the order of 5 or 6 for many elements and is near 4 for MgO and Al_2O_3.

EFFECT OF TEMPERATURE ON BULK MODULI

The pressure and the isothermal bulk modulus are volume derivatives, at constant temperature, of the free energy $F(V,T)$. The corresponding adiabatic quantities are volume derivatives of the internal energy U (V,S) at constant entropy. The equation of state of simple solids subjected to hydrostatic pressure can be written in two alternative forms. The *vibrational* formulation splits the free energy of the solid into the *lattice* energy, $U_L(V)$, which is the energy of a static solid of volume V in its electronic ground state, and a *vibrational* energy $U^*(V,T)$. The *thermal* formulation splits the free energy into a nonthermal *cohesive* energy $U_c(V)$ of the solid of volume V at 0 K and a *thermal* energy $U^*(V,T)$. Note that the lattice and cohesive energies depend only on volume and the terms with asterisks depend, in general, on both volume and temperature; in the Hildebrand approximation the thermal and vibrational energies are taken to be a function of temperature alone, this being a good approximation at high temperatures where the heat capacity at constant volume has attained its classical value. The cohesive energy is the free energy required to assemble the atoms from infinity to form the rigid lattice; it includes both static lattice and zero-point energy contributions. The total vibrational energy of the solid is the sum over all the modes of lattice vibration of all the particles. The vibrational energy $U^*(V,T)$ consists of the zero-point vibrational energy, $U^*(V,0)$, of the normal modes at $T = 0$ K plus the energy required to heat the lattice at constant volume, V, from 0 K, to T K; that is,

$$U^*(V,T) = U^*(V,0) + \int_0^T C_V \, dT$$

The Helmholtz free energy, in the Hildebrand approximation, can be written, for example,

$$F(V,T) = U_c(V) + U^*T - TS(V,T)$$

Since

$$P = -(\partial F/\partial V)_T$$

and

$$\left(\frac{\partial S}{\partial V}\right)_T = \left(\frac{\partial P}{\partial T}\right)_V = \alpha K_T$$

we have

$$P(V,T) = -\frac{\partial U_c(V)}{\partial V} + \alpha K_T T = P(V) + P^*(V,T)$$

and

$$K_T(V,T) = V\left(\frac{\partial^2 U_c(V)}{\partial V^2}\right)_T - VT\left(\frac{\partial \alpha K_T}{\partial V}\right)_T$$

$$= K_T(V) + K_T^*(V,T)$$

or

$$K_T(V,T) = K_T(V,0) + T\alpha K_T(V,T)\left[\left(\frac{\partial \ln K_T}{\partial \ln V}\right)_P - \left(\frac{\partial \ln K_T}{\partial \ln V}\right)_T\right]$$

or

$$K_T(V,T) = K_T(V,T_o) + (T - T_o)\alpha K_T \times \left[\left(\frac{\partial \ln K_T}{\partial \ln V}\right)_P - \left(\frac{\partial \ln K_T}{\partial \ln V}\right)_T\right]$$

The quantity

$$\left(\frac{\partial \ln K_T}{\partial \ln V}\right)_P - \left(\frac{\partial \ln K_T}{\partial \ln V}\right)_T$$

is of the order of -1. The quantity αK_T is of the order of 10 to 100 bar/K for elements and is between about 30 and 70 for compounds of interest in the deeper mantle. The quantity

$$\alpha K_T\left[\left(\frac{\partial \ln K_T}{\partial \ln V}\right)_P - \left(\frac{\partial \ln K_T}{\partial \ln V}\right)_T\right]$$

is of the order of -50 bar/K, and a temperature rise of some 2000 K changes the bulk modulus by about 100 kbar, which is about 10 percent of estimated values for the bulk modulus in the mantle.

The following relations are useful and serve to define the second Grüneisen parameter, δ:

$$\left(\frac{\partial \ln K_T/\partial T}{\partial \ln \rho/\partial T}\right)_P = \left(\frac{\partial \ln K_T}{\partial \ln \rho}\right)_P = \left(\frac{\rho}{K_T}\frac{\partial K_T}{\partial \rho}\right)_P$$

$$= (K_T\alpha)^{-1}\left(\frac{\partial K_T}{\partial T}\right)_P = \delta_T \quad (1)$$

$$\left[\frac{\partial \ln K_S/\partial T}{\partial \ln \rho/\partial T}\right]_P = \left(\frac{\partial \ln K_S}{\partial \ln \rho}\right)_P = \left(\frac{\rho}{K_S}\frac{\partial K_S}{\partial \rho}\right)_P$$

$$= (K_S\alpha)^{-1}\left(\frac{\partial K_S}{\partial T}\right)_P = \delta_S \quad (2)$$

The elastic moduli of a solid are affected by temperature both implicitly, through the volume, and explicitly. Thus, for example,

$$K_T = K_T(V,T)$$

and

$$d \ln K_T = (\partial \ln K_T/\partial V)_T\, dV + (\partial \ln K_T/\partial T)_v\, dT \quad (3)$$

The measured variation of K_T with temperature is, then,

$$\frac{d \ln K_T}{dT} = \left(\frac{\partial \ln K_T}{\partial V}\right)_T \frac{dV}{dT} + \left(\frac{\partial \ln K_T}{\partial T}\right)_V$$

$$\left(\frac{d \ln K_T}{d \ln V}\right)_P = \left(\frac{\partial \ln K_T}{\partial \ln V}\right)_T + \alpha^{-1}\left(\frac{\partial \ln K_T}{\partial T}\right)_V \quad (4)$$

where $(\partial \ln K_T/\partial T)_V$ is the intrinsic temperature dependence of K_T. $(\partial \ln K_S/\partial T)_V$ is positive. There is a general tendency for $(\partial \ln K_T/\partial T)_V$ to be smaller at high T/θ.

Experiments show

$$\left(\frac{\partial \ln K_S}{\partial \ln V}\right)_P < \left(\frac{\partial \ln K_S}{\partial \ln V}\right)_T$$

$$\left(\frac{\partial \ln K_S}{\partial \ln V}\right)_P < \left(\frac{\partial \ln K_T}{\partial \ln V}\right)_P$$

and

$$\left(\frac{\partial \ln K_S}{\partial \ln V}\right)_T < \left(\frac{\partial \ln K_T}{\partial \ln V}\right)_T$$

all of which are useful when trying to estimate the effects of pressure, volume, and temperature on the adiabatic bulk modulus. Note that these are all experimental and thermodynamic inequalities and are independent of the equation of state. We also note that

$$\delta_S < K' < \delta_T$$

$$\delta_T - \delta_S \approx \gamma$$

The seismic parameter Φ is simply

$$\Phi = K_S/\rho$$

so that

$$\left(\frac{\partial \ln \Phi}{\partial \ln \rho}\right)_P = \left(\frac{\partial \ln K_S}{\partial \ln \rho}\right)_P - 1 \quad (5)$$

$$\left(\frac{\partial \ln \Phi}{\partial \ln \rho}\right)_T = \left(\frac{\partial \ln K_S}{\partial \ln \rho}\right)_T - 1 \quad (6)$$

The pressure in the mantle rises to about 1500 kbar, which, for $(dK/dP)_T = 4$, corresponds to a 6000-kbar increase in the bulk modulus. Temperature can therefore be treated as a small perturbation on the general trend of bulk modulus, or Φ, with depth, at least in the deeper part of the mantle.

THERMAL EXPANSION AND ANHARMONICITY

Because the attractive and repulsive potentials have a different dependence on the separation of atoms, the thermal oscillation of atoms in their (asymmetric) potential well is anharmonic or nonsinusoidal. Thermal oscillation of an atom causes the mean position to be displaced, and thermal expansion results. (In a symmetric, or parabolic, potential well the mean positions are unchanged, atomic vibrations are harmonic, and no thermal expansion results.) The Debye model is restricted to assemblages of harmonic oscillators and, strictly speaking, cannot be used to discuss anharmonic effects such as thermal expansion. Anharmonicity causes atoms to take up new average positions of equilibrium, dependent on the amplitude of the vibrations and hence on the temperature, but the new positions of dynamic equilibrium remain nearly harmonic. At any given volume the harmonic approximation can be made so that the characteristic temperature, θ, and frequency are not explicit functions of temperature. This is called the quasi-harmonic approximation. If it is assumed that a change in volume can be adequately described by a change in θ, then the frequency of each normal mode of vibration is changed in simple proportion as the volume is changed. The Grüneisen parameter

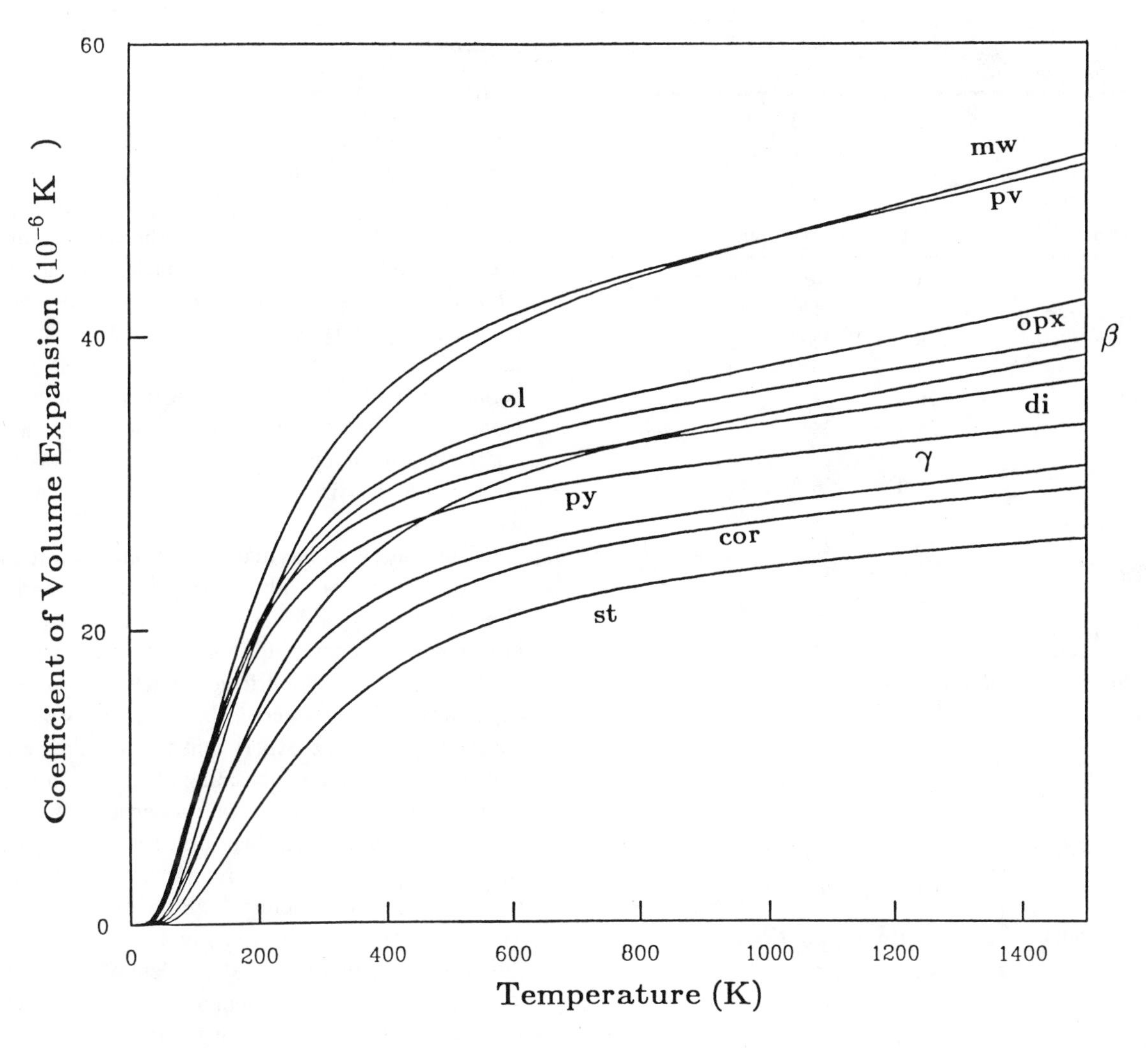

FIGURE 5-1
Coefficient of thermal expansion of mantle minerals, representing theoretical fits to available experimental data (after Duffy and Anderson, 1988).

$$\gamma = -\frac{\mathrm{d}\ln\theta}{\mathrm{d}\ln V}$$

is then a measure of anharmonicity.

From one of Maxwell's thermodynamic relations,

$$\left(\frac{\partial P}{\partial T}\right)_V = -\frac{1}{V}\left(\frac{\partial V}{\partial T}\right)\Big/\frac{1}{V}\left(\frac{\partial V}{\partial P}\right)_T = \alpha K_T$$

we have the volume coefficient of thermal expansion

$$\alpha = \frac{\gamma C_V}{K_T V} = \frac{\gamma C_P}{K_S V}$$

where C_V and C_P are molar specific heats and V is the molar volume. If the specific heats per unit volume are used, then V will not be present in these equations. The Grüneisen equation then shows that thermal expansion will only arise as a consequence of anharmonicity through the parameter γ, and if γ is itself independent of temperature, and we can ignore any explicit temperature dependence of K_T, then γ should be proportional to C_V in its temperature dependence. Since C_V is constant at high temperature (in "classical behavior"), then α should be as well; α should increase with temperature but level off at high T/θ, as shown in Table 5-4 and Figure 5-1.

The change of α with pressure is given by the thermodynamic identity

$$\left(\frac{\partial \alpha}{\partial P}\right)_T = \frac{1}{K_T^2}\left(\frac{\partial K_T}{\partial T}\right)_P$$

Thermal expansion decreases with pressure and reaches fairly low values at the base of the mantle. According to Birch (1938, 1952), α at the core-mantle boundary is only about 30 percent of its near-surface value. Birch (1968) showed that

$$\alpha/\alpha_o = 1 + \frac{P}{K}\left[\frac{1}{\alpha_o K_o}\frac{\mathrm{d}K_o}{\mathrm{d}T}\right] = 1 - \frac{\delta P}{K}$$

TABLE 5-4
Thermal Expansion of Minerals

Mineral	$\alpha = \frac{1}{V}\left(\frac{dV}{dT}\right)_P$ (10^{-6}/°C) 20°C	400°C	800°C	1500K
Quartz	34	69	−3	—
Coesite	8	11	14	—
Stishovite	16.5	22	23	24
Feldspar	12	19	24	—
Olivine	26	32	34	52
Pyroxene	24	28	32	—
Garnet	19	26	30	33
Al_2O_3	16.3	26	27	28
MgO	31.5	42	45	52
Spinel	16.2	28	29	31
β-spinel	20.6	31	34	37
γ-spinel	18.6	27	28	30
$MgSiO_3$-perovskite	—	37	42	—

Clark (1966), Jeanloz and Knittle (1986).

where the subscripts denote $P = 0$, and showed that

$$\frac{1}{K}\left(\frac{\partial K}{\partial T}\right)_P = \frac{1}{K_o}\frac{dK_o}{dT}\left[1 - \frac{P}{K}\left(\frac{\partial K}{\partial P}\right)_T\right]$$

If δ is independent of pressure $[(\delta = (\partial \ln K/\partial \ln \rho)_P)]$, then

$$\alpha/\alpha_o = (V/V_o)^\delta$$

and

$$\frac{1}{K}\left(\frac{\partial K}{\partial T}\right)_P = \frac{\alpha}{\alpha_o}\frac{1}{K_o}\frac{dK_o}{dT} = \left(\frac{V}{V_o}\right)^\delta \frac{1}{K_o}\frac{dK_o}{dT}$$

implying that $\partial K/\partial T$ is independent of pressure and $\partial K/\partial P$ is independent of temperature.

In shock-wave work it is often assumed that $\gamma C_V/V$ is independent of pressure. This gives

$$\alpha/\alpha_o = K_o/K$$

or αK is independent of pressure.

The Grüneisen theory for thermal expansion can be written

$$\alpha = (\partial E/\partial T)_P/[Q_o(1 - kE/Q_o)]^2$$

where E is the thermal or vibrational lattice energy,

$$Q_o = K_o V_o/\gamma$$

and

$$k = (1/2)(K_o' - 1)$$

where K_o, K_o' and V_o are the bulk modulus, its pressure derivative and volume, respectively. The thermal energy can be calculated from the Debye model or the Nernst-Lindemann formula,

$$E(T,\theta) = (3pR\theta/4)\,[2(e^{\theta/T} - 1)^{-1} + (e^{\theta/T} - 1)^{-1}]$$

where p is the number of atoms in the molecular formula and θ is a characteristic temperature. In fitting experimental data for α as a function of T, the parameters Q_o, k and θ can be treated, if necessary, as adjustable parameters or, if $\alpha(T)$ is to be estimated for unmeasured materials, these parameters can be estimated from other types of measurements. Both theory and experiment show that α increases rapidly with temperature and then levels off at high temperature ($T/\theta > 1$). Most of the mantle is at high temperature, but most laboratory measurements are made at relatively low temperatures. It cannot be assumed that α is constant with temperature, or varies linearly with temperature.

The theory of thermal expansion and a realistic estimate of its variation with temperature are essential in modeling the density and elastic properties of the mantle and in calculating mineral equilibria. The elastic properties of minerals have both an intrinsic and an extrinsic temperature dependence. The former is the variation of a property at constant volume, an experiment that requires a change in temperature and a compensating change in pressure. Most of the variation of the elastic properties is a result of the change in volume, and therefore it is important to understand the variation of α with temperature and to allow for this variation in modeling high-temperature phenomena. The functional forms of $\alpha(T)$ and $C_P(T)$ are related, and this is why γ is relatively independent of temperature.

In general, the coefficient of thermal expansion α is less for high-pressure phases than for low-pressure phases. Of the important mantle minerals, olivine and periclase have high thermal expansivities and γ-spinel and stishovite have relatively low coefficients. (Mg,Fe)SiO_3-perovskite violates this trend (Kuittle and others, 1986), having a relatively high α, at least for the metastable form.

There is a close relationship between lattice thermal conductivity, thermal expansion and other properties that depend intrinsically on anharmonicity of the interatomic potential. The atoms in a crystal vibrate about equilibrium positions, but the normal modes are not independent except in the idealized case of a harmonic solid. The vibrations of a crystal lattice can be resolved into interacting traveling waves that interchange energy due to anharmonic, nonlinear coupling.

In a harmonic solid:

1. There is no thermal expansion.
2. Adiabatic and isothermal elastic constants are equal.
3. The elastic constants are independent of pressure and temperature.
4. The heat capacity is constant at high temperature ($T > \theta$).

These consequences are the result of the neglect of anharmonicity (higher than quadratic terms in the interatomic displacements in the potential energy). In a real crystal the presence of one phonon, or lattice vibration of a given type, causes a periodic elastic strain that, through anharmonic interaction, modulates the elastic constants of a crystal. Other phonons are scattered by these modulations. This is a nonlinear process that does not occur in the absence of anharmonic terms.

Perhaps the simplest departure from linear or harmonic theory is to assume that the frequencies, ω_i, of lattice vibrations depend on volume. In the harmonic theory the free energy is independent of volume. The mode Grüneisen parameter expresses this volume dependence

$$\gamma_i = -(\partial \ln \omega_i / \partial \ln V)$$

and is a useful measure of anharmonicity. The crystal anharmonicity is a suitable average of all the modal γ_i. The Grüneisen approximation is that all γ_i are equal, but this is not generally true. A better approximation is to consider the longitudinal and shear modes separately, giving

$$\gamma = (1/3)(\gamma_p + 2\gamma_s)$$

where γ_p and γ_s are the longitudinal and transverse components, respectively, and all shear modes are assumed to have the same volume dependence, or, alternatively, separate averages are made of the two mode types. The above γ is sometimes called the acoustic or high-temperature γ. It is clearly dominated by the shear modes. In principle, the variation of the elastic constants with volume provides an estimate of γ or the anharmonicity and, therefore, higher order properties of the interatomic potential.

At high temperature ($T > \theta$) all phonons are excited and the acoustic γ is a weighted average of all modes. At lower temperature the value of γ is largely controlled by the lower frequency transverse waves.

According to the Mie-Grüneisen theory of the thermal expansion of solids,

$$\gamma = \frac{\alpha K_T}{C_V \rho}$$

and, in the Debye theory,

$$\gamma = -\partial \ln \theta / \partial \ln V$$

In terms of the interatomic potential function, U,

$$\gamma \approx -r^2 U''' / 3U''$$

where U'' and U''' are related to the elastic constants and their volume derivatives, respectively. Note that if $\alpha = 0$ or $U''' = 0$ (that is, no pressure dependence of elastic moduli), then $\gamma = 0$ and there is no anharmonicity. If $\gamma = 0$, the lattice thermal conductivity is infinite.

Actually, the concept of a strictly harmonic crystal is highly artificial. It implies that neighboring atoms attract one another with forces proportional to the distance between them, but such a crystal would collapse. We must distinguish between a harmonic solid in which each atom executes harmonic motions about its equilibrium position and a solid in which the forces between individual atoms obey Hooke's law. In the former case, as a solid is heated up, the atomic vibrations increase in amplitude but the mean position of each atom is unchanged. In a two- or three-dimensional lattice, the net restoring force on an individual atom, when all the nearest neighbors are considered, is not Hookean. An atom oscillating on a line between two adjacent atoms will attract the atoms on perpendicular lines, thereby contracting the lattice. Such a solid is not harmonic; in fact it has negative α and γ.

The quasi-harmonic approximation takes into account that the equilibrium positions of atoms depend on the amplitude of vibrations, and hence temperature, but that the vibrations about the new positions of dynamic equilibrium remain closely harmonic. One can then assume that at any given volume V the harmonic approximation is adequate. In the simplest quasi-harmonic theories it is assumed that the frequencies of vibration of each normal mode of lattice vibration and, hence, the vibrational spectra, the maximum frequency and the characteristic temperatures are functions of volume alone. In this approximation γ is independent of temperature at constant volume, and α has approximately the same temperature dependence as molar specific heat C_V.

ISOTHERMAL-ADIABATIC TRANSFORMATIONS

Seismic data are adiabatic in the sense that the time scale of seismic waves is short compared to the time scale required for the temperature to equilibrate between the compressed and dilated parts of the wave. To relate isothermal theories and experiments with adiabatic data, laboratory or seismic, requires isothermal-adiabatic transformations, all of which follow from

$$K_S = K_T(1 + \alpha\gamma T) \tag{7}$$

A large amount of ultrasonic data on solids at moderate pressures has accumulated in the past decades, and these transformations are also required to interpret the data in terms of isothermal equations of state.

From equation 7 we can write

$$\frac{(\partial \ln K_S/\partial T)_P}{(\partial \ln \rho/\partial T)_P} = \frac{(\partial \ln K_T/\partial T)_P}{(\partial \ln \rho/\partial T)_P} - \frac{K_T}{\alpha K_S}\left(\frac{\partial \alpha\gamma T}{\partial T}\right)_P \tag{8}$$

$$\frac{(\partial \ln K_S/\partial P)_T}{(\partial \ln \rho/\partial P)_T} = \frac{(\partial \ln K_T/\partial P)_T}{(\partial \ln \rho/\partial P)_T} + \frac{\alpha\gamma T K_T}{K_S} \times \left[\left(\frac{\partial \ln \gamma}{\partial \ln \rho}\right)_T - \left(\frac{\partial \ln K_T}{\partial \ln \rho}\right)_P\right] \tag{9}$$

The Grüneisen ratio, γ, is relatively independent of temperature, and the coefficient of volume thermal expansion α is independent of temperature at high temperatures. The second term on the right of equation 8 is, therefore, of the order of γ and is negative. The volume dependence of γ can be written:

$$\left(\frac{\partial \ln \gamma}{\partial \ln V}\right)_T = \left(\frac{\partial \ln K_T}{\partial \ln V}\right)_T - \left(\frac{\partial \ln K_T}{\partial \ln V}\right)_P + 1$$

$$= -\frac{1}{\alpha}\left(\frac{\partial \ln K_T}{\partial T}\right)_V + 1 \quad (10)$$

if we take $(\partial \ln C_V/\partial \ln V)_T = 0$, appropriate for high temperatures. Note that the multiplicative factor $(\alpha\gamma T K_T/K_S)$ in equation 9 can be written $(K_S - K_T) \div K_S$, which is a small number of the order of 0.001 for most materials at room temperature. The derivative $(\partial \ln \gamma/\partial \ln \rho)_T$ is of the order of -1 and $(\partial \ln K_T/\partial \ln \rho)_P$ is of the order of 6, so the second term on the right-hand side of equation 8 is of the order of -0.007 or about 1 percent of the first term.

The following relations are useful:

$$\frac{(\partial \ln K_T/\partial P)_T}{(\partial \ln \rho/\partial P)_T} = \left(\frac{\partial \ln K_T}{\partial \ln \rho}\right)_T = \left(\frac{\rho\, \partial K_T}{K_T\, \partial \rho}\right)_T$$

$$= \left(\frac{\partial K_T}{\partial P}\right)_T = K_T' \quad (11)$$

$$\frac{(\partial \ln K_S/\partial P)_T}{(\partial \ln \rho/\partial P)_T} = \left(\frac{\partial \ln K_S}{\partial \ln \rho}\right)_T$$

$$= \left(\frac{\rho\, \partial K_S}{K_S\, \partial \rho}\right)_T = \frac{K_T}{K_S}\left(\frac{\partial K_S}{\partial P}\right)_T \quad (12)$$

CALCULATION OF DENSITY IN THE EARTH

The variation of density ρ with radius in the Earth r can be written

$$\frac{d\rho}{dr} = \left(\frac{\partial \rho}{\partial P}\right)\frac{dP}{dr}$$

$$+ \left(\frac{\partial \rho}{\partial T}\right)\frac{dT}{dr} + \left(\frac{\partial \rho}{\partial \phi}\right)\frac{d\phi}{dr} + \left(\frac{\partial \rho}{\partial c}\right)\frac{dc}{dr} \quad (13)$$

that is, density is a function of pressure, temperature, phase (ϕ) and composition (c). For a homogeneous adiabatic self-compressed region, we have

$$\frac{d\phi}{dr} = 0, \quad \frac{dc}{dr} = 0$$

$$\frac{dP}{dr} = -g\rho, \quad g = \frac{GM(r)}{r^2}$$

In a convecting mantle the mean temperature gradient, away from thermal boundary layers, is close to adiabatic:

$$\frac{dT}{dP} = \left(\frac{\partial T}{\partial P}\right)_S = \frac{T\alpha}{\rho C_P} \quad (14)$$

It is therefore convenient to write the temperature gradient as

$$\frac{dT}{dr} = \frac{T\alpha}{\rho C_p}\frac{dP}{dr} - \tau \quad (15)$$

where τ is the superadiabatic (or subadiabatic) gradient. Adiabatic compression of a material is given by the adiabatic bulk modulus, K_S

$$K_S = \left(\rho\frac{\partial P}{\partial \rho}\right)_S$$

Seismic waves are also adiabatic, and hence we can use

$$V_P^2 - (4/3)V_S^2 = K_S/\rho = (\partial P/\partial \rho)_S = \Phi$$

to calculate the variation of density with depth in a homogeneous, adiabatic region for which we have seismic data. Making the above substitutions,

$$\frac{d\rho}{dr} = -g\rho/\Phi + \alpha\rho\tau$$

$$= -g\rho/\Phi\,(1 - \gamma C_P\tau/g) \quad (16)$$

These are the Williamson-Adams equations as modified by Birch (1938, 1952).

A useful test of homogeneity (Birch, 1952) is provided by

$$1 - g^{-1}\, d\Phi/dr = dK_S/dP + \alpha\Phi\tau/g \quad (17)$$

The Bullen parameter (Dziewonski and Anderson, 1981),

$$\eta = \frac{dK}{dP} + \frac{1}{g}\frac{d\Phi}{dr}$$

should be near unity for homogeneous regions of the mantle that do not depart too much from adiabaticity.

In the upper mantle the temperature gradients are high, decreasing from a high conductive gradient at the surface to the convective gradient in the deeper interior. There are also probably chemical, mineralogical and phase changes in the shallow mantle. The latter include partial melting and basalt-eclogite and garnet-pyroxene reactions. At greater depth the olivine-spinel, pyroxene-majorite and garnet-perovskite phase changes keep the mantle from being homogeneous in the Williamson-Adams sense. Any chemical layers also cause thermal boundary layers and superadiabatic gradients. The Bullen parameter is consequently far from unity at depths less than 670 km, and the Williamson-Adams equations cannot be used over most of the upper mantle (Butler and Anderson, 1978). The parameter dK/dP is another measure of homogeneity. It is generally close to 4 at $P = 0$ and decreases smoothly with pressure. This behavior is exhibited by the mantle below 770 km except for the region near the core-mantle boundary.

FINITE-STRAIN EQUATION OF STATE

Finite-strain theory has been applied extensively to problems in geophysics. The resulting equations are called semi-empirical because they contain parameters that have to be determined from experiment. The theory relates strain, or compression, to pressure.

The relation between strain ε and volume V or density ρ is

$$V_o/V = \rho/\rho_o = (1 - 2\varepsilon)^{3/2} = (1 + 2f)^{3/2}$$

where $f = -\varepsilon$ refers to compression, a positive quantity.

The first few terms in the Birch-Murnaghan equation of state (Birch, 1938, 1952) are

$$P = \frac{3}{2} K_o [(\rho/\rho_o)^{7/3} - (\rho/\rho_o)^{5/3}] \times \{1 - \zeta[(\rho/\rho_o)^{2/3} - 1] + \cdots\}$$

K_o is the bulk modulus at $P = 0$ and can refer to either isothermal or adiabatic conditions depending on whether an isotherm or an adiabat is to be calculated. K_o and ζ are parameters that are functions of temperature alone. In terms of strain,

$$P = 3K_o f(1 + 2f)^{5/2}(1 - 2\zeta f)$$

$$K = K_o(1 + 2f)^{5/2}[1 + 7f - 2\zeta f(2 + 9f)]$$

The term ζ can be found in terms of $(dK/dP)_o = K_o'$;

$$K_o' = 4 - \frac{4}{3}\zeta$$

This equation of state has been fitted to a large amount of shock-wave data on oxides and silicates, and K_o' is found to be generally between 2.9 and 3.6 (Anderson and Kanamori, 1968; Sammis and others, 1970; Davies and Anderson, 1971). K' generally decreases with pressure. In the lower mantle $K'(P)$ varies from about 3.8 to 3.1. Ultrasonic measurements of K_o' on minerals generally give values in the range 3.8 to 5.0. Note that for $\zeta = 0$, $K_o' = 4$, a typical value.

For quick, approximate calculations, the Murnaghan equation is useful:

$$P = \frac{3K_o}{n}\left[\left(\frac{\rho}{\rho_o}\right)^n - 1\right]$$

where $n = K_o'$. This diverges from the Birch-Murnaghan equation at high compressions but is useful at low pressures.

Finite-strain equations can also be developed for the variation of seismic velocity with pressure (Birch, 1961b; Burdick and Anderson, 1975). These have been used in the interpretation of velocity and density profiles of the mantle (Butler and Anderson, 1978; Davis and Dziewonski, 1975; Jeanloz and Knittle, 1986). The equations are:

$$V_p^2(P) = V_p^2(0)(1 - 2\varepsilon)[1 - 2\varepsilon(3K_oD_p - 1)]$$

$$V_s^2(P) = V_s^2(0)(1 - 2\varepsilon)[1 - 2\varepsilon(3K_oD_s - 1)]$$

where

$$D_{p,s} = (\partial \ln V_{p,s}/\partial P)_T$$

at $P = 0$. Pressure is calculated from

$$P = -3K_o(1 - 2\varepsilon)^{5/2}(1 - 2\varepsilon\zeta)\varepsilon.$$

The ζ parameter satisfies

$$\zeta = (9/4) - (3/2)\rho_o[V_p^2(0)D_p - (4/3)V_s^2(0)D_s]$$

The expressions to the next order in strain have been given by Davies and Dziewonski (1975). In order to apply these, the higher order pressure derivatives of K_S, V_p and V_s are required, and these are generally not available. However, the higher order terms for the lower mantle can be determined by fitting these equations to the seismic data for the lower mantle, assuming it is homogeneous and adiabatic. The zero-pressure properties of the lower mantle can therefore be estimated.

The "fourth-order" finite-strain equations can be written

$$\rho V_p^2 = (1 - 2\varepsilon)^{5/2}(L_1 + L_2\varepsilon + 1/2L_3\varepsilon^2 + \cdots)$$

$$\rho V_s^2 = (1 - 2\varepsilon)^{5/2}(M_1 + M_2\varepsilon + 1/2M_3\varepsilon^2 + \cdots)$$

$$P = -(1 - 2\varepsilon)^{5/2}(C_1\varepsilon + 1/2C_2\varepsilon^2 + 1/6C_3\varepsilon^3 + \cdots)$$

where L_i, M_i and C_i are constants.

By evaluating the above equations and their derivatives at $\varepsilon = 0$, it is possible to relate the above coefficients to the $P = 0$ values of the elastic moduli and their pressure derivatives. There is some question as to whether the finite-strain equations converge at high pressure and which order is appropriate for application to the lower mantle. There are significant differences in the inferred $P = 0$ properties of the mantle depending on whether third-order or fourth-order finite-strain equations are used.

Unfortunately, there has been little progress in determining equations of state for V_p and V_s from first principles. The bulk modulus and seismic parameter, $K_S/\rho = \Phi$, however, can be determined by simple differentiation of a wide variety of equations of state. $\Phi = \partial P/\partial\rho$ can also be determined from static-compression and shock-wave measurements. Therefore, most discussions of the composition and mineralogy of the mantle depend upon the seismic values for ρ, K_S and Φ, rather than V_p and V_s. Unfortunately, it is the velocities that can be determined most accurately.

Other potential functions in common use are the Bardeen potential:

$$U(r) = \frac{a}{r^3} + \frac{b}{r^2} - \frac{c}{r}$$

giving

$$P = X^{4/3}(X^{2/3} - 1)\left[\frac{3}{2}K_o + D(X^{2/3} - 1)\right]$$

$$K = \left(\frac{4}{3}D - 2K_o\right)X^{4/3} + (3K_o - 4D)X^2 + \frac{8}{e}DX^{8/3}$$

where $X = \rho/\rho_o$, D is an empirically determined constant and

$$K_o' = \frac{10}{3} + \frac{8}{9}\frac{D}{K_o}$$

For an exponential repulsive term in the potential function,

$$U(r) = -\frac{a}{r} + b\exp\left(-\frac{r}{c}\right)$$

giving

$$P = AX^{2/3}\exp[B(1 - X^{-1/3})] - AX^{4/3}$$

$$K = \frac{AX^{2/3}}{3}(BX^{-1/3+2})\exp[B(1 - X^{1/3})] - \frac{4}{3}AX^{4/3}$$

$$K_o = \frac{1}{3}A(B - 2)$$

and

$$K_o' = \frac{1}{9}\frac{A}{K_o}(B^2 + 3B - 12)$$

ZERO-PRESSURE VALUES OF LOWER-MANTLE SEISMIC PROPERTIES

Butler and Anderson (1978) fitted a variety of equations of state to the lower mantle in order to test for homogeneity and to obtain estimates of lower-mantle properties at zero pressure. Their results are summarized in Table 5-5. The first row gives the extrapolated zero-pressure values, based on the assumption that the lower mantle is homogeneous and adiabatic. These assumptions, for the Earth model they used, were only valid between radii of 4825–5125 km and 3850–4600 km; and I have taken the average here. In the Earth model PREM the homogeneity-adiabaticity assumption seems to hold below 5700 km radius. For comparison the Earth model PREM yields $\rho_o(T) = 3.99$–4.00 g/cm^3, $K_o(T) = 2.05$–2.23 Mbar, $K_o' = 3.8$–4.4, $G_o = 1.30$–1.35 Mbar and $G_o' = 1.5$–1.8.

In subsequent rows various temperature corrections have been made, using temperature derivatives given at the bottom of the table. In the lower part of the table, measurements or estimates of the zero-pressure, room-temperature values for various candidate lower-mantle minerals are given. Note that "perovskite" (the high-pressure form of $MgSiO_3$) and corundum (Al_2O_3) give good fits. If the mantle is homogeneous and contains abundant olivine, $(Mg,Fe)_2SiO_4$, in the upper mantle, then the lower mantle will contain substantial magnesiowüstite, (Mg,Fe)O, thereby decreasing the moduli and velocities compared to perovskite.

THE EQUATION OF STATE

The general form of an equation of state follows from considerations of elementary thermodynamics and solid-state physics. A wide variety of theoretical considerations lead to equations of state that can be expressed as

$$P = 3K_o(m - n)^{-1}[(V_o/V)^{(m+3)/3} - (V_o/V)^{(n+3)/3}] \quad (18)$$

$$K_T = K_o(m - n)^{-1}[(m + 3)(V_o/V)^{(m+3)/3} - (n + 3)(V_o/V)^{(n+3)/3}] \quad (19)$$

The choice of exponents $m = 2$, $n = 4$ leads to Birch's equation, which is based on finite-strain considerations; if $m = 1$, $n = 2$ we obtain Bardeen's equation, which was derived from quantum mechanical considerations. If $m = -3$ we obtain Murnaghan's finite-strain equation. A generalized form of the equation of state of a degenerate electron gas obeying Fermi-Dirac statistics can also be cast into this form. Equations of state based on the Mie form of the potential energy $U(r)$ of an atom in a central interatomic force field, given as

$$U(r) = -Ar^{-m} + Br^{-n}$$

where the two terms on the right correspond to an attractive and a repulsive potential and r is an interatomic distance, yield the general form of the equation of state by differentiation. The choice $m = 1$ is appropriate for electrostatic interactions, and $m = 6$, $n = 12$ is the Lennard-Jones potential, appropriate for Van der Waals or molecular crystals.

The foundations of the atomic approach were laid near the beginning of this century by Born, Van Karman, Grüneisen, Madelung, Mie and Debye. The basic premise of the theory is that ionic crystals are made up of positively charged metal atom ions and negatively charged electronegative atom ions that interact with each other according to simple central force laws. The electrostatic, or Coulomb, forces that tend to contract the crystal are balanced by repulsive forces, which, in the classical theory, are of uncertain origin. Dipole-dipole and higher order interactions, the Van der Waals forces, provide additional coupling between ions. They dominate the attraction between closed-shell atoms but are a minor part of the total attractive force in mainly ionic crystals. The Van der Waals forces are also of much shorter range than electrostatic forces.

An ionic crystal is a regular array of positive and negative ions that exert both attractive and repulsive forces on each other. The attractive force is the Coulomb or electro-

TABLE 5-5
Extrapolated Values of Lower Mantle Properties

$-\Delta T$ (°C)	ρ (q/cm³)	K_S (Mbar)	μ (Mbar)	V_P (km/s)	V_S (km/s)	K_S'	μ'
0	3.97	2.12	1.34	9.93	5.83	3.9	1.5
1400	4.14	2.48–2.62	1.67–1.72	10.7–10.9	6.3–6.4		
1600	4.16	2.53–2.68	1.71–1.76	10.8–11.0	6.4–6.5		
1800	4.18	2.57–2.74	1.74–1.80	10.8–11.1	6.5–6.6		
			Minerals				
$(Mg_{0.8}\,Fe_{0.2})O$ (mw)	4.07	1.66	1.05	8.7	5.08	4.0	2.5
Stishovite (st)	4.29	3.16	2.20	11.9	7.16	4.0*	1.1*
Al_2O_3 (cor)	3.99	2.53	1.63	10.83	6.40	4.3	1.8
Perovskite* (pv)	4.10	2.60–2.45	1.43–1.84	10.5–10.9	5.9–6.7	4.1–4.5	2.1–2.6

*Estimated.

Temperature corrections:

$$(\partial \ln K_S/\partial \ln \rho)_P = 4.0\text{–}5.5$$
$$(\partial \ln \mu/\partial \ln \rho)_P = 5.7\text{–}6.5$$
$$\alpha = 3 \times 10^{-5}/°C$$

static force between the ions, and the force that keeps the crystal from collapsing is the repulsion of filled shells. For a simple salt the attractive potential between any pair of ions with charges q_1 and q_2 is

$$U_1 = -q_1 q_2 e^2/r$$

where r is the distance between the centers of the ions and e is the electronic charge. This potential must be summed over all pairs of ions in the crystal to get the total cohesive energy. The result for the attractive potential energy of a crystal is

$$U = aU_1$$

where a (or A) is the Madelung constant, which has a characteristic value for each crystal type. The repulsive potential is much shorter range and usually involves only nearest neighbors.

The calculation of the exact form of the interatomic force law or the potential energy of an assembly of particles as a function of their separation is a difficult problem and has been treated by quantum mechanical methods for only a few cases. For many purposes it is sufficient to adopt a fictitious force law that resembles the real one in some general features and that can be made to fit it in a narrow region around the equilibrium point. The total energy U must satisfy

$$\left(\frac{dU}{dV}\right)_{V_o} = 0 \quad \text{and} \quad = \left(\frac{d^2U}{dV^2}\right)_{V_o} = \frac{K_T}{V_o}$$

which are the conditions that the crystal be in equilibrium with all forces and that the theoretical bulk modulus, K_t, should be equal to the observed value. These conditions serve to determine the constants in the fictitious force law and assure that the slope and curvature of this law are proper at the equilibrium point.

The attractive forces in a crystal are balanced by the so-called overlap repulsive forces that oppose the interpenetration of the ions. Perhaps the simplest picture is a rigid ion surrounded by a free-electron gas. The effect of hydrostatic pressure is to reduce the volume of the electron gas and to raise its kinetic energy. The kinetic energy varies as r^{-2} where r is the nearest neighbor separation. The repulsive force between ions is very small until the ions come in contact, and then it increases more rapidly than the electrostatic force. In his early work on ionic crystals, Born (1939) assumed that the repulsive forces between ions gave rise to an interaction energy of the type

$$U(r) = b/r^n$$

for the whole crystal where b and n are constants and r is the distance between nearest unlike ions. Investigations of interionic forces based on quantum mechanics indicate that a repulsive potential of this type cannot be rigorously correct, although it may be a good approximation for a small range of r. Later work has used a repulsive potential of the form

$$U(r) = be^{-r/a}$$

where b and a are constants.

Regardless of the details of the various attractive and repulsive potentials and their dependence on interatomic spacing, the Mie-Lennard-Jones potential

$$U(r) = -Ar^{-m} + Br^{-n}, \quad n > m$$

is a simple useful approximation for a restricted region of the potential energy curve and, in particular, the vicinity of the potential minimum. Constants A, B, m and n will be determined at a point in the vicinity of interest by requiring that the interatomic spacing and the bulk modulus both be appropriate for the pressure at this point.

TABLE 5-6
Thermodynamic Properties of Minerals and Metals

Substance	γ	$\left(\frac{\partial \ln K}{\partial \ln \rho}\right)_T = (\partial K/\partial P)_T$	$\left(\frac{\partial \ln K_T}{\partial \ln \rho}\right)_P = \delta_T$	$\left(\frac{\partial \ln K_S}{\partial \ln \rho}\right)_P = \delta_S$	$\left(\frac{\partial \ln K_S}{\partial T}\right)_V \times 10^5/K$	$\left(\frac{\partial \ln G}{\partial \ln \rho}\right)_T$	$\left(\frac{\partial \ln G}{\partial \ln \rho}\right)_P$	$\left(\frac{\partial \ln G}{\partial T}\right)_V \times 10^5/K$
MgO	1.53	3.89	6.18	3.61	0.842	3.13	5.58	7.72
Mg_2SiO_4	1.18	5.39	7.60	5.38	0.025	2.83	6.50	9.19
Fe_2SiO_4	1.25	5.97	6.95	5.15	2.14	—	—	—
Al_2O_3	1.32	3.99	5.14	3.41	0.945	2.73	6.85	6.72
$MgAl_2O_4$	1.13	4.19	6.14	3.97	0.389	1.31	5.89	7.42
Garnet	1.43	5.45	6.88	5.00	1.062	2.61	5.40	6.07
Cu	1.96	5.62	5.69	3.23	10.89	—	—	—
Ag	2.40	6.21	6.19	3.26	15.38	—	—	—
Au	3.03	6.50	7.03	3.92	9.64	—	—	—

Using the interatomic potential

$$U = -\frac{A}{r^m} + \frac{B}{r^n}, \quad n > m$$

and setting the molar volume of the solid $V = \overline{M}/\rho$ equal to a constant times r^3 and using the relations

$$P = -\left(\frac{\partial U}{\partial V}\right)_T$$

and

$$K_T = -\left(\frac{V\,\partial P}{\partial V}\right)_T$$

for the pressure and bulk modulus, respectively, we obtain the equations previously given as equations 18 and 19:

$$P = \frac{3K_o}{m-n}\left[\left(\frac{V_o}{V}\right)^{(m+3)/3} - \left(\frac{V_o}{V}\right)^{(n+3)/3}\right]$$

$$K = \frac{K_o}{m-n}\left[(m+3)\left(\frac{V_o}{V}\right)^{(m+3)/3} - (n+3)\left(\frac{V_o}{V}\right)^{(n+3)/3}\right]$$

where V_o and K_o are the molar volume and the bulk modulus at zero pressure.

For small compressions we can expand K about $V = V_o$ to obtain

$$\left(\frac{\partial \ln K_T}{\partial \ln V}\right)_T = -\frac{1}{3}(m+n+6) \qquad (20)$$

and can note in passing that

$$\left(\frac{\partial \ln K_T}{\partial \ln V}\right)_T = \left(\frac{\partial \ln K/\partial P}{\partial \ln V/\partial P}\right)_T = -\left(\frac{\partial K}{\partial P}\right)_T \qquad (21)$$

so, to a first approximation, the isothermal bulk modulus is a linear function of pressure and a simple power-law relationship holds between the bulk modulus and the density. Table 5-6 summarizes some pertinent experimental data on minerals and metals, which show the relationships between K_S, K_T and rigidity, G, with density and temperature.

For $m = 3$, we have the simple relation (Fürth, 1944)

$$\left(\frac{\partial \ln K}{\partial \ln V}\right)_P = \frac{(\partial \ln K/\partial T)_P}{(\partial \ln V/\partial T)_P} = \left[\frac{\partial \ln K}{\partial \ln V}\right]_T + 1$$

$$= -\frac{1}{3}(m+n+3)$$

Using Grüneisen's approximation for $(\partial \ln K_T/\partial T)_P/(\partial \ln V/\partial T)_P$, we can summarize the important results of the previous sections. For small compressions:

$$\left(\frac{\partial \ln K_T}{\partial \ln \rho}\right)_T = \frac{1}{3}(m+n+6)$$

$$\left(\frac{\partial \ln K_T}{\partial \ln \rho}\right)_P = \frac{1}{3}(m+n+9)$$

$$\left(\frac{\partial \ln K_S}{\partial \ln \rho}\right)_T = \frac{1}{3}(m+n+6) + \frac{K_T^2}{K_S}\left(\frac{\partial \alpha\gamma T}{\partial P}\right)_T$$

$$\left(\frac{\partial \ln K_S}{\partial \ln \rho}\right)_P = \frac{1}{3}(m+n+9) - \frac{K_T}{\alpha K_S}\left(\frac{\partial \alpha\gamma T}{\partial T}\right)_P$$

$$\left(\frac{\partial \ln \Phi}{\partial \ln \rho}\right)_T = \frac{1}{3}(m+n+3) + \frac{K_T^2}{K_S}\left(\frac{\partial \alpha\gamma T}{\partial T}\right)_T$$

$$\left(\frac{\partial \ln \Phi}{\partial \ln \rho}\right)_P = \frac{1}{3}(m+n+6) - \frac{K_T}{\alpha K_S}\left(\frac{\partial \alpha\gamma T}{\partial T}\right)_P$$

or

$$\left(\frac{\partial \ln K_S}{\partial \ln \rho}\right)_T = \frac{1}{3}(m+n+6) - \frac{\alpha\gamma T K_T}{K_S}\left[\left(\frac{\partial \ln K_T}{\partial \ln \rho}\right)_P - \left(\frac{\partial \ln \gamma}{\partial \ln \rho}\right)_T\right]$$

$$\left(\frac{\partial \ln K_S}{\partial \ln \rho}\right)_P = \frac{1}{3}(m + n + 9) - \frac{\gamma K_T}{K_S} + \frac{\alpha\gamma T K_T}{K_S}\left[\left(\frac{\partial \ln \gamma}{\partial \ln \rho}\right)_P + \left(\frac{\partial \ln \alpha}{\partial \ln \rho}\right)_P\right]$$

and corresponding equations for $(\partial \ln \Phi/\partial \ln \rho)$.

The Debye theory leads to a nonthermal definition of the Grüneisen ratio (Knopoff, 1963; Brillouin, 1964):

$$\gamma_D = -\frac{1}{6} - \frac{1}{2}\left(\frac{\partial \ln K_T}{\partial \ln V}\right)_T \tag{22}$$

This relation assumes that all the modes of vibration have the same volume dependence or, equivalently, that all the elastic constants depend on volume or pressure the same way. An alternative expression has been suggested by Druyvesteyn and Meyering (1941) and Dugdale and MacDonald (1953) and is hence called the DM expression:

$$\gamma_{DM} = -\frac{1}{2} - \frac{1}{2}\left(\frac{\partial \ln K_T}{\partial \ln V}\right)_T = \gamma_D - \frac{1}{3} \tag{23}$$

and this corresponds physically to a model of independent pairs of nearest neighbor atoms, in a linear chain, rather than to the Debye model of coupled atomic vibrations. This definition of the Grüneisen ratio is often used in the reduction of shock-wave data.

Taking into account the transverse oscillations of atoms leads to the free-volume γ:

$$\gamma_{FV} = -\frac{5}{6} - \frac{1}{2}\left(\frac{\partial \ln K_T}{\partial \ln V}\right)_T \tag{24}$$

This is more exact than the other derivations (Stacey, 1977; Brennan and Stacey, 1979) and is appropriate for high temperatures where the classic assumption of independent vibrations of atoms becomes a good one and where the distinction between longitudinal and transverse modes becomes fuzzy.

The γ's are related by

$$\gamma_{FV} = \gamma_{DM} - \frac{1}{3} = \gamma_D - \frac{2}{3}$$

At finite pressure,

$$\gamma_{FV} = \frac{1}{2}\left(\frac{\partial \ln K_T}{\partial \ln V} - \frac{5}{6} + \frac{2}{9}\frac{P}{K_T}\right)_T \div \left(1 - \frac{4}{3}\frac{P}{K_T}\right) \tag{25}$$

Using the above relations we can write

$$\gamma_D = \frac{1}{6}(m + n + 5)$$

and

$$\gamma_{DM} = \frac{1}{6}(m + n + 3)$$

$$\gamma_{FV} = \frac{1}{6}(m + n + 1)$$

which shows that the exponents in the equation of state are related to the anharmonic properties of the solid since the Grüneisen relation $\gamma = \alpha K_T/\rho C_V$, relating the coefficient of thermal expansion α with the specific heat C_V, is a measure of anharmonicity.

The parameter γ decreases with compression but has a tendency to be higher for the close-packed crystal structures such as face-centered cubic and hexagonal close-pack than it is for the more open structures such as diamond structure and body-centered cubic. For most materials γ is between 1 and 2, which gives a range for $m + n$ of 1 to 7 for the Lorentz-Slater theory and 3 to 9 for the DM theory. The corresponding ranges for $(-\partial \ln K_T/\partial \ln V)_T$ are 2.3 to 4.3 and 3 to 6, respectively. Many minerals have measured or inferred γ in the range 1.1 to 1.5.

In a later section I discuss the role of shear vibrations in a more accurate nonthermal definition of the Grüneisen parameter.

THE SEISMIC PARAMETER Φ

The seismic parameter Φ plays an important role in discussions of the mineralogy, composition and homogeneity of the mantle. It is more amenable to theoretical treatment than the seismic velocities and is available from static compression, shock-wave and ultrasonic experiments. Table 5-7 tabulates mean atomic weight $\overline{M}$ density ρ and Φ for a variety of important minerals and analog compounds.

The Φ for many silicates and oxides is approximately the molar average of the Φ's of the constituent oxides (Anderson, 1967b, 1969, 1970). Using the values for MgO, Al_2O_3 and SiO_2 (stishovite) in Table 5-7, we can estimate Φ for $MgAl_2O_4$ and $MgSiO_3$ as 55.3 and 60.6, respectively. Table 5-8 gives Φ calculated from the molar averaging rule; note the excellent agreement between the predictions and the measurements. This rule is useful in the estimation of Φ for compounds that have not been measured.

EFFECT OF COMPOSITION AND PHASE

We have now established the theoretical form for the expected relationship between seismic parameter Φ and density and have investigated the effect of temperature and pressure. The exponent in the power-law relationship is different for temperature and pressure, meaning that there is an intrinsic temperature effect over and above the effect of temperature on volume. We have not yet specifically allowed for composition except insofar as this information

TABLE 5-7
Mean Atomic Weight, Density and Seismic Parameter of Minerals

Mineral	$\bar{M}$	ρ (g/cm³)	Φ (km²/s²)
Albite	20.2	2.62	20.3
Nephelite	21.1	2.62	17.4
Oligoclase	20.5	2.65	24.9
Orthoclase	21.4	2.58	18.3
Microcline	21.4	2.56	20.2
Quartz	20.0	2.65	17.4
Olivine	20.1	3.22	40.1
Olivine	23.0	3.35	38.7
Orthopyroxene	21.2	3.29	32.5
Diopside	21.6	3.28	34.5
Garnets	21.9	3.73	45.1
	22.6	3.62	47.4
	—	3.64	48.7
Jadeite	22.5	3.35	40.4
Fayalite	29.1	4.14	26.0
β-spinel	20.1	3.47	50.1
γ-spinel	20.1	3.56	51.7
Majorite	20.1	3.52	49.7*
Spinel	20.3	3.58	55.1
$MgSiO_3$-perovskite	20.1	4.10	64.8
$MgFeSiO_3$-perovskite	—	4.21	61.6*
$SrTiO_3$-perovskite	36.7	5.12	34.1
TiO_2	26.6	4.26	50.6
Al_2O_3	20.4	3.99	63.2
MgO	20.2	3.58	47.4
SiO_2-stishovite	20.0	4.29	73.7
$CaSiO_3$-perovskite	24.0	4.13*	55.0*
$CaMgSi_2O_6$-perovskite	22.1	4.12*	57.8*

*Estimated.

is contained in the initial density and Φ_o. Birch (1961a) showed empirically that the mean atomic weight, $\bar{M}$, is an appropriate measure of composition although exceptions to this general rule occur. Knopoff and Uffen (1954) used a "representative atomic number" Z as a measure of composition in applying the Thomas-Fermi-Dirac (TFD) theory to compounds.

McMillan (1985) and Knopoff (1965) made semi-empirical adjustments to the Thomas-Fermi statistical model of the atom in order to obtain the proper low-pressure limit. Their equations can be put in the forms

$$-5(K_o Z^{-10/3})(ZV_o)^{-1} = d(P_{TF}Z^{-10/3})/d(ZV)_{V=V_o}$$

and

$$(K_o Z^{-10/3})(ZV_o)^{7/3} = \text{constant}$$

The latter form can be written approximately as

$$\Phi_o = \text{constant}(\rho_o/\bar{M})^{4/3}$$

Although the Thomas-Fermi model is not appropriate for pressures as low as those existing in the Earth and the extrapolation to zero-pressure conditions is not justified because of the presence of phase changes, this equation does suggest the form of the relationship between Φ_o, ρ_o and composition $\bar{M}$. At this stage the constants are more properly obtained from experiment.

Consider an equation of the general form we treated earlier:

$$P = (N - M)^{-1}K_o\left[\left(\frac{\rho}{\rho_o}\right)^N - \left(\frac{\rho}{\rho_o}\right)^M\right] \tag{26}$$

where M and N are constants, P is pressure, K_o is initial bulk modulus, ρ_o is initial density and ρ is the density at pressure P. Its derivative with respect to density gives

$$\frac{\partial P}{\partial \rho} = (N - M)^{-1}\frac{K_o}{\rho_o}\left[N\left(\frac{\rho}{\rho_o}\right)^{N-1} - M\left(\frac{\rho}{\rho_o}\right)^{M-1}\right]$$

or

$$\Phi = \Phi_o(N - M)^{-1}\left[N\left(\frac{\rho}{\rho_o}\right)^{N-1} - M\left(\frac{\rho}{\rho_o}\right)^{M-1}\right] \tag{27}$$

where the seismic parameter Φ is the ratio of the bulk modulus to the density. The adiabatic Φ for the Earth is available from seismic data. The ratio of Φ for two different densities is then

$$\frac{\Phi_1}{\Phi_2} = \frac{N(\rho_1/\rho_o)^{N-1} - M(\rho_1/\rho_o)^{M-1}}{N(\rho_2/\rho_o)^{N-1} - M(\rho_2/\rho_o)^{M-1}} \tag{28}$$

This is the seismic equation of state (Anderson, 1967a). If the total compression is small, then

$$\left(\frac{\partial \ln \Phi}{\partial \ln \rho}\right) = N + M - 1 \tag{29}$$

TABLE 5-8
Seismic Parameter $\Phi = K_S/\rho$ Calculated from Molar Average of Constituent Oxides

Mineral	Formula	Φ (km²/S²) Calculated	Measured
Spinel	$MgAl_2O_4$	54.4	54.5
Spinel	$NiFe_2O_4$	33.6	34.3
Magnetite	Fe_3O_4	33.2	36.0
Ilmenite	$FeTiO_3$	37.7	38.4
	$(Al,Cr)_2O_3$	62.5	61.7
Chromite	$MgCr_2O_3$	44.0	45.7
Titanate	$BaTiO_3$	30.7	27.0
	$SrTiO_3$	34.4	34.1
Perovskite	$MgSiO_3$	59.6*	64.8
	$CaSiO_3$	52.9*	—
Stishovite	SiO_2	—	73.7

*MgO or CaO plus SiO_2 (stishovite).

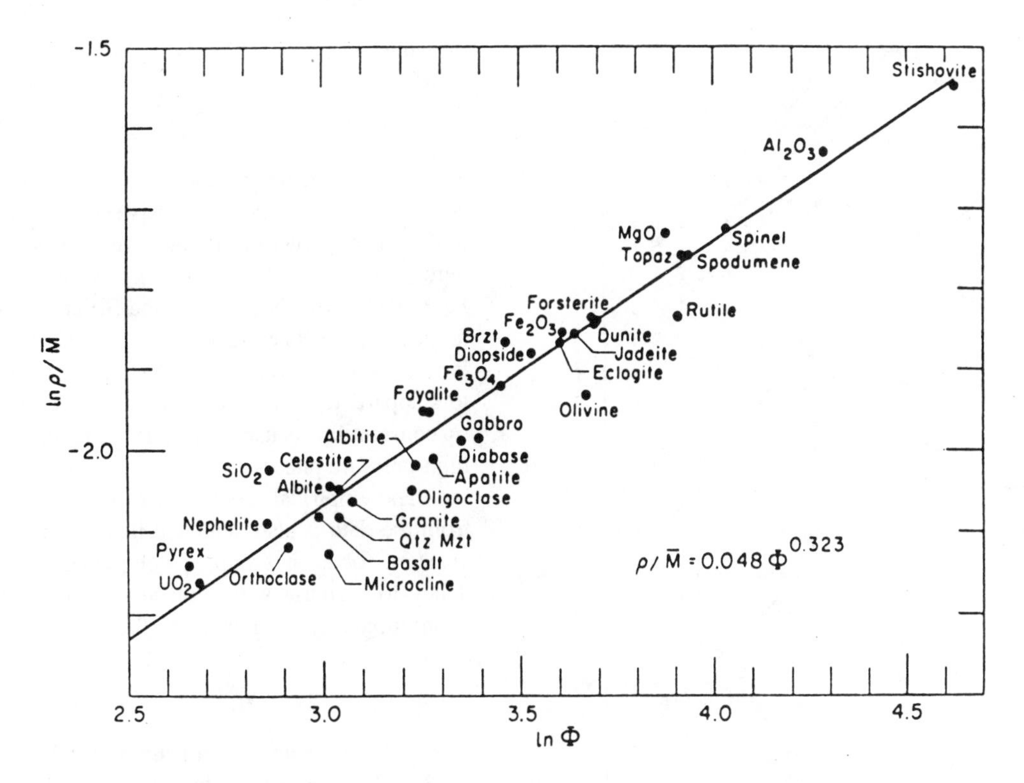

FIGURE 5-2
Seismic parameter Φ, density ρ, and mean atomic weight $\overline{M}$ for rocks, minerals and oxides (after Anderson, 1967a).

Figure 5-2 gives $\overline{M}$, ρ and Φ for selected minerals that vary in mean atomic weight. A least-square fits to these data gives

$$\frac{\rho}{\overline{M}} = 0.048\Phi^{0.323} \tag{30}$$

$$-\left(\frac{\partial \ln \Phi}{\partial \ln V}\right)_{T,P} = 3.10 \tag{31}$$

For comparison, a least-squares fit to the ultrasonic compression data on MgO and Al_2O_3 gives

$$\frac{\rho}{\overline{M}} = 0.048\Phi^{0.335}$$

$$-\left(\frac{\partial \ln \Phi}{\partial \ln V}\right)_{T} = 2.99$$

for MgO and

$$\frac{\rho}{\overline{M}} = 0.052\Phi^{0.318}$$

$$-\left(\frac{\partial \ln \Phi}{\partial \ln V}\right)_{T} = 3.15$$

for Al_2O_3. The agreement of these parameters, which are obtained from compression experiments, with those found above is remarkable. This lends support to the generalization that, as a first approximation, the bulk modulus in silicates and oxides is determined by density and mean atomic weight, or mean molar volume. The effect of changes in bulk modulus and volume at constant temperature and constant pressure are given in Table 5-6. The values for $(\partial \ln K/\partial \ln \rho)$ in that table can be compared with the above values and values computed from

$$(\partial \ln K/\partial \ln V) = (\partial \ln \Phi/\partial \ln V) - 1$$

THE REPULSIVE POTENTIAL

For any solid where the potential U can be separated into an attractive term and a repulsive term, which are functions of the interatomic separation r, we can write

$$U = U_o(r) + f(r)$$

For an ionic crystal with a Born power-law repulsive potential, it can be shown that

$$KV_o = Az_1z_2e^2(n-1)/9r_o \tag{32}$$

where A is the Madelung constant, z_i is the valence of a constituent ion, and n is the exponent in the power-law repulsive potential. If the exponential form of the repulsive potential is used, the energy per cell can be written

$$U(r) = -A/r + Be^{-r/\sigma}$$

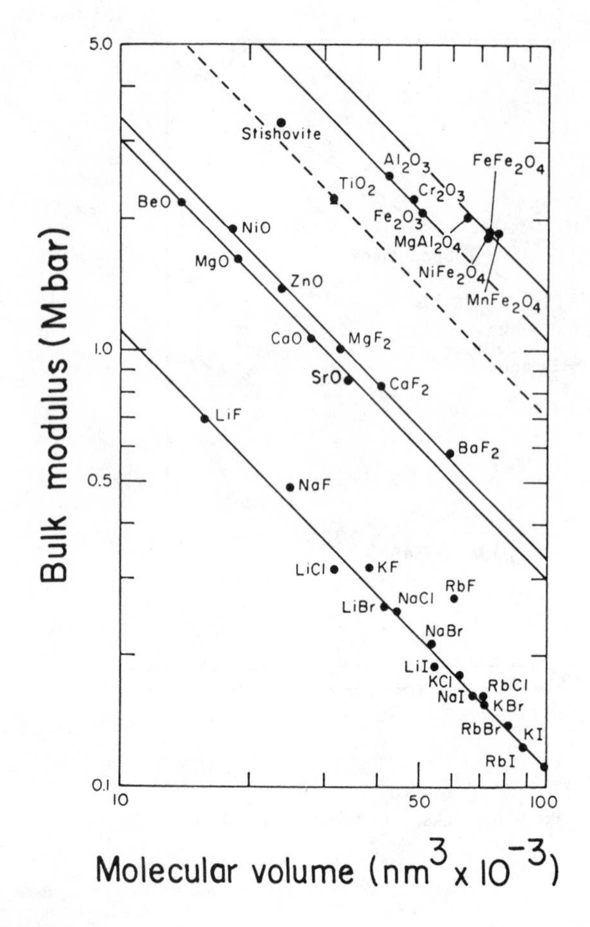

FIGURE 5-3
Bulk modulus versus molecular volume for various crystal structures (Anderson and Anderson, 1970).

where σ is a scale factor. At equilibrium the bulk modulus-volume product can be written

$$KV_o = Az_1z_2e^2(r_o/\sigma - 2)/9r_o \tag{33}$$

The parameter n in the power-law potential is simply related to σ:

$$n = (r_o/\sigma) - 1 \tag{34}$$

Data for a number of oxides are presented in Figure 5-3, which demonstrates that KV_o is a constant for a wide variety of oxide compounds. Here V_o is the specific molar volume of the formula and K is the bulk modulus at zero pressure.

The parameter $\psi = KV_o/(z_oz_ce^2)$ is tabulated in the last column of Table 5-9; z_o and z_c are, respectively, the valences of oxygen and the mean cation. Except for ZnO and TiO_2 all of the values fall in the range 0.150–0.165, and the values show no systematic behavior. ZnO and TiO_2 are anomalous in several other respects. The values of Poisson's ratio are high, the shear velocity decreases with pressure, and the oxygen coordination is anomalous for the size of the cation. If these compounds are excluded, the remaining substances for which bulk modulus data are available satisfy

$$\psi = 0.157 \pm 0.005$$

The standard deviation corresponds to an error of 3.3 percent. This is remarkable consistency when one considers that so many structures (halite, wurtzite, spinel, corundum, perovskite, and rutile) and so many cations (Mg, Be, Al, Fe^{2+}, Fe^{3+}, Mn, Ni, Cr, Sr, and Si) are involved, and that no account has been taken of structural factors (the Madelung constant) or range or repulsion parameters. These factors apparently tend to compensate for each other. The KV = constant law is useful for estimating the bulk modulus of high-pressure phases.

The empirical repulsive range parameter σ calculated from the data is also tabulated in Table 5-9. It is relatively constant for each group of compounds and shows a tendency to increase with molecular volume. It can be well approximated, as shown in Table 5-10, by the simple equation

$$\sigma_o = 0.05(1 + R + 3\Delta R) \tag{35}$$

where R is the cube root of the molecular volume and $\Delta R = R - R_c$, where R_c is the (Pauling) radius of the smallest cation.

The reduced Madelung constant A, also given in Table 5-9, is defined as

$$A_R = A/m(z_cz_o) \tag{36}$$

where m is the number of ions in the chemical formula. This is a useful parameter since it varies much less from structure to structure than the conventional Madelung constant, and it correlates well with coordination and interatomic distances. For example, it satisfies the relation

$$A_R = 0.20 + 0.45\Delta R \tag{37}$$

as shown in Table 5-10.

For crystals involving more than one cation, the valence product is defined as

$$z_cz_o = \sum^{p} x_iz_iz_o/p \tag{38}$$

where x_i is the number of cations in the formula having valence z_i, z_o is the valence of the anion (oxygen), and p is the total number of cations in the formula. z_c is the cation valence.

A check on the form of the repulsive potential is available from ultrasonic measurements of dK/dP. This quantity depends only on the parameters in the repulsive potential, that is,

$$(dK/dP)_o = (n + 7)/3 \tag{39}$$

$$(dK/dP)_o = [3(2 - r_o/\sigma)]^{-1} \times [14 - (1 + r_o/\sigma)(2 + r_o/\sigma)] \tag{40}$$

TABLE 5-9
Data for Calculation of Repulsive Range Parameter

Substance	Structure	$(z_c\ z_o)$	V (Å³)	R (Å)	A	A_R	K (Mbar)	σ (Å)	ψ
MgO	Halite	4	18.67	2.653	8.808	1.10	1.62	0.477	0.164
CaO	Halite	4	27.83	3.030	8.808	1.10	1.06	0.509	0.160
SrO	Halite	4	34.35	3.251	8.808	1.10	0.84	0.54	—
BeO	Wurtzite	4	13.77	2.397	9.504	1.19	2.20	0.482	0.164
ZnO	Wurtzite	4	23.74	2.874	9.604	1.20	1.39	0.490	0.179
$MgAl_2O_4$	Spinel	5.33	65.94	4.040	67.54	1.81	2.02	0.790	0.155
$FeFe_2O_4$	Spinel	5.33	73.85	4.195	65.48	1.75	1.87	0.769	0.161
$NiFe_2O_4$	Spinel	5.33	72.48	4.169	65.53	1.76	1.82	0.790	0.153
$MnFe_2O_4$	Spinel	5.33	76.72	4.249	(66.7)	(1.79)	1.85	0.768	0.165
$SrTiO_3$	Perovskite	6	59.56	3.905	49.51	1.65	1.79	(0.74)	0.154
Al_2O_3	Corundum	6	42.47	3.489	45.77	1.53	2.51	0.674	0.154
Fe_2O_3	Corundum	6	50.27	3.691	45.68	1.52	2.07	0.700	0.150
Cr_2O_3	Corundum	6	48.12	3.637	(45.7)	(1.52)	2.24	0.681	0.155
TiO_2	Rutile	8	31.23	3.149	30.89	1.29	2.24	0.658	0.126

Anderson and Anderson (1970).
A is the Madelung constant; A_R is the reduced Madelung constant.

for the power law and exponential forms respectively. Table 5-11 gives $(dK/dP)_o$ evaluated from equations 39 and 40 and, for comparison, the ultrasonic results. The power-law repulsive potential gives better agreement, although the measured values of $(dK/dP)_o$ are higher than computed for either potential. The exponential form gives $(dK/dP)_o$ from 0.50 to 0.58 units lower than the power-law form. Table 5-12 gives the bulk modulus calculated from the KV_o relation. Figure 5-4 shows experiments and calculations relating K_o', n and interatomic distance.

SHOCK WAVES

Pressures in the deepest parts of the Earth are beyond the reach of static-compression experiments, although pressures in diamond anvils are getting close. Explosively generated transient shock waves can be used to study material properties to pressures in excess of several megabars (1 Mbar $= 10^{11}$ Pa). The method is to fire a projectile at a target composed of the material under investigation, generating a shock wave that propagates through it at a speed,

TABLE 5-10
Comparison of Repulsive Parameter

	R_c	R	σ_c	σ	ΔR	A_c	A_R
MgO	0.65	2.65	0.48	0.48	2.00	1.10	1.10
CaO	0.99	3.03	0.51	0.51	2.04	1.12	1.10
BeO	0.31	2.40	0.48	0.48	2.09	1.14	1.19
ZnO	0.74	2.82	0.50	0.49	2.08	1.14	1.20
$MgAl_2O_4$	0.50	4.04	0.78	0.79	3.54	1.79	1.81
$FeFe_2O_4$	0.64	4.20	0.79	0.77	3.56	1.80	1.75
$NiFe_2O_4$	0.64	4.17	0.79	0.79	3.53	1.79	1.76
$MnFe_2O_4$	0.64	4.25	0.80	0.77	3.61	1.82	1.79
Al_2O_3	0.50	3.49	0.67	0.67	2.99	1.55	1.53
Fe_2O_3	0.64	3.69	0.69	0.70	3.05	1.57	1.52
Cr_2O_3	0.69	3.64	0.67	0.68	2.95	1.53	(1.52)
TiO_2	0.68	3.15	0.58	0.66	2.47	1.31	1.29
$SrTiO_2$	0.68	3.91	0.73	0.74	3.23	1.65	1.65

Anderson and Anderson (1970).
$\sigma_c = 0.05(1 + R + 3\,\Delta R)$ where $\Delta R = R - R_c$. Also given are the reduced Madelung constants calculated from $A_c = 0.20 + 0.45\,\Delta R$ and the reduced Madelung constant calculated by conventional techniques. R_c is the radius of the smallest cation.

TABLE 5-11
Repulsive Parameters and Corresponding Values of **d*K*/d*P*** for Power-Law and Exponential Repulsive Potentials

Substance	r_o	d*K*/d*P*	*n*	d*K*/d*P*	$(dK/dP)_{exp}$
MgO	5.56	3.33	4.56	3.85	3.89
CaO	5.95	3.48	4.95	3.98	5.23
BeO	4.97	3.10	3.97	3.66	5.52
ZnO	5.87	3.45	4.87	3.96	4.78
$MgAl_2O_4$	5.11	3.16	4.11	3.70	4.18
$NiFe_2O_4$	5.28	3.22	4.28	3.76	4.41
Al_2O_3	5.18	3.18	4.18	3.73	3.98
Fe_2O_3	5.27	3.22	4.27	3.76	4.53
SiO_2(st.)	5.02	3.12	4.02	3.67	7*
TiO_2	4.79	3.02	3.79	3.60	6.76

Anderson and Anderson (1970).
*Estimated.

v_s, which is faster than the following material or particle velocity, v; that is, a pressure wave travels through the solid at a speed greater than its speed of sound. The pressure rises in a thin layer to a value set up by the impact. The shock front propagates to the far side of the sample where it is reflected as a rarefaction wave. The equations of conservation of mass, momentum and energy, together with the measured shock-wave and material or particle velocities, allow one to calculate the pressure, density and internal energy of the shocked material. The sample is usually destroyed in the process. A series of shock-wave experiments, using different impact velocities and different, but hopefully similar samples then gives a relation between density, pressure and internal energy from which a shock-wave equation of state can be constructed. It is neither an adiabat nor an isotherm since the temperature and internal energy vary from point to point. Shock compression is not simply adiabatic because the compressed material acquires kinetic energy, and it is not a reversible process. The locus of points is called the Hugoniot, and a major problem is the deduction of an isothermal, or adiabatic, equation of state from this kind of data.

The basic equations are:
Conservation of mass:

$$\rho_o v_s = \rho(v_s - v) \tag{41}$$

Conservation of momentum:

$$P = \rho_o v_s v \tag{42}$$

Conservation of energy:

$$Pv = \rho_o v_s \left(\frac{1}{2} v^2 + E - E_o\right) \tag{43}$$

where $E - E_o$ is the change in internal energy per unit mass.

The rate at which material enters the shock front is $\rho_o v_s$, and it leaves at a rate $\rho(v_s - v)$ per unit area of the shock. The rate at which momentum is generated is equal to the rate of flow of material through the shock front, $\rho_o v_s$, multiplied by the velocity acquired, v, and this is equal to the difference in pressure across the front. The rate at which work is done on the material passing through the shock is equal to the rate of flow through the shock, $\rho_o v_s$, times the change of kinetic energy [$1/2\ v^2$] plus the change in internal energy, both per unit mass. Pv is the rate at which pressure does work on the material, or the rate at which kinetic en-

TABLE 5-12
Comparison of Computed (**K_c**) and Measured (**K**) Bulk Modulus

Substance	$(z_c\ z_o)$	*V*	*R*	σ_c	A_c	K_c	*K*
MgO	4	18.67	2.653	0.48	1.10	1.61	1.62
CaO	4	27.83	3.030	0.51	1.12	1.07	1.06
BeO	4	13.77	2.397	0.48	1.14	2.12	2.20
ZnO	4	23.74	2.874	0.50	1.14	1.28	1.39
$MgAl_2O_4$	5.33	65.94	4.040	0.78	1.79	2.04	2.02
$FeFe_2O_4$	5.33	73.85	4.195	0.79	1.80	1.84	1.87
$NiFe_2O_4$	5.33	72.48	4.169	0.79	1.79	1.86	1.82
$MnFe_2O_4$	5.33	76.72	4.249	0.80	1.82	1.77	1.85
$SrTiO_3$	6	59.56	3.905	0.73	1.65	1.83	1.79
Al_2O_3	6	42.47	3.489	0.67	1.55	2.58	2.52
Fe_2O_3	6	50.27	3.691	0.69	1.57	2.18	2.07
Cr_2O_3	6	48.12	3.637	0.67	1.53	2.30	2.24
SiO_2	8	23.27	2.855	0.56	1.20	3.44	3.16
TiO_2	8	31.23	3.149	0.58	1.31	2.81	2.24

Anderson and Anderson (1970).

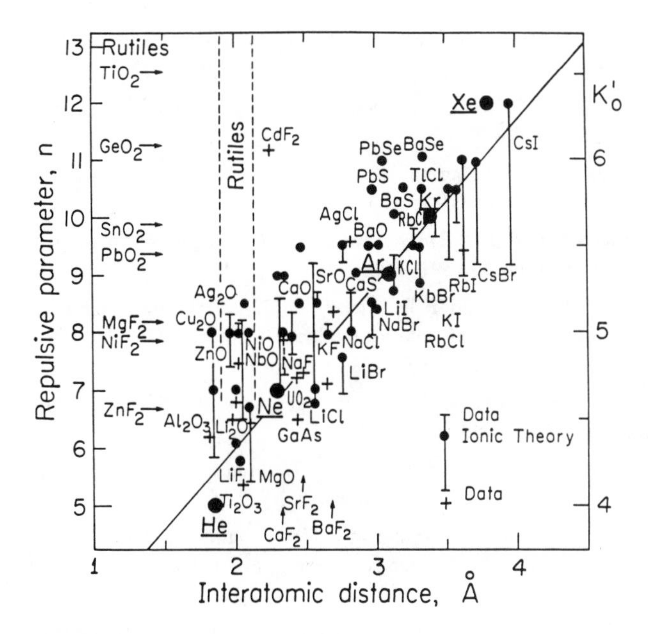

FIGURE 5-4
K'_o and repulsive parameter n versus interatomic distance. The n calculated from K using ionic theory and from compression data are shown. The line is drawn through the rare-gas solids. Ions can be treated as rare-gas atoms plus or minus electrons.

ergy plus internal energy are increased. These equations can be rewritten:

$$v_s = \frac{(P - P_o)^{1/2}}{(V_o - V)}$$

$$v = \frac{1}{2}[(P - P_o)(V_o - V)]^{1/2} \qquad (44)$$

$$E - E_o = \frac{1}{2}(P + P_o)(V_o - V)$$

where $V = 1/\rho$ and the subscript refers to the initial conditions.

In many cases the Hugoniot equations can be simplified since there is an approximately linear relation between v_s and v. In this case

$$v_s = c_o + \lambda v$$

where c_o is the bulk sound speed $(K/\rho)^{1/2}$, to which v_s reduces when the shock is weak, and λ is a constant. This gives

$$P = P_o + \frac{c_o^2(V_o - V)}{V_o - \lambda(V_o - V)^2}$$

Reduction of shock-wave data to an isotherm usually involves the Mie-Grüneisen equation, which relates the difference in pressure at fixed volume between the initial low temperature and a high-temperature state of specified thermal energy:

$$P - P_o = P^* = \alpha K_T T = \gamma\rho C_V T = \frac{\gamma}{V}(E - E_o)$$

where the total pressure, P, is the sum of an initial ambient pressure P_o plus a thermal pressure P^*. The two states (P, E) and (P_o, E_o) have the same volume. Thus, a locus of (P, V) points along a shock compression curve are reduced to a set of (P_I, V) points along an isotherm (T_o) by

$$P_I = P\left[1 - \frac{\gamma}{2}\left(\frac{V_o}{V} - 1\right)\right] - (\gamma/V)\int_{V_o}^{V}[P - \gamma\rho C_V T_o]\,dV$$

where the integral is along the shock compression curve.

The pressure, P_S, along an adiabat is

$$P_S = P[1 - (\gamma/2)(V_o/V - 1)] - (\gamma/V)\int_{V_o}^{V} P_S\,dV$$

This follows from

$$dE = -P_S\,dV$$

and

$$E_S - E_H = -\int_{V_o}^{V} P_S\,dV - \frac{1}{2}P(V_o - V)$$

the internal energy difference between the adiabat and the Hugoniot.

The pressure correction can be substantial for the higher pressure (megabar) experiments, and a reliable value of the Grüneisen ratio γ at high pressure is needed. As shown in previous sections, γ is related directly to the pressure dependence of the bulk modulus and can therefore be estimated from the shock-wave data. However, K_S and dK_S/dP require differentiation of corrected experimental data and are therefore uncertain. Usually the experimental data are fitted with theoretical or semi-empirical equations of state and the differentiations performed on these smooth functions. Shock-wave data remain our best source of information on the bulk modulus, or bulk sound speed, of rocks and minerals at high pressure, particularly high-pressure phases. These properties can be directly compared with seismic data:

$$\Phi = \left(\frac{\partial P}{\partial \rho}\right)_S = \frac{K_S}{\rho} = V_p^2 - \left(\frac{4}{3}\right)V_s^2 = c_o^2$$

Unfortunately, methods have not yet been developed for determining accurate values for the shear velocity under shock conditions.

Shock waves heat as well as compress the sample. Temperatures can be inferred from the equations already given. Temperatures are typically 1400–1700 K at pressures of the order of a megabar for materials that do not undergo phase changes, such as MgO and Al_2O_3. Silicates, which undergo shock-induced phase changes, typically end up at much higher temperature (2500–5000 K) at comparable pressures (Anderson and Kanamori, 1968). In fact, melting may occur under shock conditions.

High-speed pyrometry techniques have permitted the measurement of temperature under shock conditions (Ahrens and others, 1982). Shock-induced temperatures of 4500 to 5000 K have been measured for forsterite in the pressure range 1.5 to 1.7 Mbar. At these pressures Mg_2SiO_4 has presumably converted to MgO and $MgSiO_3$ (perovskite), and hence the temperature is due not only to compression but also due to the energy involved in phase transformation. The measured temperature is close to that calculated on the basis of the equation-of-state data (Anderson and Kanamori, 1968; Ahrens and others, 1969).

When a phase change is involved, the shock temperature T_H is calculated from

$$P(V_o - V)/2 = E_{TR} - \int_{V_o}^{V} P_a \, dV + V(P - P_a)/\gamma$$

$$T_a = T_U \exp\left[-\int_{V_o}^{V} (\gamma/V) dV \right]$$

$$V(P - P_a)/\gamma = C_V(T_H - T_a)$$

where P is the shock pressure, P_a is the isentropic pressure of the high-pressure phase, E_{TR} is the transition energy between the low- and high-pressure phases at standard conditions, and T_a is the temperature achieved on the isentrope of the high-pressure phase at volume V. The calculated temperature is therefore sensitive to E_{TR}, which is not always well known, particularly when the nature of the high-pressure phase is unknown.

References

Ahrens, T., G. Lyzenga and A. Mitchell (1982) in High-pressure research in geophysics, ed. S. Akimoto and M. Manghnani, Reidel, Dordrecht, 579–594.

Ahrens, T. J., D. L. Anderson and A. E. Ringwood (1969) Equations of state and crystal structures of high-pressure phases of shocked silicates and oxides, *Rev. Geophys.*, *7*, 667–702.

Anderson, D. L. (1967a) A seismic equation of state, *Geophys. J. Roy. Astron. Soc.*, *13*, 9–30.

Anderson, D. L. (1967b) The anelasticity of the mantle, *Geophys. J. Roy. Astron. Soc.*, *14*, 135–164.

Anderson, D. L. (1969) Bulk modulus-density systematics, *J. Geophys. Res.*, *74*, 3857–3864.

Anderson, D. L. (1970) Velocity density relations, *J. Geophys. Res.*, *75*, 1623–1624.

Anderson, D. L. and O. L. Anderson (1970) The bulk modulus-volume relationship in oxides, *J. Geophys. Res.*, *75*, 3494–3500.

Anderson, D. L. and H. Kanamori (1968) Shock-wave equations of state for rocks and minerals, *J. Geophys. Res.*, *73*, 6477–6502.

Birch, F. (1938) *J. Appl. Phys.*, *9*, 279.

Birch, F. (1952) Elasticity and constitution of the Earth's interior, *J. Geophys. Res.*, *57*, 227–286.

Birch, F. (1961a) The velocity of compressional waves in rocks to 10 kilobars, Pt. 2, *J. Geophys. Res.*, *66*, 2199–2224.

Birch, F. (1961b) Composition of the Earth's mantle, *Geophys. J.*, *4*, 295.

Birch, F. (1968) Thermal expansion at high pressures, *J. Geophys. Res.*, *73*, 817–819.

Born, Max (1939) Thermodynamics of crystals and melting, *J. Chem. Phys.*, *7*, 591–603.

Brennan, B. J. and F. D. Stacey (1979) A thermodynamically based equation of state for the lower mantle, *J. Geophys. Res.*, *84*, 5535–5539.

Brillouin, L. (1964) *Tensors in Mechanics and Elasticity*, Academic Press, New York.

Burdick, L. and D. L. Anderson (1975) Interpretation of velocity profiles of the mantle, *J. Geophys. Res.*, *80*, 1070–1074.

Butler, R. and D. L. Anderson (1978) Equation of state fits to the lower mantle and outer core, *Phys. Earth Planet. Inter.*, *17*, 147–162.

Davies, G. F. and D. L. Anderson (1971) Revised shock-wave equations of state, *J. Geophys. Res.*, *76*, 2617–2627.

Davis, G. F. and A. M. Dziewonski (1975) Homogeneity and constitution of the Earth's lower mantle and outer core, *Phys. Earth Planet. Inter.* *10*, 336.

Druyvesteyn, M. J. and J. L. Meyering (1941) *Physica*, *8*, 851.

Duffy, T. S. and D. L. Anderson, in press, (1988).

Dugdale, J. S. and D. K. C. MacDonald (1953) *Phys. Rev.*, *89*, 832.

Dziewonski, A. M. and D. L. Anderson (1981) Preliminary reference Earth model, *Phys. Earth Planet. Inter.*, *25*, 297–356.

Fürth, R. (1944) On the equation of state for solids, *Proc. R. Soc.*, *A*, *183*, 87–110.

Grüneisen, E. (1912) Theorie des festen Zustandes einatomiger Elemente, *Ann. Physik*, *39*, 257–306.

Jeanloz, R. and E. Knittle (1986) Reduction of mantle and core properties to a standard state of adiabatic decompression. In *Chemistry and Physics of Terrestrial Planets* (S. K. Saxena, ed.) 275–305, Springer-Verlag, Berlin.

Knittle, E. R., R. Jeanloz and G. L. Smith (1986) Thermal expansion of silicate perovskite and stratification of the Earth's mantle, *Nature*, *319*, 214–216.

Knopoff, L. (1963) Solids: Equations of state at moderately high pressures, Chap. 5.i. In *High Pressure Physics and Chemistry 1* (R. S. Bradley, ed.), Academic Press, New York, 444 pp.

Knopoff, L. (1965) *Phys. Rev.*, *138*, A 1445.

Knopoff, L. and R. J. Uffen (1954) The densities of compounds at high pressures and the state of the Earth's interior, *J. Geophys. Res.*, *59*, 471–484.

McMillan, W. G. (1985) *Phys. Rev.*, *111*, 479.

Sammis, C. G., D. L. Anderson and T. Jordan (1970) Application of isotropic finite strain theory to ultrasonic and seismological data, *J. Geophys. Res.*, *75*, 4478–4480.

Stacey, F. (1977) Applications of thermodynamics to fundamental Earth physics, *Geophys. Surveys*, *3*, 175–204.

6

Elasticity and Solid-State Geophysics

Mark well the various kinds of minerals, note their properties and their mode of origin.

—PETRUS SEVERINUS (1571)

The seismic properties of a material depend on composition, crystal structure, temperature, pressure and in some cases defect concentrations. Most of the Earth is made up of crystals. The elastic properties of crystals depend on orientation and frequency. Thus, the interpretation of seismic data, or the extrapolation of laboratory data, requires knowledge of crystal or mineral physics, elasticity and thermodynamics. The physics of fluids is involved in the study of magmas and the outer core. In the previous chapter we concentrated on density and the bulk modulus. In the present chapter we delve into the shear modulus and the seismic velocities. The following chapter deals with transport and nonelastic properties. Later chapters on seismic wave attenuation, anisotropy and mantle mineralogy build on this foundation.

ELASTIC CONSTANTS OF ISOTROPIC SOLIDS

The elastic behavior of an isotropic solid is completely characterized by the density ρ and two elastic constants. These are usually the bulk modulus K and the rigidity G (or μ) or the two Lamé parameters (which we will not use), λ and μ ($= G$). Young's modulus E and Poisson's ratio σ are also commonly used. There are, correspondingly, two types of elastic waves; the compressional, or primary (P), and the shear, or secondary (S), having velocities derivable from

$$\rho V_p^2 = K + 4G/3$$

$$\rho V_s^2 = G$$

The interrelations between the elastic constants and wave velocities are given in Table 6-1.

Only well-annealed glasses and similar noncrystalline materials are strictly isotropic. Crystalline material with random orientations of grains can approach isotropy, but rocks are generally anisotropic.

Laboratory measurements of mineral elastic properties, and their temperature and pressure derivatives, are an essential complement to seismic data. The elastic wave velocities are now known for hundreds of crystals. Although many of these are not directly relevant to measured seismic velocities, they are invaluable for developing rules that can be applied to the prediction of properties of unmeasured phases. Elastic properties depend on both crystal structure and composition, and the understanding of these effects, including the role of temperature and pressure, is a responsibility of a discipline called mineral physics or solid-state geophysics. Most measurements are made under conditions far from the pressure and temperature conditions in the deep crust or mantle. The measurements themselves, therefore, are just the first step in any program to predict or interpret seismic velocities.

Some information is now available on the elastic properties of all major rock-forming minerals in the mantle. On the other hand, there are insufficient data on any mineral to make assumption-free comparisons with seismic data below some 100 km depth. This is why it is essential to have a good theoretical understanding of the effects of temperature, composition and pressure on the elastic and thermal properties of minerals so that laboratory measurements can be extrapolated to mantle conditions. Laboratory results are generally given in terms of a linear dependence of the elastic moduli on temperature and pressure. The actual variation

TABLE 6-1
Connecting Identities for Elastic Constants of Isotropic Bodies

K	$G=\mu$	λ	σ
$\lambda+2\mu/3$	$3(K-\lambda)/2$	$K-2\mu/3$	$\dfrac{\lambda}{3(\lambda+\mu)}$
$\mu\dfrac{2(1+\sigma)}{3(1-2\sigma)}$	$\lambda\dfrac{1-2\sigma}{2\sigma}$	$\mu\dfrac{2\sigma}{1-2\sigma}$	$\dfrac{\lambda}{3K-\lambda}$
$\lambda\dfrac{1+\sigma}{3\sigma}$	$3K\dfrac{1-2\sigma}{2+2\sigma}$	$3K\dfrac{\sigma}{1+\sigma}$	$\dfrac{3K-2\mu}{2(3K+\mu)}$
$\rho(V_p^2-4V_s^2/3)$	ρV_s^2	$\rho(V_p^2-2V_s^2)$	—
—	—	—	$\dfrac{1}{2}\dfrac{(V_p/V_s)^2-2}{(V_p/V_s)^2-1}$

λ, μ = Lamé constants
G = Rigidity or shear modulus $= \rho V_s^2 = \mu$
K = Bulk modulus
σ = Poisson's ratio
E = Young's modulus
ρ = density
V_p, V_s = compressional and shear velocities

$$\rho V_p^2 = \lambda+2\mu = 3K-2\lambda = K+4\mu/3 = \mu\frac{4\mu-E}{3\mu-E}$$

$$= 3K\frac{3K+E}{9K-E} = \lambda\frac{1-\sigma}{\sigma} = \mu\frac{2-2\sigma}{1-2\sigma} = 3K\frac{1-\sigma}{1+\sigma}$$

of the moduli with temperature and pressure is more complex. Since the higher order and cross terms are not known in the Taylor series expansion of the moduli in terms of temperature and pressure and it is not reasonable to assume that all derivatives are independent of temperature and pressure, it is necessary to use physically based equations of state. Unfortunately, many discussions of upper-mantle mineralogy ignore the most elementary considerations of solid-state and atomic physics. For example, the functional form of $\alpha(T)$, the coefficient of thermal expansion, is closely related to the specific heat function, and the necessary theory was developed long ago by Debye, Grüneisen and Einstein. Yet α is often assumed to be independent of temperature, or linearly dependent on temperature. Likewise, interatomic potential theory shows that dK/dP must decrease with compression, yet the moduli are often assumed to increase linearly with pressure throughout the upper mantle. There are also various thermodynamic relationships that must be satisfied by any self-consistent equation of state, and certain inequalities regarding the strain dependence of anharmonic properties.

Processes within the Earth are not expected to give random orientations of the constituent anisotropic minerals. On the other hand the full elastic tensor is difficult to determine from seismic data. Seismic data usually provide some sort of average of the velocities in a given region and, in some cases, estimates of the anisotropy.

The best quality laboratory data is obtained from high-quality single crystals. The full elastic tensor can be obtained in these cases, and methods are available for computing average properties from these data. A large amount of data exists on average properties from shock-wave and static-compression experiments, and it is useful to compare these results, sometimes on polycrystals or aggregates, with the single-crystal data.

It is simpler to tabulate and discuss average properties, as I do in this section. It should be kept in mind, however, that mantle minerals are anisotropic and they tend to be readily oriented by mantle processes. Certain seismic observations in subducting slabs, for example, are best interpreted in terms of oriented crystals and a resulting seismic anisotropy. If all seismic observations are interpreted in terms of isotropy, it is possible to arrive at erroneous conclusions. The debates about the thickness of the lithosphere, the deep structure of continents, the depth of slab penetration, and the scale of mantle convection are, to some extent, debates about the anisotropy of the mantle and the interpretation of seismic data. Although it is important to understand the effects of temperature and pressure on physical properties, it is also important to realize that changes in crystal structure (solid-solid phase changes) and orientation have large effects on the seismic velocities.

Table 6-2 is a compilation of the elastic properties, measured or estimated, of most of the important mantle

TABLE 6-2
Elastic Properties of Mantle Minerals (Duffy and Anderson, 1988)

Formula (structure)	Density[1] (g/cm^3)	K_S (GPa)	G (GPa)	K_s'	G'	$-\dot{K}_S$ (GPa/K)	$-\dot{G}$ (GPa/K)
$(Mg,Fe)_2SiO_4$ (olivine)	3.222(2) + 1.182x_{Fe}	129(2)[2]	82(1) − 31(1)x_{Fe}[2]	5.1(3)[3]	1.8(1)[3]	0.016(1)[4]	0.013(1)[5]
$(Mg,Fe)_2SiO_4$ (β-spinel)	3.472(5) + 1.24(9)x_{Fe}	174(1)[6]	114(1) − 41x_{Fe}[7]	4.9	1.8	0.018	0.014
$(Mg,Fe)_2SiO_4$ (γ-spinel)	3.548(1) + 1.30x_{Fe}	184(1)[8]	119(1) − 41x_{Fe}[8]	4.8	1.8	0.017	0.014
$(Mg,Fe)SiO_3$ (orthopyroxene)	3.204(5) + 0.799(5)x_{Fe}	104(4)[9]	77(1) − 24(1)x_{Fe}[9]	5.0	2.0	0.012	0.011
$CaMgSi_2O_6$ (clinopyroxene)	3.277(5)[10]	113(1)[11]	67(2)[11]	4.5(1.8)[12]	1.7	0.013	0.010
$NaAlSi_2O_6$ (clinopyroxene)	3.32(2)[10]	143(2)[13]	84(2)[13]	4.5	1.7	0.016	0.013
$(Mg,Fe)O$ (magnesiowustite)	3.583(1) + 2.28(1)x_{Fe}	163(1) − 8(5)x_{Fe}[14]	131(1) − 77(5)x_{Fe}[14]	4.2(3)[11]	2.5(1)[11]	0.016(3)[11]	0.024(4)[11]
Al_2O_3 (corundum)	3.988(2)[10]	251(3)[11]	162(2)[11]	4.3(1)[11]	1.8(1)[11]	0.014(3)[11]	0.019(1)[11]
SiO_2 (stishovite)	4.289(3)	316(4)[11]	220(3)[11]	4.0	1.8	0.027	0.018
$(Mg,Fe)_3Al_2Si_3O_{12}$ (garnet)	3.562(2) + 0.758(3)x_{Fe}	175(1) + 1(1)x_{Fe}[11]	90(1) + 8(1)x_{Fe}[11]	4.9(5)[11]	1.4(1)[11]	0.021(2)[11]	0.010(1)[11]
$Ca_3(Al,Fe)_2Si_3O_{12}$ (garnet)	3.595(2) + 0.265(1)x_{Fe}[10]	169(2) − 11(2)x_{Fe}[15]	104(1) − 14(1)x_{Fe}[15]	4.9	1.6	0.016[16]	0.015[16]
$(Mg,Fe)SiO_3$ (ilmenite)	3.810(4) + 1.10(5)x_{Fe}	212(4)[17]	132(9) − 41x_{Fe}[17]	4.3	1.7	0.017	0.017
$(Mg,Fe)SiO_3$ (perovskite)	4.104(7) + 1.07(6)x_{Fe}	266(6)[18]	153	3.9(4)[18]	2.0	0.031	0.028
$CaSiO_3$ (perovskite)	4.13(11)[19]	227	125	3.9	1.9	0.027	0.023
$(Mg,Fe)_4Si_4O_{12}$ (majorite)	3.518(3) + 0.973(7)x_{Fe}	175 + 1x_{Fe}[20]	90 + 8x_{Fe}[20]	4.9	1.4	0.021	0.010
$Ca_2Mg_2Si_4O_{12}$ (majorite)	3.53[21]	165	104	4.9	1.6	0.016	0.015
$Na_2Al_2Si_4O_{12}$ (majorite)	4.00[21]	200	127	4.9	1.6	0.016	0.015

(1) All densities from Jeanloz and Thompson (1983) except where indicated; x_{Fe} is molar fraction of Fe endmember.
(2) Graham and others (1982), Schwab and Graham (1983), Suzuki and others (1983), Sumino and Anderson (1984), and Yeganeh-Haeri and Vaughan (1984).
(3) Schwab and Graham (1983), Sumino and Anderson (1984).
(4) Suzuki and others (1983), Sumino and Anderson (1984); $\dot{K} = \partial K/\partial T$.
(5) Schwab and Graham (1983), Suzuki and others (1983), Sumino and Anderson (1984); $\dot{G} = \partial G/\partial T$.
(6) Sawamoto and others (1984).
(7) Sawamoto and others (1984). The effect of Fe on the modulus is from Weidner and others (1984).
(8) Weidner and others (1984).
(9) Sumino and Anderson (1984), Bass and Weidner (1984), Duffy and Vaughan (1986).
(10) Robie and others (1966).
(11) Sumino and Anderson (1984).
(12) Levine and Prewitt (1981).
(13) Kandelin and Weidner (1984).
(14) Sumino and Anderson (1984).
(15) Halleck (1973), Bass (1986).
(16) Isaak and Anderson (1987).
(17) Weidner and Ito (1985).
(18) Knittle and Jeanloz (1987).
(19) Bass (1984).
(20) Inferred from Weidner and others (1987).
(21) Estimated by Bass and Anderson (1984).

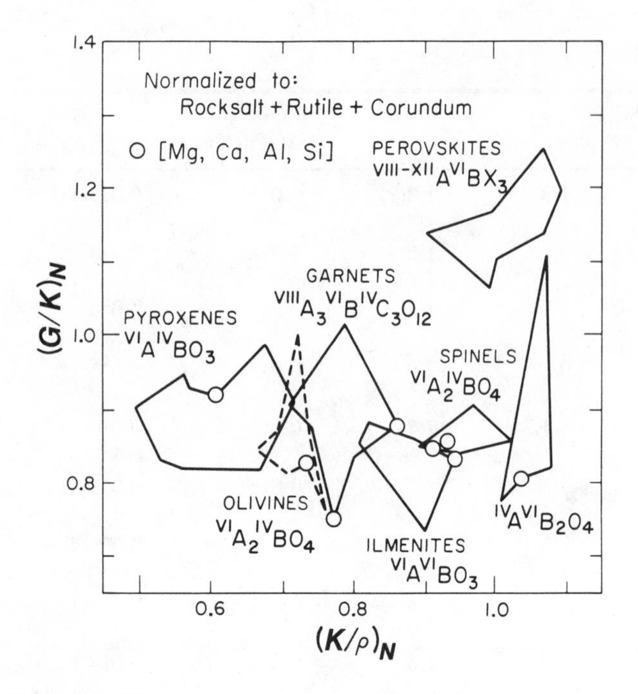

FIGURE 6-1
G/K versus K/ρ, normalized to values of constituent oxides or fluorides. Note the relatively small range of $(G/K)_N$ for Mg, Ca, Al and Si oxides. There is also little change of G/K associated with phase changes. This plot suggests a method for estimating rigidities for unmeasured phases.

minerals. Figure 6-1 relates rigidity to bulk modulus and density of major minerals.

TEMPERATURE AND PRESSURE DERIVATIVES OF ELASTIC MODULI

The pressure derivatives of the adiabatic bulk modulus, $K'_S = dK_S/dP$, for halides and oxides generally fall in the range 4.0 to 5.5. Rutiles are generally somewhat higher, 5 to 7. Oxides and silicates having ions most pertinent to major mantle minerals have a much smaller range, usually between 4.3 and 5.4. MgO has an unusually low K'_S, 3.85. The density derivative of the bulk modulus,

$$(\partial \ln K_S/\partial \ln \rho)_T = (K_T/K_S)K'_S$$

for mantle oxides and silicates that have been measured usually fall between 4.3 and 5.4 with MgO, 3.8, again being low.

The rigidity, G, has a much weaker volume or density dependence. The parameter

$$(\partial \ln G/\partial \ln \rho)_T = (K_T/G)G'$$

generally falls between about 2.5 and 2.7. For the above subset of minerals

$$(\partial \ln K_S/\partial \ln G)_T = K_S/G$$

is a very good approximation. More generally,

$$G'/K'_S = (G/K_S)^\alpha$$

where $\alpha \approx 2$ to 3 for most minerals (see Figure 6-2). These relations are useful since we know K' and K/G for many minerals for which G' is unknown.

The intrinsic temperature derivatives

$$(\partial \ln M/\alpha\partial T)_V$$

are generally negative for K_T (-4 to -1) and always negative for G (-2 to -4). The range for K_S is $+1.5$ to -2. The dimensionless intrinsic temperature derivatives are generally close to zero for K_S and about -3.4 ± 2 for G. These ranges are for most oxides, halides and silicates. The inequality

$$\left(\frac{\partial \ln K_S}{\alpha\partial T}\right)_V > \left(\frac{\partial \ln G}{\alpha\partial T}\right)_V$$

holds except for some perovskites. The magnitude of the temperature derivatives, at constant volume, indicates the intrinsic effect of temperature.

The pressure derivative of the rigidity is low for $MgAl_2O_4$-spinel. This is probably because of the low (tetrahedral) coordination of Mg. In most mantle minerals Mg exhibits 6- to 12-fold coordination, a topic considered in detail in Chapter 16.

The normalized temperature and pressure derivatives expose the intrinsic effects of temperature and show that these are generally greater for the rigidity than the bulk modulus. The normalized derivatives show a strong dependence on the constituent ions but only a weak dependence on crystal structure.

The relative changes in moduli with respect to temperature, $\partial \ln K_S/\partial T$ and $\partial \ln G/\partial T$, are of the order of

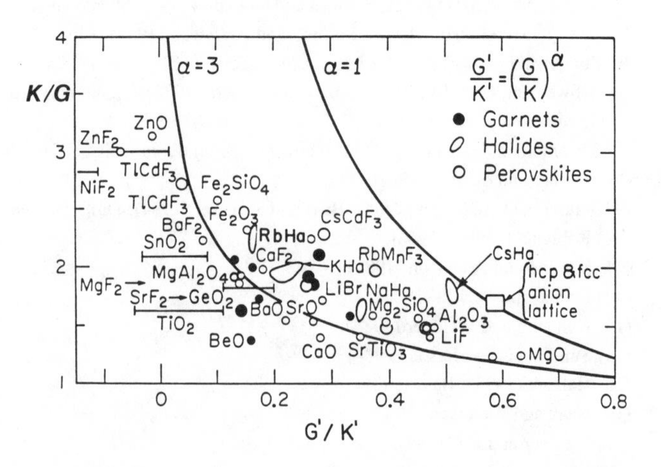

FIGURE 6-2
Correlation between G'/K' and K/G for crystals. The clustering of the alkali halides suggests cation control on these properties.

TABLE 6-3
Estimated Elastic Properties from High-pressure Experiments

Formula	Structure	ρ_0 (g/cm³)	K_0 (GPa)	K_0'	γ	Source
CaO	B2	3.76	115.0	4.9	1.8	1
	B1	3.35	112	4.8	1.5	7
FeO	B2	6.05	195.0	3.4	1.8	1
	B1	5.86	185	3.2	1.6	7
MgO	B1	3.583	163.0	4.2	1.5	1
Al_2O_3	co	3.99	253.2	3.9	1.5	6
SiO_2	st	4.29	316.0	4.0	1.5	1
$CaSiO_3$	pv	4.17	220.0	4.0	1.2	1
			263.0	4.13		3
$FeSiO_3$	pv	5.23	250.0	4.0	1.2	1
$MgSiO_3$	pv	4.10	260.0	4.0	1.2	1
			266.0	3.9	1.8	4
$CaMgSi_2O_6$	m.ox.	4.00	207.5	4.2	1.6	1
$CaMgSi_2O_6$	pv	4.12	237.7	4.0	1.2	1
$CaFeSi_2O_6$	m.ox.	4.53	214.2	4.0	1.6	1
$CaFeSi_2O_6$	pv	4.65	233.6	4.0	1.6	1
Fe_2SiO_4	ol	—	119	7	—	5
$CaMgSi_2O_6$	cpx	—	114.0	4.5	—	2
$FeSiO_4$	sp.	—	192.0	6.3	—	2
$Mg_3Al_2Si_3O_{12}$	gt	—	175.0	4.5	—	2
TiO_2	ru	—	209.0	7.3	—	2

(1) Svendsen and Ahrens (1983).
(2) Bass and others (1981).
(3) Wolf and Bukowinski (1986).
(4) Knittle and Jeanloz (1986).
(5) Yagi and others (1975).
(6) Ahrens and others (1969).
(7) Jeanloz and Ahrens (1980).

$-1.5\ (\pm 0.6) \times 10^{-4}$/°C. These are analogous to the coefficient of thermal expansion, $\alpha = \partial \ln \rho/\partial T$, but are typically five to eight times larger. These normalized moduli derivatives increase with temperature and approach a constant value at high temperature. The fully normalized derivatives,

$$(\partial \ln M/\partial \ln \rho)_P = (\partial \ln M/\alpha \partial T)_P$$

are typically 6.2 ± 0.2 for oxides and silicates. The above values are illustrative. Tables 6-3 and 6-4 give more complete compilations. The fully normalized derivatives are similar to Grüneisen parameters and might be expected to be relatively constant from material to material and as a function of temperature, at least at high temperature.

INTRINSIC AND EXTRINSIC TEMPERATURE EFFECTS

The effects of temperature on the properties of the mantle must be known for various geophysical calculations. Because the lower mantle is under simultaneous high pressure and high temperature, it is not clear that the simplifications that can be made in the classical high-temperature limit are necessarily valid. For example, the coefficient of thermal expansion, which controls many of the anharmonic properties, increases with temperature but decreases with pressure. At high temperature, the elastic properties depend mainly on the volume through the thermal expansion. At high pressure, on the other hand, the intrinsic effects of temperature may become relatively more important.

The temperature derivatives of the elastic moduli can be decomposed into extrinsic and intrinsic components:

$$(\partial \ln K_S/\partial \ln \rho)_P = (\partial \ln K_S/\partial \ln \rho)_T - \alpha^{-1}(\partial \ln K_S/\partial T)_V$$

for the adiabatic bulk modulus, K_S, and a similar expression for the rigidity, G. Extrinsic and intrinsic effects are sometimes called "volumetric" or "quasi-harmonic" and "anharmonic," respectively.

The intrinsic contribution is

$$(\partial \ln M/\partial T)_V = \alpha[(\partial \ln M/\partial \ln \rho)_T - (\partial \ln M/\partial \ln \rho)_P] \quad (1)$$

where M is K_S or G. There would be no intrinsic or an-

TABLE 6-4
Pressure Derivatives of Moduli of Cubic Crystals

Formula	C_{11}'	$(C_{11}-C_{12})'/2$	C_{44}'	K_S'	K	G
Rocksalt structures						
LiF	9.97	2.73	1.38	5.14	.70	.49
MgO	8.70	3.64	1.09	3.85	1.63	1.31
NaF	11.56	1.99	0.21	5.18	.48	.31
CaO	10.53	3.4	0.6	6.0	1.14	.81
KF	12.26	5.25	−0.43	5.26	.31	.17
SrO	11.33	4.0	−0.2	6.0	0.88	.59
RbF	12.14	4.93	−0.7	5.57	.27	.13
Perovskite structures						
$KMgF_3$	8.93	2.94	1.31	5.01	.75	.47
$RbMnF_3$	10.19	3.95	0.75	4.92	.68	.34
$SrTiO_3$	10.21	3.35	1.24	5.74	1.74	1.17

Davies (1976), Jones (1976, 1979).
K, G are moduli in Mbar.

harmonic effect if $\alpha = 0$ or if $(\partial \ln M/\partial \ln \rho)_T = (\partial \ln M/\partial \ln \rho)_P$, that is, $M(V,T) = M(V)$.

It is often assumed that αK_T is independent of pressure. This yields $\delta_T = K_T'$. When this holds there is no intrinsic temperature dependence of K_T, which is then a function of volume alone. Seismic data and most laboratory data refer to adiabatic conditions, and we will henceforth deal only with the adiabatic constants. Note that $\delta_T = K_T'$ does not imply $\delta_S = K_S'$.

The various terms in equation 1 require measurements as a function of both temperature and pressure. These are tabulated in Table 6-5 for a variety of oxides and silicates. Generally,

$$|[\partial \ln G/\partial T]_V| > |[\partial \ln K_S/\partial T]_V|$$

The intrinsic temperature effect on rigidity is greater than on the bulk modulus; G is a weaker function of volume at constant T than at constant P; and G is a weaker function of volume, at constant T, than is the bulk modulus.

The extrinsic terms are functions of pressure through the moduli and their pressure derivatives. This pressure dependence can be exposed by writing the extrinsic or quasi-harmonic terms as:

$$\alpha(K_T/K_S)(\partial K_S/\partial P)_T = -\hat{K}_{ex}$$

$$\alpha(K_T/G)(\partial G/\partial P)_T = -\hat{G}_{ex}$$

Both K_S' and G' decrease by about 30 percent from $P = 0$ to the base of the lower mantle (Dziewonski and Anderson, 1981). The coefficient of thermal expansion, at constant temperature, decreases by about 80 percent (Birch, 1968). The term K_T/G increases by about 30 percent. The net effect is an appreciable decrease for $\hat{K}_{ex}$ and a slightly smaller decrease for $\hat{G}_{ex}$. Using lower-mantle parameters from PREM, $\alpha_o = 4 \times 10^{-5}$/K and the extrapolations of Butler and Anderson (1978), we obtain -0.2×10^{-3}/deg for the high-temperature, zero-pressure value of $\hat{K}_{ex}$, similar to values for perovskites (Jones, 1979). At the core-mantle boundary (CMB) this is reduced to -0.02×10^{-3}/deg, comparable in magnitude to laboratory measurements of intrinsic temperature derivatives. The calculated zero-pressure value for $\hat{G}_{ex}$ of the lower mantle is -0.1×10^{-3}/deg decreasing to -0.02×10^{-3}/deg at the CMB. These results show that the extrinsic effects of temperature on elastic moduli become very small at lower-mantle conditions. The assumption that the adiabatic elastic moduli depend only on volume is likely to be a poor approximation. The intrinsic temperature derivatives also decrease with compression (Zharkov and Kalinin, 1971).

The inferred intrinsic temperature effect on the bulk modulus is generally small and can be either negative or positive (Table 6-5). The bulk modulus and its derivatives must be computed from differences of the directly measured moduli and therefore they have much larger errors than the shear moduli. The intrinsic temperature derivatives of the bulk modulus account for 25–40 percent of the total temperature effect for the rutiles and $SrTiO_3$-perovskite. For these structures an increase in temperature at constant volume decreases the bulk modulus. Other perovskites have positive values for $(\partial \ln K_S/\partial T)_V$ (Jones, 1979).

The intrinsic component of the temperature derivative of the rigidity is often larger than the extrinsic, or volume dependent, component and is invariably negative; an increase in temperature causes G to decrease due to the decrease in density, but a large part of this decrease would occur at constant volume. Increasing pressure decreases the total temperature effect because of the decrease of the extrinsic component and the coefficient of thermal expansion. The net effect is a reduction of the temperature derivatives, and a larger role for rigidity in controlling the temperature variation of seismic velocities and Grüneisen ratios in the lower mantle. This explains some recent data from seismic tomography. Note that α^{-1} $(\partial \ln G/\partial T)_V$ exhibits little variation for the materials in Table 6-5. It is generally larger, in absolute value, than α^{-1} $(\partial \ln K_S/\partial T)_V$ and of opposite sign. High temperature, at constant volume, makes most solids harder to compress adiabatically but easier to shear.

The total effects of temperature on the bulk modulus and on the rigidity, $(\partial \ln M/\partial \ln \rho)_P$, are comparable under laboratory conditions (Table 6-5). Therefore the compressional and shear velocities have similar temperature dependencies. On the other hand, the thermal effect on bulk modulus is largely extrinsic, that is, it depends mainly on the change in volume due to thermal expansion. The shear modulus is affected both by the volume change and a purely thermal effect at constant volume.

Although the data in Table 6-5 are not in the classical high-temperature regime ($T/\theta \gg 1$), it is still possible to separate the temperature derivatives into volume-dependent and volume-independent parts. Measurements must be made at much higher temperatures in order to test the various assumptions involved in quasi-harmonic approximations. One of the main results I have shown here is that, in general, the relative roles of intrinsic and extrinsic contributions and the relative temperature variations in bulk and shear moduli will not mimic those found in the restricted range of temperature and pressure presently available in the laboratory. The Earth can be used as a natural laboratory to extend conventional laboratory results.

It is convenient to treat thermodynamic parameters, including elastic moduli, in terms of volume-dependent and temperature-dependent parts, as in the Mie-Grüneisen equation of state. This is facilitated by the introduction of dimensionless anharmonic (DA) parameters. The Grüneisen ratio γ is such a parameter. $\partial K/\partial P = K'$ and $\partial G/\partial P = G'$ are also dimensionless anharmonic parameters, but it is useful to replace pressure, and temperature, by volume. This is done by forming logarithmic derivatives with respect to volume or density, giving Dimensionless Logarithmic An-

TABLE 6-5
Extrinsic and Intrinsic Components of $(\partial \ln M/\partial \ln \rho)_P$

Substance	Extrinsic $\left(\frac{\partial \ln K_S}{\partial \ln \rho}\right)_P$	Extrinsic $\left(\frac{\partial \ln K_S}{\partial \ln \rho}\right)_T$	Intrinsic $\frac{1}{\alpha}\left(\frac{\partial \ln K_S}{\partial T}\right)_V$	Intrinsic $\left(\frac{\partial \ln K_S}{\partial T}\right)_V$ (10^{-5}/deg)
MgO	3.0	3.8	0.8	2.3
CaO	3.9	4.8	0.9	2.6
SrO	4.7	5.1	0.4	1.6
TiO_2	10.5	6.8	−3.7	−8.7
GeO_2	10.2	6.1	−4.1	−5.6
Al_2O_3	4.3	4.3	0.0	0.0
Mg_2SiO_4	4.7	5.4	0.7	0.03
$MgAl_2O_4$	4.0	4.2	0.2	0.4
Garnet	5.2	5.5	0.3	0.5
$SrTiO_3$	7.9	5.7	−2.2	−6.2
Substance	$\left(\frac{\partial \ln G}{\partial \ln \rho}\right)_P$	$\left(\frac{\partial \ln G}{\partial \ln \rho}\right)_T$	$\frac{1}{\alpha}\left(\frac{\partial \ln G}{\partial T}\right)_V$	$\left(\frac{\partial \ln G}{\partial T}\right)_V$ (10^{-5}/deg)
MgO	5.8	3.0	−2.8	−8.6
CaO	6.3	2.4	−3.9	−11.2
SrO	4.9	2.2	−2.7	−11.3
TiO_2	8.4	1.2	−7.2	−16.9
GeO_2	5.8	2.1	−3.7	−5.1
Al_2O_3	7.5	2.6	−4.9	−16.5
Mg_2SiO_4	6.5	2.8	−3.7	−9.2
$MgAl_2O_4$	5.9	1.3	−4.6	−7.4
Garnet	5.4	2.6	−2.8	−6.1
$SrTiO_3$	7.9	3.1	−4.8	−13.4

Sumino and Anderson (1984).

TABLE 6-6
Normalized Pressure and Temperature Derivatives

Substance	$\left(\frac{\partial \ln K_S}{\partial \ln \rho}\right)_T$	$\left(\frac{\partial \ln G}{\partial \ln \rho}\right)_T$	$\left(\frac{\partial \ln K_S}{\partial \ln G}\right)_T$	$\frac{K_S}{G}$	$\left(\frac{\partial \ln K_S}{\partial \ln \rho}\right)_P$	$\left(\frac{\partial \ln G}{\partial \ln \rho}\right)_P$	$\left(\frac{\partial \ln K_S}{\alpha \partial T}\right)_V$	$\left(\frac{\partial \ln G}{\alpha \partial T}\right)_V$
MgO	3.80	3.01	1.26	1.24	3.04	5.81	0.76	−2.81
Al_2O_3	4.34	2.71	1.61	1.54	4.31	7.45	0.03	−4.74
Olivine	5.09	2.90	1.75	1.64	4.89	6.67	0.2	−3.76
Garnet	4.71	2.70	1.74	1.85	6.84	4.89	−2.1	−2.19
$MgAl_2O_4$	4.85	0.92	5.26	1.82	3.84	4.19	1.01	−3.26
$SrTiO_3$	5.67	3.92	1.45	1.49	8.77	8.70	−3.1	−4.78

harmonic (DLA) parameters. They are formed as follows:

$$(\partial \ln M/\partial \ln \rho)_T = \frac{K_T}{M}\left(\frac{\partial M}{\partial P}\right)_T = \{M\}_T$$

$$(\partial \ln M/\partial \ln \rho)_P = (\alpha M)^{-1}\left(\frac{\partial M}{\partial T}\right)_P = \{M\}_P$$

$$(\partial \ln M/\alpha \partial T)_V = \{M\}_T - \{M\}_P = \{M\}_V$$

where we use braces {} to denote DLA parameters and the subscripts T, P, V and S denote isothermal, isobaric, isovolume and adiabatic conditions, respectively. The $\{\}_V$ terms are known as intrinsic derivatives, giving the effect of temperature or pressure at constant volume. Derivatives for common mantle minerals are listed in Table 6-6. Elastic, thermal and anharmonic parameters are relatively independent of temperature at constant volume, particularly at high temperature. This simplifies temperature corrections for the elastic moduli. I use density rather than volume in order to make most of the parameters positive. By high temperature I mean $T > \theta$ where θ is the Debye or Einstein temperature. The dimensionless Grüneisen parameter γ is generally relatively constant over a broader temperature range than other thermal parameters such as α and C_P.

The DLA parameters relate the variation of the moduli to volume, or density, rather than to temperature and pressure. This is useful since the variations of density with temperature, pressure, composition and phase are fairly well understood. Furthermore, anharmonic properties tend to be independent of temperature and pressure at constant volume. The anharmonic parameter known as the thermal Grüneisen parameter γ is relatively constant from material to material as well as relatively independent of temperature.

SEISMIC CONSTRAINTS ON THERMODYNAMICS OF THE LOWER MANTLE

For most solids at normal conditions, the effect of temperature on the elastic properties is controlled mainly by the variation of volume. Volume-dependent extrinsic effects dominate at low pressure and high temperature. Under these conditions one expects that the relative changes in shear velocity, due to lateral temperature gradients in the mantle, should be similar to changes in compressional velocity. However, at high pressure, this contribution is suppressed, particularly for the bulk modulus, and variations of seismic velocities are due primarily to changes in the rigidity. Seismic data for the lower mantle can be used to estimate the Grüneisen parameters and related parameters such as the temperature and pressure derivatives of the elastic moduli and thermal expansion.

In an earlier section I showed how the bulk modulus K depended on volume through changes in temperature and pressure and how these changes, $(\partial \ln K/\partial \ln V)_T$ and $(\partial \ln K/\partial \ln V)_P$, were related to the Grüneisen parameters and the coefficient of thermal expansion α. The extrinsic, or volume-dependent, effects were shown to be greater than the intrinsic temperature effects. These considerations are now extended to the shear properties so that individual seismic wave velocities can be treated. Intrinsic effects are more important for the rigidity than for the bulk modulus. Recent geophysical results on the radial and lateral variations of velocity and density provide new constraints on high pressure–high temperature equations of state. Many of the thermodynamic properties of the lower mantle, required for equation-of-state modeling, can be determined directly from the seismic data. The effect of pressure on the coefficient of thermal expansion, the Grüneisen parameters, the lattice conductivity and the temperature derivatives of seismic wave velocities should be taken into account in the interpretation of seismic data and in convection and geoid calculations.

The lateral variation of seismic velocity is very large in the upper 200 km of the mantle but decreases rapidly below this depth. Velocity itself generally increases with depth below about 200 km. This suggests that temperature variations are more important in the shallow mantle than at greater depth. Most of the mantle is above the Debye temperature and therefore thermodynamic properties may approach their classical high-temperature limits. On the other hand, the theoretical properties of solids at simultaneous high pressure and temperature are seldom treated, and there are few precise data on variations of properties with temperature at these extreme conditions.

The effects of temperature on the bulk modulus and rigidity, that is, $(\partial \ln K/\partial T)_P$ and $(\partial \ln G/\partial T)_P$, are comparable for most minerals under normal conditions. This means that the relative variation of compressional and shear velocity with temperature should be similar. Doyle and Hales (1967), however, found that shear wave travel time anomalies, Δt_s, are about four times the compressional wave anomalies, Δt_p. This requires that lateral variations in rigidity are greater than lateral variations in compressibility or bulk modulus. The interpretation was that the anomalies are dominated by a partially molten region in the upper mantle, as partial melting affects the rigidity more than the bulk modulus. If only the rigidity changes, then

$$\Delta t_s = 3.9\ \Delta t_p$$

and

$$d \ln V_s = 2.25\ d \ln V_p$$

for Poisson's ratio $\sigma = 1/4$, or, equivalently, $(V_p/V_s)^2 = 3$ and $K = (5/3)G$. Yet a similar result has been found for the lower mantle, where the partial-melt explanation is less likely. We therefore seek a more general explanation for the apparently low lateral variation in K_s, the adiabatic bulk modulus.

The following parameters can be determined from seismology:

1. $(\partial K_S/\partial P)_S$ and $(\partial G/\partial P)_S$ from the radial variation of seismic velocities, assuming an adiabatic gradient. $(\partial K_S/\partial P)_T$ and $(\partial G/\partial P)_T$ are calculable from these.
2. $(\partial K_S/\partial T)_P$ and $(\partial G/\partial T)_P$ from the lateral changes in velocity, assuming these are due to temperature.

These derivatives are related to the Grüneisen parameters, γ and δ, and the pressure derivative of the coefficient of thermal expansion, α. The derivatives of the rigidity, G, can also be cast into Grüneisen-like parameters and used to test various assumptions in lattice dynamics, such as the assumption that longitudinal and transverse modes have the same volume dependence, or that lattice vibration frequencies depend only on volume.

We use the following relations and notation:

$$(\partial \ln K_T/\partial \ln \rho)_T = (\partial K_T/\partial P)_T = K'_T = \{K_T\}_T$$

$$(\partial \ln K_S/\partial \ln \rho)_T = (K_T/K_S)K'_S = \{K_S\}_T \tag{2}$$

$$(\partial \ln K_S/\partial \ln \rho)_P = -(\alpha K_S)^{-1}(\partial K_S/\partial T)_P = \delta_S = \{K_S\}_P \tag{3}$$

$$(\partial \ln G/\partial \ln \rho)_T = (K_T/G)(\partial G/\partial P)_T = (K_T/G)G' = G^* = \{G\}_T \tag{4}$$

$$(\partial \ln G/\partial \ln \rho)_P = -(\alpha G)^{-1}(\partial G/\partial T)_P = g = \{G\}_P \tag{5}$$

$$(\partial \ln K_T/\partial T)_P = (\partial \alpha/\partial \ln \rho)_T \tag{6}$$

$$\gamma = (1/2)K'_S - 1/6 + f(K,G,K',G') \tag{7}$$

$$-(\partial \ln \alpha/\partial \ln \rho)_T \approx \delta_S + \gamma = -\{\alpha\}_T \tag{8}$$

$$(\partial \ln K_T/\partial T)_V = \alpha[(\partial \ln K_T/\partial \ln \rho)_T - (\partial \ln K_T/\partial \ln \rho)_P] = \alpha(K'_T - \delta_T) \tag{9}$$

$$(\partial \ln G/\partial T)_V = \alpha[(\partial \ln G/\partial \ln \rho)_T - (\partial \ln G/\partial \ln \rho)_P] \tag{10}$$

$$K_S = K_T(1 + \alpha\gamma T) \tag{11}$$

$$\delta_T \approx \delta_S + \gamma \tag{12}$$

$$(\partial \ln K_S/\partial \ln \rho)_S = (\partial \ln K_S/\partial \ln \rho)_T + T\gamma(\partial \ln K_S/\partial T)_V = \{K_S\}_S \tag{13}$$

$$(\partial \ln G/\partial \ln \rho)_S = (\partial \ln G/\partial \ln \rho)_T + T\gamma(\partial \ln G/\partial T)_V = \{G\}_S \tag{14}$$

This notation stresses the volume, or density, dependence of the thermodynamic variables and is particularly useful in geophysical discussions as discussed earlier. The notation is: pressure, P, temperature, T, density, ρ, bulk modulus, K_S or K_T, rigidity, G, thermal expansion coefficient, α, and volume, V.

Values of most of the dimensionless logarithmic parameters are listed in Table 6-7 for many halides, oxides and minerals. Average values for chemical and structural classes are extracted in Table 6-8, and parameters for mineral phases of the lower mantle are presented in Table 6-9. Figures 6-3 to 6-5 show selected derivatives from Table 6-7 graphed against each other; Figure 6-6 relates some parameters to ionic radii and crystal structure.

The lateral variation of seismic velocities in the mantle can now be mapped with seismic tomographic techniques (Nakanishi and Anderson, 1982; Clayton and Comer, 1983; Dziewonski, 1984). The correlation of lower-mantle velocities with the geoid has yielded estimates of $(\partial V_p/\partial \rho)_P = 3\ (\text{km/s})/(\text{g/cm}^3)$ (Hager and others, 1985; Richards, 1986). This corresponds to $(\partial \ln V_p/\partial \ln \rho)_P = 1.2$ using lower mantle values from PREM. This is lower than laboratory values by 60 to 300 percent.

The variation of the seismic velocities can be written in terms of the moduli:

$$(\partial \ln V_s/\partial \ln \rho) = \frac{1}{2}[(\partial \ln G/\partial \ln \rho) - 1]$$

TABLE 6-7
Dimensionless Logarithmic Anharmonic Parameters

Substance	α $(10^{-6}/K)$	K_S (kbar)	G (kbar)	$\{K_S\}_T$	$\{G\}_T$	$\{K_T\}_P$	$\{K_S\}_P$	$\{G\}_P$	$\{K_T\}_V$	$\{K_S\}_V$	$\{G\}_V$	$\{K-G\}_V$	K_S/G	γ thermal	γ BR
LiF	98	723	485	4.90	3.97	4.69	2.42	6.35	0.5	2.5	−2.4	4.9	1.49	1.66	1.92
NaF	98	483	314	4.96	2.62	5.80	3.75	5.06	−0.6	1.2	−2.4	3.6	1.54	1.51	1.37
KF	99	318	164	4.81	1.97	5.05	3.18	6.17	0.0	1.6	−4.2	5.8	1.94	1.50	1.12
RbF	95	280	127	5.35	1.63	4.77	2.97	5.95	0.8	2.4	−4.3	6.7	2.20	1.43	1.05
LiCl	134	318	193	4.65	4.45	5.40	3.32	6.84	−0.4	1.3	−2.4	3.7	1.65	1.82	2.11
NaCl	118	252	148	5.10	3.00	5.45	3.74	5.07	−0.1	1.4	−2.1	3.4	1.71	1.51	1.55
KCl	105	181	93	5.10	2.01	7.48	5.67	5.54	−2.0	−0.6	−3.5	3.0	1.95	1.39	1.17
RbCl	119	163	78	5.09	1.79	5.81	4.11	5.30	−0.4	1.0	−3.5	4.5	2.10	1.44	1.09
AgCl	93	440	81	6.21	2.83	10.6	7.94	11.0	−3.8	−1.7	−8.2	6.5	5.44	2.08	1.77
NaBr	135	207	114	4.63	3.15	7.34	5.28	4.24	−2.2	−0.7	−1.1	0.4	1.81	1.72	1.59
KBr	116	150	79	5.12	2.01	5.64	3.94	4.68	−0.2	1.2	−2.7	3.8	1.90	1.45	1.16
RbBr	113	137	65	5.05	1.81	6.27	4.54	5.96	−0.9	0.5	−4.2	4.7	2.09	1.47	1.09
NaI	138	161	91	5.11	3.22	4.79	2.75	5.41	0.6	2.4	−2.2	4.5	1.76	1.74	1.65
KI	126	122	60	4.82	2.29	5.95	4.05	4.89	−0.8	0.8	−2.6	3.4	2.02	1.41	1.26
RbI	119	111	50	5.14	1.97	6.17	4.49	6.05	−0.7	0.6	−4.1	4.7	2.21	1.51	1.17
CsCl	140	182	101	5.20	4.75	6.28	3.82	6.01	−0.6	1.4	−1.3	2.6	1.80	2.04	2.30
TlCl	158	240	92	6.00	4.45	7.78	5.18	7.35	−1.0	0.8	−2.9	3.7	2.60	2.46	2.31
CsBr	138	156	88	4.97	4.53	6.33	3.86	5.90	−0.9	1.1	−1.4	2.5	1.76	1.98	2.18
TlBr	170	224	88	5.80	5.39	7.51	4.75	6.36	−0.8	1.1	−1.0	2.0	2.55	2.76	2.66
CsI	138	126	72	5.06	4.51	6.40	3.86	6.09	−0.9	1.2	−1.6	2.8	1.73	1.94	2.18
MgO	31	1628	1308	3.80	3.01	5.48	3.04	5.81	−1.6	0.8	−2.8	3.6	1.24	1.52	1.41
CaO	29	1125	810	4.78	2.42	5.45	3.92	6.29	−0.6	0.9	−3.9	4.7	1.39	1.27	1.25
SrO	42	912	587	5.07	2.25	6.99	4.68	4.98	−1.8	0.4	−2.7	3.1	1.55	1.74	1.22
BaO	38	720	367	5.42	1.95	9.46	7.34	7.88	−3.9	−1.9	−5.9	4.0	1.96	1.56	1.17
Al_2O_3	16	2512	1634	4.34	2.71	6.84	4.31	7.45	−2.5	0.0	−4.7	4.8	1.54	1.27	1.33
Ti_2O_3	17	2076	945	4.10	2.23	7.78	6.66	12.9	−3.6	−2.6	−11	8.2	2.20	1.13	1.15
Fe_2O_3	33	2066	910	4.44	1.63	5.70	3.68	3.34	−1.2	0.8	−1.7	2.5	2.27	1.99	0.95
TiO_2	24	2140	1120	6.83	1.09	12.7	10.5	8.37	−5.7	−3.7	−7.3	3.6	1.91	1.72	0.96
GeO_2	14	2589	1509	6.12	2.10	11.9	10.2	5.83	−5.7	−4.1	−3.7	−0.3	1.72	1.17	1.27
SnO_2	10	2123	1017	5.09	1.25	9.90	8.64	6.21	−4.8	−3.6	−5.0	1.4	2.09	0.88	0.85
MgF_2	38	1019	547	5.01	1.34	5.36	4.17	3.83	−0.3	0.8	−2.5	3.3	1.86	1.22	0.87
NiF_2	23	1207	459	4.98	−1.4	8.80	7.47	1.96	−3.8	−2.5	−3.4	0.9	2.63	0.88	−0.2
ZnF_2	29	1052	394	4.51	0.13	9.49	8.23	4.73	−4.9	−3.7	−4.6	0.9	2.67	0.97	0.39
CaF_2	61	845	427	4.55	2.26	5.85	3.86	4.75	−1.1	0.7	−2.5	3.2	1.98	1.83	1.21
SrF_2	47	714	350	4.67	1.66	6.02	4.75	4.94	−1.2	−0.1	−3.3	3.2	2.04	1.30	0.98
CdF_2	66	1054	330	5.77	4.05	8.36	6.04	7.11	−2.2	−0.3	−3.1	2.8	3.19	2.45	2.10
BaF_2	61	581	255	4.89	0.88	6.52	4.54	3.83	−1.4	0.3	−3.0	3.3	2.28	1.80	0.71
Opx	48	1035	747	9.26	3.18	7.26	5.43	3.34	2.2	3.8	−0.2	4.0	1.39	1.87	1.95
Olivine	25	1294	791	5.09	2.90	6.70	4.89	6.67	−1.6	0.2	−3.8	4.0	1.64	1.16	1.49
Olivine	27	1292	812	4.83	2.85	6.61	4.67	6.27	−1.7	0.2	−3.4	3.6	1.59	1.25	1.44
Olivine	25	1286	811	5.32	2.83	6.55	4.73	6.50	−1.2	0.6	−3.7	4.3	1.59	1.16	1.48
Garnets															
Fe_{16}	19	1713	927	4.71	2.70	7.89	6.84	4.89	−3.1	−2.1	−2.2	0.1	1.85	1.05	1.38
Fe_{36}	19	1682	922	4.71	2.67	7.42	5.98	5.05	−2.7	−1.3	−2.4	1.1	1.82	1.01	1.37
Fe_{54}	24	1736	955	5.38	2.52	6.79	5.52	4.81	−1.3	−0.1	−2.3	2.2	1.82	1.28	1.37
$MgAl_2O_4$	21	1969	1080	4.85	0.92	5.47	3.84	4.19	−0.6	1.0	−3.3	4.3	1.82	1.40	0.67
$SrTiO_3$	25	1741	1168	5.67	3.92	11.2	8.77	8.70	−5.3	−3.1	−4.8	1.7	1.49	1.63	1.06
$KMgF_3$	60	751	488	4.87	2.98	5.02	3.46	4.72	−0.0	1.4	−1.7	3.1	1.54	1.60	1.50
$RbMnF_3$	57	675	341	4.80	3.69	4.48	3.01	5.04	0.4	1.8	−1.4	3.1	1.98	1.49	1.80
$RbCdF_3$	40	614	257	1.09	2.15	4.12	3.06	3.27	−3.0	−2.0	−1.1	−0.8	2.39	1.06	0.80
$TlCdF_3$	49	609	228	7.43	0.95	5.09	3.86	3.14	2.4	3.6	−2.2	5.8	2.68	1.24	1.03
ZnO	15	1394	442	4.76	−2.2	9.60	6.22	3.02	−4.8	−1.5	−5.2	3.7	3.15	0.81	−0.4
BeO	18	2201	1618	5.48	1.19	5.17	3.08	4.19	0.3	2.4	−3.0	5.4	1.36	1.27	0.79
SiO_2	35	378	455	6.37	0.35	3.28	2.43	−0.3	3.1	3.9	0.7	3.3	0.85	0.67	0.40
$CaCO_3$	17	747	318	5.36	−3.5	24.0	23.1	18.5	−19	−18	−22	4.2	2.35	0.56	−1.0

TABLE 6-8
Dimensionless Logarithmic Anharmonic Derivatives

Substance	$\{K_S\}_T$	$\{G\}_T$	$\{K_S\}_P$	$\{G\}_P$	$\{K_S\}_V$	$\{G\}_V$
Averages						
Halides	5.1	2.6	4.1	5.9	0.9	−3.3
Perovskites*	4.8	2.7	4.4	5.0	0.3	−2.2
Garnets*	4.9	2.6	6.1	4.9	−1.2	−2.3
Fluorites*	5.0	2.2	4.8	5.2	0.2	−2.9
Oxides	5.3	2.0	5.7	5.8	−0.4	−3.7
Silicates	5.6	2.8	5.4	5.4	0.2	−2.6
Grand	5.1	2.5	5.0	5.7	0.1	−3.2
	(±1.0)	(±1.3)	(±1.9)	(±1.9)	(±1.9)	(±1.9)
Olivine	5.1	2.9	4.9	6.7	0.2	−3.8
Olivine	4.8	2.9	4.7	6.3	0.2	−3.4
$MgAl_2O_4$-spinel	4.9	0.9	3.8	4.2	−0.6	+1.0

$\{M\} = \partial \ln M / \partial \ln \rho$
*Structures.

$$(\partial \ln V_p/\partial \ln \rho) = \frac{1}{2}\left[\frac{3}{5}(\partial \ln K_S/\partial \ln \rho) + \frac{2}{5}(\partial \ln G/\partial \ln \rho) - 1\right]$$

where we have used $K_S = 2G$, a value appropriate for the lower mantle. For "typical" laboratory values, $\delta_S = 4$, $g = 6$, we would have $(\partial \ln V_p/\partial \ln \rho)_P = 1.9$. For comparison, if $\partial \ln V_s = 2\, \partial \ln V_p$ and $\delta_s = 4$, we would have $(\delta \ln V_p/\delta \ln \rho)_P = 4.5$, which is much greater than observed for the lower mantle.

Seismic tomography has recently established that $d \ln V_s \approx 2\, d \ln V_p$ for the lateral variations in velocity in the lower mantle (Woodhouse and Dziewonski). This implies that lateral variations in rigidity are much greater than lateral variations in bulk modulus. The value of $(\partial \ln V_p/\partial \ln \rho)_P$ can be used to establish the following bounds:

$$\{G\}_P \equiv (\partial \ln G/\partial \ln \rho)_P < 8.5$$

$$\{K_S\}_P \equiv (\partial \ln K_S/\partial \ln \rho)_P < 5.7$$

for $K_S = 2G$. If G and K contribute equally, the normalized derivatives would be about 3.4. The numerical factors in the above equations are very insensitive to K_S/G.

By combining the relations obtained from tomography and the geoid we obtain

$$\{K_S\}_P \equiv \delta_S = (\partial \ln K_S/\partial \ln \rho)_P = (K_S\alpha)^{-1}(\partial K_S/\partial T)_P = 1.8$$

$$\{G\}_P \equiv g = (\partial \ln G/\partial \ln \rho)_P = (G\alpha)^{-1}(\partial G/\partial T)_P = 5.8$$

The former is about one-fourth of normal laboratory values of δ_S taken at modest temperature and pressure, values which are dominated by the extrinsic component. A reduction of this order is consistent with the expected decrease with pressure and implies that the extrinsic component, although greatly reduced, still dominates in the lower mantle.

On the other hand, the value for g is comparable to laboratory values for most materials and within 30 percent of close-packed oxides such as Al_2O_3 and $SrTiO_3$. The intrinsic component apparently represents a large fraction of the total temperature effect. The above value for g is about twice the extrinsic contribution typical of close-packed silicates and oxides, again suggesting an important intrinsic effect.

The Grüneisen Parameters of the Lower Mantle

The thermodynamic Grüneisen parameter is defined as

$$\underline{\gamma} = \alpha K_T/\rho C_V = \alpha K_S/\rho C_P$$

This parameter generally has a value between 1 and 2 for materials of geophysical interest and is approximately constant at high temperature. On the other hand it varies with volume, but the volume dependency appears to decrease with compression (Birch, 1961). The parameter $\underline{\gamma}$ is important in temperature-dependent equations of state. For example the thermal or vibrational pressure is

$$P^* = \alpha K_T T = (C_V \underline{\gamma}/V)T$$

and the differential form of the Mie-Grüneisen equation of state is

$$(\partial P/\partial T)_V = \underline{\gamma}\rho C_V$$

Brillouin (1964) considered the vibrational pressure to be due to the radiation pressure of diffuse elastic waves. By

TABLE 6-9
Anharmonic Parameters for Oxides and Silicates and Predicated Values for Some High-pressure Phases

Mineral	$\{K_T\}_T$	$\{K_S\}_T$	$\{G\}_T$	$\{K_T\}_P$	$\{K_S\}_P$	$\{G\}_P$
α-olivine	5.0	4.8	2.9	6.6	4.7	6.5
β-spinel*	4.9	4.8	3.0	6.6	4.7	6.4
γ-spinel*	5.1	5.0	3.1	6.8	4.9	6.5
Garnet	4.8	4.7	2.7	7.9	6.8	4.9
$MgSiO_3$ (majorite)*	4.9	4.7	2.6	7.9	6.0	4.5
Al_2O_3-ilmenite	4.4	4.3	2.7	6.8	4.3	7.5
$MgSiO_3$-ilmenite*	4.7	4.5	2.7	7.0	4.3	6.0
$MgSiO_3$-perovskite*	4.5	4.5	3.5	7.5	5.5	6.5
*	4.2	4.1	2.8	6.2	4.0	5.7
SiO_2 (stishovite)*	4.5	4.4	2.4	7.5	6.4	5.0

*Predicted.

ignoring dispersion and the optical modes it is possible to estimate γ from macroscopic elastic measurements. This yields the Brillouin or acoustic Grüneisen parameter γ_{BR}.

In the Grüneisen-Debye theory of the equation of state, the Grüneisen ratio for a given mode is written

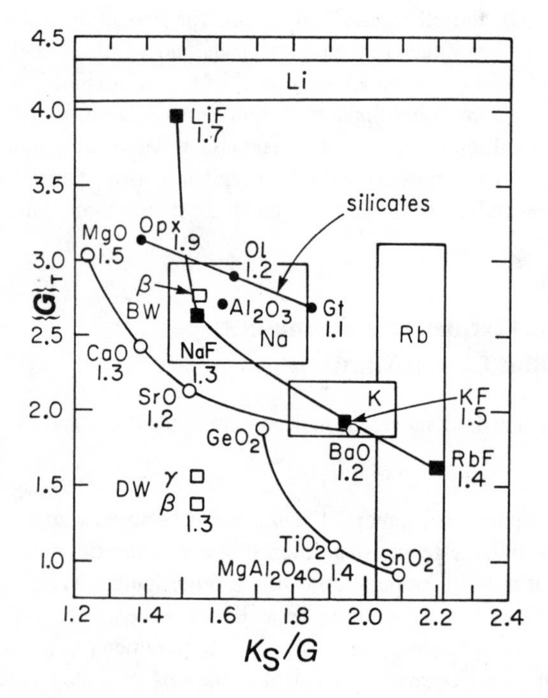

FIGURE 6-3
$\{G\}_T = (\partial \ln G/\partial \ln \rho)_T$ versus K_S/G for oxides, silicates, and halides (Opx = orthopyroxene, Ol = olivine, Gt = garnet). Boxes enclose measurements for Li, Na, K and Rb rocksalt halides. Parameter (numerals) is the Grüneisen parameter γ. Open squares are DW and BW parameters for β-spinel (from D. Weidner, 1987 and Bina and Wood, 1987, respectively). The curves connect the oxide rutiles, rocksalt oxides, silicates and the fluorides.

$$\gamma = -\frac{d \ln \theta}{d \ln V} = -\frac{d \ln \nu_m}{d \ln V}$$

where ν_m is the maximum lattice vibration frequency and θ is the Debye temperature. In an isotropic solid we have one ν_m for the longitudinal modes and a different one for the two shear modes. The acoustic or Brillouin-Grüneisen ratio (Knopoff, 1963; Brillouin, 1964) can be written

$$\gamma_{BR} = -\frac{1}{6} + \frac{1}{2}\left(\frac{d \ln K}{d \ln \rho}\right)_T - \frac{1}{3}\left(\frac{d \ln f(\sigma)}{d \ln \rho}\right)_T$$

where $f(\sigma)$ is a function of Poisson's ratio, σ, or K/G. The expression involving this term is assumed in classical theory (previous sections) to be zero, implying that all elastic constants have the same pressure or volume dependence. This is clearly not a good approximation for silicates or the Earth's mantle. In our notation the acoustic γ can be written (with $K = K_T$)

$$\gamma = \frac{1}{2}\left(\frac{\partial \ln K}{\partial \ln \rho}\right)_T - \frac{1}{6} - [(K/G) + 2][3(K/G) + 4]^{-1} \times \left[\left(\frac{\partial \ln K}{\partial \ln \rho}\right)_T - \left(\frac{\partial \ln G}{\partial \ln \rho}\right)_T\right] \quad (15)$$

which reduces to the classical form when

$$\left(\frac{\partial \ln K}{\partial \ln \rho}\right)_T = \left(\frac{\partial \ln G}{\partial \ln \rho}\right)_T$$

In the lower mantle $K \approx 2G$, and we have

$$\gamma = \frac{1}{2}\left(\frac{\partial \ln K}{\partial \ln \rho}\right)_T - \frac{1}{6} - \frac{2}{5}\left[\left(\frac{\partial \ln K}{\partial \ln \rho}\right)_T - \left(\frac{\partial \ln G}{\partial \ln \rho}\right)_T\right] \quad (16)$$

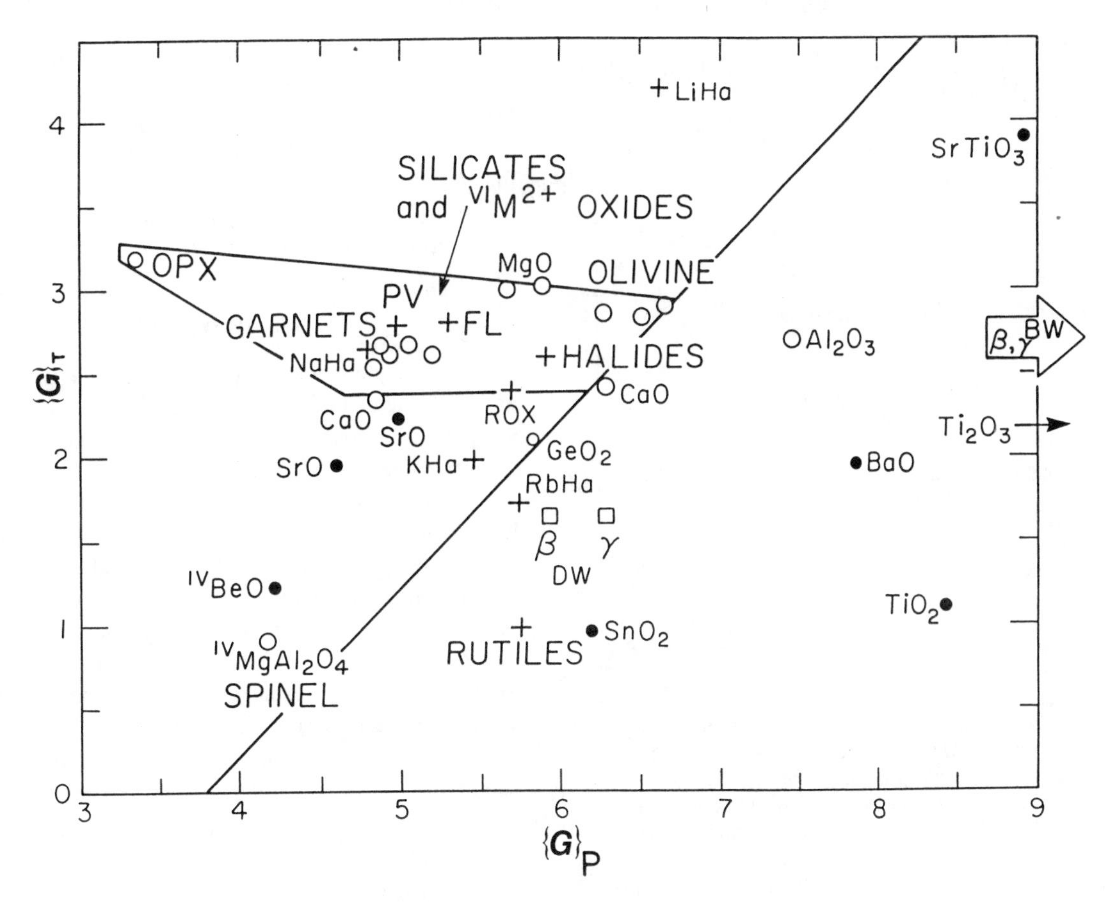

FIGURE 6-4
$\{G\}_T$ versus $\{G\}_P$ for various crystals. Symbols and sources same as for Figure 6-3.

The variation of rigidity dominates this expression. For lower-mantle values we obtain $\gamma = 1.2 \pm 0.1$ and, for extrapolated zero-pressure values, $\gamma_o = 1.4$. The numerical factor 2/5 is remarkably insensitive to variations in K/G, and equation 16 therefore has wide generality. The near cancellation of the bulk modulus terms means that the correction from adiabatic seismic properties to the high-frequency isothermal properties appropriate for lattice vibrations is very small.

The Grüneisen ratio can also be written:

$$\gamma = -(\partial \ln T/\partial \ln V)_S = K_S(\partial \ln T/\partial P)_S$$

For the adiabatic gradient at 1000 km depth we obtain $(\partial \ln T/\partial P)_S = 3.3 \times 10^{-4}$/kbar and twice this at the zero-pressure extension of the lower-mantle adiabat. The latter is 0.44°C/km for $T_o = 1700$ K. The absolute temperature along the adiabat increases by about 22 percent from zero pressure to 1000 km depth, giving T_S (1000 km) = 2080 K for $T_o = 1700$ K.

The Grüneisen parameter is related to thermal pressure, thermal expansion, and other anharmonic properties. Thermal or internal pressure is caused by diffuse elastic waves that expand the lattice. Thermally generated compressional and shear waves contribute to this internal pressure, which affects the elastic moduli at finite temperature. The Grüneisen parameter can therefore be estimated either from the thermal properties or from the volume dependence of the elastic moduli. The so-called acoustic, elastic or non-thermal Grüneisen parameter usually shows poor agreement with the thermal γ, and various attempts have been made to patch up the theory connecting these parameters to obtain better agreement. A common assumption in most theories is that the moduli or the elastic velocities have the same volume dependence at constant temperature. If so, γ can be estimated from the pressure dependence of one of the moduli. For example, the Slater value, γ_{SL}, is

$$\gamma_{SL} = \frac{1}{2}\left[\frac{\partial \ln K}{\partial \ln \rho}\right]_T - \frac{1}{6} = \frac{1}{2}\left(\frac{\partial K}{\partial P}\right)_T - \frac{1}{6} \quad (17)$$

This is derived from Debye's model of coupled atomic vibrations. If the atoms in a solid behave independently, we have the Druyvesteyn and Meyering or Dugdale and MacDonald (DM) value

$$\gamma_{DM} = \gamma_{SL} - \frac{1}{3}$$

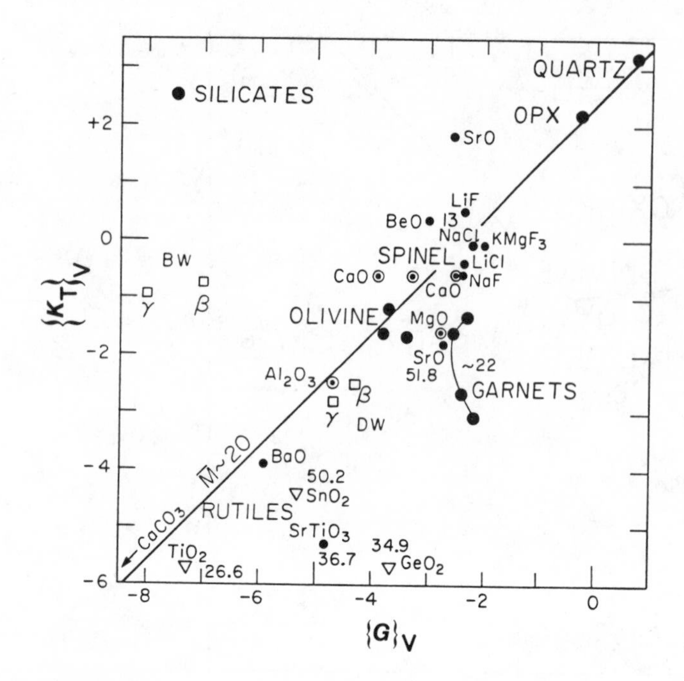

FIGURE 6-5
Intrinsic derivatives $\{K_T\}_V$ versus $\{G\}_V$. Most oxides and silicates with mean atomic weight ($\overline{M}$) near 20 ($CaCO_3$, Al_2O_3, olivine, spinel, orthopyroxene and quartz) fall near the line. Crystals with transition-element cations and high $\overline{M}$ fall below the line. The points for CaO and SrO help to indicate the scatter in presently available data. The open squares are the values adopted by BW (Bina and Wood, 1987) and DW (D. Weidner, 1987) for the spinel forms of Mg_2SiO_4. The numbers are $\overline{M}$, mean atomic weight.

The free-volume γ is, at zero pressure,

$$\gamma_{FV} = \gamma_{SL} - \frac{2}{3}$$

Equation 15 is derived with less restrictive assumptions and it shows the importance of the volume dependence of the rigidity. The rigidity, a factor in both the compressional and shear waves, plays the more important role.

The various γ's are computed for a variety of crystals and summarized in Table 6-10. The values in this table are computed from data reported in Sumino and Anderson (1984), Jones (1979) and Bohler (1982). γ_{BR} is calculated from equation 15. The factor involving K/G is nearly a constant, ranging only from 0.39 to 0.42 for K/G from 2.5 to 1.2. Most crystals relevant to geophysics exhibit an even smaller range of K/G. Therefore, a simpler form is

$$\gamma_A = \frac{1}{10}\left(\frac{\partial \ln K}{\partial \ln \rho}\right)_T + \frac{4}{10}\left(\frac{\partial \ln G}{\partial \ln \rho}\right)_T - \frac{1}{6}$$

and this is tabulated as well.

The dominance of the rigidity suggests an even simpler approximation,

$$\gamma_G = \frac{1}{2}\left(\frac{\partial \ln G}{\partial \ln \rho}\right)_T - \frac{1}{6}$$

which ignores the bulk modulus data altogether. γ_G and γ_{SL} bound $\underline{\gamma}$ for most ionic crystals. γ_G is closer to γ_{thermal} than the γ's derived from $(\partial \ln K/\partial \ln \rho)_T$ alone, even those involving elaborate refinements in the theory. The assumption that the volume dependence of the rigidity is the same as that of the bulk modulus is the most serious shortcoming of conventional lattice dynamic γ's such as the Slater and the free-volume forms. In general,

$$[\partial \ln G/\partial \ln \rho]_T < [\partial \ln K/\partial \ln \rho]_T$$

Various approximations must be made in the determination of the acoustic γ, and there are also uncertainties in the pressure derivatives of the elastic moduli. Further uncertainties are introduced by the methods of averaging the elastic properties and their derivatives when computing γ for anisotropic crystals. This makes it difficult to compare the acoustic and thermodynamic γ's and to assess the importance of high-frequency and optical modes. O. L. Anderson (1986) argued that γ nearly equals the thermodynamic γ for dense, close-packed minerals and therefore for the lower mantle. The neglect of shear-mode data and the assumption that all acoustic modes share the same volume dependence are the most serious shortcomings of most attempts to compare γ with $\underline{\gamma}$. Experimental uncertainties in

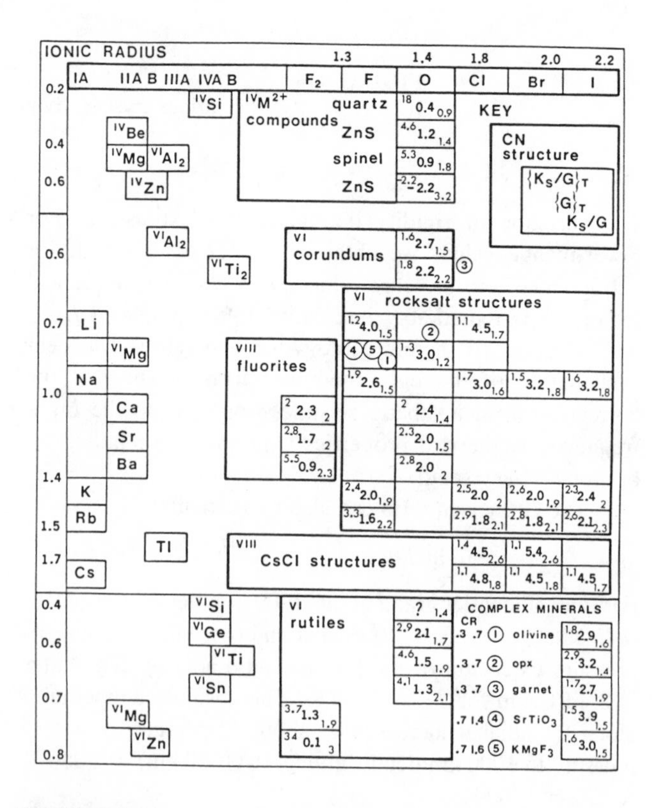

FIGURE 6-6
The shear properties of simple ionic crystals arranged by anion and cation radius. The circled numbers indicate the approximate location appropriate for the more complex ionic crystals (lower right). Note the trends with cation and anion radius. CN, cation coordination.

TABLE 6-10
Comparison of Grüneisen Parameters

Parameter	MgO	Al_2O_3	OLIVINE	GARNET	$SrTiO_3$	$KMgF_3$
$(\partial \ln K/\partial \ln \rho)_T$	3.80	4.34	4.83	4.71	5.67	4.87
$(\partial \ln G/\partial \ln \rho)_T$	3.01	2.71	2.85	2.67	3.92	2.99
K/G	1.24	1.54	1.59	1.82	1.49	1.54
γ_{SL}	1.73	2.00	2.25	2.19	2.67	2.27
γ_{DM}	1.40	1.67	1.92	1.86	2.34	1.94
γ_{FV}	1.06	1.33	1.58	1.52	2.00	1.60
γ_G	1.34	1.19	1.26	1.17	1.79	1.33
γ_{BR}	1.40	1.33	1.43	1.37	1.95	1.50
γ_A	1.40	1.35	1.46	1.37	1.97	1.52
$\gamma_{thermal}$	1.55	1.33	1.18	1.20	1.86	1.83
	1.39		1.23	1.37		
$\frac{(K/G)+2}{3(K/G)+4}$	0.42	0.41	0.41	0.40	0.41	0.41

$\gamma_{SL} = \frac{1}{2}(\partial \ln K/\partial \ln \rho)_T - 1/6$ (Slater value)

$\gamma_{DM} = \gamma_{SL} - 1/3$ (Druyvesteyn and Meyering or Dugdale and MacDonald value)

$\gamma_{FV} = \gamma_{SL} - 2/3$ (free-volume value)

$\gamma_G = \frac{1}{2}(\partial \ln G/\partial \ln \rho)_T - 1/6$ (rigidity-based approximation)

$\gamma_{BR} = \gamma_{SL} - [(K/G)+2]\,[3(K/G)+4]^{-1} \left[\left(\frac{\partial \ln K}{\partial \ln \rho}\right)_T - \left(\frac{\partial \ln G}{\partial \ln \rho}\right)_T\right]$ (Brillouin or acoustic value)

$\gamma_A = \gamma_{SL} - 2/5 \left[\left(\frac{\partial \ln K}{\partial \ln \rho}\right)_T - \left(\frac{\partial \ln G}{\partial \ln \rho}\right)_T\right]$ (simplified γ_{BR})

$\gamma_{thermal} = \alpha K_T/\rho C_V$

the pressure derivatives, particularly K_T', complicate the situation further. The acoustic γ, of course, does not include the contribution of the optical modes.

Thermodynamics of the Lower Mantle

The relative change in the variation of shear velocity in the radial (depth) direction is about 0.7 times that of the compressional velocity. On the other hand, the lateral variation of relative shear velocity is about 2 times the compressional variation. This immediately indicates that the elastic moduli are not the same function of volume and that there is an intrinsic temperature effect over and above the effect of volume. The above relations can be interpreted as

$$\left(\frac{\partial \ln V_s}{\partial \ln V_p}\right)_T = 0.7$$

$$\left(\frac{\partial \ln V_s}{\partial \ln V_p}\right)_P = 2.0$$

The relationship between V_p and V_s variations for the lower-mantle value of $K = 2G$ is

$$1 + 2(\partial \ln V_p/\partial \ln \rho)_P$$
$$= (3/5)\delta_S + (2/5)(1 + 2\,\partial \ln V_s/\partial \ln \rho)_P \quad (18)$$

Here again the numerical factors are insensitive to K_S/G. Assuming a range of $d \ln V_s = (2 \text{ to } 2.5)\, d \ln V_p$ for lateral variations and attributing these to temperature we have

$$\delta_S = 1.8 \text{ to } 1.0$$

This is much lower than the range 4 to 6 that is typical of oxides under laboratory conditions and lower than the lower-mantle K_S', which ranges from 3.0 to 3.8. Furthermore, using the value $(\partial \ln V_p/\partial \ln \rho)_P = 1.2$ inferred above from the tomography-geoid correlation (Hager and others, 1985), we have

$$(\partial \ln V_s/\partial \ln \rho)_P = 2.4 \text{ to } 3.0$$

giving

$$g = (\partial \ln G/\partial \ln \rho)_P = 5.8 \text{ to } 7.0$$

and, for completeness,

$$\delta_S = (\partial \ln K_S/\partial \ln \rho)_P = 1.8 \text{ to } 1.0$$

From PREM we have

$$(\partial \ln G/\partial \ln \rho)_S = 2.4 \text{ to } 2.6$$

$$(\partial \ln K_S/\partial \ln \rho)_S = 3.0 \text{ to } 3.8$$

This shows that the elastic constants are not strictly a function of volume and that lateral variations in rigidity are much more important than lateral variations in bulk modulus. Derivatives along the adiabat $(\partial \ln M/\partial \ln \rho)_S$ differ slightly from the corresponding isothermal derivatives. $(\partial \ln K_S/\partial \ln \rho)_T$ is within 2 percent of $(\partial \ln K_S/\partial \ln \rho)_S$. On the other hand $(\partial \ln G/\partial \ln \rho)_T$ is 4 to 16 percent higher than $(\partial \ln G/\partial \ln \rho)_S$, and this correction is important in modeling and interpreting seismic data in terms of chemistry and mineralogy.

We can also obtain some information on the coefficient of thermal expansion, α, in the lower mantle. Using the thermodynamic identity

$$\left(\frac{\partial \alpha}{\partial \ln V}\right)_T = -\left(\frac{\partial \ln K_T}{\partial T}\right)_P$$

we can write

$$\left(\frac{\partial \ln \alpha}{\partial \ln V}\right)_T = -\left(\frac{\partial \ln K_T}{\partial \ln V}\right)_P = \delta_T \qquad (19)$$

Since

$$\delta_T \approx \delta_S + \gamma$$

according to equation 12, then

$$-\left(\frac{\partial \ln \alpha}{\partial \ln \rho}\right)_T = \delta_S + \gamma$$

$$\simeq 3.1 \text{ to } 2.1$$

using values found previously. High-temperature data on NaCl and KCl (Yamamoto and Anderson, 1986) suggest an increase of about 10 percent in these values since $\delta_T - \delta_S$ is slightly greater than γ. Therefore α is inversely proportional to density to about the second or third power in the lower mantle. This causes a substantial reduction in α. From the Grüneisen relation and the extrapolated values from PREM for the lower mantle, we obtain $\alpha_o = 3.8 \times 10^{-5}$/K for zero pressure and high temperature. This is almost identical to the high-temperature value for $MgSiO_3$-perovskite (Knittle and others, 1986).

The volume dependence of γ is of both theoretical and practical importance since γ needs to be known at high pressure in order to interpret shock-wave and seismic data. The variation of thermodynamic γ with ρ at constant temperature is

$$q = \left(\frac{\partial \ln \gamma}{\partial \ln \rho}\right)_T = \left(\frac{\partial \ln \alpha}{\partial \ln \rho}\right)_T + \left(\frac{\partial \ln K_T}{\partial \ln \rho}\right)_T - \left(\frac{\partial \ln C_V}{\partial \ln \rho}\right)_T - 1 \qquad (20)$$

At this point it is common to perform some thermodynamic manipulations and then to make some approximations and to assume that one or more derivations can be set to zero. For example,

$$\left(\frac{\partial \ln \gamma}{\partial \ln \rho}\right)_T \simeq \frac{1}{\alpha}\left(\frac{\partial \ln K_S}{\partial T}\right)_V - 1 - \gamma$$

if C_V is constant, $\alpha\gamma T \ll 1$, $(\partial\alpha/\partial T)_V = 0$ and $(\partial\gamma/\partial T)_V = 0$. With the parameters derived for the lower mantle,

$$\left(\frac{\partial \ln \gamma}{\partial \ln \rho}\right)_T \simeq 0 \pm 1$$

A different approximation is

$$\left(\frac{\partial \ln \gamma}{\partial \ln \rho}\right)_T = (1 + \alpha\gamma T)^{-1}\left[\frac{1}{\alpha}\left(\frac{\partial \ln K_S}{\partial T}\right)_V - 1 - \gamma - \alpha\gamma T\left(\frac{\partial \ln K_S}{\partial \ln \rho}\right)_P\right]$$

where $(\partial\gamma/\partial T)_P$ has been set to zero (Bassett and others, 1968). This gives

$$\left(\frac{\partial \ln \gamma}{\partial \ln \rho}\right)_T \simeq -0.5 \pm 1$$

The above estimates are consistent with a constant γ or a weak variation with depth. The usual approximation

$$q = \left(\frac{\partial \ln \gamma}{\partial \ln \rho}\right)_T = -1$$

is based on the assumption that C_V is independent of volume at constant temperature and $(\partial K_T/\partial T)_V$ is zero or αK_T is independent of volume and temperature.

The approximations with the strongest experimental support, at high temperature, are

$$(\partial\alpha K_T/\partial T)_P \simeq 0$$

$$(\partial\alpha/\partial T)_V \simeq 0$$

$$(\partial K_T/\partial T)_V \simeq 0$$

Our estimate of $\frac{1}{\alpha}\left(\frac{\partial \ln K_S}{\partial T}\right)_V$ for the lower mantle is consistent with

$$\frac{1}{\alpha}\left(\frac{\partial \ln K_T}{\partial T}\right)_V \simeq 0$$

since

$$\frac{1}{\alpha}\left(\frac{\partial \ln K_S}{\partial T}\right)_V \simeq \frac{1}{\alpha}\left(\frac{\partial \ln K_T}{\partial T}\right)_V + \gamma$$

The derivative $(\partial\gamma/\partial T)_P$ is close to zero, but $(\partial\gamma/\partial T)_V$ is negative. Theoretically, $(\partial \ln C_V/\partial \ln V)_S$ should be very small.

With these relations in mind we write:

$$\left(\frac{\partial \ln \gamma}{\partial \ln \rho}\right)_S = \left(\frac{\partial \ln \alpha}{\partial \ln \rho}\right)_S + \left(\frac{\partial \ln K_T}{\partial \ln \rho}\right)_S - \left(\frac{\partial \ln C_V}{\partial \ln \rho}\right)_S - 1$$

$$= \frac{K_S}{K_T}\left[\frac{1}{\alpha}\left(\frac{\partial \ln K_T}{\partial T}\right)_V\right] + \alpha\left(\frac{\partial \ln \alpha}{\partial \ln T}\right)_P$$

$$+ \gamma\left(\frac{\partial \ln K_T}{\partial \ln T}\right)_P - \left(\frac{\partial \ln C_V}{\partial \ln \rho}\right)_S - 1$$

$$= \frac{K_S}{K_T}\left[\frac{1}{\alpha}\left(\frac{\partial \ln K_T}{\partial T}\right)_V\right] + \gamma T\left(\frac{\partial \ln \alpha}{\partial T}\right)_V$$

$$- \left(\frac{\partial \ln C_V}{\partial \ln \rho}\right)_S - 1$$

$$\simeq -1$$

Now

$$\left(\frac{\partial \ln \gamma}{\partial \ln \rho}\right)_S = \frac{K_S}{K_T}\left(\frac{\partial \ln \gamma}{\partial \ln \rho}\right)_T + \gamma\left(\frac{\partial \ln \gamma}{\partial \ln T}\right)_P$$

$$= \left(\frac{\partial \ln \gamma}{\partial \ln \rho}\right)_T + \gamma T\left(\frac{\partial \ln \gamma}{\partial T}\right)_V$$

Therefore, $|(\partial \ln \gamma/\partial \ln \rho)_S| > |(\partial \ln \gamma/\partial \ln \rho)_T|$. Since $K_S > K_T$, $(\partial \ln \gamma/\partial \ln T)_P \approx 0$ and $(\partial \ln \gamma/\partial T)_V < 0$,

$$q = \left(\frac{\partial \ln \gamma}{\partial \ln \rho}\right) = -1 + \varepsilon$$

The parameter q itself depends on compression. Model calculations show that the absolute value of q decreases with compression, meaning that γ should be only a weak function of density in the lower mantle (Hardy, 1980). The seismic data are therefore consistent with theoretical expectations.

Although extrinsic effects are suppressed by pressure, they are enhanced at high temperature. This permits us to estimate the effect of temperature on K'_S and G'. From PREM parameters for the lower mantle, we have

$$(\partial \ln M'/\partial \ln \rho) \approx -1$$

where M' is K' or G'. Assuming a mean atomic weight of 21, we derive, for the lower mantle,

$$K'_S = 0.78 V_m$$

where V_m is the mean molar volume. This relation also satisfies ultrasonic K'_S versus V_m data for olivines, garnets, $SrTiO_3$-perovskite and MgF_2-rutile and predicts K'_S to within about 10 percent for Al_2O_3, MgO and CaO. If these relations also hold for a change in temperature, then

$$\partial K'_S/K'_S = -\partial\rho/\rho = \alpha \partial T$$

and K'_S at 1700 K is predicted to be about 6 percent higher than K'_S at 300 K. This is more important than the difference between K'_S and K'_T.

Summary of Lower-Mantle Thermodynamics

Assuming that the lower mantle is adiabatic and chemically homogeneous, it is possible to obtain reliable estimates of lower-mantle properties extrapolated to zero pressure and high temperature (Butler and Anderson, 1978; Jeanloz and Knittle, 1986). (The temperature correction is described in the next section.) These extrapolations provide evidence that the lower mantle is predominantly perovskite and that it differs from the inferred composition of the upper mantle. Representative extrapolated values for the lower mantle are

$$\rho_o = 3.97\text{–}4.00\ \mathrm{g/cm^3}$$

$$K_o = 2.12\text{–}2.23\ \mathrm{kbar}$$

$$G_o = 1.30\text{–}1.35\ \mathrm{kbar}$$

$$(K'_o)_S = 3.8\text{–}4.1$$

$$(G'_o)_S = 1.5\text{–}1.8$$

The entire lower mantle is not radially homogeneous. The region between about 650 and 750 km has high gradients of seismic velocity, probably due to a mixed-phase region involving γ-spinel and majorite transforming to ilmenite and perovskite. The seismic results may also be sensing a deflection of the 650-km seismic discontinuity. The lowermost 200 to 300 km of the mantle, region D″, is also radially and laterally inhomogeneous.

The lowermost mantle is probably the dumping ground for light material exsolving from the core and ancient relict dense phases subducted while the Earth was accreting. It is therefore probably a chemically distinct layer composed of dense refractory silicates. Superposed on this is the high thermal gradient in the mantle-core thermal boundary layer. It has previously been supposed that the thickness of D″ is too great for it to be simply a thermal boundary, but the considerations in this section, particularly the decrease with rising pressure of the coefficient of thermal expansion and the large increase in the lattice thermal conductivity, show that a thick thermal boundary is likely. The low α, the high conductivity and the possibility of an intrinsic high density will tend to stabilize this layer until a large temperature gradient is built up. It may, however, accommodate small-scale convection, which will contribute to the observed lateral variability in seismic properties and, possibly, thickness. The transition zones at the top and bottom of the lower mantle are excluded from the present analysis.

The radial and lateral variation of seismic velocities in the mantle can be used to estimate important thermodynamic properties and to test some standard equation-of-state assumptions. The radial variations yield K' and G', which are related to $(\partial \ln K_S/\partial \ln \rho)_T$ and $(\partial \ln G/\partial \ln \rho)_T$ and the corresponding values along an adiabat. These in turn are related to the volume-dependent or extrinsic effects of temperature. The lateral variations, if due to temperature, yield estimates of derivatives at constant pressure, that is, δ_S and G. These derivatives are closely connected to the Grüneisen parameters γ and δ. Theoretically, the effects of temperature decrease with compression, and this is borne out by the seismic data. In the lower mantle the lateral variation of seismic velocities is dominated by the effect of temperature on the rigidity. This is similar to the situation in the upper mantle except that there it is caused by nonelastic processes, such as partial melting and dislocation relaxation, phenomena accompanied by increased attenuation. In the lower mantle the effect is caused by anharmonic phenomena and intrinsic temperature effects that are more important in shear than in compression. One cannot assume that physical properties are a function of volume alone, that classical high-temperature behavior prevails, or that shear and compressional modes exhibit similar variations. On a much more basic level, one cannot simply adopt laboratory values of temperature derivatives to estimate the effect of temperature on density and elastic-wave velocities in slabs and in the deep mantle.

From the seismic data for the lower mantle, we obtain the following estimates of in-situ values:

$(\partial \ln G/\partial \ln \rho)_P$ 5.8 to 7.0

$(\partial \ln G/\partial \ln \rho)_T$ 2.6 to 2.9

$(\partial \ln K_S/\partial \ln \rho)_P$ 1.8 to 1.0

$(\partial \ln K_S/\partial \ln \rho)_T$ 2.8 to 3.6

$(\partial \ln \gamma/\partial \ln \rho)_T$ $-1 + \varepsilon$

$-(\partial \ln \alpha/\partial \ln \rho)_T$ 3 to 2

For the intrinsic temperature terms we obtain:

$$\frac{1}{\alpha}\left[\frac{\partial \ln G}{\partial T}\right]_V \approx -3.2 \text{ to } -4.1$$

$$\frac{1}{\alpha}\left[\frac{\partial \ln K_S}{\partial T}\right]_V \approx +1 \text{ to } +2.6$$

The former is in the range of laboratory measurements. The latter is positive, indicating that an increase of temperature at constant volume causes K_S to increase. It can be compared with the value of +1.4 found for $KMgF_3$-perovskite. From lower-mantle values extrapolated to the surface, we obtain $\alpha_o = 3.8 \times 10^{-5}$/K and $\gamma_o = 1.43$. These are very close to estimates of Birch (1968).

An interesting implication of the seismic data is that the bulk modulus and rigidity are similar functions of volume at constant temperature. On the other hand, G is a stronger function of volume and K_S a much weaker function of volume at constant pressure than they are at constant temperature.

CORRECTING ELASTIC PROPERTIES FOR TEMPERATURE

The elastic properties of solids depend primarily on static lattice forces, but vibrational or thermal motions become increasingly important at high temperature. The resistance of a crystal to deformation is partially due to interionic forces and partially due to the radiation pressure of high-frequency acoustic waves, which increase in intensity as the temperature is raised. If the increase in volume associated with this radiation pressure is compensated by the application of a suitable external pressure, there still remains an intrinsic temperature effect. Thus,

$$K(V,T) = K(V,0) + K(\partial \ln K/\partial T)_V T \qquad (21)$$

$$G(V,T) = G(V,0) + G(\partial \ln G/\partial T)_V T \qquad (22)$$

These equations provide a convenient way to estimate the properties of the static lattice, that is, $K(V,0)$ and $G(V,0)$, and to correct measured values, $K(V,T)$ and $G(V,T)$, to different temperatures at constant volume. The static lattice values should be used when searching for velocity-density or modulus-volume systematics or when attempting to estimate the properties of unmeasured phases. These equations can also be used to correct laboratory or seismological data in forward or inverse modeling of mantle properties. The correction to a different or standard volume, at constant temperature, involves the extrinsic derivatives $(\partial \ln M/\partial \ln \rho)_T$.

The first step in forward modeling of the seismic properties of the mantle is to compile a table of the ambient or zero-temperature properties, including temperature and pressure derivatives, of all relevant minerals. The fully normalized extrinsic and intrinsic derivatives are then formed and, in the absence of contrary information, are assumed to be independent of temperature. The coefficient of thermal expansion, $\alpha(T)$ or $\alpha(\theta/T)$, can be used to correct the density to the temperature of interest at zero pressure. It is important to take the temperature dependence of $\alpha(T)$ into account properly since $\alpha(T)$ increases rapidly from room temperature but levels out at high T/θ. The use of the ambient α_o will underestimate the effect of temperature; the use of α_o plus the initial slope, $(\partial\alpha/\partial T)_o$ in a linear approximation, will overestimate the volume change at high temperature. Fortunately, the shape of $\alpha(T)$ is well known theoretically (a Debye function) and has been measured for many mantle minerals (see Table 6-7). The moduli can then be corrected for the volume change using $(\partial \ln M/\partial \ln \rho)_T$ and corrected for temperature using $(\partial \ln M/\partial T)_V$. The parameters $(\partial \ln M/\partial \ln \rho)_T$ or $(\partial \ln M/\partial \ln \rho)_S$ are then used in an equation of state to calculate $M(P,T)$, for example, from finite strain. The normalized form of the pressure derivatives can be assumed to be either independent of temperature or functions of $V(T)$, as discussed earlier. Theoretically, the M' are expected to increase with temperature. The expression

$$\left(\frac{\partial \ln M'}{\partial \ln \rho}\right)_P = -1$$

is suggested as a first approximation for this effect.

So far we have treated only the variation of the extrinsic or quasi-harmonic effects and showed that they decreased with compression. This means that temperature is less effective in causing variations in density and elastic properties in the deep mantle than in the shallow mantle and that relationships between these variations are different from those observed in the laboratory. Anharmonic effects also are predicted to decrease rapidly with compression (Zharkov and Kalinin, 1971; Hardy, 1980; Wolf and Jeanloz, 1985), although experimental confirmation is meager, particularly for the shear constants. Anharmonic contributions are important at high temperature and low pressure, but it is not yet clear whether temperature or pressure dominates in the lower mantle, particularly since the reference temperatures and pressures (such as the Debye, Einstein or melting temperature and the bulk modulus) also increase with compression. Hardy (1980) proposed a convenient measure of anharmonic effects relative to harmonic ones:

$$\lambda = \phi''' \bar{x}/\phi''$$

where ϕ'' and ϕ''', the second and third derivatives of the interionic potential, are the major parameters determining the size of the harmonic and cubic anharmonic effects, respectively, and $\bar{x}$ is the root-mean-square value of the ionic displacements due to thermal motion.

This measure of anharmonicity can be written

$$\lambda = (TR^n)^{1/2}$$

where R is the nearest neighbor distance, T is the absolute temperature and n is the exponent in the power-law repulsive potential, which can be estimated from

$$K_o' = \frac{n + 7}{3}$$

giving $n \sim 5$ for the lower mantle. The density increases by about 25 percent from zero pressure to the core-mantle boundary. The adiabatic temperature increases by about 33 percent. The increase of λ with temperature, therefore, is slightly greater than the decrease associated with compression, suggesting that intrinsic or anharmonic effects remain important in the lower mantle, even if they decrease with pressure. Existing laboratory data neither support nor contradict this supposition.

Model calculations by Raymond Jeanloz (written communication, 1987) suggest that anharmonic contributions are damped more rapidly with compression than quasi-harmonic and harmonic contributions. In fact, the temperature coefficients of the elastic moduli and the anharmonicity decrease exponentially with pressure (Zharkov and Kalinin, 1971). Temperature and pressure have opposing effects, so the relative roles of harmonic (or quasi-harmonic) and anharmonic contributions, particularly for the shear modes, are still uncertain for the high-temperature, high-pressure regime of the lower mantle. Although I have not treated the effect of compression on the anharmonic properties in any detail, the primary conclusions regarding the dominance of rigidity variations in the lower mantle would also follow if pressure suppresses the intrinsic temperature effects on the bulk modulus faster than on the shear modulus. It should be pointed out that the variation of temperature with pressure is different for the Earth than for shock-wave experiments, and therefore the relative role of anharmonicity may be different for these two situations.

The combination of high pressure and low temperature occurs in the vicinity of deep-focus earthquakes in subduction zones. The temperature derivatives of elastic-wave velocities should be particularly low in this situation. The large velocity anomalies associated with slabs may therefore require anisotropy or changes in mineralogy.

ELASTIC PROPERTIES OF COMPOSITE MATERIALS

Rocks are aggregates of several different anisotropic minerals that may differ widely in their properties. They often contain cracks that may be filled with various kinds of fluids. There are various ways to calculate the average properties of heterogeneous materials from the properties of the components. Most of these involve a volume average of randomly oriented components and give the effective isotropic properties.

The classical methods bound the moduli by assuming that the stress or strain is uniform throughout the aggregate:

$$M_R = \left(\sum_{i=1}^{n} v_i/M_i\right)^{-1} \leq \overline{M} \leq \sum_{i=1}^{n} v_i M_i = M_V$$

where M_i is the elastic modulus (K or G) of the ith phase, $\overline{M}$ is the actual modulus of the composite and the subscripts R and V denote the Reuss (uniform stress) and Voigt (uniform strain) averages, respectively; v_i is the volume fraction of the ith phase. The effective moduli are often taken as the arithmetic, geometric or harmonic mean of M_R and M_V. If the M_i do not differ widely, then $M_R \approx M_V$ and the method of averaging does not much matter.

For single-phase composites of anisotropic crystals, the effective isotropic moduli are found by averaging over all possible orientations:

$$K_V = \frac{1}{3}(A + 2B)$$

$$G_V = \frac{1}{5}(A - B + 3C)$$

$$A = \frac{1}{3}(C_{11} + C_{22} + C_{33})$$

$$B = \frac{1}{3}(C_{12} + C_{23} + C_{31})$$

$$C = \frac{1}{3}(C_{44} + C_{55} + C_{66})$$

The Voigt moduli depend on only nine of the single crystal moduli even in the most general case of anisotropy where there are 21 independent moduli (see Chapter 15). The Reuss moduli are a similar average over the elastic compliances. Tighter bounds can be placed on the elastic properties of composites by taking into account the interactions between grains and their shapes. The methods can also be generalized to take into account complete or partial orientation of the grains. Hashin and Shtrikman (1963) used a variational principle to bound the properties of composites from above (HS^+) and below (HS^-).

Watt and others (1976) summarized the various averaging techniques and provided a convenient formula for treating several different cases for two-phase aggregates:

$$\frac{\overline{M} - M_1}{M_2 - M_1} = v_2\left(1 + \frac{v_1(M_2 - M_1)}{M_1 + F}\right)^{-1}$$

where M is K or G, v_1 and v_2 are the volume fractions of the components with moduli M_1 and M_2. The function F is For K:

(HS^+) $$F = 4G_2/3$$

For G:

(HS^+) $$F = G_2(9K_2 + 8G_2)/6(K_2 + 2G_2)$$

for $(G_2 - G_1)(K_2 - K_1) \geq 0$.

For the lower bound (HS^-) the subscripts are interchanged. If the inequality is not satisfied, the subscripts of G_1 are interchanged for K and of K_i for G.

LIQUIDS

The liquids that are important in the deep interior of the Earth are silicate magmas and molten iron alloys. Neither of these are simple liquids. Liquids are characterized by fluidity, or low viscosity, and low or vanishing rigidity. An ideal liquid has no order and has zero rigidity. Real liquids have a short-range order and can transmit shear waves of sufficiently high frequency. Most melts are slightly less dense than their crystalline form but are more compressible, so the density difference decreases with pressure. Although liquids generally do not exhibit crystal-like long-range order, they can undergo phase changes that involve coordination changes and the packing of ions. Since solids also undergo such phase changes, the difference between the physical properties of a solid and its melt are not necessarily smoothly varying functions of temperature and pressure. Because of the buffering effect of the latent heat of fusion and the fluidity of melts, it is difficult to raise the temperature of large volumes of the Earth above the temperature of complete melting, the liquidus. Since geophysical fluids are impure, they exhibit a melting interval rather than a distinct melting temperature, and the degree of melting depends on the balance between heat input and heat removal rates and the tendency of the liquid and solid portions of the melt to separate. The largest and most rapid input of energy into the Earth was during original accretion and later large impacts. The rapid ascent of buoyant diapirs can also cause pressure-release melting. The slow process of radioactive heating is unlikely to lead to extensive melting because the processes of conduction and convection efficiently remove the heat.

Even the Earth's outer core is probably close to the melting point and may be a slurry. The inner core is solid, probably an Fe-Ni alloy that is below the melting point because of the high pressure. The outer-core–inner-core boundary, however, is not necessarily the solidus of the material in the core. It may simply represent the boundary at which the liquid fraction of the core is so small that it behaves as a solid to the transmission of seismic waves. A packing of particles with roughly 20 percent fluid-filled voids will behave as a solid. Similarly, the entire outer core may not be above the liquidus. The outer core behaves as a fluid, but it would also do so even if it contained a large fraction of suspended particles. The size of the grains that can be held in suspension depends on their density contrast, the fluid viscosity, the nature of short-range forces and grain interactions and the flow velocity of the core. Solid particles may be continually freezing and refreezing as a parcel of fluid moves around the core. If the outer core is not an ideal fluid, it may have a low but finite rigidity and be anisotropic. Liquid crystals, for example, exhibit such solid-like behavior.

The interatomic potential energy functions for liquids are similar to those for solids. One can usually assume that the total interatomic energy is the sum of the energies of atoms interacting in pairs, the pair potential. The forces between atoms interacting three or more at a time are called many-body forces. Many-body forces are far from negligible in liquid metals. The electrons are not bound to particular atoms but are delocalized over the whole of the material. The "effective pair potentials" for liquid metals are different from those of other liquids.

The interatomic energy is composed of short-range and long-range energies. The long-range energies typically vary at large interatomic distances R as R^{-n}. The Coulombic energy between charges has $n = 1$. Short-range energies are usually assumed to vary as $\exp(-kR)$ at large R. These have their origins in the overlap of orbitals of interacting atoms. The equilibrium interatomic spacing is controlled by the balance between attractive and repulsive potential energies. Elastic properties and their derivatives with pressure depend on derivatives with respect to R of the total interatomic potential energy. The repulsive energy increases rapidly as two atoms are brought close together, so we can consider that atoms and molecules have definite shapes and sizes. For packed hard spheres the repulsive energy rises sharply when R equals the sum of the radii of the spheres. The "hard sphere" model is quite useful in discussing some of the properties of liquids.

Long-range energies include electrostatic, induction and dispersion energies. The first is calculated from the distribution of charges within the molecules, which is composed of positive charges from the nuclei and a diffuse negative charge from the electrons. The second two depend on how the electrons in the molecule move under the influence of an external electric field, giving a polarization to the molecule.

The simplest pair potential for atoms is the hard-sphere function. For an atom modeled as a hard sphere of diameter d, the potential is zero for $R > d$ and infinite for $R < d$. This very simple potential, with d determined empirically, is surprisingly useful for those properties that depend primarily on repulsive forces.

Realistic potentials must contain a minimum. The most common form is the power-law potential

$$U(R) = A\left[\left(\frac{r}{R}\right)^m - \left(\frac{r}{R}\right)^n\right], \quad m > n$$

(where R is ionic radius), which is the Lennard-Jones, or L-J, potential when $m = 12$, $n = 6$. At large R the L-J potential has the correct form for the dispersion energy between closed-shell atoms and molecules. The repulsive part of the potential is steep, as required, but there is no physical significance to $m = 12$. It is mathematically convenient to have $m = 2n$. An alternative form is

$$U(R) = A\exp(-kR) - CR^{-6}$$

where the constants A and C are found from such data as average interatomic distances and compressibility.

Sound Propagation in Liquid Metals

Metals that have an open-packed crystal structure in the solid state exhibit an increase in coordination upon melting. Most metals, however, experience little change in interatomic distances or coordination when they melt, and some show little change in their physical properties at the melting point. At temperatures slightly above the melting point, metals exhibit a short-range order that becomes more random as the temperature is raised. Electronic and optical properties are well explained by assuming that the conduction electrons are nearly free and interact with the ions through a localized pseudopotential. Other physical properties are explained in terms of a "hole" model of the liquid where the holes (unoccupied electron sites) have a volume, an energy of formation and an activation energy of diffusion. The thermal expansion and compressibility are determined by a certain concentration of holes of definite volume. Properties such as the attenuation of acoustic waves depend on thermal conductivity and shear and volume viscosities. Alloys and eutectics often exhibit complex behavior, including anisotropy and inhomogeneity. At high frequencies or low temperatures, molten metals exhibit some characteristics of crystals, including the ability to transmit shear waves.

Sound velocity generally decreases with rising temperature, and the temperature coefficient is usually much greater than in the equivalent solids, although the variation is sometimes smooth through the melting point. The sound velocity depends on two competing factors, the coordination and the thermal motion of ions.

The sound velocities, for a given valency, decrease with increasing atomic weight. The sound velocity V_P, density ρ, and adiabatic compressibility are related by

$$V_P^2 = (\rho\beta_S)^{-1}$$

The isothermal compressibility is

$$\beta_T = \gamma\beta_S$$

where γ is the Grüneisen parameter. Table 6-11 summarizes some properties of liquid metals. The compressibilities of the alkali metals are an order of magnitude larger than those of the noble and polyvalent metals, indicating that the interatomic forces are smaller. The increase of isothermal compressibility through the melting point is generally between 5 and 20 percent.

A metal can be regarded as a lattice of positively charged ions immersed in a sea of conduction electrons. The electron sea about each ion screens its electric field, so that at large distances the Coulomb potential associated with a bare ion of valency z is reduced by an exponential screening factor

$$U(r) \approx (-ze^2/R)\exp(-k_s R)$$

where k_s is a screening parameter. The compressibility is predicted to scale as

$$\beta_S \approx (M/\rho)^{5/3}$$

where M is the atomic weight and M/ρ is the atomic volume.

In contrast to a solid, the binding of two atoms in a liquid is a temporary affair. One can view an atom as vibrating with a characteristic period, τ_o, in a temporary equilibrium position and having a characteristic lifetime of τ in this position, where

$$\tau = \tau_o \exp E^*/kT$$

where E^* is an activation energy. We can now distinguish several different domains. For rapid processes where $t < \tau$ and $\tau >> \tau_o$, the fluid has some solid-like characteristics since it does not have time to flow. Thus, high-frequency shear waves can be propagated and short-duration shear stresses can be maintained by an effectively high viscosity. The amplitude of the thermal vibrations of the atoms in a liquid is comparable to the interatomic distance. Therefore, the instantaneous arrangement of the atoms in a liquid possesses a high degree of disorder. However, if we look at the

TABLE 6-11
Elastic Properties and Derivatives of Solid and Liquid Metals

	β_T (Mbar^{-1})		V_p (km/s)		$\left(\frac{\partial \ln V_p}{\partial \ln \rho}\right)_P$	$\left(\frac{\partial \ln V_p}{\partial \ln \rho}\right)_T$
Metal	**Solid**	**Liquid**	**Solid**	**Liquid**	**Liquid**	**Liquid**
Na	16	18.6	3.0	2.6	0.9	1.3
Hg	3	3.8	1.7	1.5	1.7	3.7
In	2.4	3.0	2.6	2.3	1.1	2.2
Sn	1.9	2.7	3.5	2.5	0.9	2.1
Bi	3.0	4.2	2.3	1.6	0	3.3
Zn	1.7	2.5	3.8	2.9	0.7	1.7
Al	1.4	2.4	5.7	4.7	0.8	1.0
Ga	2.0	2.2		2.9	0.8	1.7
Pb	2.6	3.5	2.2	1.8	1.3	2.9
Sb	2.7	4.9	3.4	1.9	−1.1	2.4

Birch (1966), Webber and Stephens (1968).

mean positions of particles for $\tau >> t > \tau_o$, we would see a higher degree of order. The mean positions of the atoms would not change much and a quasi-crystallinity would be displayed, particularly near the melting point. The lifetime of the relative order, however, is small. It increases with pressure and decreases with temperature. For liquid metals it would be of interest to know if convection, electrical currents, magnetic fields and the presence of impurities would allow alignment of these ordered regions giving, for example, a seismic anisotropy. The inner core may be a crystalline solid or have a τ that is longer than the periods of seismic waves, acting therefore as a high-viscosity fluid or a glass.

When a liquid is sampled, or averaged, over periods greater than τ the quasi-crystallinity of the liquid disappears. Translational and other symmetries are not apparent, and the only orderliness is in the average radial distance between atoms, the radial distribution function. Experimental x-ray radial distribution functions refer to this condition rather than the instantaneous or short-time average structure.

Liquid metals exhibit a small excess sound absorption that is attributed to a bulk viscosity caused by an internal arrangement of atoms. The sound wave perturbs these arrangements and energy is thereby extracted by a relaxation process. The amplitude of an acoustic wave of frequency ω propagating in the x direction in a fluid can be written

$$u = u_o \exp - \alpha x \exp i(\omega t - kx)$$

where α is the amplitude absorption coefficient, and k is the wave number. The phase velocity $c = \omega/k$. For small attenuation

$$\alpha^* = (2/3)(\eta\omega^2/\rho c^3)$$

where η is the shear viscosity and ρ is the density. The heating associated with pressure changes in the propagating wave give rise to the flow of heat from the compressed to the dilated parts of the wave. For an ideal gas

$$\alpha^* = \frac{\omega^2 k}{2C_P\rho c^3}(\gamma - 1)$$

where k is the thermal conductivity and C_P is specific heat. Any process that removes energy from the sound wave and returns it at a later time causes dissipation of the acoustic energy. Sound propagation is still adiabatic but is no longer isentropic; that is, there is a production of entropy. The energy removed from the sound wave is transferred to vibration or rotation of atoms or to the potential energy of some structural rearrangement or chemical reaction. The transfer is not instantaneous and is characterized by a relaxation time, τ.

The velocity, $c(\omega)$, varies from the zero-frequency or relaxed value, c_o, to the high-frequency or unrelaxed value, c_∞:

$$c^2/c_o^2 = (1 + \omega^2\tau^2)/[1 + (c_o^2/c_\infty^2)\omega^2\tau^2]$$

The absorption coefficient per unit wavelength λ is

$$\lambda\alpha^* = \pi(c^2/c_o^2)\left(\frac{c_\infty^2 - c_o^2}{c_\infty^2}\right)\frac{\omega\tau}{1 + \omega^2\tau^2}$$

The quantity

$$\frac{c_\infty^2 - c_o^2}{c_\infty^2} = 1 - K_S^o/K_S^\infty$$

is the relaxation strength. These equations can be generalized to multiple relaxation mechanisms or a distribution of relaxation times.

The individual relaxation times are

$$\tau = \tau_o \exp E^*/RT$$

where E^* is the activation energy required for an atom to pass from one equilibrium position to another. The height of the potential energy barrier, E^*, is not fixed, as it is in a crystal, but depends on the changing configuration of the atoms in the neighborhood. In the absence of external forces, these barrier jumps are random and no net flow results. If an external force, such as a sound wave, is applied, then flow can result if the period of the wave is greater than τ. For short-duration forces, or waves, the flow responds as a solid. Both solid-like and liquid-like behaviors are reflected in the above equations, depending on the frequency of the sound wave or the duration of the applied stress.

The sound (phase) velocity is

$$c^2 = (K_S + 4G/3)/\rho$$

where the rigidity, G, is zero in the low-frequency or relaxed limit. The frequency dependence of the rigidity is

$$G(\omega) = G_\infty \frac{i\omega\tau_s}{1 + i\omega\tau_s}$$

giving a complex sound velocity; that is, an attenuating wave. As $\omega \rightarrow \infty$, $G \rightarrow G_\infty$. As $\omega \rightarrow 0$, $G \rightarrow G_o = 0$, and the solid-like property vanishes.

The imaginary part of the shear modulus can be interpreted as a frequency-dependent shear viscosity:

$$\eta(\omega) = \frac{G_\infty\tau_s}{1 + \omega^2\tau_s^2}$$

At low frequencies this reduces to the ordinary shear viscosity

$$\eta = G_\infty\tau_s$$

Similarly, the bulk modulus can be written

$$K = K_o + \frac{K_1\omega^2\tau_c^2}{1 + \omega^2\tau_c^2} + \frac{iK_1\omega\tau_c}{1 + \omega^2\tau_c^2}$$

where K_o is the $\omega = 0$ bulk modulus and $K_o + K_1$ is K_∞, the unrelaxed infinite frequency modulus. If the compressional relaxation time τ_c is similar to the shear relaxation time τ_s, then the attenuation coefficient for longitudinal waves is

$$\alpha^* \simeq (\omega^2/2\rho c^3)(\eta_c + 4\eta_s/3)$$

where η_c is the compressional, or bulk, viscosity.

The effects of pressure and temperature on the absorption coefficient and velocity depend on both the variation of τ and the relaxation strength. Temperature generally decreases τ and c_∞, so the net effect depends on whether $\omega\tau$ is greater than or less than 1. Vibrational relaxation frequencies are expected to increase with pressure because of the smaller interatomic distances and more chances for collisions. Structural relaxation frequencies might be expected to decrease with pressure since it is more difficult for atoms to jump from one state to another.

For simple metals, the temperature and pressure derivatives of the bulk modulus generally satisfy

$$\{K_S\}_P \equiv (\partial \ln K_S/\partial \ln \rho)_P = 2.4 \text{ to } 4.2$$

$$\{K_S\}_T \equiv (\partial \ln K_S/\partial \ln \rho)_T = 3 \text{ to } 7$$

For individual metals

$$(\partial \ln K_S/\partial \ln \rho)_P < (\partial \ln K_S/\partial \ln \rho)_T$$

which means that there is an intrinsic temperature effect at constant volume. At constant volume the bulk modulus increases with temperature.

The core is invariably treated as an ideal fluid, having zero rigidity at all frequencies. However, real fluids transmit shear waves at high frequency and, therefore, have a finite, if small, rigidity. A viscoelastic fluid can be described as a Maxwell body. The Maxwell time is

$$\tau = \eta/G$$

where η is the shear viscosity and G is the high-frequency rigidity. Such a body will flow without limit under a constant load and therefore behaves as a fluid. At high frequency ($\omega\tau > 1$) the body propagates shear waves with velocity $\sqrt{G/\rho}$. The quality factor, Q, is unity at low frequency and approaches infinity as ω for $\omega\tau > 1$. The quality factor will be discussed in great detail in a later chapter. A Q of unity means high attenuation and of infinity means no attenuation. Low-frequency shear waves have low velocities, approaching zero as $\sqrt{\omega}$, and high attenuation (low Q). Pressure generally increases τ, so $\omega\tau$ may increase with depth in the core, giving rise to an effective rigidity for a seismic wave of a given frequency. The rigidity will be lower at free-oscillation periods than at body-wave periods. The periods of normal modes, which are sensitive to properties of the core, may be affected by this rigidity.

The velocity of compressional waves will also be affected by the presence of rigidity and will start to increase with depth when the rigidity starts to have a measurable effect. The bulk modulus of fluids is also affected by relaxation effects operating through the bulk viscosity.

Properties of Oxide Melts

The volumes and bulk moduli of silicate melts can be approximated by averaging the properties of the component oxides. The bulk moduli are approximately

$$K = (K_V + K_R)/2$$

where

$$K_V = \Sigma V_i K_i$$

$$K_R = (\Sigma(V_i/K_i))^{-1}$$

and V_i are the volume fractions of components i.

The molar volumes of oxide melts are usually about 10–20 percent more than the crystalline components. The

TABLE 6-12
Molar Volumes V_0, Thermal Expansivities α and Bulk Moduli K_0 for Components in Silicate Liquids at 1400°C and Solids at 25°C

	Melts			Crystals		
Oxide	V_0 ($cm^3/gf\omega$)	α ($\times 10^5$/K)	K_0 (kbar)	V_0 ($cm^3/gf\omega$)	α ($\times 10^5$/K)	K_0 (kbar)
SiO_2	27.08	1.0	130	22.69	3.5	378
Al_2O_3	36.83	6.2	209	25.56	1.6	2544
FeO	13.31	23.6	745* 287†	12.12	3.5	1740
MgO	11.87	23.4	888* 386†	11.28	3.1	1630
CaO	16.49	24.3	49.4	16.76	3.8	1140
Na_2O	28.82	24.7	135	—	—	—
K_2O	46.05	26.1	82	—	—	—

Rivers and Carmichael (1986), Herzberg (1987).
*From fusion curve.
†From ultrasonics.

bulk moduli of the melts are considerably less than the solids. The low bulk moduli mean that the volumes of melts and crystals converge at high pressure. Table 6-12 gives the properties of components of oxide melts and solids. Details are given in the Herzberg reference.

General References

Anderson, D. L. (1988) Temperature and pressure derivatives, *J. Geophys. Res.*, *93*, 4688–4700.

Anderson, O. L., E. Schreiber, R. C. Liebermann and N. Soga (1968) Some elastic constant data on minerals relevant to geophysics, *Rev. Geophys.*, *6*, 491–524.

Bina, C. and B. Wood (1987) *J. Geophys. Res.*, *92*, 4853.

Bonzcar, L. J., E. K. Graham and H. Wang (1977) The pressure and temperature dependence of the elastic constants of pyrope garnet, *J. Geophys. Res.*, *82*, 2529–2534.

Chang, Z. P. and G. R. Barsch (1973) Pressure dependence of single-crystal elastic constants and anharmonic properties of spinel, *J. Geophys. Res.*, *78*, 2418–2433.

Chang, Z. P. and G. R. Barsch (1969) Pressure dependence of the elastic constants of single-crystalline magnesium oxide, *J. Geophys. Res.*, *74*, 3291–3294.

Chang, Z. P. and E. K. Graham (1977) Elastic properties of oxides in the NaCl-structure, *J. Phys. Chem. Solids*, *38*, 1355–1362.

Frisillo, A. L. and G. R. Barsch (1972) Measurement of single-crystal elastic constants of bronzite as a function of pressure and temperature, *J. Geophys. Res.*, *75*, 6360–6384.

Graham, E. K. and G. R. Barsch (1969) Elastic constants of single-crystal forsterite as a function of temperature and pressure, *J. Geophys. Res.*, *74*, 5949–5960.

Jones, L. E. A. (1979) Pressure and temperature dependence of the single crystal elastic moduli of the cubic perovskite $KMgF_3$, *Phys. Chem. Mineral.*, *4*, 23–42.

Jones, L. E. A. and R. C. Liebermann (1974) Elastic and thermal properties of fluoride and oxide analogues in the rocksalt, fluorite, rutile and perovskite structures, *Phys. Earth Planet. Inter.*, *9*, 101–107.

Kumazawa, M. and O. L. Anderson (1969) Elastic moduli, pressure derivatives, and temperature derivatives of single-crystal olivine and single-crystal forsterite, *J. Geophys. Res.*, *74*, 5961–5972.

Liu, H. P., R. N. Schock, and D. L. Anderson (1975) Temperature dependence of single-crystal spinel ($MgAl_2O_4$) elastic constants from 293 to 423°K measured by light-sound scattering in the Raman-Nath region, *Geophys. J. R. Astr. Soc.*, *42*, 217–250.

Manghnani, M. H. (1969) Elastic constants of single-crystal rutile under pressures to 7.5 kilobars, *J. Geophys. Res.*, *74*, 4317–4328.

Simmons, G. and H. Wang (1971) *Single-Crystal Elastic Constants and Calculated Aggregate Properties: A Handbook*, 2nd ed., The M.I.T. Press, Cambridge, MA.

Skinner, B. J., Thermal expansion, in S. P. Clark (ed.) (1966) Handbook of Physical Constants, *Geol. Soc. Am., Mem. 97, Sect. 6*, 78–96.

Soga, N. (1967) Elastic constants of garnet under pressure and temperature, *J. Geophys. Res.*, *72*, 4227–4234.

Spetzler, H. (1970) Equation of state of polycrystalline and single-crystal MgO to 8 kilobars and 800°K, *J. Geophys. Res.*, *75*, 2073–2087.

Sumino, Y. (1979) The elastic constants of Mn_2SiO_4, Fe_2SiO_4 and Co_2SiO_4 and the elastic properties of olivine group minerals at high temperature, *J. Phys. Earth.*, *27*, 209–238.

Sumino, Y., O. L. Anderson and I. Suzuki (1983) Temperature coefficients of elastic constants of single crystal MgO between 80 and 1,300 K, *Phys. Chem. Minerals, 9,* 38–47.

Sumino, Y., O. Nishizawa, T. Goto, I. Ohno and M. Ozima (1977) Temperature variation of elastic constants of single-crystal forsterite between −190 and 400°C, *J. Phys. Earth., 25,* 377–392

Watt, J. P., G. F. Davies and R. J. O'Connell (1976) The elastic properties of composite materials, *Rev. Geophys. Space Phys., 14,* 541–563.

Weidner, D. V. (1986) in Chemistry and Physics of Terrestrial Planets, ed. S. K. Saxena, Springer-Verlag, New York, 405 pp.

Weidner, D. J., J. D. Bass, A. E. Ringwood and W. Sinclair (1982) The single-crystal elastic moduli of stishovite, *J. Geophys. Res., 87,* 4740–4746.

Weidner, D. J. and N. Hamaya (1983) Elastic properties of the olivine and spinel polymorphs of Mg_2GeO_4, and evaluation of elastic analogues, *Phys. Earth Planet. Inter., 33,* 275–283.

Weidner, D. J. and E. Ito (1985) Elasticity of $MgSiO_3$ in the ilmenite phase, *Phys. Earth Planet. Int., 40,* 65–70.

Weidner, D. J., H. Sawamoto, S. Sasaki, and M. Kumazawa (1984) Single-crystal elastic properties of the spinel phase of Mg_2SiO_4, *J. Geophys. Res., 89,* 7852–7860.

References

Ahrens, T. J., D. L. Anderson and A. E. Ringwood (1969) Equations of state and crystal structures of high-pressure phases of shocked silicates and oxides, *Rev. Geophys., 7,* 667–702.

Anderson, D. L. (1987) A seismic equation of state, II. Shear and thermodynamic properties of the lower mantle, *Phys. Earth Planet. Inter., 45,* 307–323.

Anderson, O. L. (1986) Simple solid state equations for materials of terrestrial planet interiors, Proceedings NATO Conference on Physics of Planetary Interiors.

Bass et al. (1981) *Phys. Earth Planet. Int., 25,* 140, 181.

Bassett, W., T. Takahashi, H. Mao and J. Weaver (1968) Pressure induced phase transition in NaCl, *J. Appl. Phys., 39,* 319–325.

Birch, F. (1961) The velocity of compressional waves in rocks to 10 kilobars, Part 2, *J. Geophys. Res., 66,* 2199–2224.

Birch, F. (1966) *Handbook of Physical Constants,* S. P. Clark, Jr. (ed.), *Geol. Soc. Am.,* Memoir 97.

Birch, F. (1968) Thermal expansion at high pressures, *J. Geophys. Res., 73,* 817–819.

Boehler, R. (1982) Adiabats of quartz, coesite, olivine and MgO at 50 kbar and 1000 K, and the adiabatic gradient in the Earth's mantle, *J. Geophys. Res., 87,* 5501–5506.

Brillouin, L. (1964) *Tensors in Mechanics and Elasticity,* Academic Press, New York.

Butler, R. and D. L. Anderson (1978) Equation of state fits to the lower mantle and outer core, *Phys. Earth Planet. Inter., 17,* 147–162.

Clayton, R. W. and R. P. Comer (1983) A tomographic analysis of mantle heterogeneities from body wave travel times, *Eos, 62,* 776.

Davies, G. (1976) *Geophys. J. Astron. Soc., 44,* 625–648.

Doyle, H. A. and A. L. Hales (1967) An analysis of the travel times of S waves to North American stations, *Bull. Seis. Soc. Am., 57,* 761–771.

Dziewonski, A. M. (1984) Mapping the lower mantle: Determination of lateral heterogeneity in P velocity up to degree and order 6, *J. Geophys. Res., 89,* 5929–5952.

Dziewonski, A. M. and D. L. Anderson (1981) Preliminary reference Earth model, *Phys. Earth Planet. Inter., 25,* 297–356.

Hager, B. H., R. W. Clayton, M. A. Richards, R. P. Comer, and A. M. Dziewonski (1985) Lower-mantle heterogeneity, dynamic topography, and the geoid, *Nature, 313,* 541–545.

Hardy, R. J. (1980) Temperature and pressure dependence of intrinsic anharmonic and quantum corrections to the equation of state, *J. Geophys. Res., 85,* 7011–7015.

Hashin, Z. and S. Shtrickman (1963) *J. Mech. Phys. Solids, 11,* 127–140.

Herzberg, C. J. (1987) *Magmatic Processes: Physiochemical Processes,* the Geochemical Society, Special Publication.

Jeanloz, R. and T. J. Ahrens (1980) *Geophys. J. R. Astro. Soc., 62,* 505–528.

Jeanloz, R. and E. Knittle (1986) Reduction of mantle and core properties to a standard state by adiabatic decompression. In *Chemistry and Physics of Terrestrial Planets* (S. K. Saxena, ed.), 275–305, Springer-Verlag, Berlin.

Jones, L. (1976) *Phys. Earth Planet. Interiors, 13,* 105–118.

Jones, L. (1979) Pressure and temperature dependence of the single crystal elastic moduli of the cubic perovskite $KMgF_3$, *Phys. Chem. Min., 4,* 23–42.

Knittle, E., R. Jeanloz and G. L. Smith (1986) Thermal expansion of silicate perovskite and stratification of the Earth's mantle, *Nature, 319,* 214–216.

Knittle, E. R. and R. Jeanloz (1987) *Science, 235,* 668–670.

Knopoff, L. (1963) Solids: Equations of state at moderately high pressures. In *High Pressure Physics and Chemistry* (R. S. Bradley, ed.), 227–245, Academic Press, New York.

Nakanishi, I. and D. L. Anderson (1982) World-wide distribution of group velocity of mantle Rayleigh waves as determined by spherical harmonic inversion, *Bull. Seis. Soc. Am., 72,* 1185–1194.

Nataf, H.-C., I. Nakanishi and D. L. Anderson (1984) Anisotropy and shear-velocity heterogeneities in the upper mantle, *Geophys. Res. Lett., 11,* 109–112.

Nataf, H.-C., I. Nakanishi and D. L. Anderson (1986) Measurements of mantle wave velocities and inversion for lateral heterogeneities and anisotropy, Part III, Inversion, *J. Geophys. Res., 91,* 7261–7307.

Richards, M. A. (1986) Dynamical models for the Earth's geoid, Ph.D. Thesis, California Institute of Technology, Pasadena, 273 pp.

Rivers, M. and I. Carmichael (1986) *J. Geophys. Res.* (in press).

Sumino, Y. and O. L. Anderson (1984) Elastic constants of minerals. In *Handbook of Physical Properties of Rocks, v. III* (R. S. Carmichael, ed.), 39–138, CRC Press, Boca Raton, Florida.

Suzuki, I., O. L. Anderson and Y. Sumino, Elastic properties of a single-crystal forsterite Mg_2SiO_4, up to 1,200K, *Phys. Chem. Minerals, 10,* 38–46, 1983.

Svendsen and Ahrens (1983) *Geophys. Res. Lett., 10,* 501–540.

Wallace, D. C. (1972) *Thermodynamics of Crystals,* Wiley, New York.

Watt, J. P., G. F. Davies and R. J. O'Connell (1976) The elastic properties of composite materials, *Rev. Geophys. Space Phys., 14,* 541–563.

Webber, G. M. B. and R. W. B. Stephens (1968) *Physical Acoustics IVG,* W. P. Mason (ed.), Academic Press, N.Y.

Wolf and Bukowinski (1986) *High Pressure Research in Geophysics,* eds. M. Manghnani and Y. Symo.

Wolf, G. and R. Jeanloz (1985) Vibrational properties of model monatomic crystals under pressure, *Phys. Rev. B, 12,* 7798–7810.

Woodhouse, J. H. and A. M. Dziewonski (1984) Mapping the upper mantle: Three-dimensional modeling of earth structure by inversion of seismic waveforms, *J. Geophys. Res., 89,* 5983–5986.

Yagi, Ida, Sato and Akimoto (1975) *Phys. Earth Planet. Int., 10,* 348–354.

Yamamoto, S. and O. L. Anderson (1986) Elasticity and anharmonicity of potassium chloride at high temperature, *J. Phys. Chem. Minerals* (in press).

Yamamoto, S., I. Ohno and O. L. Anderson (1986) High temperature elasticity of sodium chloride, *J. Phys. Chem. Solids* (in press).

Zharkov, V. and V. Kalinin (1971) *Equations of State for Solids at High Pressures and Temperatures,* Consultants Bureau, New York, 257 pp.

7

Nonelastic and Transport Properties

Shall not every rock be removed out of his place?

—JOB 18:4

Most of the Earth is solid, and much of it is at temperatures and pressures that are difficult to achieve in the laboratory. The Earth deforms anelastically at small stresses and, over geological time, this results in large deformations. Most laboratory measurements are made at high stresses, high strain rates and low total strain. Laboratory data must therefore be extrapolated in order to be compared with geophysical data, and this requires an understanding of solid-state physics. In this chapter I discuss transport and activated processes in solids, processes that are related to rates or time. Some of these are more dependent on temperature than those treated in previous chapters. These properties give to geology the "arrow of time" and an irreversible nature.

THERMAL CONDUCTIVITY

There are three mechanisms contributing to thermal conductivity $\mathcal{K}$ in the crust and mantle. The lattice part, $\mathcal{K}_L$, is produced by diffusion of thermal vibrations in a crystalline lattice and is also called the phonon contribution. $\mathcal{K}_R$ is the radiative part, due to the transfer of heat by infrared electromagnetic waves. If the mantle is sufficiently transparent, $\mathcal{K}_R$ is significant. $\mathcal{K}_E$ is the exciton part, due to the transport of energy by quasiparticles composed of electrons and positive holes. This becomes dominant in intrinsic semiconductors as the temperature is raised. Thus, thermal conduction in solids arises partly from electronic and partly from atomic motion and, at high temperature, from radiation passing through the solid.

Introduction

Debye regarded a solid as a system of coupled oscillators transmitting thermoelastic waves. For an ideal lattice with simple harmonic motion of the atoms, the conductivity would be infinite. In a real lattice anharmonic motion couples the vibrations, reducing the mean free path and the lattice conductivity. Thermal conductivity is related to higher order terms in the potential and should be correlated with thermal expansion. Lattice conductivity can be viewed as the exchange of energy between high-frequency lattice vibrations, or elastic waves. A crude theory for the lattice conductivity, consistent with the Grüneisen approximation, gives

$$\mathcal{K}_L = a/3\gamma^2 T K_T^{3/2} \rho^{1/2}$$

where a is the lattice parameter, γ is the Grüneisen parameter, T is temperature, K_T is the isothermal bulk modulus, and ρ is density. This is valid for large T/θ. This relation predicts that the thermal conductivity decreases by about 20 percent as one traverses the mantle.

The contribution of "free carriers" such as electrons, holes and electron-hole pairs to thermal conduction can be estimated from the Wiedemann-Franz law relating electrical, $\mathcal{K}_e$, and thermal (lattice) conductivity. The mantle is a good electrical insulator, and the associated thermal conductivity is almost certainly negligible. The Wiedermann-Franz ratio is

$$\mathcal{K}_L/\mathcal{K}_e = (\pi^2/3)(k/e)^2 T$$

where k is Boltzmann's constant and e is the electronic charge. Using estimates of electrical conductivity, $\mathcal{K}_e$, of the mantle, we obtain estimates of thermal conductivity due to this mechanism that are some six orders of magnitude lower than observed lattice conductivities.

Excitons, or bound electron-hole pairs, may contribute to thermal conductivity at high temperature if the excitation energy is of the order of 1 eV. Evidence to date suggests a much higher energy in silicates and oxides, so exitonic thermal transport appears to be negligible in the mantle.

The thermal conductivities of various rock-forming minerals are given in Table 7-1. Note that the crust-forming minerals have about one-half to one-third of the conductivity of mantle minerals. This plus the cracks present at low crustal pressures means that a much higher thermal gradient is maintained in the crust, relative to the mantle, to sustain the same conducted heat flux. The gradient can be higher still in sediments that have conductivities of the order of 0.4 to 2×10^{-3} cal/cm s°C.

The thermal gradient decreases with depth in the Earth because of the increasing conductivity and the decreasing amount of radioactivity-generated heat. If the crustal radioactivity and mantle heat flow are constant and the effects of temperature are ignored, regions of thick crust should have relatively high upper-mantle temperatures.

Horai and Simmons (1970) found a good correlation between seismic velocities and lattice conductivity, as expected from lattice dynamics. They found

$$V_p = 0.17\mathcal{K}_L + 5.93$$

$$V_s = 0.09\mathcal{K}_L + 3.31$$

with $\mathcal{K}_L$ in mcal/cm s°C and $V_{p,s}$ in km/s. They also found a linear relationship between the Debye temperature, θ (in kelvins), and $\mathcal{K}_L$:

$$\theta = 25.6\mathcal{K}_L + 3.85$$

Thermal conductivity is strongly anisotropic, varying by about a factor of 2 in olivine and orthopyroxene as a function of direction. The highly conducting axes are [100] for olivine and [001] for orthopyroxene. The most conductive axis for olivine is also the direction of maximum *P*-velocity and one of the faster *S*-wave directions, whereas the most conductive axis for orthopyroxene is an intermediate axis for *P*-velocity and a fast axis for *S*-waves. In mantle rocks the fast *P*-axis of olivine tends to line up with the intermediate *P*-axis of orthopyroxene. These axes, in turn, tend to line up in the flow direction, which is in the horizontal plane in ophiolite sections. The vertical conductivity in such situations is much less than the average conductivity computed for mineral aggregates, which is about 7×10^{-3} cal/cm s°C at normal conditions. Conductivity decreases with temperature and may be only half this value at the base of the lithosphere. The implications of this anisotropy in thermal conductivity and the lower than average vertical conductivity have not been investigated. Two obvious implications are that the lithosphere can support a higher thermal gradient than generally supposed, giving higher upper-mantle temperatures, and that the thermal lithosphere grows less rapidly than previously calculated. For example, the thermal lithosphere at 80 Ma can be 100 km thick for $\mathcal{K} = 0.01$ cal/cm s deg and only 30 km thick if the appropriate $\mathcal{K}$ is 3×10^{-3} cal/cm s deg. The low lattice conductivity of the oceanic crust is usually also ignored in these calculations, but this may be counterbalanced by water circulation in the crust.

The lattice (phonon) contribution to the thermal conductivity decreases with temperature, but at high temperature radiative transfer of heat may become significant depending on the opacities of the minerals. If the opacity, ε, is independent of wavelength and temperature, then $\mathcal{K}_R$ increases strongly with temperature:

$$\mathcal{K}_R = 16n^2\sigma T^3/3\varepsilon$$

where n is the refractive index and σ is the Stefan-Boltzmann constant. If ε were constant, $\mathcal{K}_R$ would increase very rapidly with temperature and would be the dominant heat conduction mechanism in the mantle. The parameter ε^{-1} decreases from about 0.6 to 0.1 cm in the temperature range of 500 to 2000 K for olivine single crystals and approaches 0.02 cm for enstatite (Schatz and Simmons, 1972). The net result is that $\mathcal{K}_R$ is about equal to $\mathcal{K}_L$ at high temperatures and lower-mantle conditions.

Convection is the dominant mode of heat transport in the Earth's deep interior, but conduction is not irrelevant to the thermal state and history of the mantle as heat must be transported across thermal boundary layers by conduction. Thermal boundary layers exist at the surface of the Earth, at the core-mantle boundary and, possibly, at chemical interfaces internal to the mantle. Conduction is also the mechanism by which subducting slabs cool the mantle. The thicknesses and thermal time constants of boundary layers are controlled by the thermal conductivity, and these regulate the rate at which the mantle cools and the rate at which the thermal lithosphere grows.

The thermal diffusivity, κ, is defined

$$\kappa = \mathcal{K}/\rho C_V$$

and the characteristic thermal time constant of a body of dimension l is

$$\tau \approx l^2/\kappa$$

TABLE 7-1
Thermal Conductivity of Minerals

Mineral	Thermal Conductivity 10^{-3}cal/cm s°C
Albite	4.71
Anorthite	3.67
Microcline	5.90
Serpentine	7.05
Diopside	11.79
Forsterite	13.32
Bronzite	9.99
Jadeite	15.92
Grossularite	13.49
Olivine	6.7–13.6
Orthopyroxene	8.16–15.3

Horai (1971), Kobayzshigy (1974).

TABLE 7-2
Estimates of Lattice Thermal Diffusivity in the Mantle

Depth (km)	κ (cm^2/s)
50	5.9×10^{-3}
150	3.0×10^{-3}
300	2.9×10^{-3}
400	4.7×10^{-3}
650	7.5×10^{-3}
1200	7.7×10^{-3}
2400	8.1×10^{-3}
2900	8.4×10^{-3}

Horai and Simmons (1970).

The thermal diffusivities of mantle minerals are about 0.006 to 0.010 cm²/s (see Table 7-2). In order to match the observed elevation and geoid changes across oceanic fracture zones, Crough (1979) derived a diffusivity of 0.0033 cm²/s for the upper mantle. This low value may be related to the anisotropy discussed above or high temperatures.

Lattice Conductivity

Both thermal conductivity and thermal expansion depend on the anharmonicity of the interatomic potential and therefore on dimensionless measures of anharmonicity such as γ or $\alpha\gamma T$.

The lattice or phonon conductivity, $\mathcal{K}_L$, is

$$\mathcal{K}_L = \frac{1}{3} C_V \overline{V} l$$

where $\overline{V}$ is the mean sound velocity, l is the mean free path,

$$l = a/\alpha\gamma T = aK_T/\rho C_V \gamma^2 T$$

and a is the interatomic distance, and K_T is the isothermal bulk modulus. Therefore,

$$\mathcal{K}_L = (1/3)\overline{V}aK_T/\rho\gamma^2 T$$

$$= (1/3)\left(\frac{1}{V_P^3} + \frac{2}{V_S^3}\right)^{-1/3} aK_T/\rho\gamma^2 T$$

This gives

$$\frac{\partial \ln \mathcal{K}_L}{\partial \ln \rho} = \frac{\partial \ln K_T}{\partial \ln \rho} - 2\frac{\partial \ln \gamma}{\partial \ln \rho} + \gamma - \frac{4}{3}$$

where we have used the approximation

$$\partial \ln \overline{V}/\partial \ln \rho \simeq \gamma$$

For lower-mantle properties this expression is dominated by the $\partial \ln K_T/\partial \ln \rho$ term, and the variation of $\mathcal{K}_L$ is similar to the variation of K_T.

The lattice conductivity decreases approximately linearly with temperature, a well-known result, but increases rapidly with density. The temperature effect dominates in the shallow mantle, giving a decreasing $\mathcal{K}_L$, but pressure causes $\mathcal{K}_L$ to be high in the lower mantle. This has important implications regarding the properties of thermal boundary layers, the ability of the lower mantle to conduct heat from the core and into the upper mantle, and the convective mode of the lower mantle.

The general correlation of $\mathcal{K}_L$ with the elastic wave velocities suggests that pressure-induced phase changes in the transition region and lower mantle will also increase the lattice conductivity. Thermally induced velocity variations in deep slabs will be small because of the high $\mathcal{K}_L$ and low temperature derivatives of velocity. The expected increase of $\mathcal{K}_L$ with compression, with bulk modulus, with sound speed and across low pressure–high pressure phase changes has been verified by experiment (Fujisawa and others, 1968) as has the correlation of $\mathcal{K}_L$ with elastic wave velocities. We expect about a factor of 3 increase in $\mathcal{K}_L$, caused by the increase in seismic velocities, from shallow-mantle to transition-zone pressures. The erroneous use of an olivine-like $\mathcal{K}_L$ for the deep slab and a $(\partial V_P/\partial T)_P$ of about twice that of olivine (Creager and Jordan, 1984) are partly responsible for the conclusion about the persistence and extent of deep slab-related seismic anomalies. On the other hand, changes in mineralogy associated with solid-solid phase changes are much more important than temperature.

The ratio $\alpha/\mathcal{K}_L$ decreases rapidly with depth in the mantle, thereby decreasing the Rayleigh number. Pressure also increases the viscosity, an effect that further decreases the Rayleigh number of the lower mantle. The net effect of these pressure-induced changes in physical properties is to make convection more sluggish in the lower mantle, to decrease thermal-induced buoyancy and to increase the thickness of the thermal boundary layer in D″, above the core-mantle boundary.

The thermal conductivity of crystals is a complex subject even when many approximations and simplifying assumptions are made. The mechanism for transfer of thermal energy is generally well understood in terms of lattice vibrations, or high-frequency sound waves. This is not enough, however, since thermal conductivity would be infinite in an ideal harmonic crystal. We must understand, in addition, the mechanisms for scattering thermal energy and for redistributing the energy among the modes and frequencies in a crystal so that thermal equilibrium can prevail. An understanding of thermal "resistivity," therefore, requires an understanding of higher order effects, including anharmonicity.

High-frequency elastic energy is scattered by imperfections such as point defects, dislocations, grain boundaries and impurities, including isotopic differences, and nonlinearity or anharmonicity or interatomic forces. The latter effects can be viewed as nonlinear interactions of the thermal sound waves themselves, a sort of self-scattering.

The parameters that enter into a theory of lattice conductivity are fairly obvious. One expects that the temperature, specific heat and the coefficient of thermal expansion will be involved. One expects that some measure of anharmonicity, such as γ, will be involved. In addition one needs a measure of a mean free path or a mean collision time or a measure of the strength and distribution of scatterers. The velocities of the sound waves and the interatomic distances are also likely to be involved.

Debye (1912) explained the thermal conductivity of dielectric or insulating solids in the following way. The lattice vibrations can be resolved into traveling waves that carry heat. Because of anharmonicities the thermal fluctuations in density lead to local fluctuations in the velocity of lattice waves, which are therefore scattered. Simple lattice theory provides estimates of specific heat and sound velocity and how they vary with temperature and volume. The theory of attenuation of lattice waves involves an understanding of how thermal equilibrium is attained and how momentum is transferred among lattice vibrations.

The heat flow, Q, associated with a given mode type (longitudinal or transverse) can be written in terms of the energy, wave number and group velocity of the mode.

The distinction between group velocity, $\mathbf{v}_g = \partial\omega/\partial\mathbf{k}$, and phase velocity, $\omega/\mathbf{k}$, is seldom made in the theory because dispersion is generally ignored. If ω is proportional to the wave number, $\mathbf{k}$, then $\mathbf{v}$ is constant and phase and group velocities are equal. This assumption breaks down for high frequencies or large $\mathbf{k}$. At high frequencies $\mathbf{v}_g$ becomes very small, and high frequencies therefore are not efficient in transporting heat. High-frequency waves are those having wavelengths comparable to a lattice spacing.

Because of the discreteness of a lattice, the possible energy levels are quantized. We can treat the thermal properties of a lattice as a gas of phonons. A quantum of lattice vibration is called a phonon and acts as a particle of energy $\hbar\omega$, momentum $\hbar\mathbf{k}$ and velocity $\partial\omega/\partial\mathbf{k}$. There is no limit to the number of quanta in a normal mode. The phonons carry a heat current, which is the sum of the heat currents carried by all normal modes:

$$Q = \sum_{\mathbf{k}} N(\mathbf{k})\hbar\omega\ \partial\omega/\partial\mathbf{k}$$

The lattice conductivity is a sum over all wave types of the integral over all wave numbers

$$\mathcal{K}_L = \frac{1}{3}\sum_j \int d\mathbf{k}\mathbf{v}_j^2(\mathbf{k})\tau_j(\mathbf{k})C_V\mathbf{v}_j(\mathbf{k})$$

where $\mathbf{v}_j$ is the group velocity of the jth wave type and τ_j is the lifetime or collision time of this wave type.

The thermal resistance is the result of interchange of energy between lattice waves, that is, scattering. Scattering can be caused by static imperfections and anharmonicity. Static imperfections include grain boundaries, vacancies, interstitials and dislocations and their associated strain fields, which considerably broadens the defects cross-section. These "static" mechanisms generally become less important at high temperature. Elastic strains in the crystal scatter because of the strain dependence of the elastic properties, a nonlinear or anharmonic effect. An elastic strain alters the frequencies of the lattice waves.

DIFFUSION AND VISCOSITY

Diffusion and viscosity are activated processes and depend more strongly on temperature and pressure than the properties discussed up to now. The diffusion of atoms, the mobility of defects, the creep of the mantle and seismic wave attenuation are all controlled by the diffusivity.

$$D(P,T) = \zeta a^2\,\nu \exp[-G^*(P,T)/RT]$$

where G^* is the Gibbs free energy of activation, ζ is a geometric factor and ν is the attempt frequency (an atomic vibrational frequency). The Gibbs free energy is

$$G^* = E^* + PV^* - TS^*$$

where E^*, V^* and S^* are activation energy, volume and entropy, respectively. The diffusivity can therefore be written

$$D = D_o \exp -(E^* + PV^*)/RT$$

$$D_o = \zeta a^2\,\nu \exp S^*/RT$$

where S^* is generally in the range R to $5R$. The theory for the volume dependence of D_o is similar to that for thermal diffusivity, $\kappa = \mathcal{K}_L/\rho C_V$. It increases with depth but the variation is small, perhaps an order of magnitude, compared to the effect of the exponential term. The product of $\mathcal{K}_L$ times viscosity is involved in the Rayleigh number, and the above considerations show that the temperature and pressure dependence of this product depend mainly on the exponential terms.

The activation parameters are related to the derivative of the rigidity (Keyes, 1963):

$$V^*/G^* = (1/K_T)\left[\left(\frac{\partial \ln G}{\partial \ln \rho}\right)_T - 1\right]$$

The effect of pressure on D can be written

$$-\frac{RT}{K_T}\left(\frac{\partial \ln D}{\partial \ln \rho}\right)_T = V^*$$

$$= \frac{1}{K_T}\left[\left(\frac{\partial \ln G}{\partial \ln \rho}\right)_T - 1\right]G^*$$

or

$$\left(\frac{\partial \ln D}{\partial \ln \rho}\right)_T = -\frac{G^*}{RT}\left[\left(\frac{\partial \ln G}{\partial \ln \rho}\right)_T - 1\right]$$

For a typical value of 30 for G^*/RT we have V^* decreasing from 4.3 to 2.3 cm^3/mole with depth in the lower mantle, using elastic properties from the PREM Earth model. We also have

$$-(\partial \ln D/\partial \ln \rho)_T \approx 48 \text{ to } 40$$

This gives a decrease in diffusivity, and an increase in viscosity, of about a factor of 60 to 80, due to compression, across the lower mantle. In convection calculations and geoid modeling it is common practice to assume a constant viscosity for the lower mantle. This is a poor assumption.

Viscosity

A general expression for viscosity, η, is

$$\eta \approx (G/\sigma)^n$$

where n is a constant generally between 1 and 3 and σ is the nonhydrostatic stress. This gives an additional increase of viscosity over that contributed by the diffusivity, D, unless σ decreases rapidly with depth. The general tendency of η to increase with depth may be reversed in the D″ zone due to a high thermal gradient and, possibly, an increase in σ. The decrease in V^* with depth also means that compression-induced viscosity increases will be milder at the base of the mantle. A low-viscosity D″ layer could reduce the ability of mantle convection to deform the core-mantle boundary.

The change in viscosity across a chemical or phase boundary is not easily determined. The pre-exponential term in the diffusivity will increase across a boundary that involves an increase in bulk modulus and mean sound velocity, or Debye frequency. Therefore, the viscosity will *decrease* due to this factor. For the same reason lattice conductivity and the thermal diffusivity will increase. The activation volume of the low-density phase will generally be greater than that of the high-density phase. This means that the viscosity jump across a deep boundary will be negative if both phases had the same viscosity at zero pressure. The high temperature gradient in thermal boundary layers, such as D″, the lithosphere and, possibly, near the 650-km discontinuity will cause the diffusivity to increase. Therefore, viscosities will tend to *decrease* across mantle discontinuities unless the activation energies are sufficiently lower for the dense phases that the geothermal gradient can overcome the above effects or unless pre-exponential, crystal structure, defect or nonhydrostatic stress considerations play an appropriate role.

The combination of physical parameters that enters into the Rayleigh number, $\alpha/\kappa\eta$ (coefficient of expansion α, thermal diffusivity κ, and viscosity η), decreases rapidly with compression. The decrease through the mantle is of the order of 10^6 to 10^7. With parameters appropriate for the mantle, the increase due to temperature is of the order of 10^6. Therefore, there is a delicate balance between temperature and pressure. The local Rayleigh number in thermal boundary layers increases because of the dominance of the thermal gradient over the pressure gradient.

Diffusion

Diffusion of atoms is important in a large number of geochemical and geophysical problems: metamorphism, element partitioning, creep, attenuation of seismic waves, electrical conductivity and viscosity of the mantle. Diffusion means a local nonconvective flux of matter under the action of a chemical or electrochemical potential gradient.

The net flux J of atoms of one species in a solid is related to the gradient of the concentration, N, of this species

$$J = -D \text{ grad } N$$

where D is the diffusion constant or diffusivity and has the same dimensions as the thermal diffusivity. This is known as Fick's law and is analogous to the heat conduction equation.

Usually the diffusion process requires that an atom, in changing position, surmount a potential energy barrier. If the barrier is of height G^*, the atom will have sufficient energy to pass over the barrier only a fraction $\exp(-G^*/RT)$ of the time. The frequency of successes is therefore

$$\nu = \nu_o \exp(-G^*/RT)$$

where ν_o is the attempt frequency, usually taken as the atomic vibration, or Debye, frequency, which is of the order of 10^{14} Hz. The diffusivity can then be written

$$D = \zeta \nu a^2$$

where ζ is a geometric factor that depends on crystal structure or coordination and that gives the jump probability in the desired direction and a is the jump distance or interatomic spacing.

The factor ν takes into account the fact that if an atom can jump equally in m directions ($m = 6$ in a face-centered cubic lattice, for example), then the distance moved in the desired direction is different for each jump direction. Also, the atom can only jump to empty sites, that is, vacancies. The probability for this is C_v, the vacancy concentration, where

$$C_v = e^{-G_v^*/RT}$$

where G_v^* is the free energy of formation of a vacancy.

Diffusion, like most thermally activated processes, exhibits a change in activation energy between high-temperature and low-temperature regimes. At low temperatures the number of diffusing ions is independent of temperature, and therefore the energy of formation is not involved. This is the extrinsic range of temperature. Diffusion proceeds via chemical vacancies due to non-stoichiometry or impurities, which outnumber the thermally generated vacancies. At high temperature thermally generated vacancies are produced, and the energy of formation is also involved. The transition from the extrinsic to the intrinsic regime usually occurs at about 0.8 of the melting temperature. The barrier to extrinsic diffusion, G_x^*, is the maximum change in free energy due to the lattice distortion associated with the motion of the ion from its lattice site into a neighboring vacancy. D_o is generally greater than 1 cm²/s for intrinsic diffusion and much less than 1 cm²/s in the extrinsic range.

Motion of vacancies (Schottky defects) and interstitials (Frenkel defects) are important in ionic conductivity. The ionic conductivity is given by

$$\sigma = (N_o e^2 p \nu a^2/kT)e^{-E/kT}$$

where N_o is the total number of ions of the appropriate species per unit volume and p is the fraction of ions able to move.

The frequency factor ν and the activation energy can be found from diffusion experiments. They can also be found from dynamic experiments such as anelastic or attenuation measurements involving elastic waves. An absorption peak, for example, occurs when $\omega/\nu = 1$ or $\omega\tau = 1$. The activation energy is found from the shift in the absorption peak with temperature. Generally, solids exhibit a series of absorption peaks, one for each physical mechanism or diffusing species. The theories of creep, or mantle viscosity, and seismic wave attenuation are intimately related to theories of diffusion. For a mechanism to be important at seismic frequencies and mantle temperatures, the frequency factor ν must be close to the seismic frequency ω. ν can be estimated from ν_o and G^*, given the temperature, pressure, and atomic species. The diffusive properties of the common rock-forming elements (Mg, Fe, Ca, Al, Si and O) are of most relevance in studies of mantle creep and attenuation. Experimentally determined diffusion parameters in various oxides and silicates are given in Table 7-3 and 7-4. E^* is generally well determined if the temperature range is extensive enough, but D_o requires a long extrapolation. In curve-fitting diffusion data there is a trade-off between D_o and E^*.

Regions of lattice imperfections in a solid are regions of increased mobility. Dislocations are therefore high-mobility paths for diffusing species. The rate of diffusion in these regions can exceed the rate of volume or lattice diffusion. In general, the activation energy for volume diffusion is higher than for other diffusion mechanisms. At high temperature, therefore, volume diffusion can be important. In and near grain boundaries and surfaces, the jump frequencies and diffusivities are also high. The activation energy for surface diffusion is related to the enthalpy of vaporization.

The effect of pressure on diffusion is given by the activation volume, V^*:

$$V^* = RT(\partial \ln D/\partial P)_T - RT\left(\frac{\partial \ln \zeta a^2 \nu}{\partial P}\right)_T$$

TABLE 7-3
Diffusion in Silicate Minerals

Mineral	Diffusing Species	T (K)	D (m²/s)
Forsterite	Mg	298	2×10^{-18}
	Si	298	10^{-19}–10^{-21}
	O	1273	2×10^{-20}
Zn_2SiO_4	Zn	1582	3.6×10^{-15}
Zircon	O	1553	1.4×10^{-19}
Enstatite	Mg	298	10^{-20}–10^{-21}
	O	1553	6×10^{-16}
	Si	298	6.3×10^{-22}
Diopside	Al	1513	6×10^{-16}
	Ca	1573	1.5×10^{-15}
	O	1553	2.4×10^{-16}
Albite	Ca	523	10^{-14}
	Na	868	8×10^{-17}
Orthoclase	Na	1123	5×10^{-15}
	O	~1000	10^{-20}

Freer (1981).

The second term can be estimated from lattice dynamics and pressure dependence of the lattice constant and elastic moduli. This term is generally small. V^* is usually of the order of the atomic volume of the diffusing species. The activation volume is also made up of two parts, the formational part V_f^* and the migrational part V_m^*. Ordinarily the temperature and pressure dependence of a and ν are small.

For a vacancy mechanism V_f^* is simply the atomic volume since a vacancy is formed by removing an atom. This holds if there is no relaxation of the crystal about the vacancy. Inevitably there must be some relaxation of neighboring atoms inward about a vacancy and outward about an interstitial, but these effects are small. In order to move, an atom must squeeze through the lattice, and V_m^* can also be expected to about an atomic volume.

The work performed in creating a lattice imperfection can be estimated from elasticity theory if the lattice can be treated as a continuum. In this case

$$V^*/G^* = -[(\partial \ln G/\partial \ln V)_T - 1]K_T^{-1} = [1 + \{G\}_T]K_T^{-1}$$

where G is the shear modulus and K_T is the isothermal bulk modulus (Keyes, 1963). The magnitude of $\{G\}_T$, from the previous chapter, is about 3. The Grüneisen assumption, that all vibrational frequencies of the lattice depend on volume in the same way, gives

$$(\partial \ln K/\partial \ln V)_T = -(2\gamma + 1/3) = (\partial \ln G/\partial \ln V)_T$$

and

$$V^*/G^* = 2(\gamma - 1/3)K_T^{-1}$$

Borelius (1960) obtained a similar expression, but a smaller effect, by assuming that the increase in volume during the self-diffusion act is equal to the volume of the hard atomic core:

$$V^*/G^* = -1/3(K_T)^{-1}(\partial \ln K_T/\partial \ln V)_T = \{K_T\}_T/3K_T$$

The activation entropy S^* can also be estimated from a strain energy model (Keyes, 1960):

$$S^*/G^* = -\alpha[(\partial \ln G/\partial \ln V)_P + 1]$$
$$= \alpha[\{G\}_P - 1]$$

or

$$V^*/S^* = (\alpha K_T)^{-1}[(\partial \ln G/\partial \ln V)_T + 1]$$
$$\div [(\partial \ln G/\partial \ln V)_P + 1]$$
$$= (1 - \{G\}_T)/\alpha K_T(1 - \{G\}_P)$$

With the Grüneisen assumption, that the rigidity G depends on temperature and pressure only through the volume,

$$V^*/S^* \approx (\alpha K_T)^{-1}$$

and

$$S^* = 2(\gamma - 1/3)\alpha G^*$$

Some of these approximations are not necessary for upper-mantle minerals since $(\ln G/\partial \ln V)_T = \{G\}_T$ and $(\partial \ln G/\partial \ln V)_P = \{G\}_P$ are both known.

Sammis and others (1977) estimated values for V^* of about 8 to 12 cm^3/mole for oxygen self-diffusion in olivine and about 3 to 5 cm^3/mole in the lower mantle, decreasing with depth. Another estimate for activation volume of about 2 cm^3/mole for the lower mantle was based on seismic attenuation data (Anderson, 1967). Of course the V^* for attenuation may differ from the V^* for creep since different diffusing species may be involved. Nevertheless, the effect of pressure on ionic volumes leads one to expect that V^* will decrease with depth and, therefore, that activated processes became less sensitive to pressure at high pressure. Indeed, both viscosity and seismic factor Q do not appear to increase rapidly with depth in the lower mantle.

TABLE 7-4
Diffusion Parameters in Silicate Minerals

Mineral	Diffusing Species	D_0 (m^2/s)	Q ($kJ\ mol^{-1}$)
Olivine	Mg	4.1×10^{-4}	373
	Fe	4.2×10^{-10}	162
	O	5.9×10^{-8}	378
	Si	7.0×10^{-13}	173
	Fe-Mg	6.3×10^{-7}	239
Garnet	Sm	2.6×10^{-12}	140
	Fe-Mg	6.1×10^{-4}	344
Ca_2SiO_4	Ca	2.0×10^{-6}	230
$CaSiO_3$	Ca	7	468
Albite	Na	1.2×10^{-7}	149
	O	1.1×10^{-9}	140
Orthoclase	Na	8.9×10^{-4}	220
	O	4.5×10^{-12}	107
Nepheline	Na	1.2×10^{-6}	142
Glass			
Albite	Ca	3.1×10^{-5}	193
Orthoclase	Ca	2.6×10^{-6}	179
Basalt	Ca	4.0×10^{-5}	209
	Na	5×10^{-10}	41.8

Freer (1981).

HOMOLOGOUS TEMPERATURE

The ratio E^*/T_m is nearly constant for a variety of materials, though there is some dependence on valency and crystal structure. Thus, the factor E^*/T in the exponent for activated processes can be written $\lambda T_m/T$ where λ is roughly a constant and T_m is the "melting temperature." If this relation is assumed to hold at high pressure, then the effect of pressure on G^*, that is, the activation volume V^*, can be estimated from the effect of pressure on the melting point:

$$D(P,T) = D_o \exp[-\lambda T_m(P)/RT]$$

and

$$V^* = E^* \frac{dT_m}{dP} \Big/ T_m$$

which, invoking the Lindemann law, becomes

$$V^* = 2E^*(\gamma - 1/3)/K_T$$

which is similar to expressions given in the previous section. The temperature T, normalized by the "melting temperature," T_m, is known as the homologous temperature. It is often assumed that activated properties depend only on T/T_m and that the effect of pressure on these properties can be estimated from $T_m(P)$.

The melting point of a solid is related to the equilibrium between the solid and its melt and not to the properties of the solid alone. Various theories of melting have been proposed that involve lattice instabilities, critical vacancy concentrations or dislocation densities, or amplitudes of atomic motions. These are not true theories of melting since they ignore the properties of the melt phase, which must be in equilibrium with the solid at the melting point.

DISLOCATIONS

Dislocations are extended imperfections in the crystal lattice and occur in most natural crystals. They can result from the crystal growth process itself or by deformation of the crystal. They can be partially removed by annealing. Although dislocations occur in many complex forms, all can be obtained by the superposition of two basic types: the edge dislocation and the screw dislocation. These can be visualized by imagining a cut made along the axis of a cylinder, extending from the edge to the center and then shearing the cylinder so the material on the cut slides radially (edge dislocation) or longitudinally (screw dislocation). In the latter case the cylinder is subjected to a torque.

The first suggestion of dislocations was provided by observations in the nineteenth century that the plastic deformation of metals proceeded by the formation of slip bands wherein one portion of a specimen sheared with respect to another. Later the theory evolved that real crystals were composed of small crystallites, slightly misoriented with respect to one another, whose boundaries consist of arrays of dislocation lines. Theories of crystallization involve the nucleation of atoms on the ledges formed by the emergence of a dislocation at the surface of a crystal. The rapid equilibration of point defects in a crystal also involves dislocations as sources and sinks of point defects in crystals. This is also involved in diffusional and dislocation models of creep of solids.

The major impetus for the development of dislocation theory involved the consideration of the strength of a perfect crystal. The theoretical shear strength of a crystal is

$$\text{Strength} = Gb/2\pi a \approx G/5$$

where G is rigidity, b, the Burger's vector, is the magnitude of a simple lattice-translation and a is the spacing of atomic planes. The shear stress required to initiate plastic flow in metals, however, is only about 10^{-3} to 10^{-4} G. This was resolved by the introduction of lines of mismatch, or dislocations, in the lattice of crystals. Slip could then proceed by the unzippering effect of a dislocation line moving through the lattice. Work by Orowan, Polanyi, Taylor and Burgess in the 1930s to resolve the strength problem led to the development of modern dislocation theory. The theory was well developed and dislocations were almost completely understood before they were even actually observed.

An elementary property of a dislocation is the Burger's vector b. If a closed circuit is made by proceeding from one atom to another in a perfect crystal and then the same circuit is made in a part of a crystal that contains a dislocation, there will be closure error in the latter case. This closure error or closure vector is the Burger's vector and is roughly one interatomic spacing.

Consider a crystal that contains a half-cut and then is subjected to shear that displaces atoms on opposite sides of the cut by a single lattice spacing. The cut defines the slip plane, and the dislocation line, the edge of the cut, marks the boundary between the slipped and unslipped parts of the crystal. The displacement is specified by the slip or Burger's vector. Since b is a lattice vector, the atoms on opposite sides of the slip plane remain in alignment; thus the slip plane is not a discontinuity in the crystal, but there is distortion of the crystal around the dislocation line. The energy of a dislocation is proportional to b^2, so crystals tend to deform by slip with b the smallest lattice vector. The vector b is normal to the dislocation line for an edge dislocation and parallel to the dislocation line for a screw dislocation.

Dislocations can contribute to creep by climb or by glide. Dislocation climb involves the lengthening or shortening of the extra plane of atoms defining the dislocation. Dislocation glide, or slip, involves the transfer of an edge dislocation to an adjacent plane of atoms.

DISLOCATION CREEP

Creep in mantle silicates can occur by motions of vacancies by the self-diffusion mechanism (Nabarro-Herring creep) or by the motion of dislocations. The respective strain rates are

$$\dot{\varepsilon}_{NH} = K_1(D/l^2)(\sigma\Omega/kT)$$

and

$$\dot{\varepsilon}_{D} = K_2(D/b^2)(G\Omega/kT)(\sigma/G)^3$$

where l is the grain size, Ω the atomic volume, D the diffusivity, b the length of the Burger's vector of the dislocation, σ the shear stress, G the rigidity, and K_1 and K_2 are dimensionless constants. The second mechanism, called power-law creep, is commonly observed in mantle silicates at high temperature. Diffusional creep can be more important at low stress and small grain size as small grains, or subgrains, place the sources and sinks of vacancies closer together. Note that $\dot{\varepsilon}_{NH}$ depends linearly on σ and is therefore Newtonian. Power-law creep, $\dot{\varepsilon}_{D}$, is distinctly non-Newtonian.

If diffusion is preferentially along grain boundaries we have Coble creep,

$$\dot{\varepsilon}_{C} = K_1(D/l^2)(\alpha\Omega/kT)[1 + (\pi\delta/l)](D_B/D)$$

where δ is the thickness of the grain boundary ($\sim 200b$) and D_B is the grain boundary diffusivity.

Note that all these expressions contain D and therefore have the same temperature and pressure dependence as the appropriate diffusing species, presumably Si or O.

It is probable that steady-state creep in the mantle is controlled by dislocation climb and is therefore limited by the slow process of self-diffusion and, possibly, jog formation on dislocation lines. There is abundant evidence that mantle minerals contain many dislocations; most are probably associated with cell walls of subgrains. Under large stresses dislocation multiplication can take place.

The average subgrain size, L, and dislocation density, ρ_m, are observed to be related to the stress σ (Nicolas and Poirer, 1976)

$$L/b \approx G/\sigma$$

$$b^2\rho_m \approx (\sigma/G)^2$$

These relations can be used to estimate the stress in the mantle from studies of crystals derived from the mantle. These relations lead to

$$b^2\rho_m = K^2(b/L)^2$$

where the constant K is about 10 for laboratory-deformed olivine (Gittus, 1976). The creep rate can be written

$$\dot{\varepsilon} = \sigma/\tau G$$

where τ is a characteristic time often called the Maxwell time.

For diffusional creep, with stress-dependent grain or subgrain dimensions,

$$\tau = K_2(G/\sigma)^2 \frac{kTb^2 \exp E^*/RT}{GD_o\Omega}$$

where it is assumed that cell walls are efficient sources and sinks of vacancies. This differs from the usual equation for diffusional creep because of the stress dependence of the grain size. K_2 is about 20 (Minster and Anderson, 1981).

For dislocation creep the strain rate is

$$\dot{\varepsilon} = \rho_m \frac{b\Lambda}{d} V$$

where Λ is the mean free path of the dislocation and d is the distance covered at the rate-controlling speed V, usually the climb velocity. For this case

$$\tau = K_3(G/\sigma)^2 \frac{kTb^2 \exp E^*/RT}{GD_o\Omega_v}$$

where K_3 is roughly 1 and Ω_v is the volume of a vacancy.

For a polygonized network of dislocations, the climb of dislocations in cell walls is rate-limiting, but the characteristic time for creep involves both the motion of the long mobile dislocations in the cells and the shorter dislocations in the walls. Dislocations sweep across the grain, under the action of an imposed stress, and then climb and are annihilated in cell walls or grain boundaries. The Maxwell time for this mechanism is

$$\tau = (3/K^4)(G/\sigma)^2 \frac{kTb^2 \exp [(E_s^* + E_j)/RT]}{GD_o\Omega_v}$$

where K is a constant of order 10 to 50 ($\sim L\sigma/Gb$), E_s^* is the self-diffusion energy and E_j is the jog formation energy, of order 17 to 58 kcal/mole for olivine. This can be termed the cell-wall recovery model, which was formulated by Gittus (1976) and discussed by Minster and Anderson (1980, 1981). This is an efficient mechanism because of the low τ and, at high temperature, because of the high activation energy.

The dominant creep mechanism depends on such parameters as grain size, temperature, pressure and stress level. It may therefore be different in the Earth than in the laboratory. Laboratory results are generally obtained at high stress levels, high strain rates and short times. Results cannot necessarily be extrapolated to low stresses and low strain rates and, even if they can, other mechanisms may become more effective.

The low-stress creep data for olivine can support either an $\dot{\varepsilon} \approx \sigma^3$ or σ^2 interpretation, but measurements on individual samples support the high exponent. The difference can result in a several order of magnitude variation in the extrapolated creep rate at important geophysical stress levels of about 10 bars. Estimated values of τ_o at 10 bars are 10^{-11} to 5×10^{-10} s for a σ^3 law and 2×10^{-13} to 10^{-11} s for a σ^2 law.

These values for τ_o, and the associated activation energy E^*, rule out diffusional creep, involving self-diffusion of oxygen and silicon, as a mechanism to explain laboratory creep in olivine. Several other proposed mechanisms for creep can also be ruled out by this procedure. Minster and Anderson (1981) concluded that a dislocation model that involves cell-wall recovery under self-stress and silicon diffusion satisfies the olivine creep data and might therefore be an appropriate creep mechanism for the mantle. Theoretical Maxwell times, appropriate for mantle conditions, are shown in Figure 7-1.

Estimated activation volumes for O and Si diffusion in olivine are about 11 cm³/mole and 4 cm³/mole, respectively, which can be compared with the estimates from seismic data indicating a decrease from 11 to 6 cm³/mole in the upper mantle and 6 to 3 cm³/mole in the lower mantle (Anderson, 1967; Sammis and others, 1977). These considerations limit the total variation of viscosity in the mantle to about two orders of magnitude, assuming no change in the pre-exponential term.

Strain rate, $\dot{\varepsilon}$, and effective viscosity, η, are related to the characteristic time by

$$\tau = \sigma/G\dot{\varepsilon} = \eta/G$$

and therefore the viscosity can be calculated from the expressions given above. The expression for viscosity for the dislocation cell-wall recovery model is

$$\eta = (3/K^4)(G/\sigma)^2 \frac{kTb^2 \exp E^*/RT}{D_o\Omega_v}$$

where

$$E^* = E_s^* + E_j - \Delta E^*/2$$

where E_s^* is the activation energy for self-diffusion of the slowest moving species (presumably oxygen or silicon), E_j is the jog formation energy and ΔE^* is the difference between the activation energies for bulk diffusion and core diffusion, to be taken into account when E_j exceeds $\Delta E^*/2$. The ΔE^* term is only involved if dislocation core diffusion is important and there are few geometric jogs in the wall dislocations.

Figure 7-2 shows creep data for olivine. The parameter

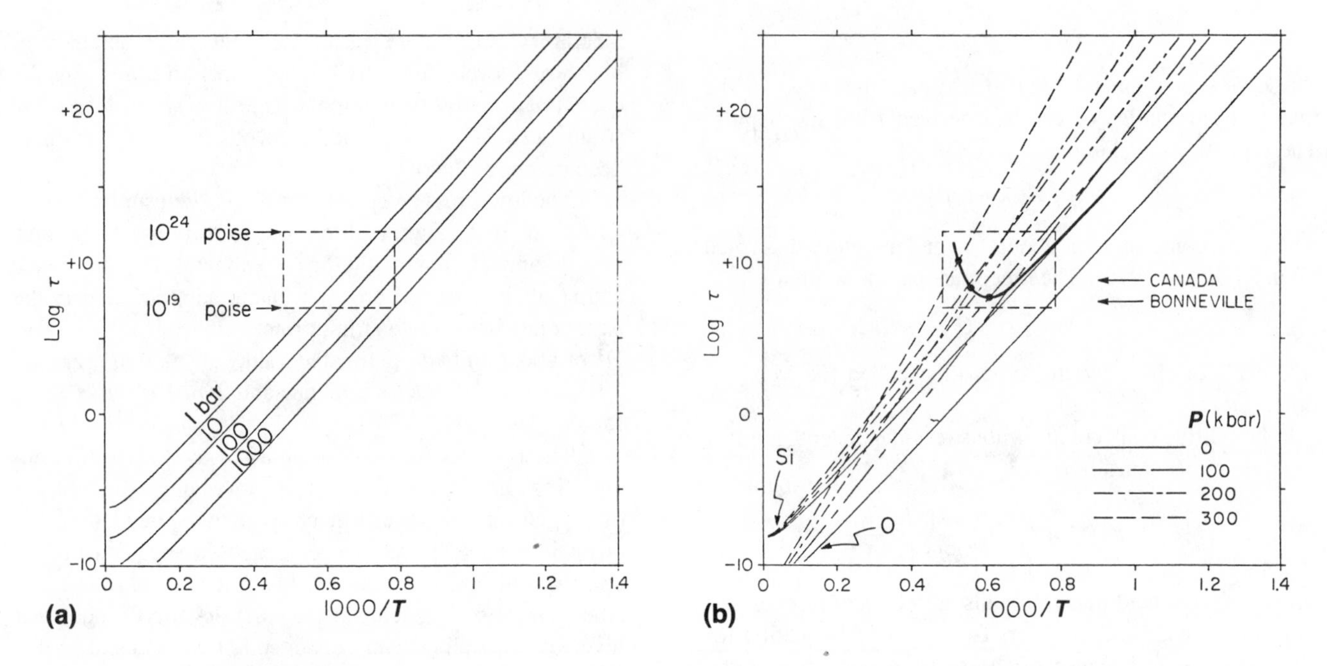

FIGURE 7-1
(a) Temperature dependence of Maxwell time τ for several values of applied load, for a creep model that satisfies laboratory observations with a σ^3 interpretation (after Minster and Anderson, 1981). **(b)** Effect of confining pressure on Maxwell times τ. At pressures greater than 200 kbar, oxygen diffusion becomes slower and controls self-diffusion. Simple thermal model of the mantle (heavy line) is compared with Crough's (1977) estimates for isostatic rebound for Lake Bonneville and the Canadian shield (after Minster and Anderson, 1981).

K is the ratio of cell diameter to the average dislocation length, including mobile and cell-wall dislocations.

Steady-state creep data for olivine for stresses between several hundred bars and several kilobars are well fit by a dislocation climb model where the strain rate is proportional to the third power of the stress (Goetze and Kohlstedt, 1973). At lower stresses the data deviate from this relationship and approach a stress-squared dependency.

THE LITHOSPHERE

The lithosphere is the cold outer shell of the Earth, which can support stresses elastically. There has been much confusion regarding the thickness of the lithosphere because the roles of time and stress have not been fully appreciated. Since mantle silicates can flow readily at high temperatures and flow more rapidly at high stresses, the lithosphere appears to be thicker at low stress levels and short times than it does for high stress levels and long times. Thus, the elastic lithosphere is thick when measured by seismic or postglacial-rebound techniques. At longer times the lower part of the instantaneous elastic lithosphere relaxes and the effective elastic thickness decreases. Thus, the elastic lithosphere is relatively thin for long-lived loads such as seamounts and topography. Estimates of the flexural thickness of the lithosphere range from 10 to 35 km for loads having durations of millions of years. The elastic lithosphere can be expected to be much thicker for short-duration loads such as ice caps. A more complete definition of the lithosphere is that part of the crust and upper mantle that deforms elastically for the load and time scale in question.

In theoretical discussions of the lithosphere, it is usually assumed that the viscosity or creep resistance of the

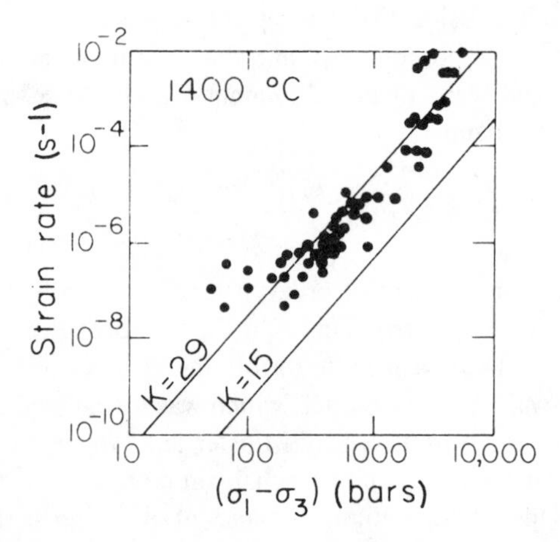

FIGURE 7-2
Strain rate versus stress difference data for olivine at 1400°C compared with the theory for two subgrain sizes (after Minster and Anderson, 1981). Data from Kohlstedt and Goetz (1974).

mantle depends only on temperature, pressure and stress. The oceanic lithosphere, in this framework, thickens with time only because it is cooling, and the thickness can be related to the depth of a given isotherm. The viscosity, or strength, of the mantle depends also on composition, mineralogy and crystal orientation. If the upper mantle is layered, then the lithosphere-asthenosphere boundary may be controlled by factors other than temperature. For example, if the subcrustal layer is olivine-rich harzburgite, it may be stronger at a given temperature than a clinopyroxene-garnet-rich layer, which has been postulated to occur at greater depth in the oceanic mantle. If the latter is weak enough, the lithosphere-asthenosphere boundary may represent a chemical boundary rather than an isotherm. Likewise, a change in the preferred orientation of the dominant crystalline species may also markedly affect the creep resistance.

The layer that translates coherently, the "plate" of plate tectonics, is often taken to be identical with the elastic lithosphere. This is probably a valid approximation if the stresses and time scales of the experiment that is used to define the flexural thickness are similar to the stresses and time scales of plate tectonics. It must be kept in mind, however, that mantle silicates are anisotropic in their flow characteristics, and that the stresses involved in plate tectonics may have different orientations than the stresses involved in surface loading experiments.

In a convecting mantle there is a thermal boundary layer through which heat must pass by conduction. The thickness of the thermal boundary layer is controlled by such parameters as conductivity and heat flow and is related in a simple way to the thickness of the elastic layer. Since temperature increases rapidly with depth in the conduction layer, and viscosity decreases rapidly with temperature, the lower part of the boundary layer probably lies below the elastic lithosphere; that is, only the upper part of the thermal boundary layer can support large and long-lived elastic stresses. Unfortunately, the conduction layer too is often referred to as the lithosphere.

Most recent models of the Earth's mantle have an upper-mantle low-velocity zone, LVZ, overlain by a layer of higher velocities, referred to as the LID (see Chapter 3). The LID is also often referred to as the lithosphere. Seismic stresses and periods are, of course, much smaller than stresses and periods of geological interest. If seismic waves measure the relaxed modulus in the LVZ and the high-frequency or unrelaxed modulus in the LID, then, in a chemically homogeneous mantle, the LID should be much thicker than the elastic lithosphere. If the LID is chemically distinct from the LVZ, then one might also expect a change in the long-term rheological behavior at the interface. If the LID and the elastic lithosphere turn out to have the same thickness, then this would be an argument for chemical or crystallographic rather than thermal control of the mechanical properties of the upper mantle.

In summary, we recognize the following "lithospheres":

1. The elastic, flexural or rheological lithosphere. This is the closest to the classical definition of a rocky, or strong, outer shell. It can be defined as that part of the crust and upper mantle that supports elastic stresses of a given size for a given period of time. The thickness of this lithosphere depends on the stress and time.
2. The plate. This is that part of the crust and upper mantle that translates coherently in the course of plate tectonics. The thickness of the plate may be controlled by chemical or buoyancy considerations or by stress, as well as by temperature.
3. The chemical lithosphere. The density and mechanical properties of the lithosphere are controlled by chemical composition and crystal structure as well as temperature. If chemistry and mineralogy dominate, then the elastic lithosphere and LID may be identical. If the lithosphere, below the crust, is mainly depleted peridotite or harzburgite, it may be buoyant relative to the underlying mantle.
4. The thermal boundary layer or conduction layer. This should not be referred to as the lithosphere, which is a mechanical concept, but if the lithospheric thickness is thermally controlled, the thickness of the lithosphere should be proportional to the thickness of the thermal boundary layer.
5. The seismic LID. This is a region of high seismic velocity that overlies the low-velocity zone. At high temperatures the seismic moduli measured by seismic waves may be relaxed, in which case they can be of the order of 10 percent less than the high-frequency or unrelaxed moduli. High-temperature dislocation relaxation and partial melting are two mechanisms that serve to decrease seismic velocities. The boundary between the LID and the LVZ would be diffuse and frequency dependent if thermal relaxation is the mechanism. A sharp interface would be evidence for a chemical or mineralogical boundary.

There are several other pieces of data regarding the nature of the lithosphere. It has recently been found that the earthquakes in some subduction zones lie along two planes about 30 km apart. These may represent the top and bottom interfaces of the downgoing slab or internal interfaces in the slab or zones of maximum stress in the slab. The separation of the two seismic zones is similar to the flexural thickness of the plate.

The best evidence for cooling of the oceanic plate, or thickening of the thermal boundary layer, comes from the deepening of bathymetry as a function of time. The simple square-root of time relationship for bathymetry falls down after 80 million years, indicating that the thermal boundary layer has reached an equilibrium thickness or that a thermal event prior to 80 Ma affected the parts of the oceanic lithosphere that have been used to calculate the bathymetry-age relation. The conductive cooling calculation apparently does not work in reverse: As the plate approaches a hotspot, it thins very rapidly to a thickness of about 30 km (Crough and Thompson, 1976). This thickness seems to be independent of plate rate or original plate thickness. Conductive heating is too slow to explain these results; mechanical thinning or delamination seems required in order to explain the heating time scale, and compositional or mineralogical control seems required in order to explain the equilibrium thickness obtained upon heating. The similarity in thickness of the heated lithosphere, the flexural lithosphere and the double seismic zone should be noted. These results suggest that there is a strong, refractory layer that extends to a depth of about 30 to 50 km. The material below this depth apparently is easily abraded, ablated or delaminated and is weaker than the subcrustal layer. This change in mechanical properties may be controlled by temperature, mineralogy, partial melting or stress.

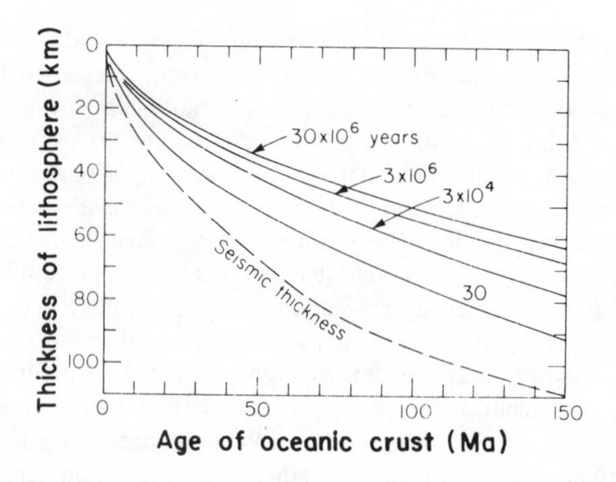

FIGURE 7-3
Relationship between thickness of "elastic" lithosphere, age of oceanic crust and duration of loading for 1 kbar stress. For a given load the elastic thickness decreases with time. The dashed line is the upper bound of seismic determinations of LID thicknesses. Compare with Figure 3-2.

EFFECTIVE ELASTIC THICKNESS OF THE LITHOSPHERE

We can now estimate the thickness of the "rheological lithosphere." Clearly it is a function of lithospheric age, the magnitude of the load, and the duration of loading. At a given depth, and therefore temperature, the rheology will appear elastic for times shorter than the characteristic relaxation time, and viscous for longer duration loads. The "thickness of the lithosphere" is therefore a more complicated concept than has been generally appreciated.

Figure 7-3 gives estimates of the thickness of the oceanic rheological lithosphere as a function of age of crust and duration of load, using oceanic geotherms and a stress of 1 kilobar. The depth to the rheological asthenosphere is defined as the depth having a characteristic time equal to the duration of the load. For example, from Figure 7-3 a 30-Ma-old load imposed on 80-Ma-old crust would yield a thickness of 45 km if the lithosphere did not subsequently cool, or if the cooling time is longer than the relaxation time. A 30-Ma-old load on currently 50-Ma-old crust would give a rheological lithosphere 34 km thick with the above qualifications. The thickness of the high seismic velocity layer overlying the low-velocity zone (the LID) is also shown in Figure 7-3. For load durations of millions of years, the rheological lithosphere is about one-half the thickness of the older estimates of the thickness of the seismological lithosphere. Note that the rheological thickness decreases only gradually for old loads. This is consistent with the observations of Watts and others (1975). On the other hand, the lithosphere appears very thick for young loads, and the apparent thickness decreases rapidly.

MELTING AND ORIGIN OF MAGMAS

There are several ways to generate melts in the mantle. Melting can occur in situ if the temperature is increasing and eventually exceeds the solidus. The main source of heating in the mantle is the slow process of radioactive decay. Heating also leads to buoyancy and convection, a relatively rapid process that serves to bring heated material toward the surface where it cools. The rapid ascent of warm material leads to decompression, another mechanism for melting due to the relative slopes of the adiabat and the melting curve. Extensive in situ melting, without adiabatic ascent, is unlikely except in deep layers that are intrinsically denser than the overlying mantle. In this case melting can progress to a point where the intrinsic density contrast is overcome by the elimination of a dense phase such as garnet. Below about 100 km the effect of pressure on the melting point is much greater than the adiabatic gradient or the geothermal gradient in homogeneous regions of the mantle. Recent work on the melting of peridotites at high pressure shows that the melting curve levels off at pressures greater than about 100 kb. This combined with chemical boundary layers at depth makes it possible to envisage the onset of melting at depths between about 300 and 400 km or deeper. Because of the high temperature gradient at the interface between chemically distinct layers, melting is most likely to initiate in chemical boundary layers. Whether melting is most extensive above or below the interface depends on the mineralogy and the amount of the lower-melting-point phase. The Earth is slowly cooling with time and therefore melting was more extensive in the past and probably extended, on average, to both shallower and greater depths than at present.

Most experimental petrology is performed at pressures less than 40 kilobars, and it is natural that most theories of petrogenesis have been shoehorned into this accessible pressure range. The extrapolation of melting gradients obtained at low pressure suggest that plausible mantle temperatures are well below the solidus at depths greater than about 200 km.

Melting most likely is the result of adiabatic ascent from greater depth, initiating where the geotherm intersects the solidus. The amount of melting depends on the temperature interval between the solidus and the liquidus, the latent heat of fusion and the ability of the melt to leave the matrix.

Fragmentary evidence at high pressure (~100 kbar) indicates that the solidus and liquidus tend to converge at high pressure and that the effect of pressure on both decreases rapidly (Ohtani et al., 1986, Takahashi, 1986). Additionally, the MgO content of the initial melt increases with pressure, and the density of the melt and residual crystals converge with depth. These factors mean that high-MgO melts such as picrites and komatiites, or even peridotites, can be generated at depth, that extensive melting can occur upon further adiabatic ascent, and that melt separation is more likely at shallow depth than at great depth. The possible increase of melt viscosity with pressure and pinching off of permeability in a rapidly rising diapir will also affect melt separation at depth.

The dependence of melting point T_m on pressure is governed by the Clausius-Clapeyron equation

$$dT_m/dP = \Delta V/\Delta S = T\Delta V/L$$

where ΔV and ΔS are the changes in volume and entropy due to melting and L is the latent heat of melting. ΔS is always positive, but ΔV can be either positive or negative. Therefore the slope of the melting curve can be either negative or positive but is generally positive. Although this equation is thermodynamically rigorous, it does not provide us with much physical insight and is not suitable for extrapolation since ΔV, ΔS and L all depend on pressure. Because of the high compressibility of liquids, ΔV decreases rapidly with pressure, and $\Delta V/V$ and ΔS probably approach limiting values at high pressure.

Lindemann proposed that melting occurs when the thermal oscillation of atoms reached a critical amplitude,

$$T_m = Am\theta^2V^{2/3}$$

where m is the mass of the atoms, V is the volume, θ is the Debye temperature and A is a constant. Gilvarry (1956) reformulated the equation in terms of the bulk modulus and volume of the solid at the melting point.

Other theories of melting propose that some critical density of dislocations or vacancies causes the crystal to melt or that a crystal becomes unstable when one of the shear moduli vanishes.

All of the above theories can be criticized because they do not involve the properties of the melt or considerations of solid-melt equilibrium. They correspond rather to an absolute stability limit of a crystal, which may differ from the crystal-liquid transition.

Stacey and Irvine (1977) were able to derive a melting relation, which resembles Lindemann's equation, from a simple adaptation of the Mie-Grüneisen equation without involving the vibration amplitude assumption. An appropriate form of their equation is

$$dT_m/T_m\,dP = 2(\gamma - 2\gamma^2\alpha T_m)/K_T$$

The Lindemann law itself can be written

$$dT_m/T_m\,dP = 2(\gamma - 1/3)/K_T$$

which gives almost identical numerical values. These can be compared with the expression for the adiabatic gradient

$$dT/T\,dP = \gamma/K_T$$

Since γ is generally about 1, the melting point gradient is steeper than the adiabatic gradient. For $\gamma < 2/3$ the reverse is true.

If the above relations apply to the mantle, the adiabat and the melting curve diverge with depth. This means that melting is a shallow-mantle phenomenon and that deep melting will only occur in thermal boundary layers in a convecting mantle. In thermal boundary layers the thermal gradient is controlled by the conduction gradient, which is typically 10–20°C/km compared to the adiabatic gradient of 0.3°C/km. A limited amount of data at very high pressure indicates that dT_m/dP decreases with pressure, and T_m may approach a constant value. The adiabatic gradient, of course, also decreases with pressure.

The Lindemann law was motivated by the observation that the product of the coefficient of thermal expansion α and the melting temperature T_m was very nearly a constant for a variety of materials. This implies that

$$dT_m/dP = -(T_m/\alpha)(\partial\alpha/\partial P)_T$$

which can be written

$$\begin{aligned} dT_m/T_m\,dP &= -(\partial \ln K_T/\partial \ln V)_P/K_T \\ &= -(1/\alpha K_T^2)(\partial K_T/\partial T)_P \end{aligned}$$

Thus, the increase of melting temperature with pressure can be estimated from the thermal and elastic properties of the solid. Typical values for silicates are T_m, ~2000K, $-(\partial \ln K_T/\partial \ln V)_P$, ~5 and K_T, ~1500 kbar, giving

$$dT_m/dP \approx 6°C/kbar$$

which is in the ballpark of observed values. For example, the mean slopes of the melting curves for albite, enstatite, diopside and pyrope at pressures below 10–30 kb range from 5 to 15°C/kbar (see Table 7-5). Metals generally have an initial melting point gradient of 3 to 9 K/kilobar.

The Simon fusion equation,

$$P_m/P_o = (T_m/T_o)^c - 1$$

TABLE 7-5
Calculated and Measured Initial Melting Point Gradients

Mineral	Calculated dT_m/dP	Measured dT_m/dP
Mg_2SiO_4	9.7	4.8
$MgSiO_3$	11.2	12.8
Diopside	9.7	13 (5 kbar)
		7.5 (50 kbar)

$(T_m)^{-1}\ dT_m/dP$ (calculated) $= -(\partial \ln K_S/\partial \ln V)_P/K_S$.

has received much attention in the geophysics literature. T_o is the melting temperature at zero pressure and P_o and c are constants. Gilvarry (1956) showed the equivalence of this to the Lindemann equation by invoking the "law of corresponding states"

$$T_m/T_o = (V_o/V_m)^{2(\gamma - 1/3)}$$

and obtained

$$c = (6\gamma + 1)/(6\gamma - 2)$$

P_o is the (negative) melting pressure at zero temperature.

Although silicates are complex multicomponent systems, the solidi of many natural systems typically have slopes of 10–15°C/km. The presence of volatiles such as water and CO_2 considerably reduces the initial melting point, or solidus, of silicates at low pressure, but the behavior at higher pressure and the behavior of the liquidus are relatively simple.

Stacey (1977), attempting to estimate the melting temperature in the mantle from seismic data using considerations similar to the above, obtained a T_m of 3157 K at the base of the mantle as shown in Table 7-6. Ohtani (1983) used measured and estimated melting information on various silicates and extrapolated these, also using solid-state physics considerations. His results for the liquidus and solidus temperature profiles are also given in the table. Stacey's

TABLE 7-6
Estimates of Temperature (T), Melting Temperature (T_m), Liquidus Temperature (T_l), Solidus Temperature (T_s) and Debye Temperature (θ) for the Mantle (All in K)

Depth (km)	T	T_m	T_l	T_s	θ	γ
600	2230	2300	2800	2300	900	0.74
1100	2400	2500	3500	3000	1100	1.00
1500	2600	2700	4200	3800	1200	0.97
1900	2700	2800	5500	4500	1250	0.95
2200	2800	2950	6000	4900	1300	0.94
2900	3200	3200	6500	5000	1400	0.91

Stacey (1977); T and T_s from Ohtani (1983).

TABLE 7-7
Selected Ionic Radii (Å)

Ion	Coordination	Radius
Al^{3+}	IV	0.39
	V	0.48
	VI	0.535
As^{3+}	VI	0.58
As^{5+}	IV	0.335
	VI	0.46
B^{3+}	III	0.01
	IV	0.11
	VI	0.27
Ba^{2+}	VI	1.35
	VII	1.38
	VIII	1.42
	IX	1.47
Ba^{2+}	X	1.52
	XI	1.57
	XII	1.61
Be^{2+}	IV	0.27
	VI	0.45
Br^-	VI	1.96
C^{4+}	III	−0.08
	IV	0.15
Ca^{2+}	VI	1.00
	VII	1.06
	VIII	1.12
	IX	1.18
	X	1.23
Ca^{2+}	XII	1.34
Cd^{2+}	VI	0.95
Ce^{4+}	VIII	0.97
Cl^-	VI	1.81
Co^{2+}(HS)*	VI	0.745
Cr^{2+}(HS)	VI	0.80
Cr^{3+}	VI	0.615
Cs^+	VI	1.67
	VIII	1.74
	XI	1.88
Eu^{+2}	VI	1.17
	VIII	1.25
F^-	IV	1.31
	VI	1.33
Fe^{2+}(HS)	IV	0.63
	IV	0.64
(HS)	VI	0.780
	VIII	0.92
Fe^{3+}	IV	0.49
(HS)	VI	0.645
Ga^{3+}	IV	0.47
Ge^{4+}	IV	0.39
	VI	0.53
H^+	I	−0.38
	II	−0.18
Hf^{4+}	VI	0.71
	VIII	0.83
Hg^{2+}	VI	1.02
I^-	VI	2.20
K^+	VI	1.38

TABLE 7-7 (continued)
Selected Ionic Radii

Ion	Coordination	Radius
	VIII	1.51
	IX	1.55
	X	1.59
	XII	1.64
Li^{+}	IV	0.59
	VI	0.76
	VIII	0.92
Mg^{2+}	IV	0.57
	VI	0.720
	VIII	0.89
Mn^{2+}(HS)	VI	0.830
	VIII	0.96
Mn^{3+}(HS)	VI	0.645
N^{3-}	IV	1.46
N^{3+}	VI	0.16
N^{5+}	III	−0.10
Na^{+}	VI	1.02
	VII	1.12
	VIII	1.18
	IX	1.24
	XII	1.39
Ni^{2+}	IV	0.55
	VI	0.690
O^{2-}	II	1.35
	III	1.36
	IV	1.38
	VI	1.40
	VIII	1.42
OH^{-}	II	1.32
	III	1.34
	IV	1.35
	VI	1.37
P^{5+}	IV	0.17
	VI	0.38
Pb^{2+}	VI	1.19
	VIII	1.29
Rb^{+}	VI	1.52
	VIII	1.61
	IX	1.63
	X	1.66
	XII	1.72
Ru^{4+}	VI	0.62
S^{2-}	VI	1.84
Se^{2-}	VI	1.98
Si^{4+}	IV	0.26
	VI	0.40
Sm^{3+}	VI	0.96
Sn^{4+}	VI	0.69
Sr^{2+}	VI	1.18
	VIII	1.26
	XII	1.44
Te^{2-}	VI	2.21
Ti^{3+}	VI	0.670
Ti^{4+}	IV	0.42
	VI	0.605
U^{4+}	VI	0.89
	VIII	1.00
V^{2+}	VI	0.79
V^{3+}	VI	0.64
V^{5+}	IV	0.355
	VI	0.54
W^{4+}	VI	0.66
Zn^{2+}	IV	0.60
	VI	0.74
Zr^{4+}	VI	0.72
	VIII	0.84

Shannon (1976).
*(HS) denotes high-spin electronic configuration.

estimated geotherm is given for comparison. The T_m calculation of Stacey is much lower than T_l of Ohtani, partly because he assumed that both the top and base of the mantle are at the melting point, not just locally but on average. Ohtani's estimated melting temperatures diverge rapidly from the geotherm with increasing depth in the lower mantle.

Also given in Table 7-6 are Stacey's estimates of the Debye temperature, θ, and the Grüneisen ratio, γ, both estimated from the seismic data. For comparison, the estimated variations of heat capacity C_P throughout the mantle is only about 1 percent, and the coefficient of thermal expansion, α, decreases by about a factor of 2 or 3.

IONIC RADII

Crystal chemists and other investigators have found the concept of ionic radii to be extremely useful. Atoms and ions are treated as hard spheres, and empirical tables of atomic and ionic radii have been computed that closely reproduce interatomic distances. Approximate additivity of atomic and ionic radii was noted by the earliest investigators. The principle of additivity of cation and anion radii in ionic crystals accurately reproduces interatomic distances if one takes into consideration the coordination number, electronic spin, covalency and repulsive forces. Ionic radii have been important to crystal chemists because structure types and coordinations are determined by cation/anion radius ratios. For example, the likely crystal structures of high-pressure phases can often be inferred from these ratios. The geochemist uses these radii to infer what ions can readily substitute for others. Thus, partition coefficients depend on ionic radii. Elasticity systematics also depend on ionic radii. Diffusivity, anelasticity and activation volumes depend on the size of the diffusing species. Table 7-7 gives ionic radii as determined by Shannon and Prewitt (1969) and Shannon

TABLE 7-8
Measured Variation of K_s with Temperature and Pressure and Intrinsic Temperature Dependence

Mineral	α (10^{-6}/K)	$(\partial \ln K_S/\partial \ln V)_T$	$(\partial \ln K_S/\partial \ln V)_P$	$\alpha^{-1}(\partial \ln K_S/\partial T)_V$
Al_2O_3	16.4	−4.3	−4.1	+0.2
Mg_2SiO_4	24.7	−4.8	−5.7	−0.9
Olivine	24.7	−5.4	−4.7	+0.7
Pyroxene	47.7	−9.5	−5.5	+4.0
$MgAl_2O_4$	22.0	−4.9	−3.7	+1.2
Quartz	36.6	−6.4	−5.8	+0.6
Garnet	21.6	−5.5	−5.2	+0.3
Garnet	19.0	−4.8	−5.9	−1.1
MgO	31.5	−3.8	−3.1	+0.7
$SrTiO_3$	28.2	−5.7	−7.7	−2.0

(measured) (measured) (computed)

$$\left(\frac{\partial \ln K}{\partial \ln V}\right)_P = \left(\frac{\partial \ln K}{\partial \ln V}\right)_T + \alpha^{-1}\left(\frac{\partial \ln K}{\partial T}\right)_V$$

(1976). The variation of the adiabatic bulk modulus with volume is given in Table 7-8 for some important crystals.

General References

Ashby, M. F. and R. A. Verrall (1978) Micromechanism of flow and fracture, and their relevance to the rheology of the upper mantle, *Phil. Trans. R. Soc. Lond. A., 288,* 59–95.

Hirth, J. P. and J. Lothe (1968) *Theory of Dislocations,* McGraw-Hill, New York, 780 pp.

Jaoul, O., M. Poumellec, C. Froidevaus and A. Havette (1981) Silicon diffusion in forsterite: A new constraint for understanding mantle deformation. In *Anelasticity in the Earth* (F. D. Stacey, M. S. Patterson, A. Nicholas, eds.), 95–100, American Geophysical Union, Washington, D.C.

Nicolas, A. and J. P. Poirer (1976) *Crystalline Plasticity and Solid State Flow in Metamorphic Rocks,* John Wiley and Sons, London, 444 pp.

Nowick, A. S. and B. S. Berry (1972) *Anelastic Relaxation in Crystalline Solids,* Academic Press, New York, 677 pp.

Stacey, F. (1977) Applications of thermodynamics to fundamental Earth physics, *Geophys. Surveys, 3,* 175–204.

References

Anderson, D. L. (1967) The anelasticity of the mantle, *Geophys. J. Roy. Astron. Soc., 14,* 135–164.

Berckhemer, H., F. Auer and J. Drisler (1979) High-temperature anelasticity and elasticity of mantle peridotite, *Phys. Earth Planet. Inter., 20,* 48–59.

Borelius, G. (1960) *Arkiv Fysik, 16,* 437.

Creager, K. and T. Jordan (1984) Slab penetration into the lower mantle, *J. Geophys. Res., 89,* 3031–3050.

Crough, S. T. (1977) Isostatic rebound and power-law flow in the asthenosphere, *Geophys. J. Roy. Astron. Soc., 50,* 723–738.

Crough, S. T. (1979) Geoid anomalies across fracture zones and the thickness of the lithosphere, *Earth Planet. Sci. Lett., 44,* 224–230.

Crough, S. T. and G. Thompson (1976) Numerical and approximate solutions for lithospheric thickening and thinning, *Earth Planet. Sci. Lett., 31,* 397–402.

Debye, P. (1912) *Ann. Phys. Lpz., 39,* 784.

Freer, R. (1981) Diffusion in silicate minerals and glasses, *Contrib. Mineral. Petrol., 76,* 440–454.

Fujisawa, H., N. Fujii, H. Mizutani, H. Kanamori and S. Akimoto (1968) Thermal diffusivity of Mg_2SiO_4, Fe_2SiO_4, and NaCl at high pressure and temperature, *J. Geophys. Res., 73,* 4727–4733.

Gilvarry, J. J. (1956) *Phys. Rev., 102,* 317.

Gittus, J. (1976) *Phil. Mag., 34,* 401–411.

Goetze, C. and D. L. Kohlstedt (1973) Laboratory study of dislocation climb and diffusion in olivine, *J. Geophys. Res., 78,* 5961–5971.

Horai, K. (1971) Thermal conductivity of rock-forming minerals, *J. Geophys. Res., 76,* 1278–1308.

Horai, K. and G. Simmons (1970) An empirical relationship between thermal conductivity and Debye temperature for silicates, *J. Geophys. Res., 75,* 678–982.

Keyes, R. (1963) Continuum models of the effect of pressure on activated processes. In *Solid Under Pressure* (W. Paul and D. M. Warschauer, eds.), 71–99, McGraw-Hill, New York.

Kobayzshigy, A. (1974) Anisotropy of thermal diffusivity in olivine, pyroxene and dunite, *J. Phys. Earth, 22,* 359–373.

Kohlstedt, D. L. and C. Goetze (1974) Low-stress high-temperature creep in olivine single crystals, *J. Geophys. Res.*, *79*, 2045–2051.

Minster, J. B. and D. L. Anderson (1980) Dislocations and non-elastic processes in the mantle, *J. Geophys. Res.*, *85*, 6347–6352.

Minster, J. B. and D. L. Anderson (1981) A model of dislocation-controlled rheology for the mantle, *Phils. Trans. R. Soc. London A*, *299*, 319–356.

Ohtani, E. (1983) Melting temperature distribution and fractionation in the lower mantle, *Phys. Earth Planet. Inter.*, *33*, 12–25.

Ohtani, E., T. Kato and H. Sawamoto (1986) *Nature*, *322*, 352–353.

Sammis, C., J. Smith, G. Schubert and D. Yuen (1977) Viscosity-depth profile of the Earth's mantle; effects of polymorphic phase transitions, *J. Geophys. Res.*, *82*, 3747–3761.

Schatz, J. F. and G. Simmons (1972) Thermal conductivities of Earth materials at high temperatures, *J. Geophys. Res.*, *77*, 6966–6983.

Shannon, R. D. (1976) *Acta Cryst.*, *A32*, 751.

Shannon, R. D. and C. T. Prewitt (1969) *Acta Cryst.*, 925.

Stacey, F. D. (1977) A thermal model of the Earth, *Phys. Earth Planet. Inter.*, *15*, 341–348.

Stacey, F. D. and R. D. Irvine (1977) *Austral. J. Phys.*, *30*, 631.

Takahashi, E. (1986) *J. Geophys. Res.*, *91*, 9367–9382.

Watts, A. B., J. R. Cochran and G. Selzer (1975), Gravity anomalies and flexures of the lithosphere: A three-dimensional study of the Great Meteor Seamount, N. E. Atlantic, *J. Geophys. Res.*, *80*, 1391–1398.

8

Chemical Composition of the Mantle

These rocks . . .
Shall yet be touched with beauty,
and reveal
the secrets of the book of earth to man.

—ALFRED NOYES

Considerations from cosmochemistry and the study of meteorites permit us to place only very broad bounds on the chemistry of the Earth's interior. These tell us little about the distribution of elements in the planet. Seismic data tell us a little more about the distribution of the major elements. General considerations suggest that the denser major elements will be toward the center of the planet and the lighter major elements, or those that readily enter melts or form light minerals, will be concentrated toward the surface. To proceed further we need detailed chemical information about crustal and mantle rocks. The bulk of the material emerging from the mantle is in the form of melts, or magmas. It is therefore important to understand the chemistry and tectonic setting of the various kinds of magmatic rocks and the kinds of sources they may have come from.

METHODS OF ESTIMATING MANTLE CHEMISTRY

The chemical composition of the mantle is one of the most important yet elusive properties of our planet. Attempts to estimate mantle composition fall into two broad categories.

Cosmochemical approaches take meteorites or mixtures of meteoritic material as the basic building blocks, and mixing ratios are adjusted to satisfy such constraints as core size, heat flow and crustal ratios of certain elements. An example is the six-component model of Morgan and Anders (1980). Cosmochemical models constrain the bulk chemistry of the Earth rather than that of the mantle alone; nevertheless they provide important input into models of the bulk chemistry of the Earth.

Petrological models begin with the reasoning that since basalts represent melts, and peridotites are thought to be residues, some mixture of these should approximate the composition of the upper mantle. With only two components this approach does not yield chondritic ratios for many key elements, most notably Si/Mg. An alternate approach is to search for the most "primitive" ultramafic rock (that is, the one with most nearly chondritic ratios of the refractory elements) and attribute its composition to the whole mantle. Unfortunately, even the most "primitive" ultramafic nodules are depleted in many of the trace elements and have nonchondritic rare-earth ratios, but theorizing has proceeded beyond the first crude models.

Both approaches utilize terrestrial and meteoritic data. The common theme is that the Earth should have an unfractionated chondritic pattern of the refractory elements. This assumption, justified by the observation that these elements occur in roughly constant proportions in the various meteorite classes, has led to the generally accepted hypothesis that the refractory elements do not suffer any preaccretional fractionation. This can be used as a formal a priori constraint in geochemical modeling of the composition of the Earth. Some recent estimates of the composition of the mantle are given in Table 8-1. Once they are in a planet, the refractory elements become fractionated by a variety of processes. The refractory siderophiles enter the core, the compatible refractories are retained in mantle silicates and the incompatible refractories preferentially enter melts and the crust along with the more volatile elements.

The volatile elements are fractionated by preaccretional processes, and they exhibit a wide range in meteorites. It is therefore difficult to estimate the volatile content of the Earth or to obtain estimates of such key volatile-to-refractory ratios as K/U, Rb/Sr, Pb/U and others. It is often assumed that ratios of this type are the same in the Earth or in primitive mantle as they are in the continental crust. The

TABLE 8-1
Estimates of Average Composition of Mantle

Oxide	(1)	(2)	(3)	(4)	(5)
SiO_2	45.23	47.9	44.58	47.3	45.1
Al_2O_3	4.19	3.9	2.43	4.1	3.9
MgO	38.39	34.1	41.18	37.9	38.1
CaO	3.36	3.2	2.08	2.8	3.1
FeO	7.82	8.9	8.27	6.8	7.9
TiO_2	—	0.20	0.15	0.2	0.2
Cr_2O_3	—	0.9	0.41	0.2	0.3
Na_2O	—	0.25	0.34	0.5	0.4
K_2O	—	—	0.11	0.2	(0.13)

(1) Jacobsen and others (1984): extrapolation of ultramafic and chondritic trends.
(2) Morgan and Anders (1980): cosmochemical model.
(3) Maaløe and Steel (1980): extrapolation of lherzolite trend.
(4) 20 percent eclogite, 80 percent garnet lherzolite (Anderson, 1980).
(5) Ringwood and Kesson (1976, Table 7): pyrolite adjusted to have chondritic CaO/Al_2O_3 ratio and Ringwood (1966) for K_2O.

crust, of course, is just one repository of the incompatible elements and is less than 0.6 percent of the mass of the mantle. The validity of this assumption is therefore not obvious and needs to be tested by an independent approach.

"Primitive mantle" as used here is the silicate fraction of the Earth, prior to differentiation and removal of the crust and any other parts of the present mantle that are the result of differentiation, or separation, processes. In some geochemical models it is assumed that large parts of the Earth escaped partial melting, or melt removal, and are therefore still "primitive." Some petrological models assume that melts being delivered to the Earth's surface are samples from previously unprocessed material. I find it difficult to believe that any part of the Earth could have escaped processing during the high-temperature accretional process. "Primitive mantle," as used here, is a hypothetical material that is the sum of the present crust and mantle. "Primitive magma" is a hypothetical magma, the parent of other magmas, which formed by a single-stage melting process of a parent rock and has not been affected by loss of material (crystal fractionation) prior to sampling.

The view that there is a single primitive mantle magma type that leaves behind a single depleted peridotite, the essence of the pyrolite model, is clearly oversimplified. There is increasing evidence that ophiolitic peridotites, for example, are not simply related to the overlying basalts. Isotopic data on basalts and nodules show that there are at least two major mantle reservoirs. The identification of a component in ocean-island tholeiites and alkali olivine basalts that is "enriched," both chemically (in LILs) and isotopically, also is not adequately accounted for in single mantle reservoir models. I will refer to this enriched component as Q.

The use of three components of primitive mantle is conventional: basalts, ultramafic rocks, and continental crust. The assumption that the crust and depleted mantle are strictly complementary and are together equivalent to the bulk Earth, however, is not consistent with isotopic results. Basalts cover a broad compositional range, from LIL-poor to LIL-rich. "Large-ion-lithophile" (LIL) is commonly, although loosely, used to refer to elements (including small high-charge elements!) that do not substitute readily for magnesium or iron and are therefore excluded from olivine and orthopyroxene. Some LILs are relatively compatible in garnet and clinopyroxene. One recent proposal is that most mantle magmas are composed of a depleted MORB component and an enriched component (Q) with high potassium, LIL, $^{87}Sr/^{86}Sr$, $^{144}Nd/^{143}Nd$ and $^{206}Pb/^{204}Pb$. Mid-ocean-ridge basalt (MORB) represents the most uniform and voluminous magma type and is an end member for LIL concentrations and isotopic ratios. This is logically taken as one of the components of the mantle. The MORB source has been depleted by removal of a component—Q—that must be rich in LIL but relatively poor in Na and the garnet-clinopyroxene-compatible elements (such as Al, Ca, Yb, Lu and Sc). Kimberlitic magmas have the required complementary relationship to MORB, and I adopt them here as the Q component. Peridotites are the main reservoirs for elements such as magnesium, chromium, cobalt, nickel, osmium and iridium. The continental crust is an important reservoir of potassium, rubidium, barium, lanthanum, uranium and thorium. Thus, each of these components plays an essential role in determining the overall chemistry of the primitive mantle. It is conventional to adopt a single lherzolite or harzburgite as the dominant silicate portion of the mantle. An orthopyroxene-rich component is also present in the mantle and is required if the Mg/Si and Ca/Al ratios of the Earth are to be chondritic. Some peridotites appear to have been enriched (metasomatized) by a kimberlite-like component.

Figure 8-1 shows representative compositions of kimberlite, crust, MORB, and ultramafic rock. For many refractory elements kimberlite and crust have a similar enrichment pattern. However, the volatile/refractory ratios are quite different, as are ratios involving strontium, hafnium, titanium, lithium, yttrium, ytterbium and lutetium. Kimberlite and MORB patterns are nearly mirror images for the refractory elements, but this is only approximately true for MORB and crust, especially for the HREE, and the small ion–high charge elements. MORB and kimberlite also represent extremes in their strontium and neodymium isotopic compositions.

An important development in recent years has been the recognition of an LIL-enriched "metasomatic" component in the mantle. The most extreme magmas from the mantle (high LIL, high LREE/HREE, high $^{87}Sr/^{86}Sr$, low $^{143}Nd/^{144}Nd$) are kimberlites and lamproites (McCulloch and others, 1982, 1983). When these are mixed with a depleted

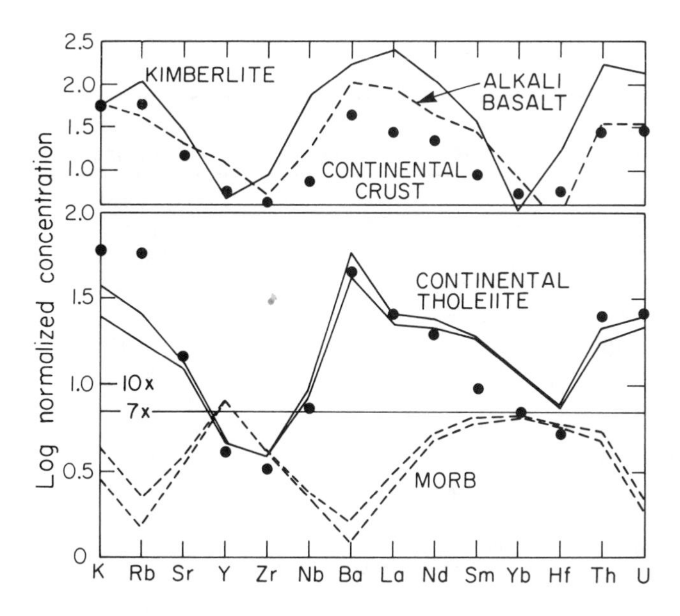

FIGURE 8-1
Trace-element concentrations in the continental crust (dots), continental basalts and mid-ocean ridge basalts (MORB), normalized to average mantle compositions derived from a chondritic model. Note the complementary relationship between depleted basalts (MORB) and the other materials. MORB and continental tholeiites are approximately symmetric about a composition of 7 × C1. This suggests that about 14 percent of the Earth may be basalt. For other estimates, see text.

magma (MORB), the resulting blend can have apparently paradoxical geochemical properties. For example, the hybrid magma can have such high La/Yb, Rb/Sr and $^{206}Pb/^{204}Pb$ ratios that derivation from an enriched source is indicated, but the $^{143}Nd/^{144}Nd$ and $^{87}Sr/^{86}Sr$ ratios imply derivation from an ancient depleted reservoir. Many ocean-island, island-arc and continental basalts have these characteristics. These apparently paradoxical results simply mean that ratios do not average as do concentrations.

In order to estimate the volatile and siderophile content of the mantle, we seek a linear combination of components that gives chondritic ratios for the refractory elements. We can then estimate such key ratios as Rb/Sr, K/U, and U/Pb. In essence, we replace the five basic building blocks of mantle chemistry (ol, opx, cpx, ga, and Q, or their high- and low-pressure equivalents) with four composites; peridotite (ol ± opx), orthopyroxenite (opx ± ol), basalt (cpx ± ga), and Q. In practice, we can use two different ultramafic rocks (UMR and OPX) with different ol/opx ratios, to decouple the ol + opx contributions. The chemistry of the components (MORB, UMR, KIMB, OPX, and crust) are given in Table 8-2.

Having measurements of m elements in n components where m far exceeds n, we can find the weight fraction x_j of each component, given the concentration C_i of the jth element in the ith component, that yields chondritic ratios of the refractory oxyphile elements. In matrix form,

$$C_{ij}x_j = kC_j \qquad (1)$$

where C_j is the chondritic abundance of element j and k is a dilution or enrichment factor, which is also to be determined. The least-squares solution is

$$x_j/k = (C^TC)^{-1}C^TC_j \qquad (2)$$

where C^T is the matrix transpose of C, with the constraints

$$\sum x_j = 1 \quad \text{and} \quad k = \text{constant} \qquad (3)$$

When x_j and k are found, equation 1 gives the mantle concentrations of the volatile and siderophile elements, elements not used in the inversion.

The mixing ratios found from equations 2 and 3 are UMR, 32.6 percent, OPX, 59.8 percent, MORB, 6.7 percent, crust, 0.555 percent, and Q, 0.11 percent. This is a model composition for primitive mantle, that is, mantle plus crust. The Q component is equivalent to a global layer 3.6 km thick. The MORB component represents about 25 percent of the upper mantle. This solution is based on 18 refractory elements and, relative to carbonaceous chondrite (C1) abundances, $k = 1.46 \pm 0.09$. The result is given in Table 8-3 under "Mantle plus crust." Concentrations normalized to C1 are also given. Note that the high-charge, small ionic radius elements (Sc, Ti, Zr, Nb, and Hf) have the highest C1-normalized ratios. Taking all the refractory oxyphile elements into account (23 elements), the C1-normalized enrichment of the refractory elements in the mantle plus crust is 1.59 ± 0.26. If six elements (Fe, S, Ni, Co, P, and O) are removed from C1 to a core of appropriate size and density, the remaining silicate fraction will be enriched in the oxyphile elements by a factor of 1.48 relative to the starting C1 composition. This factor matches the value for k determined from the inversions. Thus it appears that the Earth's mantle can be chondritic in major- and refractory-element chemistry if an appreciable amount of oxygen has entered the core (see Chapter 4).

Table 8-3 also compares these results with Morgan and Anders's cosmochemically based model. This model can be viewed as providing a first-order correction for volatile-refractory fractionation and inhomogeneous accretion. Rather than treating each element separately, Morgan and Anders estimated abundances of groups of elements: refractories, volatiles, and so on. There is strong support for unfractionated behavior of refractories prior to accretion, but the volatile elements are likely to be fractionated. Both the volatile elements and the siderophile elements are strongly depleted in the crust-mantle system relative to cosmic abundances.

In the pyrolite models, it is assumed that primitive mantle is a mix of basalt and peridotite and that one knows the average compositions of the basaltic and ultramafic components of the mantle. These are mixed in somewhat arbitrary proportions and the results are comparable with

TABLE 8-2
Chemical Composition of Mantle Components (ppm)

	MORB	Ultramafic Rocks	KIMB	Crust	Picrite	OPX	Morgan and Anders
Li	9	1.5	25	10	7	1.5	1.85
F	289	7	1900	625	219	—	14
Na	15900	2420	2030	26000	12530	—	1250
Mg*	5.89	23.10	16.00	2.11	10.19	21.00	13.90
Al*	8.48	2.00	1.89	9.50	6.86	1.21	1.41
Si*	23.20	21.80	14.70	27.10	22.85	22.60	15.12
P	390	61	3880	1050	308	—	1920
S	600	8	2000	260	452	—	14600
Cl	23	1	300	1074	17	—	20
K	660	35	17600	12500	504	10	135
Ca*	8.48	2.40	7.04	5.36	6.96	1.32	1.54
Sc	37.3	17	15	30	32	12	9.6
Ti	5500	1000	11800	4800	4375	800	820
V	210	77	120	175	177	60	82
Cr	441	2500	1100	55	956	2500	4120
Mn	1080	1010	1160	1100	1063	1010	750
Fe*	6.52	6.08	7.16	5.83	6.41	6.08	32.07
Co	53	105	77	25	66	105	840
Ni	152	2110	1050	20	642	2110	18200
Cu	77	15	80	60	62	30	31
Zn	74	60	80	52	71	20	74
Ga	18	4	10	18	14	3	3
Ge	1.5	1.1	0.5	1.5	1.4	1.1	7.6
Se	0.181	0.02	0.15	0.05	0.14	0.006	9.6
Rb	0.36	0.12	65	42	0.30	0.04	0.458
Sr	110	8.9	707	400	85	4.5	11.6
Y	23	4.6	22	22	18	0.023	2.62
Zr	70	11	250	100	55	5.5	7.2
Nb	3.3	0.9	110	11	2.7	0.45	0.5
Ag	0.019	0.0025	—	0.07	0.015	0.001	0.044
Cd	0.129	0.0255	0.07	0.2	0.103	0.01	0.016
In	0.072	0.002	0.1	0.1	0.055	—	0.002
Sn	1.36	0.52	15	2	1.15	0.52	0.39
Cs	0.007	0.006	2.3	1.7	0.01	0.002	0.0153
Ba	5	3	1000	350	4.50	2	4
La	1.38	0.57	150	19	1.18	0.044	0.379
Ce	5.2	1.71	200	38	4.33	0.096	1.01
Nd	5.61	1.31	85	16	4.54	0.036	0.69
Sm	2.08	0.43	13	3.7	1.67	0.009	0.2275
Eu	0.81	0.19	3	1.1	0.66	0.0017	0.079
Tb	0.52	0.14	1	0.64	0.43	0.0017	0.054
Yb	2.11	0.46	1.2	2.2	1.70	0.0198	0.229
Lu	0.34	0.079	0.16	0.03	0.27	0.0072	0.039
Hf	1.4	0.34	7	3	1.14	0.17	0.23
Ta	0.1	0.03	9	2	0.08	—	0.023
Re†	1.1	0.23	1	1	0.88	0.08	60
Os†	0.04	3.1	3.5	5	0.81	3.1	880
Ir†	0.0011	3.2	7	1	0.80	3.2	840
Au†	0.34	0.49	4	4	0.38	0.49	257
Tl†	0.01	0.02	0.22	0.25	0.01	—	0.00386
Pb	0.08	0.2	10	7	0.11	—	0.068
Bi	0.007	0.005	0.03	0.2	0.01	—	0.00294
Th	0.035	0.094	16	4.8	0.05	—	0.0541
U	0.014	0.026	3.1	1.25	0.02	—	0.0135

*Results in percent.
†Results in ppb.
Anderson (1983a).

TABLE 8-3
Chemical Composition of Mantle (ppm)

	C1	Morgan and Anders	UDS	Upper Mantle	Lower Mantle	Mantle + Crust	Normalized to M&A	Normalized to C1
Li	2.4	1.85	7.48	3.49	1.50	2.09	0.76	0.87
F	90	14	259	91	1	28	1.39	0.31
Na	7900	1250	13172	6004	359	2040	1.10	0.26
Mg*	14.10	13.90	9.80	18.67	21.31	20.52	1.00	1.46
Al*	1.29	1.41	6.95	3.65	1.33	2.02	0.97	1.57
Si*	15.60	15.12	23.00	22.20	22.48	22.40	1.00	1.44
P	1100	1920	387	170	9.06	57	0.02	0.05
S	19300	14600	458	158	1.19	48	0.0022	0.0025
Cl	1000	20	79	27	0.07	8	0.27	0.008
K	890	135	1356	475	14	151	0.76	0.17
Ca*	1.39	1.54	6.87	3.89	1.48	2.20	0.96	1.58
Sc	7.8	9.6	31.92	22	13	15	1.09	1.99
Ti	660	820	4477	2159	830	1225	1.01	1.86
V	62	82	176	110	63	77	0.63	1.24
Cr	3500	4120	907	1969	2500	2342	0.38	.67
Mn	2700	750	1066	1029	1010	1016	0.91	0.38
Fe*	27.20	32.07	6.39	6.18	6.08	6.11	0.13	0.22
Co	765	840	64	91	105	101	0.08	0.13
Ni	15100	18200	611	1610	2110	1961	0.07	0.13
Cu	160	31	62	31	28	29	0.62	0.18
Zn	455	74	70	63	26	37	0.34	0.08
Ga	14	3	15	7	3	4	0.94	0.31
Ge	47	7.6	1.40	1.20	1.10	1.13	0.10	0.02
Se	29	9.6	0.14	0.06	0.01	0.02	0.0016	0.0008
Rb	3.45	0.458	3.32	1.19	0.05	0.39	0.57	0.11
Sr	11.4	11.6	109	42.3	5.2	16.2	0.94	1.42
Y	2.1	2.62	18.64	9.28	0.70	3.26	0.84	1.55
Zr	5.7	7.2	60	27	6	13	1.18	2.20
Nb	0.45	0.5	4.30	2.03	0.52	0.97	1.31	2.15
Ag	0.27	0.044	0.02	0.01	0.00	0.003	0.05	0.01
Cd	0.96	0.016	0.11	0.05	0.01	0.02	1.03	0.03
In	0.12	0.002	0.06	0.02	0.0003	0.006	2.13	0.05
Sn	2.46	0.39	1.34	0.79	0.52	0.60	1.04	0.24
Cs	0.29	0.0153	0.13	0.05	0.003	0.02	0.68	0.05
Ba	3.60	4	34	13	1.72	5.22	0.88	1.45
La	0.367	0.379	3.75	1.63	0.12	0.57	1.02	1.56
Ce	0.957	1.01	8.28	3.90	0.34	1.40	0.93	1.46
Nd	0.711	0.69	6.03	2.88	0.23	1.02	1.00	1.43
Sm	0.231	0.2275	1.90	0.92	0.07	0.32	0.96	1.40
Eu	0.087	0.079	0.70	0.362	0.03	0.13	1.10	1.48
Tb	0.058	0.054	0.44	0.241	0.02	0.09	1.09	1.51
Yb	0.248	0.229	1.72	0.88	0.09	0.32	0.95	1.30
Lu	0.038	0.039	0.27	0.144	0.02	0.06	0.96	1.46
Hf	0.17	0.23	1.30	0.66	0.20	0.33	0.98	1.96
Ta	0.03	0.023	0.28	0.11	0.004	0.04	1.10	1.24
Re†	60	60	0.89	0.45	0.10	0.21	0.0023	0.0034
Os†	945	880	1.08	2.43	3.10	2.90	0.0022	0.0031
Ir†	975	840	0.88	2.43	3.20	2.97	0.0024	0.0031
Au†	255	257	0.62	0.53	0.49	0.50	0.0013	0.002
Tl†	0.22	0.0039	0.03	0.02	0.00	0.01	1.28	0.033
Pb	3.6	0.068	0.60	0.33	0.03	0.12	1.19	0.033
Bi	0.17	0.0029	0.02	0.009	0.0007	0.0033	0.75	0.019
Th	0.051	0.0541	0.48	0.224	0.014	0.0765	0.96	1.50
U	0.014	0.0135	0.12	0.057	0.004	0.0196	0.98	1.40

*Results in percent.
†Results in ppb.
Anderson (1983a).

chondritic abundances for some of the major elements. Tholeiitic basalts are thought to represent the largest degree of partial melting among common basalt types and to most nearly reflect the trace-element chemistry of their mantle source. Tholeiites, however, range in composition from depleted midocean-ridge basalts to enriched ocean-island basalts (OIB) and continental flood basalts (CFB). Enriched basalts (alkali-olivine, OIB, CFB) can be modeled as mixtures of MORB and an enriched (Q) component that experience varying degrees of crystal fractionation prior to eruption.

Ultramafic rocks from the mantle, likewise, have a large compositional range. Some appear to be crystalline residues after basalt extraction, some appear to be cumulates, and others appear to have been secondarily enriched in incompatible elements (metasomatized). This enriched component is similar to the Q component of basalts. Some ultramafic rocks have high Al_2O_3 and CaO contents and are therefore "fertile" (they can yield basalts upon partial melting), but they do not have chondritic ratios of all the refractory elements.

In the above calculation I have, in essence, decomposed the basaltic component of the mantle into a depleted (MORB) and LIL-enriched (Q) component. These can be combined and compared with the basaltic component of previous two-component models, such as pyrolite, by comparing "undepleted basalt" (MORB + 1.5 percent KIMB) with the basalts chosen by Ringwood (Hawaiian tholeiite or "primitive" oceanic tholeiite, KD11). Thus the present model's undepleted basaltic component (MORB + Q) has 914 ppm potassium, 0.06 ppm uranium, and 0.27 ppm thorium. KD11 has, for these elements, 1400, 0.152, and 0.454 ppm, respectively, and other incompatible elements are correspondingly high. Refractory ratios such as U/Ca and Th/Ca for this pyrolite are about 50 percent higher than chondritic because of the high LIL content of the chosen basalt. Hawaiian tholeiites, the basis of another pyrolite model (Ringwood, 1975), have LIL concentrations even higher than KD11. One disadvantage of pyrolite-type models is that the final results are controlled by the arbitrary choice of components, including basalt. Indeed, the various pyrolite models differ by an order of magnitude in, for example, the abundance of potassium.

By and large, there is excellent agreement with the Morgan and Anders cosmochemical model (Table 8-3). In the present model, the alkalis lithium, potassium, rubidium and cesium are somewhat more depleted, as are volatiles such as chlorine, vanadium and cadmium. The Rb/Sr and K/U ratios are correspondingly reduced. The elements that are excessively depleted (P, S, Fe, Co, Ni, Ge, Se, Ag, Re, Os, Ir, and Au) are plausibly interpreted as residing in the core. Note that the chalcophiles are not all depleted. In particular, lead is not depleted relative to other volatiles such as manganese, fluorine and chlorine, which are unlikely to be concentrated in the core, or to the alkali metals. The less

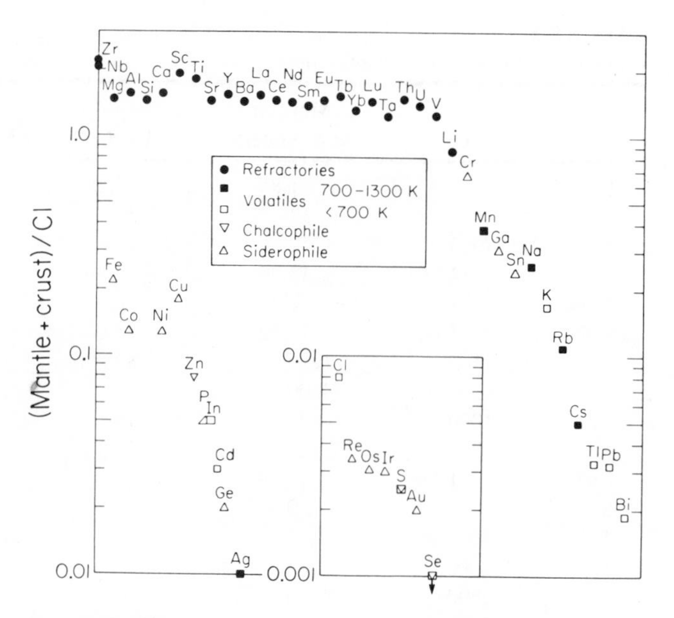

FIGURE 8-2
Abundances of elements in "primitive mantle" (mantle + crust) relative to C1, derived by mixing mantle components to obtain chondritic ratios of the refractory lithophile elements.

volatile chalcophiles (Bi, Cu, Zn) are slightly depleted. The most depleted siderophiles and chalcophiles are those that have the highest ionization potential and may, therefore, have suffered preaccretional sorting in the nebula. There is little support for the conjecture that the chalcophiles are strongly partitioned into the core.

The composition of primitive mantle, derived by the above approach, is given in Figure 8-2.

Some elements are extraordinarily concentrated into the crust. The above results give the following proportions of the total mantle-plus-crust inventory in the continental crust; rubidium, 58 percent; cesium, 53 percent; potassium, 46 percent; barium, 37 percent; thorium and uranium, 35 percent; bismuth, 34 percent; lead, 32 percent; tantalum, 30 percent; chlorine, 26 percent; lanthanum, 19 percent and strontium, 13 percent. In addition, the atmospheric argon-40 content represents 77 percent of the total produced by 151 ppm potassium over the age of the Earth. These results all point toward an extensively differentiated Earth and efficient upward concentration of the incompatible trace elements. It is difficult to imagine how these concentrations could be achieved if the bulk of the mantle is still primitive or unfractionated. If only the upper mantle provides these elements to the crust, one would require more than 100 percent removal of most of the list (U, Th, Bi, Pb, Ba, Ta, K, Rb, and Cs). More likely, the whole mantle has contributed to crustal, and upper-mantle, abundances.

The crust, MORB reservoir and the Q component account for a large fraction of the incompatible trace elements. It is likely, therefore, that the lower mantle is depleted in these elements, including the heat producers potassium, uranium and thorium.

In an alternative approach we can replace UMR and OPX by their primary constituent minerals, olivine, orthopyroxene, and clinopyroxene. The present mantle is therefore viewed as a five-component system involving olivine, orthopyroxene, clinopyroxene, MORB (cpx and ga), and Q. In this case the LIL inventory of the primitive mantle is largely contained in four components: MORB, Q, clinopyroxene, and CRUST. The results are: olivine, 33.0 percent, orthopyroxene, 48.7 percent, clinopyroxene, 3.7 percent, MORB, 14.0 percent, Q, .085 percent and CRUST, 0.555 percent. Concentrations of certain key elements are sodium, 2994 ppm, potassium, 205 ppm, rubidium, 0.53 ppm, strontium, 25 ppm, and cesium, 0.02 ppm. The alkalis are generally within 50 percent of the concentrations determined previously. Key ratios are Rb/Sr, 0.021, K/U, 4323, and K/Na, 0.07. The Rb/Sr and K/Na ratios are essentially the same as those determined previously; the K/U ratio is 44 percent lower.

In summary, a four-component (crust, basalt, peridotite and Q) model for the upper mantle can be derived that gives chondritic ratios for the refractory trace elements. The model gives predictions for volatile/refractory ratios such as K/U, Rb/Sr and Pb/U. An orthopyroxene-rich component is required in order to match chondritic ratios of the major elements. Such a component is found in the upper mantle and is implied by the seismic data for the lower mantle. The abundances in the mantle-plus-crust system are 151 ppm potassium, 0.0197 ppm uranium and 0.0766 ppm thorium, giving a steady-state heat flow of 0.9 μcal/cm^2 s. This implies that slightly more than half of the terrestrial heat flow is due to cooling of the Earth, consistent with convection calculations in a stratified Earth (McKenzie and Richter, 1981).

Primitive mantle can be viewed as a five-component system; crust, MORB, peridotite, pyroxenite and Q (quintessence, the fifth essence) or, alternatively, as olivine, orthopyroxene, garnet plus clinopyroxene (or basalt) and incompatible and alkali-rich material (crust and kimberlite or LIL-rich magmas).

THE UPPER MANTLE

The mass-balance method gives the average composition of the mantle but makes no statement about how the components are distributed between the upper and lower mantle. If we assume that the only ultramafic component of the upper mantle is UMR, we can estimate the composition of the upper and lower mantles and, as an intermediate step, the composition of the MORB source region prior to extraction of crust and Q. The lower mantle is UMR plus OPX. Orthopyroxenite, the most uncertain and to some extent arbitrary of the components, plays a minor role in the mass-balance calculations for the trace refractories and is required mainly to obtain chondritic ratios of Ca/Al and Mg/Si.

My approach to the upper mantle is similar to the conventional approach in that I consider a basaltic and an ultramafic component. However, instead of making an *a priori* selection of basalt, I have decomposed it into a depleted (MORB) and an enriched (Q) component. These represent extremes in both LIL contents and isotopic ratios. For example, fresh MORB has $^{87}Sr/^{86}Sr$ as low as 0.7023 and kimberlite usually has $^{87}Sr/^{86}Sr$ well above 0.704; alkalic basalts are intermediate in both LIL contents and isotopic ratios. The procedure is as follows. The mixing ratios of MORB, crust and Q are known from the previous section, and these ingredients are assumed to be entirely contained in the primitive upper mantle. The absolute sizes of the crustal and upper-mantle reservoirs (above 650 km) are known, so we know both the relative and absolute amounts of each component. As an intermediate step, we estimate the composition of a possible picritic parent to MORB. The relation PICRITE = 0.75 MORB + 0.25 UMR gives the results tabulated under "picrite" in Table 8-2 and 8-4.

The mixing ratios which were determined to give a chondritic pattern for the refractory elements yield

$$\text{UDS} = 0.9355\text{PICRITE} + 0.0106\text{Q} + 0.0559\text{CRUST}$$

The composition of this reconstructed Undepleted Source Region is tabulated under UDS. The fraction of the crustal component is about 10 times the crust/mantle ratio, so UDS accounts for 10 percent of the mantle. The remainder of the upper mantle is assumed to be ultramafic rocks UMR. This gives the composition tabulated under "Upper Mantle" in Tables 8-3 and 8-4. This region contains 23.4 percent basalt (MORB). The resulting upper mantle has refractory-element ratios (Table 8-5), which, in general, are in agreement with chondritic ratios. The La/Yb, Al/Ca and Si/Mg ratios, however, are too high. These are balanced by the lower mantle in the full calculation. The solution for the lower mantle is 0.145 UMR and 0.855 OPX. This gives chondritic ratios for Mg/Si and Ca/Al for the Earth as a whole. An orthopyroxene-rich lower mantle is expected for a chondritic model for the major elements, particularly if the upper mantle is olivine-rich. At low pressure olivine and orthopyroxene are refractory phases and are left behind as basalt is removed. However, at high pressure the orthopyroxene-rich phases, majorite and perovskite, are both refractory and dense. If melting during accretion extended to depths greater than about 350 km, then the melts would be olivine-rich and separation of olivine from orthopyroxene can be expected.

Figure 8-3 shows the concentrations of the lithophile elements in the various components, upper mantle and mantle-plus-crust, all normalized to the Morgan and Anders mantle equivalent concentrations. The refractory elements in the upper mantle have normalized concentrations of

TABLE 8-4
Elemental Ratios

	C1	Morgan Anders*	UDS	Upper Mantle	Lower Mantle	Mantle + Crust	Normalized M&A*	Normalized C1
Rb/Sr	0.3026	0.0395	0.0304	0.0281	0.0101	0.024	0.61	0.08
K/U	63571	10000	11429	8356	3552	7693	0.77	0.12
Sm/Nd	0.3249	0.3297	0.315	0.319	0.318	0.319	0.968	0.982
Th/U	3.64	4.01	4.08	3.94	3.62	3.98	0.97	1.07
U/Pb	0.0039	0.20	0.20	0.17	0.13	0.16	0.82	42
La/Yb	1.48	1.66	2.18	1.85	1.43	1.77	1.07	1.20
K/Na	0.11	0.11	0.10	0.08	0.04	0.07	0.69	0.66
Mg/Si	0.90	0.92	0.43	0.84	0.95	0.92	1.00	1.01
Ca/Al	1.08	1.09	0.99	1.07	1.12	1.09	1.00	1.01
Yb/Sc	0.032	0.024	0.054	0.04	0.007	0.021	0.87	0.65
Ce/Nd	1.35	1.46	1.37	1.35	1.49	1.37	0.94	1.02
Eu/Nd	0.12	0.11	0.12	0.13	0.13	0.13	1.10	1.03
Yb/Lu	6.53	5.87	6.26	6.10	4.77	5.80	0.99	0.89
Sr/Ba	3.17	2.90	3.17	3.14	2.99	3.11	1.07	0.98
U/La	0.038	0.036	0.03	0.035	0.032	0.034	0.97	0.90

*Model of Morgan and Anders (1980).

about 3; this includes the crustal contribution. Since the upper mantle is about one-third of the whole mantle, a strongly depleted lower mantle is implied. Note that the upper mantle is depleted in lithium and titanium. These elements may be in the lower mantle, or the Morgan and Anders estimates may be too high.

The upper mantle is not necessarily uniform. The basaltic fraction, as eclogite, is denser than peridotite and may form the major part of a separate layer. Seismic data are consistent with an eclogite-rich transition region and also suggest that the roots of midocean ridges extend to the transition region. The conventional view of basalt petrogenesis, however, is that the basaltic fraction of the mantle is uniformly dispersed, on a microscopic scale, in a mainly olivine matrix. The picritic fraction of the mantle corresponds to a layer about 200 km in thickness. At high pressure this would be a picritic eclogite, or piclogite. The transition region, it happens, is just over 200 km thick. The shallow mantle is probably harzburgite-rich. Basalts, or their parents, of course, must pass through the shallow mantle on their way to the surface, so parts of the shallow mantle are basalt-rich.

TABLE 8-5
Elemental Ratios in Upper-Mantle Components

	MORB	Ultramafic Rocks	KIMB	Crust	Picrite	OPX	Morgan and Anders
Rb/Sr	0.0033	0.0135	0.0919	0.105	0.0035	0.0089	0.039
K/U	47143	1346	5677	10000	29632	—	10000
Sm/Nd	0.371	0.328	0.153	0.231	0.368	0.25	0.3297
Th/U	2.50	3.62	5.16	3.84	2.94	—	4.01
U/Pb	0.18	0.13	0.31	0.18	0.15	—	0.20
La/Yb	0.65	1.24	125	8.64	0.69	2.22	1.66
K/Na	0.84	0.01	8.67	0.48	0.04	—	0.11
Mg/Si	0.25	1.06	1.09	0.08	0.45	0.93	0.92
Ca/Al	1.00	1.20	3.72	0.56	1.01	1.09	1.09
Yb/Sc	0.057	0.027	0.08	0.073	0.053	0.0017	.0239
Ce/Nd	0.93	1.31	2.35	2.38	0.95	2.67	1.46
Eu/Nd	0.14	0.15	0.04	0.07	0.14	0.05	0.11
Yb/Lu	6.21	5.82	7.50	7.33	6.18	2.75	5.87
Sr/Ba	22.00	2.97	1.18	1.14	18.83	3.00	2.90
U/La	0.01	0.46	0.021	0.066	0.014	—	0.036

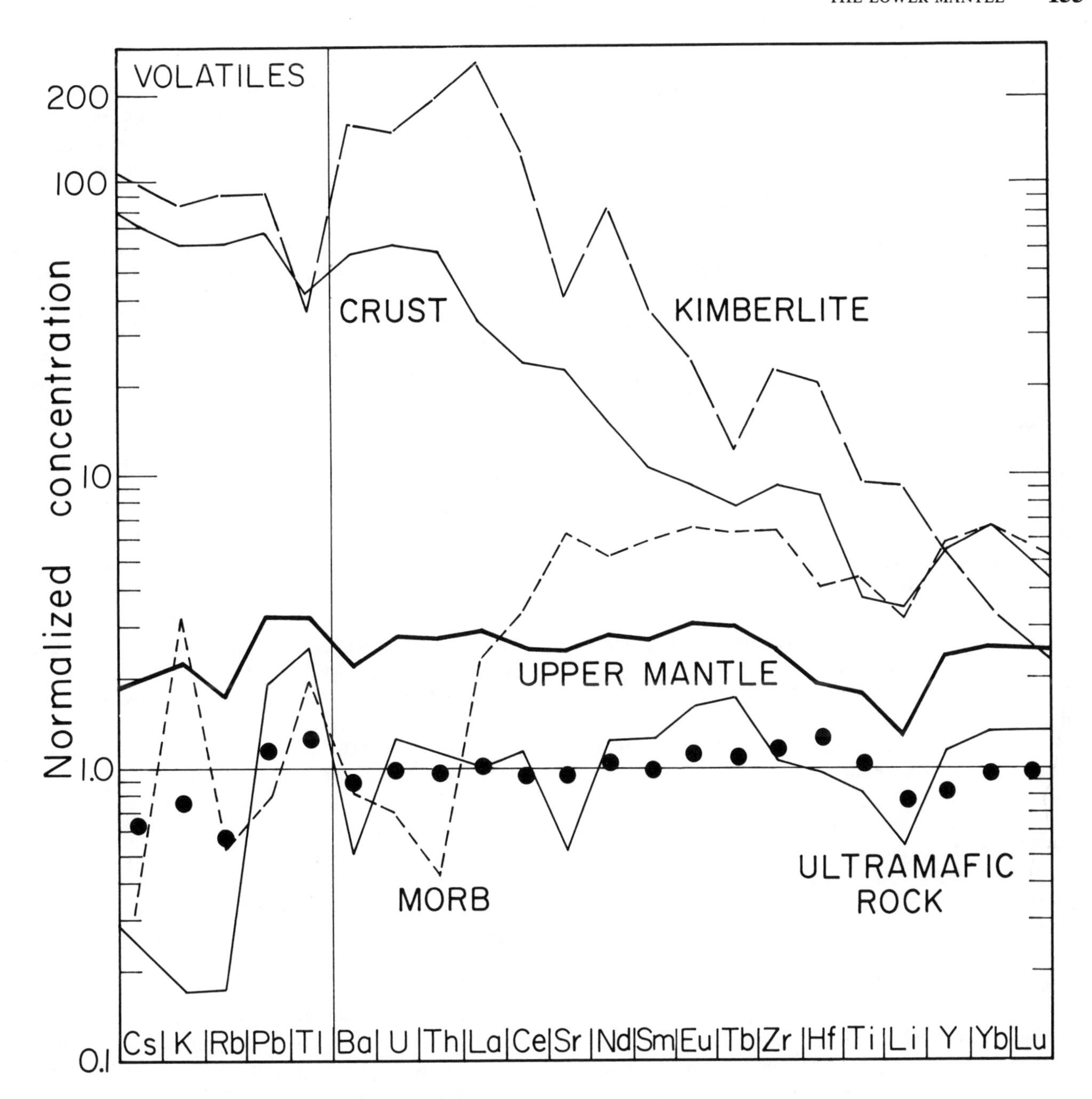

FIGURE 8-3
Lithophile elements in ultramafic rocks, MORB, continental crust and kimberlite normalized to 1.48 times the Morgan and Anders (1980) model. Derived values for upper mantle and primitive mantle (dots) are also shown (after Anderson, 1983a).

THE LOWER MANTLE

Whether the Earth is chondritic in major-element ratios, as assumed above, or is pyrolitic with a large silicon deficiency depends on the composition of the lower mantle. This section condenses the more detailed discussion in Chapter 4. Seismic data show that the lower mantle has a higher density and bulk modulus than olivine-rich peridotite or pyrolite whether in a mixed oxide or perovskite-bearing assemblage. The lower mantle is separated from the upper-mantle by a sharp discontinuity at about 650 km. The sharpness of the discontinuity and the magnitude of the density and velocity jumps suggest that this is a chemical discontinuity as well as a phase change (Burdick and Anderson, 1975; Anderson, 1977).

There has been much controversy over the composition of the lower mantle. Ringwood (1975; Ringwood and Kesson, 1977) has advocated a chemically uniform pyrolite mantle with Mg/Si of about 1.5. Others have shown, however, that the seismic data imply a chondritic or pyroxene-rich lower mantle with Mg/Si roughly 1 (Anderson, 1977; Butler and Anderson, 1978). The main seismic evidence for a high SiO_2 content for the lower mantle is the high bulk modulus, K_o.

In an attempt to resolve the density discrepancy, Ringwood (1975) suggested that enstatite might transform to $MgSiO_3$ (perovskite) with a density 3 to 7 percent greater than the isochemical mixed oxides. This transformation has subsequently been found (Liu, 1974), but the density increase is only 2 percent relative to the mixed oxides, and the density difference is negligible at high pressure. The pyrolite density discrepancy therefore still remains.

When the newer, more accurate data are used, one concludes that the zero-pressure density (4.1 g/cm^3) and bulk modulus (2.6 Mbar) of the lower mantle are coincident with a perovskite rock (ρ = 4.1 g/cm^3, K_o = 2.6 Mbar). Substantial amounts of (Mg,Fe)O, as in the pyrolite models, cannot be tolerated because of the very low K_o of (Mg,Fe)O. Thus, the methodology advocated by Ringwood, when implemented with modern data, gives support to the alternate hypothesis of a lower mantle rich in pyroxene (in the perovskite structure) with a near-chondritic Mg/Si ratio. These results, based on physical arguments, agree with the chemical model presented here. The issue is not closed since there are trade-offs between composition and temperature, and temperature corrections to physical properties are uncertain. There is certainly no reason, however, to invoke a grossly nonchondritic (major elements) lower mantle unless the recent estimates of solar composition are used for TiO_2 and CaO and the high solar Fe-content translates to a high FeO mantle content.

MAGMAS

Magmas are an important source of information about conditions in the Earth's interior. The bulk composition, trace-element chemistry, isotope geochemistry and volatile content of magmas all contain information about the source region and the processes that have affected the magmas before their eruption. Mantle fragments, or xenoliths, found in these magmas tell us about the material through which the magmas have passed on the way to the surface. Unfortunately, none of the above, even when combined with results from experimental petrology, can determine the depth of the source region, the degree of partial melting involved, or the bulk chemistry of the source region. It will prove useful to discuss the various magma types in the context of their tectonic environment. Representative compositions of various magmas are given in Table 8-6.

Three principal magma series are recognized: tholeiite, calc-alkaline and alkali. The various rock types in each series may be related by varying degrees of partial melting or crystal separation. The dominant rock type in the voluminous tholeiitic series is tholeiite, a fine-grained dark basalt containing little or no olivine. Tholeiites are found in both oceanic and intraplate settings. Those formed at midocean ridges are low-potassium and LIL-depleted and relatively high in Al_2O_3, while those found on oceanic islands and continents are generally LIL-enriched. The calc-alkaline series is characterized by andesite in island-arc volcanic terranes and granodiorite in plutonic terranes. Tholeiites and calc-alkaline basalts may be derived from a common parent and differ because of the depth of fractionation and the nature and proportions of the minerals that crystallize prior to eruption. The alkali series is characterized by alkali basalt, an olivine-bearing rock that is relatively rich in alkali elements. Tholeiites and alkali basalts are common in midplate environments. Large marginal basins have basalts similar to MORB, while smaller back-arc and fore-arc basins are more commonly calc-alkaline.

Tholeiites often contain large plagioclase crystals but seldom contain mantle fragments or xenoliths. Ridge tholeiites differ from continental and island tholeiites by their higher contents of Al_2O_3 and chromium, low contents of large-ion lithophile (LIL) elements (such as K, Rb, Cs, Sr, Ba, Zr, U, Th and REE) and low $^{87}Sr/^{86}Sr$ and $^{206}Pb/^{204}Pb$ ratios. Tholeiitic basalts have low viscosity and flow for long distances, constructing volcanic forms of large area and small slope. The major- and trace-element differences between the tholeiitic and calc-alkaline suites can be explained by varying proportions of olivine, plagioclase and pyroxene crystallizing from a basaltic parent melt.

Calc-alkaline volcanoes are commonly large, steep-sided stratovolcanoes composed of lava and fragmented material. Their eruptions are typically explosive, in contrast to the eruptions of oceanic island and continental-rift volcanoes.

Explosive volcanism is the most dramatic evidence for high temperatures in the Earth's interior. Spectacular as they are, however, the best known active volcanoes in Italy, Japan, Hawaii, Indonesia, Iceland, Alaska and Washington state represent only a small fraction of the magma that is brought to the Earth's surface. The world-encircling mid-ocean-ridge system, mostly hidden from view beneath the oceans, accounts for more than 90 percent of the material flowing out of the Earth's interior. The whole ocean floor has been renewed by this activity in less than 200 Ma. Hotspots represent less than 10 percent of the material and the heat flow.

The presence of a crust and core indicates that the Earth is a differentiated body. The age of the crust and the evidence for an ancient magnetic field indicate that differentiation was occurring prior to 3 Ga and may have been contemporaneous with the accretion of the Earth. The high concentrations of incompatible elements in the crust and the high argon-40 content of the atmosphere indicate that the differentiation has been relatively efficient. The presence of

TABLE 8-6
Average Compositions of Basalts and Andesites (Condie, 1982)

	Tholeiite					Oceanic Alkali Basalt	Cont. Rift Alkali Basalt	Andesite		
	Ridge	Arc	Continental Rift	Island	High-Al			Arc	Low-K	High-K
SiO_2	49.8	51.1	50.3	49.4	51.7	47.4	47.8	57.3	59.5	60.8
TiO_2	1.5	0.83	2.2	2.5	1.0	2.9	2.2	0.58	0.70	0.77
Al_2O_3	16.0	16.1	14.3	13.9	16.9	18.0	15.3	17.4	17.2	16.8
Fe_2O_3*	10.0	11.8	13.5	12.4	11.6	10.6	12.4	8.1	6.8	5.7
MgO	7.5	5.1	5.9	8.4	6.5	4.8	7.0	3.5	3.4	2.2
CaO	11.2	10.8	9.7	10.3	11.0	8.7	9.0	8.7	7.0	5.6
Na_2O	2.75	1.96	2.50	2.13	3.10	3.99	2.85	2.63	3.68	4.10
K_2O	0.14	0.40	0.66	0.38	0.40	1.66	1.31	0.70	1.6	3.25
Cr	300	50	160	250	40	67	400	44	56	3
Ni	100	25	85	150	25	50	100	15	18	3
Co	32	20	38	30	50	25	60	20	24	13
Rb	1	5	31	5	10	33	200	10	30	90
Cs	0.02	0.05	0.2	0.1	0.3	2	>3	~0.1	0.7	1.5
Sr	135	225	350	350	330	800	1500	215	385	620
Ba	11	50	170	100	115	500	700	100	270	400
Zr	85	60	200	125	100	330	800	90	110	170
La	3.9	3.3	33	7.2	10	17	54	3.0	12	13
Ce	12	6.7	98	26	19	50	95	7.0	24	23
Sm	3.9	2.2	8.2	4.6	4.0	5.5	9.7	2.6	2.9	4.5
Eu	1.4	0.76	2.3	1.6	1.3	1.9	3.0	1.0	1.0	1.4
Gd	5.8	4.0	8.1	5.0	4.0	6.0	8.2	4.0	3.3	4.9
Tb	1.2	0.40	1.1	0.82	0.80	0.81	2.3	1.0	0.68	1.1
Yb	4.0	1.9	4.4	1.7	2.7	1.5	1.7	2.7	1.9	3.2
U	0.10	0.15	0.4	0.18	0.2	0.75	0.5	0.4	0.7	2.2
Th	0.18	0.5	1.5	0.67	1.1	4.5	4.0	1.3	2.2	5.5
Th/U	1.8	3.3	3.8	3.7	5.9	6.0	8.0	3.2	3.1	2.5
K/Ba	105	66	32	32	12	28	16	58	49	68
K/Rb	1160	660	176	630	344	420	55	580	440	300
Rb/Sr	0.007	0.022	0.089	0.014	0.029	0.045	0.13	0.046	0.078	0.145
La/Yb	1.0	1.7	10	4.2	3.7	11	32	1.1	6.3	4.0

*Total Fe as Fe_2O_3.

helium-3 in mantle magmas shows, however, that outgassing has not been 100 percent efficient. The evidence for differentiation and a hot early Earth suggest that much of the current magmatism is a result of recycling or the processing of already processed material. The presence of helium-3 in the mantle suggests that a fraction of the magma generated remains in the mantle; that is, magma removal is inefficient.

MIDOCEAN-RIDGE BASALTS (MORB)

The floor of the ocean, under the sediments, is veneered by a layer of tholeiitic basalt that was generated at the midocean ridge systems. The pillow basalts that constitute the upper part of the oceanic crust extend to an average depth of 1 to 2 km and are underlain by sheeted dikes, gabbros and olivine-rich cumulate layers. The average composition of the crust is much more MgO-rich and lower in CaO and Al_2O_3 than the surface MORB, if ophiolites can be used as a guide. The crust rests on a depleted harzburgite layer of unknown thickness. In certain models of crustal genesis, the harzburgite layer is complementary to the crust and is therefore about 24 km thick if an average crustal thickness of 6 km and 20 percent melting are assumed. This is just a model, and it is not clear that harzburgites are genetically related to the basalts. Seismic data are consistent with a harzburgite layer in the shallow mantle, but its thickness has not been determined. Oceanic plateaus and aseismic ridges have anomalously thick crust (>30 km), and the corresponding depleted layer would be more than 120 km thick if the simple model is taken at face value. Since depleted peridotites, including harzburgites, are less dense than

Al_2O_3-rich or fertile peridotite, or eclogite, several cycles of plate tectonics, crust generation and subduction, would fill up the shallow mantle with harzburgite. Oceanic crust may in fact be deposited on ancient, not contemporaneous, harzburgite.

Obviously, midocean-ridge basalts result from the partial or complete melting of something with a basaltic component. The composition of the source region and the degree of partial melting are unknown. The smaller the degree of melting, the larger is the problem with excess residual peridotite. Also unknown are the amount and nature of phases that have crystallized from the parent material prior to the eruption of MORB. Some olivine fractionation is commonly assumed, but garnet and pyroxene crystallization may be involved if the parent magma is trapped beneath the lithosphere. At shallow depths the crystallization of plagioclase can extensively modify the composition of the remaining melt. In general, the crystallization of refractory phases increases the content of potassium, rubidium and other incompatible elements in the melt. These questions are explored in more detail in Chapter 9. Note the similarity of tholeiites from various tectonic environments as far as the major elements are concerned but the depletion of ridge tholeiites in the LIL elements (Table 8-6).

Models of MORB genesis fall into two categories. In the first, the most primitive MORB samples are considered to represent "primary" magmas formed by partial melting of the mantle beneath the midocean ridges and little modified by crystal fractionation or other processes once they segregated from this source region. In the second category it is maintained that MORB is not primary; it evolved from a parent magma by substantial amounts of fractionation, principally of olivine and perhaps other phases as well, at pressures less than those at which the primary magma separated from the mantle source region. Primary magmas would be picritic, or rich in an olivine component, and may contain about 20 percent MgO. If the parent magma ascends from great depth and is allowed to fractionate at depths of the order of 50 km or greater, it may have precipitated garnet and clinopyroxene as well as olivine.

In most models of MORB genesis, it is assumed not only that olivine and orthopyroxene are contained in the source rock but that these are the dominant phases. Petrological data alone, however, cannot rule out a source that is mainly garnet and clinopyroxene. Primitive mantle and samples from the shallow mantle contain abundant olivine and orthopyroxene. However, MORB are extremely depleted in the incompatible elements, and isotopic ratios show that the depletion of the MORB source occurred early in Earth history. Therefore, the MORB source is not primitive, nor has it been mixed with much primitive or enriched material. It may represent a deep garnet-clinopyroxene-rich cumulate layer or a peridotite layer that experienced a small amount of partial melting and melt removal. The melting in the latter case could not have been too extensive since the source is still fertile, that is, it contains clinopyroxene and garnet, the low-melting components.

The eclogite (garnet plus clinopyroxene) or olivine eclogite source hypothesis differs only in scale and melt extraction mechanism from the fertile peridotite hypothesis. In the fertile peridotite model the early melting components, garnet and clinopyroxene, are distributed as grains in a predominantly olivine-orthopyroxene rock. On a larger scale, eclogite might exist as pods or blobs in a peridotite mantle. Since eclogite is denser than peridotite, these blobs would tend to sink and coalesce in a hot, low-viscosity mantle. In the extreme case an isolated eclogite-rich layer might form below the lighter peridotite layer. Such a layer could also form by subduction or by crystal settling in an ancient magma ocean. The isotopic evidence for isolated reservoirs and the geophysical evidence for gross layering suggest that differentiation and chemical stratification may be more important in the long run than mixing and homogenization. This is not to deny that magma mixing occurs in the processes leading up to eruption.

The important geophysical questions regarding the MORB source region and MORB genesis are (1) depth of source, (2) amount of olivine and orthopyroxene in the source, (3) the extent of melting, (4) the depth of melt-crystal separation (or final equilibration), (5) the amount of crystal fractionation subsequent to melt separation from the source but prior to eruption, and (6) the amount of contamination. The answers to most of these questions are assumed in many models of basalt genesis. It is also generally assumed that the upper mantle is homogeneous and that non-MORB magmas must originate elsewhere. In fact, the upper mantle is neither radially nor laterally homogeneous.

HOTSPOTS

Most of the world's visible volcanoes occur at plate boundaries and are associated with plate creation and consumption. There are about 50 other volcanic centers, scattered about the globe, that are either remote from plate boundaries or that are anomalous in the volume or chemistry of eruptive material (Figure 8-4). The numerous linear island chains in the central Pacific are perhaps the most dramatic manifestations of long-lived hotspot activity. Hawaii and Tahiti are examples of hotspots in the Pacific; Iceland, the Azores, Tristan de Cunha and the Canaries are Atlantic examples. Réunion and Kerguelen are hotspots in the Indian Ocean, and tiny Bouvet Island is in the Southern Ocean.

J. Tuzo Wilson (1963) and W. Jason Morgan (1971) proposed an important additional element to plate tectonics. The concepts of seafloor spreading and continental drift leave unexplained many aspects of seafloor and continental geology and geochemistry; not the least of these are the driving forces for plate tectonics and the requirement for

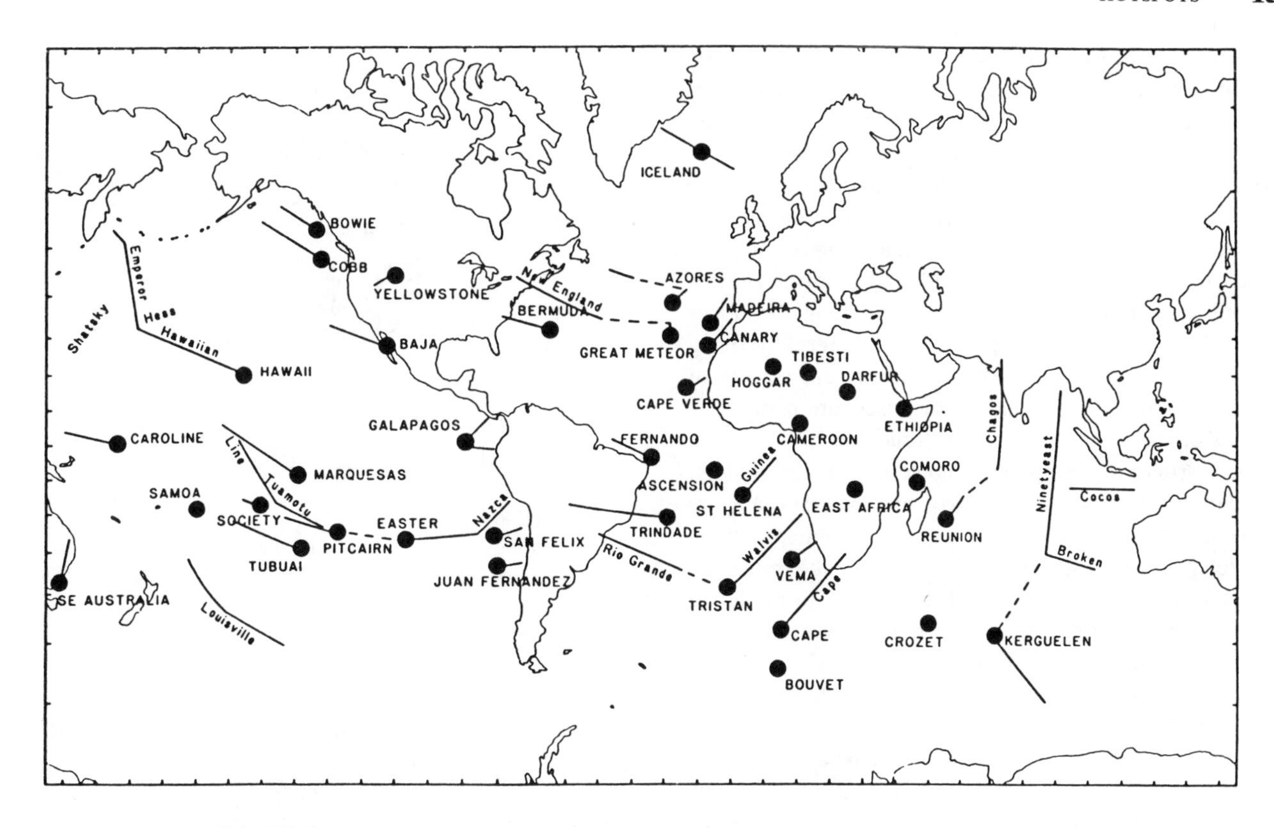

FIGURE 8-4
Global distribution of hotspots and hotspot tracks (after Crough, 1983).

multiple magma sources in the mantle. Wilson proposed that hotspots, or melting spots in the mantle, could be used as absolute reference points. Morgan (1982) used hotspots as fixed reference points in his continental reconstructions to about 200 million years ago. Morgan (1972) further proposed that hotspots had a deep-mantle, thermal, origin and that they drove plate motions. He interpreted hotspots in terms of thermal convection, a concentrated upwelling of hot material from the deep mantle.

The Wilson-Morgan hypothesis attributes linear volcanic chains to thermal plumes that originate deep in the mantle. A relatively fixed mantle source of heat seems to be required by the age progression data, but a deep mantle source for the material is not required by either the geochemical or geophysical data. For this reason others have proposed that localities such as Hawaii and Iceland be termed "hotspots" or "melting spots" rather than "plumes," a term that has deep mantle implications. Alternative mechanisms include propagating fractures, shear melting, bumps on the asthenosphere and gravitational anchors. These hypotheses do not directly address the problem of why melts, tears or bumps occur in the first place.

In 1975 I introduced the concept of "chemical plumes" and proposed that plumes differ in intrinsic chemistry from ordinary mantle, rather than being just thermal perturbations (Anderson, 1975). Chemical inhomogeneity of the lower mantle, for example, could be caused by upwelling of material from a chemically distinct layer at the base of the mantle, D″, which is also the location of the mantle-core thermal boundary layer. If this material is intrinsically denser than normal lower mantle it would rise only when hot, much in the manner of "lava lamps" where hot wax blobs rise through oil, cool and then sink. Even if lower-mantle material cannot cross into the upper mantle, the locations of hot regions may fix the locations of melting anomalies in the upper mantle. The composition of lower mantle plumes, thermal or chemical, may be unrelated to hotspot magmas. If the upper mantle is chemically stratified, upwellings from the deeper layer may contribute material different in chemistry from melts from the shallow mantle. Plumes in the upper mantle could originate near or above the 650-km discontinuity or where melting is most extensive.

There is now abundant evidence that hotspots provide magmas that are chemically distinct from MORB. The thermal plume hypothesis must therefore be abandoned. Hotspot magmas, however, are similar to MORB in major elements, and it may not be necessary to invoke a distinct reservoir (in bulk chemistry or location). The trace-element and isotopic characteristics of hotspot magmas may have been acquired during ascent, or during evolution in the shallow mantle.

The concept of relatively fixed deep mantle plumes was initially proposed to explain the age progression of linear volcanic chains in the Atlantic and Pacific. Before the introduction of the plume concept, many geophysicists believed that convection in the mantle should be restricted to the low-velocity layer or asthenosphere. This seemed reasonable since the combined effects of temperature and pressure on melting and viscosity would lead to lower viscosities and greater degrees of partial melting in the upper mantle. In this view, most of the material presently rising to the Earth's surface is recycled, and present centers of volcanic activity, such as midoceanic rises, are passive reflections of where cracks form. If "fundamental" volcanic centers, variously called plumes, hotspots or melting spots, are relatively motionless or slowly moving, then a more profound fixed source of heat or melting is implied.

Hotspots appear to define a relatively stable reference frame that is sometimes assumed to be fixed to the lower mantle. However, if the upper mantle and lower mantle are convecting, either as a single system or separately, hotspots should wander relative to one another. The return flow associated with plate tectonics would be expected to disrupt the pattern or, at least, move cells in one hemisphere relative to another. However, this motion is expected to be much smaller than plate velocities. For example, if the slabs are 50 km thick and the return flow occurs over the upper 500 km of the mantle, the relative motions of hotspots will be only one-tenth of the relative motions of the plates. This is consistent with observations. Return flow may also be below the upper-mantle convection cells, down in the transition zone or the lower mantle.

Although it is generally agreed that the lower mantle is convecting, it is likely to be a much more sluggish convection than occurs in the upper mantle. Convection rates depend on viscosity and temperature contrasts. Pressure serves to increase viscosity. High stress decreases viscosity. Near-surface temperatures range from near the melting point of rocks to below 0°C. Such extremes cannot occur in the deeper mantle. The transition zone or the lower mantle is therefore a better candidate than the lithosphere, the spin axis or the liquid core for a stable, or at least slowly varying reference system. A convecting lower mantle will have relatively stable uprising and descending limbs, which may show up in the geoid. Convection deforms the core-mantle and the upper mantle–lower mantle boundaries. This deformation will influence convection in the core and the upper mantle, even if no material transfer takes place. For example, hotspots may be preferentially located over the hot upcurrents in the lower mantle, and then hotspots would indeed appear fixed to the lower mantle even if hotspot magmas originate in the upper mantle. On the other side of the system, downwellings in the core occur preferentially under colder parts of the lower mantle, these being where heat is most effectively removed.

Both continental and oceanic geology give ample evidence of uplift and subsidence throughout the recognizable geologic record. The concepts of mobile plates combined with hotspots or mantle upwellings give a natural explanation for these phenomena. A plume, either thermal or chemical, will generate "bumps" at the base of the upper mantle, the asthenosphere or the lithosphere that can contribute to uplift and, ultimately, fracture and magmatism in the overlying lithosphere. Crustal uplift may be due to lithosphere overriding mantle upwellings. Thermal expansion, partial melting and upwelling all contribute to lithospheric uplift and extension.

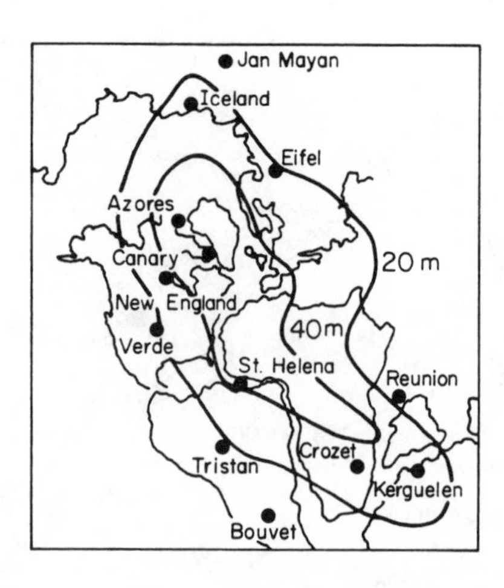

FIGURE 8-5
Distribution of hotspots relative to Pangaea and the Atlantic-Africa geoid high (contours) at 200 Ma (after Anderson, 1982c).

The distribution of hotspots on the surface of the Earth is not random. Most hotspots occur in the long-wavelength geoid highs centered in the central Pacific and Africa (Figure 8-5); few occur in geoid lows. These geoid highs correlate well with low seismic velocity regions in the lower mantle (Chapter 15). In a general sense the broad-scale distribution of hotspots seems to be controlled by hotter than average regions in the lower mantle. This does not imply that material is actually transferred from the lower mantle to the surface. The presumed high viscosity of the lower mantle makes convection sluggish, and hot areas of the lower mantle appear more permanent than surface features. Hot areas in the lower mantle may be due to upwellings from the thermal boundary layer at the core-mantle boundary, region D″. This, in turn, may be chemically distinct from most of the lower mantle.

The hotspot and plume concepts were developed from geometric and kinematic observations involving the relative fixity and large volume of certain melting spots. It later became evident that these hotspots differ in trace-element and isotopic chemistry from ocean-ridge volcanism. Hotspot magmas are richer in volatiles and LIL, and have high $^{87}Sr/^{86}Sr$, $^{144}Nd/^{143}Nd$, $^{206}Pb/^{204}Pb$, $^{207}Pb/^{204}Pb$ and $^{208}Pb/^{204}Pb$ ratios and, in some cases, high $^{3}He/^{4}He$ ratios. Their source is therefore chemically distinct, and has been for some time, from that providing the more voluminous MORB.

Hotspot magmas, including tholeiites and alkali basalts, are intermediate in LIL and isotope chemistry between MORB, the most depleted magmas, and kimberlites, in many ways the most enriched magmas. The bulk chemistry of ocean-island and continental flood basalts, two manifestations of hotspot magmatism, are similar to midocean-ridge basalts (Table 8-6), suggesting that the geochemical signatures of hotspot magmas may reflect contamination of MORB, or a magma mixture that may involve only a small fraction of an enriched component, such as Q. Although there are subtle chemical and isotopic differences between ocean-island (hotspot) and ocean-ridge basalts, most of the material coming out of both regions probably originates in the upper mantle, some or all of which has possibly been recycled. The dilution of plume material by "normal" upper-mantle melts (MORB) makes it difficult to unravel the chemistry of plumes.

Large oceanic islands, such as Iceland, Hawaii, Kerguelen and Tahiti, smaller islands such as Tristan da Cunha, St. Helena, Canaries and Ascension, and continental flood basalt and alkali basalt provinces are considered to be manifestations of hotspot volcanism. Carbonatites and kimberlites may be early continental manifestations of hotspot activity. Carbonatites and kimberlites are exotic assemblages that apparently result from anomalous chemical or thermal conditions of the upper mantle. The spatial, petrologic and geochemical relationship between kimberlites and carbonatites is well established. These suites commonly occur in stable continental areas and form trends from continent to continent. Carbonatites or related rock types occur on many of the proposed Atlantic and Indian Ocean hotspots or on their tracks. In several cases continental carbonatites, kimberlites or alkali-rich provinces abut proposed hotspots when the continents are reconstructed. The many geochemically similar seamounts on the ocean floor and evidence for simultaneity of magmatism over large areas, once thought to be hotspot tracks, makes the concept of a relatively small number of fixed deep mantle plumes less attractive than when originally proposed. Island-arc and back-arc basin basalts have many of the same geochemical properties as do hotspots, suggesting that the distinctive geochemical signatures of hotspot magmas may be acquired at shallow depth. In these regions a cold slab intervenes between the source region and the deep mantle.

The similarity in major-element chemistry of ocean floor, ocean island and continental rift basalts, or their inferred parents, suggests that all may share a common source, differing only in their trace elements, isotopes and cooling/fractionation history. For example, hotspot magmas may be contaminated MORB, intruded at depth and cooled prior to eruption. In fact, most basalts can be treated as mixtures of MORB and an enriched component.

An important and controversial problem in geochemistry and petrology is the source of the hotspot reservoir. This is actually three separate problems, usually treated as one:

1. The source of the thermal anomaly.
2. The source of the magma.
3. The source of the distinctive trace-element, volatile and isotopic signature.

That these are not necessarily the same question can be seen from the following possible scenario: Convection currents in the lower mantle provide nonuniform heating to the base of the upper mantle. If the mantle is chemically stratified, no mass transfer takes place from the lower to the upper mantle. This nonuniform heating localizes sites of upwelling and melting in the transition region, which provides depleted MORB-type magmas. These magmas rise through the LIL- and volatile-rich peridotitic shallow mantle and pick up the geochemical signature of the enriched low-velocity zone or lower lithosphere. Enriched alkali or plume-type magmas form by fractionation of this hybrid magma at sublithospheric depths. Ancient magmatic events probably involved a component of crustal and lithospheric underplating and intrusion as well as extrusion. Modern magmatism through these regions will acquire a hotspot geochemical signature even if the initial melt is MORB, particularly the early stages.

Ocean-Island Basalts

Oceanic islands are composed of both alkali and tholeiitic basalts. Alkali basalts are subordinate, but they appear to dominate the early and waning stages of volcanism. Ocean-island basalts, or OIB, are LIL-rich and have enriched isotopic ratios relative to MORB. Their source region is therefore enriched, or depleted parent magmas suffer contamination en route to the surface. The larger oceanic islands such as Iceland and Hawaii generally have less enriched magmas than smaller islands and seamounts. A notable exception is Kerguelen, which has very high $^{87}Sr/^{86}Sr$ ratios and low $^{143}Nd/^{144}Nd$ ratios (Dosso and Murthy, 1980).

The newest submarine mountain in the Hawaii chain, Loihi, is alkalic. Intermediate age islands are tholeiitic, and the latest stage of volcanism is again alkalic. The volcanism in the Canary Islands in the Atlantic also changes from alkalic to tholeiitic. The sequence suggests that the hotspot over which the islands are progressing is supplying voluminous tholeiitic magmas and the alkalics are being provided on the edges, perhaps from the wings of a magma chamber.

Trace-element and isotopic patterns of OIB overlap continental flood basalts (CFB), and a common source region and fractionation pattern can be inferred. Island-arc basalts (IAB) also share many of the same geochemical characteristics, suggesting that the enriched character of these basalts may be derived from the characteristics of the shallow mantle. This is not the usual interpretation. The geochemistry of CFB is usually attributed to continental contamination, that of IAB to sediment and subducted crust involvement, and that of OIB to "primitive" lower mantle. In all three cases the basalts have evolved beneath thick crust or lithosphere, giving more opportunity for crystal fractionation and contamination by crust, lithosphere or shallow mantle (the low-velocity zone) prior to eruption.

Kimberlite

Kimberlite is a rare igneous rock that is volumetrically insignificant compared to other igneous rocks such as basalt. Kimberlite provinces themselves, however, cover very broad areas and occur in most of the world's stable craton, or shield, areas. Kimberlites are best known as the source rock for diamonds, which must have crystallized at pressures greater than about 50 kilobars. They carry other samples from the upper mantle that are the only direct samples of mantle material below about 100 km. Some kimberlites appear to have exploded from depths as great as 200 km or more, ripping off samples of the upper mantle and lower crust in transit. The fragments, or xenoliths, provide samples unavailable in any other way. Kimberlite itself is an important rock type that provides important clues as to the evolution of the mantle. It contains high concentrations of lithophile elements (Table 8-7) and higher concentrations of the incompatible trace elements than any other ultrabasic rock. Although rare, kimberlites or related rocks have been found on most continents including Africa, Asia, Europe, North America, South America, Australia, India and Greenland. Most kimberlites are found in ancient cratons and relatively few are found in circumcratonic foldbelts. The most important areas for diamond production are in South Africa, Siberia and northwest Australia, although diamonds are also found in North America and Brazil. Diamond-bearing kimberlites are usually close to the craton's core, where the lithosphere may be thickest. Barren kimberlites are usually on the edges of the tectonically stable areas. Kimberlites range in age from Precambrian to Cretaceous. Some areas have been subjected to kimberlite intrusion over long periods of geological time.

TABLE 8-7
Kimberlite Composition Compared with Ultrabasic and Ultramafic Rocks

Oxide	Average Kimberlite	Average Ultrabasic	Average Ultramafic
SiO_2	35.2	40.6	43.4
TiO_2	2.32	0.05	0.13
Al_2O_3	4.4	0.85	2.70
Fe_2O_3	6.8	—	—
FeO	2.7	12.6	8.34
MnO	0.11	0.19	0.13
MgO	27.9	42.9	41.1
CaO	7.6	1.0	3.8
Na_2O	0.32	0.77	0.3
K_2O	0.98	0.04	0.06
H_2O	7.4	—	—
CO_2	3.3	0.04	—
P_2O_5	0.7	0.04	0.05

Wederpohl and Muramatsu (1979).
Dawson (1980).

Compared to other ultrabasic rocks such as lherzolite, dunite or harzburgite, kimberlites contain unusually high amounts of K_2O, Al_2O_3, TiO_2, FeO, CaO, CO_2, P_2O_5 and H_2O. For most incompatible trace elements kimberlites are the most enriched rock type; important exceptions are elements that are retained by garnet and clinopyroxene.

Since kimberlites are extremely enriched in the incompatible elements, they are important in discussions of the trace-element inventory of the Earth's mantle. Such extreme enrichment implies that kimberlites represent a small degree of partial melting of a mantle silicate or a late-stage residual fluid of a crystallizing cumulate layer. The LIL elements in kimberlite show that it has been in equilibrium with a garnet-clinopyroxene-rich source region, possibly an eclogite cumulate. The LIL contents of kimberlite and MORB are complementary. Removal of a kimberlite-like fluid from an eclogite cumulate gives a crystalline residue that has the required geochemical characteristics of the depleted source region that provides MORB.

Carbonatites and other ultramafic alkaline rocks, closely related to kimberlites, are widespread. In spite of their rarity, kimberlites provide us with a sample of magma that probably originated below about 200 km and, as such, contain information about the chemistry and mineralogy of the mantle in and below the continental lithosphere. Kimberlites are anomalous with respect to other trace-element-enriched magmas, such as nephelinites and alkali basaslts, in their trace-element chemistry. They are enriched in the very incompatible elements such as rubidium, thorium and LREE, consistent with their representing a small degree of partial melting or the final concentrate of a crystallizing liquid. However, they are relatively depleted in the elements (Sc, Ti, V, Mn, Zn, Y, Sr and the HREE) that are retained by garnet and clinopyroxene. They are also low in silicon and aluminum, as well as other elements (Na, Ga, Ge) that are geochemically coherent with silicon and aluminum. This suggests that kimberlite fluid has been in equilibrium with an eclogite residue. Kimberlites are also rich in cobalt and nickel.

Despite their comparative rarity, disproportionately high numbers of eclogite xenoliths have been found to contain diamonds. Diamond is extremely rare in peridotitic xenoliths. Eclogitic garnets inside diamonds imply a depth of origin of about 200–300 km if mantle temperatures in this depth range are of the order of 1400–1600°C (Cohen and Rosenfeld, 1979). Seismic velocities between 400 and 670 km depth in the mantle are consistent with picritic eclogite (Bass and Anderson, 1984). The bulk modulus in the transition zone for olivine in its spinel forms is higher than is consistent with the seismic data. Eclogites may originate in the transition region or they may be subducted oceanic crust.

It is obvious that if a fluid similar to kimberlite is the enriching or metasomatic fluid that is responsible for the trace-element pattern of alkali basalts, and other enriched continental and ocean-island magmas, then these magmas will appear to have been in equilibrium with a garnet-rich residue, even if no garnet is present in their immediate parent reservoir. That is to say, a magma with a garnet-equilibrium signature in its trace-element pattern, for instance HREE depletion, does not necessarily imply that it is a result of small degrees of partial melting of a garnet-bearing parent, leaving behind a garnet-rich residue. The signature could be inherited from the parent reservoir of the enriching fluid, and probably has been if kimberlitic fluids are the enriching fluids. Similar comments apply if the immediate parent of continental basalts had experienced a prior episode of eclogite extraction, as in a deep magma chamber or a deep magma ocean.

Alkali basalts have LIL concentrations that are intermediate to MORB and kimberlite. Although kimberlite pipes are rare, there may be a kimberlite-like component (Q) dispersed throughout the shallow mantle. Indeed, alkali basalts can be modeled as mixtures of a depleted magma (MORB) and an enriched magma (kimberlite), as shown in Table 8-8. Peridotites with evidence of secondary enrichment may also contain a kimberlite-like component.

Kimberlites commonly contain calcium zeolites and anorthite and contain higher concentrations of such refractory trace elements as barium, strontium, lanthanum, zirconium and niobium than would be expected from rocks originating in normal mantle. There are many other minor elements in excess (including Li, B, Ca, Ga, Rb, Cs, Ta and Pb). Xenoliths in kimberlites commonly contain rutile, kyanite, garnet, corundum, spinel, ilmenite, perovskite, nepheline, anatase, strontianite and barite and are sometimes associated with flows of olivine basalt unusually rich in barium, titanium, zirconium, niobium, P_2O_5 and H_2O. The $^{87}Sr/^{86}Sr$ ratio for fresh kimberlites is often high. A rock often associated with kimberlite is melilite basalt. There is a complete chemical gradation between kimberlites and carbonatites; both are enriched in barium, phosphorus, lanthanum and niobium. In addition carbonatites are com-

TABLE 8-8
Trace-Element Chemistry of MORB, Kimberlite and Intermediate (Alkali) Basalts (ppm)

		Alkali Basalts					Kimberlite	
Element	MORB	EPR*	Australia	Hawaii	BCR†	Theoretical‡	Average	Max
Sc	37	27	25	33	34–37	35	30	—
V	210	297	260	170	399	197–214	120	250
Co	53	41	60	56	38	57–61	77	130
Ni	152	113	220	364	15	224–297	1050	1600
Rb	0.36	13	7	24	47	10–45	65	444
Sr	110	354	590	543	330	200–289	707	1900
Y	23	37	20	27	37	23–28	22	75
Zr	70	316	111	152	90	97–133	250	700
Nb	3.3	—	24	24	14	19–48	110	450
Ba	5	303	390	350	700	150–580	1000	5740
La	1.38	21	24	23	26	16–31	150	302
Ce	5.2	46	47	49	54	31–57	200	522
Nd	5.6	28	24	23	29	16–26	85	208
Sm	2.1	6.7	6.8	5.5	6.7	3.1–4.8	13	29
Eu	0.8	2.1	2.1	2.0	2.0	1.1–1.4	3	6.5
Tb	0.5	1.2	0.92	0.87	1.0	0.6–0.7	1	2.1
Yb	2.1	3.4	1.6	1.8	3.4	2.0–2.1	1.2	2.0
Lu	0.34	0.49	0.28	0.23	0.55	0.32–0.34	0.16	0.26
Hf	1.4	5.3	2.8	3.9	4.7	2.2–4.3	7	30
Ta	0.1	9	1.8	2.1	0.9	1.3–2.5	9	24
Pb	0.08	—	4.0	5	18	1.6–5.1	10	50
Th	0.35	2.7	2.0	2.9	6.8	2.4–5.4	16	54
K	660	8900	4700	8800	1200	3200–4770	17,600	41,800
U	0.014	0.51	0.4	0.6	1.8	0.48–1.84	3.1	18.3

*EPR = East Pacific Rise.
†BCR = Basalt, Columbia River (USGS Standard Rock).
‡Mixture ranging from 85 percent MORB plus 15 percent average kimberlite to 90 percent MORB plus 20 percent maximum kimberlite.

monly enriched in uranium, thorium, molybdenum and tungsten, and kimberlites are enriched in boron, vanadium and tantalum.

Alkali ultrabasic rocks in ring complexes usually occur in stable or fractured continental regions; they contain nepheline, melilite-rich rocks, carbonatites, and kimberlite dikes. Kimberlites are mostly confined to interiors and margins of stable continental areas. The distribution appears to be related to deep-seated tectonics with linear trends. There appear to be genetic links between kimberlites, carbonatites and alkalic ultrabasic complexes.

The intrusion of kimberlites takes place, during uplift of platform areas, along deep-seated fractures either bounding or cutting across the uplifted areas. In South Africa, Brazil and the USSR, kimberlite eruptions closely followed widespread extrusions of flood basalt.

Extensive magmatic and metamorphic events in continental interiors may occur when sublithospheric heat sources are focused for long periods of time, many millions of years, on the same part of the lithosphere. The localization of this effect, rather than being a continent-wide phenomenon, also suggests a rather localized source. During the past 800 Ma Africa has been the site of three main thermal episodes. In some cases uplift and fracturing preceded volcanism.

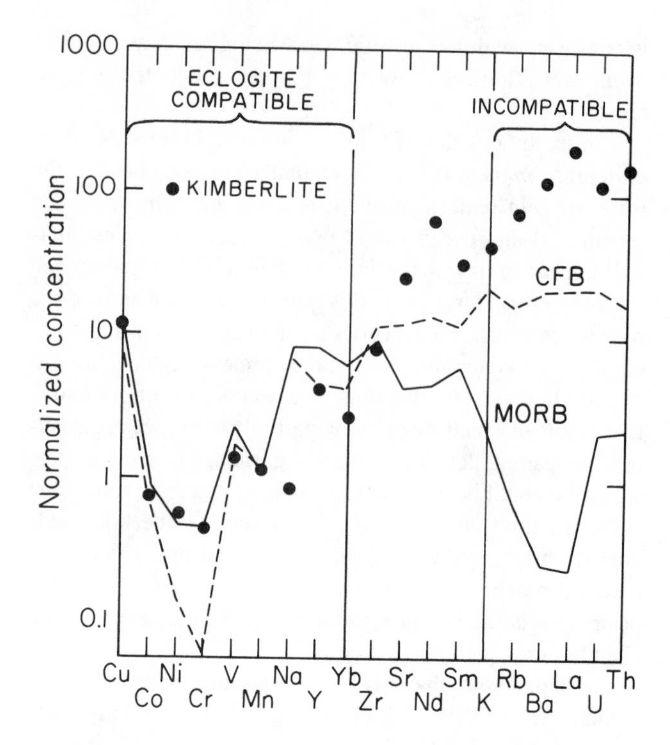

FIGURE 8-6
Trace-element concentrations in MORB, continental flood basalts (CFB) and kimberlites. The elements to the right are incompatible in all major mantle phases (olivine, pyroxene and garnet) while those to the left are retained in the eclogite minerals (clinopyroxene and garnet). Note the complementary pattern between MORB and kimberlite and the intermediate position of CFB. Concentrations are normalized to estimates of mantle composition.

The Kimberlite-MORB Connection

Kimberlites are enriched in the LIL elements, especially those that are most depleted in MORB. Figure 8-6 gives the composition of kimberlites, MORB and continental tholeiites. The complementary pattern of kimberlite and MORB is well illustrated as is the intermediate position of continental tholeiites (CFB). Note that kimberlite is not enriched in the elements that are retained by the eclogite minerals, garnet and clinopyroxene. This is consistent with kimberlite having been a fluid in equilibrium with subducted oceanic crust or an eclogite cumulate. If a residual cumulate is the MORB reservoir, the ratio of an incompatible element in kimberlite relative to the same element in MORB should be the same as the liquid/solid partition coefficient for that element. This is illustrated in Figure 8-7. The solid line is a profile of the MORB/kimberlite ratio for a series of incompatible elements. The vertical lines bracket the solid/liquid partition coefficients for garnet and clinopyroxene. Although MORB is generally regarded as an LIL-depleted magma and kimberlite is ultra-enriched in most of the incompatible elements, MORB is enriched relative to KIMB in yttrium, ytterbium and scandium, elements that have a high solid/melt partition coefficient for an eclogite residue. The trend of the MORB/KIMB ratio is the same as the partition coefficients, giving credence to the idea that kimberlite and MORB are genetically related. MORB is now generally regarded as descended from a more picritic magma by olivine fractionation. The dashed line shows the LIL ratios of MORB plus 50 percent olivine, relative to KIMB. Its fit with the partition coefficients indicates that the MORB parent magma could have lost up to about 50 percent olivine, presumably in shallow upper-mantle magma chambers.

These results support the joint hypotheses that (1) the MORB reservoir is an eclogite-rich cumulate and (2) kimberlite was in equilibrium with eclogite and could be the late-stage melt of a deep crystallizing cumulate layer. A kimberlite-like fluid may be partially responsible for the LIL enrichment of the shallow mantle as well as the depletion of the MORB reservoir. An eclogite-rich cumulate could be the result of crystallization of a magma ocean. Such crystallization would tend to separate olivine and orthopyroxene, the shallow-mantle minerals, from garnet and clinopyroxene.

The LIL content of continental tholeiites and alkali-olivine basalts suggest that they are mixtures of MORB and a melt from a more enriched part of the mantle. If diapirs initiate in a deep depleted layer, they must traverse the shallow mantle during ascent, and cross-contamination seems unavoidable. The more usual model is that the whole upper

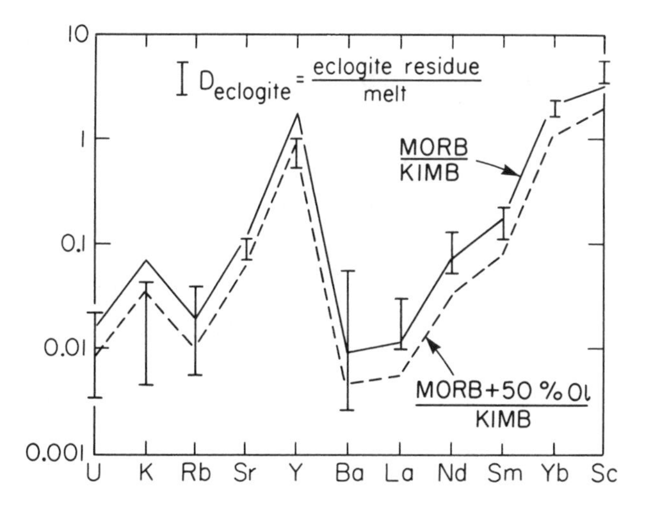

FIGURE 8-7
Solid line is ratio of concentrations in MORB and kimberlite (KIMB). Vertical bars are solid/liquid partition coefficients for garnet plus clinopyroxene. The dashed line gives the ratio for MORB plus 50 percent olivine, a possible picrite parent magma for MORB. If the MORB or picrite source region is the crystalline residue remaining after removal of a kimberlitic fluid, the ratio of concentrations, MORB/KIMB or picrite/KIMB, should equal the solid/liquid partition coefficient D_{xl}, where D_{xl} depends on the crystalline (xl) phases in the residue. D_{ecl} is the solid/liquid partition coefficient for an eclogite residue (after Anderson, 1983b).

mantle is homogeneous and provides depleted (MORB) magmas. Enriched magmas, in this scenario, must come from the lower mantle or from the crust or continental lithosphere. The alternative is that the upper mantle is both radially and laterally inhomogeneous, an alternative adopted here. It should be stressed that there is no evidence that any magma or magma component originates in the lower mantle. This applies to both direct and indirect, or mass balance, evidence.

The Kimberlite-KREEP Relation

KREEP is a lunar material having very high concentrations of incompatible elements (K, REE, P, U, Th) and so on (see Chapter 2). It is thought to represent the residual liquid of a crystallizing magma ocean (Warren and Wasson, 1979). Given proposals of a similar origin for kimberlite (Anderson, 1982a, 1982b), it is of interest to compare the composition of these two materials. An element by element comparison gives the remarkable result that for many elements (K, Cs, P, S, Fe, Ca, Ti, Nb, Ta, Th, U, Ba and the LREE) kimberlite is almost identical (within 40 percent) to the composition of KREEP (Figure 8-8 and 8-9). This list includes compatible and incompatible elements, major, minor and trace elements, and volatiles as well as refractories. KREEP is relatively depleted in strontium and europium, elements that have been removed from the KREEP source region by plagioclase fractionation. Kimberlite is depleted in sodium, HREE, hafnium, zirconium and yttrium, elements that are removed by garnet-plus-clinopyroxene fractionation. It appears that the differences between KIMB and KREEP are due to difference in pressure between the Earth and the Moon; garnet is not stable at pressures occurring in the lunar mantle. The plagioclase and garnet signatures show through such effects as the different volatile-refractory ratios of the two bodies and expected differences in degrees of partial melting, extent of fractional crystallization and other features. The similarity in composition extends to metals of varying geochemical properties and volatilities such as iron, chromium, manganese as well as phosphorus. KREEP is depleted in other metals (V, Co, Ni, Cu and Zn), the greater depletions occurring in the more volatile metals and the metals that are partitioned strongly into olivine. This suggests that olivine has been more important in the evolution of KREEP than in the evolution of kimberlite, or that cobalt, copper and nickel are more effectively partitioned into MgO-rich fluid such as kimberlite.

It appears from Figure 8-8 that KREEP and kimberlite differ from chondritic abundances in similar ways. Both are depleted in volatiles relative to refractories, presumably due to preaccretional processes. Strontium is less enriched than the other refractories, although this is much more pronounced for KREEP. The pronounced europium and strontium anomalies for KREEP are consistent with extensive plagioclase removal. The HREE, yttrium, zirconium and hafnium are relatively depleted in kimberlite, suggesting eclogite fractionation or a garnet-rich source region for kimberlite. The depletion of scandium simply indicates that olivine, pyroxene or garnet have been in contact with both KREEP and KIMB. The depletion of sodium in KIMB is also consistent with the involvement of eclogite in its history. The depletion of sodium in KREEP, relative to the other alkalis, is presumably due to the removal of feldspar, and the greater relative depletion of sodium in kimberlite therefore requires another process, such as removal of a jadeite component by eclogite fractionation.

Chromium, manganese and iron are approximately one-third the chondritic level in both KREEP and kimberlite. This is about the level in the mantles of these bodies. Nickel and cobalt in kimberlite are about the same as estimated for the Earth's mantle. These elements are extremely depleted in KREEP, indicating that they have been removed by olivine or iron extraction from the source region.

The similarity of kimberlite and KREEP is shown in Figure 8-9. For many elements (such as Ca, Ba, Nd, Eu, Nb, Th, U, Ti, Li and P) the concentrations are identical within 50 percent. Kimberlite is enriched in the volatiles rubidium, potassium and sulfur, reflecting the higher volatile content of the Earth. Kimberlite is also relatively enriched in strontium and europium, consistent with a prior extraction of plagioclase from the KREEP source region. Kimberlite is depleted in the HREEs and sodium, consistent

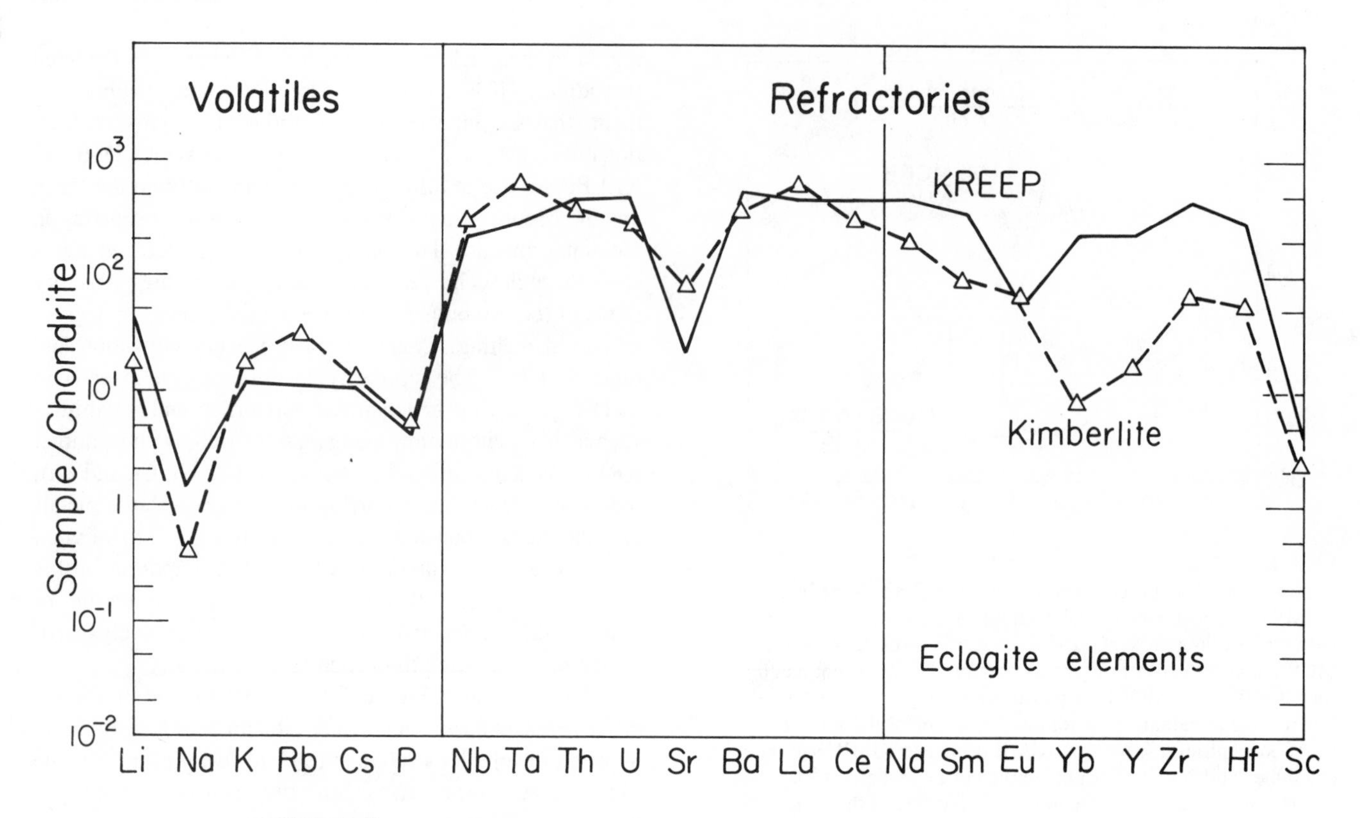

FIGURE 8-8
Chondrite-normalized trace-element compositions for kimberlite and lunar KREEP. The strontium and europium anomalies for KREEP are due to plagioclase extraction from the lunar source region. The relative depletion of kimberlite in the eclogite elements and sodium may reflect eclogite and jadeite extraction from the terrestrial source region. Lithium is plotted with the alkali metals although it is a refractory.

with an eclogite source region for kimberlite. This difference in controlling mineralogy is simply a reflection of the greater pressures in the Earth.

Kimberlite and lunar KREEP are remarkably similar in their minor- and trace-element chemistry. The main differences can be attributed to plagioclase fractionation in the case of KREEP and eclogite fractionation in the case of kimberlite. KREEP has been interpreted as the residual fluid of a crystallizing magma ocean. In a small body the Al_2O_3 content of a crystallizing melt is reduced by plagioclase crystallization and flotation. In a magma ocean on a large body, such as the Earth, the Al_2O_3 is removed by the sinking garnet. Kimberlite, in fact, is depleted in eclogite elements including the HREE and sodium. This suggests that kimberlite may represent the late-stage residual fluid of a crystallizing terrestrial magma ocean. A buried eclogite-rich cumulate layer is the terrestrial equivalent of the lunar anorthositic crust.

We have established that removal of a kimberlite-like fluid from a garnet-clinopyroxene-rich source region gives a crystalline residue that has the appropriate trace-element chemistry to be the reservoir for LIL-depleted magmas such as MORB. It is usually assumed that the continental crust serves the role that we attribute to kimberlite, and in fact the crust may also be involved. I propose, however, that enriched fluid permeates the shallow mantle and is responsible for the LIL-enrichment of island-arc and oceanic island basalts. Accretion of arcs, islands and plateaus would be the mechanism for continental growth. The continental lithosphere would therefore also be enriched in incompatible elements even if depleted in a basaltic fraction. Clearly, the LIL-enrichment of the shallow mantle, including the continental lithosphere, can affect the chemistry of depleted magmas that traverse it.

ALKALI BASALT

Continental alkaline magmatism may persist over very long periods of time in the same region and may recur along lines of structural weakness after a long hiatus. The age and thickness of the lithosphere play an important role, presumably by controlling the depth and extent of crystal fractionation and the ease by which the magma can rise to the surface. Alkali magmatism occurs over broad areas and is not necessarily associated with rifts but is often accompanied

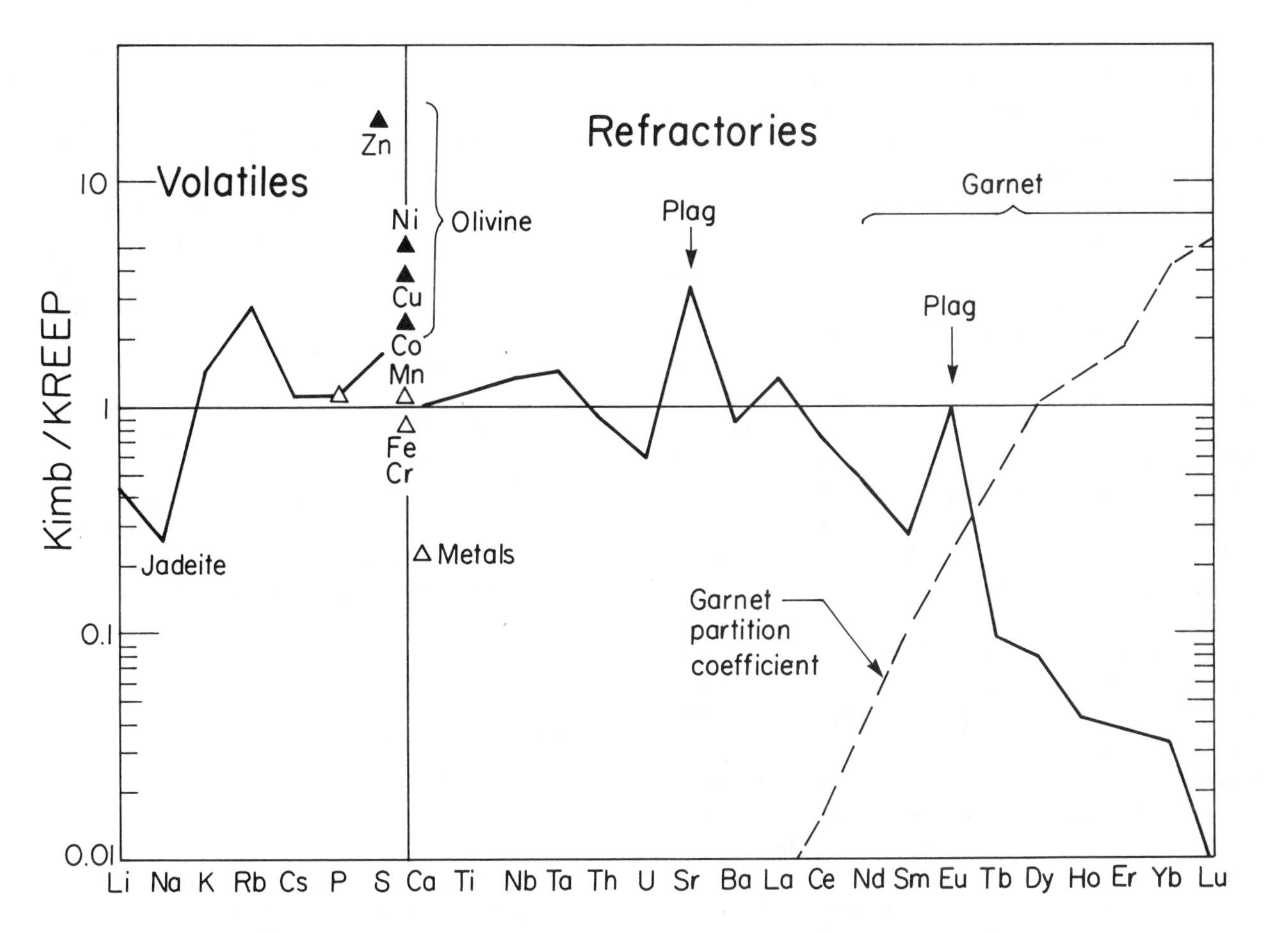

FIGURE 8-9
Trace-element concentrations in kimberlite relative to KREEP. Elements that are fractionated into olivine, plagioclase, garnet and jadeite are noted. The solid/liquid partition coefficient for garnet is shown. If kimberlite has been in equilibrium with garnet, or an eclogite cumulate, the concentrations in kimberlite will be proportional to the reciprocal of the garnet partition coefficient.

by crustal doming. Some of the more important alkaline provinces are in Africa, Siberia, the Baltic shield, Scotland, the Oslo graben, southwest Greenland, the Appalachians, the Basin and Range, the White Mountains (New England), and the St. Lawrence and Ottawa grabens. Many of these are associated with grabens and transcurrent faults and with the opening of ocean basins. Alkali magmatism precedes and follows the opening of the Atlantic. Alkali magmatism can be traced along a small circle, presumably representing a transform fault, from provinces in North America to the New England seamounts and the Canary Islands to the Moroccan alkaline province (Sykes, 1978). Likewise alkaline provinces in South America can be linked with those in west-central Africa through the Cape Verde Islands. Oceanic transform faults in general have long been recognized as the loci of alkali basalts.

The non-random distribution of alkaline provinces has been interpreted as hotspot tracks and structural weaknesses in the lithosphere. Sykes (1978) suggested that hotspots are passive centers of volcanism whose location is determined by pre-existing zones of weakness, rather than manifestations of deep mantle plumes.

Examples are known of alkaline basalts preceding and following abundant tholeiitic volcanism. In general alkaline basalts occur beneath thicker lithosphere and are early when the lithosphere is thinning (rifting) and late when the lithosphere is thickening (flanks of ridges, downstream from midplate volcanic centers).

On Kerguelen, the third largest oceanic island, and on Ascension Island, the youngest basalts are alkalic and overlie a tholeiitic base. In Iceland, tholeiites dominate in the rift zone and grade to alkali basalts along the western and southern shores. Thus, there appears to be a relation between the nature of the magmatism and the stage of evolution. The early basalts are tholeiitic, change to transitional and mixed basalts and terminate with alkaline compositions. It was once thought that Hawaii also followed this sequence, but the youngest volcano, Loihi, is alkalic and probably represents the earliest stage of Hawaiian volcanism. It could easily be covered up and lost from view by a major tholeiitic shield-building stage. Thus, oceanic islands may start and end with an alkalic stage.

In continental rifts, such as the Afar trough, the Red Sea, the Baikal rift and the Oslo graben, the magmas are alkalic until the rift is mature and they are then replaced by tholeiities, as in a widening oceanic rift. A similar sequence occurs in the Canary Islands and the early stages of Hawaii. The return of alkalics in the Hawaiian chain occurs when the islands drift off the hotspot. They, and seamounts on the flanks of midocean ridges, may be tapping the flanks of a mantle upwelling, where crystal fractionation and contamination by shallow mantle are most prevalent. In all cases lithospheric thickness appears to play a major role. This plus the long duration of alkalic volcanism, its simultaneity over large areas and its recurrence in the same parts of the crust argue against the simple plume concept.

There are two prevailing types of models to explain the diversity of basalts. In the primary magma models a parent magma evolves to a wide variety of derivative magmas depending on the depth of fractionation and, therefore, the nature of the fractionation crystals: olivine, plagioclase, orthopyroxene, clinopyroxene, garnet, ilmenite, magnetite and so on. Thus, alkalics can evolve from tholeiites by garnet and clinopyroxene fractionation. Isotopic data show that crystal fractionation is not enough; several isotopically distinct reservoirs must be involved. In the partial melt models the different basalt types are a result of varying degrees of partial melting at different depths and depend on the nature of the residual crystals in the solid rocks that are left behind. In this case alkalics represent small degrees of partial melting, small enough so that the low melting point crystals, clinopyroxene and garnet, are left behind in the residue. Tholeiites would represent a greater degree of melting, enough to eliminate the clinopyroxene and garnet. Another component is required to explain the isotopic ratios. Basalts probably have a complex history involving mixing of materials representing variable degrees of partial melting from different depths, followed by variable amounts of crystal fractionation and interaction with, or contamination by, shallower parts of the mantle and crust.

Highly alkaline products are characteristic of the later stages of many areas of extensive basaltic magmatism. These often occur as dikes, sills and caps associated with late tholeiitic layers. They constitute rock types such as basanites, melilites, nephelinites, lamprophyres and trachybasalts. Similar lavas occur in the early stages of continental rifting and on some oceanic islands in the early stages of volcanism. The distinctive chemical character of these magmas is often attributed to low degrees of partial melting of a deep, previously untapped mantle source, or melting of mantle that has been metasomatized by a CO_2, H_2O-rich fluid. An alternative mechanism involves crystal fractionation from a tholeiitic melt or melting of a shallow reservoir caused by ascending tholeiitic magma. In these models the origin of tholeiitic and alkali magmas are closely related, but melts from several reservoirs may be involved. The HREE depletion of alkali magmas can be explained by eclogite fractionation at sublithospheric depths.

Figure 8-10 shows that alkali-rich basalts generally have trace-element concentrations intermediate between MORB and highly alkalic basalts such as nephelinite. In fact, they can be treated as mixtures of MORB, or a picritic MORB-parent, and nephelinite.

Alkaline rocks are the main transporters of mantle inclusions. They include basanites, nephelinites, hawaiites, alkali olivine basalts, phonolites, kimberlites and lamproites. All of these magmas or melts are known to transport spinal- or garnet-bearing peridotites or high-pressure crystals such as diamond, sapphire, garnet and so on to the surface. The presence of these inclusions demonstrates that these melts are from the mantle and that part of the mantle is enriched in LIL and isotopic ratios.

CONTINENTAL FLOOD BASALTS

Continental flood basalts, the most copious effusives on land, are mainly tholeiitic flows that can cover very large areas and are typically 3 to 9 km in thickness. These are also called plateau basalts and fissure basalts. Examples are the Deccan traps in India, the Columbia River province in the United States, the Parana basin in Brazil, the Karoo province in South Africa, the Siberian traps, and extensive flows in western Australia, Tasmania, Greenland and Antarctica. They have great similarity around the world. Related basalts are found in recent rifts such as East Africa and the Basin and Range province. Rift and flood basalts may both form during an early stage of continental rifting. In some cases alkali-rich basalts were erupted first, occurring on the edge of rifts while tholeiites occur on the floor of the rift. Continental alkali basalts and tholeiites are highly enriched in LIL elements and have high $^{87}Sr/^{86}Sr$ ratios. There are many similarities between continental and ocean-island basalts. Experiments indicate that fractional crystallization or varying degrees of partial melting can produce various members of the alkali series (Green and Ringwood, 1967; Frey and others, 1978). The large range of isotopic ratios in such basalts indicates that their source region is inhomogeneous or that mixing between at least two magma types is involved. The similarity between many continental basalts and ocean-island basalts indicates that involvement of the continental crust is not necessary to explain all continental flood basalts; high $^{87}Sr/^{86}Sr$ ratios, for example, are not necessarily indicative of contamination by the crust.

Continental flood basalts have been the subject of

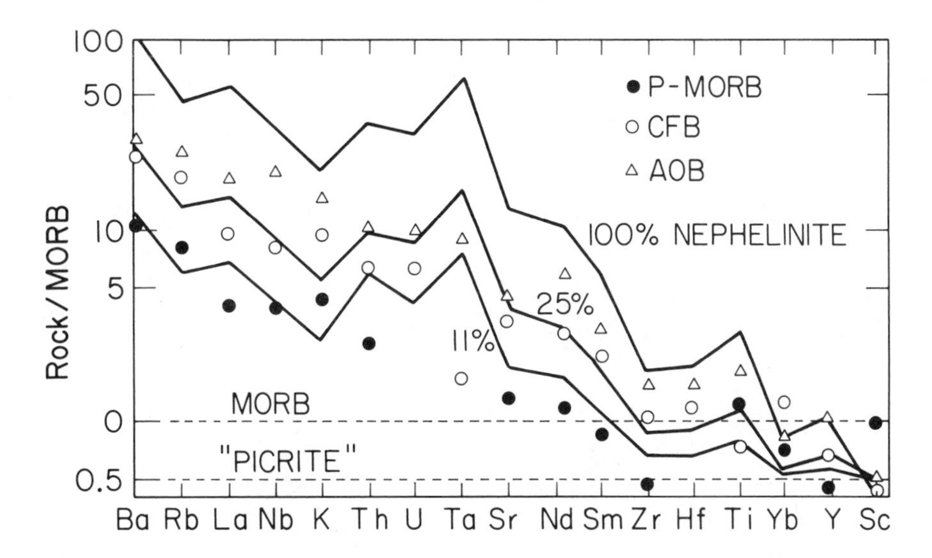

FIGURE 8-10
Trace-element concentrations of "enriched" magmas relative to MORB. P-MORB are "plume-type" MORB, found along midocean ridges; CFB are continental flood basalts; AOB are alkali olivine basalts, found in continents and on oceanic islands. Lines are mixing curves. This suggests that enriched magmas may be contaminated MORB, rather than melts from a different source region or varying degrees of melting from a common source region.

much debate. Their trace-element and isotopic differences from depleted midocean-ridge basalts have been attributed to their derivation from enriched mantle, primitive mantle, magma mixtures or from a parent basalt that experienced contamination by the continental crust. In some areas these flood basalts appear to be related to the early stages of continental rifting or to the proximity to an oceanic spreading center. In other regions there appears to be a connection with subduction, and in these cases they may be analogous to back-arc basins. In most isotopic and trace-element characteristics they overlap basalts from oceanic islands and island arcs.

Chemical characteristics considered typical of continental tholeiites are high SiO_2 and incompatible-element contents, and low MgO/FeO and compatible-element concentrations. The very incompatible elements such as cesium and barium can be much higher than in typical ocean-island basalts.

The Columbia River flood basalts are probably the best studied, but they are not necessarily typical. They appear to have been influenced by the subduction of the Juan de Fuca plate beneath the North American plate and to also have involved interaction with old enriched subcontinental mantle (Carlson and others, 1980). The idea that continental flood basalts are derived from a primordial "chondritic" source (DePaolo and Wasserburg, 1979) does not hold up when tested against all available trace-element and isotopic data, even though some of these basalts have neodymium isotopic ratios expected from a primitive mantle.

Many of the characteristics of continental and ocean-island basalts can be explained by sublithospheric fractionation of garnet and clinopyroxene from a MORB parent, combined with contamination by a high-LIL component. Superimposed on these effects are shallow-mantle or crustal fractionation and, possibly, crustal assimilation. It does not appear necessary to invoke two distinct, isolated fertile source regions such as the low-velocity zone of the upper mantle and some part of the lower mantle, for example, to explain MORB and hotspot magmas. Common basalts with different histories of cooling, fractionation and contamination can exhibit the observed chemical differences.

The relationships between some continental basalts and oceanic volcanism are shown in Figure 8-11. The basalts in the hatched areas are believed to result from hotspot volcanism. The dashed lines show boundaries that apparently represent sutures between fragments of continents. Most hotspots are inside the two noted geoid highs. Table 8-9 gives some flood basalt provinces and the possibly related hotspot. Also shown are some kimberlite provinces and their ages.

ANDESITE

Andesites are associated primarily with island arcs and other convergent plate boundaries and have bulk compositions similar to the continental crust. The accretion of island arcs onto the edges of continents is probably the primary mechanism by which continents grow or, at least, maintain their area. The "average" andesite contains about 60 per-

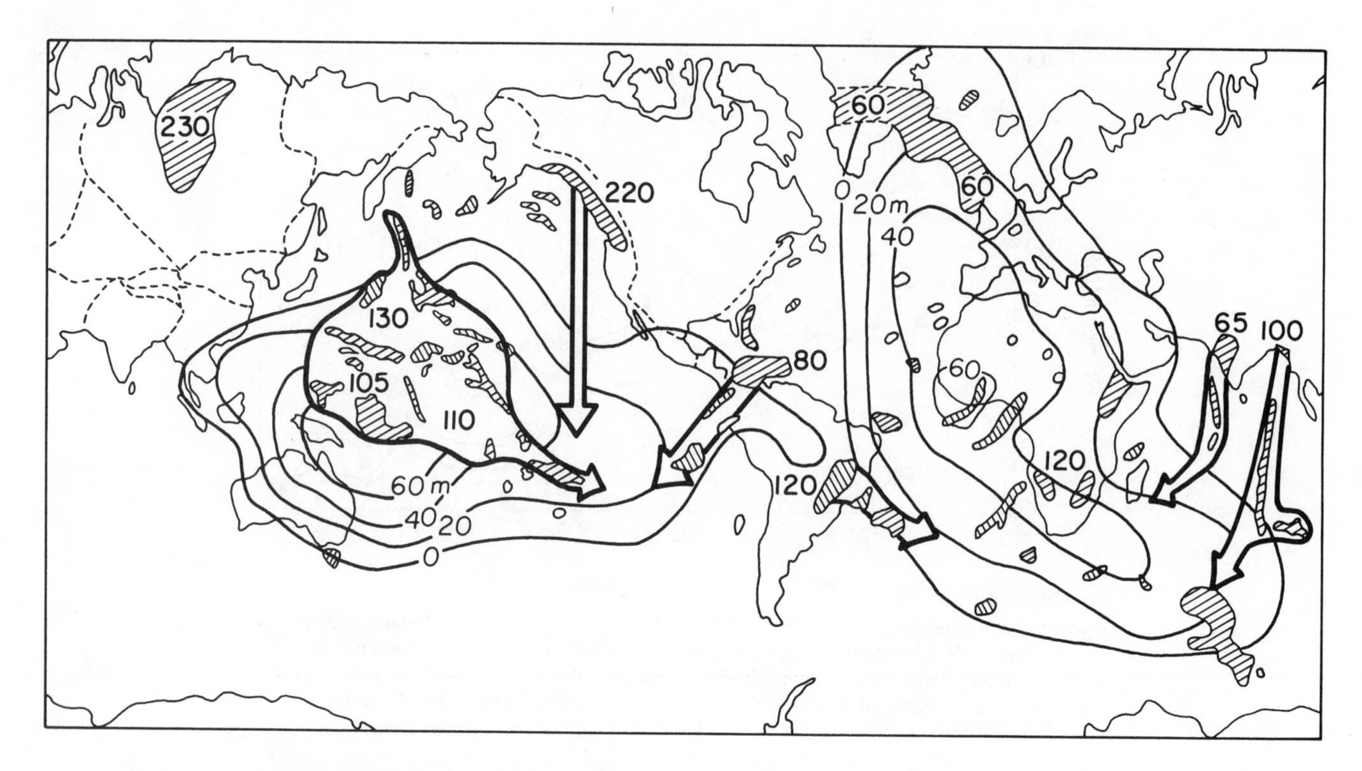

FIGURE 8-11
Map of continental flood basalts, selected hotspot islands, and oceanic plateaus, with ages in millions of years. The period between 130 and 65 Ma was one of extensive volcanism and most of this occurred inside geoid highs (contours) that, in general, are areas of high elevation. Some of the basalt areas have drifted away from their point of origin (arrows). Geoid highs are probably due to hotter than average mantle.

cent SiO_2, 17 percent Al_2O_3, 3 percent MgO, 7 percent CaO, 3.3 percent Na_2O, 1.6 percent K_2O and 1.2 percent H_2O. However, the range in SiO_2, by convention, is from 53 to 63 percent. The K_2O content is variable and seems to correlate with the depth to the underlying slab. "Calc-alkaline" is often used as an adjective for andesite. Tholeiites are also found in island arcs.

Andesites can, in principle, originate from melting of continental crust, contamination of basalt by continental crust, partial fusion of hydrous peridotite, partial fusion of subducted oceanic crust, differentiation of basalt by crystal or vapor fractionation, or by magma mixing. Melting of hydrous peridotite or low-pressure differentiation of basalt by crystal fractionation seem to be the preferred mechanisms of most petrologists, although involvement of subducted material and crust, and magma mixing, are sometimes indicated. Andesites probably originate in or above the subducted slab and, therefore, at relatively shallow depths in the mantle. The thick crust in most andesitic environments means that shallow-level crystal fractionation and crustal contamination are likely. In their incompatible-element and isotopic chemistry, andesites are similar to ocean-island basalts and continental flood basalts. They differ in their bulk chemistry, a difference that can be explained by shallower crystal fractionation.

Volcanic or island arcs typically occur 100 to 250 km from convergent plate boundaries and are of the order of 50 to 200 km wide. The volcanoes are generally about 100 to 200 km above the earthquake foci of the dipping seismic zone. Crustal thicknesses at island arcs are generally 20 to

TABLE 8-9
Continental Flood Basalts and Possibly Related Hotspots and Kimberlites

Region	Hotspot	Age (Ma)
Southeast Greenland	Iceland	60
India	Reunion	65
Columbia River	Yellowstone	20
Southern Brazil	Tristan	120
Southern Africa	Crozet	190
Siberia	Jan Mayen (?)	240
Wrangellia	Hawaii (?)	210
Central Japan	—	280
Kimberlite Provinces		
India		70
Southern Africa		194
Brazil		120
Siberia		148–159

50 km. Andesitic volcanism appears to initiate when a subducting slab has reached a depth of about 70 km and changes to basaltic volcanism when the slab is no longer present or when compressional tectonics changes to extensional tectonics.

Gill (1982) proposed that two conditions are required for voluminous oceanic andesite volcanism: (1) an underlying asthenosphere modified by subduction shortly before the volcanism and (2) compressional stress, or sialic crust greater than about 25 km thick. The latter apparently are required to permit extensive fractionation of the magma prior to eruption.

Although melting of subducted oceanic crust is not necessarily the source of andesitic magmas, the dehydration of subducted sediments and crust may trigger melting in the overlying mantle wedge. Alternatively, the subducted slab may displace material from the MORB source, which then ascends, becomes contaminated, and fractionates prior to eruption. The crust of back-arc basins is more like typical MORB and may represent less impeded ascent of MORB magmas from a deep (>200 km) source.

The overwhelming majority of andesitic volcanoes are located above the intermediate- and deep-focus earthquakes that define the "deep seismic zone." This zone typically dips toward the continent at angles between 15° and near vertical. The dip generally steepens with depth and often bends most noticeably just below the volcanic arc. In most regions the earthquakes cease between about 200 and 300 km. The deep seismic zone is often called the Benioff zone, but it was first described in Japan, and if a name must be applied it should be called the Wadati or Wadati-Benioff zone. Earthquakes occur below 300 km in regions where old (>50 Ma) oceanic lithosphere has subducted. This could mean that old, cold lithosphere has a higher negative buoyancy than young lithosphere, and can therefore sink deeper into the mantle, or that it takes longer for the thicker lithosphere to warm up and lose its brittleness. The crust and depleted peridotite parts of the slab are buoyant relative to the underlying mantle and can only sink when they are very cold. Although the crustal part of the slab can convert to dense eclogite at depths of the order of 50 km, it experiences high temperatures and high stresses as it enters the mantle and may well remain in the shallow mantle. The large number of earthquakes at shallow depths certainly indicates that the top of the slab is extensively fractured before it reaches 50 to 100 km. Even if subducted crust does not contribute directly to the overlying andesites, it probably contributes to the high H_2O and LIL content of the shallow mantle, including the mantle wedge. Oceanic crust becomes extremely altered by hydrothermal activity as it ages and is not a suitable candidate, by remelting, to provide depleted MORB, unless the high temperatures and dehydration experienced during subduction can undo the alteration. If subducted crust can escape melting and converts to eclogite, the relatively small fragments would have difficulty sinking very far into the mantle unless dragged down by the underlying lithosphere.

The mantle immediately below the crust is probably olivine-rich harzburgite or dunite. This is consistent with oceanic P_n velocities and anisotropy and with the mantle section at the base of ophiolites. The thickness of this layer, however, is unknown. In some models the harzburgite is considered the depleted complement to the overlying basalts and is therefore expected to be 25 to 30 km thick for 20 percent melting of a pyrolite source. Since it is depleted in the basaltic component, it is poor in Al_2O_3 and, at high pressure, in garnet. It is therefore buoyant and can only subduct, temporarily, if it is very cold. Over the long run, therefore, it stays in the shallow mantle or returns to the shallow mantle after heating.

The seismic data, however, do not require such a thick olivine-rich layer in the oceanic lithosphere. Probably 5–10 km would be thick enough to explain the P_n observations. There are also petrological and geochemical data suggesting that the peridotites in oceanic sections are not genetically related to the overlying basalts. They may, in fact, be the refractory residues from ancient melting processes. Since depleted peridotites are buoyant relative to fertile mantle, and since melting has been occurring over all of Earth history, there should be harzburgites and dunites in the shallow mantle that date back to early Earth history. The cold lithosphere under stable shields may be the best place to find ancient material.

If the oceanic crust represents only part of the molten material that attempts to rise to the surface of the Earth, then there may be basaltic or picritic material elsewhere in the shallow mantle that eventually intrudes or underplates the oceanic lithosphere. Part of the lower oceanic lithosphere may therefore convert to eclogite as it ages. High-velocity regions in the oceanic lithosphere imply a large garnet content, consistent with eclogite. The negative buoyancy of old lithosphere may therefore be due to chemical rather than temperature effects, and the deep subduction of old oceanic lithosphere can be understood.

The most effective way to cool the mantle is to subduct old, cold lithosphere. The cooling works in two ways: Hot mantle is replaced or displaced by the cold subducted material, and as the slab warms up the adjacent mantle is cooled. The latter process is very slow because of the long conduction time constant. The overall temperature of the slab and adjacent mantle system stays roughly constant since the heating of one is counterbalanced by the cooling of the other.

Subduction tends to occur in oceanic lithosphere of all ages. There does not seem to be a critical age at which subduction can initiate, and therefore a thermal instability, by itself, is not sufficient for spontaneous subduction. It may be that subduction always initiates at continental boundaries or at regions of thick crust, such as oceanic plateaus, or at fracture zones. A novel possibility is that sub-

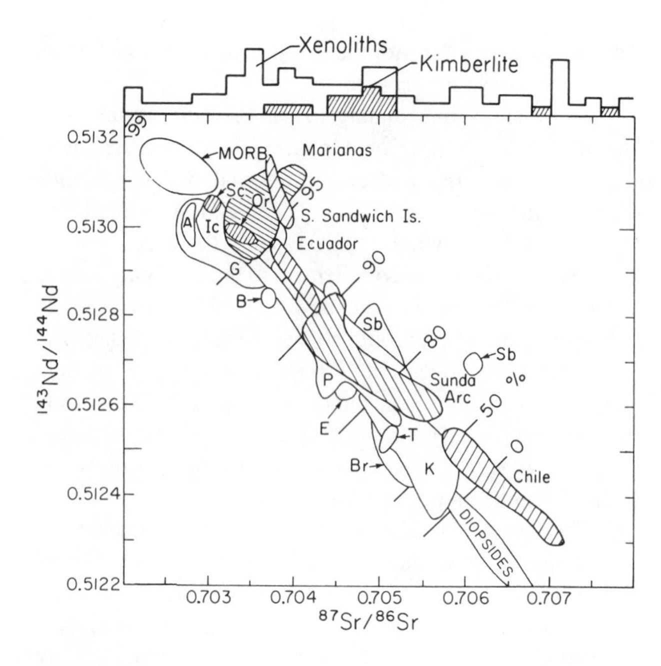

FIGURE 8-12
Neodymium and strontium isotopic ratios for oceanic, ocean-island, continental and island-arc (hatched) basalts and diopside inclusions from kimberlites. The numbers are the percentage of the depleted end-member if this array is taken as a mixing line between depleted and enriched end members. Mixing lines are flat hyperbolas that are approximately straight lines for reasonable choices of parameters. The fields of MORB and CFB correspond to >97 percent and 70–95 percent, respectively, of the depleted end-member. The enriched end-member has been arbitrarily taken as near the enriched end of Kerguelen (K) basalts. The most enriched magmas are from Kerguelen, Tristan da Cunha (T) and Brazil (Br). Other abbreviations are Sc (Scotia Sea), A (Ascension), Ic (Iceland), H (Hawaii), G (Gouch), Or (Oregon), Bo (Bouvet), Sb (Siberia), Co (Columbia River), P (Patagonia) and E (Eiffel). Along top are strontium isotopic data for xenoliths and kimberlites.

duction can occur, not only in normal collision situations, but also when the oceanic lithosphere overrides abnormally hot and light mantle. The Tonga-Fiji arc, for example, seems to be embedded in such a region, with abnormally slow mantle velocities to great depth.

The "mantle wedge" between the base of the crust and the top of the slab has low seismic velocities and high attenuation, consistent with partial melting. Below about 300 km the mantle near subduction zones is faster than average, presumably due to the insertion of cold material.

Figure 8-12 shows that island-arc basalts (hatched) overlap ocean-island and continental basalts in isotopic characteristics. A similar mantle source is indicated.

MARGINAL BASINS

Marginal basins (see Table 8-10) are small oceanic basins, underlain by oceanic crust, that represent trapped and isolated fragments of oceanic crust (Bering Sea, south Fiji Basin, Caribbean) or formed by processes akin to continental rifting (South China Sea, Tasman Basin, Coral Sea) or seafloor spreading (Gulf of California, Lau Basin, Andaman Sea). Some are related to subduction and are called back-arc basins. Some of these are actively spreading (Mariana Trough, Ryukyu Trough, Lau Basin, Scotia Sea), and oth-

TABLE 8-10
Classification and Ages of Trench–Arc–Back-Arc–Marginal Basin Systems

Classification	System
Continental Arc	Peru-Chile Andes Cordilleran
	Middle America Cordilleran
	North America Basin and Range
	Alaska Cordilleran
	Sumatra-Java-Sunda Shelf
Trapped Ocean Remnants	Bering Sea–Aleutian Basin (117–132 Ma)
	West Philippine Sea? (39–50 Ma)
	South Fiji Basin? (25–34 Ma)
	Caribbean Plate (100 Ma)
	Emerald Basin? (South of New Zealand)
Active Back-Arc Spreading	Mariana Trough (Late Miocene)
	Ryukyu Trough
	Lau Basin (Tonga) (0–4 Ma)
	Scotia Sea (0–9 Ma)
	New Hebrides? (41–55 Ma)
	Bonin?
Leaky Transform	Gulf of California
	Andaman Sea (0–11 Ma)
	Bismark Sea (New Guinea) (0–3.5 Ma)
	Fiji Plateau (New Hebrides) (0–10 Ma)
Inactive Back-Arc	Grenada Trough (Antilles)
	Kamchatka Basin
	Kurile Basin
	Japan Basin
	Tsushima Basin (Japan)
	Shikoku Basin? (Izu-Bonin) (17–26 Ma)
	Parece Vela Basin (Mariana) (17–30 Ma)
Small Ocean Basin (Continental rift related)	Coral Sea (56–64 Ma)
	South China Sea (17–32 Ma)
	Tasman Sea (56–77 Ma)
Uncertain Origin	Banda Sea
	Caroline Basin (28–36 Ma)
	Celebes Basin (42–48 Ma)
	New Caledonia Basin
	Solomon Basin
	West Scotia Sea (16–30 Ma)

Brooks and others (1984).

TABLE 8-11
Composition of Marginal Basin Basalts Compared with Other Basalts

Oxide	MORB	MORB	Mariana Trough	Scotia Sea	Lau Basin	Bonin Rift	North Fiji Plateau	Arc Tholeiites		Island Tholeiite
SiO_2	49.21	50.47	50.56	51.69	51.33	51.11	49.5	48.7	51.1	49.4
TiO_2	1.39	1.58	1.21	1.41	1.67	1.10	1.2	0.63	0.83	2.5
Al_2O_3	15.81	15.31	16.53	16.23	17.22	18.20	15.5	16.5	16.1	13.9
FeO	7.19	10.42	8.26	8.28	5.22	8.18	6.2	8.4	—	—
Fe_2O_3	2.21	—	—	—	2.70	—	3.9	3.4	11.8	12.4
MnO	0.16	—	0.10	0.17	0.15	0.17	0.1	0.29	0.2	—
MgO	8.53	7.46	7.25	6.98	5.23	6.25	6.7	8.2	5.1	8.4
CaO	11.14	11.48	11.59	11.23	9.95	11.18	11.3	12.2	10.8	10.3
Na_2O	2.71	2.64	2.86	3.09	3.16	3.12	2.7	1.2	1.96	2.13
K_2O	0.26	0.16	0.23	0.27	1.07	0.38	0.3	0.23	0.40	0.38
P_2O_5	0.15	0.13	0.15	0.18	0.39	0.18	0.1	0.10	0.14	—
H_2O	—	—	1.6	—	1.18	—	1.4	—	—	—
$^{87}Sr/^{86}Sr$	0.7023	0.7028	0.7028	0.7030	0.7036	—	—	—	—	—

Gill (1976), Hawkins (1977).

ers are inactive (Kamchatka and Kurile Basins, Japan Basin, Shikoku Basin). Some have an uncertain origin (Banda Sea, New Hebrides–Fiji Plateau, west Philippine Sea).

The crustal structure, heat flow and magnetic lineation patterns of back-arc and major ocean basins are similar, implying that the origins are similar. Although back-arc spreading appears to be related to subduction, most arc-trench systems do not have actively spreading back-arc basins. Back-arc basin basalts (BABB) are similar to mid-ocean-ridge basalts but in many respects they are transitional toward island-arc tholeiites and ocean-island basalts (Tables 8-11, 8-12 and 8-13). They are LIL-enriched compared to normal MORB and generally have LREE enrichment. High $^3He/^4He$ ratios have been found in some BABB, and in this and other respects they are similar to, or gradational toward, ocean-island basalts and other hotspot magmas.

Back-arc spreading may be initiated above subducted lithosphere, but with time the axis of spreading migrates and no longer lies above the earthquakes of the Wadati-Benioff zone. If present the seismic zone is deeper than 200–300 km below the spreading center. Spreading centers in the Bismarck Sea, east Scotia Sea, Mariana trough and Coriolis trough (New Hebrides) are beyond the deepest seismic extent of the slab.

If the geometric and temporal relationships can be used as a guide, the relatively enriched island-arc tholeiites come from a depth of about 100 km and BABB generally comes from deeper than 200 km. The seismic velocities and seismic Q's are low in the mantle wedge above subducting plates. This probably reflects partial melting.

It is not yet clear whether back-arc basins are formed as a result of global plate motions or as a direct result of subduction. Many back-arc regions exhibit cordilleran-type

TABLE 8-12
Representative Properties of Basalts and Kimberlites

Ratio	Midocean Ridge Basalt	Back-Arc Basin Basalt	Island-Arc Basalt	Ocean-Island Basalt	Alkali-Olivine Basalt	Kimberlite
Rb/Sr	0.007	0.019–0.043	0.01–0.04	0.04	0.05	0.08
K/Rb	800–2000	550–860	300–690	250–750	320–430	130
Sm/Nd	0.34–0.70	0.31	0.28–0.34	0.15–0.35	0.37	0.18
U/Th	0.11–0.71	0.21	0.19–0.77	0.15–0.35	0.27	0.24
K/Ba	110–120	33–97	7.4–36	28–30	13.4	7.5
Ba/Rb	12–24	9–25	10–20	14–20	15	15
La/Yb	0.6	2.5	2–9	7	14	125
Zr/Hf	30–44	43	38–56	44	41	36
Zr/Nb	35–40	4.8	20–30	14.7	2.7	2.3
$^{87}Sr/^{86}Sr$	0.7023	0.7027–0.705	0.703–0.706	0.7032–0.706	—	0.71
$^{206}Pb/^{204}Pb$	17.5–18.5	18.2–19.1	18.3–18.8	18.0–20.0	—	17.6–20.0

TABLE 8-13
Representative Values of Large-Ion Lithophile and High-Field-Strength Elements in Basalts (ppm)

Element	Midocean-Ridge Basalt	Back-Arc Basin Basalt	Island-Arc Basalt	Ocean-Island Basalt
Rb	0.2	4.5–7.1	5–32	5
Sr	50	146–195	200	350
Ba	4	40–174	75	100
Pb	0.08	1.6–3	9.3	4
Zr	35	121	74	125
Hf	1.2	2.8	1.7	3
Nb	1	25	2.7	8
Ce	3	32	6.7–32.1	35
Th	0.03	1–1.9	0.79	0.67
U	0.02	0.4	0.19	1.18

tectonics or nonspreading back-arc basins. In the Marianas, back-arc spreading alternates with arc volcanism. Extension, of course, is associated with spreading, whether as cause or effect, and arcs and trenches are basically convergent or compressional phenomena. Arcs with active back-arc spreading do not have the giant earthquakes associated with the cordilleran-style arcs. This is probably a result of the local coupling between plates.

In most respects active back-arc basins appear to be miniature versions of the major oceans, and the spreading process and crustal structure and composition are similar to those occurring at midocean ridges. The age-depth relationships of marginal seas, taken as a group, are indistinguishable from those of the major oceans. To first order, then, the water depth in marginal sea basins is controlled by the cooling, contraction and subsidence of the oceanic lithosphere, and the density of the mantle under the spreading centers is the same. The marginal basins of Southeast Asia, however, tend to be deeper than similar age oceanic crust elsewhere, by one-half to one kilometer, and to deepen faster with age. The presence of cold subducted material underneath the basins may explain why they are deeper than average but not why they sink faster. The low-density material under the ridge axis may be more confined under the major midocean ridges. On average the upper 200–300 km of the mantle in the vicinity of island arcs and marginal basins has slower than average seismic velocities and the deeper mantle is faster than average, probably reflecting the presence of cold subducted material. The depth of back-arc basins is an integrated effect of the thickness of the crust and lithosphere, the low-density shallow mantle and the presumed denser underlying subducted material. It is perhaps surprising then that, on average, the depths of marginal basins are so similar to equivalent age oceans elsewhere. The main difference is the presence of the underlying deep subducting material, which would be expected to depress the seafloor in back-arc basins.

Basalts in marginal basins with a long history of spreading are essentially similar to MORB, while basalts generated in the early stages of back-arc spreading have more LIL-enriched characteristics. A similar sequence is found in continental rifts (Red Sea, Afar) and some oceanic islands, which suggests a vertical zonation in the mantle, with the LIL-rich zone being shallower than the depleted zone. This is relevant to the plume hypothesis, which assumes that enriched magmas rise from deep in the mantle. The early stages of back-arc magmatism are the most LIL-enriched, and this is the stage at which the effect of hypothetical deep mantle plumes would be cut off by the presence of the subducting slab. Continental basalts, such as the Columbia River basalt province, are also most enriched when the presence of a slab in the shallow mantle under western North America is indicated. The similarity of the isotopic and trace-element geochemistry of island-arc basalts, continental flood basalts and ocean-island basalts and the slightly enriched nature of the back-arc basin basalts all suggest that the enrichment occurs at shallow depths, perhaps by contamination of MORB rising from a deeper layer. The high $^3He/^4He$ of Lau Basin basalts also suggests a shallow origin for "primitive" gases.

Both back-arc basins and midocean ridges have shallow seafloor, high heat flow and thin sedimentary cover. The upper mantle in both environments has low seismic velocity and high attenuation. The crusts are typically oceanic and basement rocks are tholeiitic. The spreading center in the back-arc basins, however, is much less well defined. Magnetic anomalies are less coherent, seismicity is diffuse and the ridge crests appear to jump around. Although extension is undoubtedly occurring, it may be oblique to the arc and diffuse and have a large shear component. The existence of similar basalts at midocean ridges and back-arc basins and the subtle differences in trace-element and isotopic ratios provide clues as to the composition and depth of the MORB reservoir.

TABLE 8-14
Major-Element Analyses of Komatiites and Other High-MgO Magmas (percent)

Major Element	Komatiite Munro Township	Komatiite Barberton	Picritic Tholeiite (Baffin Is.)	Bonin Volcanic Province
SiO_2	45.8	45.2	44.0	59.2
Al_2O_3	7.30	3.66	8.3	11.25
TiO_2	0.30	0.20	0.58	0.22
Fe_2O_3	—	—	2.2	—
FeO	11.2	11.0	8.8	8.86
MnO	0.21	0.22	0.19	0.16
MgO	26.1	32.2	26.0	11.4
CaO	7.64	5.28	7.3	6.55
Na_2O	0.69	0.44	0.90	1.78
K_2O	0.10	0.17	0.06	0.43
Cr_2O_3	0.24	—	0.36	0.17

BVP (1980).

KOMATIITES

Komatiites are ultrabasic melts that occur mainly in Archean rocks (Table 8-14). The peridotitic variety (MgO > 18 percent) apparently require temperatures of the order of 1450–1500°C and degrees of partial melting greater than 60–70 percent in order to form. Although they have been identified in the Mesozoic, they are extremely rare in the Phanerozoic. Cawthorn and Strong (1974) estimated that they must originate below 300 km if the melting is due to adiabatic ascent, which seems likely.

The great depth of origin implied for high-temperature magmas such as komatiites means that there is little experimental petrological control on their conditions of formation. At low pressure, olivine is the most refractory phase, and melts are commonly much less MgO-rich than the parent. At high pressure it is possible to generate MgO-rich melts with smaller degrees of partial melting. It is even possible that olivine is replaced as the liquidus phase by the high-pressure majorite phase of orthopyroxene, again giving high-MgO melts. Komatiites may therefore represent large degrees of melting of a shallow olivine-rich parent, small degrees of melting of a deep peridotite source (Herzberg, 1984) or melting of a rock under conditions such that olivine is not the liquidus phase. At depths greater than about 200 km, the initial melts may be denser than the residual crystals (Rigden and others, 1984). This may imply that large degrees of partial melting are possible and, in fact, are required before the melts, or the source region, become buoyant enough to rise. The high CaO/Al_2O_3 ratios of some komatiites suggest that garnet has been left behind in their source region or that high-pressure garnet fractionation occurred prior to eruption. Much has been made of the high temperatures and large degrees of melting implied for komatiite formation, particularly with regard to high upper-mantle temperatures in the Precambrian. Their existence appears to refute the common claim that melt-crystal separation must occur at relatively small degrees of partial melting, about 20–25 percent in ascending diapirs. The rarity of komatiites since Precambrian times could mean that the mantle has cooled, but it could also mean that a suitable peridotite parent no longer exists at about 300 km depth, or that the currently relatively thick lithosphere prevents their ascent to the surface. Diapirs ascending rapidly from about 300 km from an eclogite-rich source region would be almost totally molten by the time they reach shallow depths, and picrite and basaltic magmas would predominate over komatiitic magma. If peridotites only extend to a depth of 200–400 km in the present mantle, then the absence of recent komatiites could be readily understood. Picritic magmas, the currently popular precursor to basalts, are also rare, presumably because they are too dense to rise through the crust. Cooling and olivine fractionation causes the density to decrease, and tholeiites are now generally viewed as fractionated picrites. The MgO content of magmas also increases with depth of melting. Komatiites may therefore be the result of deep melting. By cooling and crystal fractionation komatiitic melts can evolve to less dense picritic and tholeiitic melts. Low-density melts, of course, are more eruptable. It may be that komatiitic melts exist at present, just as they did in the Precambrian, but, because of the colder shallow mantle they can and must cool and fractionate more, prior to eruption. In any case komatiites provide important information about the physics and chemistry of the upper mantle.

The existence of komatiitic melts seems to contradict the petrological prejudice that large amounts of melting are impossible. However, if melts are to drain from a source region, the surrounding matrix must deform to fill the space. Marsh (1984) argued that melting must exceed 45 to 55 volume percent before eruption can occur.

SUMMARY

Most models of mantle chemistry and evolution based on magmas alone tend to relate the various magma types by varying degrees of partial melting or crystal fractionation. Depth of melt-crystal separation is another parameter. The actual depth of the source region, its composition and the amount of partial melting cannot be tightly constrained by observational or experimental petrology. The simplest petrological models tend to view the mantle as homogeneous throughout and capable of providing basalt, for the first time, by partial melting. When the petrological data are combined with isotopic and geophysical data, and with considerations from accretional calculations, a more complex evolution is required. Similarly, simple evolutionary models have been constructed from isotopic data alone that conflict with the broader data base.

General References

Anderson, D. L. (1979) The upper mantle transition region: Eclogite? *Geophys. Res. Lett.*, *6*, 433–436.

Anderson, D. L. (1981) Hotspots, basalts and the evolution of the mantle, *Science*, *213*, 82–89.

Anderson, D. L. (1982a) Isotopic evolution of the mantle, *Earth Planet. Sci. Lett.*, *57*, 1–24.

Anderson, D. L. (1982b) Chemical composition and evolution of the mantle. In *High-Pressure Research in Geophysics* (S. Akimoto and M. Manghnani, eds.), 301–318, D. Reidel, Dordrecht.

Anderson, D. L. (1983a) Chemical composition of the mantle, *J. Geophys. Res.*, *88*, B41–B52.

Anderson, D. L. (1983b) Kimberlite and the evolution of the mantle, *Proc. Third Intl. Kimberlite Conference,* Clermont-Ferrand, France.

Anderson, D. L. and J. Regan (1983) Uppermantle anisotropy and the oceanic lithosphere, *Geophys. Res. Lett.*, *10*, 841–844.

Dawson, J. B. (1980) *Kimberlites and their Xenoliths*, Springer-Verlag, Berlin, 252 pp.

Taylor, S. (1982) Lunar and terrestrial crusts, *Phys. Earth Planet. Inter.*, *29*, 233.

Wedepohl, K. H. and Y. Muramatsu (1979) The chemical composition of kimberlites compared with the average composition of three basaltic magma types. In *Kimberlite, Diatremes and Diamonds* (F. R. Boyd and H. O. Meyer, eds.), 300–312, American Geophysical Union, Washington, D.C.

References

Anderson, D. L. (1975) Chemical plumes in the mantle, *Geol. Soc. Am. Bull.*, *86*, 1593–1600.

Anderson, D. L. (1977) Composition in the mantle and core, *Ann. Rev. Earth Planet. Sci.*, *5*, 179.

Anderson, D. L. (1980) *Bull. Volc.*, 663.

Anderson, D. L. (1982a) Isotopic evolution of the mantle, *Earth Planet. Sci. Lett.*, *57*, 1–24.

Anderson, D. L. (1982b) Chemical composition and evolution of the mantle. In *High-Pressure Research in Geophysics* (S. Akimoto and M. H. Manghnani, eds.), 301–318, D. Reidel, Dordrecht.

Anderson, D. L. (1982c) Hotspots, polar wander, mesozoic convection and the geoid, *Nature*, *297*, 391–393.

BVP, Basaltic Volcanism Study Project (1980) *Basaltic Volcanism on the Terrestrial Planets,* Pergamon, New York, 1286 pp.

Bass, J. D. and D. L. Anderson (1984) *Geophys. Res. Lett*, *11*, 237–240.

Brooks, D. A., R. L. Carlson, D. L. Harry, P. J. Melia, R. P. Moore, J. E. Rayhorn and S. G. Tubb (1984) Characteristics of back-arc regions, *Tectonophys.*, *102*, 1–16.

Burdick, L. and D. L. Anderson (1975) Interpretation of velocity profiles of the mantle, *J. Geophys. Res.*, *80*, 1070–1074.

Butler, R. and D. L. Anderson (1978) Equation of state fits to the lower mantle and outer core, *Phys. Earth Planet. Inter.*, *17*, 147–162.

Carlson, R. W., G. W. Lugmair and J. D. Macdougall (1980) Crustal influence in the generation of continental flood basalts, *Nature*, *289*, 160–162.

Cawthorn, R., R. Grant and D. Strong (1974) The petrogenesis of komatiites and related rocks as evidence for a layered upper mantle, *Earth Planet. Sci. Lett.*, *23*, 369.

Cohen, L. and J. Rosenfeld (1979) Diamond; depth of crystallization inferred from compressed included garnet, *J. Geol.*, *87*, 330–340.

Crough, S. T. (1983) Hotspot swells, *Ann. Rev. Earth Planet. Sci.*, *11*, 165.

DePaolo, D. J. and G. J. Wasserburg (1979) Neodymium isotopes in flood basalts from the Siberian Platform and inferences about their mantle sources, *Proc. Nat. Acad. Sci.*, *76*, 3056–3060.

Dosso, L. and V. R. Murthy (1980) A Nd isotopic study of the Kerguelen Islands; inferences on enriched oceanic mantle sources, *Earth Planet. Sci. Lett.*, *48*, 268–276.

Frey, F. A., D. H. Green and S. D. Roy (1978) Integrated models of basalts petrogenesis: A study of quartz tholeiites to olivine melilitites from southeastern Australia utilizing geochemical and experimental petrological data, *J. Petrol. 19*, 463–513.

Gill, J. (1981) *Orogenic Andesites and Plate Tectonics,* Springer-Verlag, New York, 390 pp.

Gill, J. B. (1976) Composition and age of Lau basin and ridge volcanic rocks; implications for evolution of an interarc basin and remnant arc, *Geol. Soc. Am. Bull.*, *87*, 1384–1395.

Green, D. H. and A. E. Ringwood (1967) The stability fields of aluminous pyroxene peridotite and garnet peridotite and their relevance in upper mantle structure, *Earth Planet. Sci. Lett.*, *3*, 151–160.

Hawkins, J. W. (1977), in Island Arcs, Deep Sea Trenches, and Back-Arc Basins, M. Talwani and W. C. Pittman, eds., AGU, Washington, D. C., 367–377.

Herzberg, C. T. (1984) Chemical stratification in the silicate Earth, *Earth Planet. Sci. Lett.*, *67*, 249–260.

Jacobsen, S. B., J. E. Quick and G. J. Wasserburg (1984) A Nd and Sr isotopic study of the Trinity Peridotite; implications for mantle evolution, *Earth Planet. Sci. Lett.*, *68*, 361–378.

Liu, L. (1974) Silicate perovskite from phase transformation of pyrope garnet, *Geophys. Res. Lett.*, *1*, 277–280.

Maaløe, S. and R. Steel (1980) Mantle composition derived from the composition of lherzolites, *Nature*, *285*, 321–322.

Marsh, Bruce D. (1984). In *Explosive Volcanism.* (F. R. Boyd, Jr., ed.) National Academy Press, Washington, D.C.

McCulloch, M. T., A. Jaques, D. Nelson and J. Lewis (1983) Nd and Sr isotopes in kimberlites and lamproites from Western Australia: An enriched mantle origin, *Nature*, *302*, 400.

McCulloch, M. T., R. Arculus, B. Anappell and J. Ferguson (1982) Isotopic and geochemical studies of nodules in kimberlites, *Nature*, *300*, 166.

McKenzie, D. and F. Richter (1981) Parameterized thermal convection in a layered region, *J. Geophys. Res., 86,* 11,677.

Morgan, J. W. and E. Anders (1980) Chemical composition of the Earth, Venus and Mercury, *Proc. Natl. Acad. Sci., 77,* 6973.

Morgan, W. J. (1971) Convection plumes in the lower mantle, *Nature, 230,* 42–43.

Morgan, W. J. (1972) Plate motions and deep mantle convection. In *Studies in Earth and Space Sciences* (Geol. Soc. Am. Mem. 132), 7–22.

Morgan, W. J. (1981) Hotspot tracks and the opening of the Atlantic and Indian Oceans. In *The Oceanic Lithosphere* (C. Emiliani, ed.), 443–387, Wiley, New York.

Rigden, S. S., T. J. Ahrens and E. M. Stolper (1984) Densities of liquid silicates at high pressures, *Science, 226,* 1071–1074.

Ringwood, A. E. (1966) Mineralogy of the mantle. In *Advances in Earth Science,* 357–399, MIT Press, Cambridge, Mass.

Ringwood, A. E. (1975) *Composition and Petrology of the Earth's Mantle,* McGraw-Hill, New York, 618 pp.

Ringwood, A. E. and S. Kesson (1976) A dynamic model for mare basalt petrogenesis, *Proc. Lunar Sci. Conf., 7,* 1697–1722.

Ringwood, A. E. and S. Kesson (1977) Siderophile and volatile elements in Moon, Earth and chondrites, *Moon, 16,* 425.

Sykes, L. R. (1978) Intraplate seismicity, reactivation of preexisting zones of weakness, alkaline magmatism, and other tectonism postdating continental fragmentation, *Rev. Geophys. Space Phys., 16,* 621–688.

Walck, M. (1982) Models of upper mantle compressional velocity under a spreading center, *Eos Trans. AGU, 63,* 1036.

Warren, P. H. and J. T. Wasson (1979) The origin of KREEP, *Rev. Geophys. Space Phys., 17,* 73–88.

Wilson, J. T. (1963) Evidence from islands on the spreading of the ocean floor, *Nature, 197,* 536–538.

9

The Source Region

Long is the way
And hard, that out of hell leads up to light.

—JOHN MILTON

Magmas probably represent partial, rather than complete, melts of the basalt source region. Much or most of the source region is left behind when a melt is extracted from a partially molten source region. The composition, mineralogy and depth of the source region are all controversial. Little progress can be made without a simultaneous consideration of field and experimental petrology, isotopes and geophysics. Many plausible scenarios can be constructed from parts of the data, but most have little merit when tested against other sets of data. Magmas are just part of the story. The denser and more refractory parts of the mantle are the subject of this chapter.

BACKGROUND

The rocks of the mantle are compounds involving predominately MgO, SiO_2 and FeO. These oxides account for more than 90 percent of the mantle. Intuitively, one might expect that the bulk properties of the mantle, such as density and seismic velocity and the locations of phase change discontinuities, are controlled only by compounds such as $(Mg,Fe)SiO_3$ and $(Mg,Fe)_2SiO_4$ and their high-pressure phases. Most discussions of mantle chemistry are based on this assumption. CaO and Al_2O_3 are minor constituents in the Earth as a whole, accounting for less than 8 percent of the mantle; however, they have an importance far beyond their abundance. In an alumina-poor mantle the mineralogy is dominated by olivine and pyroxene and their high-pressure forms such as β- and γ-spinel, SiO_2 (stishovite) and majorite. However, a small amount of Al_2O_3 completely changes the mineralogy. (Mg,Fe)O, CaO and SiO_2 combine with Al_2O_3 to form garnet, which may be the dominant mineral between 400 and 650 km. Generally, changes in mineralogy have a greater effect on density and seismic velocities than small changes in composition of a given phase. CaO and Al_2O_3 change the melting point as well as the mineralogy and depths of phase changes. They therefore influence the evolution, stratigraphy and seismic profiles of the mantle. Some regions of the mantle may, in fact, be CaO- and Al_2O_3-rich.

The major shallow-mantle minerals, $MgSiO_3$ (pyroxene) and Mg_2SiO_4 (olivine), are unstable at high pressure and transform to denser phases in the spinel, garnet (majorite), ilmenite, perovskite, periclase and rutile structures. The small amount of FeO in the upper mantle readily substitutes for MgO and serves to increase the density, decrease the seismic velocities and lower the transition pressure for mantle phase changes. CaO and Al_2O_3 drastically alter the phase assemblages and the densities in the mantle. They also are major constituents of the basaltic or low-melting fraction of the mantle and serve to decrease the melting point. Al_2O_3 stabilizes the garnet structure, which means that the basaltic fraction of the mantle is denser than "normal" or Al_2O_3-poor upper mantle, but less dense than the perovskite-rich lower mantle in the pressure interval over which garnet is stable. Early differentiation of the Earth tends to separate the Al_2O_3-rich (basalt, eclogite) and Al_2O_3-poor silicates (peridotite), and the density differential is such that there should be little remixing. It is probable that the major geochemical reservoirs of the mantle, which have remained isolated for a large fraction of Earth history, also differ in Al_2O_3 content and intrinsic density. Reservoirs rich in Al_2O_3 and CaO are called "fertile."

We do not know the composition or mineralogy of the MORB reservoir. At the depth of MORB genesis, basalts would crystallize as eclogite. Eclogite is a dense clinopyroxene-garnet-rich rock that is the high-pressure form of MORB and picrite, the postulated direct parent of MORB. The parent reservoir, in addition, probably contains orthopyroxene and olivine, the other major minerals of the mantle. The composition of the MORB reservoir can therefore range from eclogite to garnet peridotite, depending on depth and the extent of partial melting and crystal fractionation occurring prior to magma separation. The bulk of the material erupting at oceanic islands and as continental flood basalts may also come from either a garnet peridotite or eclogite source region. The source region for so-called hotspot magmas and midocean-ridge basalts may, in fact, be the same if hotspot magmas are contaminated in the shallow mantle prior to eruption.

The generally accepted view that basalts represent various degrees of partial melting, up to about 20 percent, of a garnet peridotite source region is based on several lines of evidence and assumptions (Yoder, 1976). Peridotites are a common rock type and are found as xenoliths in kimberlite and magma, in obducted sections of oceanic lithosphere and in dredge hauls at fracture zones. The seismic properties of the shallow mantle are generally consistent with peridotite. Garnet peridotites can form basalts by partial melting, and some peridotites appear to be the refractory residue remaining after basalt extraction. Other potential source materials, such as eclogite or pyroxenite, are not only rarer but are thought (erroneously) to require more extensive melting in order to provide basaltic magmas. Melt-crystal separation is believed to occur before such extensive melting can occur. These latter arguments only hold if the source region is identical in chemistry to the basalts it produces and if melt-crystal separation is more effective at the depth of initial melting than source buoyancy. Eclogites cover a wide range of compositions and they melt in a eutectic-like fashion, thus basalts can form from eclogites over a wide range of partial melting conditions.

There are several developments that have reopened the question of the nature and depth of the basalt source region. Seismic evidence strongly suggests that the source of midocean-ridge basalts is in the transition region, well below the low-velocity zone. The properties of the shallow mantle are, therefore, not relevant. The low-velocity zone itself is highly variable in depth, thickness and velocity. It locally extends below 400 kilometers. Midocean-ridge basalts are possibly derivative from picrites after extensive crystal fractionation and may, therefore, require more extensive melting than previously thought. Komatiites either require more than 60 percent melting, indicating that large amounts of melting are possible prior to melt separation, or that they formed at great depth; in either case high temperatures are implied. High melt densities make melt-crystal separation more difficult at great depth. If diapiric ascent is rapid, extensive melting can occur because of adiabiatic decompression. Melt-crystal separation, or crystal fractionation, may be restricted to depths less than 100 km, where diapiric ascent is slowed by high viscosities, high-strength lithosphere or buoyancy considerations. Finally, the elastic properties of the transition zone (see Chapter 4) are compatible with an eclogite-rich composition.

Basalts are chemically equivalent to garnet plus clinopyroxene (diopside and jadeite) and melting of these minerals is involved in basalt genesis. Experimental petrology, however, cannot tell us if the garnet and clinopyroxene are dispersed, as in peridotite, or how much olivine and orthopyroxene are in the source region. The basalt fraction of pyrolite, a hypothetical source peridotite, is arbitrary. Experimental petrology cannot constrain the bulk composition of the source region. The existence of meteorites rich in olivine has provided support for the argument that peridotite is the parental basalt material in the Earth. On the other hand, cosmochemical models of the Earth have more pyroxene and less olivine than most peridotite models for the mantle, and partial melting of meteorites does not yield liquids with appropriate iron-magnesium ratios: Meteorite olivines are generally much higher in FeO than mantle olivines. There is little doubt that there is abundant olivine in the mantle. The issues are whether it, and its high-pressure phases, are distributed uniformly throughout the mantle and whether it is the dominant mineral in the basalt source region.

Observations that have been used in support of an olivine-rich (>50 percent) source region for basalts—garnet peridotites are stable in the upper mantle, their bulk compositions are consistent with the materials forming the Earth, they are prominent among the recovered deep-seated samples, they are capable of yielding material of basaltic composition (Yoder, 1976)—deal with possible mineral assemblages and bulk compositions somewhere in the mantle rather than dealing directly with the basalt source region. The "conclusion" that garnet peridotite is the immediate parent of basaltic or picritic magma is actually a working hypothesis, not well established (Yoder, 1976). Melting of garnet and clinopyroxene, whether embedded in a peridotite or eclogite matrix, will form basalt. The eclogite itself may well have been derived from partial melting or crystal fractionation of peridotite at an earlier stage of Earth evolution. Yoder (1976) and Carmichael and others (1974) have given well-balanced treatments of the various possible source rocks. A source region with less than 50 percent olivine cannot be ruled out.

The effects of melt compressibility and matrix stiffness (reviewed in Chapter 7) are such that separation of melt is difficult at high pressure (Rigden and others, 1984). When sufficient melting occurs at depth, the whole source region may become unstable, bringing melt plus matrix to shallow levels where melt separation can occur. In a chemically stratified mantle the high temperature gradient in a thermal boundary layer causes melting to initiate in the deeper layer. Melt can be retained at depth if the overlying mantle has high viscosity and low permeability. The high density, and possible high viscosity, of melts at depth also makes melt extraction more difficult than at shallower levels. The increased buoyancy of a partially molten source region may be more important than the density contrast between melts and residual crystals. Garnet exsolution at high temperature can also cause a deeper layer to become buoyant relative to a colder shallow peridotite layer. These effects serve to accentuate the instability of thermal boundary layers in a chemically stratified upper mantle. Such boundary layers may occur at the base of the upper mantle (near 650 km) and near 400 km. Of course, the instability breaks the strict layering, and such a mantle will exhibit both radial and lateral inhomogeneity, as suggested by the seismic results discussed in Chapter 10.

The previous chapter discussed the properties of the various types of magmas that emerge from the mantle. These magmas provide clues about the chemistry of some parts of the mantle but do not provide a unique composition or mineralogy. We do not even know what fraction of the mantle the basalts represent or from what depth they originated. Presumably, the parts that are left behind are denser and, possibly, more refractory. It is not clear that we have samples of the original source rock or even a representative sample of the residual material. Various attempts have been made to infer the properties of the basalt source region from rocks that are exposed at the surface, and we now turn our attention to observed and hypothetical rock types that may be important in the upper mantle.

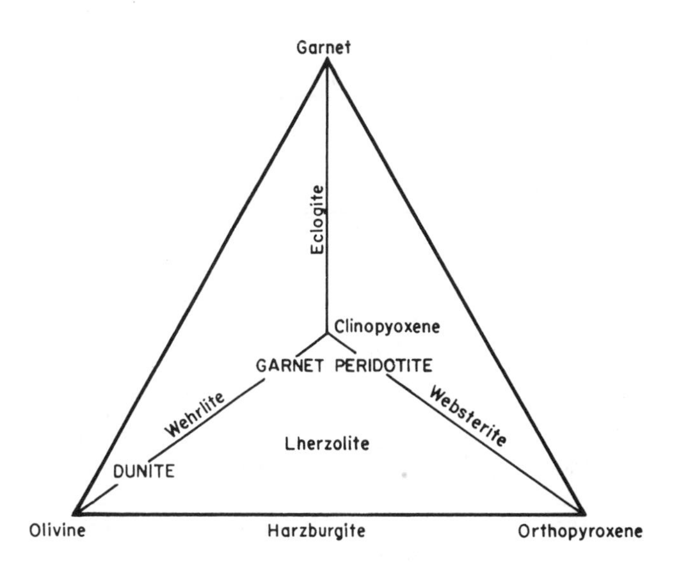

FIGURE 9-1
Nomenclature tetrahedron for assemblages of olivine, clinopyroxene, orthopyroxene and garnet. Dunites and garnet peridotites lie within the tetrahedron.

ULTRAMAFIC ROCKS

Ultramafic rocks are composed chiefly of ferromagnesian minerals and have a low silicon content. The term is often used interchangeably with "ultrabasic," although pyroxene-rich rocks are ultramafic but not ultrabasic because of their high SiO_2 content. Peridotites, lherzolite, dunite and harzburgite are specific names applied to ultramafic rocks that are chiefly composed of olivine, orthopyroxene, clinopyroxene and an aluminous phase such as plagioclase, spinel or garnet. Ultramafic rocks are dense and mainly composed of refractory minerals with high seismic velocities. Most of the shallow mantle is probably ultramafic in composition. Basic rocks, such as basalts, become dense at high pressure (for example, eclogite) and can have properties comparable to the more refractory peridotites. The main subdivisions of ultramafic rocks, along with the dominant minerals, are garnet peridotite (olivine, orthopyroxene, clinopyroxene, garnet), lherzolite (olivine, orthopyroxene, clinopyroxene), harzburgite (olivine, orthopyroxene), wehrlite (olivine, clinopyroxene), dunite (olivine ± clinopyroxene), websterite (orthopyroxene, clinopyroxene) and eclogite (clinopyroxene, garnet). Some eclogites overlap basalts in their bulk chemistry. The relationships between these rocks are shown in Figure 9-1.

Peridotites can represent

1. The refractory residue left after basalt extraction
2. Cumulates formed by the crystallization of a magma
3. Primitive mantle that can yield basalts by partial melting
4. Cumulates or residues that have been intruded by basalt
5. High-pressure or high-temperature melts (Herzberg, 1984, 1986)

Peridotites are divided into fertile or infertile (or barren) depending on their Al_2O_3, CaO and Na_2O content. Fertile peridotites can be viewed as having an appreciable basaltic component. The terms "enriched" and "depleted" are often used interchangeably with "fertile" and "infertile" but have trace-element and isotopic connotations that are often inconsistent with the major-element chemistry. I use "infertile" and "barren" as attributes of a rock poor or very poor in CaO and Al_2O_3 and "depleted" for rocks poor in the incompatible trace elements and having low ratios of Rb/Sr, LREE/HREE and so on. These distinctions are necessary since some rocks are fertile yet depleted, for instance, the source region for midocean-ridge basalts. Table 9-1 gives compositions for representative ultramafic rocks.

Garnet lherzolites are composed mainly of olivine and orthopyroxene (Table 9-2). Olivine is generally in the range of 60 to 70 volume percent and orthopyroxene 30 to 50 percent. The average clinopyroxene and garnet proportions

TABLE 9-1
Compositions of Spinel and Garnet Lherzolites

	Spinel Lherzolite		
Oxide	Continental (avg. of 301)	Oceanic (avg. of 83)	Garnet Lherzolite
SiO_2	44.15	44.40	44.90
Al_2O_3	1.96	2.38	1.40
FeO	8.28	8.31	7.89
MgO	42.25	42.06	42.60
CaO	2.08	1.34	0.82
Na_2O	0.18	0.27	0.11
K_2O	0.05	0.09	0.04
MnO	0.12	0.17	0.11
TiO_2	0.07	0.13	0.06
P_2O_5	0.02	0.06	—
NiO	0.27	0.31	0.26
Cr_2O_3	0.44	0.44	0.32

Maaløe and Aoki (1977).

are about 5 percent and 2 percent, respectively (Maaløe and Aoki, 1977).

The major oxides in peridotites and lherzolites generally correlate well (Figure 9-2 and Table 9-3). An increase in MgO correlates with decreases in SiO_2, Al_2O_3, CaO, TiO_2 and Na_2O and an increase in NiO. CaO, Al_2O_3 and Na_2O all go to approximately zero at an MgO content of about 48 weight percent. Cr_2O_3, MnO and FeO are roughly constant. The lherzolite trend can be explained by variable amounts of clinopyroxene and garnet.

Olivine- and orthopyroxene-rich rocks, presumably from the mantle, are found in foldbelts, ophiolite sections, oceanic fracture zones and, as xenoliths, in kimberlites and alkali-rich magmas. They are rare in less viscous magmas such as tholeiites. Olivine and orthopyroxene in varying proportion are the most abundant minerals in peridotites. These are dense refractory minerals, and peridotites are therefore generally thought to be the residue after melt extractions. Some peridotites are shallow cumulates deposited from cooling basalts and are therefore not direct samples of the mantle. Alumina in peridotites is distributed among the pyroxenes and accessory minerals such as plagioclase, spinel and garnet. At higher pressure most of the Al_2O_3 would be in garnet. Garnet-rich peridotite, or pyrolite, is the commonly assumed parent of mantle basalts. This variety is fertile peridotite since it can provide basalt by partial melting. Most peridotites, however, have relatively low Al_2O_3 and can be termed barren. These are commonly thought to be residual after melt extraction. Al_2O_3-poor peridotites are less dense than the fertile variety and should concentrate in the shallow mantle. Given sufficient water at crustal and shallow mantle temperatures, peridotite may be converted to serpentinite with a large reduction in density and seismic velocity. Hydrated upper mantle may therefore be seismically indistinguishable from lower crustal minerals.

Lherzolites typically contain 60 to 80 percent olivine, 20 to 40 percent orthopyroxene, less than 14 percent clinopyroxene and 1 to 10 percent of an aluminous phase such as spinel or garnet. Spinel lherzolites, the lower-pressure assemblages, dredged from the ocean bottom are similar in composition to those found in alkali basalts and kimberlites on oceanic islands and continents. Garnet lherzolites, the higher-pressure assemblages, have lower Al_2O_3, CaO and FeO and are therefore denser than spinel lherzolites only when they contain appreciable garnet. They would become less dense at higher temperature, lower pressure or if partially molten.

The major-element chemistries of lherzolites vary in a systematic fashion. Most of the oxides vary linearly with MgO content. SiO_2, Al_2O_3, CaO, Na_2O and TiO_2 decrease with an increase in MgO. These trends are generally consistent with variable amounts of a basaltic component. However, the basaltic component, especially for the MgO–SiO_2 trend, is not tholeiitic or MORB. If lherzolites represent olivine-orthopyroxene-rich rocks with variable amounts of melt extraction or addition, this melt component is andesitic in major elements.

The major-element trends of lherzolites may also be controlled by melt-crystal equilibration at various depths in

TABLE 9-2
Mineralogy of Spinel and Garnet Lherzolites

	Spinel Lherzolite		Garnet Lherzolite	
Mineral	Average (wt. pct.)	Range (vol. pct.)	Average (vol. pct.)	Range (vol. pct.)
Olivine	66.7	65–90	62.6	60–80
Orthopyroxene	23.7	5–20	30	20–40
Clinopyroxene	7.8	3–14	2	0–5
Spinel	1.7	0.2–3	—	—
Garnet	—	—	5	3–10
Phlogopite	—	—	0.4	0–0.5

Maaløe and Aoki (1977).

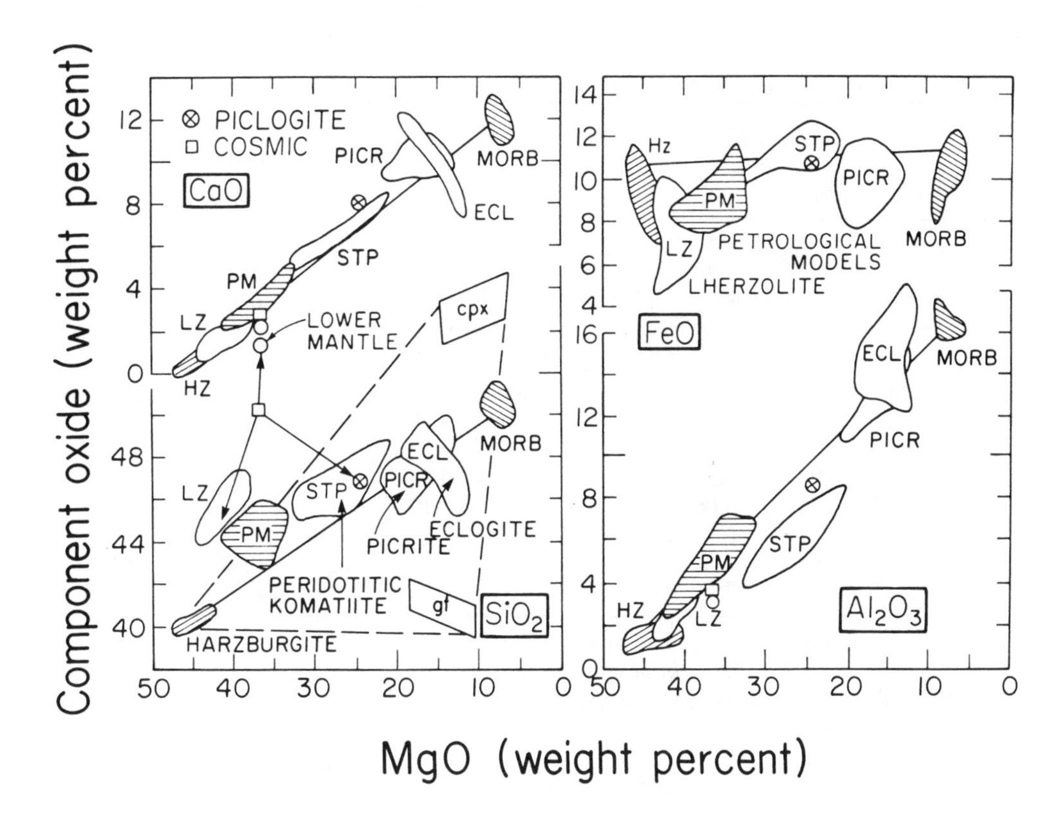

FIGURE 9-2
SiO_2, CaO, Al_2O_3 and FeO versus MgO for igneous rocks. The basalt source region probably has a composition intermediate between basalt (MORB) and harzburgite. Most petrological models (PM) of the major-element chemistry of the source region favor a small basalt fraction. STP (spinifex textured peridotites) are high-temperature MgO-rich magmas. Picrites (PICR) are intermediate in composition between STP and MORB and may evolve to MORB by olivine separation. Picrites and eclogites (ECL) overlap in composition. Lherzolites (LZ) contain an orthopyroxene component, but the other rock types are mainly clinopyroxene + garnet ± olivine. Squares represent estimates of primitive mantle composition based on a chondritic model. If the upper mantle is primarily lherzolite, basalt and harzburgite, the lower mantle (open dot) will be primarily orthopyroxene. The composition of the MORB source (piclogite model) probably falls between PICR and PM or STP.

the mantle. Lherzolites, and most other ultramafic rocks, are generally thought to be the refractory residue complementary to melts presently being extracted from the mantle. They differ, however, from primitive mantle compositions. In particular they contain more olivine and less orthopyroxene than would be appropriate for a chondritic or "cosmic" mantle. Upper-mantle lherzolites and basalts may be complementary to the lower mantle, representing melts from the original, accretional differentiation of the mantle. The MgO content of melts increases with temperature and with depth of melting. At great depth (>200 km) relatively low-MgO phases, such as orthopyroxene and garnet-majorite may remain behind, giving olivine-rich melts. The major-element trends in lherzolites may therefore represent trends in high-pressure melts (Herzberg, 1984; Herzberg and others, 1986).

Unserpentinized peridotites have seismic velocities and anisotropies appropriate for the shallow mantle. This situation is often generalized to the whole mantle, but seismic data for depths greater than 400 km are not in agreement with that hypothesis. It is not even clear that peridotite has the proper seismic properties for the lower lithosphere. In the depth interval 200–400 km both eclogite and peridotite can satisfy the seismic data.

Garnet pyroxenites and eclogites are also found among the rocks thought to have been brought up from the mantle, and they have physical properties that overlap those of the ultramafic rocks. Garnet-rich pyroxenites and eclogites are denser than peridotites and should therefore occur deeper. The extrapolation of the properties of peridotites to the deep upper mantle, much less the whole mantle, should be done with caution. Not only do other rock types emerge from the mantle, but there is reason to believe that peridotites will be concentrated in the shallow mantle. The common belief that seismic data *require* a peridotite mantle is simply false. The V_p and V_s velocities in the upper 150 km of the mantle under stable shield areas are consistent with olivine-rich perido-

TABLE 9-3
Compositions of Peridotites and Pyroxenite

	Lherzolites				Pyrox-			
	Spinel		Garnet	Dunite	enite	Peridotites		
Oxide	(1)	(2)	(3)	(4)	(5)	(6)	(7)	(8)
SiO_2	44.15	44.40	44.90	41.20	48.60	44.14	46.36	42.1
Al_2O_3	1.96	2.38	1.40	1.31	4.30	1.57	0.98	
FeO	8.28	8.31	7.89	11.0	10.0	8.31	6.56	7.10
MgO	42.25	42.06	42.60	43.44	19.10	43.87	44.58	48.3
CaO	2.08	1.34	0.82	0.80	13.60	1.40	0.92	
Na_2O	0.18	0.27	0.11	0.08	0.71	0.15	0.11	
K_2O	0.05	0.09	0.04	0.016	0.28	—	—	
MnO	0.12	0.17	0.11	0.15	0.18	0.11	0.11	
TiO_2	0.07	0.13	0.06	0.06	0.83	0.13	0.05	
P_2O_5	0.02	0.06	—	0.10	0.10	—	—	
NiO	0.27	0.31	0.26	—	—	—	—	
Cr_2O_3	0.44	0.44	0.32	—	—	0.34	0.33	
H_2O	—	—	—	0.50	0.90	—	—	

(1) Average of 301 continental spinel lherzolites (Maaløe and Aoki, 1977).
(2) Average of 83 oceanic spinel lherzolites (Maaløe and Aoki, 1977).
(3) Average garnet lherzolite (Maaløe and Aoki, 1977).
(4) Dunite (Beus, 1976).
(5) Pyroxenite (Beus, 1976).
(6) High-T peridotites, South Africa (Boyd, 1986).
(7) Low-T peridotites, South Africa (Boyd, 1986).
(8) Extrapolated lherzolite trend (~0 percent Al_2O_3, CaO, Na_2O, etc.).

tite. Elsewhere in the shallow mantle both eclogite and peridotite are consistent with the data.

It should be kept in mind that the most abundant rock type by far that emerges from the mantle is tholeiite (equivalent to garnet plus clinopyroxene). A garnet peridotite with suitable compositions of garnet and clinopyroxene, particularly Na_2O and TiO_2 content, can certainly yield basalts by partial melting. The amount of melting, however, and the amount of residual olivine and orthopyroxene are unknown. A suitable natural garnet peridotite (pyrolite) has not yet been found that can provide the major- and trace-element abundances found in, for example, midocean-ridge tholeiites. Such a peridotite would have to have high Al_2O_3, Na_2O, CaO and TiO_2 and low concentrations of the incompatible elements. Pyrolite (next section) therefore remains a hypothetical mantle rock.

If picrites are the parent for tholeiitic basalts, then roughly 30 percent melting is implied for generation from a shallow peridotitic parent. If the parent is eclogitic, then similar temperatures would cause more extensive melting. However, generation of basaltic magmas from an eclogitic parent does not require extensive melting. Melts of basaltic composition are provided over a large range of partial melting.

Basalts and peridotites are two of the results of mantle differentiation. They both occur near the surface of the Earth and may not represent the whole story. They are also not necessarily the result of a single-stage differentiation process. In the next chapter I discuss the possible evolution of the mantle, from its high-temperature birth to the formation of the geochemical reservoirs that we sample today. This complements the standard approach that attempts to take the observed rocks and work backward in time.

PYROLITE

The general idea of a peridotitic mantle as a source for basaltic magmas by partial melting goes back at least as far as Bowen (1928). Ringwood (1962) formalized the concept in a series of papers starting in 1962. He proposed the existence of a primitive mantle material that was defined by the property that on fractional melting it would yield a typical basaltic magma and leave behind a residual refractory dunite-peridotite. This he called "pyrolite" (pyroxene-olivine-rock). The initially proposed composition was one part basalt and four parts dunite, although this is arbitrary. Green and Ringwood (1967) considered that 20–40 percent melting will be necessary before liquid segregates and begins an independent existence, and this mechanical idea influenced their choice of basalt-to-dunite ratios. The stable phase assemblages in pyrolite, and the disposition of Al_2O_3, depend on temperature and pressure. Garnet pyrolite is essentially identical with garnet peridotite but is more fertile than most natural samples (Ringwood, 1962). Ringwood earlier argued that garnet pyrolite would not yield, by direct partial melting, basaltic magma and proposed that partial melting of plagioclase pyrolite must be the main source of basalts. He attributed the low-velocity zone to the presence of plagioclase and assembled many arguments in favor of these conjectures. Plagioclase is not in fact stable at upper-mantle pressures, and Ringwood revised his pyrolite model and ideas about the basalt source region many times in the subsequent years.

In an early model garnet pyrolite was considered to be free of orthopyroxene. In 1963 Green and Ringwood calculated a pyrolite composition based on the assumption that it was three parts dunite plus one part of the averages of tholeiitic and alkali olivine basalt; it therefore contained some orthopyroxene. In 1966 Ringwood assumed that pyrolite was a three-to-one mix of Alpine-type peridotite and a Hawaiian olivine-tholeiite. Table 9-4 gives compositions of some of these pyrolite models.

The cornerstones of the pyrolite hypothesis were that the basaltic magmas that were thought to represent the largest degrees of partial melting, tholeiites, were primitive, unfractionated melts and that they were in equilibrium with the residual peridotite at the depth of melt separation. O'Hara (1968) argued that extensive crystal fractionation, at depth, operated on the parent magma and, therefore, basalts including tholeiites were not primary magmas. He suggested that partial melting and magma segregation occur at pressures greater than 30 kbar, rather than the 5–25 kbar in the Green-Ringwood models. Under these conditions the parent magma is picritic, and extensive olivine crystallization at low pressure produces tholeiitic magmas. He also proposed that extensive eclogite fractionation at depths of order 100 km occurred. Ringwood (1975) argued strongly against O'Hara's model.

Jaques and Green (1980) on the basis of an extensive series of melting experiments on synthetic peridotites abandoned the earlier model of Green and Ringwood (1967) and Green (1971), which involved the segregation of olivine tholeiite magmas at about 30 km depth. Studies by Green and others (1979) and Stolper (1980) have shown that primary picritic magmas are likely parents for midocean-ridge basalts by separation of about 15 percent olivine at about 60–70 km depth and temperatures of about 1400–1450°C. Picritic and komatiitic magmas require about 40–60 percent melting of peridotite if the melting initiates at such shallow depths. Green and others showed that there can be no simple genetic relationship between the primary tholeiitic picritic magmas that are parental to midocean-ridge basalts and the residual harzburgite or accumulate peridotite of ophiolite complexes. Thus, these studies effectively pull the rug out from under the basic pyrolite assumptions. Nevertheless, the same authors proposed yet more pyrolite models that are mixtures of picrite and varying, arbitrary proportions of olivine (50–69 percent), orthopyroxene (0–25 percent) and spinel (0.5 to 1 percent). The new pyrolites are richer in Al_2O_3 and CaO than previous pyrolite models and less rich in TiO_2, Na_2O and K_2O. Their preferred model generates the primary midocean-ridge basalts by 24 percent partial melting.

At this point one is reminded of Karl Popper's (1972) "conventionalist stratagem": "Some genuinely testable theories, when found to be false, are still upheld by their admirers by introducing ad hoc some auxiliary assumptions

TABLE 9-4
Compositions of Pyrolites and Garnet Peridotites (Weight Percent)

	Pyrolite				Garnet Peridotite	
Oxide	**(1)**	**(2)**	**(3)**	**(4)**	**(5)**	**(6)**
SiO_2	45.1	42.7	46.1	45.0	42.5	46.8
TiO_2	0.2	0.5	0.2	0.2	0.1	0.0
Al_2O_3	3.3	3.3	4.3	4.4	0.8	1.5
Cr_2O_3	0.4	0.5	—	0.5	—	—
MgO	38.1	41.4	37.6	38.8	44.4	42.0
FeO	8.0	6.5	8.2	7.6	3.8	4.3
MnO	0.15	—	—	0.11	0.10	0.11
CaO	3.1	2.1	3.1	3.4	0.5	0.7
Na_2O	0.4	0.5	0.4	0.4	0.1	0.1
K_2O	0.03	0.18	0.03	0.003	0.22	0.02

(1) Ringwood (1979), p. 7.
(2) Green and Ringwood (1963).
(3) Ringwood (1975).
(4) Green and others (1979).
(5) Boyd and Mertzman (1987).
(6) Boyd and Mertzman (1987).

or by re-interpreting the theory ad hoc in such a way that it escapes refutation. Such a procedure is always possible, but it rescues the theory from refutation only at the price of destroying, or at least lowering its scientific status."

Jaques and Green (1980), in common with most petrologists, have restricted their attention to rocks that can form basalts by a relatively small degree of partial melting since high degrees of partial melting may not be geologically feasible. They point to laboratory measurements showing strong crystal settling of olivine when melting exceeds 40 percent, implying that conditions in a convecting mantle are the same as in a static laboratory experiment. The real Earth has a cold, high-viscosity, buoyant, strong and relatively impermeable lithosphere. An important question, then, is how much melting can occur in the underlying convecting mantle that cannot easily lose its melt fraction to the surface? What are the likely products of partial melting in a convecting, differentiating mantle?

In a garnet-clinopyroxene-rich rock there is a small temperature differential between the onset of melting and the temperature of extreme melting. Adiabatic ascent of such a rock can lead to extensive melting. In contrast, in a garnet peridotite, the eclogite fraction is melted near the solidus but the bulk of the rock does not melt until much higher temperatures, near the liquidus. Therefore, garnet peridotites require a large temperature increase in order to melt extensively.

The pyrolite hypothesis was based entirely on major elements and on several arbitrary assumptions regarding allowable amounts of basalt and melting in the source region. It does not satisfy trace-element or isotopic data on basalts, and it violates chondritic abundances and evidence for mantle heterogeneity.

ECLOGITES

Häuy (1822) introduced the term "eclogite" for rocks composed of omphacite (diopside plus jadeite) and garnet, occasionally accompanied by kyanite, zoisite, amphibole, quartz and pyrrhotite. Natural eclogites have a variety of associations, chemistries, mineralogies and origins, and many names have been introduced to categorize these subtleties. Nevertheless, the term "eclogite" implies different things to different workers. To some eclogites mean metamorphic crustal rocks, and to others the term implies bimineralic kimberlite xenoliths. The chemical similarity of some eclogites to basalts prompted early investigators to consider eclogite as a possible source of basalts but more recently has been taken as evidence that these eclogites are simply subducted oceanic crust or basaltic melts that have crystallized at high pressure. Some eclogites are demonstrably metamorphosed basalts, while others appear to be igneous rocks ranging from melts to cumulates. The trend in recent years has been toward the splitters rather than the lumpers, and toward explanations that emphasize the derivative and secondary nature of eclogite rather than the possible importance of eclogite as a source rock for basaltic magmas.

Pyroxene-garnet rocks with jadeite-bearing pyroxenes are found as inclusions in alkali basalt flows as layers in ultramafic intrusions, as inclusions in kimberlite pipes, as tectonic inclusions in metamorphic terranes associated with gneiss and schist, and as inclusions in glaucophane schist areas. Jadeite-poor garnet clinopyroxenites are abundant in Salt Lake Crater, Hawaii. Only the eclogites from kimberlite pipes are demonstrably from great depth, that is, well into the upper mantle. Some of these contain diamonds.

Peridotites and lherzolites predominate over all other rock types as inclusions in diamond-bearing kimberlites. The presence of diamond indicates origin depths of at least 130 km, and other petrological indicators suggest depths even greater. In a few kimberlite pipes eclogites form the majority of inclusions. The overwhelming majority of peridotites and lherzolites are infertile, that is, very low in aluminum, sodium and calcium. If the distribution of rock types in kimberlite inclusions is representative of the source region, the majority of basaltic components in the upper mantle resides in eclogites rather than in an olivine-rich rock (Smyth and others, 1984).

Table 9-5 shows typical clinopyroxene and garnet compositions of eclogites and peridotites, a synthetic two-mineral eclogite and for comparison, a typical MORB. Note that, in general, diopside plus garnet from peridotite does not approximate the composition of MORB. In most cases, however, omphacite and garnet from eclogite bracket MORB compositions, and therefore eclogite is a more appropriate source rock. Table 9-6 gives comparisons of the bulk chemistry of some South African eclogites and MORB and an estimate of the average composition of the oceanic crust. There is a close correspondence between the composition of kimberlite eclogites and the material in the oceanic crust.

Many kimberlite eclogites show signs of garnet exsolution from clinopyroxene, implying that there can be substantial changes in the density of eclogites as a function of temperature in the subsolidus region. Clinopyroxenes in eclogites have exsolved 20 percent or more garnet, implying a substantial increase in pressure or decrease in temperature (Smyth and Caporuscio, 1984). A pressure increase is unlikely. A representative eclogite can increase in density by 2.5 percent by cooling from 1350°C to 950°C at 30 to 50 kbar. The reverse process can happen as garnet plus clinopyroxene is elevated into a lower pressure regime along an adiabat. For example, the density of an eclogite can decrease by about 3 percent simply by rising 50 km. Thus, garnet exsolution caused by pressure release can accomplish more than a 1000°C rise in temperature, by thermal expansion, all in the subsolidus regime. This plus the low

TABLE 9-5
Typical Trace Element Concentrations (ppm) and Ratios in Eclogite and Peridotite Minerals

	Eclogites		Synthetic	Peridotites		
	Omphacite	Garnet	Eclogite	Diopside	Garnet	MORB
K	1164	337	820	615	296	700
Rb	0.565	1.14	0.7	2.1	1.45	0.4
Sr	249	8.25	95	337	5.50	110
Na	52244	74	12700	1332	2420	17300
Ti	4856	899	2500	659	959	5500
Zn	106	15	55	28	69	80
Rb/Sr	0.002	0.14	—	0.006	0.26	0.004
Sm/Nd	0.206	0.522	—	0.211	0.590	0.335
Rb/K	4.9*	33.8*	8.5*	34.1*	49.0*	5.7*

Basu and Tatsumoto (1982), Wedepohl & Muramatsu (1979).
$* \times 10^{-4}$.

melting point of eclogite means that rising eclogite-rich convection currents obtain considerable buoyancy.

The MORB source region has been depleted by removal of either a small melt fraction or a late-stage intercumulus fluid. The abundances of uranium, thorium and potassium in various mantle samples (given in Table 9-7) show that eclogite xenoliths found in kimberlite are often less rich in these elements than peridotites found in the same pipe, although both may have been contaminated by the kimberlite matrix. These elements are, nevertheless, higher than in primitive mantle. The $^{206}Pb/^{204}Pb$ ratios of eclogites are generally low, implying long-term depletion.

At one time eclogite was considered a possible parent material for basalt. In recent years the consensus has swung strongly to the view that partial (~20 percent) melting of garnet peridotite provides the basaltic magmas. Cosmochemical and seismic data have been used in support of the latter hypothesis. However, the differentiation of the Earth into crust, depleted and enriched reservoirs and, in particular, the evidence for a depleted MORB reservoir indicate that a chondritic source is not appropriate for all, if any, mantle reservoirs, even if the average composition of the mantle is chondritic. An eclogite cumulate layer is one possible product of a differentiated chondritic Earth and a possible source for basaltic magmas.

The other arguments for a peridotite source have to do with the amount of melting that is considered plausible. Melt-crystal separation can occur easily at relatively shal-

TABLE 9-6
Comparison of Kimberlite Eclogites and Some Other Rock Types

Oxide (percent)	Eclogite (1)	Eclogite (2)	Picrite (3)	MORB (4)	Ocean Crust (5)
SiO_2	45.2	47.2	44.4	47.2	47.8
TiO_2	0.5	0.6	1.18	0.7	0.6
Al_2O_3	17.8	13.9	10.2	15.0	12.1
Cr_2O_3	0.4	—	0.22	—	—
Fe_2O_3	—	—	—	3.4	—
FeO	11.2	11.0	10.92	6.6	9.1
MgO	13.1	14.3	18.6	10.5	17.8
MnO	0.3	—	0.17	0.1	—
CaO	9.6	10.1	9.7	11.4	11.2
Na_2O	1.6	1.6	1.37	2.3	1.3
K_2O	0.03	0.8	0.13	0.1	0.03

(1) Bobbejaan eclogite (Smyth and Caporuscio, 1984).
(2) Roberts Victor eclogite (Smyth and Caporuscio, 1984).
(3) Picrite, Svartenhuk (Clarke, 1970).
(4) Oceanic tholeiite (MORB).
(5) Average oceanic crust (Elthon, 1979).

TABLE 9-7
Heat-Producing Elements in Mantle Rocks (ppm)

Rocks	Uranium	Thorium	Potassium
Eclogites			
Roberts Victor	0.04–0.8	0.29–1.25	83–167
	0.04–0.05	1.3	1833
Zagadochnaya	0.024		
Jagersfontein	0.07	0.17	1000
Bultfontein	0.07	0.31	1300
MORB	0.014	0.035	660
Garnet peridotite	0.22	0.97	663
Ultramafic rock	0.26	0.094	35
Primitive Mantle	0.02	0.077	151

Note: Measured values of $^{206}Pb/^{204}Pb$ for eclogites are 15.6–19.1; for primitive mantle it is ~17.5.

low depths in stationary magma chambers but is less likely at depth or in rising decompressing diapirs. In particular, partial melts at depth can be trapped if the overlying layer is subsolidus and impermeable or if the melt density increases rapidly with pressure. More likely Rayleigh-Taylor instabilities occur in the deeper layer, and large diapirs with entrained melt rise through the shallow mantle. Pressure-release melting then can cause extensive melting in the diapir. The instability occurs when melting in the deeper layer is extensive enough that, by the elimination of garnet, it becomes buoyant relative to the shallow mantle. The top of an eclogite-rich zone underlying a peridotite region will be in a thermal boundary layer and therefore 100–200°C warmer than the shallower mantle. A density reversal would set in for about 10–20 percent melting in eclogite if the overlying material is peridotite. We refer to a basalt or eclogite-rich rock as a piclogite.

PICLOGITE

Peridotites and pyrolite are rock types composed primarily of the refractory crystals olivine and orthopyroxene. Olivine is generally more than 60 percent, and both minerals together typically constitute more than 80 percent of the rock. Clinopyroxene and aluminous phases such as spinel, garnet and jadeite—the basalt assemblage—are minor constituents. At the other extreme are rocks such as clinopyroxenites and eclogites, which are low in orthopyroxene and olivine. Intermediate are rocks such as picrites, olivine eclogites and komatiites. Rocks that have less than 50 percent olivine plus orthopyroxene have been given the general name *piclogite*. In major elements they fall between dunites and basalts and between lherzolites and picrites. They contain a larger basaltic fraction than peridotites, although they contain the same minerals: olivine, orthopyroxene, diopside-jadeite and garnet. At high pressure they are denser than lherzolite, but at high temperature they can become less dense. Piclogites can represent garnet-rich cumulates or frozen high-pressure melts and can generate basaltic melts over a wide range of temperatures. They are not compositionally equivalent to basalts and do not require large degrees of partial melting to generate basaltic magmas. However, if piclogites are the source reservoir for mid-ocean-ridge basalts, the garnet must be mainly eliminated in order to satisfy the HREE data and in order to decrease the density of the reservoir so that it can rise into the shallow mantle. Therefore, clinopyroxene probably exceeds garnet in any reservoir with clinopyroxene rather than garnet as a near-liquidus phase.

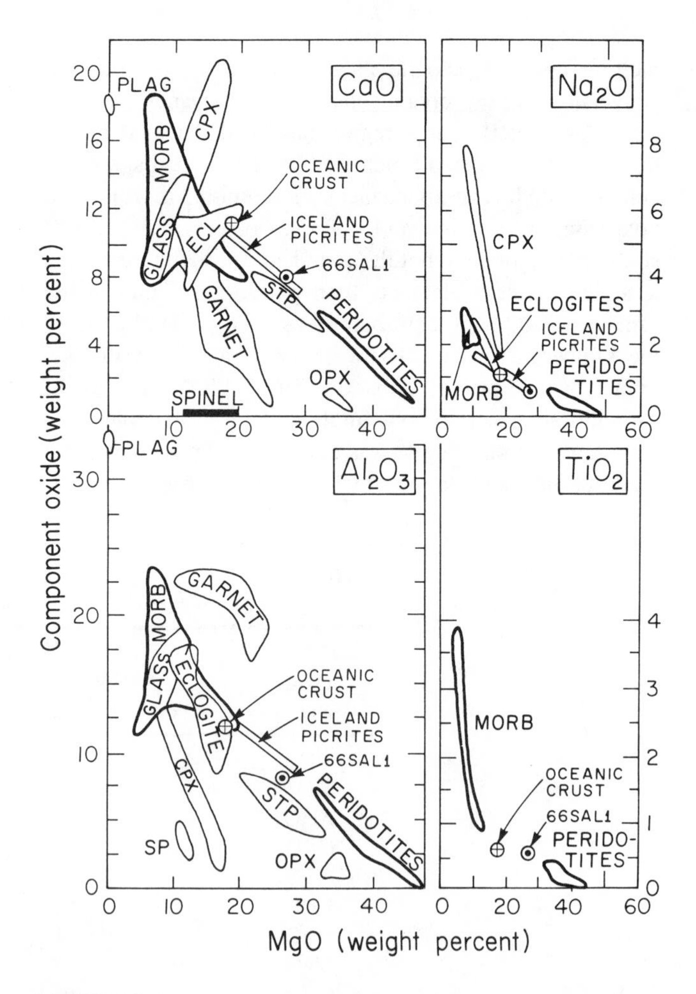

FIGURE 9-3
Major-element oxides of igneous rocks and rock-forming minerals. If peridotites represent residues after basalt extraction and MORB and picrites represent melts, and if these are genetically related, the MgO content of the basalt source is probably between about 20 and 30 percent, much less than most lherzolites or peridotites. 66 SAL-1 is a garnet pyroxenite from Salt Lake Crater, Hawaii, and falls in the composition gap between basalts and peridotites. It is a piclogite.

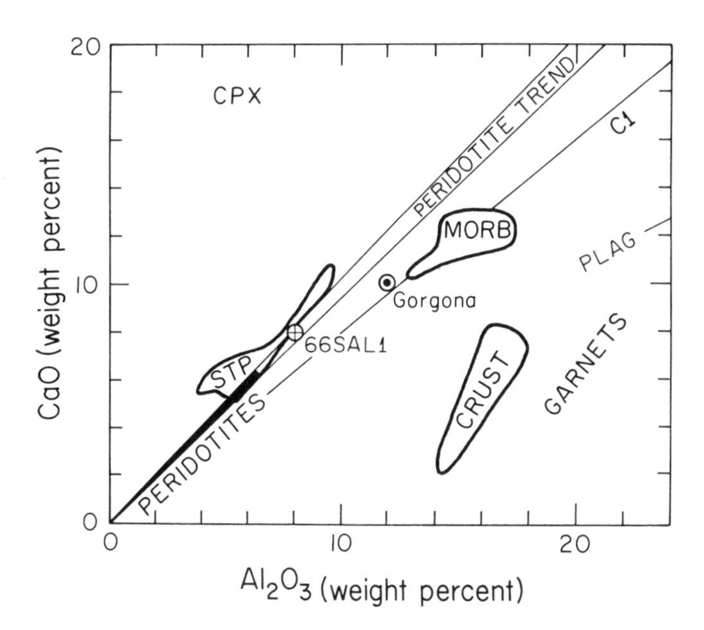

FIGURE 9-4
CaO versus Al_2O_3 for igneous rocks and minerals. Note that peridotites and STP are CaO-rich compared to C1 chondrites, and meteorites in general, while the continental crust is Al_2O_3-rich. MORB falls on the C1 trend, possibly suggesting that it is an extensive melt of primitive mantle. Gorgona is an island in the Pacific that has komatiitic magmas. The solar CaO/Al_2O_3 ratio is slightly higher than peridotites (Chapter 1).

Peridotitic komatiites (STP, for spinifex-texture peridotites) have major-element chemistries about midway between harzburgites and tholeiites and occupy the compositional gap between lherzolites and picrites (Figures 9-3 and 9-4). The present rarity of komatiitic magmas may be partially due to their high density and the thick lithosphere. Komatiites evolve to picrites and then to tholeiites as they crystallize and lose olivine, becoming less dense all the while. In a later chapter I show that the properties of the transition region (400–650 km depth) are consistent with less than 50 percent olivine and low orthopyroxene content, that is, piclogite.

Compositions of picrites, komatiites and a garnet pyroxenite (66 SAL-1) from Salt Lake Center, Hawaii are given in Table 9-8. These are all examples of rocks that fall between basalts and peridotites in their major-element chemistry. In terms of mineralogy at modest pressure they would be olivine eclogites or piclogites. They are all suitable source rocks for basalts.

DEPTHS OF MANTLE RESERVOIRS

Petrological data suggest that basalts separated from their parent material at depths of about 60 to 90 km. This is not necessarily the depth of the source region, although it is generally assumed that melting initiates in the low-velocity zone and that this part of the mantle is in fact the source region. More likely, melting is caused by adiabatic ascent from greater depth, and the shallower depths are where cooling, fractionation and melt-crystal separation occur. Recent seismic results show that the low velocities under the East Pacific Rise extend to a depth of 400 km (Walck, 1984). Surface-wave tomographic results (Nataf and others, 1986) are consistent with this and indicate even greater depths under some ridge systems. In a convecting mantle, with ridge migration, the roots of midocean ridges are displaced relative to their surface expression, and this is shown clearly in the tomographic results. It is likely, therefore, that the ultimate source region of midocean-ridge basalts is in the transition region rather than in the low-velocity zone. The low-velocity zone is the location of final crystal-melt equilibration and MORB, or picrite, separation.

The V_p/V_s ratio, closely related to Poisson's ratio, generally increases with temperature, pressure and partial melting. The high V_p/V_s of the ridge structure, particularly between 90 and 150 km depth, suggests high temperature and, for the highest values, partial melting. The peculiar reversal of V_p/V_s under the shield LID at 150 km, associated with a decrease in K/ρ, can be explained by either a high orthopyroxene content or by anisotropy. In olivine aggregates the slow direction has low V_p/V_s and K/ρ. This zone has been sampled by kimberlites and seems to be predominantly highly deformed peridotite. This is probably the decoupling zone between the shield lithosphere and the underlying mantle. The rapid increase in velocity near 200 km implies a change in petrology.

In an accreting planet the upper mantle is above the solidus throughout most of the growth phase and the planet ends up zone refined, with melts and incompatible elements concentrated near the surface and dense, depleted residual crystals concentrated toward the interior (see Chapter 2). Even if a planet can avoid differentiation and gravitational separation during accretion, the relative slopes of the adiabat and the melting point assure that upper-mantle melting will occur in the rising limbs of convection calls. In olivine-rich peridotites it is unlikely that the liquidus is exceeded, but the separation of melt and crystals serves to remove the high-melting components so that a large part of the ancient upper mantle may actually have been completely molten. Freezing of the molten layer proceeds both from the top and the bottom, giving, on a Moon-size planet, plagioclase crystals that float to the surface and, in the case of an Earth-size planet, dense aluminous phases, such as garnet, which sink. In an accreting or crystallizing planet, refractory residual phases form the base of the system. Nearer the surface are layers of cumulate crystals, which represent the early crystallization products of the melt, and the late-stage residual melts, which contain most of the incompatible elements. The expulsion of late-stage intercumulus melts from deep cumulate layers is an alternative to small degrees of

TABLE 9-8
Compositions of Picrites, Komatiites and Garnet Pyroxenite

	Group II[1]		Komatiite		
Oxide	Svartenhuk	Baffin Is.	S. Africa[2]	Gorgona[3]	66SAL-1[4]
SiO_2	44.4	45.1	45.7	45.3	44.8
TiO_2	1.18	0.76	0.91	0.60	0.52
Al_2O_3	10.2	10.8	7.74	10.6	8.21
Cr_2O_3	0.22	0.27	—	—	—
FeO	10.92	10.26	—	10.9	9.77
MnO	0.17	0.18	0.23	0.18	
MgO	18.6	19.7	19.43	21.9	26.53
CaO	9.7	9.2	8.04	9.25	8.12
Na_2O	1.37	1.04	0.61	1.04	0.89
K_2O	0.13	0.08	—	0.02	0.03
P_2O_5	0.14	0.09	0.12	—	0.04
Fe_2O_3	—	—	15.44	—	—
NiO	0.13	0.12	—	—	—
H_2O	2.33	2.14	3.46	—	—

[1] Clarke (1970).
[2] Jahn and others (1980).
[3] Echeverria (1980).
[4] Frey (1980).

partial melting as a mechanism for depleting part of the mantle and forming the continental crust. In this scenario the concentration of volatiles and incompatible elements would decrease from the crust to the upper mantle to the transition region to the lower mantle, and each region would differ in major-element chemistry and mineralogy as well.

The origin of the oceanic crust and lithosphere is undergoing a dramatic paradigm shift. The conventional (pyrolite) theory is that midocean-ridge tholeiites are primary unmodified melts from the mantle, which form the oceanic crust, and the immediately underlying mantle is residual peridotite. The presumed parent is reconstructed by combining MORB and peridotite, and this is taken to be the composition of primitive mantle.

The average composition of the oceanic crust now appears, however, to be picritic, and tholeiites represent only the upper portions. The harzburgites underlying the oceanic crust in ophiolite sections appear not to be genetically related to the overlying basalts. The source region for MORB is not primitive but has experienced an ancient fractionation event that stripped it of its incompatible elements. A picritic magma is more refractory than tholeiite, and therefore the degree of partial melting and the required upper-mantle temperatures are much higher than previously thought. The source region for oceanic crust is remarkably homogeneous and has remained isolated from other parts of the mantle for a large fraction of the age of the Earth. This is known from isotopic ratios that show that this reservoir has been depleted in uranium, rubidium and samarium for over 10^9 years. These observations all suggest that the MORB reservoir is large and deep, otherwise it would be quickly contaminated by subducted sediments, altered oceanic crust and the metasomatism that has affected the shallow mantle. The upper 200 km of the mantle is extremely heterogeneous, and enriched mantle appears to underlie both continents and oceanic islands. If the enriched source is deep, it must provide magmas that penetrate the depleted reservoir, without contaminating it, and emerge at such diverse regions as continental rifts and flood basalt provinces (CFB), oceanic islands (OIB) and island arcs (IAB). Figure 9-5 shows a possible configuration for the various reservoirs.

The source of the material that provides the distinctive trace-element and isotopic signature of hotspot magmas is controversial. In most discussions the source of energy and the source of material are not differentiated. It is possible that material rising from one depth level advects high temperatures to shallow levels, causing melting and the production of hotspot magmas from the material in the shallow mantle. It is not necessary that material melt itself in situ. In a stratified mantle, without transfer of material between layers, convection in a lower layer can control the location of hotspots in the overlying layer. In fact, the high temperature gradient in the thermal boundary layer between layers is the preferred source of magma genesis since the melting gradient is larger than the adiabatic gradient in a homogeneous convecting fluid (Figure 9-5). In a homogeneous mantle, if the melting point is not exceeded in the shallow mantle, then it is unlikely to be exceeded at greater depth since the melting point and the adiabat diverge with depth.

Modern midocean-ridge basalts are extremely depleted in the large-ion lithophile elements (LIL) such as rubidium,

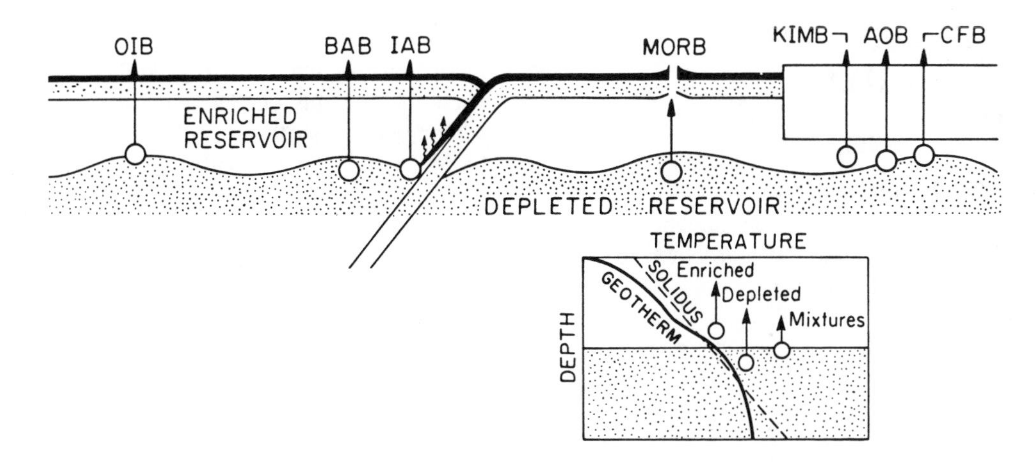

FIGURE 9-5
A possible configuration of the enriched and depleted reservoirs; a continent, midocean ridge and subduction zone are shown schematically. In this model MORBs are relatively uncontaminated melts from a deep depleted reservoir, one that may become buoyant when hot or partially molten, and rise to the shallow mantle. Other basalts from the depleted reservoir are trapped by thick crust or lithosphere and interact with the enriched reservoir. Inset shows the effects of the thermal boundary between the reservoirs.

strontium, LREE and uranium. The low values of isotopic ratios such as $^{87}Sr/^{86}Sr$ show that the MORB source cannot be primitive; it must have been depleted in the LIL for at least 10^9 years. On the other hand the helium-3 in these basalts suggests the presence of a "primitive" component. The $^{206}Pb/^{204}Pb$ ratios are low compared to other basalts but still imply a high U/Pb compared to primitive or undifferentiated mantle. U/Pb should be decreased, just as Rb/Sr, in the processes that led to depletion of the MORB reservoir. These paradoxes can be understood if (1) partial melting processes are not 100 percent efficient in removing volatiles such as helium-3 from the mantle and (2) MORBs are hybrids or blends of magmas from a depleted reservoir and an enriched reservoir. In a chemically stratified mantle one reservoir is of necessity deeper than the other, and mixing or contamination upon ascent seems inevitable. Even with this mixing it is clear from the isotopes that there are at least two ancient magma reservoirs in the mantle, reservoirs that have had an independent history for much of the age of the Earth.

Although island-arc magmas and oceanic-island magmas differ in their petrography and inferred pre-eruption crystal fractionation, they overlap almost completely in the trace elements and isotopic signatures that are characteristic of the source region. It is almost certain that island-arc magmas originate above the descending slab, although it is uncertain as to the relative contribution of the slab and the intervening mantle wedge. The consensus is that water from the descending slab may be important, but most of the material is from the overlying "normal" mantle. It is curious that the essentially identical range of geochemical signatures found in oceanic-island and continental flood basalts have generated a completely different set of hypotheses, generally involving a "primitive" lower-mantle source. If island-arc magmas originate in the shallowest mantle, it is likely that island and continental basalts do as well. Subduction may, in fact, be the primary cause of shallow mantle enrichment.

The main diagnostic differentiating some islands from most arcs is the high $^3He/^4He$ ratio of the former. High ratios, greater than atmospheric, are referred to as "primitive," meaning that the helium-3 content is left over from the accretion of the Earth and that the mantle is not fully outgassed. The word "primitive" has introduced semantic problems since high $^3He/^4He$ ratios are assumed to mean that the part of the mantle sampled has not experienced partial melting. This is not correct since it assumes that (1) helium is strongly concentrated into a melt, (2) the melt is able to efficiently outgas, (3) mantle that has experienced partial melting contains no helium and (4) helium is not retained in or recycled back into the mantle once it has been in a melt. As a matter of fact it is difficult to outgas a melt. Magmas must rise to relatively shallow depths before they vesiculate, and even under these circumstances gases are trapped in rapidly quenched glass. Although most gases and other volatiles are probably concentrated into the upper mantle by magmatic processes, only a fraction manages to get close enough to the surface to outgas and enter the atmosphere. High helium-3 contents relative to helium-4, however, require that the gases evolved in a relatively low-uranium-thorium reservoir for most of the age of the Earth. Shallow depleted reservoirs may be the traps for helium.

MANTLE METASOMATISM AND THE ENRICHED RESERVOIR

Kimberlites and alkali basalts are high in potassium, titanium, volatile components and large-ion-lithophile (LIL) elements. Several mechanisms have been proposed for the generation of highly enriched basalts:

1. They could represent small degrees of melting (2 percent or less) of mantle peridotite.
2. If the peridotite is already enriched in the LIL, then a larger degree of melting is allowed, and separation of very small melt fractions is not necessary. The enrichment event, however, must involve strongly fractionated fluids (high Rb/Sr, La/Sm and so on), and these must be the result of small degrees of partial melting or be a fractionated vapor phase.
3. The enriched fluid component could be the late-stage fluids, the last crystallizing fraction of a cumulate layer.
4. The enriched component could be introduced into the mantle by the subduction of sediments or altered oceanic crust. This cannot be the whole story since many enriched magmas from oceanic islands have high $^3He/^4He$ ratios.

The isotopic ratios of some enriched magmas indicate that the enriched component is ancient, although it could have been introduced into the magma, or the source region, relatively recently. A precursory metasomatic event is a currently popular hypothesis, but the source of the metasomatic fluid is seldom addressed. This mechanism is hard to distinguish from a mechanism that involves contamination or hybridization of a rising depleted magma with melts from an enriched source region.

Mantle-derived xenoliths from depths shallower than 170 km show evidence that titanium, alkalis, volatiles and LIL elements were introduced into the mantle, converting peridotite to assemblages bearing phlogopite, amphibole and clinopyroxene. These regions cannot be the source of LIL-depleted MORB.

Neodymium and strontium isotopic data indicate that significant isotopic heterogeneity occurs in the source regions of kimberlites and alkali basalts or that these basalts represent variable degrees of mixing between an enriched and a depleted magma. The combination of variable degrees of mixing and subsequent crystal fractionation can give suites of magmas that may not appear to be simple mixtures. The residual or cumulate assemblages will also display LIL and isotopic heterogeneity.

Metasomatism introduces ingredients such as iron, titanium and H_2O, which lower the solidus temperature, and also introduces the heat-producing elements potassium, uranium and thorium. Retention of interstitial fluids by a crystallizing cumulate also introduces phases that are rich in these elements and that will yield them up early in a subsequent melting episode. These phases will probably be concentrated in the shallow mantle. Enriched material will also be concentrated in the shallow mantle by subduction.

The time sequence of eruption of alkali and tholeiitic basalts suggests that the enriched source is shallower than the depleted source. Extremely depleted MORB only occurs at mature continental and oceanic rifts, implying that the earlier basalts are contaminated. The shallow mantle may provide the "metasomatic" trace-element and isotopic signature, and the deeper source may provide the bulk of the magma.

The refractory rocks contained in magmas and kimberlites and exposed in mountain belts are harzburgites, dunites, orthopyroxenites and lherzolites, with harzburgites constituting a large fraction of the volume. Many of these, particularly the alpine peridotites, are depleted in the large-ion-lithophile elements, making them both infertile and depleted, as expected for refractory residues or cumulates. Most, however, are enriched in the incompatible elements, suggesting that they have been invaded by the upward migration of silicate melts or water-rich fluids. These melts introduce a kimberlite-like component into an otherwise refractory rock. Ultramafic rock masses exhibit a wide range of isotopic ratios, indicating a variety of ages, including ancient, for the enrichment events. Basalts passing through such peridotites pick up these LIL and isotopic signatures.

Table 9-9 shows that trace-element-enriched peridotites have about the same patterns for some elements as kimberlite. The enriched (Q) component of both basalts and peridotites is similar to kimberlite.

RISE OF DEEP DIAPIRS

Melt extraction and crystal settling are important processes in igneous petrology. Basic melts apparently separate from magma chambers, or rising diapirs, at depths as great as 90 km. The basic law governing crystal settling, Stokes' Law, expresses a balance between gravitational and viscous forces,

$$V = \frac{2g\alpha^2\Delta\rho}{9\eta}$$

where α is the radius of a spherical particle, $\Delta\rho$ is the solid-liquid density contrast, η is viscosity of the melt and V is the terminal settling velocity. The same equation can be applied to the rise of diapirs through a mantle with Newtonian viscosity, with modifications to take into account nonspherical objects and non-Newtonian viscosity. Additional complications are introduced by turbulence in the magma chamber and the possibility of a finite yield strength of the magma. Both serve to keep small particles in suspension.

TABLE 9-9
Trace-Element Contents of Enriched and Depleted Peridotites and Kimberlite

	Peridotites		
	Enriched*	Depleted	Kimberlite
Ba	345	2.3	1000
Sr	132	0.5	740
Rb	19	0.05	65
La	7	0.18	150
Ce	20	0.69	200
Nd	11	0.56	85
Nb	16	0.4	110
Sc	10	14	15
Sm	1.9	0.27	13
Yb	0.42	0.37	1.2
Th	0.55	0.25	11.4
Ta	1.13	0.03	9
Hf	0.75	0.15	7

*RSI Mica parqusite lherzolite, Germany (Menzies and others, 1987).

Diapirs are usually treated as isolated spheres or cylinders rising adiabatically through a static mantle. Because of the relative slopes of the adiabat and the melting curve, diapirs become more molten as they rise. At some point, because of the increased viscosity or decreased density contrast, ascent is slowed and cooling, crystallization and crystal settling can occur. The lithosphere serves as a viscosity or strength barrier, and the crust serves as a density barrier. Melt separation can therefore be expected to occur in magma chambers at shallow depths.

In a convecting mantle the actual temperatures (adiabatic) diverge from the melting point as depth increases. Melting can therefore only occur in the upper parts of the rising limbs of convection cells or in thermal boundary layers. The additional buoyancy provided by melting contributes to the buoyancy of the ascending limbs. Although melts will attempt to rise relative to the adjacent solid matrix, they are embedded in a system that is itself rising and melting further. If broad-scale vertical convection is fast enough, diapirs can melt extensively without fractionating. The stresses and temperatures in the vicinity of rising plumes or diapirs are high, and these serve to decrease the mantle viscosity; thus rapid ascent is possible. In order to achieve observed equilibration temperatures and large degrees of partial melting, allowing for specific and latent heats, melting must initiate at depths of at least 200 km. The solidus temperature at this depth is at least 1900 K. The questions then are how fast does material rise between about 200 and 90 km, and is the material at 90 km representative of the deeper mantle source region or has it been fractionated upon ascent?

Viscosities in silicates are very stress- and temperature-dependent, and diapirs occur in regions of the mantle that have higher than normal stresses and temperatures. Diapiric emplacement itself is a high-stress process and occurs in regions where mantle convection may have oriented crystals along flow lines. Diapirs may rise rapidly through such low-viscosity material. I have estimated that a 50-km partially molten diapir at a depth of 200 km could rise at a rate of about 40 cm/s (Anderson, 1981). Kimberlites travel an order of magnitude faster still. Crystal settling velocities are of the order of 10^{-2} cm/s. It appears therefore that deep diapirs can rise rapidly enough to entrain both melt and crystals. If the magma behaves as a Bingham fluid with interacting crystals and a finite yield strength or if the magma is turbulent, the Stokes settling velocities are upper bounds. Crystals in excess of 10 cm can be held in suspension with measured values of yield strength of basaltic melts.

In addition, the density contrast between melt and crystals, $\Delta\rho$, is very small at great depth because of the change in composition and the high compressibility of the melt. At depth the melt content and the permeability are also low, and melt segregation is therefore very slow. The importance of crystal settling is not yet clear even in shallow, static cooling magma chambers. The situation is far more complex in a rising, melting and perhaps turbulent diapir.

In a chemically stratified mantle, for example residual peridotite over piclogite or fertile peridotite, there is a conductive thermal boundary between the convecting layers. In such a region the thermal gradient is in excess of the melting gradient, and melting is most likely to initiate at this depth. Partial melting in the deeper, denser layer causes a reduction in density, and a Rayleigh-Taylor instability can develop. Material can be lifted out of the piclogite layer by such a mechanism and extensive melting occurs during ascent to the shallower mantle. If the peridotite is hydrous, it can also melt and magma mixing is likely, particularly if the diapir is trapped beneath thick lithosphere. Unimpeded ascent of a picritic diapir to the surface yields relatively unfractionated, uncontaminated picritic-tholeiitic melts, as at midocean ridges. In midplate environments, such as Hawaii and other midplate hotspots, a similar melt will cool, fractionate and mix prior to eruption. Such a mechanism seems capable of explaining the difference between hotspot (ocean island, continental flood) basalts and MORB.

Material can leave a deep, dense source region by several mechanisms;

1. Melting in the thermal boundary at the bottom or the top of the layer because of the high thermal gradient in boundary layers compared to melting point gradients.
2. Melting, or phase changes, due to adiabatic ascent of hotter regions of layer and crossing of phase boundary.
3. Entrainment of material by adjacent convecting layer.

Some of these mechanisms are illustrated in Figure 9-6.

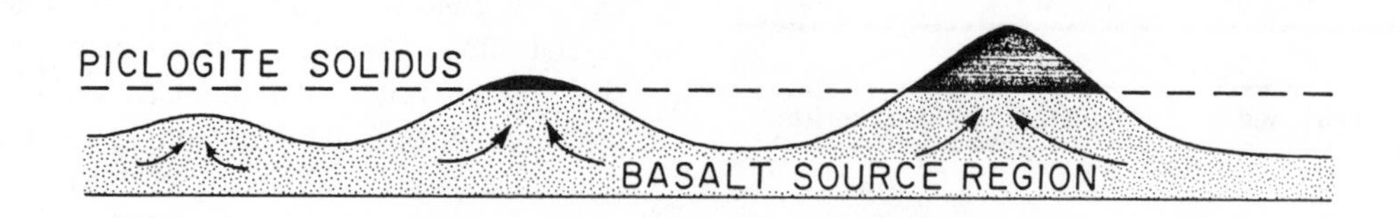

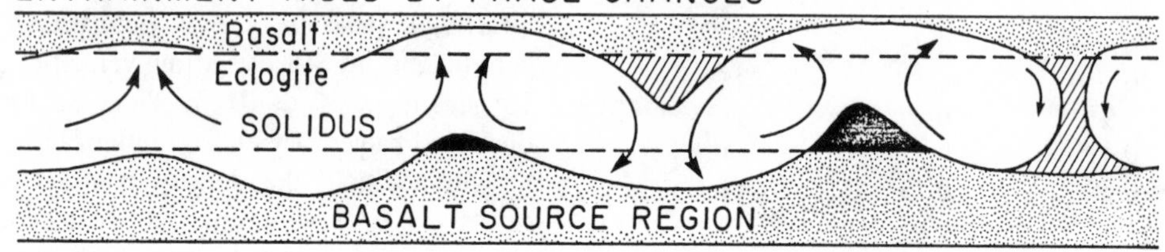

FIGURE 9-6
Methods for removing material from a deep, dense source region. In a layered mantle the interface between chemically distinct layers deforms due to convection and thermal expansion. This may induce phase changes, or partial melting, in the deeper layer, causing it to become buoyant. Material from the deeper layer may also be entrained.

TRACE-ELEMENT MODELING

The trace-element contents of basalts contain information about the composition, mineralogy and depth of their source regions.

When a solid partially melts, the various elements composing the solid are distributed between the melt and the remaining crystalline phases. The equation for equilibrium partial melting is simply a statement of mass balance:

$$C_m = \frac{C_o}{D(1 - F) + F} = \frac{C_r}{D}$$

where C_m, C_o and C_r are the concentrations of the element in the melt, the original solid and the residual solid, respectively; D is the bulk distribution coefficient for the minerals left in the residue, and F is the fraction of melting. Each element has its own D that depends not on the initial mineralogy but on the residual minerals, and, in some cases, the bulk composition of the melt. For the very incompatible elements ($D \ll 1$), essentially all of the element goes into the first melt that forms. The so-called compatible elements ($D \gg 1$) stay in the crystalline residue, and the solid residual is similar in composition to the original unmelted material. The above equation is for equilibrium partial melting, also called batch melting. The reverse is equilibrium crystal fractionation, in which a melt crystallizes and the crystals remain in equilibrium with the melt. The same equations apply to both these situations. The effective partition coefficient is a weighted average of the mineral partition coefficients.

The Rayleigh fractionation law

$$C_m/C_o = F^{(D-1)}$$

applies to the case of instantaneous equilibrium precipitation of an infinitesimally small amount of crystal, which immediately settles out of the melt and is removed from further equilibration with the evolving melt. The reverse situation is called fractional fusion.

Nickel and cobalt are affected by olivine and orthopyroxene fractionation since D is much greater than 1 for these minerals. These are called compatible elements. Vanadium and scandium are moderately compatible. Rare-earth patterns are particularly diagnostic of the extent of garnet involvement in the residual solid or in the fractionating crystals. Note the high partition coefficients for the heavy REE, yttrium and scandium for garnet (Table 9-10). Melts in equilibrium with garnet would be expected to be low in these elements. MORB is high in these elements; kimberlite is relatively low.

Melt-crystal equilibrium is likely to be much more complicated than the above equations imply. As semi-molten diapirs or melt packets rise in the mantle, the equilibrium phase assemblages change. The actual composition of a melt depends on its entire previous history, including shallow crystal fractionation. Melts on the wings of a magma chamber may represent smaller degrees of partial

TABLE 9-10
Crystal/Melt Partition Coefficients

	cpx	opx	ol	gt	plag	plag
La	0.02	0.0005	0.0005	0.004	—	0.02–0.11
Ce	0.04	0.0009	0.0008	0.021	0.10	0.02–0.11
Nd	0.09	0.0019	0.0013	0.087	0.05	0.02–0.07
Sm	0.14	0.0028	0.0019	0.217	0.05	0.02–0.04
Eu	0.16	0.0036	0.0019	0.320	1.50	0.14–0.39
Tb	0.19	0.0059	0.0019	0.70	—	—
Ho	0.195	0.0089	0.0020	1.4	—	—
Yb	0.20	0.0286	0.0040	4.03	0.04	0.006–0.03
Lu	0.19	0.038	0.0048	5.7	—	0.006–0.037
Sc	3.1	1.1	0.25	6.5	—	—
V	1.5	0.3	0.09	0.27	—	—
Co	1.2	2	1.3–6.5	2	—	—
Ni	2–4	3–5	3.8–35	0.8	—	—
Rb	0.05	0.02	0.01	0.02	0.04	0.019–0.36
Sr	0.165	0.016	0.016	0.014	1–4.4	1.2
Y	0.20	0.009	0.002	1.4	—	—
Ba	0.001	0.001	0.0001	0.0015	0.15–0.4	—
	0.08	0.01	0.01	0.04	0.35	—
U	0.04	0.006	0.0025	0.04	—	—
K	0.002	0.001	0.0002	0.001	0.1	0.18
He	—	—	<0.07	—	—	—
Ne	0.5	—	0.006–0.08	—	—	—
Ar	0.7	—	0.05–0.15	—	—	—
Kr	0.14	—	<0.15	—	—	—
Xe	<0.5	—	<0.3	—	—	—

Estimated Rock/Melt Partition Coefficients

Th	0.0004	P	0.015
Ta	0.001	Hf	0.015
Nb	0.0015	Zr	0.015

Frey and others (1978), Wedepohl and Muramatsu (1979), Chen and Frey (1985), Hiyagon and Ozima (1986).

melting than those on the axis. Melts representing a variety of melting conditions may mix prior to eruption. It is therefore difficult to state that a melt represents a given degree of partial melting from a given depth. Nevertheless, extremely fractionated and enriched melts must, at some point in their history, be the result of very small degrees of partial melting or large degrees of crystal fractionation.

Solid trace elements form ionic or covalent bonds with the surrounding ions forming the crystalline lattice. Solid elements can be substituted into a lattice if they have similar ionic charge and radius as the ions in the lattice. For a given crystal the partition coefficient is a strong function of ionic radius; garnet and clinopyroxene, for example, are able to retain elements with ionic radii close to Al and Ca while olivine and orthopyroxene are not.

Noble gases are likely to be accommodated in a crystal as neutral ions, and therefore the mechanism of noble gas retention is quite different from that for solid elements. Neutral ions are likely to be trapped in dislocations, grain boundaries, vacancies and interstitial sites. These ions are subject to the electric field of surrounding ions and undergo electric polarization. The heavier noble gases have a large electronic polarizability and may be held more readily in a lattice, in spite of their larger size. Measured partition coefficients between olivine and basalt melt are $D(\mathrm{He}) \lesssim 0.07$, $D(\mathrm{Ne}) = 0.006–0.08$, $D(\mathrm{Ar}) = 0.05–0.15$, $D(\mathrm{Kr}) \lesssim 0.15$ and $D(\mathrm{Xe}) \lesssim 0.3$ (Hiyagon and Ozima, 1986). The partition coefficients for diopside are up to an order of magnitude higher, although data are lacking for helium. No data are available for garnet, which might be an important trap for helium in the upper mantle. Note that these partition coefficients for the rare gases are larger than for some of the incompatible solid trace elements.

The so-called incompatible elements are those that have low ($< \sim 0.1$) crystal-liquid partition coefficients in olivine and orthopyroxene. Most of the elements listed in Table 9-10 are incompatible elements. Nickel and cobalt are compatible elements, being retained by the major mantle minerals ($D > 1$). Scandium is a compatible element for all pyroxenes and garnet and moderately incompatible for olivine. Note that the heavy rare-earth elements (HREE) are compatible in garnet and moderately compatible in clinopyroxene. This is why eclogite fractionation, or retention of garnet in the source rock, is effective in fractionating the REE pattern.

References

Anderson, D. L. (1981) Rise of deep diapirs, *Geology, 9,* 7–9.

Basu, A. R. and Tatsumoto, M. (1982) Nd isotopes in kimberlites and mantle evolution, *Terra Cog.,* 2, 214.

Beus, A. A. (1976) *Geochemistry of the Lithosphere,* MIR Publications, Moscow, 366 pp.

Bowen, N. L. (1928) *The Evolution of the Igneous Rocks,* Princeton University Press, 332 pp.

Boyd, F. R. (1986) In *Mantle Xenoliths* (P. Nixon, ed.) Wiley, New York.

Boyd, F. R. and S. A. Mertzman (1987) In *Magmatic Processes* (B. O. Mysen, ed.) The Geochemical Society, University Park, Pennsylvania, 500 pp.

Carmichael, I. S. E., F. Turner and J. Verhoogen (1974) *Igneous Petrology,* McGraw-Hill, New York, 739 pp.

Chen, C. and F. A. Frey (1983) Origin of Hawaiian theiite and alkali basalt, *Nature, 302,* 785.

Clarke, D. B. (1970) Tertiary basalts of Baffin Bay; possible primary magma from the mantle, *Contrib. Mineral. Petrol. 25,* 203–224.

Echeverria, L. M. (1980) *Tertiary Komatiites of Gorgona Island,* Colombia, Carnegie Instn. Wash. Ybk., *79,* 340–344.

Elthon, D. (1979) High magnesia liquids as the parental magma for ocean floor basalts, *Nature, 278,* 514–518.

Frey, F. A. (1980) The origin of pyroxenites and garnet pyroxenites from Salt Crater, Oahu, Hawaii, trace element evidence, *Am. J. Sci., 280-A,* 427–449.

Frey, F. A., D. H. Green and S. D. Roy (1978) Integrated models of basalts petrogenesis: A study of quartz tholeiites to olivine melilitites from southeastern Australia utilizing geochemical and experimental petrological data, *J. Petrol., 19,* 463–513.

Green, D. H. (1971) Composition of basaltic magmas as indicators of conditions of origin; application to oceanic volcanism, *Phil. Trans. R. Soc., A268,* 707–725.

Green, D. H., W. O. Hibberson and A. L. Jaques (1979) Petrogenesis of mid-ocean ridge basalts. In *The Earth: Its Origin, Structure and Evolution* (M. W. McElhinny, ed.), 265–295, Academic Press, New York.

Green, D. H. and A. E. Ringwood (1963) Mineral assemblages in a model mantle composition, *J. Geophys. Res., 68,* 937–945.

Green, D. H. and A. E. Ringwood (1967) The genesis of basaltic magmas, *Contrib. Mineral. Petrol., 15,* 103–190.

Häuy, R. J. (1822) *Minéralogie, 4,* 548.

Herzberg, C. T. (1984) Chemical stratification in the silicate Earth, *Earth Planet. Sci. Lett., 67,* 249–260.

Herzberg, C. T., (1986). *Eos, Trans. Am. Geophy. Union 67,* 408.

Hiyagon, H. and M. Ozima (1986) *Geochim. Cosmochim. Acta, 50,* 2045.

Jaques, A. and D. Green (1980) Anhydrous melting of peridotite at 0-15 Kb pressure and the genesis of tholeitic basalts, *Contrib. Mineral. Petrol. 73,* 287–310.

Maaløe, S. and K. Aoki (1977) The major element composition of the upper mantle estimated from the composition of Iherzolites, *Contrib. Mineral. Petrol. 63,* 161–173.

Menzies, M., N. Rogers, A. Zindle and C. Hawkesworth (1987) In *Mantle Metasomatism* (M. A. Menzies, ed.) Academic Press, New York.

Nataf, H.-C., I. Nakanishi and D. L. Anderson (1986) Measurements of mantle wave velocities and inversion for lateral heterogeneities and anisotropy, Part III, Inversion, *J. Geophys. Res., 91,* 7261–7307.

O'Hara, M. J. (1968) The bearing of phase equilibria studies in synthetic and natural systems on the origin and evolution of basic and ultrabasic rocks, *Earth Sci. Rev., 4,* 69–133.

Popper, K. (1972) *Conjectures and Refutations,* 4th ed., Routledge and Kegan Paul, London.

Rigden, S. S., T. J. Ahrens and E. M. Stolper (1984) Densities of liquid silicates at high pressures, *Science, 226,* 1071–1074.

Ringwood, A. E. (1962) Mineralogical constitution of the deep mantle, *J. Geophys. Res., 67,* 4005–4010.

Ringwood, A. E. (1966) Mineralogy of the mantle. In *Advances in Earth Science,* 357–399, MIT Press, Cambridge, Mass.

Ringwood, A. E. (1975) *Composition and Petrology of the Earth's Mantle,* McGraw-Hill, New York, 618 pp.

Ringwood, A. E. (1979) *Origin of the Earth and Moon,* Springer-Verlag, New York, 295 pp.

Smyth, J. R. and F. A. Caporuscio (1984) Petrology of a suite of eclogite inclusions from the Bobbejaan Kimberlite; II, Primary phase compositions and origin. In *Kimberlites* (J. Kornprobst, ed.), 121–131, Elsevier, Amsterdam.

Smyth, J. R., T. C. McCormick and F. A. Caporuscio (1984) Petrology of a suite of eclogite inclusions from the Bobbejaan Kimberlite; I, Two unusual corundum-bearing kyanite eclogites. In *Kimberlites* (J. Kornprobst, ed.), 109–119, Elsevier, Amsterdam.

Stolper, E. (1980) *Contrib. Mineral. Petrol. 74,* 13.

Walck, M. C. (1984) The P-wave upper mantle structure beneath an active spreading center; the Gulf of California, *Geophys. J. Roy. Astron. Soc., 76,* 697–723.

Wedepohl, K. H. and Y. Muramatsu (1979) The chemical composition of kimberlites compared with the average composition of three basaltic magma types. In *Kimberlites, Diatremes, and Diamonds* (F. R. Boyd and H. O. Meyer, eds.), 300–312, American Geophysical Union, Washington, D.C.

Yoder, H. S., Jr. (1976) *Generation of Basaltic Magma,* National Academy of Science, Washington, D.C., 265 pp.

Zindler, A. and E. Jagoutz, *Geochim. Cosmochim. Acta* (in press).

10

Isotopes

The Earth has a spirit of growth.

—LEONARDO DA VINCI

The various chemical elements have different properties and can therefore be readily separated from each other by igneous processes. The various isotopes of a given element are not so easily separated. The abundances of the radioactive isotopes and their decay products are not constant in time. Therefore, the information conveyed by the study of isotopes is different in kind than that provided by the elements. Each isotopic system contains unique information, and the various radioactive isotopes allow dating of various processes in a planet's history. The unstable isotopes most useful in geochemistry have a wide range of decay constants, or half-lives, and can be used to infer processes occurring over the entire age of the Earth (Table 10-1). In addition, isotopes can be used as tracers and in this regard they complement the major- and trace-element chemistry of rocks and magmas.

Studies of isotope ratios have played an important role in constraining mantle and crustal evolution, convective mixing and the long-time isolation of mantle reservoirs. Isotope studies derive their power from the existence of suitable pairs of isotopes of a given element, one a "primordial" isotope present in the Earth since its formation, the other a radiogenic daughter isotope produced by radioactive decay at a known rate throughout geological time. The isotopic composition of these isotope pairs in different terrestrial reservoirs—for example, the atmosphere, the ocean, and the different parts of the crust and mantle—are a function of the transport and mixing of parent and daughter elements between the reservoirs. In some cases the parent and daughter have similar geochemical characteristics and are difficult to separate in geological processes. In other cases the parent and daughter have quite different properties, and isotopic ratios contain information that is no longer available from studies of the elements themselves. For example the $^{87}Sr/^{86}Sr$ ratio gives information about the time-integrated Rb/Sr ratio of the rock or its reservoirs. Since rubidium is a volatile element and separates from strontium both in preaccretional and magmatic processes, the isotopic ratios of strontium in various products of mantle differentiation, combined with mass-balance calculations, are our best guide to the rubidium content, and volatile content, of the Earth. Lead isotopes can be similarly used to constrain the U/Pb ratio, a refractory/volatile pair, of the mantle. In other cases, such as the neodymium-samarium pair, the elements in question are both refractory, have similar geochemical characteristics and are probably in the Earth in chondritic ratios. The neodymium isotopes can therefore be used to infer ages of reservoirs and to discuss whether these reservoirs are enriched or depleted, in terms of Nd/Sm, relative to chondritic or undifferentiated reservoirs. The Rb/Sr and Nd/Sm ratios of a reservoir are changed when melt is removed or added or if it is mixed with or contaminated by sediment, crust or seawater. With time, the isotopic ratios of such reservoirs diverge. The isotopic ratios of the crust and various magmas show that mantle differentiation is ancient and that remixing and homogenization is secondary in importance to separation and isolation. On the other hand we know that these isolated reservoirs can readily provide material to the surface. Although isotopes cannot tell us where these reservoirs are, or their bulk chemistry, their long-term isolation and lack of homogenization plus the temporal and spatial proximity of their products suggests that, on average, they occur at different depths. This in turn suggests that the reservoirs differ in intrinsic density and therefore in bulk chemistry and mineralogy. Melts from the reservoirs, however, are buoyant

TABLE 10-1
Radioactive Nuclides and their Decay Products

Radioactive Parent	Decay Product	Half-life (billion years)
^{238}U	^{206}Pb	4.468
^{232}Th	^{208}Pb	14.01
^{176}Lu	^{176}Hf	35.7
^{147}Sm	^{143}Nd	106.0
^{87}Rb	^{87}Sr	48.8
^{235}U	^{207}Pb	0.7038
^{40}K	^{40}Ar, ^{40}Ca	1.250
^{129}I	^{129}Xe	0.016
^{26}Al	^{26}Mg	8.8×10^{-4}

relative to the shallow mantle. It may be that basalts originate primarily from only one of these reservoirs and that the trace-element and isotopic diversity is acquired when they traverse the shallow mantle on their way to the surface.

The crust is extremely enriched in many of the so-called incompatible elements, generally the large ionic radius or high-charge elements that do not readily fit into the lattices of the "major mantle minerals," olivine and orthopyroxene. The crust is not particularly enriched in elements of moderate charge having ionic radii between those of Ca^{2+} and Al^{3+}. This suggests that the mantle has retained elements that can be accommodated in the garnet and clinopyroxene structures. The crust is also not excessively enriched in lithium, sodium, lead, bismuth and helium.

Some mantle rocks and magmas have high concentrations of the incompatible elements and isotopic ratios that reflect long-term enrichment of an appropriate incompatible-element parent. The common geochemical prejudice is that the crust is somehow involved in the evolution of these magmas, either by crustal contamination prior to or during eruption or by recycling of continent-derived sediments. Usually, however, it is impossible to distinguish this possibility from the alternative that all potential crust-forming material has not been removed from the upper mantle or that the crust formation process, efficient as it seems to be, is not 100 percent efficient in removing the incompatible elements from the mantle. Ironically, a parallel geochemical prejudice is that some magmas represent melts from "primitive" mantle, which has survived from the accretion of the Earth without any melting or melt extraction, the apparent reasoning being that part of the mantle provided the present crust, with 100 percent efficiency, and the rest of the mantle has been isolated, again with 100 percent efficiency. In this scenario, "depleted" magmas are derived from a reservoir, complementary to the continental crust, which has experienced a two-stage history (stage one involves an ancient removal of a small melt fraction, the crust; stage two involves a recent extensive melting process, which generates MORB). "Enriched" magmas (also called "primitive," "more primitive," "less depleted," "hotspot" or "plume" magmas) are single-stage melts from a "primitive" reservoir or depleted magmas that have experienced some type of crustal contamination.

There is no room in these models for an ancient enriched mantle reservoir. This "box model" of the Earth contains three boxes: the continental crust, the "depleted mantle" (which is equated to the upper mantle or MORB reservoir) and "primitive mantle" (which is equated to the lower mantle) with the constraint that primitive mantle is the sum of continental crust and depleted mantle. With these simple rules many amusing games can be played with crustal recycling rates and mean age of the crust. When contradictions appear they are explained by hiding material in the lower crust, the continental lithosphere or the core, or by storing subducted material somewhere in the mantle for long periods of time. The products of mantle differentiation are viewed as readily and efficiently separable but, at the same time, storable for long periods of time in a hot, convecting mantle. Magmas having similar geochemical characteristics are given a variety of origins. "Enriched" magmas are variously attributed to continental contamination, recycling and a "lower-mantle primitive reservoir."

Isotopes are extremely useful as probes of planetary processes, and they are even more useful when used in conjunction with other petrological, geochemical and geophysical data. They are also much more interesting as tracers of the Earth's dynamics than is implied by static box models involving crust, mantle and core—the main features of the Earth acknowledged by isotope geochemists.

Isotopes make it possible to rule out conventional ideas that relate various basalts by different degrees of partial melting at different depths (the pyrolite hypothesis) or by crystal fractionation (the parent magma hypothesis). These mechanisms may still be important, but they must be combined with contamination or mixing of materials from isotopically distinct reservoirs.

LEAD ISOTOPES

Lead has a unique position among the radioactive nuclides. Two isotopes, lead-206 and lead-207, are produced from radioactive parent isotopes of the same element, uranium-238 and uranium-235. The simultaneous use of coupled parent-daughter systems allows one to avoid some of the ambiguities associated with a single parent-daughter system.

In discussing the uranium-lead system, it is convenient to normalize all isotopic abundances to that of lead-204, a stable nonradiogenic lead isotope. The total amount of lead-204 in the Earth has been constant since the Earth was formed; the uranium parents have been decreasing by radioactive decay while lead-206 and lead-207 have been increasing. The U/Pb ratio in various parts of the Earth changes by chemical fractionation and by radioactive decay. The $^{238}U/^{204}Pb$ ratio, calculated as of the present, can be used to remove the decay effect in order to study the chemical fractionation of various reservoirs. If no chemical separation of uranium from lead occurs, the $^{238}U/^{204}Pb$ ratio for the system remains constant. This ratio is called the μ of the system.

The decay schemes are

$$(^{206}Pb/^{204}Pb) = (^{238}U/^{204}Pb)(e^{\lambda t_o} - 1) + (^{206}Pb/^{204}Pb)_o$$

$$(^{207}Pb/^{204}Pb) = (^{235}U/^{204}Pb)(e^{\lambda' t_o} - 1) + (^{207}Pb/^{204}Pb)_o$$

where λ and λ' are the decay constants and the second terms on the right are the initial ratios (at $t = t_o$).

By combining these equations we can write an expression for t_o in terms of isotopic ratios of lead and uranium: $^{207}Pb/^{204}Pb$, $^{206}Pb/^{204}Pb$ and $^{235}U/^{238}U$. This is the basis for determining the lead-lead age of the Earth. Use of this equation does not involve measurements of the absolute concentration of either the uranium or the lead in the rock, nor does t_o depend on the absolute concentrations of these elements. The ratio $^{235}U/^{238}U$ is very nearly constant (1/137.9) in natural uranium. For uranium-rich minerals, $^{207}Pb/^{204}Pb$ and $^{206}Pb/^{204}Pb$ are high, and the equations do not much depend on the initial ratios. These equations were used by Patterson (1956) and Houtermans (1947) to calculate the age of the Earth, using initial ratios inferred from meteoritic data.

The coupled equations can be used to calculate $^{206}Pb/^{204}Pb$ and $^{207}Pb/^{204}Pb$ for a given $^{238}U/^{204}Pb$ ratio, as a function of time. If the initial ratios and μ are known, or assumed from meteorite data, then the locus of points so calculated is called the *geochron*. The lead-isotopic ratios can therefore be calculated for any time, including the present, for unfractionated reservoirs. For $\mu = 8$ and $t_o = 4.6$ Ga, values thought to be appropriate for the Earth, $^{206}Pb/^{204}Pb$ is about 17.6 and $^{207}Pb/^{204}Pb$ is about 15.45. If the Earth fractionated at any time, giving high U/Pb in the melts and low U/Pb in the residue, then present day lead-isotopic ratios will be greater and less, respectively, than present-day ratios inferred for primitive mantle.

Most lead-isotopic results can be interpreted as growth in a primitive reservoir for a certain period of time and then growth in reservoir with a different μ-value from that time to the present. By measuring the isotopic ratios of lead and uranium in a rock, the time at which the lead ratios were the same as inferred for the primitive reservoir can be determined, thus giving the lead-lead age of the rock. This dates the age of the uranium-lead fractionation event, assuming a two-stage growth model. In some cases multistage or continuous differentiation models are used.

A melt removed from the primitive reservoir at t_o will crystallize to a rock composed of minerals with different μ values. If these minerals can be treated as closed systems, then they will have distinctive lead ratios that plot as a straight line on a $^{207}Pb/^{204}Pb$–$^{206}Pb/^{204}Pb$ plot (Figure 10-1). This line is an *isochron* because it is the locus of points all of which experienced fractionation at the same time to form minerals with differing U/Pb ratios. The residual rock will also plot on this line, on the other side of the geochron. The time at which the rock was fractionated can be calculated from the slope of the isochron. Mixing lines between genetically unrelated magmas will also be straight lines, in which case the age will be spurious unless both magmas formed at the same time.

In the uranium-lead decay system, the curve representing the growth of radiogenic lead in a closed system has marked curvature. This is because uranium-238 has a half-life (4.47 Ga) comparable to the age of the Earth, whereas uranium-235 has a much shorter half-life (0.704 Ga). In early Earth history lead-207, the daughter of uranium-235, is formed at a higher rate than lead-206. For a late fractionation event $^{207}Pb/^{204}Pb$ changes slowly with time.

For isotopic systems with very long half-lives, such as samarium-142 (106 Ga) and rubidium-87 (48.8 Ga), the analogous closed-system geochrons will be nearly straight lines. On the other hand, isochrons and mixing lines for other systems, in general, are not straight lines. They are straight in the uranium-lead system because $^{238}U/^{204}Pb$ and $^{235}U/^{204}Pb$ have identical fractionation factors, and mixing lines for ratios are linear if the ratios have the same denominator.

The initial lead-isotopic composition in iron meteorites can be obtained since these bodies are essentially free of uranium. Galenas are also high in lead and low in uranium and therefore nearly preserve the lead-isotopic ratios of their parent at the time of their birth. Galenas of various ages fall close to a unique single-stage growth curve. The small departures can be interpreted as further fractionation events.

The equations describing the evolution of a given set of lead-isotope compositions from a single common composition in systems that may lose lead or gain or lose uranium are

$$^{206}Pb/^{204}Pb = (^{206}Pb/^{204}Pb)_{t_o} + \int_o^{t_o} \mu(t)e^{\lambda t}\lambda \, dt$$

$$^{207}Pb/^{204}Pb = (^{207}Pb/^{204}Pb)_{t_o} + (^{235}U/^{238}U) \int_o^{t_o} \mu(t)e^{\lambda' t} \, dt$$

If μ changes discontinuously at various times, then the above equations can be written as the appropriate sums.

The μ values for basaltic magmas are usually quite high, 15–45, compared to primitive mantle. Their lead-isotopic ratios will therefore grow more rapidly with time than the primitive mantle, and the $^{206}Pb/^{204}Pb$ and $^{207}Pb/^{204}Pb$ ratios of such magmas are high. Oceanic islands have such high lead-isotopic ratios that they must have come from ancient enriched reservoirs or contain, as a component, ancient enriched material. MORB thought to come from an ancient depleted reservoir, also have ratios in excess of the geochron. This suggests either the mantle (or

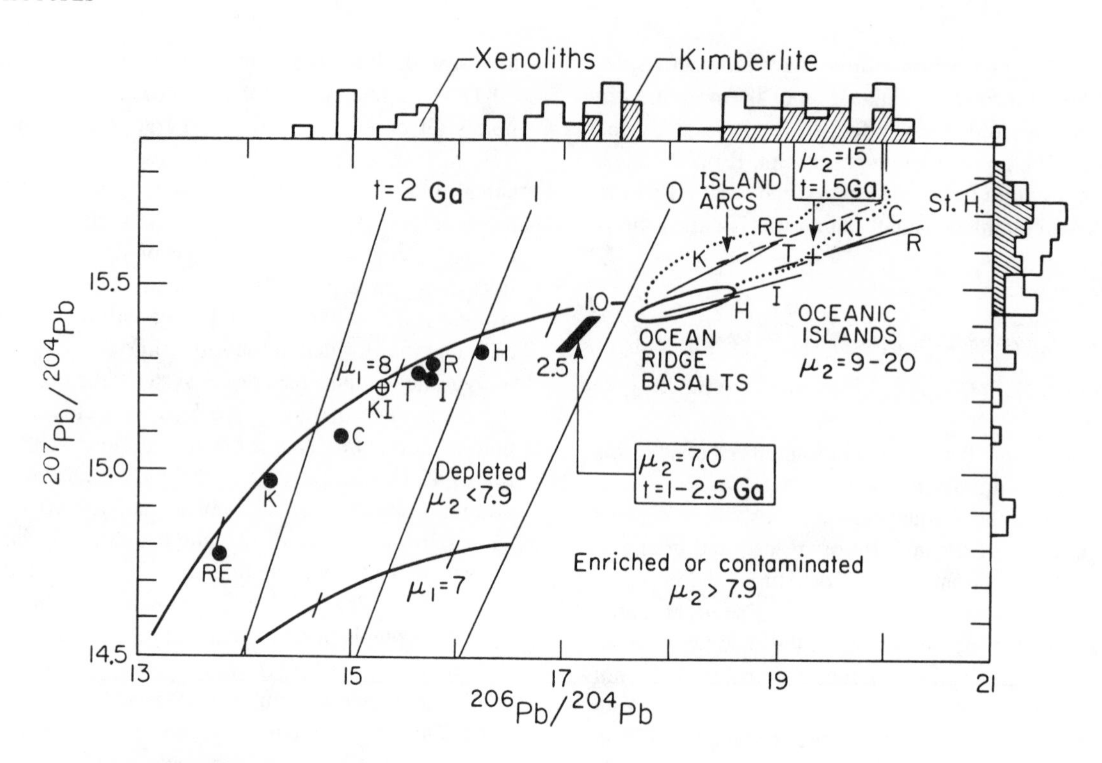

FIGURE 10-1
Lead isotope diagram. Age of Earth is taken as 4.57 Ga. Straight lines labeled with letters are values for oceanic islands. Black dots are the inferred primary isotopic ratios if the island data are interpreted as secondary isochrons. Growth curves for $^{238}U/^{204}Pb$ (μ_1) values of 7.0 and 8.0 and primary isochrons at 1 Ga intervals are shown. The primary mantle reservoir appears to have a μ_1 of 7.9. Oceanic-island basalts appear to have evolved in enriched reservoirs ranging in age from 1 to 2.5 Ga with μ_2 (the second-stage μ value) values ranging from 9 to 20. A point is shown for a two-stage model with $\mu_1 = 7.9$ before 1.5 Ga and $\mu_2 = 15$ subsequently. The black bar represents the range of values for depleted reservoirs with $\mu_2 = 7.0$ and a range of depletion ages from 1 to 2.5 Ga. The range for mid-ocean-ridge basalts could be due to growth in an enriched reservoir or due to contamination by enriched magmas. Isotopic ratios for xenoliths and kimberlites are shown along the axes. Xenoliths are primarily from the shallow mantle and many are enriched. KI is kimberlite. Diagram is modified from Chase (1981).

upper mantle) has continuously lost lead, relative to uranium, to the lower crust or core (or lower mantle), or that MORB basalts have been contaminated by material with high isotopic ratios, prior to eruption.

Oldenberg (1984) performed a detailed analysis of lead-isotopic data from conformable ore deposits to obtain information about the movement of uranium, thorium and lead into or out of the mantle source region. He showed that $\mu(t) = {}^{238}U(t)/{}^{204}Pb$ increased from 8.03 ± 0.3 in the first billion years after Earth formation to 9.8 ± 0.3 between 2.0 and 3.0 Ga ago. The ratio μ apparently decreased in the time interval 2 to 1 Ga ago, perhaps representing removal of uranium and lead from the mantle to the continental crust or the preferential sampling of a relatively depleted reservoir during that period of time. The ratio $\xi = {}^{232}Th/{}^{204}Pb$ also increased with time from about 35 to 38, also with a dip at 2–1 Ga ago. The ratio $k(t) = \xi(t)/\mu(t)$ thus maintained a value near 4.0 since the Earth formed.

In a cooling, crystallizing mantle the μ of the residual melt will increase with time, assuming that solid silicates and sulfides retain lead more effectively than uranium. Modeling shows that most of the mantle had solidified prior to 3.8 Ga, close to the age of the oldest known rock. Oldenburg estimated that the residual, depleted solid mantle might have μ of about 6.0. Basalts from oceanic islands have apparently experienced secondary growth in reservoirs with μ from about 10 to 20, after a long period of growth in a more "primitive" reservoir ($\mu \approx 7.9$).

Leads from basaltic suites in many oceanic islands form linear areas on $^{206}Pb/^{204}Pb$ versus $^{207}Pb/^{204}Pb$ diagrams (Figure 10-1). These could represent either mixing lines or secondary isochrons. Two-stage histories indicate that the leads from each island were derived from a common primary reservoir (μ = 7.9) at different times from 2.5 to 1.0 Ga ago. Alternatively, the magmas from each island could represent mixtures between an enriched end-member and a less enriched or depleted end-member. In either case the ocean-island basalts involve a source region with ancient U/Pb enrichment. One mechanism for such enrichment is removal of a melt from a primitive reservoir to another part of the mantle that subsequently provides melts to the oceanic islands or contaminates MORB. The most logical storage place for such a melt is the shallow mantle. The enrichment event must have been ancient, older than 1 Ga. To explain the various trends of the individual islands by mixing, the enriched end-member must come from parts of the mantle that were enriched at different times or that have different time-integrated U/Pb ratios. In a crystallizing cumulate or magma ocean, the U/Pb ratio of the remaining fluid probably increases with time, and regions of the mantle that were enriched by this melt would have variable μ depending on when and how often they were enriched. If the enriched reservoir is global, as indicated by the global distribution of enriched magmas, it is plausible that different parts of it were enriched at different times.

STRONTIUM AND NEODYMIUM ISOTOPES

Strontium-isotope measurements are now almost as routine as trace-element measurements and are used in similar ways in petrological investigations. Magmas that are genetically related have the same isotopic ratios. Midocean-ridge basalts have $^{87}Sr/^{86}Sr$ less than 0.703, and "pure" MORB may have values of 0.702 or less. Ocean-island, island-arc and continental flood basalts are generally much higher than 0.703, commonly higher than 0.71. Primitive mantle values of $^{87}Sr/^{86}Sr$ are unknown because the mantle Rb/Sr is unknown, but it probably falls between 0.704 and 0.705. Basalts with high $^{87}Sr/^{86}Sr$ are often considered to be contaminated by crust or recycled material or seawater. They may also come from ancient enriched mantle reservoirs, but other isotope and trace-element data are required to remove the ambiguities. Attributing the properties of MORB to "normal mantle" and, more recently, to the whole upper mantle, leaves crustal contamination, recycling or the lower mantle as the only alternatives to explain ocean-island and other "plume" or "hotspot" basalts. However, the upper mantle itself is probably inhomogeneous. In some cases continental, sediment or seawater contamination can be ruled out. There is no evidence that any magma comes from a primordial reservoir or from the lower mantle.

The radioactive isotope samarium-147 decays to neodymium-143 with a half-life of 106 Ga. $^{143}Nd/^{144}Nd$ ratios are expressed in terms of deviations, in parts per 10^4, from the value in a reservoir that has had chondritic ratios of Sm/Nd for all time,

$$\varepsilon_{Nd} = \left[\frac{(^{143}Nd/^{144}Nd)}{(^{143}Nd/^{144}Nd)_c} - 1 \right] \times 10^4$$

Clearly, a chondritic unfractionated reservoir has $\varepsilon_{Nd} = 0$ at all times. Samarium and neodymium are both refractory rare-earth elements and should be in the Earth in chondritic ratios. However, they are separated by magmatic process and thus record the magmatic or fractionation history of the Earth. Samarium has a higher crystal-melt partition coefficient than neodymium, and thus the Sm/Nd ratio is smaller in melts than in the original rock. The $^{143}Nd/^{144}Nd$ ratio, normalized as above, will therefore be positive in reservoirs from which a melt has been extracted and negative in the melts or regions of the mantle that have been infiltrated by melts. The Sm/Nd ratio depends on the extent of melting and the nature of the residual phases, and ε_{Nd} depends on the Sm/Nd and the age of the fractionation event.

Since neodymium and samarium are geochemically similar and are both refractory elements, the neodymium-isotope system has certain advantages over the strontium and lead systems:

1. Rocks are less sensitive to alteration, in particular seawater alteration.
2. The Sm/Nd ratio of the Earth is probably well known (chondritic).
3. There is probably little separation of samarium and neodymium once a rock has been formed.

Nevertheless, there is generally good correlation between neodymium and strontium isotopes. Positive ε_{Nd} correlates with low $^{87}Sr/^{86}Sr$ and vice versa. Midocean-ridge basalts have high ε_{Nd} and low $^{87}Sr/^{86}Sr$, indicating time-integrated depletions of Nd/Sm and Rb/Sr. The isotopic ratios are so extreme that the depletion must have occurred in the MORB reservoir more than 1 Ga ago, probably more than 2 Ga ago. The depletion may have occurred at the time the continental crust formed or even during the accretion of the Earth. The measured Sm/Nd and Rb/Sr ratios in MORB generally would not support such ancient ages, but the depletion may have been progressive and MORB may be mixtures of a more depleted and an enriched material.

Incompatible-element ratios such as Rb/Sr and Nd/Sm are increased in partial melts. However, for large fractions of partial melting the ratios are similar to the original rock. Since elements with D much less than 1 (such as Rb, Sr, Nd and Sm) are not retained effectively by residual crystals, it is difficult to change their ratio in melts, but the residual crystals, although low in these elements, have highly fractionated ratios. Partial melts representing large degrees of partial melting from primitive mantle will also have near-primitive ratios, as will regions of the mantle invaded by

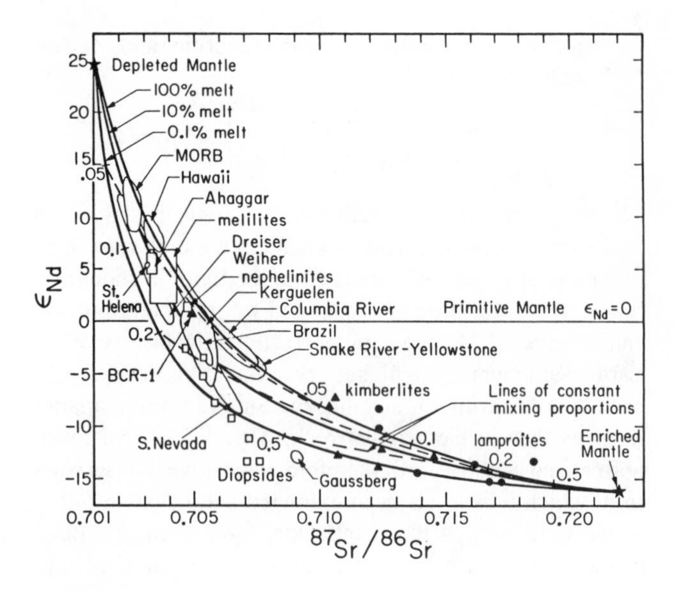

FIGURE 10-2

ε_{Nd} versus $^{87}Sr/^{86}Sr$ for mixtures involving a depleted magma or residual fluids from such a magma after crystal fractionation, and an enriched component (EM) (after Anderson, 1985).

these melts. If the melt cools and crystallizes, with refractory crystals being removed and isolated, the Sm/Nd ratio changes. Thus, it is dangerous to infer that a melt came from a primitive reservoir simply because the $^{143}Nd/^{144}Nd$ ratio appears primitive. Similarly, magmas with ε_{Nd} near 0 can result from mixtures of melts, with positive and negative ε_{Nd}.

Figure 10-2 shows the ε_{Nd} versus strontium isotope correlation for a variety of materials.

THE LEAD PARADOX

Both uranium and lead are incompatible elements in silicates, and uranium apparently enters the melt more readily than lead. The U/Pb ratio should therefore increase in melts and decrease in the solid residue. One would expect, therefore, that the MORB reservoir should be depleted in U/Pb as well as Rb/Sr and Nd/Sm. A time-average depletion would give $^{206}Pb/^{204}Pb$ and $^{207}Pb/^{204}Pb$ ratios that fall to the left of the primary geochron and below the mantle growth curve. Figure 10-1 shows, however, that both MORB and ocean-island tholeiites appear enriched relative to the primary growth curve. This implies that MORB has been contaminated by high-uranium or high-U/Pb material before being sampled, or that lead has been lost from the MORB reservoir. Early lead loss to the core, in sulfides, is possible, but the isotopic results, if interpreted in terms of lead removal, also require lead extraction over the period 3–1 Ga. Contamination of MORB is a possible explanation, particularly if the depleted reservoir is deeper than the enriched reservoir.

The lead reservoirs for oceanic islands are heterogeneous but may have been derived from a fairly uniform reservoir with $\mu = 7.9$ at times ranging from 2.5 to 10 Ga ago (Chase, 1981). There is some indication that the more recently formed reservoirs are more enriched in U/Pb. This is consistent with progressive enrichment of the ocean-island basalt (OIB) reservoir over time. Kimberlites from South Africa appear to have evolved in a similar reservoir prior to 1.5 Ga and in an enriched reservoir subsequent to that time (Kramers, 1977). The uniform mantle reservoir would have $^{206}Pb/^{204}Pb$ of about 17.5 if sampled today. After depletion, this reservoir would evolve in a low-μ environment ($\mu <$ 7.9) and would generate magmas with $^{206}Pb/^{204}Pb$ values less than 17.5. If midocean-ridge basalts originate in the depleted reservoir and are contaminated by magmas from the complementary enriched reservoirs, then the field of MORB on a lead-lead plot should fall on a mixing line between OIB and the uncontaminated magmas. This constrains the field of these magmas on a lead-lead evolution diagram. Assuming that the depleted and enriched reservoirs have the same range, 1.0–2.5 Ga, one can calculate that μ of the depleted reservoir should be about 7 or less. Magmas from depleted reservoirs that formed over the above time interval would have $^{206}Pb/^{204}Pb$ ratios of 17.0 to 17.4. In the MORB system itself variations of U/Pb, Rb/Sr and Nd/Sm are correlated.

TABLE 10-2

Parameters Adopted for Uncontaminated Ocean Ridge Basalts (1) and Contaminant (2)

Parameter	Pure MORB (1)	Enriched Contaminant (2)
Pb	0.08 ppm	2 ppm
U	0.0085 ppm	0.9 ppm
U/Pb	0.10	0.45
$^{238}U/^{204}Pb$	7.0	30.0
$^{206}Pb/^{204}Pb$	17.2	19.3, 21
Rb	0.15 ppm	28 ppm
Sr	50 ppm	350 ppm
$^{87}Sr/^{86}Sr$	0.7020	0.7060
Sm	2 ppm	7.2 ppm
Nd	5 ppm	30 ppm
$^{143}Nd/^{144}Nd$	0.5134	0.5124

Enrichment Factors*

Pb	24.6	U/Pb	4.5
Sr	7.0	Rb/Sr	26.7
Nd	6.0	Sm/Nd	0.60

(1) Assumed composition of uncontaminated midocean-ridge basalts.

(2) Assumed composition of contaminant. This is usually near the extreme end of the range of oceanic-island basalts.

*Ratio of concentration in two end-members.

In order to test if contamination is a viable explanation for the location of the field of MORB on lead-lead isotopic diagrams, we must estimate the lead content of uncontaminated depleted magmas and the lead and lead-isotopic ratios of the contaminant. Table 10-2 lists the parameters we shall use.

The lead content appropriate for a $^{206}Pb/^{204}Pb$ range of 17.0 to 17.2 is 0.07 to 0.09 ppm (Sun and others, 1975). Let us adopt 0.08 ppm. The U/Pb ratio for basalt glasses is about 0.08 to 0.19. Adopting 0.11, we derive a uranium content for pure MORB basalts of 0.0085 ppm. For the contaminant, we adopt the values in Table 8-4. These are close to the extreme values measured on ocean-island tholeiites. For the contaminant $^{206}Pb/^{204}Pb$ ratio we adopt 19.3, corresponding to 1.5 Ga of secondary growth in a reservoir with $\mu = 15$. This is a conservative value since ratios as high as 21 are observed on oceanic islands. The effect of a contaminant with this value is also shown in Figure 10-3.

The results of mixing calculations are shown in Figure 10-3 for the $^{143}Nd/^{144}Nd$, $^{87}Sr/^{86}Sr$ and $^{206}Pb/^{204}Pb$ systems. The differences between the lead and other systems is striking. A small amount of contamination, less than 0.5 percent, pushes MORB compositions into the enriched field for lead but not for neodymium or strontium. In terms of single-stage evolution, both observed (contaminated) MORB and oceanic-island basalt will appear to have future ages on a lead-lead geochron diagram. The neodymium and strontium isotopic ratios are not affected as much, and contaminated MORB will appear to come from depleted reservoirs.

The large increase of $^{206}Pb/^{204}Pb$ caused by a small amount of contamination is mainly a function of the choice of the ratios Pb(contaminant)/Pb(MORB) and $^{206}Pb/^{204}Pb$ of the contaminant. The lowest measured lead concentration in MORB is about 0.2 ppm, a factor of 2.5 times greater than the value adopted for uncontaminated MORB on the basis of uranium-lead systematics. On the other hand, the value adopted for the contaminant, 2 ppm, is five to eight times less than values commonly observed for oceanic-island basalts. The lead enrichment adopted for the mixing calculation, 24.6, therefore is conservative. The enrichment factors for strontium and neodymium are much less, and this is the main reason for the greater sensitivity of MORB to lead contamination. If the enriched and depleted reservoirs are chemically distinct layers in the mantle, they will experience convection with scale lengths of the order of one of the layer thicknesses. The individual cells may experience enrichment/depletion events at different times. Thus, the reservoirs, even if global in extent, need not be homogeneous in trace-element geochemistry. In the mantle differentiation scheme I present here, the depleted source region is below the enriched source region. It seems likely that MORB evolves in shallow mantle magma chambers prior to eruption. Therefore, contamination of MORB is probably unavoidable, at least at the low levels discussed here. The isotopes alone, of course, do not constrain the locations of the depleted and enriched reservoirs. It is the time sequence of eruption of enriched and depleted basalts, the seismic data for the depth of the MORB reservoir, the tendency of melts to migrate upward and the vulnerability of the shallow mantle to contamination by subduction and trapped melts that suggest the stratification favored here.

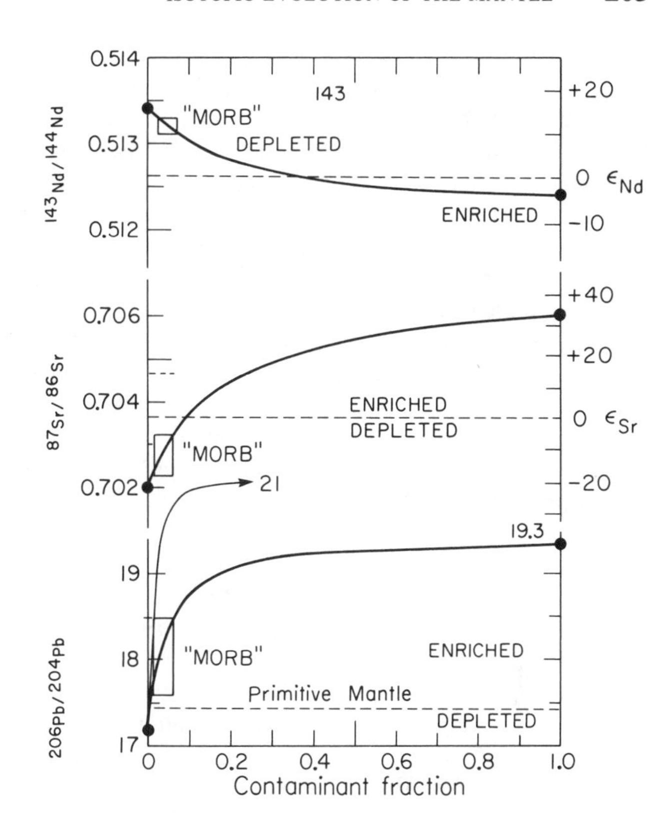

FIGURE 10-3
Isotopic ratios versus contamination. Note that a small amount of contamination has a large effect on the lead system. Enriched magmas and slightly contaminated depleted magmas will both fall in the "enriched" field relative to primitive mantle and will both give "future" ages on a single-stage Pb-Pb evolution diagram. Slight contamination has less effect on ε_{Nd} and ε_{Sr}, and MORB will still appear depleted. Curves for two values for $^{206}Pb/^{204}Pb$ for the contaminant are shown. The correlation line cannot be used to estimate the primitive value for $^{87}Sr/^{86}Sr$ if basalts are mixtures (e.g. Fig. 10-2).

ISOTOPIC EVOLUTION OF THE MANTLE

Neodymium-isotope results were initially interpreted in terms of simple one- and two-stage differentiation schemes, with a primary stage characterized by growth in a primitive reservoir followed by secondary growth in a reservoir with constant Sm/Nd ratio. In this scheme the change in Sm/Nd ratio of the mantle source of MORB was attributed to extraction of the continental crust. Continental tholeiites were regarded as coming from a primitive unfractionated reservoir and MORB from the depleted reservoir. Enriched magmas were considered to be contaminated by the continental crust or to have come from a mantle reservoir that had been metasomatized prior to being sampled.

Lead isotopes, on the other hand, indicate the presence of ancient enriched reservoirs in the mantle and, in many cases, require episodic or continuous differentiation rather than simple two-stage growth. If the U/Pb ratios of the mantle reservoirs have changed with time, it is likely that the Rb/Sr and Sm/Nd ratios have changed as well. Lead isotopes also show that MORB are enriched in U/Pb relative to primitive mantle. Accretional calculations suggest that extensive melting and differentiation were processes that were occurring during most of the accretion of the Earth. Geochemical models for the evolution of the mantle, however, usually assume the survival of a large primitive unfractionated reservoir and that ancient fractionation events involved only very small degrees (1–2 percent) of partial melting, based on the assumption that the continental crust is the only enriched reservoir and that it forms by partial melting of primitive mantle and 100 percent efficient extraction of the crustal melt.

The present levels of Rb/Sr and Nd/Sm in oceanic-island and continental basalts are often too high to explain their strontium and neodymium isotopic ratios by one- or two-stage differentiation schemes. Episodic, multistage or continuous differentiation models are required to interpret lead-isotopic data, and similar models should be applicable to the strontium and neodymium systems. Lead-isotope data identify a single relatively homogeneous reservoir from about 4.5 to 3.8 Ga ago. This reservoir is not necessarily the whole mantle, nor need it have chondritic or terrestrial ratios of the key elements used in isotopic studies, such as U/Th, U/Pb, Rb/Sr and Sm/Nd. It may itself represent the product of early core-mantle or upper mantle–lower mantle separation. The latter may be a result of partial melting during accretion and concentration of aluminum, calcium, and the incompatible trace elements into the upper mantle. If the partial melt fraction, F, is appreciable, say 15–25 percent, then the incompatible elements will be more or less uniformly enriched by a factor of $1/F$ into the upper mantle, and such ratios as Rb/Sr and Sm/Nd will be nearly unfractionated but slightly higher and lower, respectively, than primitive mantle values because of the differences in partition coefficients.

The existence of complementary depleted and enriched (relative to each other) reservoirs is evident in most isotopic systems by about 2 Ga ago or earlier. The process depleting one is presumably enriching the other. For example, transfer of fluids (such as terrestrial KREEP) from one level in the primary reservoir to a different level will generate complementary depleted and enriched reservoirs. The present levels of enrichment (such as high values of Rb/Sr and U/Pb) inferred for the present source region of many alkali basalts, continental tholeiites and other enriched magmas are so high that these reservoirs must have been enriched very recently, or they have reached their present level of enrichment gradually over time, or they are mixtures of enriched and depleted magmas. The depletion/enrichment process could have started as early as 3.8 Ga ago, but the fractionation may have been gradual. It is useful to calculate the isotopic evolution of a continuously differentiating mantle.

The trend of oceanic-island leads on lead-lead diagrams supports a model of episodic and heterogeneous secondary enrichment in uranium relative to lead of the mantle source regions for oceanic-island basalt (Oldenburg, 1984; Chase, 1981). At the same time, episodic or continuous enrichment of rubidium relative to strontium and neodymium relative to samarium may explain the trace-element and isotopic data for these systems.

It is natural to assume that an enriched reservoir in the mantle, along with the continental crust, is the complement to the depleted reservoir and that both formed over the same time interval. Because of the upward migration of fluids and subduction of sediments and hydrothermally altered oceanic crust, it is also natural to assume that the enriched layer is in the uppermost mantle. Because of the various contributions to enrichment (metasomatism, subduction), it is difficult to determine an appropriate functional form for the time history of enrichment. Because of the isotopic evidence for slow enrichment in the first 1 Ga after continental formation and rapid enrichment in the last billion years, an enrichment of the form $\exp(t/\tau)$ may be appropriate for the early stages of the process, where t is measured from the start of the enrichment/depletion process and τ is a time constant. The time constant of differentiation appears to be several billion years. We therefore adopt

$$f(t) = (\mathrm{Rb/Sr})_N - 1 = ae^{t/\tau} = ae^{t/2}$$

where $(\mathrm{Rb/Sr})_N$ is the mantle-normalized ratio, and similar equations for $f(t)_{\mathrm{Sm/Nd}}$ and U/Pb. Results are shown in Figure 10-4.

If we assume that the fractionation factors $(\mathrm{Rb/Sr})_N - 1$ and $(\mathrm{Sm/Nd})_N - 1$ have grown exponentially with time, with $\tau = 2$ Ga, from 3.8 Ga ago, the magnitude and spread of the ε values are much reduced compared to simple two-stage models. Also the ages of the more enriched reservoirs agree with the ages of the more depleted reservoirs and with the age spread of the continental crust. Although the depleted reservoir may also deplete progressively, successive depletion is difficult, because a large fraction of the incompatible elements is removed with the first melt increment. These results are shown in Figure 10-4. Progressive enrichment and magma mixing are two of the ways that apparently contradictory isotopic and parent/daughter ratios in a magma can be reconciled. It is not necessary, with these

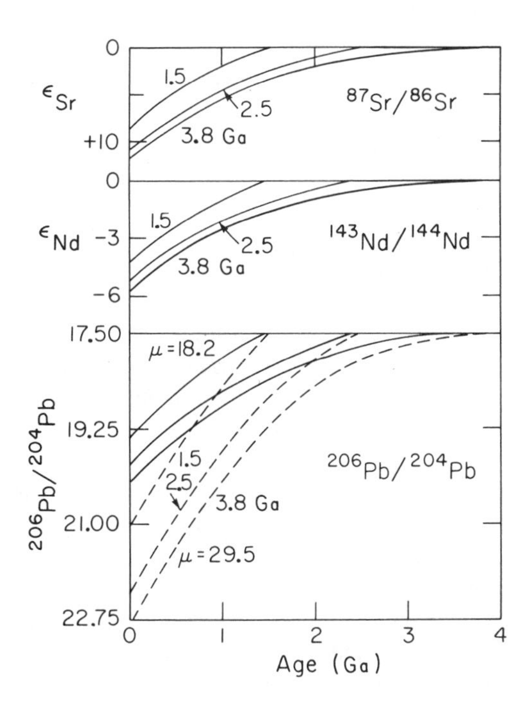

FIGURE 10-4
Evolution of ε_{Sr}, ε_{Nd} and $^{206}Pb/^{204}Pb$ in enriched reservoirs ranging in age from 3.8 to 1.5 Ga. The fractionation factor is assumed to vary as $f(t) = a(e^{t/\tau} - 1)$ where t is measured to the left from the start of differentiation, τ is the characteristic time associated with the fractionation and a is calculated from the present enrichment. Adopted values are $f_{Rb/Sr} = 0.67$, $f_{Sm/Nd} = -0.17$ and $^{238}U/^{204}Pb$ (μ) = 29.5 (dashed lines) and 18.2 (solid lines). These are determined from the inferred present composition of the source region for oceanic-island and continental flood basalts. τ is assumed to be 2 Ga. Upper curve in each series is for a differentiation initiation age of 1.5 Ga; central and lower curves are for 2.5 and 3.8 Ga respectively. This type of progressive enrichment can result from, for example, progressive crystallization of the donor reservoir, giving more enriched late-stage fluids. Islands showing positive ε_{Nd} values may represent mixtures of depleted MORB magmas and enriched (hotspot) magmas. On this basis Tristan and Kerguelen magmas appear to be the most enriched, or least diluted by MORB.

mechanisms, that the Rb/Sr, Sm/Nd and U/Pb ratios, for example, of magmas represent the time-integrated values in the reservoir.

OXYGEN ISOTOPES

The analysis of $^{18}O/^{16}O$ ratios is a powerful geochemical tool because of the large difference between crustal and mantle rocks. It has recently become evident that various mantle basalts also exhibit differences in this ratio. Rocks that have reacted with the atmosphere or hydrosphere are typically richer in oxygen-18, through mass fractionation, than those from the mantle. Crustal contamination and reflux of crustal material into the mantle serve to increase the $^{18}O/^{16}O$ ratio of basalts affected by these processes.

Oxygen is not a trace element; it comprises about 50 weight percent of most common rocks. There are three stable isotopes of oxygen, of mass 16, 17 and 18, which occur roughly in the proportions of 99.76 percent, 0.038 percent and 0.20 percent. Variations in $^{18}O/^{16}O$ are expressed as

$$\delta^{18}O(‰) = \left[\frac{(^{18}O/^{16}O)_m - (^{18}O/^{16}O)SMOW}{(^{18}O/^{16}O)SMOW}\right] \times 1000$$

where SMOW is Standard Mean Ocean Water, a convenient standard.

Oxygen isotopic variations in rocks are chiefly the result of fractionation between water and minerals. The fractionation between minerals is small. $\delta^{18}O$ generally decreases in the order quartz–feldspar–pyroxene–garnet–ilmenite, with olivine and spinel showing large variations. Metasomatism increases $\delta^{18}O$ as well as the LIL elements (McCulloch and others, 1981). Appropriate fluids can be derived from subducted oceanic crust.

Most unaltered MORB has $\delta^{18}O$ near 5.7 per mil whereas alkali basalts tend to be more ^{18}O-rich. Carbonatites, potassic lavas and andesites have $\delta^{18}O$ of 6.0 to 8.5 per mil (Kyser and others, 1982). The maximum $\delta^{18}O$ variations in rocks known to be derived from the mantle are those recorded in eclogite, 1.5 to 9.0 per mil (Garlich, 1971). Peridotites are generally 5.2 to 7.0, but the olivines in the same rocks range from 4.4 to 7.5. The $\delta^{18}O$ range in pyroxenes are very much smaller, +5.3 to 6.5 per mil. The upper parts of the subducted slab are extremely ^{18}O-rich ($\delta^{18}O$ = 10–25 per mil; Gregory and Taylor, 1981), and subduction is a plausible way of increasing $\delta^{18}O$ of the shallow mantle. Gregory and Taylor (1981) proposed that oxygen-18 and LIL enrichment go hand in hand and suggested that both subducted material and kimberlitic magmas have the appropriate characteristics.

Although $\delta^{18}O$ values are generally used to investigate crustal contamination, the recognition that mantle magmas also vary in $\delta^{18}O$ provides an additional tool for characterizing the various mantle reservoirs, and for unraveling the nature of the enrichment process. Whether $\delta^{18}O$ enrichment is caused by upward migration of kimberlitic fluids or subduction of oceanic crust, it is probable that the enriched layer is shallow. Most islands are high in $^3He/^4He$ (that is, "primitive"), suggesting that subducted material, or crustal recycling, is not the sole source of enrichment.

Continental rocks are generally enriched in $\delta^{18}O$ (6 to 9 per mil) relative to MORB. It is usually stated that the continental crust is enriched relative to the mantle, but some crustal rocks are low, due to interaction with rain water, and some mantle magmas have high $\delta^{18}O$, as previously mentioned.

Contamination of mantle-derived magmas by the shallow mantle, lithosphere or crust may be caused by bulk assimilation of solid rock, by isotopic and trace-element exchange between magma and wallrock, or by magma mixing between the original melt and melts from the wallrock. Isotopic and trace-element exchange between magma and solid rock are likely to be too inefficient to be important because of the very low diffusivities. Diffusion distances, and therefore equilibration distances, are only a few centimeters per million years. Bulk assimilation or isotope exchange during partial melting are probably the most efficient means of magma contamination, and Taylor (1980) discussed these processes in detail. This is not a simple two-component mixing process. It involves three end-members, the magma, the contaminant and a cumulate phase, which crystallizes to provide the heat required to partially melt the wallrock or dissolve the assimilated material.

Most ocean-ridge tholeiites have $\delta^{18}O$ near 5.7 per mil. Potassic lavas have $\delta^{18}O$ values of 6.0 to 8.5, and continental tholeiites range up to 7.0. Oceanic alkalic basalts go as high as 10.7. Kimberlites and carbonatites have values up to 26. Alkalic basalts apparently form in a region of the mantle that is more ^{18}O-rich than the MORB source region. At high temperature clinopyroxene and garnet have lower $\delta^{18}O$ values than olivine. It is possible, therefore, from the ^{18}O evidence that tholeiites originate in a garnet-clinopyroxene-rich reservoir and alkalic basalts have more olivine in their source regions. In addition, olivine fractionation at low temperature increases the $\delta^{18}O$ of residual melts.

A sample mixing calculation is shown in Figure 10-5. In order to match the higher $\delta^{18}O$ values found for oceanic islands and have reasonable amounts of fractionation and contamination, a $\delta^{18}O$ of about 17 per mil is implied for the enriched component. Oxygen fractionation is ignored in this calculation.

Pillow lavas in the upper parts of ophiolites have $\delta^{18}O$ values of over 12 per mil (Gregory and Taylor, 1981). Subduction of this material may contribute to the high $\delta^{18}O$ of the enriched reservoir. Regardless of the origin of the distinct $\delta^{18}O$ reservoirs, mixing of magmas from these reservoirs can explain the spread of values over the $^{87}Sr/^{86}Sr$–$\delta^{18}O$ plane.

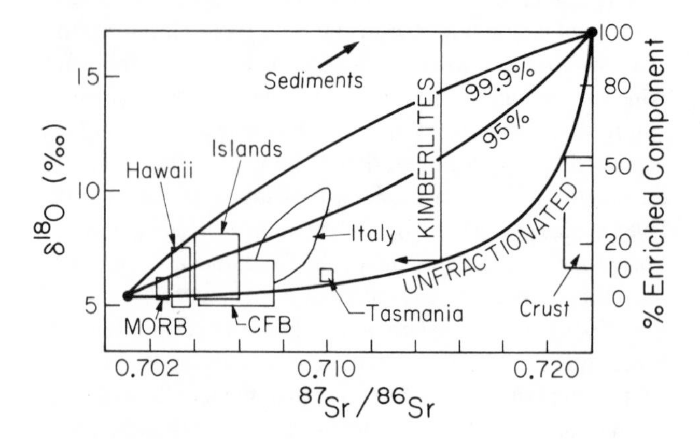

FIGURE 10-5
$\delta^{18}O$ versus $^{87}Sr/^{86}Sr$ for magmas, oceanic sediments, kimberlites and continental crust. Mixing curves are shown for various degrees of crystal removal from the depleted end-member.

RARE GASES

Rare-gas isotope pairs place constraints both on the evolution of the atmosphere and on transport processes in the mantle. The rare-gas radiogenic/primordial isotope pairs are ^4He–^3He, ^{40}Ar–^{36}Ar, ^{129}Xe–^{130}Xe and ^{136}Xe–^{130}Xe. The radiogenic isotopes are produced from α-decay of uranium and thorium, β-decay of potassium-40, β-decay of iodine-129 and spontaneous fission of plutonium-244 and uranium-238. Iodine-129 and plutonium-244 are referred to as extinct since they have respective half-lives of 17 and 76 Ma; they therefore provide constraints on the earliest stages of outgassing of the atmosphere. Due to the volatility of the noble gases and the involatility of the parent isotopes (^{40}K, ^{238}U, ^{235}U, ^{232}Th, ^{244}Pu), the isotope evolution is strongly affected by degassing. This is not the case for the samarium-neodymium and rubidium-strontium systems. Results to date show that the depleted MORB reservoir was outgassed in earliest Earth history. Midocean-ridge basalts, which are low in potassium, uranium, thorium and rubidium compared to ocean-island basalts, are high in $^{40}Ar/^{36}Ar$, $^3He/^4He$ and $^{129}Xe/^{130}Xe$. Ocean-island basalts are higher in the "primordial" components. This does not necessarily mean that they come from a primordial or undegassed reservoir. The presence of primordial gases in the mantle shows that outgassing of the mantle has not been 100 percent efficient. This in turn implies that the crust extraction process has left "crustal elements" behind in the mantle. These are important points since the assumptions that the upper mantle was totally and efficiently depleted by extraction of the continental crust and that the presence of primitive gases implies primitive, undegassed mantle have led to hypotheses about lower-mantle reservoirs and deep-mantle plumes.

Helium Isotopes

Helium has two stable isotopes of mass 3 and mass 4, which vary in their ratio by a factor of more than 10^3 in terrestrial materials. Helium-4, the most abundant isotope, produced by radioactive decay of uranium and thorium, was discovered in the atmosphere of the Sun, from which it takes its name. The cosmic abundance of helium is about 10 percent, making it second only to hydrogen in the composition of stars, such as the Sun, and presumably the preplanetary solar nebula. In the Sun helium is about 100 times more abundant than oxygen and 3000 times more abundant than silicon. The Earth has clearly not received its solar complement of this element, but the small amount that it has

TABLE 10-3
Argon and Helium Abundances

	Units	Atmosphere	Crust	MORB	Plume
^{36}Ar	$10^{19}cm^3$	12.46	0.04	—	—
	$10^{-8}cm^3/g$	2	<3	0.005–0.013	3.3–5.9
^{40}Ar	$10^{21}cm^3$	36.78	4.6–14	—	—
	$10^{-5}cm^3/g$	—	—	0.2–0.4	2
$^{40}Ar/^{36}Ar$	10^3	0.2955	1	24.5	0.35–0.50
^3He	$10^{-10}cm^3/g$	—	—	0.2–0.7	10–36
^4He	$10^{-6}cm^3/g$	—	—	2–6	28–170
$^3He/^4He$	10^{-5}	0.14	10^{-2}–10^{-3}	1.2	>6
$^4He/^{40}Ar$	—	10^{-3}	—	2–10	1.4–1.7
$^4He/^{36}Ar$	10^4	10^{-4}	—	1.6–2.9	0.05–0.09
$^3He/^{36}Ar$	—	—	—	0.16–1.4	0.03–0.06
$(^3He/^{36}Ar)_N$	—	—	—	20±6	1
$(^{36}Ar)_N$	—	—	—	1	250–1100
$(^3He)_N$	—	—	—	1	5
$^3He/U$	—	—	—	$\sim10^{-5}$	10^{-3}

Hart and others (1985).

trapped, as primordial helium-3, provides important clues regarding differentiation and outgassing of the Earth.

The atmospheric $^3He/^4He$ ratio is about 1.4×10^{-6} (Table 10-3). A small amount of helium-3 is produced by nuclear reactions involving lithium, but most of the helium-3 in the mantle is thought to be original or primordial. The high $^3He/^4He$ ratio of mantle gases indicates that the Earth is not completely outgassed. Both isotopes escape from the atmosphere because of their low masses. The residence time of the helium in the atmosphere is about 10^6 years, thus a geochemical mass balance cannot be attempted, but the mantle flux rate can be estimated.

The $^3He/^4He$ ratio in gas-rich meteorites is about 3×10^{-4}, about 200 times greater than the atmospheric ratio. About half of the production of helium-4 in the Earth takes place in the continental crust. The $^3He/^4He$ ratio in helium from the mantle is at least 100 times greater than the ratio produced in the crust.

The $^3He/^4He$ ratio, *R*, of terrestrial samples varies from less than 10^{-8} to more than 10^{-3} with a general increase from old, stable terranes to currently active regions. These bracket the current atmospheric value of 1.4×10^{-6} and are generally much less than the presumed "cosmic" value of about 10^{-4}. The highest values are found in regions of recent volcanism such as Hawaii, Iceland and Yellowstone and in diamond inclusions, which are probably armored from the addition of radiogenic helium-4. Some hotspot islands and other plume-like basalts have low *R* values, suggesting the addition of a low-*R* component, such as sediments or seawater, or a decrease of *R* with time due to helium-4 production by uranium and thorium. Studies in Hawaii suggest that the earliest magmas to emerge in a given region are the most enriched in the "primitive" helium-3 component, and *R* decreases as volcanism proceeds. *R* is higher in tholeiites than in alkalics and is relatively low in olivine inclusions.

Midocean-ridge tholeiites average about 8.5 times the atmospheric ratio for $^3He/^4He$, and this ratio is remarkably constant. The MORB reservoir, therefore, must be extremely uniform. Basalts and phenocrysts from Hawaii range up to 37 times atmospheric. The $^3He/^4He$ ratio in natural diamond crystals from kimberlites ranges up to 280 times atmospheric, close to the solar value. $^3He/^4He$ values found in Iceland and Hawaii and along the Reykjanes Ridge can be modeled as two-component mixtures with MORB and Loihi being the end-members. Other islands have lower $^3He/^4He$ ratios than MORB, suggesting a third component, or a variably outgassed enriched reservoir. The $^3He/^4He$ ratio for oceanic-island basalts is highly variable.

Values of the $^3He/^4He$ ratio in some natural diamond crystals range up to 3.2×10^{-4} (Ozima and Zashu, 1983), suggesting that the primitive ratio has been preserved, that these samples have been isolated from uranium and thorium and that crystals can hold helium for long periods of time. These values are close to the solar value ($\sim4 \times 10^{-4}$) and exceed the value commonly attributed to meteorites and planets. Some of these diamonds appear to be nearly as old as the Earth itself. They appear to have retained high $^3He/^4He$ values by being completely isolated from the uranium in the surrounding mantle. This may be evidence that a cold and thick continental lithosphere (refractory and buoyant assemblages of olivine) was an early result of terrestrial differentiation. Cold proto-continental rafts moving away from the hotter regions of the mantle offer a possible explanation for the helium data and other evidence for ancient mantle under some cratons.

The $^3He/^4He$ ratio in some oceanic-island basalts is very high. The correlation of $^3He/^4He$ with $^{87}Sr/^{86}Sr$ on oceanic islands and along ridge segments is not simple, leading to the speculation that helium and the other incompatible elements are decoupled. Primordial helium-3 is often assumed to come from primordial or undifferentiated mantle, but it can be extracted from primitive mantle in a melt phase and stored elsewhere. If the melt is high in uranium and thorium, the $^3He/^4He$ ratio will, of course, decrease with time.

Large oceanic islands such as Hawaii and Iceland not only have some magmas with high $^3He/^4He$ but also exhibit a very large variation in $^3He/^4He$ for relatively constant $^{87}Sr/^{86}Sr$. The wide range of $\delta^{18}O$ values of basalts from these islands, as well as the low $^3He/^4He$ values, indicate involvement of the atmosphere, meteoric water and older altered basaltic crust. Low $^3He/^4He$ ratios in island basalts are not necessarily representative of the mantle source region. In Hawaii, the highest $^3He/^4He$ ratios are found for the youngest volcanoes and the smallest edifices. This suggests that the gases in the earliest stage of volcanism are representative of the mantle source and gases at later stages are more representative of the lithosphere (Kurz and others, 1983) or that shallow mantle gases are flushed out by the earliest volcanism.

The low $^3He/^4He$ and high helium-4 content of olivine xenocrysts in Hawaiian basalts indicate that helium can be trapped in significant quantities in crystals and that this gas can contribute to low measured $^3He/^4He$ values of some alkali basalts (Kurz and others, 1983). These observations also indicate that rare gases can be trapped in the upper mantle. The high helium content of xenocrysts relative to glasses also shows that the effective crystal-melt partition coefficient for helium can be high relative to other incompatible elements. The high $^3He/^4He$ and low $^{87}Sr/^{86}Sr$ ratios of some Hawaiian basalts show that these basalts are not directly derived from "primitive" mantle. The relation $^3He/^4He$ versus $^{40}Ar/^{39}Ar$ of hotspot magmas suggests an atmospheric component (Kaneoka and others, 1983).

The $^3He/^4He$ ratio of a melt or basalt increases with time at a rate determined by the abundance of uranium and thorium (Craig and Lupton, 1981):

$$J = 0.2355 \times 10^{-12}\,(\mathrm{U}) \times [1 + 0.123(\mathrm{Th/U} - 4)]$$

where J is the growth of helium-4 in units of cm^3/g year and U is in ppm. The total $^3He/^4He$ is therefore

$$^3\mathrm{He}/^4\mathrm{He} = \frac{(^3\mathrm{He}/^4\mathrm{He})_o \times \dfrac{[^4\mathrm{He}]_o}{[^4\mathrm{He}]_R} + \left(\dfrac{^3\mathrm{He}}{^4\mathrm{He}}\right)_R}{\dfrac{[^4\mathrm{He}]_o}{[^4\mathrm{He}]_R} + 1}$$

where $(^3He/^4He)_R$ is the production ratio of $^3He/^4He$, about 10^{-8} for basalt. Assimilation of deeply buried older basalts by new magma will decrease both $^3He/^4He$ and $\delta^{18}O$.

Mixing calculations are shown in Figure 10-6 with two assumptions about the $^{87}Sr/^{86}Sr$ ratio of the depleted end-member. The highest $^3He/^4He$ samples from Loihi can be matched with about 18 percent of the enriched end-member

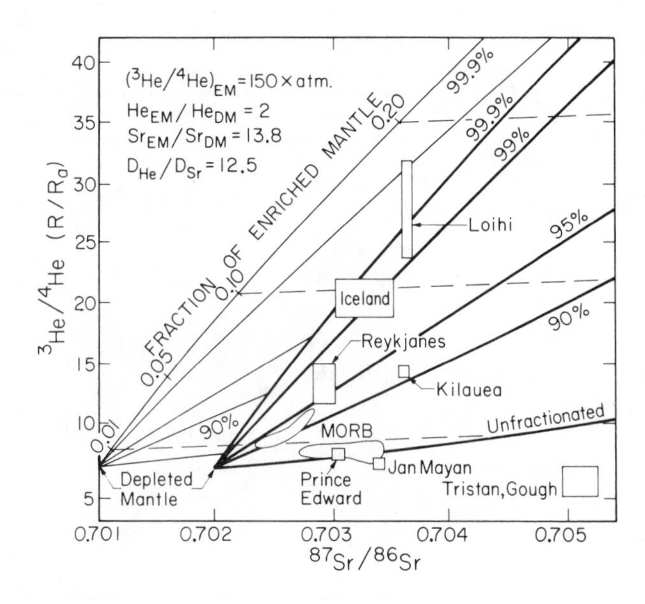

FIGURE 10-6
Mixing curves for $^3He/^4He$ versus $^{87}Sr/^{86}Sr$. Two choices for the depleted end-member (DM) are shown. Hawaiian basalts imply 90 to over 99 percent crystal fractionation and 5–18 percent contamination from enriched mantle (EM) (Anderson, 1985).

($^3He/^4He$ = 150 × atmospheric) combined with 99.9 percent crystal fractionation of the depleted magma. Islands such as Prince Edward (in the Southern Ocean) and Jan Mayen can be modeled as mixtures involving nearly unfractionated MORB. Tristan da Cunha and Gough may require a third component, such as subducted sediments, or may have experienced previous outgassing. In order to cover the entire range by two-component mixing and fractionation, it is necessary to assume that the helium abundance in the enriched reservoir is about twice the abundance in the depleted reservoir and that the partition coefficient for helium is much greater than for strontium. The partition coefficient for helium in basalt is unknown, but it is not obvious that it will behave as a classic incompatible element. The total helium abundances in some diamonds and xenoliths are higher than in some basalts, suggesting that appreciable helium can be trapped in crystals.

The clinopyroxene-melt partition coefficients for argon and krypton are 0.2 to 0.5, and there is some indication that the value increases with decreasing atomic weight (Hiyagon and Ozima, 1982). These surprisingly high values for the rare gases, compared to the values for other incompatible elements, are of the order of those required to explain the helium and strontium isotopic results by the fractionation mixing, two-components model.

The olivine-basalt partition coefficients for helium, neon and argon are $\lesssim$ 0.07, 0.006–0.08 and 0.05–0.15, respectively (Hiyagon and Ozima, 1986). Clinopyroxene values may be an order of magnitude higher. In the eclogite fractionation model the important phases are clinopyroxene and garnet, but the partitioning of helium into either is unknown. Garnet might be an important accommodater for the rare gases.

The distribution of $^3He/^4He$ in the enriched reservoir is likely to be uneven because it depends on the amount of trapped primordial helium-3, as well as the uranium and thorium content, and age of the enrichment. It therefore depends on the outgassing history as well as the magmatic and metasomatic history of the reservoir. If part of the LIL enrichment is due to reinjection of oceanic crust and sediments, this part of the enriched reservoir will have a low $^3He/^4He$ ratio.

Nevertheless, combined crystal fractionation and contamination, utilizing only two end-members, can explain much of the range and several of the trends observed in oceanic basalts. For example, the MORB–Loihi and Kilauea–Loihi trends can be due to increasing fractionation and contamination. The MORB trends require only small differences in the amount of contamination.

The $^3He/^4He$ ratio of hotspot magmas may help constrain the enrichment process. Subducted material, a possible candidate for enrichment in LIL, $^{87}Sr/^{86}Sr$ and $\delta^{18}O$, has low $^3He/^4He$, and the ratio decreases with time. Ancient subducted crust or sediment, therefore, cannot be the source of high-$^3He/^4He$ basalts. Some oceanic islands are low in $^3He/^4He$, and these may have a subducted component. Except for the high $^3He/^4He$ found in kimberlitic diamonds, there are few data on kimberlitic material, so this source cannot be evaluated.

If we admit the possibility that helium-3, uranium and thorium can all be concentrated in the enriched source region by upward migrating melts or fluids, then the present-day $^3He/^4He$ ratio depends on the $^3He/(Th+U)$ ratio and the age of enrichment events and, of course, the degree of outgassing.

The study of $^3He/^4He$ has led to several important conclusions:

1. Primordial helium-3 is present; hence, the Earth is not completely outgassed.
2. The results from submarine basalts indicate a uniform, probably well-mixed reservoir. The similarity in results from the Atlantic and the Pacific suggests that the depleted reservoir is global. This suggests that the depleted reservoir is deep.
3. The variability in island basalts shows that the whole mantle does not share this homogeneity. Either the enriched source region is heterogeneous or melts from the enriched source region are variably mixed with depleted basalts.

The Earth is an extensively differentiated and probably well-outgassed planet. The crust, for example, contains 35–58 percent of the estimated mantle-plus-crustal abundances for various incompatibles (Rb, Cs, K, Ba, Th and U). The amount of argon-40 in the atmosphere represents 77 percent of the total produced by 152 ppm potassium that is estimated to reside in the mantle-plus-crust. Since argon-40 is produced slowly over time, its presence in the atmosphere cannot be the result of an early catastrophic outgassing. Helium-3 is probably more outgassed than argon-40, but we do not know what fraction of the primordial component still resides in the mantle.

O'Nions and Oxburgh (1983) concluded that the uranium and thorium abundance in the mantle required to support the observed radiogenic helium flux into the atmosphere would only provide about 5 percent of the mantle heat flux. That can be raised to 10 percent if half the mantle heat flow is due to secular cooling. This implies that vastly more helium is being generated in the Earth than is being outgassed, or that the time constants for the loss of heat and loss of helium differ greatly. Helium brought into the shallow mantle can only escape at ridges and hotspots and only then if magmas are brought close enough to the surface to vesiculate. By contrast heat is continuously lost from the mantle as the seafloor spreads and as continents cool from their last thermal event. Magmas solidifying at depth can even recycle helium back into the mantle. The diffusion distances for helium are relatively short, and therefore helium can be trapped at any depth, not just the lower mantle.

If outgassing was more efficient in early Earth history, then a larger fraction of the mantle argon-40 may reside in the atmosphere than is the case for helium-4 because of the short half-life of potassium-40. It also appears that potassium is more efficiently concentrated in the crust than uranium or thorium and therefore can more readily dump its decay products into the atmosphere. The present rate of outgassing also may not be typical of that integrated over the past 200 Ma. If oceanic islands, plateaus and continental flood basalts are any indication, there may have been more hotspot activity and outgassing during the breakup of Pangaea. Seafloor spreading, and therefore ridge-crest volcanism, was much greater in the recent past than at present.

Two sources of mantle helium can clearly be identified, MORB helium and a high $^3He/^4He$ source. The latter is, unfortunately, often called undepleted, undegassed or primitive mantle. The range of isotopic abundances, and correlations, or lack thereof, with other isotopes can be produced by mixing, crystal fractionation, degassing, and subsequent addition of radiogenic ^4He. The location of the high $^3He/^4He$ source is not known. It may be inside of crystals, such as garnet or diamond or in a separate part of the mantle, for example, in depleted harzburgites. The shallow mantle may be the best storage place since it is relatively cold and probably buoyant. Entrainment of upward migrating helium by previously depleted harzburgite would make the harzburgite a secondary, rather than primitive, source.

Argon and Helium

The argon-40 in the atmosphere is produced from the decay of potassium-40 in the crust and mantle via the decay equation

$$^{40}Ar = \lambda_1 {}^{40}K(e^{\lambda\tau} - 1)/\lambda$$

where λ_1 is the decay constant for decay of potassium-40 to argon-40 (0.581×10^{-10}/yr) and λ is the total decay constant of potassium-40 to calcium-40 and argon-40 (0.5543×10^{-9}/yr). The $^{40}K/K$ ratio is about 1.17×10^{-4}. The total argon-40 in the atmosphere is then

$$^{40}Ar = {}^{40}Ar_o + 1.39 \times 10^{-4}\ K$$

where $\tau = 4.5 \times 10^9$ yr and $^{40}Ar_o$ is the initial argon-40 content, which is probably negligible compared to that produced and outgassed over the age of the Earth.

The decay equation for helium is

$$^{4}He = 8\ {}^{238}U(e^{\lambda_8\tau} - 1) + 7\ {}^{235}U(e^{\lambda_5\tau} - 1) + 6\ {}^{232}Th(e^{\lambda_2\tau} - 1)$$

where the decay constants are $\lambda_8 = 1.551 \times 10^{-10}$/yr, $\lambda_5 = 9.85 \times 10^{-10}$/yr and $\lambda_2 = 4.948 \times 10^{-10}$/yr. The amount of helium-4 generated in 4.5×10^9 yr is approximately

$$^{4}He = 16.64\ U$$

which gives

$$^{4}He/^{40}Ar = 1.22 \times 10^5\ (U/K)$$

The predicted bulk Earth value for $^4He/^{40}Ar$ is about 1.5 ± 0.2, which is about 10^5 higher than the atmospheric ratio. The difference is due to the rapid escape of helium from the atmosphere. The bulk Earth value, however, is in the range of both MORB and hotspot magmas.

On the basis of $^3He/^4He$ versus $^{40}Ar/^{36}Ar$ systematics, there are four discrete reservoirs: atmosphere, continental crust, MORB and "plume." Most basalts appear to be mixtures of material from these reservoirs. Hotspot or "plume," magmas have $^{40}Ar/^{36}Ar$ ratios similar to atmospheric values but have $^3He/^4He$ up to two orders of magnitude higher. This suggests that hotspot basalts or their source region are contaminated with atmospheric, seawater or sedimentary components or that the present atmosphere is mainly generated by outgassing of the plume source with differential escape of helium-3 relative to helium-4. In any event the mixing of plume magmas with atmosphere, seawater or sediments does not change the $^{40}Ar/^{36}Ar$ ratio. Plume magmas therefore appear to be homogeneous in $^{40}Ar/^{36}Ar$ just as MORB appears to homogeneous with respect to $^3He/^4He$. If most magmas are mixtures, then He/Ar for the MORB reservoir is greater than He/Ar for the plume source (Hart and others, 1979).

ISOTOPIC CONSTRAINTS ON MAGMA GENESIS

Midocean-ridge basalts (MORB) are the most depleted (low concentrations of LILs, low values of Rb/Sr, Nd/Sm, $^{87}Sr/^{86}Sr$, $^{144}Nd/^{143}Nd$, $^{206}Pb/^{204}Pb$) and the most voluminous magma type. They erupt through thin lithosphere and have apparently experienced some crystal fractionation prior to eruption. Isotopic ratios show that the MORB reservoir was outgassed and depleted more than 10^9 years ago. There is growing evidence that there is also an ancient enriched reservoir in the mantle and that the continental crust is not the only complement to the MORB reservoir (Menzies and Murthy, 1980; Anderson, 1981, 1983; McCulloch and others, 1983). The most enriched mantle magmas are lamproites, kimberlites, and basalts from islands such as Kerguelen, St. Helena and Tristan da Cunha. There is a complete spectrum of basalts lying between these extremes. The presence of helium-3 in MORB and the fact that its $^3He/^4He$ is higher than atmospheric (Hart and others, 1979; Kaneoka, 1983) suggests that outgassing was not 100 percent efficient, or that MORB is contaminated. Lead isotopes suggest that MORB has experienced some contamination prior to eruption (Anderson, 1982). Therefore, MORB itself may be a hybrid magma.

The geochemistry of various magma types requires at least two distinct reservoirs in the mantle. The relatively uniform depleted mantle suffered an ancient depletion event, presumably by removal of a residual fluid. The other reservoir, enriched mantle, is less homogeneous and is presumed to provide the enriched signature of hotspot magmas. Correlated LIL and isotopic variations in mantle magmas are often explained by mixing of material from the two reservoirs.

Basalts with high Rb/Sr, La/Yb, and Nd/Sm generally have high $^{87}Sr/^{86}Sr$ and low $^{143}Nd/^{144}Nd$. These can be explained by binary mixing of depleted and enriched magmas or by mixtures of a depleted magma and a component representing varying degrees of melting of an enriched reservoir (Anderson, 1982). The variable LIL ratios of such an enriched component generates a range of mixing hyperbolas or "scatter" about a binary mixing curve, even if there are only two isotopically distinct end-members. Thus a model with two isotopically distinct reservoirs can generate an infinite variety of mixing lines (Anderson, 1982). In some regions, however, the inverse relationship between LIL and isotopic ratios cannot be explained by binary mixing (Lipman and Mehnert, 1975; Dunlop and Fitton, 1979; Barberi and others, 1980; Menzies and Murthy, 1980b, Zhou and Armstrong, 1982; Chen and Frey, 1983). These regions are all in midplate or thick lithosphere environments, and sublithospheric crystal fractionation involving garnet and clinopyroxene might be expected prior to eruption.

The wide range of isotopic ratios in island and continental basalts has led to the view that the mantle is heterogeneous on all scales. The failure of simple binary mixing to explain the spread of isotopic and trace-element ratios, and the presence of LIL-enriched basalts with time-integrated depleted isotopic ratios, has also given rise to the concept that recent mantle metasomatism has affected the source region and may even be a prerequisite for magmatism (Boettcher and O'Neil, 1980).

Some of the observations that have been interpreted in terms of metasomatism, or an extremely heterogeneous mantle, are (1) scatter of $^{143}Nd/^{144}Nd$–$^{87}Sr/^{86}Sr$ values about the mantle array, (2) lack of correlation between $^{87}Sr/^{86}Sr$ and $^{206}Pb/^{204}Pb$ or $^{3}He/^{4}He$, and (3) inverse correlations between La/Ce or Rb/Sr and $^{87}Sr/^{86}Sr$. However, if one or both end-members of a hybrid is a fractionated melt, or represents varying degrees of partial melting, then these observations do not necessarily require multiple sources. If melts from one reservoir must traverse the other reservoir on their way to the surface, there is no need for a "plum pudding" mantle or recent metasomatism of the source.

The cases where a depleted magma causes variable degrees of melting of an enriched reservoir or where an undepleted magma partially melts a depleted layer have already been treated (Anderson, 1982; Chen and Frey, 1983). We now treat a fractionating depleted magma interacting with an enriched reservoir. To fix ideas, consider the depleted magma from the depleted-mantle reservoir to be the parent of MORB. If this magma is brought to a near-surface environment, it may fractionate olivine, plagioclase and orthopyroxene. If arrested by thick lithosphere it will fractionate garnet and clinopyroxene. Assume that the fractionating crystals and melt are in equilibrium. I emphasize varying degrees of crystal fractionation, but these results are applicable to the case where the depleted component represents varying degrees of partial melting of a garnet- and clinopyroxene-rich region, such as a garnet pyroxenite lower oceanic lithosphere or an eclogite slab (Chen and Frey, 1983).

There are two situations that have particular relevance to mantle-derived magmas. Consider a "normal" depleted mantle magma. If it can rise unimpeded from its source to the surface, such as at a rapidly spreading ridge, it yields a relatively unfractionated, uncontaminated melt—MORB. Suppose now that the magma rises in a midplate environment, and its ascent is impeded by thick lithosphere. The magma will cool and crystallize, simultaneously partially melting or reacting with the surrounding shallow mantle. Thus, crystal fractionation and mixing occur together, and the composition of the hybrid melt changes with time and with the extent of fractionation. Fractionation of garnet and clinopyroxene from a tholeiitic or picritic magma at sublithospheric depths (>50 km) can generate alkalic magmas with enriched and fractionated LIL patterns. Let us investigate the effects of combined eclogite fractionation (equal parts of garnet and clinopyroxene) and "contamination" on melts from depleted mantle. "Contamination" is modeled by mixing an enriched component with the fractionating depleted magma. This component is viewed as a partial melt generated by the latent heat associated with the crystal fractionation. The assumed geochemical properties of the end-members, the enrichment factors of the elements in question and the partition coefficients, D, assumed in the modeling are given in Table 10-4. D for six elements (Rb, Sr, Sm, Nd, La and Ce) are the mean for clinopyroxène and garnet (Frey and others, 1978). The figures that follow show various ratios for mixes of a fractionating depleted melt and an enriched component. We assume equilibrium crystal fractionation, as appropriate for a turbulent or permeable magma body, and constant D.

La/Ce Versus $^{87}Sr/^{86}Sr$

La/Ce is high in melts relative to crystalline residues containing garnet and clinopyroxene. It is low in depleted reservoirs and high in enriched reservoirs. Low and high values of $^{87}Sr/^{86}Sr$ are characteristics of time-integrated depleted and enriched magmas, respectively. Since high $^{87}Sr/$

TABLE 10-4
Parameters of End-Members

Parameter	Depleted Mantle (1)	Enriched Mantle (2)	Enrichment Ratio (2)/(1)	
$^{87}Sr/^{86}Sr$	0.701–0.702	0.722	Sr_2/Sr_1	13.8
ε_{Nd}	24.6	−16.4	Nd_2/Nd_1	73.8
$^{206}Pb/^{204}Pb$	16.5–17.0	26.5	Pb_2/Pb_1	62.5
$^{3}He/^{4}He$ (R_a)	6.5	150	He_2/He_1	2.0
$\delta^{18}O$ (permil)	5.4	17	O_2/O_1	1.0
La/Ce	0.265	0.50	Ce_2/Ce_1	165.0
Sm/Nd	0.50–0.375	0.09–0.39		

Partition Coefficients

Rb	0.02	Sr	0.04	La	0.012	Ce	0.03
Sm	0.18	Pb	0, 0.002	He	0.50	Nd	0.09

^{86}Sr implies a time-integrated enrichment of Rb/Sr, there is generally a positive correlation of Rb/Sr and La/Ce with ^{87}Sr/^{86}Sr. Some magmas, however, exhibit high La/Ce and low ^{87}Sr/^{86}Sr.[9] This cannot be explained by binary mixing of two homogeneous magmas.

On a theoretical La/Ce versus ^{87}Sr/^{86}Sr plot (Figure 10-7), the mixing lines between the crystallizing MORB and enriched mantle components reverse slope when MORB has experienced slightly more than 99 percent crystal fractionation. The relationships for equilibrium partial melting and equilibrium, or batch, crystallization are, of course, the same. Therefore, large degrees of crystallization of a MORB-like melt or small amounts of partial melting of a MORB layer are implied by an inverse relationship between La/Ce (or Rb/Sr, La/Yb, Nd/Sm, and so on) and ^{87}Sr/^{86}Sr (or ^{143}Nd/^{144}Nd) such as observed at many midplate environments. The apparently contradictory behavior of magmas with evidence for current enrichment and long-term depletion is often used as evidence for "recent mantle metasomatism." Figure 10-7 illustrates an alternative explanation. Note that, with the parameters chosen, the Hawaiian alkalics have up to 10 percent contamination by the enriched component.

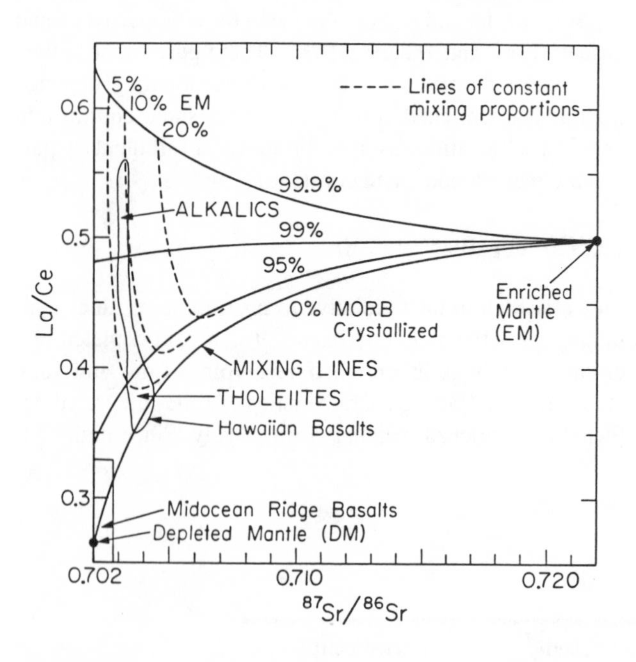

FIGURE 10-7
La/Ce versus ^{87}Sr/^{86}Sr for the fractionation-contamination model, compared with Hawaiian basalts. Unfractionated MORB (DM) has La/Ce = 0.265. La/Ce of the depleted end-member increases as crystal fractionation proceeds. The enriched end-member (EM) has La/Ce = 0.5, in the range of kimberlitic magmas. The Hawaiian tholeiites can be modeled as mixes ranging from pure MORB plus 2–7 percent enriched component to melts representing residuals after 95 percent clinopyroxene-plus-garnet crystal fractionation and 5–8 percent enriched component. Alkali basalts involve more crystal fractionation and more contamination. In this and subsequent figures solid curves are mixing lines between EM and melts representing fractionating depleted magmas. Dashed curves are trajectories of constant mixing proportions. Data from Chen and Frey (1983).

Neodymium Isotopes Versus ^{87}Sr/^{86}Sr

Isotopic ratios for ocean-island and continental basalts are compared with mixing curves (Figure 10-2), these basalts can be interpreted as mixes between a fractionating depleted magma and an enriched component. The value for primitive mantle (ε_{Nd} = 0) is also shown. The primitive mantle value of ^{87}Sr/^{86}Sr is unknown and cannot be inferred from basalts that are themselves mixtures. For example, the basalts cover a field extending from 0 to 99.9 percent crystal fractionation, and the corresponding ^{87}Sr/^{86}Sr ratios at ε_{Nd} = 0 range from 0.7043 to 0.7066. The extreme situations, constant degree of fractionation combined with variable contamination, and constant mixing proportions combined with variable fractionation, can each explain the trend of the data. Together, they can explain most of the dispersion of the data.

Neodymium Isotopes Versus Sm/Nd

The mixing-fractionation curves for the system ε_{Nd} versus Sm/Nd are shown in Figure 10-8. High-Sm/Nd basalts from Iceland, Hawaii, Siberia, Kerguelen, and Brazil all fall near the curve for unfractionated MORB with 1 to 5 percent contamination. Alkalics from large oceanic islands with thick crust (Hawaii, Iceland, and Kerguelen) are consistent with large amounts of crystal fractionation and moderate (5–10 percent) amounts of contamination.

The interpretation is that the more voluminous tholeiites are slightly fractionated and contaminated MORB, while the alkalics have experienced sublithospheric crystal fractionation and contamination prior to eruption. MORB itself has about 1 percent contamination, similar to that required to explain lead isotopes (Anderson, 1982). The Hawaiian tholeiites can be modeled as MORB that has experienced variable degrees of deep crystal fractionation (0 to 95 percent), mixed with 1–5 percent of an enriched component. The alkali basalts represent greater extents of crystal fractionation and contamination. The Columbia River basalts can be interpreted as MORB that has experienced 80 percent crystal fractionation and 5 percent contamination.

Lead and Helium Isotopes

Rubidium, strontium, and the light REEs are classic incompatible elements, and the effects of partial melting, fractionation, and mixing can be explored with some confidence. Relations between these elements and their isotopes

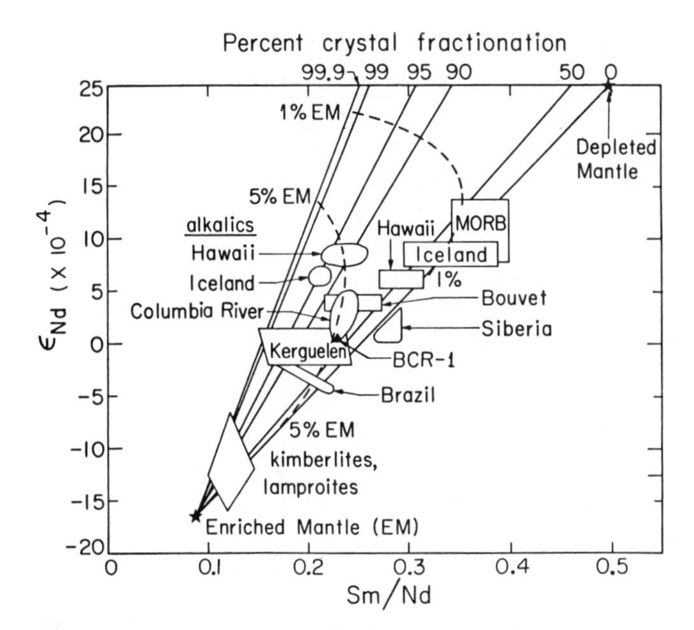

FIGURE 10-8
$^{143}Nd/^{144}Nd$ (relative to primitive mantle) versus Sm/Nd for mixtures of enriched mantle (EM) and residual melts resulting from high-pressure crystal fractionation of MORB-like depleted magma (Anderson, 1985).

should be fairly coherent. Some enriched magmas also have high $^3He/^4He$, $\delta^{18}O$, and $^{206}Pb/^{204}Pb$. These isotopes provide important constraints on mantle evolution, but they may be decoupled from the LIL variations and are usually assumed to be so. $^3He/^4He$ depends on the outgassing and magmatic history of the depleted reservoir and on the uranium and thorium content and age of the enriched component. $^{206}Pb/^{204}Pb$ and $^3He/^4He$ are sensitive to the age of various events affecting a reservoir because of the short half-life of uranium. The U/Pb ratio may be controlled by sulfides and metals as well as silicates. Nevertheless, it is instructive to investigate the effects of contamination and fractionation on these systems to ferret out what other effects or components are involved.

A minimum of three components is required to satisfy the combined strontium and lead isotopic systems when only simple mixing is considered. St. Helena has high $^{206}Pb/^{204}Pb$–low $^{87}Sr/^{86}Sr$; Kerguelen has high $^{87}Sr/^{86}Sr$–low $^{206}Pb/^{204}Pb$; MORB has low $^{87}Sr/^{86}Sr$–low $^{206}Pb/^{204}Pb$. The lead isotopes in MORB are higher than expected for a primitive or depleted reservoir. A mixing-fractionation curve for $^{87}Sr/^{86}Sr$ and $^{206}Pb/^{204}Pb$ (Figure 10-9) shows that the data can be explained with only two isotopically distinct mantle reservoirs. However, even if the enriched mantle reservoir is global, it can have variable $^{206}Pb/^{204}Pb$ and $^{207}Pb/^{204}Pb$ since these are sensitive to the U/Pb ratio and the age of enrichment events. U/Pb and $^3He/U$ of the enriched mantle will change locally by the subduction of sediments and oceanic crust and by the insertion of basalts that do not make their way to the surface. Isotopic ratios of lead and helium, therefore, are variable and change rapidly with time. The various contributions to the enrichment process (subduction, trapped magmas) make it likely that the enriched reservoir is in the shallow mantle and is laterally inhomogeneous. In this sense the mantle contains multiple reservoirs; their dimensions are likely to be of the order of the size of convection cells.

A MODEL FOR MAGMA GENESIS AND EVOLUTION IN THE MANTLE

Chen and Frey (1983) proposed a model for Hawaiian volcanism that involves melting of an enriched mantle plume from the lower ("primitive") mantle. This upwelling plume traverses a depleted (MORB) region, which partially melts (0.05–2.0 percent); these melts mix in varying proportions to form the spectrum of basalts found in Hawaiian volcanoes. With their choice of end-members the MORB component must be small, less than 2 percent.

The results just described suggest an alternative explanation. Melts from the depleted reservoir rise to the base of the lithosphere, fractionate and mix with melts from the enriched reservoir. At midocean ridges the lithosphere is thin, crystal fractionation occurs at shallow depths, and eruption occurs with little contamination. Midplate volcanism involves crystal fractionation at greater depth, deeper than 60 km. Because of slower cooling at depth, there is more chance for mixing with melts from the asthenosphere and lithosphere prior to eruption. Thus, diapirs from a deep depleted reservoir may be the precursors for both midocean and midplate volcanism, but the trace-element and isotopic spectra of basalts depend on the thickness of the lithosphere or crust. Material from the depleted source, which initiates melting in the shallow enriched reservoir, dominates the mix. The choice of a single enriched component is clearly simplistic, but it serves to illustrate the mechanism. On their ascent MORB diapirs can interact with asthenosphere, oceanic or continental lithosphere, sediments, seawater or subducted material.

In the Chen-Frey model the amount of the MORB component decreases as the inferred degree of partial melting of the MORB source increases, the reverse of what might be expected. The relative proportions of the MORB and enriched components are also the reverse of what the $^3He/^4He$ data indicate. Kaneoka (1983) suggested that the MORB component of Hawaiian basalts is more than 50–70 percent. In the present model the amount of contamination increases as crystal fractionation proceeds.

Kaneoka (1983) favored a model in which LIL- and ^3He-rich material from a deep primitive mantle reservoir rises through and mixes with a shallow, depleted region. He argued that if the enriched layer is shallow, plumes should

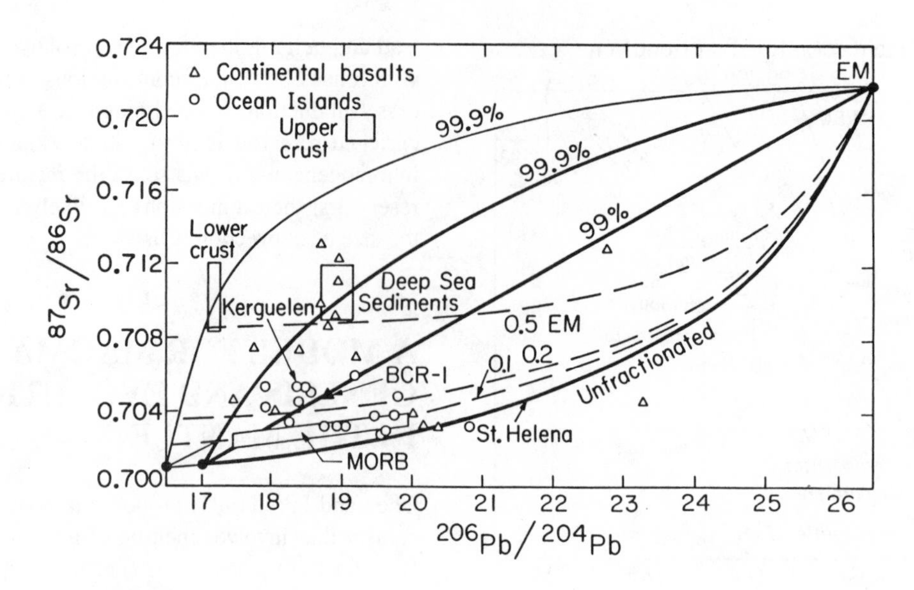

FIGURE 10-9

$^{87}Sr/^{86}Sr$ versus $^{206}Pb/^{204}Pb$ trajectories for mixtures of a fractionating depleted MORB-like magma (at two different initial $^{206}Pb/^{204}Pb$) and an enriched component. Mixing hyperbolas are solid lines. Hawaiian basalts fall in the region of 50–99 percent crystal fractionation (garnet plus clinopyroxene) and 10–30 percent contamination by EM (dashed lines).

originate at shallow depth, and the only overlying depleted layer is the oceanic lithosphere. However, the distinctive geochemical signature of hotspot basalts may be acquired at shallow depths, and the majority of the erupted material may be depleted basalts rising from greater depth. Continental contamination is an oft-invoked mechanism of this type, although it clearly does not work for the rare gases. An LIL- and ^{3}He-rich shallow mantle does not imply that plumes originate at shallow depth. In discussions of hotspots it should be recognized that the source of heat, the source of the majority of the magma and the source of the isotopic signatures may be at different depths. Hot regions of the lower mantle, for example, can localize melting in the transition region (400–650 km depth), and depleted melts from this region may become contaminated as they rise. Melts trapped in or rising through the shallow mantle may be responsible for the distintive physical properties of the low-velocity zone and asthenosphere. Most workers assume that noble gases are strongly partitioned into melts, which then efficiently lose these gases to the atmosphere. Outgassing, however, probably occurs only at relatively shallow depths, where the major gases, such as CO_2 and H_2O, can exsolve (Kaneoka, 1983). If melts are trapped at depth, the noble gases will be as well. The presence of primordial helium-3 and argon-36 in magmas does not require the existence of a previously unmelted or primitive reservoir. The $^{3}He/^{4}He$ characteristics of hotspot magmas therefore do not require an explanation separate from that for the other isotopes, except that $^{3}He/^{4}He$ can decrease rapidly with time in a high-uranium reservoir.

The magnitude of tholeiitic volcanism, the dominance of the MORB component in mantle magmas, and the close temporal and spatial relationship of depleted and more enriched and evolved alkalic magmas support a controlling role for melts from the depleted reservoir in mantle magma genesis. Oceanic ridges, islands and seamounts may share a common source and mantle magmas may differ primarily because of different degrees of sublithospheric crystal fractionation and contamination, and because of the nature of the (minor) enriched component. Shallow fractionation and crustal contamination, alteration and recycling will be superimposed on the effects considered here. There is also the possibility that the enriched region is inhomogeneously enriched or veined. Lead isotopes, for example, require more than two end-members (Stille and others, 1983). These may differ in age or intrinsic chemistry or both.

The location of the enriched reservoir, the source of the contamination discussed here, cannot be settled by petrological and geochemical arguments alone. It has been a common, but unfortunate, practice in the geochemical literature to beg or prejudice the issue by using the terms lower mantle, primitive mantle, undegassed mantle, plume reservoir and oceanic-island reservoir interchangeably. The similarity in trace-element and isotopic geochemistry between large islands, seamounts, continental rifts, and island arcs suggests that the enriched source is widespread and shallow. The most enriched magmas, kimberlites, appear to originate at a depth of about 200 km. Midocean ridges can be traced by seismic means to depths greater than 200 km, in some cases greater than 400 km (Walck, 1984; Nataf and others, 1984). The ascent phase of a deep diapir is characterized by melting. Crystallization is caused by entry of the diapir into the cold surface boundary layer. The heat of crystallization causes melting, or wallrock reactions, and hence crystal fractionation and contamination are concurrent processes. The adoption of a single-stage process involving two isotopically distinct end-members, only one of which is allowed to fractionate, is clearly an oversimplification. The enriched component may represent variable degrees of partial melting of an enriched region that is probably inhomogeneous. The shallow mantle is periodically replenished with subducted material and trapped magmas. Some of these processes serve to diminish the extreme amount of crystal fractionation implied by the present calculations, as can non-Henry's Law behavior (Harrison and Wood, 1980; Harrison, 1981). The possibility of large amounts of eclogite or clinopyroxene fractionation has been discussed by O'Hara (1973) and by many others (Frey and others, 1974; O'Hara and others, 1975; Clague and Bunch, 1976; Maaløe and Peterson, 1976; O'Nions and others, 1976; Leeman and others, 1977; de Argolo and Schilling, 1978; Bender and others, 1978).

A thin, shallow layer is unlikely to be globally homogenized by plate tectonic processes. The large-scale isotope anomaly in the southern hemisphere mantle (Hart, 1984) may have been the locus of ancient subduction, just as the circum-Pacific belt is today or the circum-Pangaea belt was in the Cenozoic. In this case the southern hemisphere anomaly would be due to an ancient subducted component in the shallow mantle (high ^{87}Sr, ^{206}Pb, ^{4}He). Such an anomaly would be smeared out in time, in the direction of mantle flow, but would diffuse only slowly across the upwellings and downwellings that separate upper-mantle convection cells, particularly if it is embedded in buoyant material such as harzburgite.

This model, which has been approached from various lines of evidence in this book, contradicts conventional views of hotspot petrogenesis. The enriched reservoir is shallow rather than deep, and it may be relatively infertile, that is, low in CaO and Al_2O_3. Being shallow it can collect volatiles and LIL by subduction and by trapping of fluids rising from below. Although global, it is unlikely to be homogeneous. It protects the deeper, depleted but fertile (low LIL, high CaO and Al_2O_3) reservoir from contamination. It is usually assumed that the source of heat, the source of the basalts and the source of the hotspot geochemical signatures are all the same. The relative stationarity of hotspots, the associated swells and high geoid may be due to increased melting in the upper mantle caused by locally high lower-mantle temperatures rather than transfer of material from the lower mantle to the surface. It is the thickness of the lithosphere that controls the geochemical signature of intraplate volcanics, not the ease of communication to the lower mantle. The lower mantle, other than being a source of heat, may not be involved at all in petrogenesis.

References

Anderson, D. L. (1981) Hotspots, basalts, and the evolution of the mantle, *Science, 213,* 82–89.

Anderson, D. L. (1982) Isotopic evolution of the mantle, *Earth Planet. Sci. Lett., 57,* 1–24.

Anderson, D. L. (1983) *J. Geophys. Res., 88,* B41–B52.

Anderson, D. L. (1985) *Nature, 318,* 145–149.

Barberi, F., L. Civetta and J. Varet (1980) Sr isotopic composition of Afar volcanics and its implication for mantle evolution, *Earth Planet. Sci. Lett., 50,* 247–259.

Bender, J. F., F. N. Hodges and A. E. Bence (1978) Petrogenesis of basalts from the FAMOUS area; experimental study from 0 to 15 kilobars, *Earth Planet. Sci. Lett., 41,* 277–302.

Boettcher, A. L. and O'Neil, J. R. (1980) *Am. J. Sci., 280A,* 594–621.

Chase, C. G. (1981) Oceanic island Pb; two-stage histories and mantle evolution, *Earth Planet. Sci. Lett., 52,* 277–284.

Chen, C.-Y. and Frey, F. A. (1983) *Nature, 302,* 785–789.

Clague, D. and Bunch, T. E. (1976) *J. Geophys. Res., 81,* 4247–4256.

Craig, H. and Lupton, J. (1981) In *Sea 1,* (C. Emiliani, ed.) 391–428, Wiley, New York.

de Argollo, R. M. and Schilling, J. G. (1978) *Nature, 276,* 24–28.

Dunlop, H. M. and Fitton, J. G. (1979) *Contr. Miner. Petrol., 71,* 125–131.

Frey, F. A., W. B. Bryan and G. Thomson (1974) Atlantic Ocean floor; geochemistry and petrology of basalts from Legs 2 and 3 of the Deep Sea Drilling Project, *J. Geophys. Res., 79,* 5507–5527.

Frey, F. A., D. H. Green and S. D. Roy (1978) Integrated models of basalts petrogenesis: A study of quartz tholeiites to olivine melilitites from southeastern Australia utilizing geochemical and experimental petrological data, *J. Petrol., 19,* 463–513.

Garlick, G., I. MacGregor, and D. Vogel (1971) Oxygen isotope ratios in eclogites from kimberlites, *Science, 171,* 1025–1027.

Gregory, R. T. and H. P. Taylor, Jr. (1981) An oxygen isotope profile in a section of Cretaceous oceanic crust, Samail ophiolite, Oman, *J. Geophys. Res., 86,* 2737–2755.

Harrison, W. J. (1981) *Geochim. Cosmochim. Acta, 45,* 1529–1544.

Harrison, W. J. and B. J. Wood (1980) An experimental investigation of the partitioning of REE between garnet and liquid, *Contrib. Mineral. Petrol., 72,* 145–155.

Hart, R., J. Dymond and L. Hogan (1979) Preferential formation of the atmospheric-sialic crust system from the upper mantle, *Nature 278,* 156–159.

Hart, R., L. Hogan and J. Dymond (1985) The closed-system approximation for evolution of argon and helium in the mantle, crust and atmosphere, *Chem. Geol., 52,* 45–73.

Hart, S. R. (1984) A large-scale isotope anomaly in the Southern Hemisphere mantle, *Nature, 309,* 753–757.

Hiyagon, H. and M. Ozima (1982) Noble gas distribution between basalt melt and crystals, *Earth Planet. Sci. Lett., 58,* 255–264.

Hiyagon, H. and M. Ozima (1986) *Geochim. Cosmochim. Acta, 50,* 2045.

Houtermans, F. G. (1947) *Z. Naturforsch., 2a,* 322.

Kaneoka, I. (1983) Noble gas constraints on the layered structure of the mantle, *Nature, 302,* 698–700.

Kaneoka, I., N. Takaoka and J. D. Clague (1983) Noble gas systematics for coexisting glass and olivine crystals in basalts and dunite xenoliths from Loihi Seamount, *Earth Planet. Sci. Lett., 66,* 427–437.

Kramers, J. D. (1977) Lead and strontium isotopes in Cretaceous kimberlites and mantle-derived xenoliths from southern Africa, *Earth Planet. Sci. Lett., 34,* 419–431.

Kurz, M. D., Jenkins, W. J., Hart, S. R. and Clague, D. A. (1983) Helium isotopic variations in volcanic rocks from Loihi Seamount and the Island of Hawaii, *Earth Planet. Sci. Lett., 66,* 388–406.

Kyser, T. K., J. R. O'Neil and I. S. E. Carmichael (1982) Genetic relations among basic lavas and ultramafic inclusions; evidence from oxygen isotope compositions, *Contrib. Mineral. Petrol. 81,* 88–102.

Leeman, W. P., A. V. Murali, M.-S. Ma and R. A. Schmitt (1977) Mineral constitution of mantle source regions for Hawaiian basalts; rare earth element evidence for mantle heterogeneity, *Oregon Dept. Geol. Minerals. Indus. Bull., 96,* 169–183.

Lipman, P. W. and H. H. Mehnert (1975) Late Cenozoic basaltic volcanism and development of the Rio Grande Depression in the southern Rocky Mountains, *Geol. Soc. Am. Mem., 144,* 119–154.

Maaløe, S. and Peterson, T. S. (1976) *Lithos 9,* 243–252.

McCulloch, M. T., R. Arculus, B. Anappell and J. Ferguson (1982) Isotopic and geochemical studies of nodules in kimberlites, *Nature 300,* 166.

McCulloch, M. T., R. T. Gregory, G. J. Wasserburg and H. P. Taylor, Jr. (1981) Sm-Nd, Rb-Sr and O-18/O-16 isotopic systematics in an oceanic section; evidence from the Samail ophiolite, *J. Geophys. Res., 86,* 2721–2735.

Menzies, M. and V. R. Murthy (1980a) Enriched mantle; Nd and Sr isotopes in diopsides from kimberlite nodules, *Nature 283,* 634–636.

Menzies, M. and V. R. Murthy (1980b) Nd and Sr isotope geochemistry of hydrous mantle nodules and their host alkali basalts; implications for local heterogeneities in metasomatically veined mantle, *Earth Planet. Sci. Lett., 46,* 323–334.

Nataf, H.-C., Nakanishi, I. and Anderson, D. L. (1984) Anisotropy and shear velocity heterogeneities in the upper mantle, *Geophys. Res. Lett., 11,* 109–112.

O'Hara, M. J. (1973) Non-primary magmas and dubious mantle plume beneath Iceland, *Nature, 243,* 507–508.

O'Hara, M. J., M. J. Saunders and E. L. P. Mercy (1975) Garnet-peridotite, primary ultrabasic magma and eclogite; interpretation of upper mantle processes in kimberlite, *Phys. Chem. Earth, 9,* 571–604.

Oldenburg, D. W. (1984) The Inversion of Lead Isotope Data, *Geophys. J.R. Astron. Soc., 78,* 139–158.

O'Nions, R. K. and Oxburgh, E. R. (1983) Heat and helium in the Earth, *Nature, 306,* 429–431.

O'Nions, R. K., R. J. Pankhurst and K. Gronvod (1976) Nature and development of basalt magma sources beneath Iceland and the Reykjanes Ridge, *J. Petrol., 17,* 315–338.

Ozima, M. and S. Zashu (1983) Primitive helium in diamonds, *Science, 219,* 1067–1068.

Ozima, M., S. Zashu, D. P. Matley and C. T. Pillinger (1985) Helium, argon and carbon isotopic compositions in diamonds and their implications in mantle evolution, *Geochim. J., 19,* 127–134.

Patterson, C. (1956) Age of meteorites and the Earth, *Geochim. Cosmochim. Acta, 10,* 230–237.

Stille, P., Unruh, D. M. and Tatsumoto, M. (1983) Pb, Sr, Nd and Hf isotopic evidence of multiple sources for Oahu, Hawaii basalts, *Nature 304,* 25–29.

Sun, S., M. Tatsumoto and J. G. Schilling (1975) Mantle plume mixing along the Reykjanes Ridge axis; lead isotopic evidence, *Science, 190,* 143–147.

Taylor, H. P., Jr. (1980) The effects of assimilation of country rocks by magmas on $^{18}O/^{16}O$ systematics in igneous rocks, *Earth Planet. Sci. Lett., 47,* 243–254.

Walck, M. C. (1984) The P-wave upper mantle structure beneath an active spreading center; the Gulf of California, *Geophys. J. Roy. Astron. Soc., 76,* 697–723.

Zhou, X. and R. L. Armstrong (1982) Cenozoic volcanic rocks of eastern China; secular and geographic trends in chemistry and isotopic composition, *Earth Planet. Sci. Lett., 58,* 301–329.

11

Evolution of the Mantle

Rocks, like everything else, are subject to change and so also are our views on them.

—FRANZ Y. LOEWINSON-LESSING

I have now discussed the various kinds of magmas and refractory rocks that are important in the mantle, or at least the upper mantle, and some isotopic and seismic properties of the mantle. The visible rocks are the end, or present, product of mantle evolution. If these were our only source of information, we could come up with a fairly simple scheme of magma genesis, perhaps involving single-stage melting of a homogeneous, even primitive, mantle. It now seems unlikely that we will find the "Rosetta Stone"—a rock fragment that represents "original Earth" or even the parent or grandparent of other rocks. Rocks and magmas represent products of complex multistage processes, just as the crust and mantle do. As we delve deeper into the Earth and further back in time, we depend more and more on isotopes and on modeling of planetary accretion and mantle processes.

A variety of studies have lent support to the concept of a chemically inhomogeneous mantle. The mantle contains at least two distinct reservoirs that differ in their large-ion-lithophile (LIL) contents and that have maintained their separate identities for at least 10^9 years. One reservoir, itself heterogeneous, has high values of Rb/Sr, Nd/Sm, U/Pb and, possibly, ^{3}He/U. A variety of LIL-enriched basalts ranging from basanites, nephelinites and alkali basalts to tholeiites either form or evolve in, or become contaminated by, this reservoir. There may be several enriched reservoirs and several mechanisms of enrichment (subduction, trapping of melts or gas). The simplest hypothesis, however, is that there is one enriched reservoir that may be radially and laterally inhomogeneous, having been enriched at various times and by various mechanisms. The other major reservoir provides fairly uniformly depleted tholeiites (MORB) to the midocean ridges. The incompatible-element patterns for MORB and enriched or alkali-rich basalts are complementary, and their ratios parallel the partition coefficient trend of garnet. This suggests that neither reservoir is primitive and that they are complementary products of a more primitive reservoir. The LIL-depleted reservoir appears to be garnet-rich, or fertile, relative to the enriched reservoir. The high density of garnet suggests that the depleted reservoir is the deeper one, and that ancient depletion/enrichment processes involved the upward transport of fluids that have been in equilibrium with garnet. The recycling of sediments and hydrothermally altered oceanic crust, too, has probably been important throughout most of Earth history. This material may contribute to the chemical and isotopic heterogeneity of the enriched source region, or it may be preferentially concentrated in the shallow mantle and act as a separate source.

From a geophysical and tectonic point of view the following possible sources can be identified. Some may be fertile sources of basalts, while others may simply interact with magmas generated elsewhere.

1. The continental shield lithosphere.
2. The shallow mantle, sometimes called the asthenosphere.
3. The transition region.
4. Subducted sediments.
5. Subducted oceanic crust.

In addition, magmas may interact with the continental crust, the oceanic crust, the oceanic lithosphere, seawater and the atmosphere. Until all of these sources are evaluated, it seems premature to invoke such remote sources as the lower mantle and D″.

MODELS OF THE MANTLE PETROGENESIS

Most models of petrogenesis assume that basic and ultrabasic magmas are formed by varying degrees of partial melting of fertile peridotite in the upper mantle. In these models the degree of partial melting and crystal fractionation and the depth of origin are the main variables in controlling the composition of the melt. Variations in volatile content and mineralogy of the source region, and crustal contamination are additional variables that have been used to explain the full range of magma compositions. Tholeiites are generally regarded as representing large degrees of partial melting of a shallow source, and the more K_2O-rich magmas, such as alkali basalts and nephelinites, are regarded as resulting from relatively small amounts of melting at deeper levels. The incompatible-element-enriched magmas with high LREE content, such as melilites, nephelinites, basanites and kimberlites, are assumed to result from extremely small degrees of partial melting from a deep garnet-rich source region.

Isotopic studies require at least two source regions in the mantle. The most voluminous magma type, MORB has low LIL contents and low $^{87}Sr/^{86}Sr$, $^{143}Nd/^{144}Nd$ and $^{207}Pb/^{204}Pb$ ratios. The Rb/Sr, Nd/Sm and U/Pb ratios in this reservoir have been low for at least 2.5 Ga, and perhaps were set by accretional differentiation. Since these ratios are high in melts and low in residual crystals, the implication is that the MORB reservoir is a cumulate or a crystalline residue remaining after the removal of a melt fraction. The enriched melt fraction is commonly assumed to efficiently enter the continental crust. The continental crust is therefore regarded as the only complement to the depleted MORB source. Mass-balance calculations based on this premise indicate that most of the mantle is undepleted or primitive, and the depleted reservoir is assumed to occupy most or all of the upper mantle. Since the continental crust is the only enriched reservoir in this model, magmas that have high LIL contents and $^{87}Sr/^{86}Sr$ ratios are assumed to be contaminated by the continental crust, or to contain a recycled ancient crustal component.

However, enriched magmas such as alkali-olivine basalts (AOB), oceanic-island basalts (OIB) and nephelinites also occur on oceanic islands and have similar LIL and isotopic ratios to continental flood basalts (CFB) and continental alkali basalts. Continental contamination is unlikely in these cases. Veins in mantle peridotites and xenoliths contained in alkali basalts and kimberlites are also commonly enriched and, again, crustal contamination is unlikely. In many respects these enriched magmas and xenoliths are also complementary to MORB (in LIL contents and isotopic ratios), suggesting that there is ancient enriched material in the mantle. Island-arc basalts are also high in LIL, $^{87}Sr/^{86}Sr$, $^{143}Nd/^{144}Nd$ and $^{206}Pb/^{204}Pb$, suggesting that there is a shallow and global enriched reservoir. Back-arc basin basalts (BABB) are closer to MORB in composition and, if the depth of the low-velocity zone and the depths of earthquakes can be used as a guide, tap a source deeper than 150 km. Many BABBs are intermediate in chemistry to MORB and OIB. This and other evidence indicates that the enriched reservoir is shallow and the depleted MORB reservoir is deeper.

Since the seismic velocities and anisotropy of the shallow mantle, above 220 km, are consistent with an olivine-rich aggregate, and since most mantle xenoliths are olivine-rich, it is natural to assume that the enriched reservoir is a peridotite. This peridotite can be infertile (low in Al_2O_3) yet enriched in LIL. Since the depleted reservoir has already lost a melt fraction, it should be depleted in garnet, unless it is an eclogite cumulate. Peridotites depleted in basalt have less Al_2O_3 and garnet than nondepleted or fertile peridotites.

The trace-element signatures of enriched magmas are consistent with derivation from a reservoir that has experienced eclogite fractionation or metasomatism by melts from an eclogite-rich source. Magmas from this reservoir will therefore be LREE-enriched, not because there is garnet in the residue but because the reservoir itself had experienced a prior stage of garnet removal or secondary enrichment by a fluid from a garnet-rich reservoir. This eliminates one of the arguments for derivation of LIL- and LREE-enriched magmas, such as alkali olivine basalts, by small degrees of partial melting from a deep garnet-rich peridotite layer. In general, diapirs from great depth will experience a greater degree of partial melting than diapirs originating at shallow depths. This is another argument against a shallow source for tholeiites. Extensive melting probably requires adiabatic ascent from a deep thermal boundary layer.

Kimberlites are among the most enriched magmas. Although they are rare, the identification of a kimberlite-like component in enriched magmas means that they may be volumetrically more important than generally appreciated.

The eclogite-fractionation, magma-mixing hypothesis for the evolution of the mantle (Anderson, 1982a,b,c) differs from conventional petrological assumptions in several ways: (1) The depleted source region is an eclogite-rich cumulate or a piclogite rather than a garnet peridotite; (2) enriched magmas are blends or hybrids of a depleted magma (MORB or picrite) and a melt from a shallow enriched peridotite reservoir; (3) the LIL pattern of enriched magmas, in particular LREE enrichment, is due to small degrees of partial melting of a garnet peridotite, or garnet-clinopyroxene fractionation from a picritic or tholeiitic magma. This pattern may also have been introduced into the source at some earlier time by metasomatic fluids.

There is general consensus that the depleted, or MORB, reservoir is depleted in the incompatible elements such as rubidium, neodymium, uranium and LREEs and that the very incompatible elements are depleted more than the less incompatible elements, thereby giving low Rb/Sr, Nd/Sm and U/Pb ratios. The corresponding isotopic ratios indicate that the depletion event was ancient, perhaps dating back to the formation of the continental crust. The complementary enrichment events would likewise have been ancient. We use "depleted" to describe basalts and reservoirs that have low LIL contents and low LREE/HREE, Rb/Sr, Nd/Sm, U/Pb, $^{87}Sr/^{86}Sr$, $^{143}Nd/^{144}Nd$, $^{206}Pb/^{204}Pb$ and other ratios. A depleted reservoir can still be fertile, as it can provide basalts by partial melting. A garnet-clinopyroxenite cumulate, for example, can be depleted but fertile. Similarly, an enriched reservoir can be infertile, being low in CaO, Al_2O_3, Na_2O and so on.

What are not agreed upon are the following:

1. The location of the MORB reservoir. It has been variously placed within the low-velocity zone, at the bottom of the low-velocity zone, and in the transition region, that is, anywhere between 220 km and 650 km. The homogeneity of MORB, on a global basis, the arguments that attribute it to large degrees of partial melting, the tendency of enriched interstitial fluids to migrate upward, and the possible contamination of the shallow mantle by subducted sediments and altered oceanic crust all argue for a relatively deep origin for the MORB reservoir. The time sequence of erupted magmas at island arcs, continental rifts and oceanic islands is consistent with a shallow enriched reservoir and a deeper LIL-depleted reservoir.
2. The composition of the MORB reservoir. The conventional petrological view is that MORBs result from large degrees of partial melting of a garnet peridotite. It now appears that midocean-ridge tholeiites are not primitive magmas but are the result of extensive crystal fractionation of a more picritic parental melt of a nonchondritic reservoir. Picrites imply that the parent peridotite be melted by 30 percent or more. The high degree of partial melting is required in order to generate high-MgO picritic magmas at shallow depths. The alternative point of view is that the depleted source region is a deep garnet-rich pyroxenite cumulate, such as a piclogite (olivine eclogite).
3. The nature of the depletion process. All workers agree that the source region of midocean-ridge tholeiites is depleted in the LIL elements and that this depletion was an ancient event. In the conventional view this depletion was the result of a small amount of partial melting and melt removal to the continental crust. Since the MORB source is obviously still fertile, in the sense that it can provide basalts, it must be garnet-rich in spite of having lost a basaltic component. The MORB source, in this model, must have remained homogeneous and must have escaped the early high temperatures implied by thermal history calculations. Alternatively, the depletion of the MORB source in a cooling Earth could be explained if it is a garnet-pyroxenite cumulate. The late-stage residual fluids in equilibrium with such a layer would be LIL- and LREE-enriched and, in this respect, kimberlitic. Such fluids, if buoyant, could form the proto-crust.

Fluids and melts are LIL-enriched, and they tend to migrate upward. Sediments and altered ocean crust, also LIL-enriched, re-enter the upper mantle at subduction zones. Thus there are several reasons to believe that the shallow mantle serves as a scavenger of incompatible elements, including the radioactive elements (U, Th and K) and the key tracers (Rb, Sr, Nd, Sm and, possibly, Pb and ^3He). The continental crust and lithosphere are commonly assumed to be the main repositories of the incompatible elements, but oceanic islands, island arcs and deep-seated kimberlites also bring LIL-enriched material to the surface. This is one reason for invoking a global upper-mantle enriched layer and for investigating reservoirs other than the continental crust in compiling inventories of the incompatible elements. Even a moderate amount of LIL in the upper mantle will destroy the arguments for a primitive lower mantle.

Recent metasomatism of an upper-mantle source region is sometimes invoked to explain LIL-enriched magmas, particularly those with low $^{87}Sr/^{86}Sr$ ratios. The source of the enriching fluid is seldom addressed. Recycling of sediments or remelting of subducted oceanic crust may explain some enriched magmas. These mechanisms do not explain the high $^3He/^4He$ ratios found in some oceanic-island basalts.

The continuum in trace elements and isotopic ratios between such enriched magmas as nephelinites and alkali-olivine basalts and depleted magmas such as MORB suggests that most mantle basalts represent mixtures between MORB, or its picritic parent, and partial melts from a shallow enriched reservoir. Mixing or contamination upon ascent is an alternative to recent metasomatism of the source reservoir.

The study of basalts combined with experimental petrology allow one to estimate the temperatures and pressures at which basalts might have separated from their source region. The amount of melting involved and the composition of the source region, however, cannot be determined by these means. The emphasis on olivine-rich and peridotite source regions is based on the following arguments:

1. Peridotite is consistent with seismic velocities for the shallow mantle, and basalts come from the mantle. The seismic arguments for an olivine-rich shallow mantle include both the velocities and the anisotropy. However, peridotites rarely have either the major-element or trace-element chemistry necessary to provide basalts of the required composition. Most of these rocks are refractory

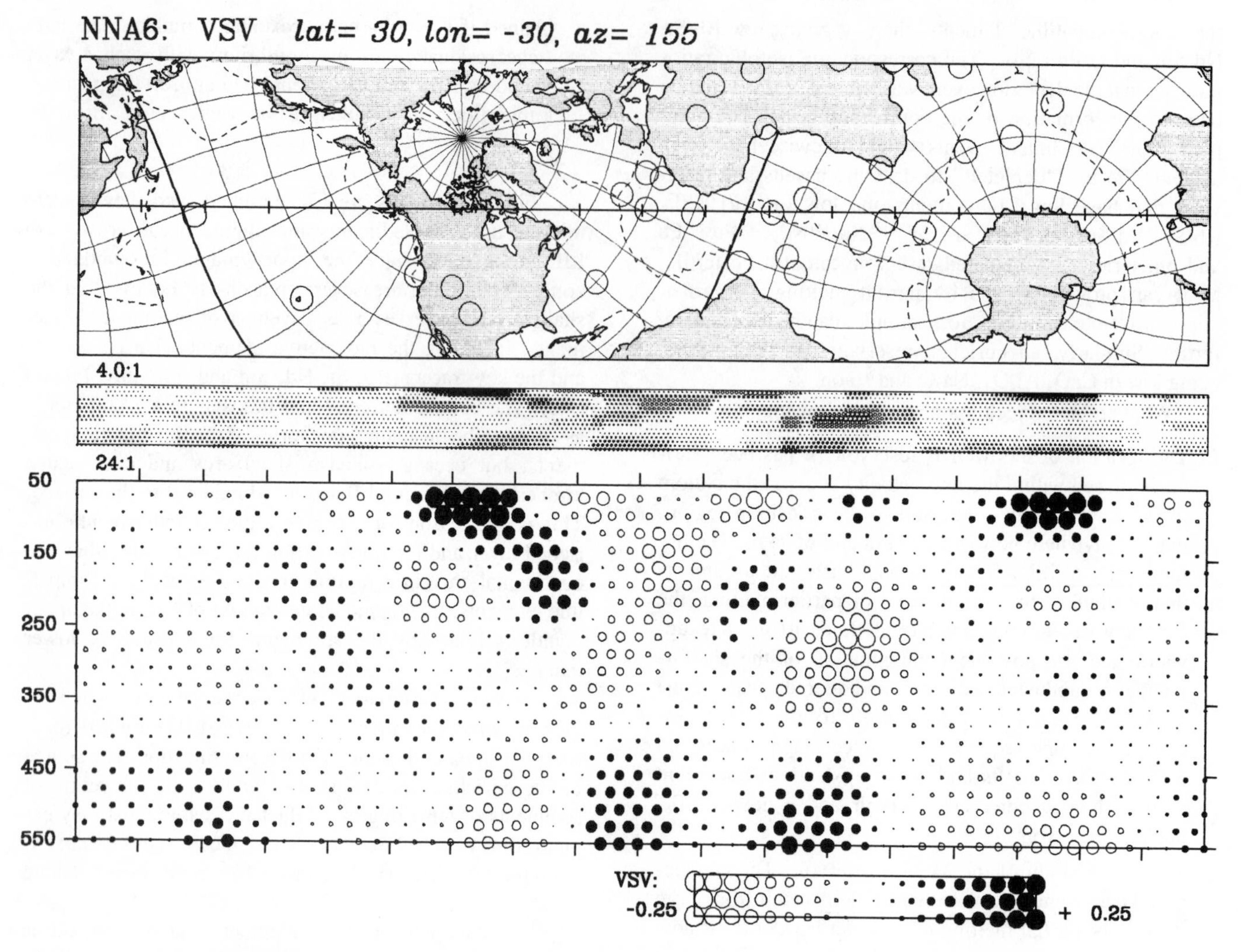

FIGURE 11-1
Seismic shear velocities from 50 to 550 km depth (bottom) along great-circle path shown in center of top panel. Low velocities are open circles, fast velocities are filled circles. Vertical exaggeration is 24:1 in bottom panel and 4:1 in center panel. Note the slow velocities in the North and South Atlantic, where the cross section cuts across the Mid-Atlantic Ridge. Dashed lines are plate boundaries. Circles in upper panel are hotspots. (Model from Nataf and others, 1986.)

residuals or cumulates. In regions of very low seismic velocity, such as under tectonic regions and midocean ridges, the velocities are much lower than in peridotites. Partial melting or high-temperature grain boundary relaxation can reduce the velocities considerably, but when this occurs one can have almost any major-element chemistry for the matrix; that is, seismic velocities are no longer a constraint. The slow velocities associated with midocean-ridge upper mantle now appear to extend much deeper than 200 km, and, in some cases, into the transition region (Figure 11-1). Velocities in the transition region do not seem to be appropriate for peridotites. Thus, although arguments can be made that some parts of the shallow mantle have seismic velocities appropriate for peridotite, the connection between these regions and the source regions for basalts has not been made.

2. The occurrences of garnet peridotite are appropriate to deep-seated environments. Garnet peridotite is stable in the upper mantle but, again, this does not prove that it occurs in the basalt source region or is the immediate parent for basalts. Most natural peridotites, in fact, are buoyant relative to fertile peridotites, eclogites and piclogite.

3. Garnet peridotites have close compositional relationships to meteorites. This is an argument that the average Earth or even the bulk composition of the mantle can be inferred from meteoritic abundances. Midocean-ridge

basalts clearly do not come from a chondritic source; the LIL elements are depleted in MORB relative to chondritic abundances. The MORB source is at least one generation removed from a chondritic ancestor. This argument, and many like it, confuse what may be in the mantle or what may be the average composition of the mantle with what is actually required of the immediate parent of basalts. A single-stage basalt forming process is implied whereas isotopic and trace-element data clearly require a multistage process.

4. Partial melts of natural samples of garnet peridotite at high pressure have basaltic compositions. Actually, partial melting of natural peridotites has not provided magmas with MORB compositions, particularly in the trace elements. Synthetic peridotitic aggregates (minus olivine) come close to matching *inferred* compositions of parental picrites, except for K_2O (Green and others, 1979). Melting of eclogites, of course, also gives basaltic composition melts, so the above argument, at best, is permissive rather than persuasive.
5. Melting of eclogites would have to be very extensive, and melt-crystal segregation would occur before such extensive melting can be achieved. Actually, eclogites provide basaltic melts over a wide range of melting, and large amounts of melting may be required for melt separation.

There is no doubt that garnet peridotites can and do come from the shallow mantle, and some regions of the upper mantle are probably mostly garnet peridotite. Some garnet peridotites can provide basalts by partial melting, as can pyrolite by definition. The average composition of the Earth is probably close to chondritic in major-element chemistry, and the mantle therefore contains abundant, although not necessarily predominant, olivine. By the same reasoning the mantle contains even more pyroxene and garnet. These arguments do not prove that the source region of the most abundant basalt type, midocean-ridge basalts, is garnet peridotite or that the regions of the mantle that appear to be peridotitic, on the basis of seismic velocities, are the regions where midocean basalts are generated.

Although some of the older ideas about the source regions, such as melting of a glassy or basaltic shallow source, can be ruled out, the possibility that a deep eclogite, or picritic eclogite, is the basalt source region cannot be ruled out. Olivine eclogite, or piclogite, the inferred composition of the transition region, is also a candidate source rock.

PETROLOGICAL EVOLUTION OF THE MANTLE

Most models of petrogenesis assume that so-called primary magmas are the result of varying degrees of partial melting of peridotite. Abyssal tholeiites, because of their relative depletions in the LIL elements, have had a more complex history, although it seems likely that partial melting of a peridotite was involved at an early stage of their evolution. The trace-element inhomogeneity of the mantle (Chapter 13) plus the long-term isolation of the major reservoirs suggests that differentiation has been more effective in the long run than mixing. Mixing can be avoided in a chemically stratified mantle if the layers have a large intrinsic density contrast. Garnet has the highest density of any abundant upper-mantle mineral and therefore plays a role in determining the density of various regions of the mantle.

The conventional model for the origin of magmas might well be designated PM^3 for "primitive mantle partial melt pyrolite model." In this model, tholeiitic basalts are considered to be primary, unfractionated melts resulting from about 20 percent melting of fertile garnet peridotite. In recent variations, oceanic tholeiites are treated as partial melts from a reservoir that has been depleted by removal of the continental crust. Continental and oceanic-island basalts (OIB) are assumed to be melts from a primitive undifferentiated lower mantle. The oceanic lithosphere is modeled as 6 km of basalt, the primary melt, and 24 km of depleted residual harzburgite (pyrolite = 1 part basalt, 4 parts depleted peridotite). Upon subduction, the lithosphere sinks to the core-mantle boundary (Ringwood, 1975). Except for a layer of depleted peridotite in the upper mantle, with perhaps some isolated blobs of eclogites, the mantle is uniform in composition and composed of pyrolite, which, by definition, can yield basaltic magmas by partial melting.

There are several problems with this model. There is increasing evidence that midocean-ridge tholeiites are not primary magmas but are the result of olivine fractionation from a more MgO-rich picritic parent. The harzburgite residue after removal of partial melt from a garnet peridotite is less dense than the parent and will remain in the upper mantle. Several billion years of seafloor spreading will fill up the entire upper mantle with this depleted residue. In addition, harzburgites do not appear to have been in equilibrium with MORB (Green and others, 1979).

The Moon differentiated early in its history and evolved into a series of cumulate layers. It is likely that the Earth did as well. Even if a large fraction of the heat of accretion was radiated away or was convected efficiently to the surface, it is difficult to construct geotherms that remain below the solidus of silicates during most of the accretion of the Earth (Chapter 1). Gravitational separation of the melt will concentrate the LIL elements and Al_2O_3, CaO and SiO_2 into the upper mantle. In contrast to the conventional model, the "primitive mantle," which can yield basalt by partial melting, has already been processed into a depleted lower mantle and an upper mantle that on average is enriched. This processing occurs near the surface where material delivered by accretion, or by whole-mantle convection, crosses the melting zone. If olivine and orthopyroxene are the main residual phases, the incompatible-element en-

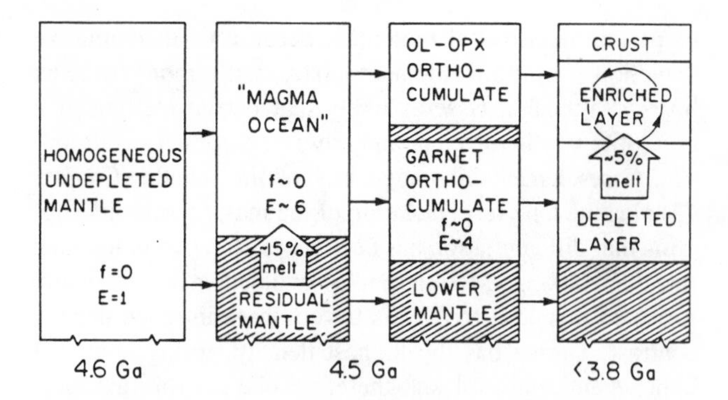

FIGURE 11-2
Differentiation of a planet during accretion and early high-temperature evolution. *E* is the enrichment of incompatible elements, relative to the starting materials. These elements have low crystal-melt partition coefficients and therefore readily enter the melt fraction. *f* is the fractionation factor and gives the ratio of two incompatible elements in the melt, expressed as the difference from the starting material. Very incompatible elements occur in the same ratio in melts as in the original, or primitive, material. Isotopic ratios of these elements will evolve at the same rate as in primitive material. A magma ocean will therefore be enriched but unfractionated. As the magma ocean crystallizes, the fractionating crystals will either float or sink, leaving behind an enriched, fractionated residual liquid layer. This fluid may permeate the shallow mantle, giving an enriched geochemical signature to this region, and to the continental crust. If melting and melt extraction are efficient in early Earth history, most regions of the planet will be depleted in incompatible elements. The large difference in crystallization temperature and density of olivine-orthopyroxene (ol-opx) and garnet means that mineralogically distinct regions can form in early Earth history.

richment *E* of the upper mantle will be relatively uniform at $1/F$ where F is the melt fraction (Figure 11-2). That is, for elements with low partition coefficients (such as Sm, Nd, Rb and Sr), the "primitive" upper mantle, a product of whole-mantle differentiation, will appear to be unfractionated (Chapter 8). Crystallization of a melt layer or magma ocean will lead to a series of cumulate layers, and fractionation of the LIL.

Although it is unlikely, and unnecessary, that the liquidus of peridotite is exceeded during accretion, the separation of melt from residual crystals in effect lowers the liquidus temperature from that of peridotite to that of basalt at low pressure and eclogite at high pressure. In the near-surface chill layer of a magma ocean, olivine and plagioclase are the low-pressure liquidus phases. At greater depth, clinopyroxene crystallizes. Since the liquidus of basalt and eclogite increases with depth faster than the adiabat, crystallization will also occur at the base of a magma ocean. At depths greater than 60 km, garnet and clinopyroxene crystallize from the melt. A deep eclogite cumulate layer is therefore one of the first products of crystallization of a deep (>50 km) magma ocean. The remaining fluid is depleted in Al_2O_3, CaO, SiO_2 and Na_2O. Garnet and clinopyroxene between them retain FeO, MnO, yttrium, ytterbium, titanium, zirconium, and the heavy rare-earth elements (HREE). Olivine crystallizing at shallow levels sinks but will react with the silica-undersaturated fluid to form clinopyroxenes and garnet. It is buoyant relative to melt at high pressure. Olivine, therefore, is primarily a shallow mantle phase. The removal of eclogite from the melt means that there will not be much Al_2O_3 in the melt for formation of a thick, lunar-type anorthosite layer at the surface. The eclogite will probably be mixed with some olivine, which is also denser than the melt, at least in the shallow mantle.

The cumulate layers originally contain interstitial fluids that hold most of the incompatible elements. As crystallization proceeds, these melts migrate upward. Melts from an eclogite or olivine eclogite cumulate will be depleted in Al_2O_3, CaO, Na_2O, yttrium, ytterbium, manganese, and the HREE. These are the characteristics of kimberlites. Removal of late-stage (kimberlite) intercumulus fluids from an eclogite-rich cumulate layer will deplete it and enrich the overlying olivine-rich layer. The enrichment, however, will be selective. It will be uniform in the very incompatible elements, giving primitive ratios of Rb/Sr, Sm/Nd and such, but will impart a pattern of depletion in the HREE, yttrium, sodium, manganese and so on since these are the eclogite-compatible elements. Partial melts from a shallow enriched reservoir will therefore appear to have a garnet-residual pattern, even if this reservoir contains no garnet. This pattern can be transferred to any MORB magmas interacting with this layer.

An eclogite-rich layer such as piclogite is normally denser than peridotite; therefore, the enriched, fertile and depleted, less fertile or infertile, reservoirs remain separate and isolated. However, convection in a chemically stratified system causes lateral variations in temperature, thermal boundary layers and deformation of the interfaces because of the buoyancy of the uprising currents. If this deformation raises a chemical boundary across the solidus, or if the temperature is perturbed by, for example, continental insulation, then partial melting can generate a buoyant diapir. Subsolidus reactions between garnet and clinopyroxene also occur at high temperature. This results in a temperature-induced density decrease much greater than can be achieved by thermal expansion. Adiabatic ascent of a diapir from a buried eclogite or piclogite layer (see Chapter 9) can lead to extensive melting because of the small amount of refractory component compared to olivine-rich (>60 percent) peridotites. Crystal settling apparently can be avoided in a rapidly rising diapir because of the high temperatures, temperature gradients and stresses (see Chapter 9). Depleted picritic melt, delivered to the shallow mantle, can evolve to tholeiites by shallow-level olivine fractionation and to alkali basalts by deeper clinopyroxene and garnet fractionation. Some of the melt may remain in the shallow mantle.

Ringwood's (1975) petrogenesis model assumes that tholeiites are primary magmas representing 20–30 percent partial melting of pyrolite. Most of the mantle is assumed to be still capable of yielding a variety of basalts by varying degrees of partial melting. In the other extreme under the model presented here, all of the mantle has already been processed, and the melt products and their cumulates are concentrated in the upper mantle. The crystalline residue of the early differentiation, about 75 percent of the mantle, is now the lower mantle. Separation of upper and lower mantle may be irreversible. Transfer of material between the lower mantle and the upper mantle will be prohibited if the 650-km discontinuity is a chemical boundary with a sufficiently high intrinsic density contrast. The high melting temperature of the lower mantle and the small difference between melt density and crystal density at high pressure makes it unlikely that lower-mantle material can overcome a compositional density barrier by partial melting. The high FeO-content of the lower mantle, inferred from its density and from solar abundances (Chapter 1) is presumably a result of accretional fractionation and makes the Earth irreversibly differentiated.

With the recognition that tholeiites are not primary magmas, the temperatures and extent (or depth) of partial melting in the upper mantle have had to be revised upward. The maximum amount of melting of a mantle silicate is buffered by the latent heat of fusion, the specific heat and, most importantly, by the segregation of melt from crystal at high degrees of partial melting. High degrees of partial melting are probably restricted to the rising limbs of convection cells. The deep-mantle adiabat in these regions extends into the upper mantle, and temperatures well in excess of the solidus are feasible. With the higher heat productivity in the past, the heat is most efficiently removed if there are more regions of upwelling or more rapid transfer of material through the upper-mantle melting zone. In either case, it is likely that large volumes of melts were generated in early Earth history, and melt/crystal separation is likely.

The process of planetary accretion and melting during accretion is akin to a zone-refining process. The surface of the planet, where the kinetic energy of accretion is turned into heat, acts as the furnace and refractory, "purified" material is fed into the planet. The incompatible elements and melts are preferentially retained near the surface. A deep magma ocean at any one time is not required, or even desirable. It is not desirable since if the surface melt layer is in equilibrium with dense phases such as perovskite, there should be anomalies in the trace-element patterns of upper-mantle materials. Large amounts of garnet extraction (~10 percent) and, presumably, majorite are allowed.

The crystallization behavior of a partial melt removed from a garnet peridotite has been discussed in detail by M. J. O'Hara and his coworkers (O'Hara and Yoder, 1967; O'Hara, 1968; O'Hara and others, 1975). Table 11-1 illustrates their scheme for generating tholeiites from primitive mantle. Early melting is likely to be extensive since large amounts of melt, 15–25 percent, occur in small temperature range just above the solidus. Eclogite extraction at moderate depth leaves a residual fluid deficient in CaO, Al_2O_3 and Na_2O with a high MgO and FeO content. Crystallization at shallow pressures gives an olivine cumulate and a residual fluid enriched in CaO, Al_2O_3, TiO_2, Na_2O and K_2O and low in MgO.

Table 11-2 gives a more detailed comparison of the possible products of mantle differentiation. The cosmochemical mantle model of Ganapathy and Anders (1974),

TABLE 11-1
Effect of Eclogite and Olivine Fractionation on Primitive Magma

Magma	SiO_2	Al_2O_3	FeO	MgO	CaO	TiO_2	Na_2O	K_2O
1. Primitive	46.2	11.1	10.8	20.2	9.4	0.77	1.06	0.08
2. Extract	46.2	13.9	9.3	16.3	11.9	0.81	1.29	0.02
3. Picrite	46.2	8.3	12.3	24.1	6.9	0.74	0.83	0.14
Tholeiites								
4. Model	50.0	13.8	12.4	8.5	11.5	1.23	1.38	0.23
5. Hawaiian	50.0	14.1	11.4	8.6	10.4	2.53	2.16	0.39
6. Continental	50.6	13.6	10.0	8.5	10.0	1.95	2.90	0.54
7. Average oceanic	50.7	15.6	9.9	7.7	11.4	1.49	2.66	0.17

1. Possible primitive magma. The partial melt product of primitive mantle differentiation (O'Hara and others, 1975).
2. Eclogite extract (O'Hara and others, 1975).
3. Residual liquid after 50 percent eclogite (2) removal from primitive magma (1). This is a model picritic primary magma.
4. Residual liquid after a further removal of 40 percent olivine ($Fo_{87.5}$) from liquid (3).
5. Average Hawaiian parental tholeiite.
6. Continental tholeiite (Tasmania) (Frey and others, 1978).
7. Average oceanic tholeiite glass (Elthon, 1979).

TABLE 11-2
Composition of Mantle, Upper Mantle, Possible Picritic Parent Magmas and Eclogites

Material	SiO_2	Al_2O_3	FeO	MgO	CaO	TiO_2	Na_2O	K_2O
Mantle and Upper Mantle Compositions								
1. Bulk mantle	48.0	5.2	7.9	34.3	4.2	0.27	0.33	
2. Residual mantle	48.3	3.7	7.1	37.7	2.9	0.15	0.15	
3. Pyrolite	45.1	3.3	8.0	38.1	3.1	0.2	0.4	
Possible Picritic Parent Magmas								
4. Eclogite extract	46.2	13.9	9.3	16.3	11.9	0.81	1.29	0.02
5. Oceanic crust	47.8	12.1	9.0	17.8	11.2	0.59	1.31	0.03
6. Tortuga dikes	47.3	13.6	9.8	17.6	9.6	0.79	0.89	0.06
7. High-MgO	46.2	12.6	11.0	16.6	10.5	0.69	1.18	0.02
tholeiites	46.3	13.0	11.3	15.5	10.9	0.71	1.26	0.03
Kimberlite Eclogites								
8. Average	47.2	13.9	11.0	14.3	10.1	0.60	1.55	0.84
9. Roberts Victor	46.5	11.9	10.0	14.5	9.9	0.42	1.55	0.85

1. Bulk mantle composition (Ganapathy and Anders, 1974).
2. Residual after 20 percent extraction of primitive magma (line 1, Table 11-1).
3. This is an estimate of shallow mantle composition (Ringwood, 1975).
4. Possible eclogite extract from primary magma (O'Hara and others, 1975).
5. Average composition of oceanic crust (Elthon, 1979).
6. High magnesia Tortuga dike NT-23 (Elthon, 1979).
7. High magnesia-tholeiites.
8. Average bimineralic eclogite in kimberlite.
9. Eclogite, Roberts Victor 11061 (O'Hara and others, 1975).

on line 1 of the table, uses the iron content (an element that condenses at high temperature) of the Earth and the observed heat flow to estimate the abundances of the refractory elements (the heat flow constrains the abundances of U and Th, two of the refractory elements). The relative abundance of the refractory elements is chondritic. The composition of pyrolite is given in line 3. Various candidates for picritic parent magmas are given in lines 4 to 7. The composition of eclogites in kimberlites are also given. These are possible cumulates from the primary melt and also possible parents for MORB. Note the similarity between the eclogite and picrite compositions.

These tables illustrate the plausibility of a large eclogite-rich cumulate layer in the mantle that has properties similar to those inferred for the picritic parent magma of MORB. An eclogite or basalt layer representing about 10–15 percent of the mantle can reconcile the major-element compositions inferred by cosmochemical, geophysical and petrological techniques.

Table 11-3 delineates a self-consistent petrological and geochemical model of the mantle and crust as it traces the enrichment and fractionation of the incompatible elements during the various stages of mantle differentiation. All results are given relative to the starting composition. The fractionation factors, for Rb/Sr and Sm/Nd, are also given for the products at various stages. The level of enrichment varies from stage to stage, but fractionation of rubidium relative to strontium and samarium relative to neodymium is only significant in the solid residues of partial melting and in melts that represent only a small degree of partial melting (or late-stage residual fluids). Thus, an "enriched" reservoir can have unfractionated Rb/Sr and Sm/Nd ratios and can give "primitive" ratios of $^{87}Sr/^{86}Sr$ and $^{143}Nd/^{144}Nd$. This fairly obvious point needs to be stressed because it has been overlooked by advocates of an undifferentiated primitive mantle.

In addition to demonstrating the effects of partial melting, crystal fractionation and redistribution of an enriched fluid from a crystallizing cumulate layer (mantle metasomatism), Table 11-3 also compares the composition of MORB, continental flood basalts, continental crust and kimberlites with the various products of this differentiation. In order to make this comparison I have normalized to the whole-mantle concentrations of Ganapathy and Anders (1974). Most other recent estimates of the abundances of refractory trace elements in the Earth are within 15 percent of these values and volatile-refractory ratios are within a factor of 2, generally much closer. The main assumption common to all estimates is that the large-ion incompatible refractories occur in the Earth in the same relative proportions as they occur in chondrites. This seems to be valid since these elements apparently do not undergo significant preaccretional, nonmagmatic fractionation.

These calculations indicate that the source reservoir for midocean-ridge basalts may be an eclogite-rich cumulate which has been depleted by removal of a kimberlite-like fluid. A deep eclogitic cumulate layer can become unstable at depth as it warms up, due to garnet-clinopyroxene

TABLE 11-3
Effect of Partial Melting and Eclogite and Peridotite Fractionation from Primary Melt, Relative to Primitive Mantle

Material	K	Rb	Sr	La	Sm	Nd	Yb	$f_{Rb/Sr}$	$f_{Sm/Nd}$
1. Primitive mantle	1	1	1	1	1	1	1	0	0
2. Primitive melt	3.98	3.97	3.86	3.93	3.64	3.82	2.33	0.03	−.05
3. Primary eclogite cumulate	3.33	3.32	3.30	3.30	3.25	3.27	2.78	.01	−.01
4. Primary peridotite residual	6.58	6.31	6.10	6.45	5.28	6.02	0.53	.03	−.12
5. Depleted eclogite	1.06	1.10	2.45	1.77	2.84	2.57	2.83	−.55	.11
6. Model MORB	2.12	2.20	4.85	3.52	5.59	5.09	5.46	−.55	.10
7. Average MORB	2.9	1.5	3.4	2.5	6.1	4.9	6.5	−.56	.24
8. Enriched extract from eclogite	60.8	59.6	24.9	42.0	13.2	21.0	1.43	1.39	−.37
9. Continental crust	59.5	58.1	14.9	26.8	9.7	20.5	5.12	2.90	−.53
10. Kimberlite	41.3	75.6	27.5	211	34.2	65.9	2.79	1.75	−.48
11. Enriched reservoir	6.03	5.84	4.11	5.08	3.19	3.89	2.76	0.42	−.18
12. Model enriched tholeiite	30	29	20	25	14	18	7	0.45	−.22
13. Average continental tholeiite	25	18	13	22	18	22	12	0.38	−.18

Modified from Anderson (1982a).

1. All values in table are normalized to primitive mantle. All calculations based on equilibrium partial melting or equilibrium fractional crystallization. Partition coefficients from Frey and others (1978). f is (Rb/Sr)−1 or (Sm/Nd)−1.

2. 25 percent melt, garnet peridotite residual.

3. 50 percent garnet; 50 percent interstitial fluid. Note that this is nearly unfractionated. This layer and its fluids evolve as nearly primitive material until melt extraction occurs. This is model MORB reservoir prior to depletion. This garnet-rich reservoir becomes depleted as it further crystallizes and expels the intercumulus liquids. The fluid at this point is enriched but not very fractionated.

4. Residual fluid after removal of eclogite (3) from primitive melt (2). This evolves into a shallow peridotite source region (above the garnet cumulate layer). It is somewhat enriched and becomes more enriched as fluid is added to it from the underlying crystallizing garnet cumulate layer.

5. Remove 75 percent of final 5 percent melt fraction from garnet cumulate (3). This is a model for the depleted (MORB) reservoir. The late-stage fluid is both enriched and fractionated.

6. Remove 50 percent olivine plus orthopyroxene from depleted eclogite (5).

7. Midocean-ridge tholeiite (MORB); combined range of uncertainties in MORB and mantle compositions is greater than ±50 percent.

8. 5 percent melt extract from garnet cumulate (3). This is used to enrich the peridotite layer (4) and form continental crust (9). This enriched fluid can be compared with continental crust (9) and kimberlite (10).

9. Continental crust.

10. Kimberlite.

11. Enriched source region. Two-thirds of melt extract from garnet cumulative layer used to enrich the primary residual fluid (4); one-third removed to continental crust. Composition of enriched reservoir; 5 parts (4), 0.5 part (8), 5 parts peridotite.

12. 20 percent melt of (11). This is model theoleiite from enriched source region (continental tholeiite, ocean-island tholeiite). Alkali basalts and other continental basalts are generated from this reservoir by smaller degrees of partial melting. Alternatively, these basalts are mixtures of MORB and a melt from 11.

13. Continental tholeiite.

reactions or to partial melting. The seismic properties of the transition region are consistent with a picritic eclogite (piclogite) composition (Anderson and Bass, 1986). Furthermore, the seismic velocity anomalies under ridges can be traced to 400 km or deeper (Figure 11-1).

These calculations show that extensive quantities of garnet can be removed from the shallow mantle or, put another way, that the upper mantle could have been molten and in equilibrium with a large amount of garnet. Majorite, a high-pressure garnet-like form of pyroxene, probably has similar partition coefficients. Therefore a deep magma ocean, say 300 km deep, is not inconsistent with trace-element data. Actually, the lower mantle can be depleted in LIL during accretion, by zone refining, even if the magma ocean is much shallower.

CHEMICAL STRATIFICATION OF THE MANTLE

Chemical stratification resulting from early differentiation of the mantle, upward removal of the melt and fractionation via crystal settling of the resulting magma is one way to explain the presence of separate and chemically distinct reservoirs. In the first stage, probably during accretion, the incompatible elements (including Rb, Sr, Nd, Sm, and U) are concentrated into melts (zone refined) and the upper mantle is more or less uniformly enriched in these elements. As the upper-mantle magma layer cools, a garnet-rich cumulate, containing intercumulus fluids, forms at the base. Peridotitic cumulates or melts concentrate at shallower depths. The deep garnet-rich cumulate layer can have near-primitive ratios of Rb/Sr and Sm/Nd if it contains a moderate amount of interstitial fluid. Transfer of a late-stage melt (KREEP or kimberlite) is one mechanism by which this layer may become depleted and the complementary region enriched. For this type of model, the isotopic ratios will be a function of the crystallization (fractionation) history of the upper mantle and the history of redistribution of LIL-enriched fluids. It is possible that even the most depleted MORB is contaminated by magmas from the enriched reservoir. The MORB reservoir may therefore be more depleted than generally assumed.

To calculate the isotopic evolution of a reservoir, relative to primitive mantle, we need only know how Rb/Sr and Sm/Nd vary relative to primitive ratios. The entries in Table 11-3 are therefore all relative to primitive, undifferentiated mantle. Since the composition of the melt and the crystals are complementary, the composition of a cumulate of crystals and interstitial fluids will appear less fractionated than either component. Intercumulus material typically constitutes 35–50 percent of an orthocumulate (crystals plus frozen liquids).

The calculations displayed in Table 11-3 show that relatively "primitive" ratios of incompatible elements and, therefore, isotopic ratios, do not necessarily imply a primitive reservoir. In particular, orthocumulates are less fractionated than either of the components. The highly fractionated kimberlites may represent the late-stage fluids of an apparently primitive crystallizing cumulate layer. KREEP probably formed on the Moon by a similar process. Other evidence indicates that the kimberlite source region, which is probably below 200 km depth, is garnet- and clinopyroxene-rich. Reaction relations between olivine, orthopyroxene and the melt also favor a garnet-clinopyroxene-rich cumulate at depth. A garnet-clinopyroxene-rich residue, even after extensive partial melting, is possible at high pressure.

What is the fate of eclogite in the mantle? Ringwood's (1975) model assumes that it sinks to the core-mantle boundary and is removed from the system. Estimated densities as a function of depth for eclogite and garnet peridotite are shown in Figure 11-3. Eclogite and garnetite (garnet solid-solution) are denser than peridotite to depths at least as great as 500 km. On the other hand the post-spinel phases of olivine and the perovskite form of orthopyroxene are denser than garnetite or ilmenite eclogite. Eclogite-rich cumulates, or subducted eclogitic lithosphere, are therefore unlikely to sink into the lower mantle. Whether eclogite can sink below 500 km depends on temperature and the compressibility and thermal expansivity relative to peridotite. The abrupt termination of earthquakes at 670 km suggests that oceanic lithosphere can penetrate to this depth. The sharpness of the discontinuity near 650 km is consistent with a chemical interface. This may be the lower boundary of the eclogite-rich layer. If eclogite represents 10 percent of the mantle, this would imply a thickness of about 200 km. The top of the eclogite layer would therefore be at a depth of 400 km but could rise to the shallow mantle when partially molten. Below some 400 km pyroxene dissolves in garnet to form a garnet solid solution (S.S.). This is

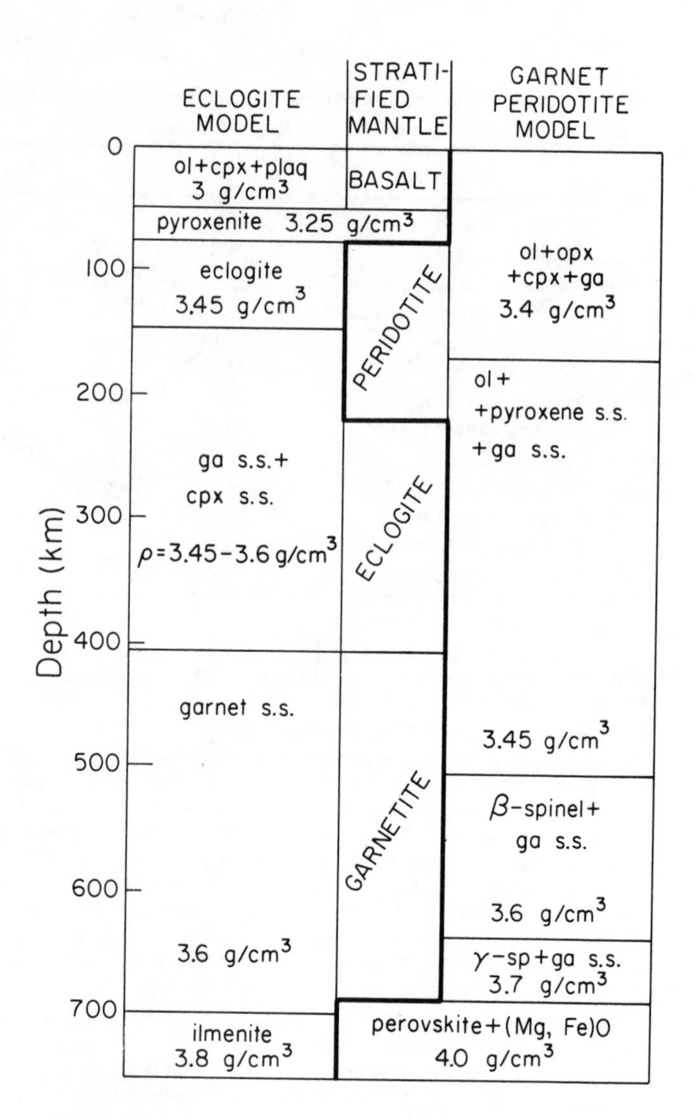

FIGURE 11-3
Approximate densities of basalt/eclogite (left) and garnet peridotite (right). Eclogite (subducted oceanic lithosphere or a cumulate in a deep magma ocean) is denser than peridotite until olivine converts to β-spinel. Below some 400 km the garnet and clinopyroxene in eclogite convert to garnet solid solution. This is stable to very high pressure, giving the mineralogical model shown in the center, the gravitationally stable configuration. The heavy line indicates that the bulk chemistry varies with depth (eclogite or peridotite).

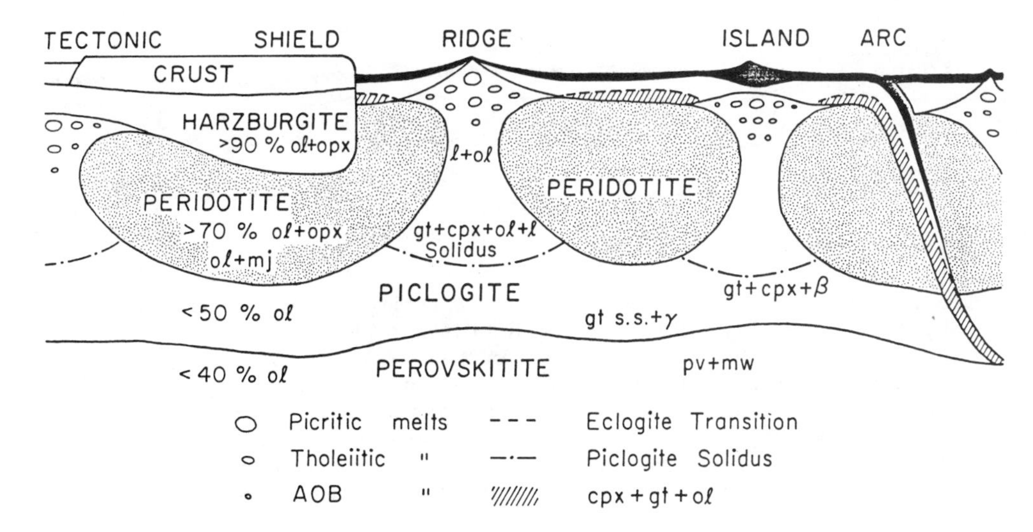

FIGURE 11-4
A possible configuration of the major rock types in the mantle. Peridotite accumulates in the shallow mantle because of its low density, particularly residual or infertile peridotite. Fertile peridotite, eclogite, olivine eclogite and piclogite are denser and are, on average, deeper. Partial melting, however, can reverse the density contrast. Upwellings from a fertile piclogite layer are shown under ridges, islands and tectonic continents. The isotopic and trace-element contents of the magmas are controlled by tectonic and shallow-mantle and crustal processes. The continental lithosphere is stable and is dominantly infertile harzburgite. It may have been secondarily enriched by upward migration of fluids and melts. The shallow mantle is predominantly infertile peridotite (stippled). This is underlain by piclogite, which, however, can rise into the shallow mantle if it becomes hot, partially molten or entrained. Melts from the piclogite layer also underplate the lithosphere, giving a dense, eclogite-rich base to the oceanic lithosphere, which can recycle into the transition region. The top parts of the subducted slab rejuvenate the shallow mantle with LIL. The slab is confined to the upper mantle, being too buoyant to sink into the lower mantle.

termed garnetite, the high-pressure form of eclogite. Piclogite contains olivine as well as eclogite and a piclogite layer can be thicker than an eclogite layer, for a given garnet content of the mantle.

The spatial relationship of the geochemical reservoirs is uncertain. If these were generated by petrological processes such as partial melting and crystal fractionation, the melts, residues, and cumulates would differ in bulk chemistry as well as in trace-element chemistry. The intrinsic density increases in the order basalt, picrite, depleted peridotite, fertile peridotite, eclogite. Basalts and picrites crystallizing or recrystallizing below about 50 km transform to eclogite, which is the densest upper-mantle assemblage. Garnet-poor and olivine-rich residues or cumulates are likely to remain at the top of the upper mantle since they are less dense than parental peridotites and do not undergo phase changes in the upper 300 km of the mantle. Eclogites are about 15 percent and 3 percent denser than basalts and fertile peridotites, respectively. With a coefficient of thermal expansion of $3 \times 10^{-5}/^{0}C$, it would require temperature differences of 1000–5000°C to generate similar density contrasts by thermal effects alone, as in normal thermal convection, or to overcome the density contrasts in these assemblages.

Simple Stokes' Law calculations show that inhomogeneities having density contrasts of 0.1 to 0.4 g/cm^3 and dimensions of 10 km will separate from the surrounding mantle at velocities of 0.5 to 2.5 m/yr in a mantle of viscosity 10^{20} poises. This is orders of magnitude greater than average convective velocities. Inhomogeneities of that magnitude will be generated by partial melting as material is brought across the solidus in the normal course of mantle convection. The higher mantle temperatures in the past make partial melting in rising convection cells even more likely and the lowered viscosity makes separation even more efficient. It seems unlikely, therefore, that chemical inhomogeneities can survive as blobs entrained in mantle flow for the long periods of time indicated by the isotopic data. Gravitational separation is more likely, and this leads to a chemically stratified mantle like that shown in Figures 11-4 and 11-5. The unlikely alternative is that the reservoirs differ in trace elements but not major elements. Small differences in bulk chemistry change the mineralogy and therefore the intrinsic density, and, in general, mineralogy is more important than temperature in generating density inhomogeneities.

The density differences among basalt, depleted peridotite, fertile peridotite, and eclogite are such that they can-

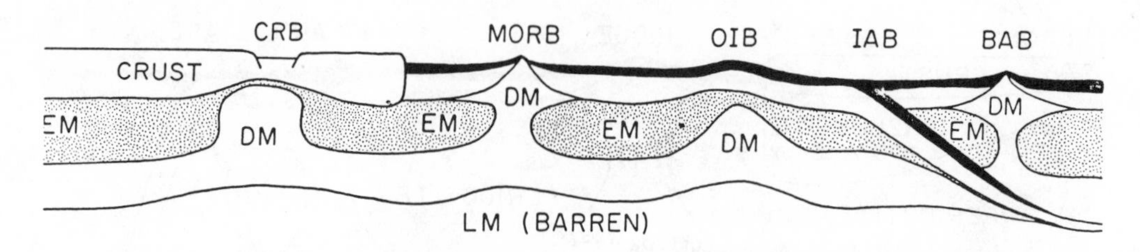

FIGURE 11-5
Possible configurations of geochemical reservoirs (DM, Depleted Mantle; EM, Enriched Mantle; LM, Lower Mantle; CRB, Continental Rift Basalts; MORB, Midocean-Ridge Basalts; OIB, Oceanic-Island Basalts; IAB, Island-Arc Basalts; BAB, Back-Arc Basalts). In this model EM is heterogeneous and probably not continuous. It is isotopically heterogeneous because it has been enriched at various times. LM does not participate in plate tectonic or hotspot volcanism. Contrast this with the model of Wasserburg and DePaolo (1979), which has OIB and CFB coming from a primitive lower mantle, and that of Allegré and others (1982), which has OIB coming from the lower mantle and a uniform upper mantle providing MORB.

not be reversed by the kinds of temperature differences normally encountered in mantle convection. However, phase changes such as partial melting and basalt–eclogite involve large density changes. A picritic or pyroxenitic lithosphere, for example, will be less dense than fertile peridotite at depths shallower than about 50 km where it is in the plagioclase or spinel stability field. As the lithosphere thickens and cools, it becomes denser at its base than the underlying mantle and a potential instability develops. Similarly, if the temperature in a deep garnet-rich layer exceeds the solidus, the density may become less than the overlying layer. The large density changes associated with partial melting and the basalt–eclogite phase change may be more important in driving mantle convection than thermal expansion.

NATURE OF THE ENRICHMENT/DEPLETION PROCESS

Detailed modeling of the isotopic and trace-element characteristics of the enriched and depleted source regions of the mantle show that these regions are genetically related. The continental crust is also approximately complementary to the MORB source, but it was probably not derived by a single-stage partial melting event operating on primitive mantle. In detail, crust plus depleted mantle does not equal primitive or chondritic mantle. Although continental crust recycling via sediment subduction can increase the LIL and certain isotopic ratios in the shallow mantle, it cannot explain the high $^3He/^4He$ ratios of some enriched magmas and other characteristics of the enriched source region.

The melting of ancient subducted oceanic crust has also been proposed as the source of enriched magmas. The subducted crust, being at the top of the slab, may be exposed to high temperatures and, as evidenced by the seismic activity, high stresses, and is probably fractured and possibly melted shortly after subduction. It is not clear how or where ancient subducted crust might be stored. It is doubtful that it would sink into the lower mantle, or even into the transition region. Partial melting of depleted MORB, in the eclogite stability field, gives magmas that are more LIL-enriched than MORB and, for small degrees of melting, can have LREE enrichment. The very high LIL contents and high $^3He/^4He$ ratios of enriched magmas cannot be explained by this mechanism. A source enriched in LIL, LREE, and $^{87}Sr/^{86}Sr$ as well as $^3He/^4He$ is required. The top part of the oceanic crust is altered by hydrothermal activity, and this is a more promising source if the altered part of the crust can survive the subduction process. However, the isotopic data require that any subducted oceanic crust contribution to the enriched reservoir must be ancient.

The emphasis on recycling and continental contamination as mechanisms of enrichment is based on an all-or-nothing philosophy such that when primitive mantle is melted it only forms continental crust and a complementary depleted (MORB) reservoir. In this philosophy present melts from the mantle are either primitive, from previously unfractionated mantle, or depleted. Most mantle magmas are, in fact, in the depleted quadrant of the mantle array (high ε_{Nd}, low $^{87}Sr/^{86}Sr$). However, MORBs are enriched in the lead isotopes and have $^3He/^4He$ ratios higher than atmospheric. This suggests that mantle magmas are mixtures of a depleted and enriched component, with the depleted component dominant.

Although the high LIL content of the crust and the argon-40 content of the atmosphere suggest efficient upward transport of volatiles and the incompatible elements, the high $^3He/^4He$ ratio of the mantle indicates that this process has not gone to completion. The process that put the LIL into the crust can also be expected to enrich the shallow mantle.

Semantics has played a subtle part in most petrological

models. It has long been recognized that midocean-ridge basalts are low in LIL and $^{87}Sr/^{86}Sr$ compared to continental magmas. Since MORB do not pass through continental crust, it has been common practice to refer to depleted basalts as mantle magmas and to assume that low LIL and $^{87}Sr/^{86}Sr$ is a characteristic of the mantle. High-LIL and high-$^{87}Sr/^{86}Sr$ basalts (and high $\delta^{18}O$) therefore are deemed crustal or crustal-contaminated. When enriched magmas were found on oceanic islands, it was difficult to accept these as mantle melts since, by definition, the mantle was depleted. Thus, some islands were assumed to rest on continental crust or to represent melting of recycled crustal material. As the number of enriched islands and seamounts increased, this became an untenable position, and the idea became popular that deep mantle plumes provided melts to the islands. Some of the early $^{143}Nd/^{144}Nd$ results for islands, kimberlites and continental flood basalts fell near the chondritic growth curve, and it was presumed then that hotspots were the result of deep mantle plumes from primitive, unfractionated mantle. This ignored the lead-isotopic evidence. As data accumulated it became clear that island basalts ranged from very depleted to very enriched in $^{143}Nd/^{144}Nd$ and $^{87}Sr/^{86}Sr$ (and always enriched in $^{206}Pb/^{204}Pb$ and $^{207}Pb/^{204}Pb$). Some geochemists still maintained that the most enriched islands were continental fragments in spite of geophysical evidence to the contrary. Tristan da Cunha, for example, is one of the most enriched islands but lies close to the axis of the Mid-Atlantic Ridge. It is also low in $^{3}He/^{4}He$.

Although there are a variety of ways to enrich the shallow mantle by present tectonic processes, the isotopic evidence points toward ancient enrichment as well. We also, of course, must start forming the oceanic crust early on.

A crystallizing cumulate layer contains interstitial melt that becomes progressively more enriched and fractionated as crystallization proceeds. Fluids from such a layer may be the source of the original crust and the metasomatic fluids that have invaded the upper mantle at various times and may account for the progressive enrichment of the source region for oceanic-island and continental basalts.

A depleted reservoir can represent either a cumulate or a region of the mantle that has been subjected to melt extraction. Because of packing considerations and the fact that crystallization at depth is slower than crystal settling, an early cumulate probably retains a large amount of interstitial fluid. Removal of a late-stage melt is therefore also required in this case to generate a very depleted reservoir. The composition of the residual fluid in equilibrium with eclogite, as a function of crystallization, is shown in Figure 11-6. I have chosen 50 percent garnet and 50 percent clinopyroxene for the residue in this calculation. For both trace-element systems the fractionations increase rapidly as the residual melt fraction drops below about 20 percent (that is, above 80 percent crystallization). Kimberlites may represent such late-stage fluids. They appear to have been in

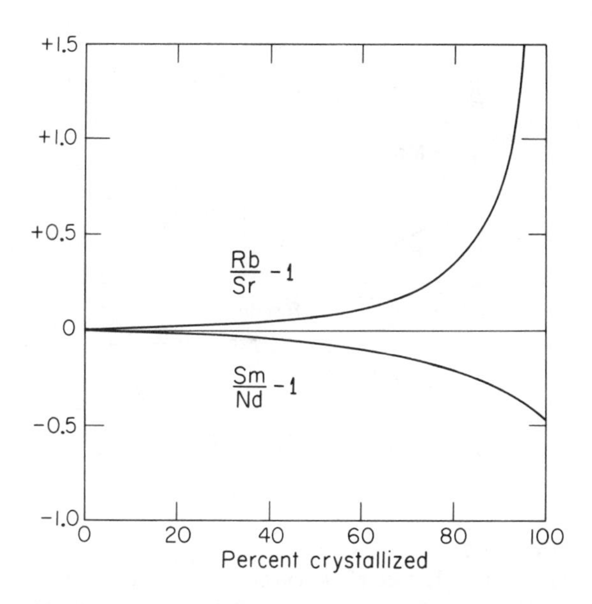

FIGURE 11-6
Variation of the normalized Rb/Sr and Sm/Nd ratios in the melt fraction of a crystallizing eclogite cumulate. Equilibrium crystallization is assumed. The fractionation factor in the melt increases as freezing progresses. If melt extracts from this layer are the enriching fluids for upper-mantle metasomatism, then enrichment will increase with time. Early melt extracts have nearly primitive ratios. The crystalline residue is 50 percent garnet and 50 percent clinopyroxene.

equilibrium with eclogite and often contain eclogite xenoliths. They also have LIL patterns that are complementary to MORB. From Figures 11-6 and 11-7 we see that progressive crystallization of an eclogite cumulate and removal of the melt at various stages is not efficient in fractionating Rb/Sr and Sm/Nd until the late stages. Isotopic evolution in the region enriched by the expelled fluids will start to deviate significantly from the primitive-mantle growth curve only at relatively late times. The more recently enriched reservoirs will have higher Rb/Sr and Nd/Sm ratios than ancient enriched and subsequently isolated reservoirs (Figure 11-7). The partition coefficients for lead are not well known, but the U/Pb ratios appears to behave similarly to the Rb/Sr ratio unless sulfides are involved. That is, the U/Pb ratio in the melt also should increase as crystallization proceeds. For a simple crystallization history of the depleting reservoir, the fractionation factor of the melt increases rapidly with time, for example as an exponential or power law of time (Figure 11-6).

The melt content of a cooling garnet- and clinopyroxene-rich silicate decreases rapidly within a few tens of degrees of the solidus. The average temperature of the mantle has decreased about 150°C in the past 3 billion years (Turcotte, 1980), and the cooling is roughly exponential with a characteristic time of about 2 Ga. Complete crystallization of the upper mantle therefore occurs on a time scale comparable to the age of the Earth, and extreme fractionation,

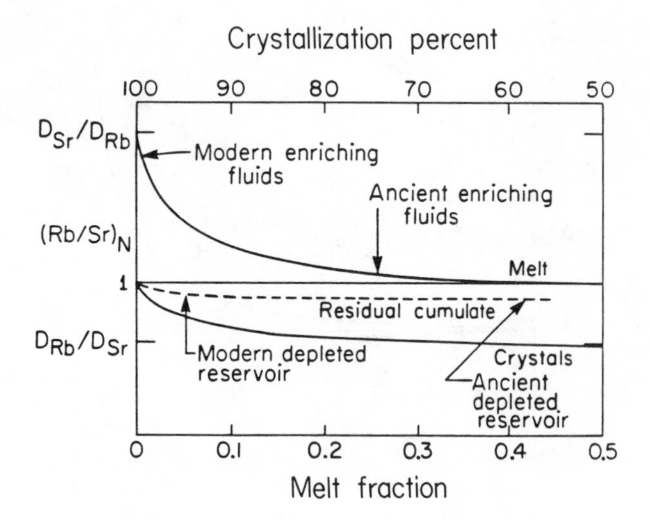

FIGURE 11-7
Composition of the melt phase and the residual crystals as a function of melt fraction for a crystallizing eclogite cumulate. A hypothetical residual cumulate is also shown. The residual cumulate is formed by extracting half of the melt fraction. This shows that the composition of the depleted reservoir is nearly constant with time. Therefore, progressive depletion is less important an effect than progressive enrichment. D_{Sr} and D_{Rb} are partition coefficients.

at least on a global scale, is restricted to the past 1 to 2 Ga. The distribution of oceanic ridges and hotspots is evidence that a large part of the upper mantle is still above or close to the solidus.

The normalized Rb/Sr ratios of a melt and residual crystals can be written

$$(\mathrm{Rb/Sr})_m = \frac{F + (1 - F)D_{Sr}}{F + (1 - F)D_{Rb}}$$

$$= (\mathrm{Rb/Sr})_{res}\,(D_{Sr}/D_{Rb}) = f_m + 1$$

where the D are solid-melt partition coefficients and F is the melt fraction. The range of $(\mathrm{Rb/Sr})_m$ is from 1 for large F to D_{Sr}/D_{Rb} as F approaches zero. For an eclogite residue D_{Sr}/D_{Rb} is about 15, so the fractionation factor of the melt f_m, which is taken to be the enriching fluid, varies from zero for large degrees of melting to 14 for the final melt fraction. On the other hand, the total range of f for the residual crystals is only from 0 to -0.8. Considering that the residual cumulate may consist of interstitial melt as well as crystals, the total range is even smaller. As the cumulate freezes, continuously or episodically losing its fluids to the overlying layer it is enriching, it contains less of a more enriched fluid. The net result is a nearly constant f for the cumulate as it evolves. This may explain why depleted magmas require only a simple two-stage model to describe the evolution of their source reservoir and why depleted magmas (MORB) appear to come from a uniform reservoir. Most of the fractionation that a crystallizing reservoir experiences occurs upon the removal of the first batch.

The depleted reservoir becomes depleted by the removal of enriched fluids representing late-stage interstitial fluids or small degrees of partial melting (Figure 11-2). The enriched fluid is not necessarily removed to the continental crust; it can also serve to enrich the uppermost mantle. The enriched and depleted layers may also differ in major elements and mineralogy, possibly the result of crystallization and gravitational separation. This chemical stratification—that is, the difference in intrinsic density—makes it possible for the two source regions to maintain their isotopic identity. If the enriching fluid comes from a garnet-rich layer, it will impart an HREE-depleted pattern to the enriched layer. Since bulk chemistry controls the density and location of the source regions, it is not necessary to assume that one grows at the expense of the other except for a small change in thickness caused by the exchange of a small amount of melt. There also is no need to invoke a large primitive reservoir. Since the enrichment increases with time, there is no inconsistency between a currently highly enriched reservoir and isotopic ratios that depart only slightly from primitive mantle. Isotopic ratios reflect the time-integrated concentrations of the appropriate elements. The mantle need not be primitive even up to the times of the enrichment/depletion events. Partial melting of primitive mantle followed by crystallization and gravity separation gives upper-mantle source regions that, at least initially, have LIL ratios, including Rb/Sr and Sm/Nd ratios, similar to primitive mantle. Residual fluids in a cooling Earth become more fractionated with time.

The lead, neodymium and strontium isotopic data are consistent with progressive enrichment of the reservoir providing the enriched continental and oceanic-island magmas. The time scale of enrichment is of the order of 2 Ga. Such time-dependent enrichment can be accomplished if the enriching fluid is a late-stage residual melt of a deeper, slowly crystallizing cumulate layer. The trace-element pattern of the enriching fluid suggests that it was in equilibrium with garnet.

Partial melting of the mantle during accretion followed by melt separation, crystal fractionation and formation of upper-mantle cumulate layers is one model that can explain the observations. The layers evolve into the major upper-mantle reservoirs. The upward transfer of KREEP-like or kimberlitic material can explain the progressive depletion and enrichment of these reservoirs. Similar scenarios have been developed for the Moon.

Garnet precipitation below 50 km and olivine and orthopyroxene reaction relations with the melt lead to the formation of an eclogite-rich cumulate layer and a residual silica-undersaturated and alumina-poor melt that, upon further cooling, forms a shallower, less garnet-rich, peridotite cumulate layer. The lower mantle is formed of depleted residual crystals and is relatively cold because most of the radioactive elements are in the crust and upper mantle. If the lower mantle is "primitive" or undepleted in radioactives, then temperatures there will be much greater, and temperatures in the upper mantle will be lower. In the primitive-lower-mantle scenarios of geochemistry, the large amount of heat produced by radioactivity gives a large increase of temperature across the upper mantle–lower mantle boundary and, therefore, a significant thermal boundary layer. In the depleted-lower-mantle scenario the temperature rise is much less.

THE ROLE OF MAGMA MIXING

When two magmas are mixed, the composition of the mix, or hybrid, is

$$xC_i^1 + (1 - x)C_i^2 = C_i^{\text{mix}}$$

where x is the weight fraction of magma 1, and C_i^1, C_i^2 and C_i^{mix} are, respectively, the concentration of the ith element in magma 1, magma 2 and the mix. Mixing relations for elements are therefore linear.

The mixing relations for ratios of elements or isotopic ratios are more complicated. For example,

$$(\text{Rb/Sr})_{\text{mix}} = \frac{x(\text{Rb/Sr})_1 + (1 - x)(\text{Rb/Sr})_2(\text{Sr}_2/\text{Sr}_1)}{x + (1 - x)(\text{Sr}_2/\text{Sr}_1)}$$

$$({}^{87}\text{Sr}/{}^{86}\text{Sr})_{\text{mix}} = \frac{x({}^{87}\text{Sr}/{}^{86}\text{Sr})_1 + (1 - x)({}^{87}\text{Sr}/{}^{86}\text{Sr})_2(\text{Sr}_2/\text{Sr}_1)}{x + (1 - x)(\text{Sr}_2/\text{Sr}_1)}$$

These are hyperbolas, and the shape or curvature of the hyperbola depends on the enrichment factor E, in this case Sr_2/Sr_1. Depleted magmas when mixed with an enriched magma can appear to be still depleted for some elemental and isotopic ratios, undepleted or "primitive" for others, and enriched for others, depending on the appropriate E. This simple observation can explain a variety of geochemical paradoxes. For example, many basalts are clearly enriched, relative to primitive mantle, in such ratios as La/Yb and ${}^{206}\text{Pb}/{}^{204}\text{Pb}$ but depleted in ${}^{87}\text{Sr}/{}^{86}\text{Sr}$.

Trace-element and isotopic data for magmas from the two major mantle reservoirs sometimes appear to be inconsistent. The incompatible elements and strontium and neodymium isotopes show that abyssal tholeiites (MORB) are from a reservoir that has current and time-integrated depletions of the elements that are fractionated into a melt. MORBs, however, have ${}^{206}\text{Pb}/{}^{204}\text{Pb}$ and ${}^{207}\text{Pb}/{}^{204}\text{Pb}$ ratios suggesting long-term enrichment in U/Pb as shown in Chapter 10. Alakli basalts and tholeiites from continents and oceanic islands are derived from LIL- and U/Pb-enriched reservoirs. Strontium and neodymium isotopic ratios, however, appear to indicate that some of these basalts are derived from unfractionated reservoirs and others from reservoirs with time-integrated depletions.

These inconsistencies can be reconciled by treating oceanic and continental basalts as mixtures of magmas from depleted and enriched reservoirs. MORBs are slightly contaminated, depleted magmas, while oceanic-island and continental basalts are mixtures of MORB, or a depleted picritic parent magma, and an enriched end-member having trace-element patterns similar to potassic magmas such as kimberlites or nephelinites. The mixing relations are such that mixtures can be enriched in U/Pb, Rb/Sr, Nd/Sm or ${}^{206}\text{Pb}/{}^{204}\text{Pb}$ relative to primitive mantle, yet appear to have time-integrated depletions in ${}^{143}\text{Nd}/{}^{144}\text{Nd}$ and ${}^{87}\text{Sr}/{}^{86}\text{Sr}$.

A small amount of contamination by material from an enriched reservoir can explain the lead results for MORB (the "lead paradox"; see Chapter 10). Depleted basalts are more sensitive to lead than to rubidium or neodymium contamination. The ${}^{238}\text{U}/{}^{204}\text{Pb}$ of uncontaminated MORB may be about 7 compared to 7.9 for the primary growth curve and above 10 for the enriched reservoirs. Similarly, continental and oceanic-island basalts may represent mixtures of enriched and depleted magmas.

In previous sections I proposed that both of the major mantle magma sources are in the upper mantle and that both are global in extent. The shallower one is inhomogeneous, having been enriched (metasomatized) at various times by fluids from the deeper depleting layer. It may also be the sink of subducted sediments and hydrothermally altered oceanic crust. The deeper source, 400–670 km depth, is more homogeneous and more garnet-rich and provides depleted magmas that, however, become contaminated as they rise through the shallow enriched layer. Partial melting and diapiric ascent originate in the thermal boundary layers between chemically distinct layers. Magma mixing, prior to eruption, is an inevitable consequence of a chemically layered mantle with depleted magmas rising through an enriched uppermost mantle.

Abyssal tholeiites have ${}^{87}\text{Sr}/{}^{86}\text{Sr}$ and ${}^{143}\text{Nd}/{}^{144}\text{Nd}$ ratios that require that the time-averaged Rb/Sr and Sm/Nd ratios of their sources be lower and higher, respectively, than those estimated for the bulk Earth. These results can be explained by a general depletion in large-ion lithophile elements of the MORB reservoir, such as would be caused by extraction of a partial melt or a late-stage residual fluid from a crystallizing cumulate layer.

The lead isotopes for ocean islands can be interpreted in terms of a two-stage evolutionary model with a high ${}^{238}\text{U}/{}^{204}\text{Pb}$ ratio (μ ratio) for the second stage. The radiogenic nature of MORB, which by other measures appears to be depleted, has been attributed to continuous extraction of lead from the reservoir to the core, the lower crust or lower mantle. An alternative is that abyssal tholeiites are slightly contaminated by overlying high-μ material. This would require that enriched mantle underlies both oceans and continents.

Trace-element and isotopic data indicate that many basalts come from a source that is more enriched in LIL than primitive mantle. The presence of a large enriched reservoir in the mantle, perhaps forming contemporaneously with the continental crust, would materially affect models that as-

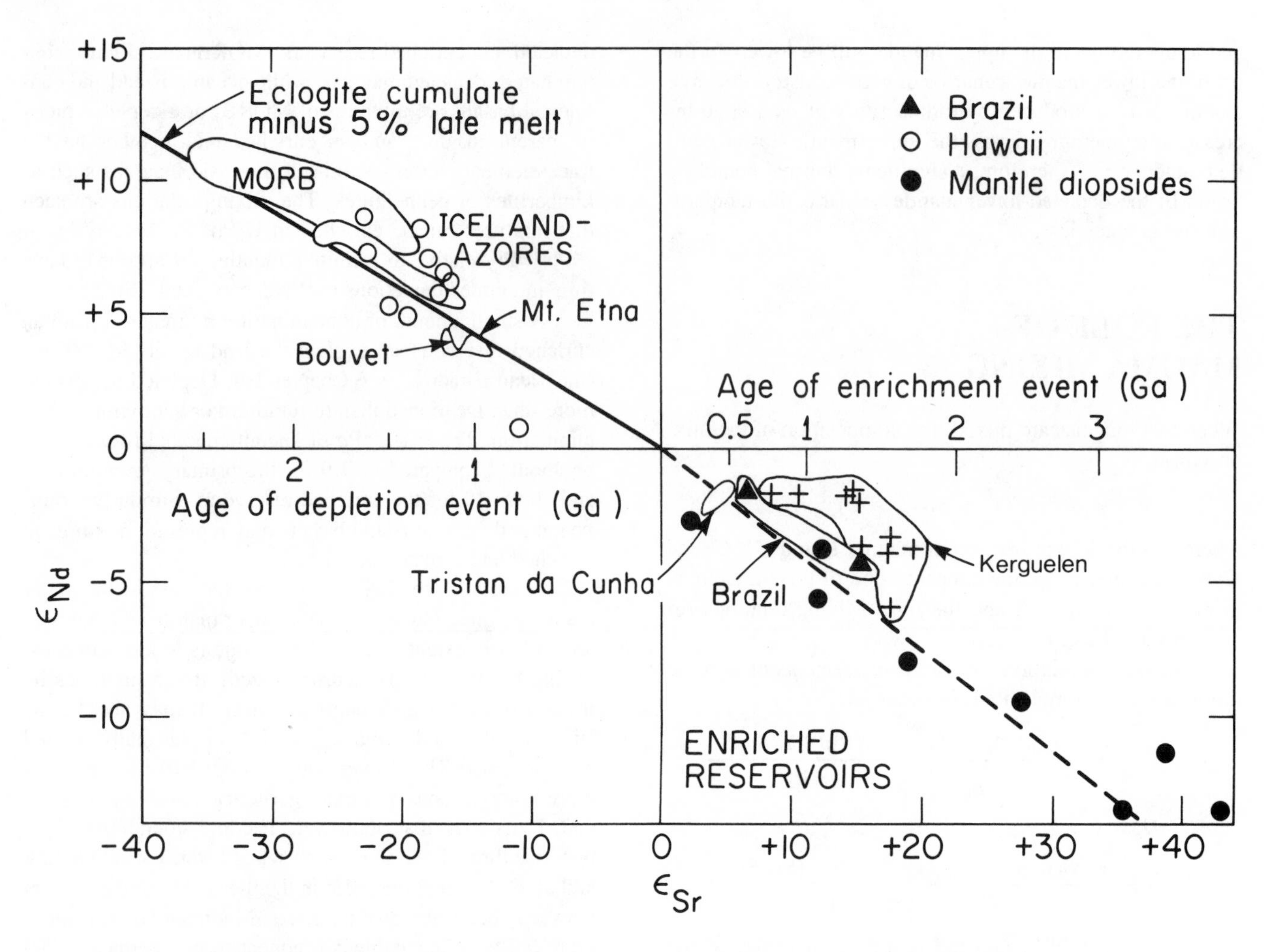

FIGURE 11-8
Neodymium and strontium fractionation trends for an eclogite cumulate (solid line) that has been depleted at various times by removal of a melt fraction. ε_{Nd} and ε_{Sr} are the $^{143}Nd/^{144}Nd$ and $^{87}Sr/^{86}Sr$ ratios expressed as fractional deviations in parts in 10^4 from those in a primitive undifferentiated reference reservoir. The dashed line is the complementary reservoir that has been enriched at various times by the melt extract from the eclogite layer. Reservoirs are unfractionated until the enrichment/depletion event and uniform thereafter. If present values of enrichment/depletion have been reached gradually over time or if these magmas are mixtures, then ages shown are lower bounds. "Depleted" OIB (upper left quadrant) may represent mixtures of depleted and enriched magmas.

sume that the continental crust is the only complement to the depleted reservoir. The presence, size, location and time history of enrichment of this reservoir are critical issues in discussions of mantle evolution. If this reservoir is shallow and global, it may be expected that MORB and partial melts from the enriched layer may be common.

Figure 11-8 shows the correlation in differentiation for MORB, oceanic islands, and some continental basalts and mantle diopsides. The depletion and enrichment ages are calculated for a simple two-stage model for the development of the enriched and depleted reservoirs. The Rb/Sr and Sm/Nd ratios are assumed to be unfractionated up to the age shown and then fractionated to values appropriate for the depleted and enriched reservoirs. Subsequent isotopic evolution occurs in these fractionated reservoirs. For this kind of model the MORB reservoirs were apparently depleted and isolated at times ranging from 1.5 to 2.5 Ga, and the enriched reservoirs (giving magmas in the lower right quadrant) were enriched between 0.5 and 1.8 Ga. If the enrichment has been progressive, that is, Rb/Sr and Nd/Sm increasing with time, the start of enrichment could have been much earlier. Progressive enrichment, of the type often discussed for the uranium-lead system, could explain the low ε_{Nd}–ε_{Sr} values found for continental and oceanic-island basalts with high Rb/Sr and Nd/Sm ratios.

The data shown in Figure 11-8 may also be interpreted

in terms of mixtures of magmas from depleted and enriched reservoirs. In fact, the compositions of alkali olivine basalts, basanites and continental tholeiites are bracketed by MORB and potassium-rich magmas such as nephelinites for most of the major and minor elements as well as for the isotopes. This supports the possibility that many continental and oceanic-island basalt types are mixtures.

Contamination of MORB seems to be a plausible way of reconciling the lead isotopic results with those from the other isotopic systems. Mixing alone, however, cannot explain the large spread of lead and U/Pb in MORB that accompanies the relatively restricted range in $^{206}Pb/^{204}Pb$. Crystal fractionation and hydrothermal alteration operating at the ridge-crest environment may also be involved.

It appears that, in general terms, the isotopic and trace-element results for continental and oceanic-island basalts can be understood in terms of magma mixing. A conceptual model is as follows. Consider a stratified mantle with a shallow enriched layer and a deeper (>200 km) depleted layer. Assume that partial melting in the thermal boundary layer of the depleted layer initiates diapiric ascent. The upward advection of hot isotherms initiates melting in the shallower layer. The initially erupted magmas will be the most enriched, and subsequent magmas will have more of the depleted component. The final eruptive product is MORB, which as it wanes is increasingly contaminated by its passage through and evolution in the enriched upper mantle. In this model, magma mixing is unavoidable and there is a specific temporal sequence of eruption; enriched magmas occur in the initial and waning stages. Enriched magmas would also be expected on the flanks of the surface expression of the diapir. Active seamounts and islands away from oceanic ridges would be expected to be the most enriched unless the depleted oceanic lithosphere is remelted and contributes to the erupted magmas. This model is illustrated in Figure 11-9.

The mixing parameters of Table 11-4 (see also Figure 10-4) indicate how several apparent geochemical contradictions can be resolved. If we adopt the point of view that all oceanic-island basalts are contaminated MORBs, then mixtures involving more than about 60 to 75 percent MORB will appear to represent magmas having time-integrated depleted ratios of Nd/Sm and Rb/Sr, respectively, that is, depleted ε_{Nd} and ε_{Sr}. On the other hand, the current Sm/Nd and Rb/Sr ratios of the mixture will appear enriched until the mixture is more than 78 to 94 percent of the depleted end-member. Therefore, contaminated MORB can exhibit LREE and Rb/Sr enrichment but have apparent time-integrated depletions, as is the case for most oceanic-island and continental flood basalts. With the parameters given, only magmas with more than about 94 percent MORB will appear depleted by all the above measures, and even purer MORB is required to exhibit $^{206}Pb/^{204}Pb$ depletion. On the other hand, magmas must contain less than about 60 percent MORB component in order to appear enriched by all measures. Magmas containing 20 to 60 percent MORB component have major- and trace-element compositions characteristic of the basanite–alkali olivine basalt suite.

The isotopic ratios of end-member MORB are greatly affected by small degrees of contamination, contamination that is probably unavoidable if MORB rises through, or evolves in, continental lithosphere or enriched upper mantle. Basalts at anomalous ridge segments show clear signs of contamination, as in T- and P-MORB (transitional- and plume-type MORB), and normal MORB may simply show less obvious signs of contamination. Since relatively depleted magmas are so sensitive to contamination, it seems unlikely that they can be used to define secondary isochrons. On the other hand, enriched magmas such as oceanic-island basalts are less sensitive to "contamination." The fact that MORB contains high $^{207}Pb/^{204}Pb$ and $^{206}Pb/^{204}Pb$ ratios, in spite of being a "depleted" magma, is called the "lead paradox" (see Chapter 10 and Figure 10-4 in particular).

The dichotomy between high Nd/Sm in some alkali basalts coupled with depleted $^{143}Nd/^{144}Nd$ ratios ($\varepsilon_{Nd} > 0$) requires that the Nd/Sm enrichment be recent. This implies either recent metasomatism of the depleted source region or recent magma mixing. If deep diapirs from the depleted reservoir cause the melting in the enriched layer, then magma mixing would be an expected prelude to eruption, and enriched magmas exhibiting isotopic evidence for long-term depletion could be easily understood. Magmas containing 10–20 percent contaminant will still appear isotopically depleted for Nd and Sr. Using values from Table 11-4, the Rb/Sr ratio of such mixtures is 0.037 to 0.052, enriched relative to primitive mantle and similar to ratios in oceanic-island and continental basalts. The Sm/Nd ratio is 0.30 to 0.34, also enriched compared to MORB and similar to primitive ratios. Such mixtures will appear to exhibit long-term enrichment in the lead isotopic systems.

MELTS FROM THE ENRICHED RESERVOIRS

So far I have considered a single enriched end-member that may represent a small degree of partial melting from an enriched refractory peridotitic layer in the upper mantle. I have speculated that melting in this layer is initiated by the rise of a diapir from a deeper depleted layer. Even if the enriched layer is homogeneous, the partial melts from this region will have variable LIL contents and ratios such as Rb/Sr, Sm/Nd, and U/Pb, which depend on the extent of partial melting. Magma mixtures, therefore, may appear to require a range of enriched end-members. A plot of an isotopic ratio versus a ratio such as Rb/Sr, Sm/Nd or La/Sm may exhibit considerable scatter about a two-component mixing line even if the end-members are isotopically homogeneous. An example is shown in Figure 11-10. The en-

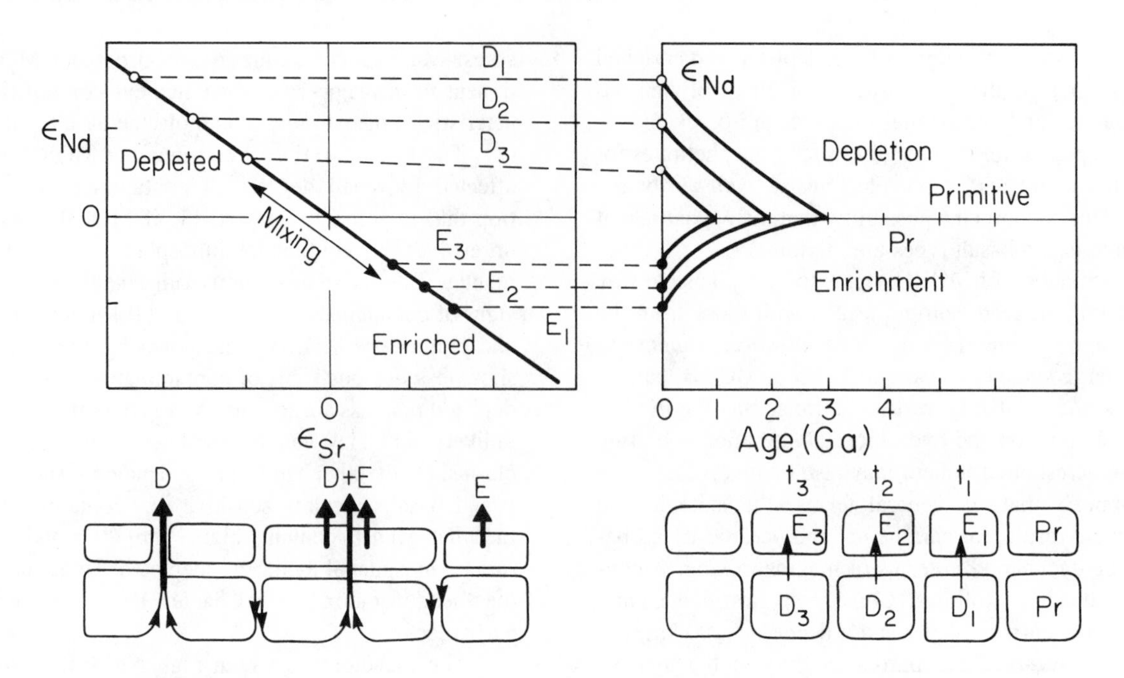

FIGURE 11-9
Illustration of isotopic growth in a two-layer mantle. The lower layer is formed of heavy cumulates, perhaps at the base of a magma ocean. As it freezes it expels enriched fluids to the shallow layer (E), thereby becoming depleted (D). As time goes on, and crystallization proceeds, the melts become more enriched and more fractionated. Isotopic growth is more rapid for the parts of the shallow mantle enriched at later times, but the earlier enriched reservoirs (E_1) have had more time for isotopic growth. The mantle array can be interpreted as the locus of points representing magmas from different aged reservoirs or as a mixing array between products of enriched and depleted reservoirs, or some combination. Melts from D may be contaminated by E if they cannot proceed directly to the surface. If E is a trace-element-enriched, but infertile, peridotite, then D may be the main basalt source region, and enriched basalts may simply represent contaminated MORB. The enriched component in E may be kimberlitic. The deeper layer transfers its LIL upward, forming a depleted layer (D) and a complementary enriched layer (E). The growth of ε_{Nd} in the depleted and enriched cells (upper right) combined with similar diagrams for ε_{Sr} generates the mantle array (upper left). Magma mixing reduces the spread of ε-values, decreasing the apparent ages of the depletion/enrichment events. Layer D may be the transition region.

riched source has ε_{Nd} and Rb/Sr as shown on the lower dashed line. The Rb/Sr of partial melts from this reservoir are also shown. The solid lines are mixing lines between these melts and a melt from the depleted reservoir having properties estimated for "pure MORB." The dashed lines are labeled by the fraction of MORB in the mixture. The data points are representative compositions of various basalts; most fall in the field representing 50 to 95 percent "MORB" and an enriched component representing 2 to 20 percent melt from the enriched reservoir. The Rb/Sr ratio may also be affected by crystal fractionation and true heterogeneity of one or both of the two source regions.

The hypothesis that oceanic and continental magmas represent mixtures of melts from ancient enriched and depleted reservoirs in the mantle appears capable of explaining a variety of geochemical data and resolving some isotopic paradoxes. The hypothesis explains the lead paradox and other apparently contradictory trace-element and isotopic evidence for enrichment and depletion of the various mantle reservoirs. LIL-enrichment associated with time-integrated depletion, usually attributed to mantle metasomatism, is a simple consequence of mixing relations. Most oceanic-island and continental flood basalts may be "contaminated" MORB. If these basalts are mixtures, then the ε_{Nd} ages of their parent reservoirs will be underestimated, that is, the sources are more enriched or depleted in their isotopic ratios than the hybrid magmas. This may reconcile the geological evidence for ancient and rapid continental growth with the apparently conflicting isotopic evidence for the history of mantle reservoirs. The close proximity in time and space of enriched and depleted magmas in all tectonic environments—continental rifts, oceanic islands, fracture zones, midoceanic rifts and arcs—supports the concept of a stratified mantle. In a chemically stratified mantle the in-

TABLE 11-4
Mixing Model for Basalts

Parameter	MORB	Contaminant*
Pb	0.08 ppm	2 ppm
U	0.0085 ppm	0.9 ppm
U/Pb	0.10	0.45
$^{238}U/^{204}Pb$	7.0	30.0
$^{206}Pb/^{204}Pb$	17.2	19.3, 21
Rb	0.15 ppm	28 ppm
Sr	50 ppm	350 ppm
$^{87}Sr/^{86}Sr$	0.7020	0.7060
Sm	2 ppm	7.2 ppm
Nd	5 ppm	30 ppm
$^{143}Nd/^{144}Nd$	0.5134	0.5124

Enrichment Factors†

Pb	25.0	U/Pb	4.5
Sr	7.0	Rb/Sr	26.7
Nd	6.0	Sm/Nd	0.60

*Assumed composition of contaminant. This is usually near the extreme end of the range of oceanic-island basalts.

†Ratio of concentration in two end-members.

dividual reservoirs can maintain their isotopic identity. In a layered, convecting mantle the average temperature gradient is adiabatic except in the thermal boundary layer between reservoirs. Melting is most likely to initiate in these regions of high temperature gradient, permitting the ascent and mixing of magmas from previously isolated reservoirs. Partial melting in the deeper layer, caused by deformation of the boundary or continental insulation, may be the trigger for diapiric ascent. If the solidus temperatures in the two reservoirs are similar, the geotherm will cross the solidus first in the lower part of the thermal boundary layer, that is, in the deeper layer (see Figure 9-5).

The geochemical homogeneity of MORB is one reason why it is desirable to place it deep, so it can be homogenized by convection and isolated from subducted sediments and altered oceanic crust. The heterogeneity of OIB and CFB and the requirement of several enriched end-members to satisfy all the various kinds of data is one reason why this reservoir appears to be shallow. The enrichment and heterogeneity of the OIB reservoir is also intermediate between the crust and the MORB reservoir.

MAGMA OCEAN

The separation of the mantle into distinct geochemical reservoirs probably occurred during accretion and possibly via a magma ocean. A magma is a molten or extensively molten rock that may contain suspended crystals. It behaves as a high-viscosity fluid. An ocean is a liquid that covers extensive areas of a planet. When the surface is below the freezing point, the ocean may be covered by a solid cap. As an ocean freezes the components separate according to their densities and crystallization temperatures. The stability of a freezing ocean depends on the variation of temperature and composition with depth. Thus, a layer of molten rock containing suspended, settling or rising crystals and overlain by a cap of the lighter crystals can be called a magma ocean. The freezing of a magma ocean results in a layered stack of crystals, sorted according to their crystallization temperatures and densities.

A magma ocean is transient because the surface or atmospheric temperatures are undoubtedly less than the interior temperature of a planet. On the other hand, during the accretional stage of planetary evolution, energy can be delivered to the surface and interior much faster than it can be removed by convection and conduction. The extreme case is a large impact that can melt a large fraction of the mantle as well as eject material from the Earth. It is likely that the Earth experienced many giant impacts, and therefore the mantle was partially molten many times and at all stages in its evolution.

Since the settling of crystals in a freezing magma ocean is likely to be much faster than the freezing of the ocean itself, the result is a chemically stratified planet, during and after accretion. The high melting points of olivine and orthopyroxene, the buffering provided by the latent heats of melting, the high density of these refractory crystals relative to melts in the outer part of a planet, and the increase of melting temperature with pressure mean that low-pressure melting and melt-solid separation will dominate and that separation of basaltic material from peridotitic material is likely to be an efficient process. The effect of pressure on crystallization temperatures means that freezing in a adiabatic magma ocean will proceed principally from

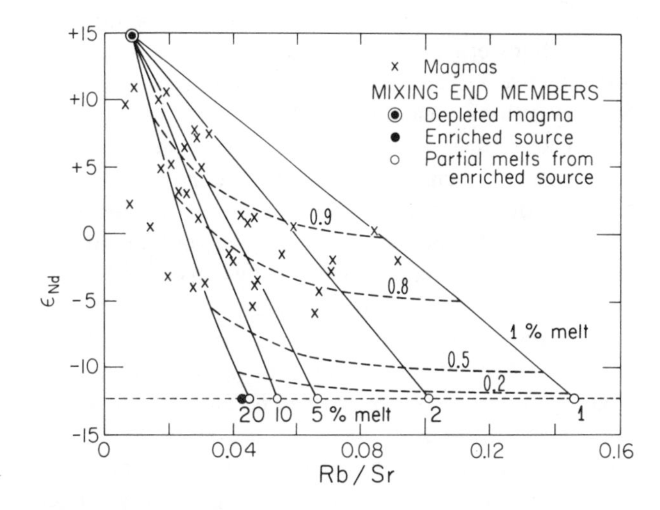

FIGURE 11-10
Mixing relations for a depleted magma and partial melts from an enriched peridotite reservoir. The solid lines are mixing lines, and dashed lines give the fraction of the depleted component.

the bottom, forcing residual melts and their incompatible elements toward the surface. Thus, the minor elements as well as the major elements are redistributed during accretion and the subsequent freezing of the magma ocean. It is difficult to imagine a scenario that will avoid this chemical stratification or allow a large portion of the mantle to retain cosmic abundances of the elements.

The transient nature of the ocean and the possibility that it may be largely buried or discontinuous should not divert attention away from the extensive and widespread melting that must occur in the accretion of a terrestrial planet.

The extreme concentration of trace elements in both the lunar and terrestrial crusts and the thick anorthositic crust on the Moon provide strong geochemical and petrological support for the concept of a magma ocean. The absence of a correspondingly thick crust on the Earth is simply a reflection of the higher pressures on the Earth. Plagioclase, the chief component of anorthosite, floats on magma but it crystallizes only at low pressure.

At low pressure there is a very large melting interval, more than 500°C, between the solidus and the liquidus of four-phase rocks consisting of olivine, orthopyroxene, clinopyroxene and an aluminous phase such as plagioclase, spinel or garnet. Olivine is the most refractory component and the calcic, aluminous and alkali-bearing phases are the most fusible. The large melting interval plus the large difference in density between melts and the near-liquidus phases, olivine and orthopyroxene, means that the basaltic and refractory components will have time to separate during accretion and the subsequent crystallization phase of Earth evolution. If melting ever extends deeper than about 200 km, melts may become denser than the refractory crystals and settle downward. Orthopyroxene transforms to denser phases, spinel plus stishovite or majorite, depending on the Al_2O_3 content, below about 300 km. This plus the olivine–β-spinel transformation at 400 km restricts a deep melt zone to between 200 and 300 or 400 km. This deep melt zone crystallizes much more slowly than near-surface melts. Garnet may settle as crystallization proceeds, because of its high density, but garnet and clinopyroxene coprecipitate over a narrow temperature interval and are more likely to stay together than melt and olivine-orthopyroxene, which differ substantially both in density and crystallization temperature.

The high density of garnet and majorite relative to mafic and ultramafic melts and olivine means that the transition region and lower mantle will be enriched in Si and have a higher Si/Mg ratio than the shallow mantle. If the pressure at the base of the magma ocean was greater than 100 kbar, the molten zone would act as a gravitational filter with orthopyroxene, as majorite, sinking to the base and olivine rising. As the planet grows, the majorite transforms to perovskite, thus explaining the seismic evidence for a perovskite-rich lower mantle. If the magma ocean is deeper than about 400 km, any residual olivine that has not already floated to shallow depths will transform to β- and γ-spinel, phases that are probably denser than the melt at these depths. With this scenario most of the material in the lower mantle is the refractory residue remaining after melt-crystal separation and is therefore orthopyroxene-rich and depleted in the incompatible elements. Although garnet-majorite and perovskite can probably accommodate a variety of trace elements, it is the partitioning of elements between the phases and melts at lower pressures that sets the trace-element contents of the phases that presently exist in the high-pressure forms. It is likely therefore that the transition region and lower mantle are depleted in the incompatible elements.

The elastic properties of the lower mantle depend mainly on the Mg/Si ratio. For iron-free systems this ranges from 2 for olivine stoichiometry to 1 for orthopyroxene and garnet stoichiometries and 1/2 for diopside. For a chondritic mantle the ratio is slightly greater than 1. This reflects both the low olivine content, relative to upper-mantle peridotites, and the high FeO content. The high Mg/Si ratio of the upper mantle is partially due to the removal of basalt ($CaMgSi_2O_6$ plus $Mg_3Al_2Si_3O_{12}$). This also reduces the CaO and Al_2O_3 content of peridotite that has experienced basalt extraction. The amount of basalt at the Earth's surface is not enough to account for the difference in Mg/Si between peridotite and cosmic abundances, and the amount of CaO and Al_2O_3 in chondrites is not adequate to balance the Mg/Si if observed peridotites are the complement to basalt. The conclusion is that there is abundant basalt and orthopyroxene buried at depth in the mantle. One way to separate orthopyroxene from olivine is at the base of a deep (>300 km) magma ocean, where majorite, the liquidus phase, is denser than olivine.

A globe-encircling magma ocean has interesting implications regarding the early evolution of the mantle. Such an ocean probably existed throughout most of the accretional phase of planetary formation and may even have been regenerated by large late impacts. The rate of freezing of such an ocean depends on the temperature of the early atmosphere, which may have been quite dense and hot, and the rate at which heat is conducted through the base of the ocean. The temperatures in the ocean will be nearly adiabatic. The melting curve is much steeper. The ocean will therefore experience pressure freezing at its base and will crystallize upwards. The earliest crystals are olivine, orthopyroxene-majorite or orthopyroxene-perovskite, depending on depth. The composition of the residual fluid therefore also changes with depth. Deep liquids below 400 km will be MgO-rich, while shallower liquids will be SiO_2-rich. In general, these crystals are denser than the residual melts except for a possible density crossover between olivine and melt at depths between some 200 and 300 km. The liquids will also be enriched in CaO, Al_2O_3 and the incompatible elements. The processing of infalling material by a magma ocean is an efficient mechanism for separation of elements on the basis of the melting temperatures and densities of their host materials. A growing Earth with a deep magma ocean will result in an olivine-rich upper mantle and a depleted, infertile, orthopyroxene-rich lower mantle. Continued freezing will deposit clinopyroxene and garnet at depth and feldspar in the upper layers. The lighter SiO_2-rich and alkali-rich melts may even form a separate liquid layer at the surface, which eventually freezes to form the protocrust. Such fluids would form by the crystallization of an olivine cumulate layer and, if lighter than the bulk of the magma ocean, will rise to the surface.

The end result is a chemically stratified mantle with the incompatible elements strongly enriched toward the surface and the olivine-rich melts and crystals concentrated in the upper mantle. The deeper mantle would be enriched in orthopyroxene.

The chemical stratification of the magma ocean and the sorting of crystals and melts results in a stratified mantle that cools much more slowly than a homogeneous fluid. Deeply buried melt and partial melt layers will cool and crystallize extremely slowly. Convection cools a planet rapidly only if all the material in the interior can be brought to the near surface to lose its heat or if surface cooled material is circulated back into the interior. This can only happen in a homogeneous material or in a stratified system if the density contrasts between strata can be overcome by temperature or phase changes. Otherwise, cooling at depth is slow, even in convecting regions, because the convected heat must be removed by conduction through each thermal boundary layer. There is the interesting possibility that truly primitive melts may still exist somewhere in the deep interior.

SUMMARY OF CONSTRAINTS ON MANTLE EVOLUTION

Many models of petrogenesis assume that large parts of the mantle have never experienced differentiation. A primitive, or undepleted or pyrolite lower mantle is the cornerstone of many models. The alternative point of view I have developed in this chapter is that the Earth, including the mantle, is extensively differentiated, the result primarily of high temperatures early in Earth history. The following observations are particularly relevant and have been adopted as constraints on mantle evolution models:

1. A large part of the upper mantle is close to or above the solidus as evidenced by the distribution of oceanic ridges, hotspots, and island arcs and seismic velocities.
2. The upper mantle has cooled appreciably in the past 4.5 Ga, indicating that extensive partial melting was even more important in early Earth history.
3. The most voluminous magma type is tholeiite, which probably represents large (>15–20 percent) degrees of partial melting; picrites and komatiites represent even larger degrees of melting.
4. Although enriched magmas such as nephelinites and alkali-olivine basalts are considerably less abundant than oceanic and continental tholeiites, they often occur in close proximity to tholeiites in continental interiors, oceanic islands, and along ridge crests. Magmas exhibiting extreme LIL-enrichment, implying small degrees of partial melting, are rare now and may have been rarer in the Precambrian.
5. Inhomogeneities in the mantle having density contrasts of greater than about 1 percent and dimensions greater than about 10 km have a lifetime much less than the convective overturn of the mantle or the age of the mantle reservoirs.
6. Bodies much smaller than the Earth have apparently experienced extensive partial melting and differentiation early in their history.
7. Since the melting gradient increases faster with depth than the adiabatic gradient, melting in the mantle must begin in thermal boundary layers, found near chemical interfaces in the mantle.
8. The intrinsic density differences between the various products of mantle differentiation, such as basalts, depleted peridotite, and eclogites, are much greater than density variations associated with thermal expansion for reasonable variations in temperature.
9. Komatiites imply that high temperatures existed in the Archean mantle and that deep, or extensive, melting was common.
10. The isotopic compositions of MORB and the atmosphere indicate that the Earth was efficiently (but not 100 percent) outgassed early in its history and that subsequent outgassing (such as ^{40}Ar) has also been relatively efficient.

A mantle evolution model involving early and extensive partial melting, cooling, crystal fractionation and gravitational stratification seems to have the potential of satisfying a variety of geochemical data. An eclogite-rich cumulate layer at depth is the terrestrial equivalent of the lunar anorthosite layer. High-temperature magmas and magmas representing relatively large degrees of partial melting are currently being delivered to the Earth's surface. Convection calculations indicate that the mantle has cooled appreciably in the past 3 Ga. Extensive melting and crystal fractionation were therefore probably important processes in early Earth history. Isotopic data show that differentiation, or separation, rather than homogenization has been the rule throughout Earth history.

References

Allègre, C. J. (1982) Chemical geodynamics, *Tectonophysics, 81,* 109–132.

Anderson, D. L. (1982a) Chemical composition and evolution of the mantle. In *High-Pressure Research in Geophysics* (S. Akimoto and M. Manghnani, eds.), 301–318, D. Reidel, Dordrecht.

Anderson, D. L. (1982b) Isotopic evolution of the mantle: The role of magma mixing, *Earth Planet. Sci. Lett., 57,* 1–12.

Anderson, D. L. (1982c) Isotopic evolution of the mantle: A model, *Earth Planet. Sci. Lett., 57,* 13–24.

Anderson, D. L. (1985) Hotspot magmas can form by fractionation and contamination of MORB, *Nature, 318,* 145–149.

Anderson, D. L. and J. Bass (1986) Transition region of the Earth's upper mantle, *Nature, 320,* 321–328.

DePaolo, D. J. and G. J. Wasserburg (1979) Neodymium isotopes in flood basalts from the Siberian Platform and inferences about their mantle sources, *Proc. Natl. Acad. Sci., 76,* 3056.

Elthon, D. (1979) High magnesia liquids as the parental magma for ocean ridge basalts, *Nature, 278,* 514–518.

Frey, F. A., D. H. Green and S. D. Roy (1978) Integrated models of basalts petrogenesis: A study of quartz tholeiites to olivine melilitites from southeastern Australia utilizing geochemical and experimental petrological data, *J. Petrol., 19,* 463–513.

Ganapathy, R. and E. Anders (1974) Bulk compositions of the Moon and Earth estimated from meteorites, *Proc. Lunar Sci. Conf., 5,* 1181–1206.

Green, D. H., W. Hibberson and A. L. Jaques (1979) Petrogenesis of mid-ocean ridge basalts. In *The Earth: Its Origin, Structure and Evolution* (M. W. McElhinny, ed.), 265–295, Academic Press, New York.

Nataf, H.-C., I. Nakanishi and D. L. Anderson (1984) Anisotropy and shear-velocity heterogeneities in the upper mantle, *Geophys. Res. Lett., 11,* 109–112.

Nataf, H.-C., I. Nakanishi and D. L. Anderson (1986) Measurements of mantle wave velocities and inversion for lateral heterogeneities and anisotropy, Part III. Inversion, *J. Geophys. Res., 91,* 7261–7307.

O'Hara, M. J. (1968) The bearing of phase equilibria studies in synthetic and natural systems on the origin and evolution of basic and ultrabasic rocks, *Earth Sci. Rev., 4,* 69–133.

O'Hara, M. J., M. J. Saunders and E. L. P. Mercy (1975) Garnet-peridotite, primary ultrabasic magma and eclogite; interpretation of upper mantle processes in kimberlite, *Phys. Chem. Earth, 9,* 571–604.

O'Hara, M. J. and H. S. Yoder (1967) Formation and fractionation of basic magmas at high pressures, *Scot. J. Geol., 3,* 67–117.

Ringwood, A. E. (1975) *Composition and Petrology of the Earth's Mantle,* McGraw-Hill, New York, 618 pp.

Turcotte, D. L. (1980) On the thermal evolution of the Earth, *Earth Planet. Sci. Lett., 48,* 53–58.

12

The Shape of the Earth, Heat Flow and Convection

When Galileo let his balls run down an inclined plane with a gravity which he had chosen himself. . . then a light dawned upon all natural philosophers.

—I. KANT

We now come to the more global properties of the Earth. The shape of the Earth and its heat flow are both manifestations of convection in the interior and conductive cooling of the outer layers. The style of convection, however, is unknown. There are various hypotheses in this field that parallel those in petrology and geochemistry. The end-members are whole-mantle convection in a chemically uniform mantle and layered convection with little or no interchange of material between layers. Layered schemes have several variants involving a primitive lower mantle or a depleted lower mantle. In a convecting Earth we lose all of our reference systems, including the axis of rotation.

TOPOGRAPHY

The topography of the Earth's surface is now generally well known, though some areas in Tibet, central Africa and the southern oceans remain poorly surveyed. The distribution of elevations is distinctly bimodal, with a peak near 0.1 km representing the mean elevation of continents and a peak near -4.7 km corresponding to the mean depth of the oceans. This bimodal character contrasts with that of the other terrestrial planets. The spherical harmonic spectrum of the Earth's topography shows a strong peak for $n = 1$, corresponding to the distribution of most continents in one hemisphere, and a regular decrease with increasing n. The topography spectrum is similar to that of the other terrestrial planets. There are small peaks in the spectrum at $n = 3$ and $n = 9$–10, the latter apparently corresponding to the distribution of large oceanic islands and hotspots.

In general, the most recent orogenic belts such as the Alpine and Himalaya are associated with high relief, up to 5 km, while older orogenic belts such as the Appalachian and Caledonian, because of erosion, are associated with low relief, less than 1 km. Regional changes in the topography of the continents are generally accompanied by changes in mean crustal thickness. Continents stand high because of thick, low density crust, compared to oceans.

The long-wavelength topography of the ocean floor exhibits a simple relationship to crustal age. The systematic increase in the depth of the ocean floor away from the mid-ocean ridges can be explained by simple thermal models for the evolution of the oceanic lithosphere. Parsons and Sclater (1977) using data from the western North Atlantic and central Pacific Oceans showed that, for seafloor ages from 0 to 70 Ma, topography is described by

$$d(t) = 2500 + 350t^{1/2}$$

where t is age in Ma and $d(t)$ is the depth in meters. Older seafloor does not follow this simple relationship, being shallower than predicted. There are large portions of the ocean floor whose depth cannot be explained by simple thermal models; these include oceanic islands, hotspot swells, aseismic ridges and oceanic plateaus as well as other areas where the effects of surface tectonics and crustal structure are not readily apparent. Simple cooling models assume that the underlying mantle is uniform and that all of the variation in bathymetry is due to cooling of a thermal boundary layer. The North Atlantic is generally too shallow for its age, and the Indian Ocean between Australia and Antarctica is too deep. The residual depth anomaly is the departure of the depth of the ocean from the value expected for its age and is given by

$$\Delta d = d(t) + \frac{S(\rho_s - \rho_w)}{(\rho_m - \rho_s)} - d$$

where $d(t)$ is the expected depth based on age, S is the sediment thickness, ρ_s, ρ_w and ρ_m are the densities of sediments, water and mantle, respectively, and d is the observed depth. Residual depth anomalies observed in the ocean have dimensions of order 2000 km and amplitudes greater than 1 km. Part of the residual anomalies are due to regional changes in crustal thickness. This cannot explain all of the anomalies. Positive depth anomalies are generally associated with volcanic regions such as Bermuda, Hawaii, the Azores and the Cape Verde Islands. These might be due to thinning of the lithosphere or the presence of abnormally hot upper mantle. Hotspot swells generally have a higher heat flow than appropriate for the age of the surrounding oceanic crust.

Menard and Dorman (1977) noted that ridge-crest depths, a measure of the depth of the ocean floor at zero age, were greatest in the equatorial regions. They expanded the depths in the Pacific Ocean as a function of age and latitude:

$$d(t) = Kt^{1/2} + \sum_{n=0} C_n P_n (\sin \phi)$$

and obtained

$$K = 0.313 \text{ km (Ma)}^{-1/2}$$

$$C_0 = 2.64 \text{ km}$$

$$C_1 = 0.09 \text{ km}$$

$$C_2 = -0.26 \text{ km}$$

$$C_3 = -0.253 \text{ km}$$

Thus, the mean depth of the ocean at zero age is 2.64 km, and it deepens at an average rate of 0.313 km/(million years)$^{1/2}$. In addition to these age- and latitude-dependent effects, they found other depth anomalies having wavelengths of 4000 km and amplitudes of +0.8 to −0.5 km and a crest-to-trough relief of 0.7 to 1.0 km. The anomalies under the Pacific plate trend northwest-southeast, subparallel to the Hawaiian hotspot track, and those under the Nazca plate trend east-west. A large fraction of recent hotspots are associated with shallow anomalies. The Nazca plate is abnormally shallow.

Although cooling of the oceanic upper mantle is the first-order control of oceanic bathymetry, many large bathymetric features are not related to standard cooling and subsidence. Crough (1978, 1979) summarized these "depth anomalies" and, in particular, oceanic hotspot swells.

A few places are markedly deep, notably the seafloor between Australia and Antarctica and the Argentine Basin of the South Atlantic. Other deep regions occur in the central Atlantic and the eastern Pacific and others, most notably south of India, are not so obvious because of deep sedimentary fill. Most of the negative areas are less than 400 m below the expected depth, and they comprise a relatively small fraction of the seafloor area.

Shallow areas often exceed 1200 m in height and occupy almost the entire North Atlantic and most of the western Pacific that has been mapped. Almost every volcanic island, seamount or seamount chain surmounts a broad topographic swell. The swells generally occur directly beneath the volcanic centers and extend away from them in the downstream direction of plate-hotspot motion. Some extend a short distance in the upstream direction. Small regions of anomalously shallow depth occur in the northwestern Indian Ocean south of Pakistan, in the western North Atlantic near the Caribbean, in the Labrador Sea and in the southernmost South Pacific. They are not associated with volcanism but are slow regions of the upper mantle as determined from seismic tomography. Shallow regions probably associated with plate flexure border the Kurile Trench, the Aleutian Trench and the Chile Trench. Major volcanic lineaments without swells include the northern end of the Emperor Seamount chain, the Cobb Seamounts off the west coast of North America and the Easter Island trace on the East Pacific Rise. Bermuda and Vema, in the southeast Atlantic, are isolated swells with no associated volcanic trace. For most of the swells explanations based on sediment or crustal thickness and plate flexure can be ruled out. They seem instead to be due to variations in lithospheric composition or thickness, or abnormal upper mantle. Underplating the lithosphere by basalt or depleted peridotite, serpentinization of the lithosphere, delamination, or reheating and thinning the lithosphere are mechanisms that can decrease the density or thickness of the lithosphere and cause uplift of the seafloor. A higher temperature asthenosphere, greater amounts of partial melt, chemical inhomogeneity of the asthenosphere and upwelling of the asthenosphere are possible sublithospheric mechanisms. It would seem that the presence of a hotspot requires anomalous upper mantle, and it is likely that both lithospheric and asthenospheric properties contribute to the swell. Surface-wave tomography, although presently having low resolving capability, suggests that many swells do not have abnormally thin lithosphere but are associated with slower than average upper-mantle shear-wave velocities.

In general, smaller swells are located on younger seafloor, and the larger ones (Cape Verde, Hawaii, Great Meteor, Bermuda, Reunion) are on older crust. Within the scatter of the data, the midplate swells reach an approximately uniform depth below sea level, approximately 4250 m, or the swell height is proportional to the square root of crustal age (Crough, 1978, 1979; Menard and McNutt, 1982). Swells on ridge crests have heights that are inversely proportional to the local spreading rate; Iceland, the largest swell, is on a slow-opening ridge and Easter Island, one of the smallest, is on a fast-spreading ridge. Vogt (1975) showed that this is consistent with relatively shallow flow along the ridge axis. The widths of the subridge asthenospheric channel is proportional to the local spreading rate, so that flow along a fast-spreading ridge encounters less viscous resistance than flow along a slower spreading ridge. The pressure gradient necessary to drive the lateral flow manifests itself as a topographic gradient. Slow-spreading ridges require a large pressure gradient, and they therefore support high swells. The persistence of swells after they leave the ridge crest implies an anomalous lithosphere or additional shallow flow in the spreading direction. An alternative explanation for the relation between swell height and spreading rate is constant flux of material from the hotspot.

The great highlands of Africa including the Ethiopian and East African plateaus and the Hoggar and Tibesti massifs, the northern Rocky Mountains around Yellowstone and the Brazilian highlands are examples of possible hotspot-related continental swells. In shape and size these swells are similar to their oceanic counterparts. Western North America and northeast Africa are also associated with slow upper-mantle seismic anomalies.

The departure of the bathymetry-age relationship from a simple cooling law for Cretaceous lithosphere may reflect the extensive igneous activity that was occurring during this period. Many of the seamounts and plateaus in the western Pacific were formed in the Cretaceous.

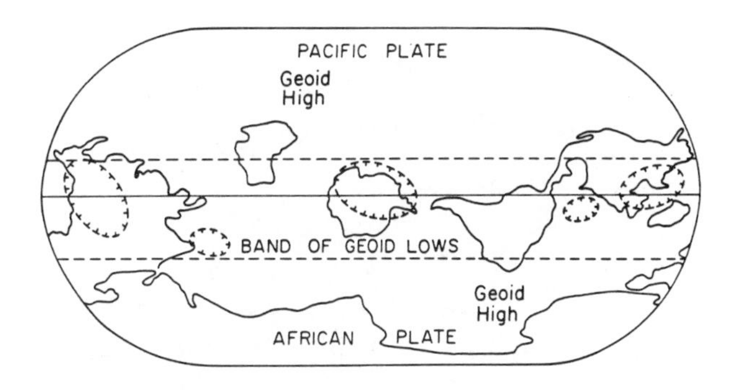

FIGURE 12-1
Geoid lows are concentrated in a narrow polar band passing through Antarctica, the Canadian Shield and Siberia. Most of the continents and smaller tectonic plates are in this band. Long-wavelength geoid highs and the larger plates (Africa, Pacific) are antipodal and are centered on the equator. The geoid highs control the location of the axis of rotation.

THE GEOID

Although the Earth is not flat or egg-shaped, as previously believed at various times, neither is it precisely a sphere or even an ellipsoid of revolution. Although mountains, ocean basins and variations in crustal thickness contribute to the observed irregular shape and gravity field of the Earth, they cannot explain the long-wavelength departures from a hydrostatic figure.

The centrifugal effect of the Earth's rotation causes an equatorial bulge, the principal departure of the Earth's surface from a spherical shape. If the Earth were covered by oceans then, apart from winds and internal currents, the surface would reflect the forces due to rotation and the gravitational attraction of external bodies, such as the Sun and the Moon, and effects arising from the interior. When tidal effects are removed, the shape of the surface is due to density variations in the interior. Mean sea level is an equipotential surface called the *geoid* or *figure* of the Earth. Crustal features, continents, mountain ranges and mid-oceanic ridges represent departures of the actual surface from the geoid, but mass compensation at depth, *isostasy,* minimizes the influence of surface features on the geoid. To first order, near-surface mass anomalies that are compensated at shallow depth have no effect on the geoid.

The shape of the geoid is now known fairly well, particularly in oceanic regions, because of the contributions from satellite geodesy. Apart from the geoid highs associated with subduction zones, there is little correlation of the long-wavelength geoid with such features as continents and midocean ridges. The geoid reflects temperature and density variations in the interior, but these are not simply related to the surface expressions of plate tectonics.

The largest departures of the geoid from a radially symmetric rotating spheroid are the equatorial and antipodal geoid highs centered on the central Pacific and Africa (Figure 12-1). The complementary pattern of geoid lows lie in a polar band that contains most of the large shield regions of the world. The largest geoid highs of intermediate scale are associated with subduction zones. The most notable geoid high is centered on the subduction zones of the southwest Pacific near New Guinea, again near the equator. The equatorial location of geoid highs is not accidental; mass anomalies in the mantle control the moments of inertia of the Earth and, therefore, the location of the spin axis and the equator. The largest intermediate-wavelength geoid lows are found south of India, near Antarctica (south of New Zealand) and south of Australia. The locations of the mass anomalies responsible for these lows are probably in the lower mantle. Many shield areas are in or near geoid lows, some of which are the result of deglaciation and incomplete rebound. The thick continental crust would, by itself, raise the center of gravity of continents relative to oceans and cause slight geoid highs. The thick lithosphere ($\sim$150 km) under continental shields is cold, but the seismic velocities and xenoliths from kimberlite pipes suggest that it is olivine-rich and garnet-poor; the temperature and petrology have compensating effects on density. The long-term stability of shields indicates that, on average, the crust plus its underlying lithosphere is buoyant. Midocean ridges show mild intermediate-wavelength geoid highs, but they occur on the edges of long-wavelength highs. Hotspots, too, are associated with geoid highs. The long-wavelength features of the geoid are probably due to density variations in the lower mantle and the resulting deformations of the core-

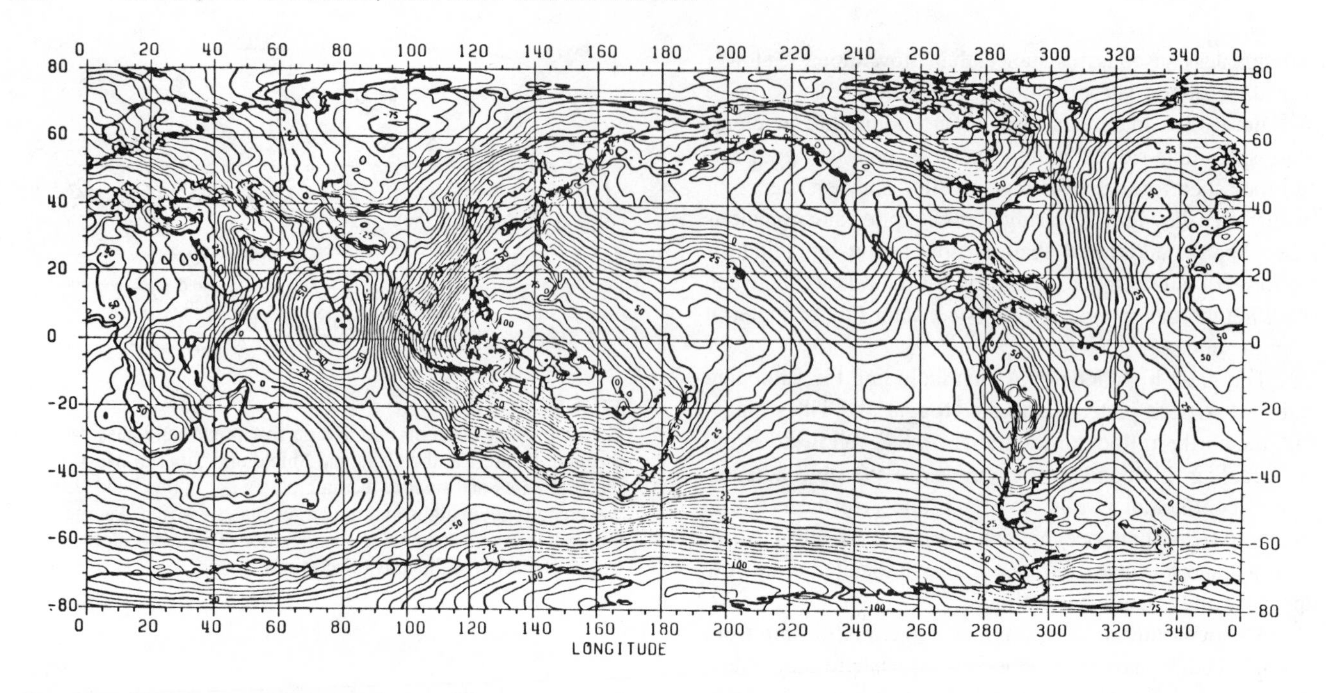

FIGURE 12-2
Geoid undulations (to degree 180) referred to a hydrostatic flattening of $1/299.638$. Contour interval is 5m (after Rapp, 1981).

mantle boundary and other boundaries in the mantle (Richards and Hager, 1984).

Geoid anomalies are expressed as the difference in elevation between the measured geoid and some reference shape. The reference shape is usually either a spheroid with the observed flattening or the theoretical hydrostatic flattening associated with the Earth's rotation. The latter, used in Figure 12-2, is the appropriate geoid for geophysical purposes and is known as the nonhydrostatic geoid. By referencing the shape of the Earth's equipotential to the observed flattening by subtracting the C_{20} terms from the observed shape, one is throwing away some potentially important geophysical data. The observed C_{20} is not in fact the same as one obtains for an Earth in hydrostatic equilibrium, and this fact is telling us about the mass distribution in the mantle. By convention the origin is taken as the center of mass, so there is no C_{10} term. The geometric flattening of the Earth is $1/298.26$. The hydrostatic flattening is $1/299.64$.

The C_{ij} terms are spherical harmonic coefficients. C_{m0} are zonal harmonics, with boundaries between positive and negative features being small circles parallel to the equatorial plane. For a rotating homogeneous sphere the C_{20} term is the major term in a spherical harmonic expansion of the shape. This is sometimes referred to as the equatorial bulge. The C_{m1} are sectoral harmonics, and the general case C_{ml}, $0 < m < l$, are known as tesseral harmonics. If the origin of the coordinate system is not at the center of the Earth, then there will be a C_{10} term. North-south-running bands of high or low geoid anomalies will show up as strong sectoral components in a spherical harmonic expansion.

The maximum geoid anomalies are of the order of 100 m. This can be compared with the 21-km difference between the equatorial and polar radii. To a good approximation the net mass of all columns of the crust and mantle are equal when averaged over dimensions of a few hundred kilometers. This is one definition of isostasy. Smaller-scale anomalies can be supported by the strength of the crust and lithosphere. The geoid anomaly is nonzero in such cases and depends on the distribution of mass. It depends on the dipole moment of the density distribution

$$\int_0^h \Delta\rho(z) z \, dz$$

where z is the radial direction, h is the depth of the anomalous density distribution and $\Delta\rho(z)$ is the density anomaly. Clearly, if the depth extent of the anomalous mass is shallow, the geoid anomaly is small. This is why continents do not show up well in the geoid.

The geoid anomaly due to a long-wavelength isostatic density distribution is

$$N = \frac{-2\pi G}{g} \int_0^h z\Delta\rho(z) \, dz$$

The dipole moment $\Delta\rho(z)$ is nonzero, and the first moment of the density anomaly, that is the net mass anomaly, is zero for isostatic density distributions.

The elevation anomaly, Δh, associated with the density anomaly is

$$\int_0^h \Delta\rho(z)\, dz = 0$$

or

$$\frac{\Delta h}{h} = -\frac{\Delta\rho}{\rho}$$

A negative $\Delta\rho$, caused for example by thermal expansion, will cause the elevation of the surface to increase (Δh = positive) and gives a positive geoid anomaly because the center of mass is closer to the Earth's surface. The mass deficiency of the anomalous material is more than canceled out by the excess elevation.

All major subduction zones are characterized either by geoid highs (Tonga and Java through Japan, Central and South America) or by local maxima (Kuriles through Aleutians). The long-wavelength part of the geoid is about that expected for the excess mass of the cold slab. The shorter-wavelength geoid anomalies, however, are less, indicating that the excess mass is not simply rigidly supported. Hager (1984) showed that there is an excellent correlation between the $l = 4$–9 geoid and the theoretical slab geoid and showed that this could be explained if the viscosity of the mantle increased with depth by about a factor of 30. The high viscosity of the mantle at the lower end of the slab partially supports the excess load. A chemical boundary near 650 km depth could also support the slab. The thick crust of island arcs and the high temperatures and partial melting of the "mantle wedge" above the slab may also contribute to the geoid highs at subduction zones. The deep trenches represent a mass deficiency, and this effect alone would give a geoid low. Other geoid highs, unrelated to slabs, appear to be associated with hotter than normal mantle, such as hotspot volcanism and low seismic velocities in the upper mantle.

The ocean floor in back-arc basins is often lower than equivalent age normal ocean, suggesting that the mass excess associated with the slab is pulling down the surface. This is not the only possibility. A thinner-than-average crust or a colder or denser shallow mantle could also depress the seafloor.

Cooling and thermal contraction of the oceanic lithosphere causes a depression of the seafloor with age and a decrease in the geoid height. Cooling of the lithosphere causes the geoid height to decrease uniformly with increasing age, symmetrically away from the ridge crest. The change is typically 5–10 m over distances of 1000 to 2000 km. The elevation and geoid offset across fracture zones is due to the age differences of the crust and lithosphere. Assuming a plate model with a fixed thickness and lower boundary temperature, a plate thickness of 66 km has been estimated for plates less than 30 Ma in age and 92 km for older plates (Cazenave, 1984; Cazenave and others, 1986). This implies a broad upwarping of geotherms in the younger parts of the Pacific Ocean.

The variation of geoid height with age is

$$\Delta h/\Delta t \approx 2\pi G/g\rho_m \alpha kT$$

for a cooling half-space model and at young ages in the plate model (where G is gravitational constant, g is surface gravity, ρ_m is upper-mantle density, Δh is change in geoid height, Δt is difference in age and T is boundary temperature). Observations are more consistent with a plate model in which lithospheric evolution slows down with time and plate buoyancy approaches an asymptotic value. However, it is also likely that lateral heterogeneity in the mantle beneath the plates, and extensive volcanism at certain periods of time, notably the Cretaceous, affect the geoid-age relation.

The long-wavelength topographic highs in the oceans generally correlate with positive geoid anomalies, giving 6–8 meters of geoid per kilometer of relief (Cazenave, 1984; Cazenave and others, 1986).

Tanimoto and Anderson (1985) showed that there was good correlation between intermediate-wavelength geoid anomalies and seismic velocities in the upper mantle; slow regions were geoid highs and vice versa. Subduction zones are slow in the shallow mantle, presumably due to the hot, partially molten mantle wedge under back-arc basins.

In the subduction regions the total geoid anomaly is the sum of the positive effect of the dense sinker and the negative effects caused by boundary deformations (Hager, 1984). For a layer of uniform viscosity, the net dynamic geoid anomaly caused by a dense sinker is negative; the effects from the deformed boundaries overwhelm the effect from the sinker itself. For an increase is viscosity with depth, the deformation of the upper boundary is less and the net geoid anomaly is positive.

For a given density contrast, the magnitude and sign of the resulting geoid anomaly in a dynamic Earth depends on the viscosity structure and the chemical stratification. Observation of the gravitational field of the Earth thus provides a null experiment, where the net result is a small number determined by the difference of large effects (Richards and Hager, 1984). The sign of the result depends on which of the effects is dominant. The anomaly also depends on the depth of the convecting system, with deep systems leading to larger geoid anomalies for a given density anomaly. Observations of the geoid in conjunction with observations of seismic velocity heterogeneities place constraints upon the variation of mantle viscosity and the depth of mantle convection.

Interpreting the Geoid

The geoid bears little relation to present tectonic features of the Earth other than trenches. The Mesozoic supercontinent of Pangaea, however, apparently occupied a central position in the Atlantic-African geoid high (Anderson, 1982). This and the equatorial Pacific geoid high contain most of the world's hotspots. The plateaus and rises in the western Pacific formed in the Pacific geoid high, and this may have been the early Mesozoic position of Pacifica, the fragments of which are now the Pacific rim portions of the continents. Geoid highs that are unrelated to present subduction zones may be the former sites of continental aggregations and mantle insulation and, therefore, hotter than normal mantle. The pent-up heat causes rifts and hotspots and results in extensive uplift, magmatism, fragmentation and dispersal of the continents, and the subsequent formation of plateaus, aseismic ridges and seamount chains. Convection in the upper mantle would then be due to lateral temperature gradients as well as heating from below and would be intrinsically episodic.

The Earth's largest positive geoid anomalies are associated with subduction zones and hotspots and have no simple relationship to other elevated regions such as continents and ridges. When the subduction-related geoid highs are removed from the observed field, the residual geoid shows broad highs over the central Pacific and the eastern Atlantic-African regions (see Figures 8-5 and 8-11). Like the total geoid, the residual geoid does not reflect the distribution of continents and oceans and shows little trace of the ocean ridge system. Residual geoid highs, however, correlate with regions of anomalously shallow ocean floor and sites of extensive Cretaceous volcanism.

The Atlantic-African geoid high extends from Iceland through the north Atlantic and Africa to the Kerguelen plateau and from the middle of the Atlantic to the Arabian Peninsula and western Europe (Figure 8-5). Most of the Atlantic, Indian Ocean, African and European hotspots are inside this anomaly. The hotspots Iceland, Trindade, Tristan, Kerguelen, Reunion, Afar, Eiffel and Jan Mayen form the 20-m boundary of the anomaly and appear to control its shape. The Azores, Canaries, New England seamounts, St. Helena, Crozet and the African hotspots are interior to the anomaly.

Although the geoid high cuts across present-day ridges and continents, there is a remarkable correspondence of the predrift assemblage of continents with both the geoid anomaly and hotspots. Reconstruction of the mid-Mesozoic configuration of the continents reveals, in addition, that virtually all of the large shield areas of the world are contained inside the geoid high (Figure 8-5). These include the shield areas of Canada, Greenland, Fennoscandia, India, Africa, Antarctica and Brazil. Most of the Phanerozoic platforms are also in this area. In contrast, today's shields and platforms are concentrated near geoid lows. They may have drifted into, and come to rest over, these geoid lows.

The area inside the geoid high is also characterized by higher-than-normal elevations, for example in Africa, the North Atlantic and the Indian Ocean southeast of Africa. This holds true also for the axial depth of oceanic ridges. Most of the continental areas were above sea level from the Carboniferous and Permian through the Triassic, at which time there was subsidence in eastern North and South America, central and southern Africa, Europe and Arabia. The widespread uplift, magmatism, breakup and initial dispersal of the Pangaean landmass apparently occurred while the continents were centrally located with respect to the present geoid anomaly. The subsequent motions of the plates, by and large, were and are directed away from the anomaly. This suggests that the residual geoid high, hotspots, the distribution of continents during the late Paleozoic and early Mesozoic, and their uplift and subsequent dispersal and subsidence are all related. The shields are regions of abnormally thick lithosphere. The thickest lithosphere is in eastern and central North America, northeastern South America, northwestern and central Africa and northern Siberia. These regions were all within the Atlantic-African geoid high at 200 Ma and, possibly, at 350 Ma as well, assuming the geoid high was present at those times and fixed relative to hotspots.

This area experienced exceptional magmatism during the Mesozoic. The great flood basalts of Siberia and South Africa were formed during the Triassic and Jurassic, possibly at the sites of the Jan Mayen and Crozet hotspots. The plateau basalt provinces of southeast Greenland and Brazil were formed during the Cretaceous, possibly at the sites of the Iceland and Tristan or Trindade hotspots. The Deccan Traps in India were also formed in the Cretaceous, presumably at the Reunion hotspot. The Walvis Ridge and the Rio Grande Rise are mainly on Mesozoic crust. A large part of the Pacific also experienced extensive on- and off-ridge volcanism in the Cretaceous. This extensive ridge and hotspot volcanism apparently reflects itself in a rapid rise in sea level during the Jurassic and Cretaceous. If sea level can be used as a guide to the volume of the ocean basins, then the end of the Cretaceous to the end of the Oligocene was a period of less intense oceanic volcanism and subsidence of the oceanic and continental crust. Sea-level variations may therefore indicate that the thermal and geoid anomalies formed in the Paleozoic have attenuated since the early Mesozoic.

Much of the Pacific rim appears to be accreted terrane that originated in the Pacific. The possibility of a continent centrally located in the Pacific—Pacifica—has been discussed for some time, but its location has been an enigma and its size uncertain. The central Pacific geoid anomaly may mark the early Mesozoic location of the anomalous circumPacific terrane and the site of extensive Cretaceous ridge-crest and midplate volcanism. It is underlain by abnormally low seismic velocities throughout most of the upper mantle (Nataf et al., 1984).

Paleomagnetic and other data indicate that various blocks of Asia such as Kolyma, Sikhote Alin, Sino-Korea, Yangtze, Southeast Asia and Japan have moved northward by up to 32° since the Permian. It is unlikely that they were in the vicinity of Australia or associated with Gondwana,

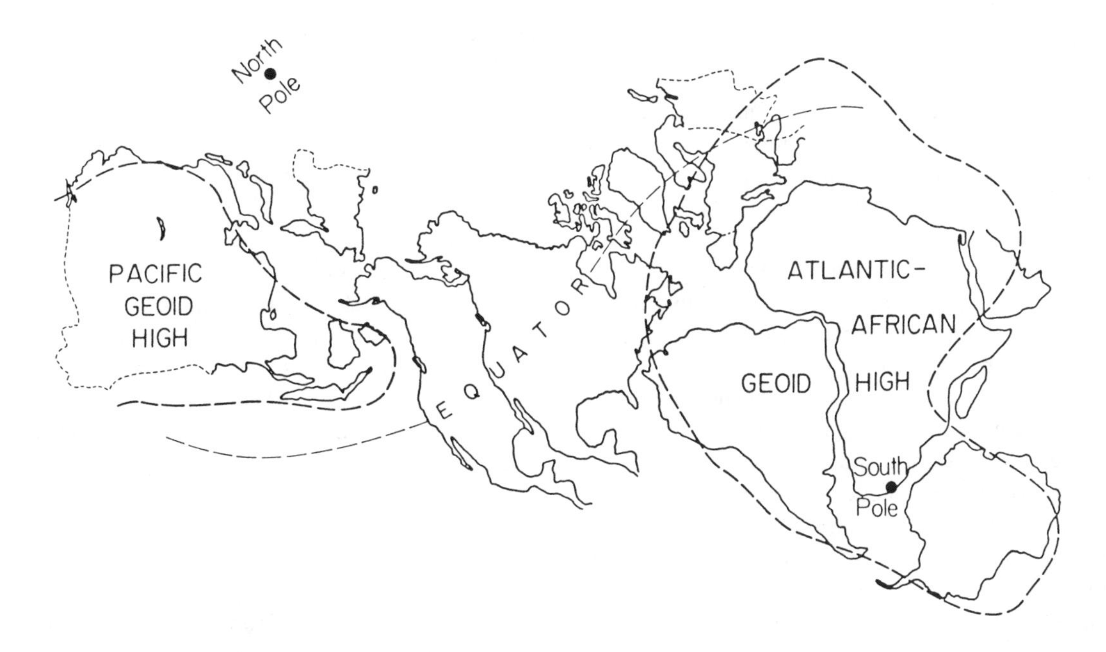

FIGURE 12-3
A possible configuration of the continents in the Triassic relative to present geoid highs. Both continental drift and polar wander are required to bring the continents to their present positions relative to the geoid and the rotation axis.

and a central Pacific location is likely. Part or all of Alaska and northwestern North America were also far south of their present positions in the Permian. The same may be true of California, Mexico, Central America and other accreted terranes in the Pacific rim continents.

The central Pacific residual (corrected for slabs) geoid high (> 20 m) extends from Australia to the East Pacific Rise and from Hawaii to New Zealand. Its western part overlaps the subduction zones of the southwest Pacific and may be eliminated by a different slab model. The actual residual geoid anomaly may therefore be more equidimensional and is perhaps centered over the Polynesian plume province in the central Pacific. In any case it encompasses most of the Pacific hotspots and is approximately antipodal to the Atlantic-African anomaly. Figure 12-3 shows present-day geoid anomalies superposed on a hypothetical Triassic assemblage of the continents.

The western Pacific contains numerous plateaus, ridges, rises and seamounts that have been carried far to the northwest from their point of origin (see Figure 8-11). The Ontong-Java plateau, for example, presently on the equator, was formed about 4000 km further south in the mid-Cretaceous. The Hess rise, Line Islands ridge and Necker ridge were formed in the Cretaceous on the Pacific, Farallon and Phoenix ridge crests that were, at the time, in the eastern Pacific in the vicinity of the Polynesian seamount province. The ridge-crest volcanism was accompanied by extensive deep-water volcanism and, possibly, rapid seafloor spreading. The Caribbean and Bering Seas may have been formed at the same location and subsequently carried northward on the Farallon and Kula plates. It is significant that the Caribbean and the anomalous regions in the Pacific have similar geophysical and geochemical characteristics.

The Pacific geoid high also has anomalously shallow bathymetry, at least in the eastern part. This shallow bathymetry extends from the East Pacific Rise to the northwest Pacific in the direction of plate motion, and includes the central and southwest Pacific hotspots. The extensive volcanism in the central western Pacific between about 120 and 70 Ma occurred about 60° to 100° to the southeast, in the hotspot frame, which would place the event in the vicinity of the southeastern Pacific hotspots and the eastern part of the geoid and bathymetry anomaly. This suggests that the anomaly dates back to at least the Early Cretaceous. A similar thermal event in the Caribbean may mean that it was also in this region in the Cretaceous, particularly since the basalts in the Caribbean are similar to those in the western Pacific. The flattening of oceanic bathymetry after about 80 Ma may be related to extensive volcanism ending at about that time rather than due simply to the passage of time. It should be noted that there is an extensive upper-mantle low-velocity anomaly in the central Pacific.

Going back even further are the Triassic basalts in Wrangellia of northwestern North America and the Permian greenstones in Japan, both of which formed in equatorial latitudes and subsequently drifted to the north and northwest respectively. They may record the initial rift stages of continental crust. They also may have formed in the Pacific geoid high.

Episodicity in continental drift, polar wander, sea-level variations and magmatism may be due to the effect of thick continental lithosphere on convection in the mantle. The supercontinent of Pangaea insulated a large part of the mantle for more than 150 Ma. The upper mantle under Pangaea could cool, by subduction, around the edges, but the center of the continent was shielded from subduction and could only lose heat by conduction through a thick lithosphere. The excess heat caused uplift by thermal expansion and partial melting and, eventually, breakup and dispersal of Pangaea. A large geoid high is generated by this expansion, and, if this is the dominant feature of the geoid, the spin axis of the Earth would change so as to center the high on the equator, much as it is today. Surface plates experience membrane stresses during this shift, which may also contribute to breakup. Active subduction zones are also geoid highs. The continents stand high while they are within the geoid high and subside as they drift off.

At 100 Ma Europe, North America and Africa were relatively high-standing continents. This was after breakup commenced in the North Atlantic but before significant dispersal from the pre-breakup position. North America suffered widespread submergence during the Late Cretaceous while Africa remained high. Europe started to subside at about 100 Ma. This is consistent with North America and Europe drifting away from the center of the geoid high while Africa remained near its center, as it does today.

The association of the Atlantic-Africa geoid high with the former position of the continents and the plateau basalt provinces with currently active hotspots suggests that Mesozoic convection patterns in the mantle still exist. The Mid-Atlantic and Atlantic-Indian Ridges and the Atlantic, African and Indian Ocean hotspots and the East African rift are in the Atlantic-Africa geoid high. If geoid anomalies and hotspots are due to a long period of continental insulation and if these regions had higher-than-normal heat flow and magmatism for the past 200 Ma, then we might expect that these features will wane with time.

Horizontal temperature gradients can drive continental drift. The velocities decrease as the distance increases away from the heat source and as the thermal anomaly decays. The geoid high should also decay for the same reasons. Thick continental lithosphere then insulates a new part of the mantle, and the cycle repeats. Periods of rapid polar motion and continental rift follow periods of continental stability and mantle insulation.

The relatively slow motion of the continents or the pole during the Permian and the relative stability of sea level during this period suggests that the geoid anomalies had essentially developed at this time. Continental drift velocities and sea level changed rapidly during the Triassic and Jurassic. This I interpret as the start of extensive rift and hotspot magmatism and the start of the decay of the anomaly.

The rotation axis has shifted with respect to the hotspot reference frame by about 20° in the last 200 Ma (Gordon and Cape, 1981; Morgan, 1981; Gordon, 1982; Harrison and Lindh, 1982; Andrews, 1987). This shift could represent a growth of the equatorial regions of the geoid highs, a decay of the South Pacific or North Atlantic portions of the high due to extensive ridge-axis and hotspot volcanism, or a reconfiguration of the world's subduction zones.

Chase (1979) concluded that the lack of correlation of the large geoid anomalies to plate boundaries requires that they reflect a deep convective flow regime in the mantle that is unrelated to plate tectonics. However, the correspondence of the Atlantic-African anomaly with the Mesozoic continental assemblage and of the central Pacific anomaly with extensive Cretaceous volcanism in the Pacific is remarkable. Surface-wave tomography shows a good correlation of intermediate-wavelength geoid highs ($l \approx 6$) and hot regions of the upper mantle. However, the long-wavelength components of the geoid ($l = 2,3$) correlate best with the lower mantle. Most of the present continents, except Africa, and most of the present subduction zones (except Tonga-Fiji) are in long-wavelength ($l = 2-4$) geoid lows and therefore overlie colder than average lower mantle.

With the scenario sketched above, the relationship between surface tectonic features and the geoid changes with time. Supercontinents periodically form and insulate the underlying mantle and also control the locations of mantle cooling (subduction zones). When the continent breaks up, the individual fragments move away from the hot part of the mantle, and the geoid high, and came to rest over cold mantle, in geoid lows.

The locations of hot and cold regions in the upper mantle may be controlled by convection in the lower mantle. The lower mantle contribution to the geoid is probably long-lived. Empirically, subduction zones and continents are primarily in long-wavelength geoid lows and over long-wavelength fast seismic regions of the lower mantle. This can be understood if continents come to rest in geoid lows and if subduction zones, on average, are controlled by the advancing edges of continents. Midocean ridges tend to fall between the long-wavelength highs and lows. By long wavelength, we mean features having dimensions of thousands of kilometers.

Figure 12-4 shows the approximate locations of the continents just after breakup of Pangaea commenced. The hatched regions show oceanic lithosphere that has been overridden by the advancing continents. These are labeled "fast" because these are seismically fast regions of the transition region, where cold lithosphere may have cooled off the mantle. The arrows represent the motions of the continents over the past 110 Ma. Most of the hatched regions are also geoid lows.

Involvement of the Lower Mantle

Body-wave tomographic techniques can be applied to the problem of lateral heterogeneity of the lower mantle. Tomographic results (Dziewonski, 1984; Clayton and Comer,

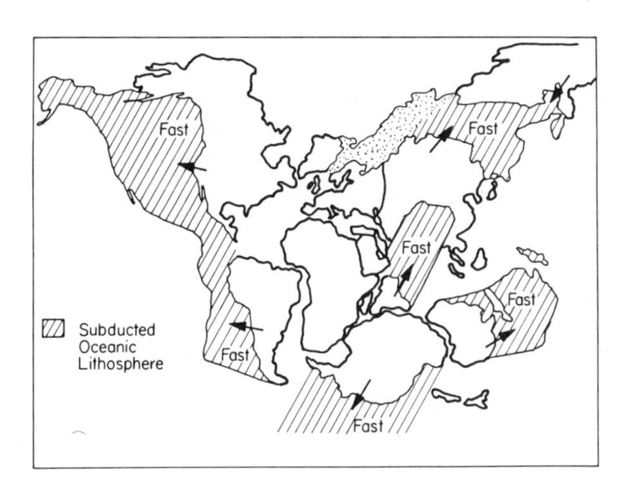

FIGURE 12-4
Reconstruction of the continents at about 110 Ma ago. The hatched areas represent former oceanic lithosphere. These regions, in general, have high seismic velocities in the transition region, consistent with the presence of cold subducted lithosphere. They are also, in general, geoid lows. Dots represent possible convergence areas.

1988) give long-wavelength velocity anomalies that correlate well with the $l = 2,3$ geoid (Hager and others, 1985). Since one expects density to correlate with velocity, there is a strong indication that the long-wavelength geoid originates in the lower mantle. Phenomena such as tides, Chandler wobble, polar wander and the orientation of the Earth's spin axis depend on the $l = 2$ component of the geoid, so the lower mantle appears to be important in these areas. By contrast the upper mantle has only a weak correlation with the $l = 2,3$ geoid and is the reverse of the correlation with shorter wavelengths $l = 4$–6.

Regions of high velocity, and presumably high density, in the lower mantle correlate with geoid lows—the same sense of correlation found for shorter-wavelength features in the upper mantle. This is the relation expected for an isostatic mantle, with equal mass in equal columns. The deformation of the density interfaces counteracts the effect of high density. Low-density regions ride high. The $l = 2,3$ correlation of the geoid with surface-wave or upper-mantle velocities is weak and in the opposite sense.

Since hotspots occur preferentially in the long-wavelength geoid highs, as defined by the $l = 2,3$ geoid, they also occur over regions of the lower mantle that have lower than average velocities, averaged over tens of thousands of kilometers. The correlation breaks down at shorter wavelengths. This suggests that, in general terms, the locations of hot regions of the upper mantle are controlled by the locations of hot regions in the lower mantle, but upper-mantle and plate processes impose their own shorter length scales. Subduction of cold slabs and continental insulation may also influence the style of convection in the lower mantle, even if no transfer of material takes place. Subduction confined to the upper mantle can influence lower-mantle convection by deforming the boundary, cooling the lower mantle from above or by shear coupling. A thermal boundary layer and a large contrast in viscosity would favor thermal coupling over shear coupling.

HEAT FLOW

The average heat flow through the surface of a planet is one of its fundamental properties. It is controlled by the radioactivity in the interior and the secular cooling of the planet, and it is a constraint on temperatures and styles of heat transport in the interior. Most of the heat transport in the deep interior is by convection, but heat convected toward the surface must ultimately be conducted through a cold conductive layer at the surface.

The terrestrial heat flow is defined as the quantity of heat escaping per unit time across each unit area of the Earth's solid surface. This quantity varies from place to place over the Earth's surface and with time throughout Earth history. The total heat being lost from the Earth at a given time is the integral of the heat flow over the Earth's surface. The heat arriving from the interior is a small fraction of the heat arriving from the Sun but the latter, of course, is not significant in controlling the internal temperature. The solar input, the surface albedo and atmospheric properties control the mean surface temperature of the Earth, which is about 0°C.

The heat flow through the crust and lithosphere is mainly by conduction and is governed by Fourier's Law,

$$q = -\mathcal{K} \text{ grad } T$$

where q is the heat flux vector, $\mathcal{K}$ is the thermal conductivity of the material through which the heat is being conducted and grad T is the local temperature gradient. Heat flow is usually measured in deep boreholes to get below the effects of seasonal and longer variations. One must measure both the thermal gradient and the rock conductivity. Temperatures in shallow holes are perturbed to varying degrees by ground-water circulation, climatic variations, irregularities of local topography, vegetation patterns, slope orientation and Sun angle. Even deep holes can be affected by local geological structures, erosion, tectonic uplift and conductivity variations that distort the isothermal surfaces.

The thermal diffusivity, κ, of a material is given by $\kappa = \mathcal{K}/\rho C_P$ where $\mathcal{K}$ is the thermal conductivity, ρ the density and C_P the specific heat. Rocks have relatively low conductivities so that they respond slowly to any change in ambient temperature. The conductive time constant is approximately

$$\tau = l^2/\kappa$$

where l is the characteristic length over which heat must be transferred. This relation shows that over the age of the Earth, conductive cooling can only extend to a depth of about 400 km. The long time constant means that as crustal rocks are buried or eroded, they tend to carry their thermal structure with them as they move. This is a form of convective transport of heat.

The present heat loss of the Earth is 10^{13} cal/cm²/s (4.2 × 10^{13} W). The heat loss through the surface is greatest near the midocean ridges and in tectonic regions and is least through continental Precambrian shields. Table 12-1 provides a summary of various components of the terrestrial heat-flow budget.

The mean heat flow through the ocean basins, based on a theoretical fit to oceanic heat-flow observations, decreases from about 8 μcal/cm²/s at the ridge crest, with a high variability, to a relatively constant value of about 1.2 μcal/cm²/s (50 mW/m²) for old ocean. The unit μcal/cm²/s or 10^{-6} cal/cm²/s is called a Heat Flow Unit or HFU.

On continents there is a rough correlation between heat flow and age, or age of last tectonic or magmatic event (Sclater and others, 1980, 1981). In young regions, heat flow has a high mean and large scatter. The heat flow in active or recently active areas ranges from about 0.5 to over 3 HFU. The mean heat flow decreases from 1.8 HFU (77 mW/m²) in the youngest provinces to a constant value of 1.1 HFU after about 800 Ma. A few continental heat-flow values are below 0.6 HFU, and strangely, most of these are in young areas. The scatter in older areas seems to be due to variations in crustal radioactivity.

It has been discovered that there is a simple linear relationship between the heat flow and surface radioactivity in specific areas, termed "heat-flow provinces":

$$q_o = q_r + DA_o$$

where q_o and A_o are the heat flow and the radioactivity at the surface, q_r is the "reduced" heat flow and D is a depth scale for the vertical distribution of radioactivity (Birch and others, 1968). This depth scale is probably different for potassium, thorium and uranium. The reduced heat flow, due to heat flow primarily from the lower crust and mantle (Sclater and others, 1980, 1981), ranges from about 0.5 to 0.67 HFU. (See Table 12-2.) This can be compared with the estimated equilibrium flux through the base of the oceanic crust, 0.9 HFU.

The steady-state temperature at moderate depth z can be estimated from

$$T(z) = T_o + q \int_0^z \frac{dz}{K}$$

The temperature gradient decreases with depth. If heat production due to radioactivity is significant over the temperature interval of interest, then this must be allowed for. Table 12-3 gives the heat productivity of various rock types and of the continental crust.

Velocity of Love waves shows a high degree of correlation with global heat-flow maps (Nakanishi and Anderson, 1984; Tanimoto and Anderson, 1984, Nataf and others, 1986). Love waves are sensitive to shear velocity in the shallow mantle. This correlation makes it possible to identify regions of possibly anomalous heat flow, which is useful since global heat-flow coverage is far from complete. Low Love-wave velocities and, therefore, regions of high upper-mantle temperature occur along the East Pacific Rise, western North America and Central America, Red Sea–East African Rift–Gulf of Aden, eastern Asia, southeast Asia, the Indian Ocean triple junction, the Tasman Sea, the south-central Pacific, the far North Atlantic, southern Europe and the Tristan da Cunha–Rio Grande Rise. Fast regions, and regions of presumed low heat flow are most con-

TABLE 12-1
Heat Loss of the Earth

Component	Value
Heat loss through continents	2.8 × 10^{12} cal/s
Heat loss through oceans	7.3 × 10^{12} cal/s
Total heat loss through surface	10^{13} cal/s
Heat loss by hydrothermal circulation	2.4 × 10^{12} cal/s
Heat loss by plate creation	6.3 × 10^{12} cal/s
Mean heat flow	2.0 × 10^{-6} cal/cm²/s
Continents	1.4 × 10^{-6} cal/cm²/s
Oceans	2.4 × 10^{-6} cal/cm²/s
Radioactive decay in crust	~17 percent of heat loss
Convective heat transport by surface plates	~65 percent of heat loss

Sclater and others (1981).

TABLE 12-2
Measured and Reduced Heat Flow

Province	Heat Flow (mW/m²) Measured	Reduced
Shields		
Superior	34 ± 1.0	22 ± 2.0
Brazil	56 ± 1.5	29 ± 3.5
Baltic	33 ± 1.4	22 ± 2.0
Ukraine	37 ± 0.4	24 ± 0.7
Western Australia	39 ± 1.2	30 ± 1.8
Mean	27 ± 4.0	
Other		
Basin and Range	77 ± 1.8	63 ± 5.3
Southeast Appalachians	49 ± 0.0	28 ± 0.1
Sierra Nevada	36 ± 0.7	15 ± 1.3
Eastern United States	64 ± 0.7	33 ± 1.3
England	69 ± 3.1	24 ± 5.4
India	71 ± 1.2	38 ± 4.4
Upper Crust	12 to 56	

Pujol and others (1985).

TABLE 12-3
Heat Production of Common Rocks

Rock type	Uranium (ppm)	Thorium (ppm)	Potassium (pct.)	Heat Production (10^{-6} W/m^3)
Granite	3.9	16.0	3.6	2.5
Granodiorite	2.3	9.0	2.6	1.5
Andesite	1.7	7.0	1.1	1.1
Basalt	0.5	1.6	0.4	0.3
Peridotite	0.02	0.06	0.006	0.01
Dunite	0.003	0.01	0.0009	0.002
Continental crust	1.25	4.8	1.25	0.8

Pollack (1980).

tinental shield areas, the North Pacific, south and northwest Africa, the eastern Indian Ocean and much of the Atlantic.

Figure 12-5 shows the low-order geoid and heat-flow maps.

ROTATION OF THE EARTH AND POLAR WANDER

The axis of rotation of the Earth is not fixed, either in space or relative to the body of the Earth, nor is the speed of rotation constant. The movement of the axis of rotation with respect to the crust is called polar wander or polar motion. Variation in the speed of rotation about the instantaneous axis causes changes in the length of day. *Precession* is the motion of the rotation axis with respect to inertial space—an oscillation of the rotation axis about the pole of the ecliptic with a period of 26,000 years. In addition, there are shorter period nutations. Various geophysical phenomena cause the pole to wander and the Earth to slow down.

Astronomical data, including records of ancient eclipses, show that the Earth has been slowing down at a rate of about 5×10^{22} rad/s, meaning that the length of the day is increasing at a rate of 1.5 ms per century. To balance the change in angular momentum, the Moon is receding from the Earth at a rate of about 5 cm per year. Paleontological evidence suggests that these accelerations have persisted for at least the past 500 Ma. The rotation rate of the Earth is also controlled by the internal rearrangement of mass.

The tidal deceleration of the Earth's rotation is due to the attraction of the Moon on the Earth's tidal bulge, which itself is primarily due to the Moon's gravitational pull. Since the Earth is not perfectly elastic, the tidal bulge is not directly beneath the Moon but is carried beyond the Earth–Moon line by the Earth's rotation. The attraction of the Moon on this displaced bulge then serves to brake the Earth's rotation and speed up the Moon in its orbit.

The eccentricity of the lunar orbit increases with time. The inclination of the lunar orbit also evolves and was greater in the past. The history of the lunar orbit is such that there was a minimum Earth–Moon distance at some time in the past, the inclination of the lunar orbit was substantially greater in the past, and the eccentricity of the lunar orbit increases as the Moon spirals away from the Earth. The tidal dissipation in the Earth occurs mainly in the oceans, especially shallow seas and along north-south coastlines. This effect is not constant with time, and it is therefore difficult to extrapolate backwards. In some calculations the Moon approaches the Roche limit, 2.8 Earth radii, early in Earth history. Within this limit the self-gravitation of the Moon is exceeded by the gravitational attraction of the Earth, and the Moon cannot cohere. The high tides raised on the Earth would cause extensive melting of the crust. The geological effect of close approach on both the Earth and the Moon would rival that of the early bombardments of these bodies.

The spin axis of an irregular, rotating body such as the Earth is controlled by the integrated mass distribution in the interior. For an evolving, dynamic planet, or for a planet being bombarded from outside, the spin axis is not fixed to any particular part of the planet but changes with time. It is no accident that the largest mass anomalies, those associated with subduction zones, are symmetric about the equator. The largest positive mass anomalies are centered over New Guinea–Borneo and the northern Andes, and these seem to control the present axis of rotation. However, if subduction in these regions were to cease and old oceanic lithosphere continued to pile up under the Aleutians and Tonga, these regions would eventually define the equator. The processes that change the mass distribution are slow, limited by the thermal time constant of the mantle and by the reconfiguration of continents and subduction zones. The reorientation of the spin axis, however, can be relatively fast, limited by the viscosity of the mantle, if the magnitudes of the principal moments of inertia interchange their order (Figure 12-6).

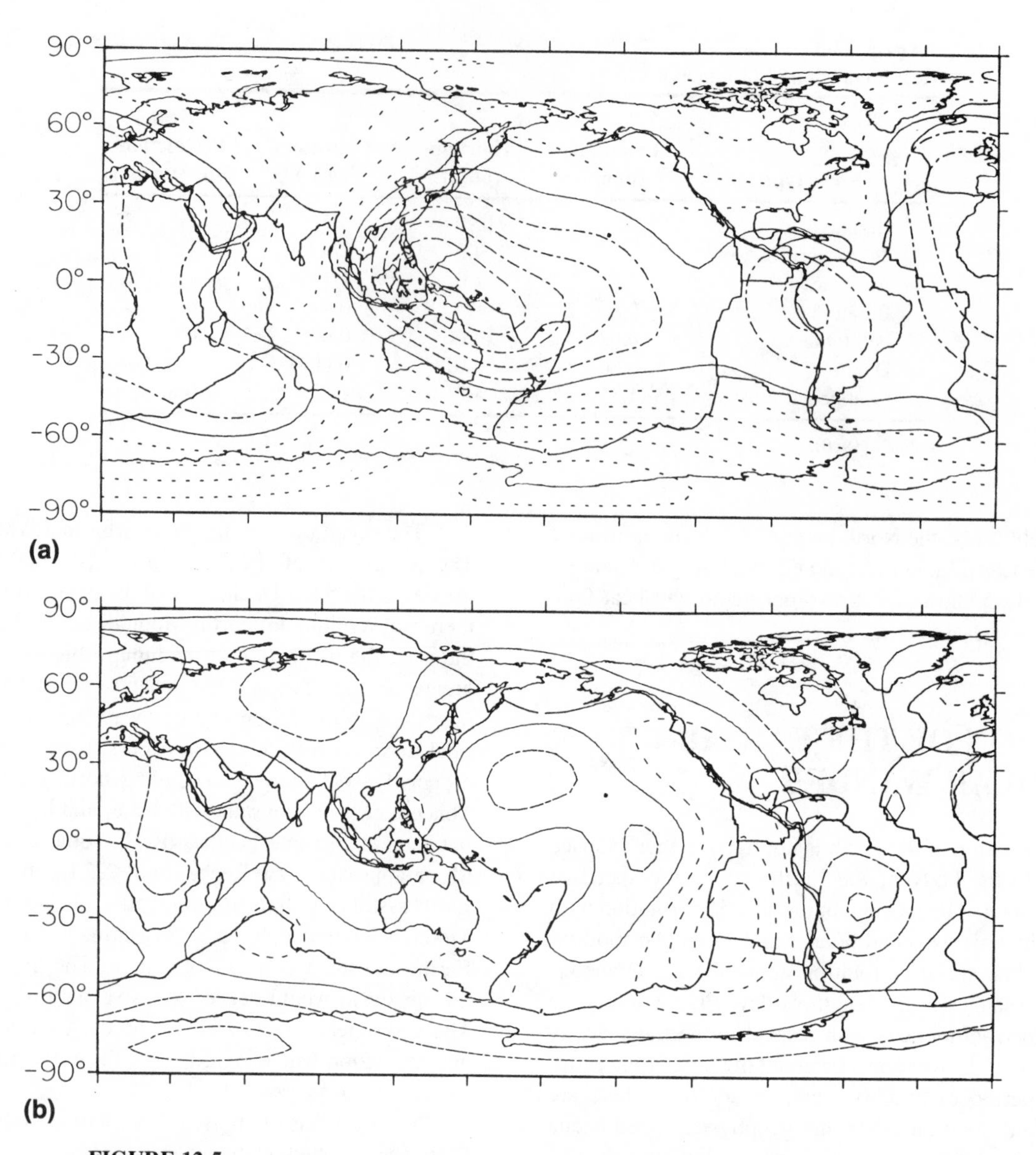

FIGURE 12-5
(a) The low-order nonhydrostatic geoid (l = 2–6). Contour interval 20 m. Long-wavelength geoid highs are centered on the equator near New Guinea and central Africa. **(b)** Heat flow map of Chapman and Pollack (see Pollack, 1980), l = 2–6. Contour interval 10 mW/m^2. Dashed lines are higher than average heat flow (after Nakanishi and Anderson, 1984).

The axis of rotation of the Earth is currently moving at a rate of 9 ± 2 cm/yr toward eastern Canada (75°–85° W). This is primarily due to isostatic rebound associated with ice and water redistribution following the last ice age and therefore contains information about the viscosity of the mantle. True polar wander must be distinguished from the apparent polar wander that is the result of plate motions; both phenomena are occurring, and it is sometimes difficult to separate the two effects. The concepts of mobile plates and a convecting mantle are now well engrained, but the concept of true polar wander, the physics of which is better understood (e.g. Goldreich and Toomre, 1969), has received less attention.

Imagine a rotating homogeneous sphere. Now place a dense mass anywhere on the surface or in the interior. This mass will eventually end up in the equatorial plane since this minimizes the kinetic energy of rotation. Although the mass is fixed to the sphere, it ends up on the equator by tilting the whole sphere. Now paint a dot on the north pole and consider a mass that can move around the interior or the surface. If it moves slowly, with respect to the time scale of viscous deformation of the sphere, it will never leave the equatorial plane. The sphere will tilt, with the point defining the north pole moving toward the mass while the rotation axis remains fixed in space. The mass will always be

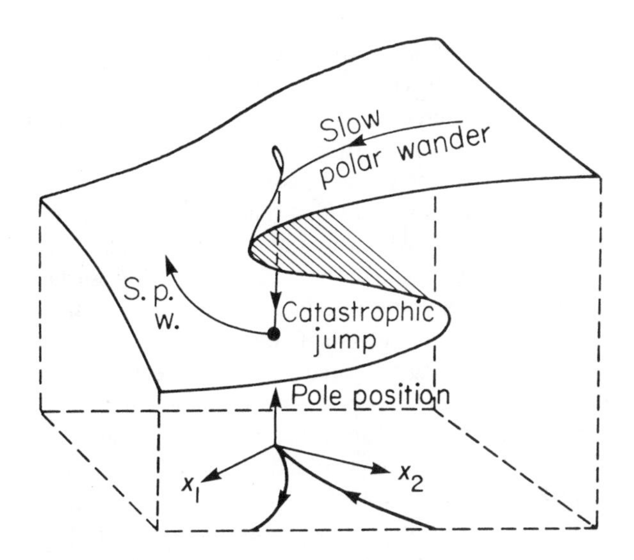

FIGURE 12-6
The principal moments of inertia shown on a cusp catastrophe diagram. As the moments of inertia vary, due to convective processes in the interior, the pole will slowly wander unless the ratios of the moments x_1 and x_2 pass through unity, at which point a catastrophe will occur, leading to a rapid change in the rotation axis.

at the farthest point from the spin axis. If there are many such masses, they will, on average, define the equator. The masses do not have to physically move. They can change their density by, for example, cooling or melting. The rotation axis of the Earth is determined by the integrated effect. The dominant effect on the present Earth appears to be the large amount of subduction of old oceanic lithosphere that has occurred in the western Pacific, and the broad upwellings in the lower mantle under the central Pacific and Africa.

Because the Earth is a dynamic body with a constantly shifting surface, it is impossible to define a permanent internal reference frame. There are three in common use: the rotation axis, the geomagnetic reference frame and the hotspot reference frame. The rotational frame is a function of the size of the mass anomalies and their distance from the axis of rotation. Upper-mantle effects are important because lateral heterogeneity is greater than lower-mantle or core heterogeneity and because they are farthest from the center of the Earth. The lower mantle is also important because of its large volume, but a given mass anomaly has a greater effect in the upper mantle. The location of the magnetic pole is controlled by convection in the core, which in turn is influenced by the rotation of the Earth and the temperatures at the base of the mantle. On average the rotational pole and the magnetic pole are close together, but the instantaneous poles can be quite far apart. The hotspot frame is based on the observation that hotspots seem to move relative to one another less rapidly than plates move relative to one another. The 40 or so long-lived hotspots have been used with some success as reference markers to track the motions of surface plates. This has led to the view that hotspots are anchored deep in the mantle and may reflect a different kind of convection than that which is responsible for large-scale convection in the mantle. Actually, the return flow associated with the motion of thin plates is spread out over a larger volume than the plates themselves, and the average velocities required to maintain mass balance are correspondingly lower. For example, if plates are 100 km thick and the return flow is spread out over the rest of the upper mantle, the average flow velocity beneath the plates is 4.5 times less than the plate velocity. Such a small relative motion beneath hotspots cannot be ruled out. A variety of evidence suggests that the lower mantle has a higher viscosity than the upper mantle. Convection in the lower mantle is therefore sluggish, and any embedded hot areas will tend to heat the upper mantle, giving rise to hotspots even if no material exits the lower mantle. Thus, there are at least two possible explanations for the small motion between hotspots that do not require a source of magma in the lower mantle, rapid radial ascent of plumes from the core-mantle boundary, or whole-mantle convection. On the other hand, material from a deep plume source must traverse a convecting mantle, and some relative motion is expected between plumes.

A relative displacement of 22° ± 10° has been deduced between the hotspot frame and the geomagnetic frame over the past 180 Ma (Andrews, 1987). The motion has not been smooth. Episodes of rapid motion (~1 cm/yr) include the present, 65–50 Ma, 115–85 Ma and 180–160 Ma.

The rotation axis has apparently wandered about 8° in the past 60 Ma and 20° in the past 200 Ma (Gordon and Cape, 1981; Morgan, 1981; Harrison and Lindh, 1982; Gordon, 1982). During these periods there have been major changes in the configurations of continents and subduction zones. These apparently had minor effects on the principal moments of inertia, suggesting a relatively stable and slowly evolving geoid. Even a slowly changing geoid, however, can cause a rapid shift in the whole mantle relative to the spin axis if the moment of inertia along this axis becomes less than some other axis of inertia. This may have happened sometime between 450 and 200 Ma ago, when Gondwana experienced a large latitude shift.

On the familiar Mercator projection the continents appear to be more or less randomly disposed over the surface of the globe. On other projections, however, there is a high degree of symmetry (for example, Figure 12-1). The centers of the largest (African and Pacific) plates are antipodal, and the other smaller plates, containing most of the large shield areas, occupy a relatively narrow polar band. The geoid lows in Siberia, the Indian Ocean, Antarctica, the Caribbean, and the Canadian shield also fall in this band. Long-wavelength geoid highs are centered on the large antipodal plates with hot spots, and it appears that shield areas have emigrated from these thermal and geoid highs. The polar band of discontinuous geoid lows correlates with areas of Cretaceous subduction and subsequent tertiary subsidence and areas in the upper mantle where seismic velocities are fast. Isostatically compensated cold mantle generates geoid lows. The geoid lows found in the wake of the Americas, India, and Australia exhibit fast surface-wave velocities, indicating that the anomalies are in the upper mantle, at least in part.

Density inhomogeneities in the mantle grow and subside, depending on the locations of continents and subduction zones. The resulting geoid highs reorient the mantle relative to the spin axis. Whenever there was a major continental assemblage in the polar region surrounded by subduction, as was the case during the Devonian through the Carboniferous, the stage was set for a major episode of true polar wandering.

The outer layers of the mantle, including the brittle lithosphere, do not fit properly on a reoriented Earth. Membrane stresses generated as the rotational bugle shifts may be responsible for the breakup and dispersal of Pangaea as it moves toward the equator. In this scenario, true polar wandering and continental drift are intimately related. A long period of continental stability allows thermal and geoid anomalies to develop. A shift of the axis of rotation causes plates to split, and the horizontal temperature gradient causes continents to drift away from the thermal anomalies that they caused. The continents drift toward cold parts of the mantle and, in fact, make the mantle cold as they override oceanic lithosphere.

Polar wandering can occur on two distinct time scales. In a slowly evolving mantle the rotation axis continuously adjusts to changes in the moments of inertia. This will continue to be the case as long as the major axis of inertia remains close to the rotation axis. If one of the other axes becomes larger, the rotation vector swings quickly to the new major axis (Figure 12-5). The generation and decay of thermal perturbations in the mantle are relatively gradual, and continuous small-scale polar wandering can be expected. The interchange of moments of inertia, however, occurs more quickly, and a large-scale 90° shift can occur on a time scale limited only by the relaxation time of the rotational bulge. The rate of polar wandering at present is much greater than the average rate of relative plate motion, and it would have been faster still during an interchange event. The relative stability of the rotation axis for the past 200 Ma suggests that the geoid highs related to hotspots have existed for at least this long. On the other hand, the rapid polar wandering that started 500 Ma ago may indicate that the Atlantic-African geoid high was forming under Gondwana at the time and had become the principal axis of inertia. With this mechanism a polar continental assemblage can be physically rotated to the equator as the Earth tumbles.

The southern continents all underwent a large northward displacement beginning sometime in the Permian or Carboniferous (280 Ma ago) and continuing to the Triassic (190 Ma). During this time the southern periphery of Gondwana was a convergence zone, and a spreading center is inferred along the northern boundary. One would expect that this configuration would be consistent with a stationary or a southern migration of Gondwana, unless a geoid high centered on or near Africa was rotating the whole assemblage toward the equator. The areas of very low surface-wave velocities in northeast Africa and the western Indian Ocean may be the former site of Gondwana.

Thus, expanding the paradigm of continental drift and plate tectonics to include continental insulation and true polar wandering may explain the paradoxes of synchronous global tectonic and magmatic activity, rapid breakup and dispersal of continents following long periods of continental stability, periods of static pole positions separated by periods of rapid polar wandering, sudden changes in the paths of the wandering poles, the migration of rifting and subduction, initiation of melting, the symmetry of ridges and fracture zones with respect to the rotation axis, and correlation of tectonic activity and polar wandering with magnetic reversals. Tumbling of the mantle presumably affects convection in the core and orientation of the inner core and offers a link between tectonic and magnetic field variations.

The largest known positive gravity anomaly on any planet is associated with the Tharsis volcanic province on Mars. Both geologic and gravity data suggest that the positive mass anomaly associated with the Tharsis volcanoes reoriented the planet with respect to the spin axis, placing the Tharsis region on the equator (Melosh, 1975a,b, 1980). There is also evidence that magmatism associated with large impacts reoriented the Moon. The largest mass anomaly on Earth is centered over New Guinea, and it is also almost precisely on the equator. The long-wavelength part of the geoid correlates well with subduction zones and hotspot provinces, and these in turn appear to control the orientation of the mantle relative to the spin axis. Thus, we have the possibility of a feedback relation between geologic processes and the rotational dynamics of a planet. Volcanism and convergence cause mass excesses to be placed near the surface. These reorient the planet, causing large stresses that initiate rifting and faulting, which in turn affect volcanism and subduction. Curiously, Earth scientists have been more reluctant to accept the inevitability of true polar wandering than to accept continental drift, even though the physics of the former is better understood.

The lower mantle plays an important role in the dynamics of the Earth. The seismically fast regions of the lower mantle are presumably cold, downwelling regions. Subduction zones (except Tonga-Fiji) and continents (except Africa) are located over these features. The colder parts of the lowermost mantle will have the highest thermal gradients in D″ and heat will be removed from the core most efficiently in these places. Downwellings in the core will therefore occur under cold parts of the lower mantle. Since, in general, hot regions of the lower mantle will be associated with geoid highs and mass excesses, when boundary deformation is taken into account, they will tend to lie along the equator, putting the colder regions of the lower mantle into high latitudes. Thus, convection in the core will have a preferential alignment relative to the spin axis. This accentuates the symmetry caused directly by rotation of the Earth on convection. Details of convection in the core may be controlled by the thermal structure in the lower mantle.

CONVECTION

The Earth's mantle transmits short-period seismic shear waves and responds as an elastic solid to tidal stresses. However, to long sustained loads it behaves as a viscous fluid. In a viscous fluid, stresses relax with time, and heat is transported by convection as well as by conduction. Both plate tectonics and the thermal history of the Earth are controlled by the fluid-like behavior of the mantle over long periods of time. A fluid will convect if heated from below or within, or cooled from above, or driven by loading at the boundaries or in the interior. Fluid-like behavior is not only a rheology-dependent process but is also a size-dependent characteristic: Large objects are intrinsically weak.

The slow uplift of the surface of the Earth in response to the removal of a load such as an ice cap or drainage of a large lake is not only proof of the fluid-like behavior of the mantle but also provides the data for estimating its viscosity. In contrast to everyday experience, however, mantle convection has some unusual characteristics. The container has spherical geometry. The "fluid" has stress-, pressure- and temperature-dependent properties. It is heated from within and from below. The boundary conditions and heat sources change with time. Melting and phase changes contribute to the buoyancy and provide additional heat sources and sinks. Convection may be driven partly by plate motions and partly by sinking and rising chemical differences. The boundaries are deformable rather than rigid. None of these characteristics are fully treated in numerical calculations, and we are therefore woefully ignorant of the style of convection to be expected in the mantle.

The theory of convection in the mantle has, by and large, been decoupled from the theory of solids and petrological constraints. The non-Newtonian rheology, the pressure and temperature sensitivity of viscosity, thermal expansion, and thermal conductivity, and the effects of phase changes and compressibility make it dangerous to rely too much on the intuition provided by oversimplified calculations. There are, however, some general characteristics of convection that transcend these details.

The energy for convection is provided by the decay of the radioactive isotopes of uranium, thorium and potassium and the cooling and crystallization of the Earth. This heat is removed from the interior by the upwelling of hot, buoyant material to the top of the system where it is lost by conduction and radiation. The buoyancy is provided by thermal expansion and phase changes including partial melting. In a chemically layered Earth the heat is transferred by convection to internal thermal boundary layers across which the heat is transferred by the slow process of conduction. Thermal boundary layers at the top of a layer cool the interior by becoming unstable and sinking, displacing new material toward the surface. Thermal boundary layers at the base of convecting systems warm up and can also become unstable, generating hot upwellings or plumes. Adiabatically ascending hot plumes are likely to cross the solidus in the upper mantle, at least, thereby buffering the temperature rise and magnifying the buoyancy. Because of the divergence, with pressure, of melting curves and the adiabat, deep plumes are probably subsolidus, at least initially.

In the simplest model of convection in a homogeneous fluid heated from below, hot material rises in relatively thin sheets and spreads out at the surface where it cools by conduction, forming a cold surface thermal boundary layer that thickens with time. Eventually the material achieves enough negative buoyancy to sink back to the base of the system where it travels along the bottom, heating up with time until it achieves enough buoyancy to rise. This gives the classical cellular convection with most of the motion and temperature change occurring in thin boundary layers, which surround the nearly isothermal or adiabatic cores. It has not escaped the attention of geophysicists that midocean ridges and subducting slabs resemble the edges of a convection cell and that the oceanic lithosphere thickens as a surface boundary layer cooling by conduction.

The planform of the convection depends on the Rayleigh number. At high Rayleigh number three-dimensional patterns result that resemble hexagons or spokes in plan view. Upwellings and downwellings can be different in shape.

In an internally heated fluid the heat cannot be completely removed by narrow upwellings. The whole fluid is heating up and becoming buoyant, so very broad upwellings result. On the other hand, if the fluid has a stress-dependent rheology, or a component of buoyancy due to phase changes such as partial melting, then the boundary layers can become thinner. A temperature-dependent rheology can force the length scales of the surface boundary to be larger than the bottom boundary layer. The lower boundary layer, having a higher temperature and, possibly, undergoing phase changes to lighter phases, can go unstable and provide upwellings with a smaller spacing than the downwellings. The effect of pressure and phase changes on the rheology may reinforce or reverse this tendency. Finally, the presence of accumulations of light material at the surface—continents—can affect the underlying motion. Subduction, for example, depends on more than the age of the oceanic lithosphere. Convection in the mantle is therefore unlikely to be a steady-state phenomenon. Collection of dense material at the base of the mantle, or light material at the top of the core, may help explain the unusual characteristics of D″.

Although convection in the mantle can be described in

general terms as thermal convection, it differs considerably from convection in a homogeneous Newtonian fluid with constant viscosity and thermal expansion. The temperature dependence of viscosity gives a "strong" cold surface layer. This layer must break or fold in order to return to the interior. When it does, it can drag the attached "plate" with it, a sort of surface tension that is generally not important in normal convection. In addition, light crust and depleted lithosphere serve to decrease the average density of the cold thermal boundary layer, helping to keep it at the surface. Buoyant continents and their attached, probably also buoyant, lithospheric roots move about and affect the underlying convection. The stress dependence of the strain rate in solids gives a stress-dependent viscosity. This concentrates the flow in highly stressed regions; regions of low stress flow slowly. Mantle minerals are anisotropic, tending to recrystallize with a preferred orientation dictated by the local stresses. This in turn gives an anisotropic viscosity, probably with the easy flow direction lined up with the actual flow direction. The viscosity controlling convection may therefore be different from the viscosity controlling postglacial rebound.

One of the major controversies regarding mantle convection is the radial scale length. There are proponents of whole-mantle convection and stratified convection. A relatively small chemical contrast can prevent whole-mantle convection. For example, if the lower mantle has a higher SiO_2/MgO content than subducting material, the density at a given depth will be different. This density contrast must be overcome by thermal expansion in order for material to puncture the boundary. For a temperature contrast of 10^3 K and a coefficient of thermal expansion of 3×10^{-6}/K, an intrinsic density contrast at a discontinuity of 3 percent will prevent circulation of material through the boundary. Differences in FeO, CaO and Al_2O_3 content also affect intrinsic density. The problem is somewhat more complicated than this since the boundary will be warped by the convection and the cold upper material can be pushed down to greater depths and pressures. If substantial enough phase changes occur as a result of this increased density, without a large increase in temperature associated with latent heats, then subducted material may be able to overcome the intrinsic density contrast. Convecting material in one layer may also entrain material from an adjacent layer (Figure 9-6).

The upper mantle itself may be chemically stratified. In this case, temperatures are more likely to be close to the solidus, and partial melting in the deeper layer provides additional buoyancy. Buried layers that become hot can become unstable.

Although the viscosity of the mantle increases with depth, because of pressure, and although the lower mantle may also have a higher viscosity, because the mineralogy is different and stresses may be lower, this is not sufficient to prevent whole-mantle convection. The actual viscosity structure of the mantle is uncertain because of the low resolving power of the data and the ambiguities regarding the contribution of non-Newtonian and transient rheology. If the viscosity of the upper mantle is less than that of the lower mantle, circulation will be faster and there is more opportunity to recirculate crustal and upper-mantle material through the shallow melting zone. Differentiation processes would therefore change the composition of the upper mantle, even if the mantle were chemically homogeneous initially. The separation of light and dense material, and low melting point and high melting point material, however, probably occurred during accretion and the early high-temperature history of the Earth. The differentiation will be irreversible if the recycled products of this differentiation (basalt, eclogite, depleted peridotite, continental crust) are unable to achieve the densities of the lower mantle or the parts of the upper mantle through which they must pass. A small radial difference in the intrinsic density of the mantle can prevent whole-mantle convection. The mean atomic weight, a measure of FeO content, is not an adequate measure of composition, and this parameter is poorly resolved from seismology. It is not clear if there is a difference in mean atomic weight between the upper and lower mantles, but even if this parameter were constant, there could be significant chemical, and intrinsic density, differences. MgO, SiO_2 and Al_2O_3 and minerals composed of these oxides have nearly the same mean atomic weight, but small variations in the proportions of these components change the stable mineral assemblages and phase transition boundaries and, therefore, the intrinsic densities. For example, if the lower mantle is orthopyroxene-rich, then perovskite will be the major mineral. An upper mantle richer in MgO and Al_2O_3 will have more olivine and garnet, and this assemblage will be unable to subduct into the lower mantle even when it has transformed to high-pressure assemblages and even if it has a higher FeO content than the lower mantle. The plausibility of whole-mantle convection then depends on the plausibility of avoiding separation of mantle components during accretion and the early high-temperature history of the Earth.

Considering these uncertainties in even what calculations to perform and in the parameters, much less the uncertainties in the pressure, temperature and stress dependence of the parameters, observational constraints will continue to be the most important diagnostics of the style of mantle convection. These constraints include surface topography, heat flow, gravity field, lateral variations in seismic velocities and depths of mantle discontinuities, and the distribution of earthquakes. Lateral variations in density and seismic velocity are due to temperature and temperature-induced phase changes. In general, phase changes cause larger variations than temperature alone.

There are several variants of the stratified mantle convection idea. In the hot accretion–magma ocean scenario, the mantle is not homogeneous in either the trace elements or the major elements. The incompatible elements are concentrated in the crust and shallow mantle, and this includes the radioactive elements. The lower mantle is composed of the denser, refractory silicates and is depeleted in radioactives. The upper mantle itself may be chemically zoned. In this model there are several thermal boundary layers, and most of the mantle cools very slowly.

Some geochemical models assume that the crust formed only from the upper mantle and that the lower mantle is still primitive, somehow surviving the accretional melting and differentiation and retaining primitive abundances of the heat-producing elements.

Other models combine whole-mantle circulation with embedded inhomogeneities to provide isotropic heterogeneity, or assume that the mantle is homogeneous except for the D″ region, which is chemically distinct.

The whole-mantle convection models have received the most attention from the convection community. The various layered models, particularly with deformable boundaries and pressure- and temperature-dependent properties and phase changes, have received little attention. It is not yet even clear if layers are primarily coupled by thermal or mechanical phenomena.

I have discussed in previous chapters a three-layer mantle: shallow mantle, transition region and lower mantle with the transition region being potentially unstable (buoyant) at high temperature. The oceanic and continental lithospheres and D″ are additional layers that may be chemically distinct. The surface layer is partially chemically buoyant (shields) and partially potentially unstable (oceans). It moves around relative to underlying features and both affects and is affected by the temperature in the upper mantle. This type of model needs to be tested by numerical calculation.

Dimensionless Numbers

The theory of convection is littered with dimensionless numbers named after prominent dead physicists. The relative importance of conduction and convection is given by the Péclet number

$$Pe = \frac{vl}{\kappa}$$

where v is a characteristic velocity, l a characteristic length, and κ the thermal diffusivity,

$$\kappa = \frac{\mathcal{K}}{\rho C_P}$$

expressed in terms of conductivity $\mathcal{K}$, density and specific heat at constant pressure. The Péclet number gives the ratio of convected to conducted heat transport. For the Earth Pe is about 10^3 and convection dominates conduction. For a much smaller body, conduction would dominate; this is an example of the scale as well as the physical properties being important in the physics. There are regions of the Earth, however, where conduction dominates, such as at the surface where the vertical velocity vanishes.

The Rayleigh number

$$Ra = \frac{g\alpha\rho d^3\Delta T}{\kappa\eta}$$

is a measure of the vigor of convection due to thermally induced density variations, $\alpha\Delta T$, in a fluid of viscosity η operating in a gravity field g. This is for a uniform fluid layer of thickness d with a superadiabatic temperature difference of ΔT maintained between the top and the bottom. If the fluid is heated internally, the ΔT term is replaced by $d^2A/\mathcal{K}$ where A is the volumetric heat production. Convection will occur if Ra exceeds a critical value of the order of 10^3. For large Ra the convection and heat transport are rapid. The Rayleigh number of the mantle is thought to be of the order of 10^5 to 10^7. This depends on the scale of convection as well as the physical properties.

The Nusselt number gives the relative importance of convective heat transport:

Nu = total heat flux across the layer/conducted heat flux in the absence of convection

$$= (Ra/Ra_c)^{1/3}$$

For an internally heated layer Nu is the ratio of the temperature drops across the layer with and without convection or equivalently, the ratio of the half-depth of the layer to the thermal boundary layer thickness.

The Prandtl number Pr

$$Pr = \eta/\kappa$$

for the mantle is about 10^{24}, which means that the viscous response to a perturbation is instantaneous relative to the thermal response.

The Reynolds number, Re, is

$$Re = vl/\eta = Pe/Pr$$

For $v \approx 10^{-7}$cm/s, a typical plate tectonic rate and $l \approx 10^8$ cm, the dimension of the mantle, $Re \approx 10^{-21}$. For $Re \ll 1$ inertial effects are negligible, and this is certainly true for the mantle.

Convection can be driven by heating from below or within, or by cooling from above. The usual case treated is where the convection is initiated by a vertical temperature gradient. When the vertical increase of temperature is great enough to overcome the pressure effect on density, the deeper material becomes buoyant and rises. An adiabatic gradient simply expresses the condition that the parcel of fluid retains the same density contrast as it rises. Horizontal temperature gradient can also initiate convection. These can be caused by the presence of continents or variations in lithosphere thickness such as at fracture zones or between oceans and continents.

The stability of a fluid layer heated from below is a

classic problem. In the simplest, Boussinesq, approximation, the fluid is assumed to be elastically incompressible, changing volume only by thermal expansion, and the onset of convection is controlled entirely by the Rayleigh number. In a thin layer, or on a small planet, the temperature gradient in the Rayleigh number is the actual gradient. In the presence of gravity the gradient is the excess over the adiabatic gradient, which for the upper mantle is about 0.3°C/km. An adiabatic gradient is neutrally stable, and only deviations from it can provide buoyancy. For a uniform fluid, convection is initiated for $Ra > 10^3$ and is characterized by a horizontal wavelength comparable to the thickness of the convecting layer. This is the same as saying that a parcel of fluid traveling along the bottom and becoming warmer becomes buoyant enough to rise when it has traveled a distance about equal to the depth of the layer. Likewise, a parcel moving along the top, becoming colder and denser, becomes unstable after traveling a comparable distance. If the surface layer is prevented from sinking because of its high strength or viscosity, then it may travel greater distances. The spacing of upwelling blobs may therefore be less than the spacing between downwellings. The lower part of the surface boundary may also detach and sink. In a system with a strongly temperature-dependent viscosity, there may therefore be several scales of convection. Midplate volcanism may be a manifestation of the second scale length.

For high Ra and Pr most of the temperature contrast ΔT occurs across boundary layers of thickness δ. In a boundary layer the horizontal convection of heat is balanced by vertical diffusion,

$$\frac{u \Delta T \delta}{H} \approx \kappa \frac{\Delta T}{\delta}$$

where u is the horizontal velocity and H is the cell width. The boundary layer thickness can be written

$$\delta \approx H(Ra_c/Ra)^{1/3}$$

The convective heat flux $\dot{Q}$ is

$$\dot{Q} \simeq \frac{\rho C_P v \Delta T \delta}{H}$$

where $v \approx u$.

Scale of Mantle Convection

The dimensions of the major lithospheric plates are of the order of 10^3 to 10^4 km. This has sometimes been taken as evidence for whole-mantle convection. Plates, however, are strong and have high viscosity and because of the crustal and depleted lithospheric component have some nonthermal buoyancy. They, therefore, can impose their own scale length on the problem. Both the lower mantle and the upper mantle are convecting, but the important question is how they interact with each other. The lower mantle certainly differs in mineralogy and pressure from the upper mantle and therefore has different, probably higher, viscosity and thermal conductivity. The non-Newtonian (stress-dependent) rheology of solids also affects the viscosity profile. Convection is probably much faster in the shallow mantle than in the deeper mantle. The geoid and seismicity both provide evidence that subducting slabs encounter a barrier near 650 km. This could be due to an increase in viscosity or intrinsic density due to chemical differences.

The mantle contains two prominent discontinuities near 400 and 650 km. If these are equilibrium phase boundaries in a homogeneous mantle, they may inhibit but will not necessarily prevent whole-mantle convection. Cold downwelling material will transform to the denser assemblage unless it is unable to reach the appropriate pressure. If it does get deep enough to transform, it should have no trouble sinking through the lower mantle unless the latent heat associated with the phase change causes it to warm up to temperatures in excess of the surrounding mantle. Flow through a phase boundary is governed by two competing effects, density change and latent heat. The olivine-spinel transition is exothermic: The transition pressure decreases with a decrease in temperature. Cold subducted olivine-rich material transforms to the denser phase while the surrounding hotter mantle is still in the less dense phase. The slab therefore has an additional contribution to its negative density, which is only partially removed by the latent heating. The olivine-spinel transition, therefore, does not inhibit convection unless the low slab temperatures prevent the phase change from happening because of slow reaction rates. Even in this case, the slab will eventually warm up and be able to sink through the phase boundary.

In the case of an endothermic reaction, with a negative Clapeyron slope, the slab must push its way into the deeper mantle before it can reach pressures great enough to transform it to the denser phase. When it does it will cool off and may become denser than the surrounding mantle. Some of the phase changes predicted to occur near 650 km have negative Clapeyron slopes. Whether convection is precluded across the interface in this case depends not only on the volume change and latent heat but also on the total deformation of the boundary. In a chemically layered mantle the interface is depressed under cold downwellings in the upper layer by an amount that depends on the integrated density contrast across the upper mantle. If cold, chemically distinct, upper-mantle material is to penetrate the boundary, the boundary deformation must be enough to allow the phase change in the colder material, and the density increase must be enough to provide a negative buoyancy with respect to the lower mantle, otherwise upper-mantle material will be confined to the upper mantle.

The upper and lower mantles will not behave completely independently even if there is a chemical stratification. Hot upwellings in the lower mantle will elevate the boundary and heat the upper mantle. Cold downwellings in the upper layer will depress the boundary and cool the top of the lower mantle. The presence of a conductive boundary layer will decrease the viscosity near the boundary and reduce the shear coupling across it. A large viscosity contrast across the boundary will also inhibit shear coupling.

Three-dimensional tomographic studies of the upper mantle in the northwest Pacific show that the high-velocity slabs become nearly horizontal at depths just beneath the deepest earthquakes (Zhou and Clayton, 1988). This is consistent with slab confinement to the upper mantle.

COOLING OF THE EARTH

The principal contributions to the surface heat flow of the Earth are cooling of the Earth and the heating due to the decay of radioactive isotopes. In 1856 Lord Kelvin used the surface heat flow and a conduction calculation to estimate the age of the Earth as 20–400 Ma. This ignored radioactive heating and convection. Later, it was noted that the present heat flow is about that expected from the radioactivity of chondritic meteorites, and it was suggested that the Earth is in steady-state heat balance between heat generation and heat loss. However, since heat productivity decreases with time, it is likely that convective velocities and temperatures have also decreased with time. The cooling contributes a specific heat component to the observed heat flow, the term treated by Lord Kelvin, and this is not negligible. For example, a cooling rate of only 200 K in 10^9 years gives a surface heat flow of more than 1.5 $\mu cal/cm^2/s$.

For a fluid layer of thickness h heated from within a dimensionless temperature, Θ, can be written

$$\Theta = 2\mathcal{K}\overline{T}/\rho Q h^2$$

where $\overline{T}$ is the mean temperature, ρ is the density and Q is the heating rate per unit mass. For a fluid heated from below at rate q,

$$\Theta = \mathcal{K}\overline{T}/qh$$

Turcotte (1980) takes the empirical scaling

$$\Theta \sim Ra^{-n},\ n \approx 1/4$$

and an average heat productivity

$$Q = Q_o \exp(t \ln 2/\tau)$$

or

$$q = q_o \exp(t \ln 2/\tau)$$

and a temperature-dependent viscosity

$$\eta = \eta_o \exp(E^*/RT)$$

to deduce a cooling rate of

$$\frac{dT}{dt} = -(\ln 2)\left(\frac{1-n}{n}\right)\frac{RT_o^2}{E\tau}$$

where τ is an average radioactive half-life, t is time backwards from the present and T_o is the present mean mantle temperature ($t = 0$). The calculated cooling rate varies from 90 K/Ga at 3 Ga ago to 36 K/Ga at present. The average temperature of the Earth has therefore decreased at least several hundred degrees since the Earth stopped accreting, and this cooling contributes about 20 percent to the current surface heat flow. (For thermal modeling of accretion itself see Turcotte and Pflugrath, 1985.) The thickness of the surface thermal boundary has increased with time since a lower average thermal gradient suffices to conduct out the internal heat. This means that the temperature of the upper mantle has decreased more than the deeper temperatures. In a chemically stratified mantle, the deeper layers cool more slowly. Upper-mantle melting was more extensive in the past, and higher temperature melts could rise closer to the surface. Today's thick lithosphere prevents high-temperature and high-density melts (such as komatiites) from rising directly to the surface. Some cooling and fractionation in the shallow mantle is probably inevitable. Komatiites and other olivine-rich magmas may, in fact, still be forming in the upper mantle by adiabatic ascent from deeper layers but would be unable to penetrate the present crust and lithosphere.

Estimates of the cooling rate of the Earth, such as the above, are highly uncertain. Cooling may contribute somewhere between 20 and 50 percent of the present heat flow, and this uncertainty makes it difficult to constrain the radioactive abundance of the mantle from the observed heat flow. A chemically layered mantle will cool more slowly than a homogeneous mantle because of the necessity of conducting heat through internal thermal boundary layers. The heat loss from the Earth may also be episodic. The present rate of heat loss may not be typical.

References

Anderson, D. L. (1982) Hotspots, polar wander, Mesozoic convection and the geoid, *Nature, 297,* 391–393.

Andrews, J. (1987) True polar wander, *J. Geophys. Res., 90,* 7737.

Birch, F., R. F. Roy and E. R. Decker (1968) Heat flow and thermal history in New England and New York. In *Studies of Appalachian Geology* (E. Anzen, ed.), 437–451, Interscience, New York.

Cazenave, A. (1984) Thermal cooling of the oceanic lithosphere; new constraints from geoid height data, *Earth Planet. Sci. Lett., 70,* 395–406.

Cazenave, A., C. Rosemberg-Borot and M. Rabinowicz (1986) Geoid lows over deep-sea trenches, *J. Geophys. Res., 91,* 1989–2003.

Chase, C. G. (1979) Subduction, the geoid and lower mantle convection, *Nature, 282,* 464–468.

Clayton, R. W. and R. P. Comer (1988) Reconstruction of mantle heterogeneity by iterative back-projection of travel times: 2. Results for P-waves, to be submitted to *J. Geophys. Res.*, 1988.

Crough, S. T. (1978) Thermal origin of mid-plate hotspot swells, *Geophys. J. R. Astron. Soc.*, *55*, 451.

Crough, S. T. (1979) Hotspot epeirogeny, *Tectonophysics*, *61*, 321–333.

Dziewonski, A. M. (1984) Mapping the lower mantle: Determination of lateral heterogeneity in P-velocity up to degree and order 6, *J. Geophys. Res.*, *89*, 5929–5952.

Goldreich, P. and A. Toomre (1969) *J. Geophys. Res.*, *74*, 2555–2567.

Gordon, R. G. (1982) The late Maastrichtian paleomagnetic pole of the Pacific Plate, *Geophys. J. R. Astr. Soc.*, *70*, 129–140.

Gordon, R. G. and C. D. Cape (1981) Cenozoic latitudinal shift of the Hawaiian hotspot and its implications for true polar wander, *Earth Planet. Sci. Lett.*, *55*, 37–47.

Hager, B. H. (1984) Subducted slabs and the geoid; constraints on mantle rheology and flow, *J. Geophys. Res.*, *89*, 6003–6015.

Hager, B. H., R. W. Clayton, M. A. Richards, R. P. Comer and A. M. Dziewonski (1985) Lower mantle heterogeneity, dynamic topography and the geoid, *Nature*, *313*, 541–545.

Harrison, C. G. A. and T. Lindh (1982) Comparison between the hotspot and geomagnetic field reference frames, *Nature*, *300*, 251–252.

Melosh, J. (1975a) *Earth Planet. Sci. Lett.*, *25*, 322.

Melosh, J. (1975b) *Earth Planet. Sci. Lett.*, *26*, 353.

Melosh, J. (1980) *Icarus*, *44*, 745.

Menard, H. W. and L. M. Dorman (1977) Dependence of depth anomalies upon latitude and plate motion, *J. Geophys. Res.*, *82*, 5329–5335.

Menard, H. W. and M. K. McNutt (1982) Evidence for and consequences of thermal rejuvenation, *J. Geophys. Res.*, *87*, 8570–8580.

Morgan, W. J. (1981) Hotspot tracks and the opening of the Atlantic and Indian Oceans. In *The Oceanic Lithosphere* (C. Emiliani, ed.), 443–487, Wiley, New York.

Nakanishi, I. and D. L. Anderson (1984) Measurements of mantle wave velocities and inversion for lateral heterogeneity and anisotropy, Part II: Analysis by the single-station method, *Geophys. J. R. Astron. Soc.*, *78*, 573–617.

Nataf, H.-C., I. Nakanishi and D. L. Anderson (1986) Measurements of mantle wave velocities and inversion for lateral heterogeneities and anisotropy, Part III: Inversion, *J. Geophys. Res.*, *91*, 7261–7307.

Parsons, B. and J. G. Sclater (1977) An analysis of the variation of ocean floor bathymetry and heat flow with age, *J. Geophys. Res.*, *82*, 803–827.

Pollack, H. N. (1980) The heat flow from the earth: A review. In *Mechanisms of Continental Drift and Plate Tectonics* (P. A. Davies and S. K. Runcorn, eds.), 183–192, Academic, London.

Pujol, J. and D. M. Fountain (1985) Statistical analysis of the mean heat flow/reduced heat flow relationship for continents and its tectonophysical implications, *J. Geophys. Res.*, *90*, 11,335–11,344.

Rapp, R. H. (1981) The Earth's gravity field to degree and order 180 using Seaset altimeter data, terrestrial gravity data, and other data, *Report 322, Dept. Geodetic. Sci. and Surv.*, Ohio State University, Columbus.

Richards, M. A. and B. H. Hager (1984) Geoid anomalies in a dynamic Earth, *J. Geophys. Res.*, *89*, 5987–6002.

Sclater, J., B. Parsons and C. Jaupart (1981) Oceans and continents: similarities and differences in the mechanism of heat transport, *J. Geophys. Res.*, *86*, 11,535–11,552.

Sclater, J. G., C. Jaupart and D. Galson (1980) The heat flow through oceanic and continental crust and the heat loss of the earth, *Rev. Geophys. Space Phys.*, *18*, 269–311.

Tanimoto, T. and D. L. Anderson (1984) Mapping convection in the mantle, *Geophys. Res. Lett.*, *11*, 287–290.

Tanimoto, T. and D. L. Anderson (1985) Lateral heterogeneity and azimuthal anisotropy of the upper mantle: Love and Rayleigh waves 100–250 sec., *J. Geophys. Res.*, *90*, 1842–1858.

Turcotte, D. L. (1980) On the thermal evolution of the Earth, *Earth Planet. Sci. Lett.*, *48*, 53–58.

Turcotte, D. L. and J. C. Pflugrath (1985) Thermal structure of the accreting Earth, *J. Geophys. Res.*, *90*, C541–C544.

Vogt, P. R. (1975) Changes in geomagnetic reversal frequency at times of tectonic change; evidence for coupling between core and upper mantle processes, *Earth Planet. Sci. Lett.*, *25*, 313–321.

Zhou, H.-W. and R. Clayton (1988) preprint.

13

Heterogeneity of the Mantle

"You all do know this mantle."

—SHAKESPEARE, JULIUS CAESAR

We must now admit that the Earth is not like an onion. The lateral variations of seismic velocity and density are as important as the radial variations. The shape of the Earth tells us this directly but provides little depth resolution. The long-wavelength geoid tells us that lateral density variations extend to great depth. Lateral variations in the mantle affect the orientation of Earth in space and convection in the core; this property of the Earth is known as asphericity. In the following chapters we further admit that the Earth is not elastic or isotropic.

UPPER-MANTLE HETEROGENEITY FROM SURFACE-WAVE VELOCITIES

The most complete maps of seismic heterogeneity of the upper mantle are obtained from surface waves. By studying the velocities of Love and Rayleigh waves, of different periods, over many great circles, small arcs, and long arcs, it is possible to reconstruct both the radial and lateral velocity variations. Although global coverage is possible, the limitations imposed by the locations of long-period seismic stations and of earthquakes limit the spatial resolution. Features having half-wavelengths of about 2000 km, as a global average, can be resolved with presently available data. The raw data consist of average group and/or phase velocity over many arcs. These averages can be converted to images using techniques similar to medical tomography. Body waves have better resolution, but coverage, particularly for the upper mantle, is poor.

Even the early surface-wave studies indicated that the upper mantle was extremely inhomogeneous. Shield paths were fast, oceanic paths were slow, and tectonic regions were also slow (Toksöz and Anderson, 1966; Anderson, 1967). The most pronounced differences are in the upper 200 km, but substantial differences between regions extend to about 400 km. A high-velocity region was inferred to extend to depths of 120–150 km under stable continental shields. This thick LID, or seismic lithosphere, probably represents the thickness of the plate. Body-wave results also give a shield LID thickness of about 150 km. On average, the shield mantle is also faster than average oceanic or tectonic mantle down to 400 km, but the differences below 200 km are much less than above this depth. Some shield-bearing continents have overriden oceanic lithosphere in the past 50–100 million years, and others were bordered by subduction zones prior to the breakup of Pangaea. Old oceanic lithosphere has a long thermal time constant and serves to cool off adjacent mantle. Below 200 km the regions of higher than average mantle velocity may represent the cooling effect of overriden oceanic lithosphere and/or the absence of a partial melt phase. The rapid decrease in velocity under shields below about 150 km suggests that this is the depth of decoupling. The suggestion that continental roots extend deeper than 400 km (Jordan, 1975) rather than the 150–200 km of earlier studies was based on the low-resolution ScS phase rather than the higher resolution P-waves and surface waves (Jordan, 1975; Jordan and Sipkin, 1976).

There are two basic approaches for interpreting global surface-wave data. The regionalized approach introduced by Anderson (1967) and Toksöz and Anderson (1966) divides the Earth into tectonic provinces and solves for the velocity of each. Applications of this technique yielded fast shield velocities at short periods or shallow depths but showed that convergence regions were faster at long periods, or greater depth, suggesting that cold subducted material was being sampled (Nakanishi and Anderson, 1983, 1984a,b). Convergence regions, on average, are slow at short periods, due to high temperatures and melting at shallow mantle depths. The regionalization approach is necessary when the data are limited or when only complete great-circle data are available. In the latter case the velocity anomalies cannot be well isolated.

If short-arc and long-arc data are available, a complete spherical harmonic expansion can be performed with no assumptions required about tectonic regionalizations. Studies of this type have been reported by Nakanishi and Anderson (1983, 1984a,b), Tanimoto and Anderson (1984, 1985) and Woodhouse and Dziewonski (1984). If only complete great-circle data, or free-oscillation data, are available, then only the even-order harmonics can be determined.

In the regionalized models it is assumed that all regions of a given tectonic classification have the same velocity. This is clearly oversimplified. It is useful, however, to have such maps in order to find anomalous regions. In fact, all shields are not the same, and velocity does not increase monotonically with age with the ocean. The region around Hawaii, for example, is faster than equivalent age ocean elsewhere at shallow depth and slower at greater depth. There is also a slow patch in the south-central Pacific. The differences between Love wave and Rayleigh wave maps are primarily due to differences in penetration depth and anisotropy between vertical and horizontal *S*-waves. The early surface-wave results showed that the shallow mantle was slow, and presumably hot, under young oceans, tectonic regions and subduction zones. Most of the slow regions are in geoid highs. There is a good correspondence of surface-wave velocity maps with heat-flow maps but generally poor correlation with the geoid. The geoid, apparently, has little sensitivity to upper-mantle density variations.

GLOBAL SURFACE-WAVE TOMOGRAPHY

A global view of the lateral variation of seismic velocities in the mantle can now be obtained with surface-wave tomography. In the regionalization approach one assumes that the velocities of surface waves are linearly dependent on the fraction of time spent in various tectonic provinces. The inverse problem then states that the velocity profile depends only on the tectonic classification. For example, all shields are assumed to be identical at any given depth. This assumption appears to be approximately valid for the shallow structure of the mantle but becomes increasingly tenuous for depths greater than 200 km. However, it probably provides a maximum estimate of the depth of tectonic features, and it also provides a useful standard model with which other kinds of results can be compared.

The second approach subdivides the Earth into cells or blocks or by some smooth function such as spherical harmonics. Nataf and others (1984, 1986) used spherical harmonics for the lateral variation and a series of smooth functions joined at mantle discontinuities for the radial variation. In this approach no a priori tectonic information is built in.

In both of these approaches the number of parameters that one would like to estimate far exceeds the information content (the number of independent data points) of the data. It is therefore necessary to decide which parameters are best resolved by the data, which is the resolution, or averaging length, which parameters to hold constant and how the model should be parameterized (for example as layers or smooth functions, isotropic or anisotropic). In addition, there are a variety of corrections that might be made (such as crustal thickness, water depth, elevation, ellipticity, attenuation). The resulting models are as dependent on these assumptions and corrections as they are on the quality and quantity of the data. This is not unusual in science: Data must always be interpreted in a framework of assumptions, and the data are always, to some extent, incomplete and inaccurate. In the seismological problem the relationship between the solution, or the model, including uncertainties, and the data can be expressed formally. The effects of the assumptions and parameterizations, however, are more obscure, but these also influence the solution. The hidden assumptions are the most dangerous. For example, most seismic modeling assumes perfect elasticity, isotropy, geometric optics and linearity. To some extent all of these assumptions are wrong, and their likely effects must be kept in mind.

Nataf and others (1986) made an attempt to evaluate the resolving power of their global surface-wave dataset and invoked physical a priori constraints in order to reduce the number of independent parameters that needed to be estimated from the data. For example, the density, compressional velocity and shear velocity are independent parameters, but their variation with temperature, pressure and composition show a high degree of correlation; that is, they are coupled parameters. Similarly, the fact that temperature variations in the mantle are not abrupt means that lateral and radial variations of physical properties will generally be smooth except in the vicinity of phase boundaries, including partial melting. Changes in the orientation of crystals in the mantle will lead to changes in both the shear-wave and compressional velocity anisotropies. These kinds of physical considerations can be used in lieu of the standard seismological assumptions, which are generally made for mathematical convenience rather than physical plausibility.

The studies of Woodhouse and Dziewonski (1984) and Nataf and others (1984, 1986) give upper-mantle models that are based on quite different assumptions and analytical techniques. Woodhouse and Dziewonski inverted for

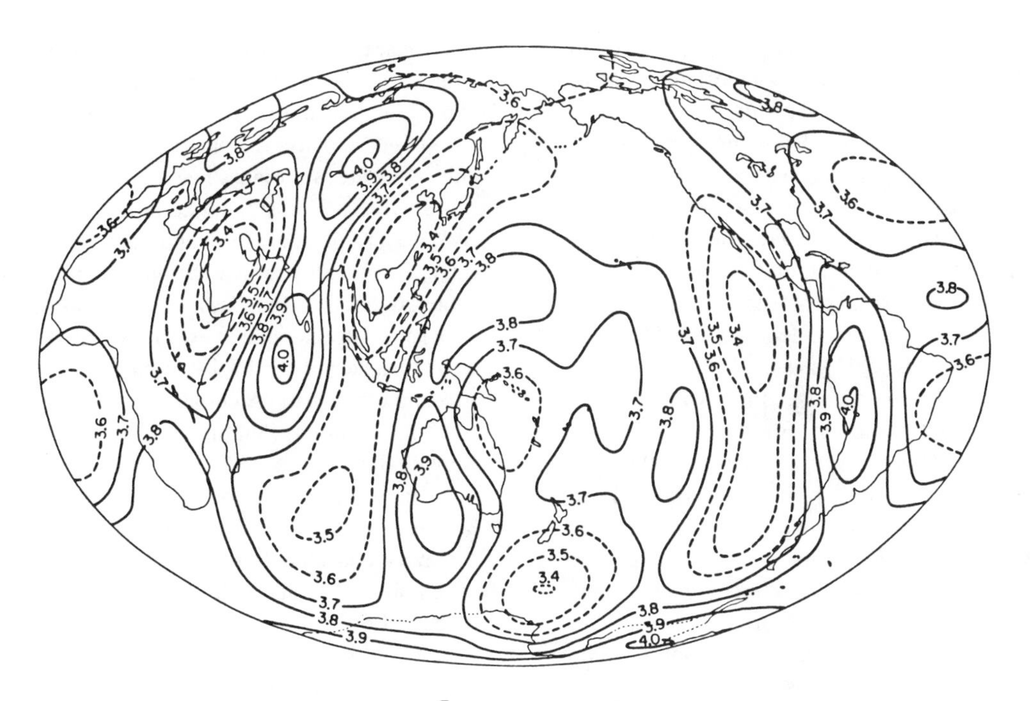

FIGURE 13-1
Group velocity of 152-s Rayleigh waves (km/s). Tectonic and young oceanic areas are slow (dashed), and continental shields and older oceanic areas are fast. High temperatures and partial melting are responsible for low velocities. These waves are sensitive to the upper several hundred kilometers of the mantle (after Nakanishi and Anderson, 1984a).

shear velocity, keeping the density, compressional velocity and anisotropy fixed. They also used a very smooth radial perturbation function that ignores the presence of mantle discontinuities and tends to smear out anomalies in the vertical direction. They corrected for near-surface effects by assuming a bimodal crustal thickness, continental and oceanic.

Nataf and others corrected for elevation, water depth, shallow-mantle velocities and measured or inferred crustal thickness. They inverted for shear velocity and anisotropy, the dominant parameters, but included physically plausible accompanying changes in density, compressional velocity and anisotropy. Corrections were also made for anelasticity. The radial perturbation functions were allowed to change rapidly across mantle discontinuities, if required by the data. In spite of these differences, the resulting models are remarkably similar above about 300 km. The main differences occur below 400 km and seem to arise from differences in the assumptions and parameterizations (crustal corrections, radial smoothing functions) rather than the data. The choice of an a priori radial perturbation function can degrade the vertical resolution intrinsic to the dataset. The solution, in this case, is overdamped or oversmoothed.

Before discussing the inversion results—the earth structures themselves—I will briefly describe the distribution of Love and Rayleigh wave velocities (Nakanishi and Anderson, 1983, 1984a,b).

Love waves are sensitive to the *SH* velocity of the shallow mantle, above about 300 km for the periods considered here. The slowest regions are at plate boundaries, particularly triple junctions. Slow velocities extend around the Pacific plate and include the East Pacific Rise, western North America, Alaska-Aleutian arcs, Southeast Asia and the Pacific-Antarctic Rise. Parts of the Mid-Atlantic Rise and the Indian Ocean Rise are also slow. The Red Sea–Gulf of Aden–East Africa Rift (Afar triple junction) is one of the slowest regions. The upper-mantle velocity anomaly in this slowly spreading region is as pronounced as under the rapidly spreading East Pacific Rise. Since it also shows up for long-period Rayleigh waves, this is a substantial and deep-seated anomaly. Shields are generally fast, particularly the Brazilian, Australian and South African shields. The fastest oceanic regions are the north-central Pacific, centered near Hawaii, and the eastern Indian Ocean.

Rayleigh waves are sensitive to shallow *P*-velocities and *SV* velocity from about 100 to 600 km. The fastest regions (see Figure 13-1) are the western Pacific, western Africa and the South Atlantic. Western North America, the Red Sea area, Southeast Asia and the North Atlantic are the slowest regions. Slower than average Rayleigh-wave phase velocities are obtained at long period for the stable continental areas. The velocities in the South Atlantic and the Philippine Sea plate are faster than shields.

Regions of convergence, or descending mantle flow, are relatively fast for long-period Rayleigh waves, particularly near New Guinea, Sumatra, the western Pacific and northwestern Pacific. These regions have large positive geoid anomalies. Dense, cold material in the upper mantle can explain these results. To explain the surface-wave results, the fast material must be below the depth sampled by 250-s Love waves and 150-s Rayleigh waves, which is about 300 km depth, and must occupy a considerable volume of the upper mantle. Conductive cooling of the oceanic lithosphere probably extends no deeper than about 150 km. It is doubtful that the subduction of such a thin high-velocity plate could explain these results. Either slabs preferentially subduct in cold regions of the mantle or they pile up in the upper mantle. At shorter periods the velocity of the mantle beneath subduction zones is average or slower than average, presumably reflecting the extensional tectonics and hot shallow mantle under back-arc basins.

A map synthesized from the $l = 2$ coefficients at 250 s is characterized by a broad closed high-velocity region centered in the western Pacific, northeast of New Guinea, and a low-velocity region centered on the East Pacific Rise. Because of symmetry the antipodal regions are identical. The $l = 2$ coefficients are just part of the complete set that is required to fully describe the Earth's asphericity and need no special explanation. The $l = 2$ pattern of the associated density field does have special significance because it is involved in the Chandler wobble, tidal friction, the orientation of the Earth's spin axis and polar wander. Tectonic models based on surface tectonics have a strong $l = 2$ component, because of the ocean-continent configuration, the locations of island arcs, and the thermal structure of the seafloor. In particular, the velocity highs encompass most of the world's old oceanic lithosphere and convergence regions, and the lows include most of the young oceanic lithosphere and many of the world's hotspots. The correlation between surface-wave velocity and heat flow and tectonics, and between velocity and geoid is relatively high for $l = 2$, particularly at long periods. If the rotation axis of the Earth is controlled by density anomalies in the upper mantle, and if density correlates with velocity, then the pattern would be symmetric about the equator.

The overall pattern of the Love-wave phase velocity variations shows a general correlation with surface tectonics. These waves are most sensitive to the *SH* velocity in the upper 200 to 300 km of the mantle.

The lowest velocity regions are located in regions of extension or active volcanism: the southeastern Pacific, western North America, northeast Africa centered on the Afar region, the central Atlantic, the central Indian Ocean, and the marginal seas in the western Pacific. A high-velocity region is located in the north central Pacific, centered near the Hawaiian swell. Love-wave velocities are low along parts of the Mid-Atlantic Ridge, especially near the triple junctions in the North and South Atlantic.

High-velocity regions in continents generally coincide with Precambrian shields and Phanerozoic platforms (northwestern Eurasia, western and southern parts of Africa, eastern parts of North and South America, and Antarctica). Tectonically active regions, such as the Middle East centered on the Red Sea, eastern and southern Eurasia, eastern Australia, and western North America are slow, as are island arcs or back-arc basins such as the southern Alaskan margin, the Aleutian, Kurile, Japanese, Izu-Bonin, Mariana, Ryukyu, Philippine, Fiji, Tonga, Kermadec, and New Zealand arcs.

Rayleigh-wave velocity also correlates well with surface tectonics, particularly at short periods. The periods generally considered are most sensitive to *SV* velocity between about 100 and 400–600 km and thus sample deeper than the Love waves.

The slowest regions, for Rayleigh waves, are centered on the Red Sea–Afar region, Kerguelen–Indian Ocean triple junction, western North America centered on the Gulf of California, the northeast Atlantic, and Tasman Sea–New Zealand–Campbell Plateau. The fastest regions are the western Pacific, New Guinea–western Australia–eastern Indian Ocean, west Africa, northern Europe and the South Atlantic.

One difference between Love waves and Rayleigh waves is evident in the island arcs along the northwestern margins of the Pacific Ocean, such as the Aleutian, Kurile, Japanese, Izu-Bonin-Mariana, Ryukyu, and Philippine regions. These regions are characterized by high velocity for Rayleigh waves. The difference between Love and Rayleigh wave results is partially explained by the difference in penetration depth. The Love-wave results indicate that the shallow mantle, in the vicinity of island arcs and back-arc basins, is slow. Rayleigh waves sample the fast material that has subducted beneath the island arcs. Because of the large wavelengths, the lateral extent of the fast material must be considerable. Anisotropy may also be involved: In regions of convergence and divergence, the mantle flow can be expected to be generally vertical, whereas in the centers of convection cells, the flow is mainly horizontal. The sense of anisotropy will change at ridges and subduction zones.

For Rayleigh waves the Atlantic Ocean is generally high velocity except for the northern part. The low velocities associated with the midoceanic ridge system, which appear in the Love-wave maps, are not evident in the long-period Rayleigh-wave maps. This may indicate that many ridge segments are passive or that *SV* exceeds *SH* in regions of ascending flow.

Maps of surface-wave velocity (Nakanishi and Anderson, 1983, 1984a,b) provide, perhaps, the most direct display possible of the lateral heterogeneity of the mantle. The phase and group velocities can be obtained with high precision and with relatively few assumptions. In general, the shorter period waves, which sample only the crust and shallow mantle, correlate well and as expected with surface tectonics. The longer period waves, which penetrate into the transition region (400–650 km), correlate less well with surface tectonics. The inversion of these results for Earth structure involves many more assumptions and approximations.

REGIONALIZED INVERSION RESULTS

Figure 13-2 shows vertical shear-velocity profiles, expressed as differences from the average Earth, using the regionalization approach (Nataf and others, 1986). Young oceans, region D, have slower than average velocities throughout the upper mantle and are particularly slow between 80 and 200 km, in agreement with the higher resolution body-wave studies. Old oceans, region A, are fast throughout the upper mantle. Intermediate-age oceans (B and C) are intermediate in velocity at all depths. Most of the oldest oceans are adjacent to subduction zones, and the subduction of cold material may be partially responsible for the fast velocities at depth. Notice that velocities converge toward 400 km but that differences still remain below this depth. The continuity of the low velocities beneath young oceans, which include midocean ridges, suggests that the ultimate source region for MORB is below 400 km. Shields (S) are faster than all other tectonic provinces except old ocean from 100 to 250 km. Below 220 km the velocities under shields decrease, relative to average Earth, and below 400 km shields are among the slowest regions. At all depths beneath shields the velocities can be accounted for by reasonable mineralogies and temperatures without any need to invoke partial melting. Trench and marginal sea regions (T), on the other hand, are relatively slow above 200 km, probably indicating the presence of a partial melt, and fast below 400 km, probably indicating the presence of cold subducted lithosphere. The large size of the tectonic regions and the long wavelengths of surface waves require that the anomalous regions at depth are much broader than the sizes of slabs or the active volcanic regions at the surface. This suggests very broad upwellings under young oceans and abundant piling up of slabs under trench and old ocean regions. The latter is evidence for layered mantle convection and the cycling of oceanic plates into the transition region.

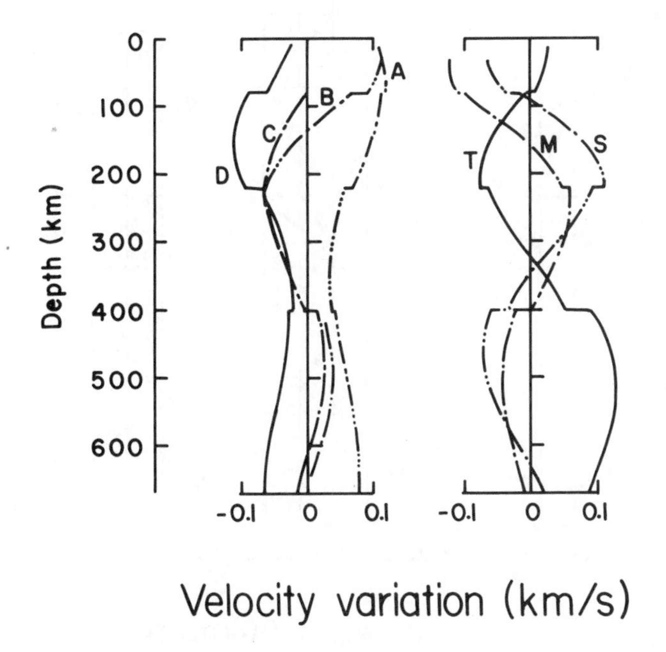

FIGURE 13-2
Variation of the *SV* velocity with depth for various tectonic provinces. A–D, oceanic age provinces ranging from old, A, to young, D; S, continental shields; M, mountainous areas; T, trench and island-arc regions. These are regionalized results (after Nataf and others, 1986).

Shields and young oceans are still evident at 250 km. At 350 km the velocity variations are much suppressed. Below 400 km, most of the correlation with surface tectonics has disappeared, in spite of the regionalization, because shields and young oceans are both slow, and trench and old ocean regions are both fast. Most of the oceanic regions have similar velocities at depth. This is a severe test of the continental tectosphere hypothesis of Jordan (1975), described later in this chapter. Shields do not have higher velocities than some other tectonic regions below 250 km and definitely do not have "roots" extending throughout the upper mantle or even below 400 km. Results for other depths are given by Nataf and others (1984, 1986). In high-resolution body-wave studies, subshield velocities drop rapidly at 150 km depth, although velocities remain relatively high to about 390 km. These high velocities could represent "roots" physically attached to the shield lithosphere, overridden cold oceanic lithosphere or simply "normal" convecting mantle weakly coupled to the overlying shield lithosphere via a boundary layer at 150–200 km depth. The velocities below 200 km can be explained by an adiabatic temperature gradient and therefore probably represent normal convecting mantle. Therefore, it is the slow mantle under ridges and tectonic regions that is anomalous, and, if anything, these are the regions with the roots. If the mantle under shields is convectively stagnant, as implied by the deep tectosphere hypothesis, a high thermal gradient would extend over a large depth interval. This could lead to partial melting and a depression of the olivine-spinel phase boundary under shields. I therefore prefer the 150-km-thick plate hypothesis, that is, a correspondence of the thickness of the plate with the seismic high-velocity layer. The slightly higher than average subshield mantle between about 200 km and 350–390 km may be a boundary layer or could represent colder than average mantle over which the continents have drifted.

SPHERICAL HARMONIC INVERSION RESULTS

An alternative way to analyze the surface-wave data is through a spherical harmonic expansion that ignores the surface tectonics (Nataf and others, 1986). This provides a less biased way to assess the depth extent of tectonic features.

The mantle is assumed to have the same average properties, including anisotropy, as the PREM Earth model given in the Appendix. Perturbations are assumed to be smooth between the discontinuities in PREM (60, 220, 400 and 670 km) and to be loosely coupled across these discontinuities. Thus, the radial variation in perturbation can change rapidly at physical discontinuities. Woodhouse and Dziewonski (1984) invoked a smooth radial perturbation throughout the upper mantle in their inversions. The data set is not yet complete enough to favor one parameterization over another. In the parameterization of Nataf and others (1984, 1986), the variation across the discontinuities is continuous unless the data require otherwise. In general, the perturbations across discontinuities are highly correlated. In both of these models the character of the perturbations changes at 220 and 400 km. In particular, the amplitudes of the perturbations decrease with depth.

At shallow depths we have little resolution because global maps for short-period surface waves have not yet been prepared. Nevertheless, surface waves sense the shallow structure, and some information is available. At a depth of 50 km the major tectonic features correlate well with the shear velocity. Shields and old oceans are fast. Young oceanic regions and tectonic regions are slow. The slowest regions are centered near the midocean ridges, some back-arc basins and the Red Sea. The hotspot province in the south Pacific is slow at shallow depths, but the shallow mantle in the north-central Pacific, including Hawaii, is fast. The shields are particularly fast. At 150 km all the major shield

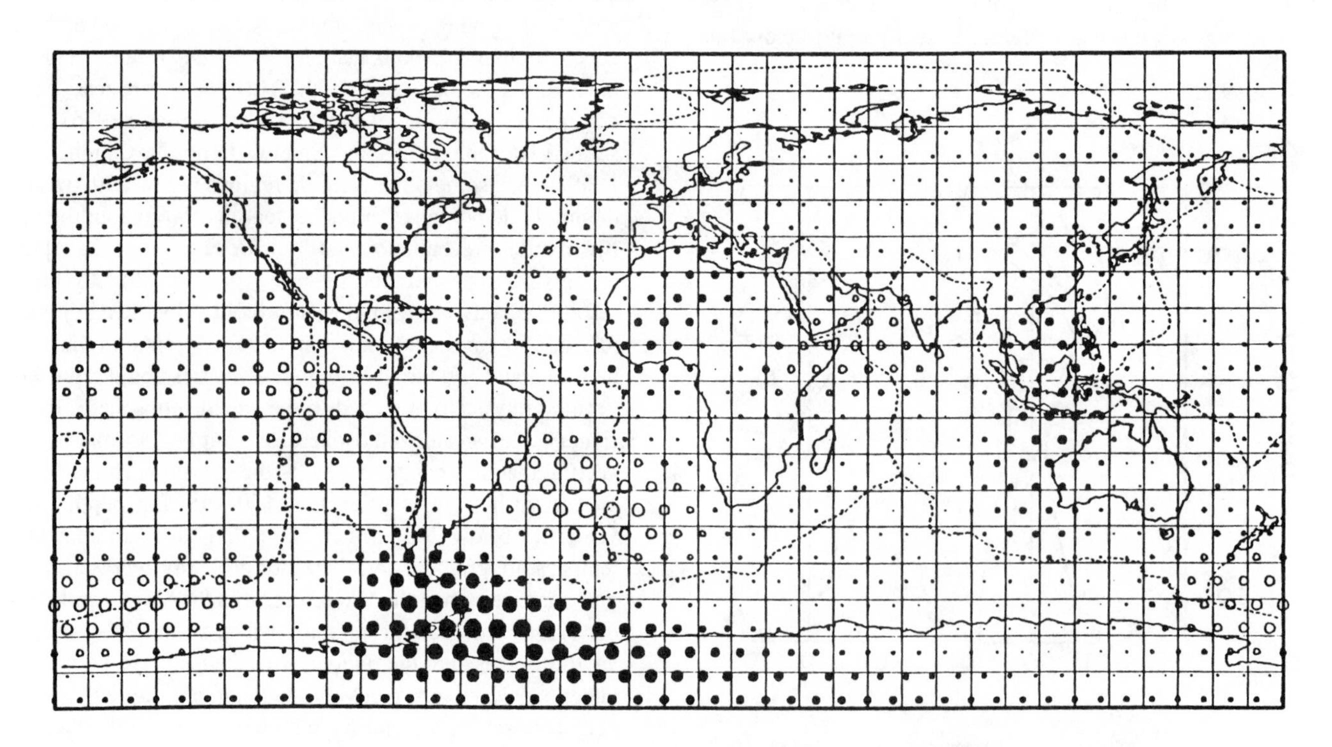

FIGURE 13-3
Map of *SV* velocity at 250 km depth from sixth-order spherical harmonic representation of Nataf and others (1984). Note the slow regions associated with the midocean ridges. The fastest regions are in the far south Atlantic, some subduction areas and northwest Africa.

areas fall near the centers of high-velocity anomalies, and the ridges are in low-velocity regions. The main differences at this depth, compared to the regionalized models, are the very low velocities in eastern Asia, the Red Sea region, and New Zealand. The central Pacific and the northeastern Indian Ocean are fast. The Arctic Ocean between Canada and Siberia is slow. At 250 km (see Figure 13-3), the shields are less evident than at shallower depths and than in the regionalized model. On the other hand, the areas containing ridges are more pronounced low-velocity regions. The central Pacific and the Red Sea are also very slow. The prominent low-velocity region in the central Pacific roughly bounded by Hawaii, Tahiti, Samoa and the Caroline Islands is the Polynesian Anomaly. This feature may be related to the extensive volcanism that occurred in the western Pacific in the Cretaceous when the Pacific plate was over this anomaly. The northern and southern Atlantic are also slow. Northern Europe and Antarctica are mainly fast. The highest velocity anomalies are in the far south Atlantic, northwest Africa to southern Europe and the eastern Indian Ocean to southeast Asia and are not confined to the older continental areas. At 340 km (Figure 13-4), most ridges are still evident although the slow region of the eastern Pacific has shifted off the surface expression of the East Pacific Rise. A central Pacific slow-velocity region persists throughout the upper mantle. The shield areas are no longer evident. Some ridges have average or faster than average velocities at this depth, particularly the Australian-Antarctic ridge and the central Atlantic ridge. These ridge segments are in relative geoid lows. This may indicate that parts of the ridge system involve lateral transport at shallow depth. Generally, however, the source for midocean ridges appears to extend below 350 km depth. Below 400 km there is little correlation with surface tectonics, and in many areas the velocity anomalies are of opposite sign from those in the shallow mantle, in agreement with the regionalized results. The net result is that shields, on average, have very high velocities to 150–200 km, and ridges, on average, have low

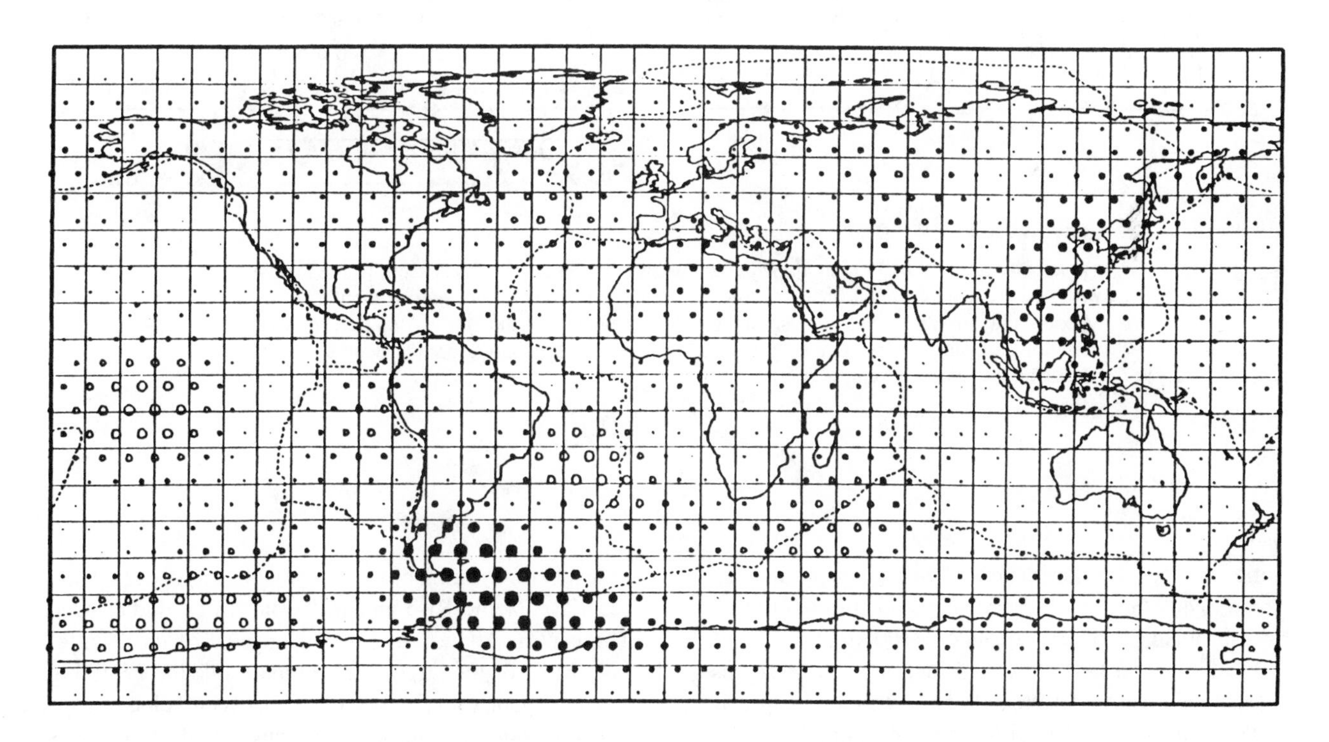

FIGURE 13-4
Map of *SV* velocity at 340 km depth. A prominent low-velocity anomaly shows up in the central Pacific (the Polynesian Anomaly). The fast anomalies under eastern Asia, northern Africa and the South Atlantic may represent mantle that has been cooled by subduction.

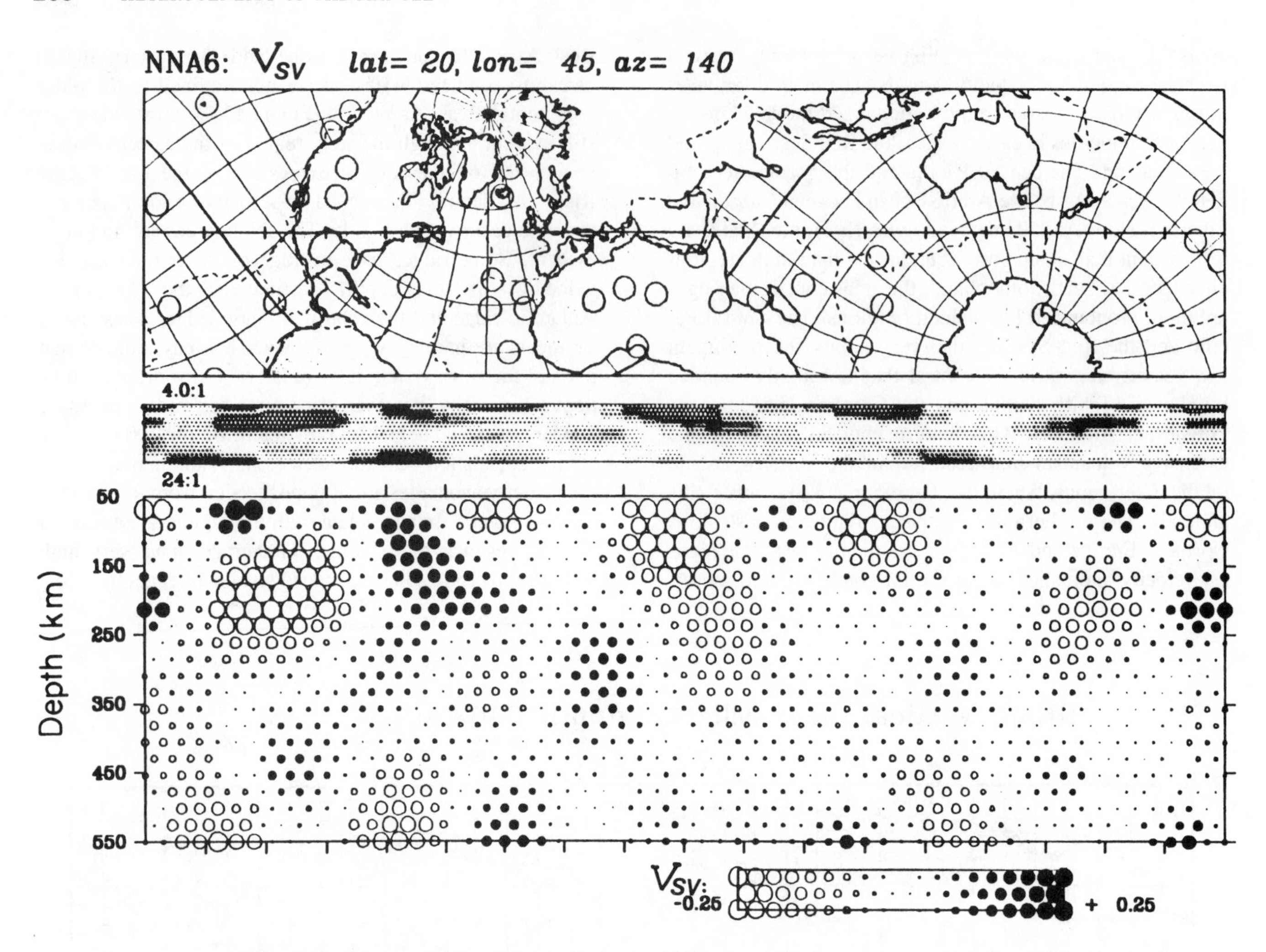

FIGURE 13-5
SV velocity from 50 to 550 km along the great-circle path shown. Cross-sections are shown with two vertical exaggerations. Note the low-velocity regions in the shallow mantle below Mexico, the Afar and south of Australia and the asymmetry of the North Atlantic. Velocity variations are much more extreme at depths less than 250 km than at greater depths. The circles on the map represent hotspots.

velocities to 350–400 km. The Red Sea anomaly appears to extend to 400 km, but the very slow velocities associated with western North America die out by 300 km.

By 400 or 450 km the pattern seen at shallow depths changes dramatically. The mantle beneath most ridges is fast. The Polynesian Anomaly, although shifted, is still present. Eastern North America and/or the western North Atlantic are slow. Most of South America, the South Atlantic and Africa are fast. The north-central Pacific is slow. Most hotspots are above faster-than-average parts of the mantle at this depth. The fast regions under the Atlantic, western North America, the western Pacific and south of Australia may be sites of subducted or overriden oceanic lithosphere. Prominent slow anomalies are under the northern East Pacific Rise and in the northwest Indian Ocean. The central Atlantic and the older parts of the Pacific are fast. Most of the large slow anomalies define geoid highs. There is poor overall correlation between velocities in the upper mantle and the geoid because subduction zones in general are associated with geoid highs and regions of fast velocity below 300 km. Hot and slow regions of the upper mantle generate geoid highs because of thermal expansion and uplift of the surface. Cold and fast regions of the mantle (subduction zones) generate geoid highs because the excess mass is held up by increasing intrinsic density or viscosity with depth (that is, it is uncompensated).

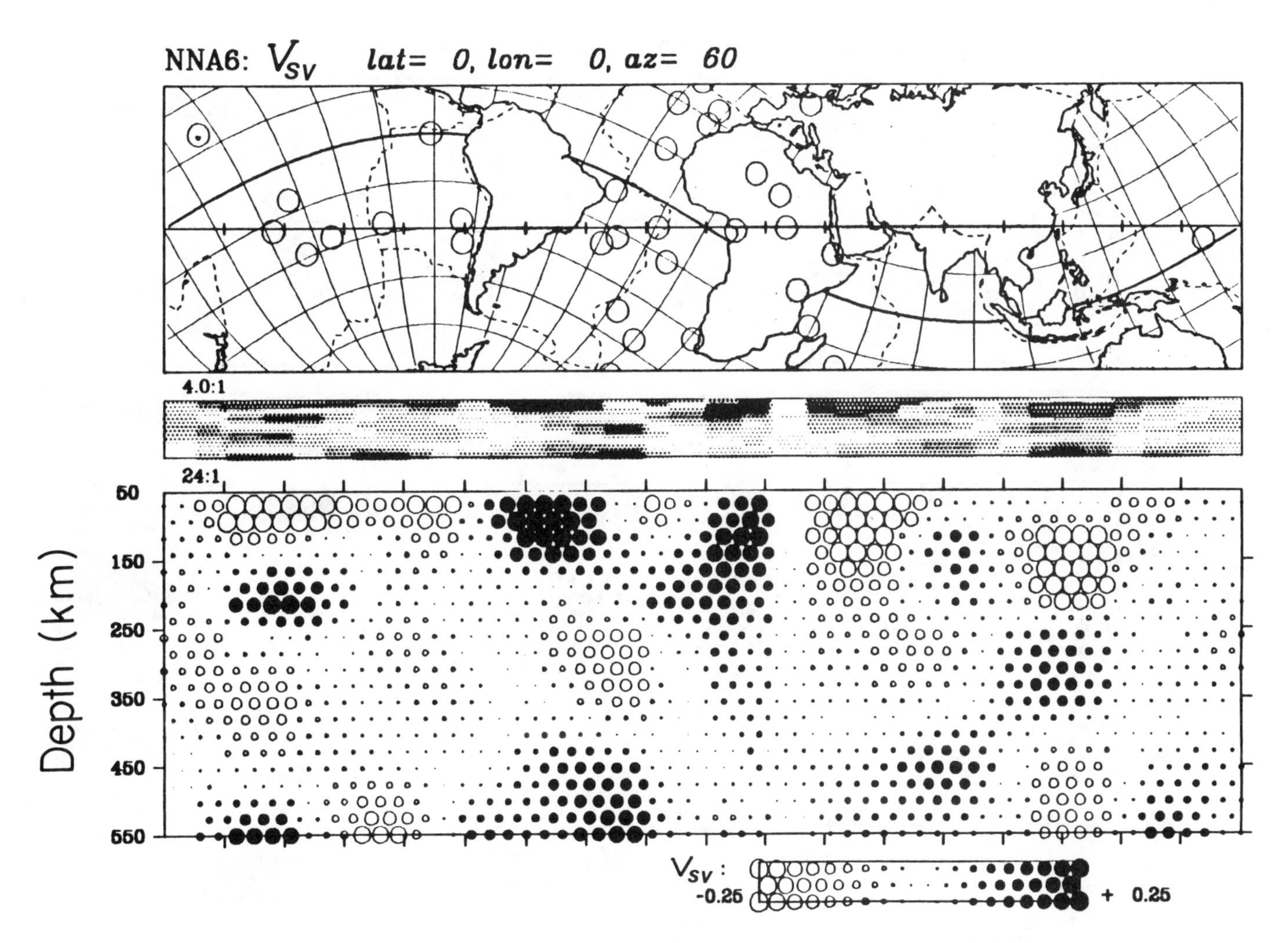

FIGURE 13-6
SV velocity in the upper mantle along the cross-section shown. Note low velocities at shallow depth under the western Pacific, replaced by high velocities at greater depth. The eastern Pacific is slow at all depths. The Atlantic is fast below 400 km.

The upper-mantle shear velocities along three great-circle paths are shown in Figures 13-5 to 13-7. The open circles are slower than average regions of the mantle.

CORRELATION WITH HEAT FLOW AND GEOID

Heat-flow and geoid maps are generally presented in terms of spherical harmonics, and comparison with surface-wave results is straightforward. Love-wave velocities are well correlated with heat flow and, therefore, surface tectonics. The geoid has a weaker correlation with the surface-wave velocities. The heat flow has positive correlation with phase slowness (that is, reciprocal velocity) for degrees from 1 to 6. The geoid shows negative correlation at $l = 2$ (Love and Rayleigh) and 3 (Rayleigh) and positive correlation at $l = 4$ to 6.

Shields are areas of low heat flow and exhibit high Love-wave velocities, in spite of the thick low-velocity crust. From the Love-wave data one would predict that southeast Asia and the Afar region are characterized by high heat flow. These regions have few heat-flow observations. The Canadian Shield and Siberian Platform anomalies are less evident in the Love-wave maps than in heat-flow maps.

The correlation between surface-wave velocities and the geoid is weak. Geoid highs are associated with both subduction and hotspots. Ridges are not evident in either the long-wavelength geoid or Rayleigh-wave maps. The

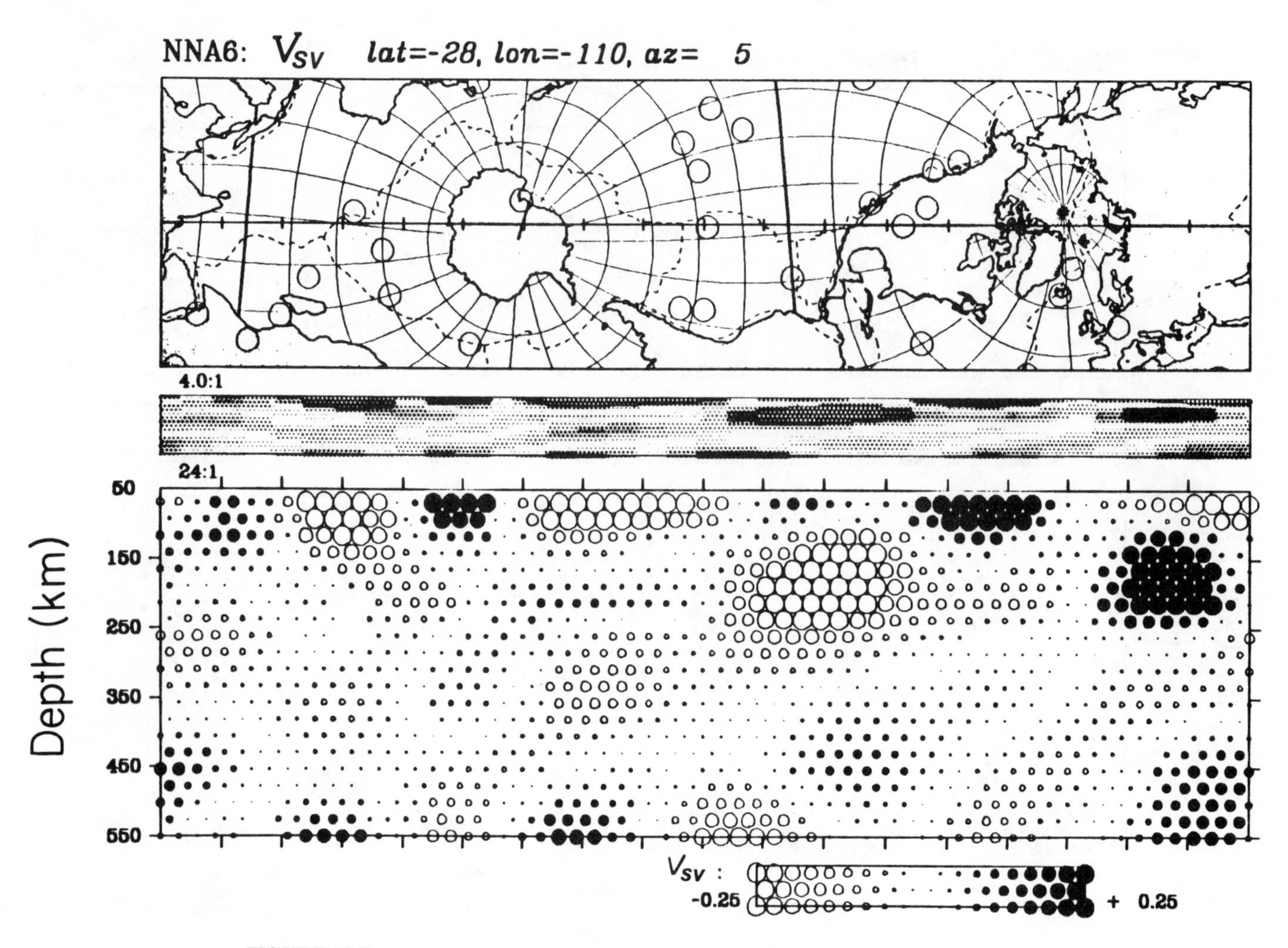

FIGURE 13-7
Cross-section illustrating the low upper-mantle shear velocities under midocean ridges and western North America. Stable regions are relatively fast in the shallow mantle.

$l = 2$ correlation indicates that the long-wavelength part of the geoid has a negative correlation with slowness; that is, $l = 2$ geoid highs correlate with $l = 2$ velocity highs. For shorter wavelength features the reverse is the case.

Surface-wave studies show (1) low-velocity anomalies centered on southeast Asia, Afar and the Gulf of California, the latter two persisting to long periods thus indicating a deep upper mantle origin; (2) an association of high velocities, at long periods, with the western Pacific subduction zones, indicating a large volume of fast material in the upper mantle; and (3) high velocities in the South Atlantic that do not have an obvious explanation in terms of present plate configurations. The South Atlantic may overlie ancient subducted Pacific Ocean lithosphere.

Regions of generally faster than average velocity occur in the western Pacific, the western part of the African plate, Australia–southern Indian Ocean, part of the south Atlantic, northeastern North America–western North Atlantic, and northern Europe. These are all geoid lows in an $l = 4$–9 expansion of the geoid.

Dense regions of the mantle that are in isostatic equilibrium generate geoid lows (Haxby and Turcotte, 1978; Hager, 1983). High density and high velocity are both consistent with cold mantle. The above regions may be underlain by cold subducted material. Geoid highs occur near Tonga-Fiji, the Andes, Borneo, the Red Sea, Alaska, the northern Atlantic, and the southern Indian Ocean. These are generally slow regions of the mantle and are therefore presumably hot. The upward deformation of boundaries counteracts the low density associated with the buoyant material, and for uniform viscosity the net result is a geoid high (Hager, 1983). An isostatically compensated column of low-density material also generates a geoid high because of the elevation of the surface.

Correlation coefficients between phase velocity and the geoid show that Love waves generally have a negative correlation (geoid highs correlate with slow velocity) for $l = 4-6$. Rayleigh waves generally correlate well with the geoid only for $l = 6$ and, at some periods, with $l = 5$. The $l = 6$ correlation is particularly strong for both Love waves and Rayleigh waves.

Hager (1983) found an excellent correlation between the observed geoid and the "slab geoid" for $l = 4$, 5, and 7–9 and weak correlation at $l = 6$. There is a high degree of correlation of subduction zones with geoid highs, particularly those centered on Tonga-Fiji, Borneo, Alaska and Peru, and associated lows such as the Nazca plate, the western Pacific, and western Australia. The slab geoid does not explain the geoid lows in Siberia, Canadian shield–western Atlantic, and the South Atlantic or the geoid highs in the south Indian Ocean, the Red Sea, and the North Atlantic. These are high-velocity and low-velocity regions, respectively. The slab geoid predicts higher than observed anomalies in Tonga-Fiji, Japan and the Kuriles. The slab-geoid correlation can be understood in terms of a large amount of dynamically supported dense material either in the upper mantle or in both the upper and lower mantles in the vicinity of active subduction zones (Hager, 1983). The surface wave–geoid correlation, which apparently explains the missing $l = 6$ component of the slab geoid, however, must be due to an anomalous mass distribution in the shallow mantle.

These results suggest that features of the geoid having wavelengths of about 4000 to 10,000 km are generated in the upper mantle. Geoid anomalies of this wavelength generally have an amplitude of about 20 to 30 m. An isostatically compensated density anomaly of 0.5 percent spread over the upper mantle would give geoid anomalies of this size. It therefore appears that a combination of slabs and broad thermal anomalies in the upper mantle can explain the major features of the $l = 4-9$ geoid. The longer wavelength part of the geoid, $l = 2$ and 3, correlates with seismic velocities in the deeper part of the mantle. The shorter wavelength ($l > 9$) part of the geoid correlates with topography. Topography and geoid also correlate moderately well at $l = 4$ and 6, but crustal thickness variations and the distribution of the continents, with shallow compensation, do not explain the magnitude of the effect. Figure 13-8 shows the actual distribution of Love-wave phase velocities and that computed from the geoid assuming a linear relationship between velocity and geoid height. Most subduction regions are slow at short periods, presumably because of the presence of back-arc basins and hot, upwelling material above the slab.

Slabs are colder than normal mantle and therefore they are denser. Denser minerals also occur in the slab because of temperature-dependent phase changes (Chapter 16). The phase change effect leverages the role of temperature with the result that slabs confined to the upper mantle can explain the magnitude of the slab-related geoid.

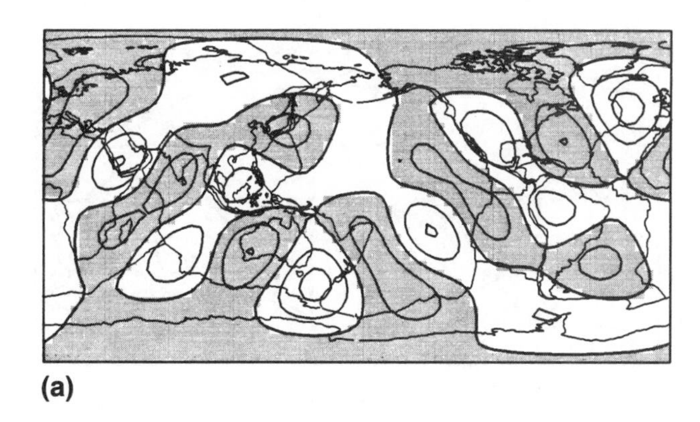

(a)

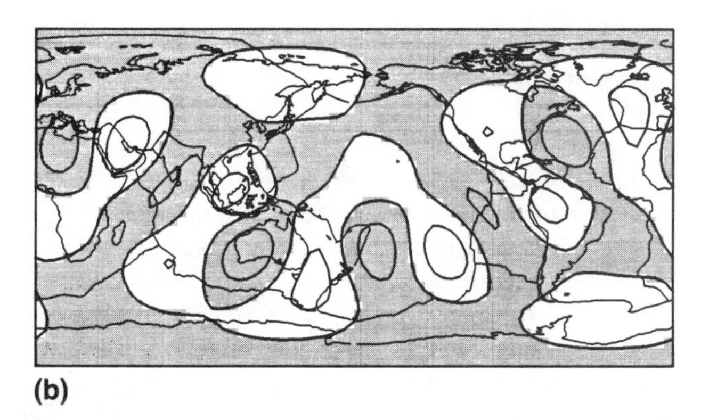

(b)

FIGURE 13-8
The intermediate wavelength ($l \approx 6$) geoid is controlled by processes in the upper mantle (slabs, hot asthenosphere). **(a)** The $l = 6$ component of a global spherical harmonic expansion of Love-wave phase velocities. These are sensitive to shear velocity in the upper several hundred kilometers of the mantle. Note that most shields are fast (gray areas), and oceanic and tectonic regions are slow (white areas). Hot regions of the upper mantle, in general, cause geoid highs because of thermal expansion and uplift of the surface and internal boundaries. Tectonic and young oceanic areas are generally elevated over the surrounding terrain. **(b)** Phase velocity computed from the $l = 6$ geoid, assuming a linear relationship between geoid height and phase velocity. Note the agreement between these two measures of upper-mantle properties (Tanimoto and Anderson, 1985).

AZIMUTHAL ANISOTROPY

The velocities of surface waves depend on position and on the direction of travel. If an adequately dense global coverage of surface-wave paths is available, then azimuthal anisotropy as well as lateral heterogeneity can be studied. Preliminary maps of azimuthal anisotropy have been prepared by Tanimoto and Anderson (1984, 1985). Regional surface-wave studies also give azimuthal variations (Forsyth, 1975).

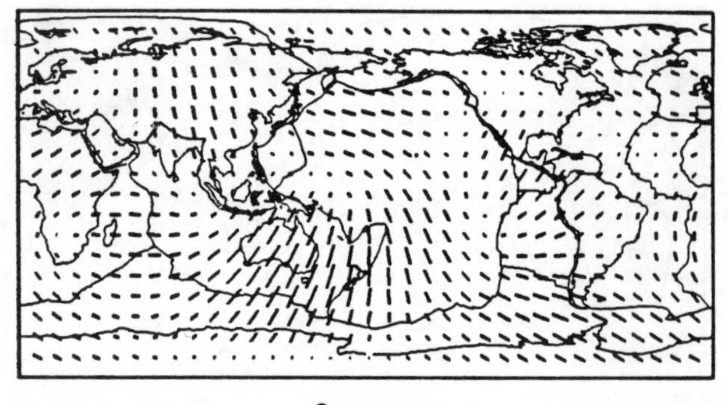

FIGURE 13-9
Azimuthal variation of phase velocities of 200-s Rayleigh waves (expanded up to $l = m = 3$) (after Tanimoto and Anderson, 1984, 1985).

Azimuthal anisotropy can be caused by oriented crystals or a consistent fabric caused by, for example, dikes or convective rolls in the shallow mantle. In either case, the azimuthal variation of seismic-wave velocity is telling us something about convection in the mantle. Since long-period surface waves see beneath the plates, it may be possible to map the direction of flow in the asthenosphere and thus discuss the nature of lithosphere-asthenosphere coupling, style of mantle convection and the viscosity structure of the mantle.

Figure 13-9 is a map of the azimuthal results for 200-s Rayleigh waves expanded up to $l = m = 3$. The lines are oriented in the maximum velocity direction, and the length of the lines is proportional to the anisotropy. The azimuthal variation is low under North America and the central Atlantic, between Borneo and Japan, and in East Antarctica. Maximum velocities are oriented approximately northeast-southwest under Australia, the eastern Indian Ocean, and northern South America and east-west under the central Indian Ocean; they vary under the Pacific Ocean from north-south in the southern central region to more northwest-southeast in the northwest portion. The fast direction is generally perpendicular to plate boundaries. There is little correlation with plate motion directions, and little is expected since 200-s Rayleigh waves are sampling the mantle beneath the lithosphere. Pn velocity correlates well with spreading direction (Morris and others, 1969). The lack of correlation for long-period Rayleigh waves implies a low-viscosity asthenosphere.

LATERAL HETEROGENEITY FROM BODY WAVES

Seismic waves recorded at various seismic observatories on the Earth's surface arrive with a delay or advance relative to average travel-time tables. These are called station residuals, or statics, and give information about the velocity in the shallow mantle under the station. In general, tectonic regions are slow and shield areas are fast. Likewise, different source regions have different anomalies; these are called source residuals. Because of the irregular distribution of stations and events, one cannot determine these anomalies on a global basis. In contrast to surface-wave studies, however, the anomalies can be fairly well localized geographically although the depth extent is ambiguous.

Station anomalies cover the range from about $+1$ to -1 s and are too large to be caused by crustal variations alone. In general, the fastest regions are stable shields and platforms; the slowest regions are tectonic and oceanic-island stations. This is consistent with surface-wave results.

The highest resolution body-wave studies, involving the use of travel times, apparent velocities and waveform fitting, have provided details about upper-mantle velocity structures in several tectonic regions. Figure 13-10 shows some of these results. Note that low velocities extend to depths of about 390 km for the tectonic and oceanic structures. These regional studies confirm the general features of the earlier global surface-wave studies.

Shields have extremely high shear velocities extending to 150 km depth. It is natural to assume that this is the

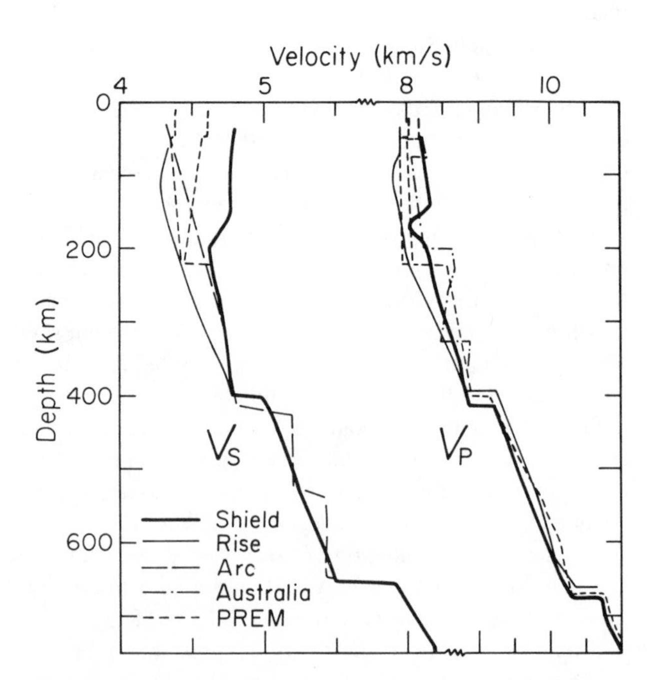

FIGURE 13-10
V_s and V_p in various tectonic provinces. Note the large lateral variations above 200 km and the moderate variations between 200 and 400 km. The reversal in velocity between 150 and 200 km under the shield area may indicate that this is the thickness of the stable shield plate. Models from Grand and Helmberger (1984a,b) and Walck (1984).

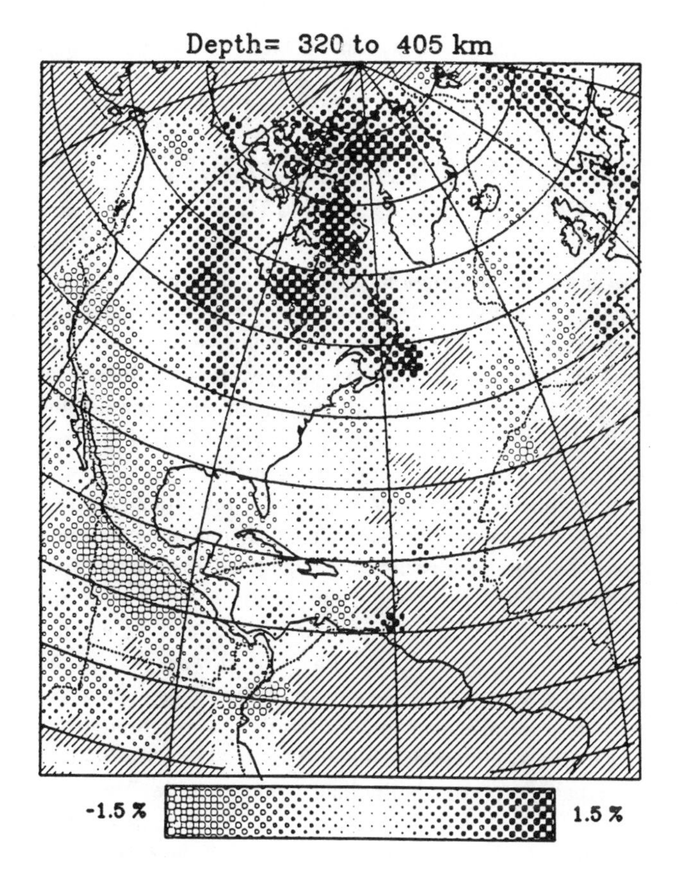

FIGURE 13-11
SH velocities in the upper mantle at depths of 320 to 405 km (after Grand, 1986).

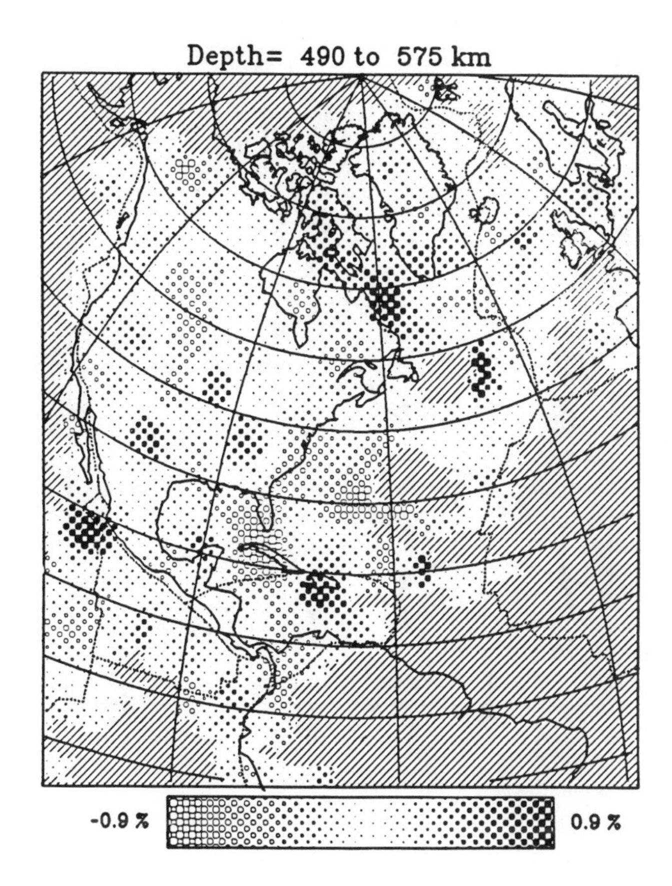

FIGURE 13-12
SH velocities at depths of 490 to 575 km (after Grand, 1986).

thickness of the stable continental plate and that the underlying mantle is free to deform and convect. The seismic velocities in the shield lithosphere are consistent with cold depleted peridotite (harzburgite or dunite). This is buoyant relative to fertile peridotite. The high-velocity layer, or LID, under tectonic and oceanic regions is much thinner, of the order of 30 to 50 km, and the shear velocities of the underlying mantle are much lower than under shields, implying higher temperatures and, possibly, the presence of a partial melt phase. The implication is that oceanic plates are much thinner and possibly more mobile than continental plates. Jordan (1975) and Sipkin and Jordan (1976) made a radically different proposal. They suggested that the high seismic velocity associated with shields extended to depths in excess of 400 km and perhaps to 700 km and that the continental plates are equally thick. Jordan called this hypothetical deep continental root the tectosphere. But others showed that the large differences in oceanic and continental ScS times (shear waves that reflect off the core), the data used in the development of the continental tectosphere hypothesis, were mainly caused by differences shallower than 200 km (Okal and Anderson, 1975; Anderson, 1979). These waves have very little depth resolution and can only resolve differences below 400 km if the shallower mantle is independently constrained.

Although the largest variations (of the order of 10 percent) in seismic velocity occur in the upper 200 km of the mantle, the velocities from 200 to about 400 km under oceanic and tectonic regions are slightly less (on the order of 4 percent on average) than under shields. The question then arises, what is the cause of these deeper velocity variations? Is the continental plate 400 km thick or are the velocities between 150–200 and 400 km beneath shields appropriate for "normal" subsolidus convecting mantle?

Grand (1986) performed a shear-wave tomographic study of the North American plate and adjacent regions. His results (Figures 13-11 to 13-14) are similar to the surface-wave study of Nataf and others (1986). Oceanic and tectonic regions, including western North America, are slow down to 400 km. The range of velocities is greater than 6 percent above 140 km, about 3 percent to 400 km and less than 2 percent below 400 km. A narrow planar zone of high velocity appears at the top of the lower mantle under eastern North America and the Caribbean. The most dramatic fea-

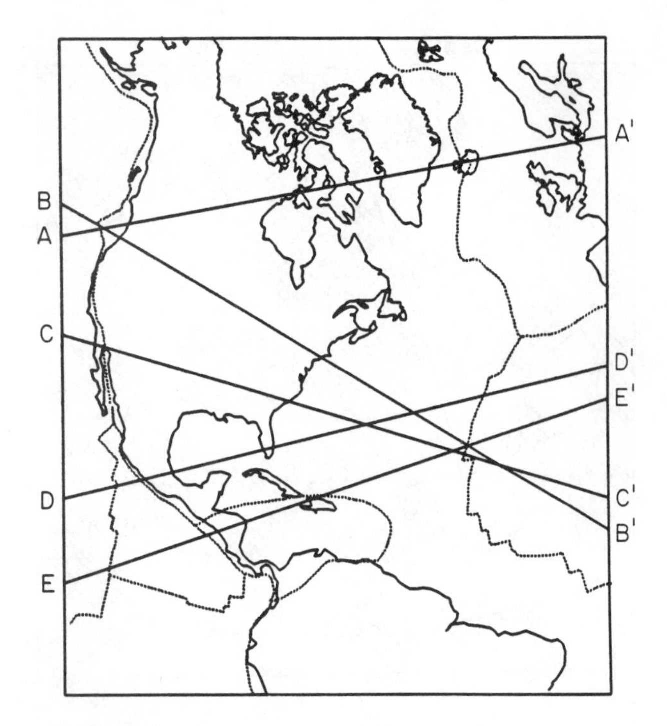

FIGURE 13-13
Locations of cross-sections (after Grand, 1986).

ture of Grand's model is the large decrease in the velocity variation below 400 km depth. This is clearly an important tectonic boundary.

Graves and Helmberger (1987) studied the shear velocity along a profile from Tonga across the East Pacific Rise (EPR) to Canada (Figure 13-15). The low-velocity zone under the EPR is very pronounced, being both shallow and of great lateral extent, some 6000 km. Midocean ridges are not the narrow features often assumed.

BODY-WAVE TOMOGRAPHY OF THE LOWER MANTLE

The large lateral variations of seismic velocity in the upper mantle make it difficult to detect the smaller variations in the lower mantle. On the other hand, long-wavelength density variations in the lower mantle have more influence on the geoid and orientation of the Earth than comparable variations in the upper mantle. Both types of study are relevant to the style of convection in the mantle.

In spherical harmonic terms, the low-order and low-degree components (long-wavelength features) of the compressional velocity distribution in the lower mantle are similar to the low-order components of the geoid (Hager and Clayton, 1986). The polar regions are fast, and the equatorial regions, in general, are slow. The slowest regions are centered near the long-wavelength geoid highs, which occupy the central Pacific and the North Atlantic through Africa to the southwestern Indian Ocean. The slow regions presumably represent hotter than average mantle. The range of velocity is about ± 1 percent, much less than in the shallow mantle. The largest variations are near the core-mantle boundary (CMB), with the maximum slowness beneath the southern tip of Africa.

Slow regions of the lower mantle occur under the Mid-Indian Ocean Ridge, the East Pacific Rise, western North America and South Africa. Fast regions occur south of Australia, China, South America and northern Pacific.

Generally, there is lack of radial continuity in the lower mantle. Large-scale convection-like features are not evi-

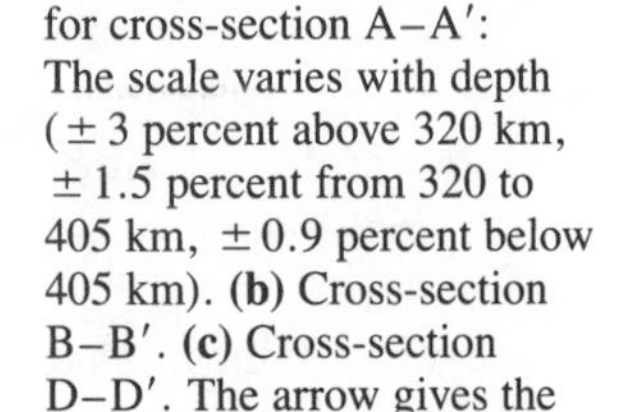

FIGURE 13-14
(a) *SH* velocity versus depth for cross-section A–A′: The scale varies with depth (± 3 percent above 320 km, ± 1.5 percent from 320 to 405 km, ± 0.9 percent below 405 km). **(b)** Cross-section B–B′. **(c)** Cross-section D–D′. The arrow gives the location of the trench off Mexico–Central America. **(d)** Cross-section E–E′ (after Grand, 1986).

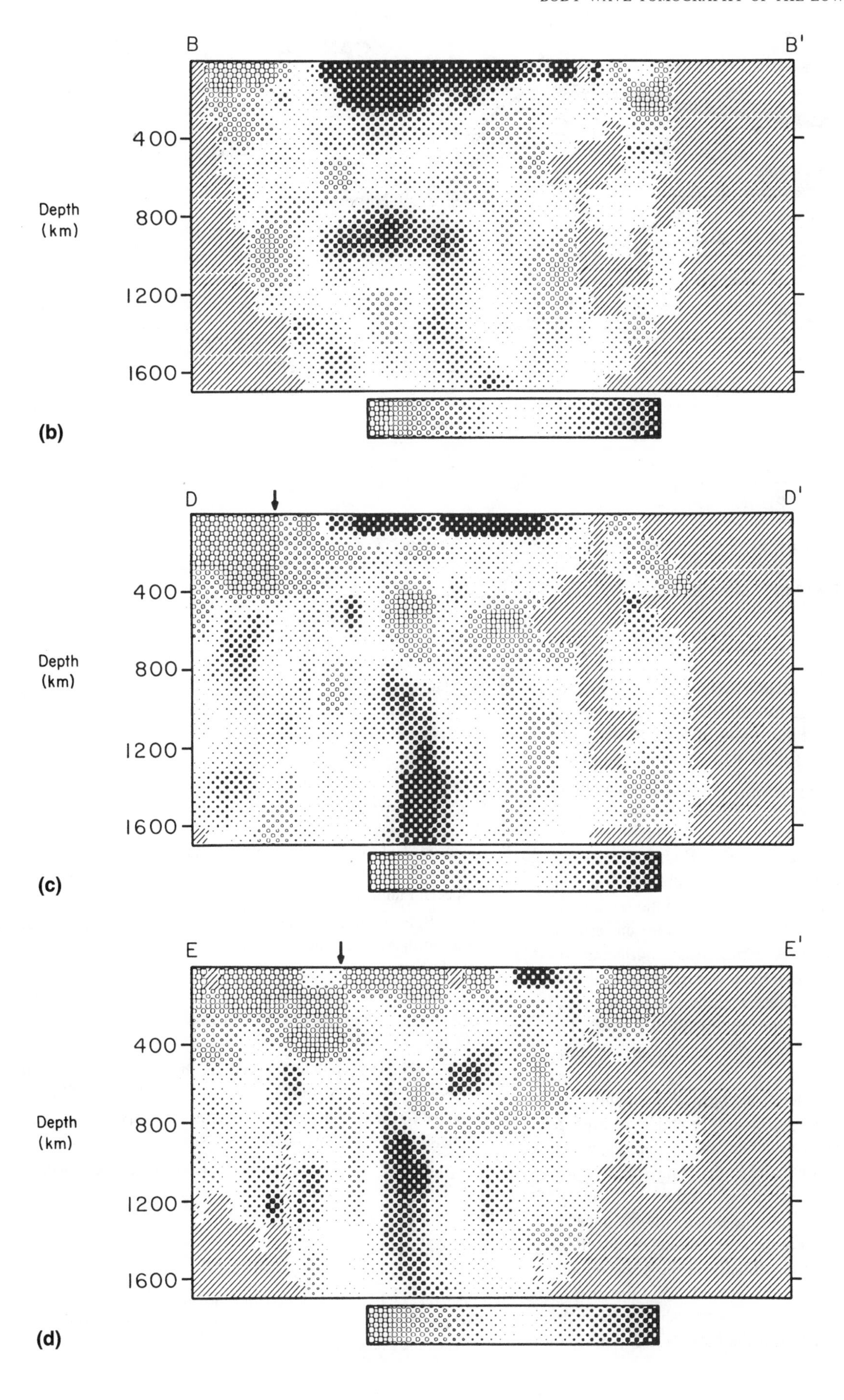
B
B'
400
800
1200
1600
Depth (km)
(b)
D
D'
400
800
1200
1600
Depth (km)
(c)
E
E'
400
800
1200
1600
Depth (km)
(d)

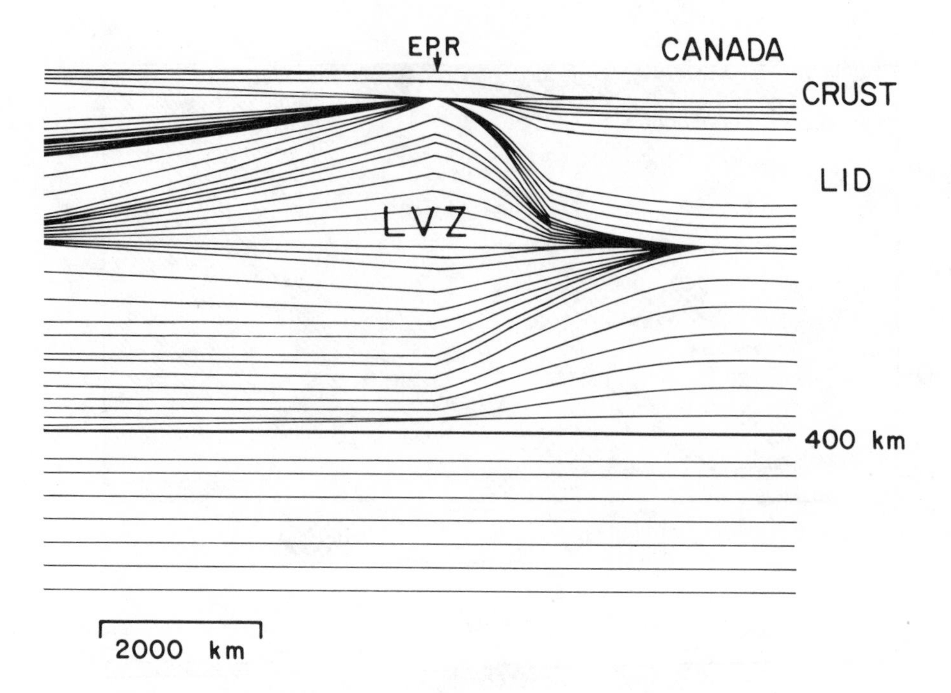

FIGURE 13-15
Shear-wave velocities along a profile from Tonga to Canada (after Graves and Helmberger, 1987). EPR, East Pacific Rise; LVZ, low-velocity zone.

dent. In the top part of the lower mantle (Figure 13-16), the prominent low-velocity regions are under the Indian Ocean, the western Pacific and China, south of New Zealand, the central Atlantic, the East Pacific Rise and the central Pacific. The fastest regions are Siberia, south Africa, south of South America and the northeast Pacific. At mid-to-lower mantle depths (Figures 13-17 and 13-18), the slowest regions are southeast Africa and adjacent Indian Ocean, the Cape Verde–Canaries–Azores region of the Atlantic, and the equatorial Pacific. The fast regions are in the north polar regions, North America, the southern Pacific, eastern Indian Ocean and the South Atlantic. Most continental regions are fast. Near the base of the lower mantle (Figure 13-19), the prominent low-velocity regions are southern and northern Africa, Brazil and the south-central Pacific. The fast regions are Asia, the North Atlantic, the northern Pacific and Antarctica. The locations of hotspots do not correlate with the slower regions at the base of the mantle.

The slow regions are presumably hotter than average and therefore probably represent buoyant upwellings, which in turn are responsible for geoid highs. They will tend to reorient the Earth so as to lie in low latitudes. This brings colder than average mantle into the polar regions. This, in turn, will affect convection in the core—cold downwellings in the core should occur preferentially in the polar regions and beneath the other fast regions at the base of the mantle. The colder regions of the lower mantle should have the highest temperature gradients at the base of the mantle, in D″. These are the regions where core heat is most readily removed. Heat flow from the core is probably far from uniform.

General References

Anderson, D. L. (1987) The depths of mantle reservoirs. In *Magmatic Processes* (B. O. Mysen, ed.), Geochemical Society, Special Publication No. 1.

Grand, S. P. (1986) Shear velocity structure of the mantle beneath the North American plate, Ph.D. Thesis, California Institute of Technology.

Hager, B. H. (1983) Global isostatic geoid anomalies for plate and boundary layer models of the lithosphere, *Earth Planet. Sci. Lett.*, *63*, 97–109.

Hager, B. H. (1984) Subducted slabs and the geoid; constraints on mantle rheology and flow, *J. Geophys. Res.*, *89*, 6003–6015.

Hager, B. H. and R. J. O'Connell (1979) Kinematic models of large-scale mantle flow, *J. Geophys. Res.*, *84*, 1031–1048.

Haxby, W. and D. Turcotte (1978) On isostatic geoid anomalies, *J. Geophys. Res.*, *83*, 5473.

References

Anderson, D. L. (1967) Latest information from seismic observations. In *The Earth's Mantle* (T. F. Gaskell, ed.), 355–420, Academic Press, New York.

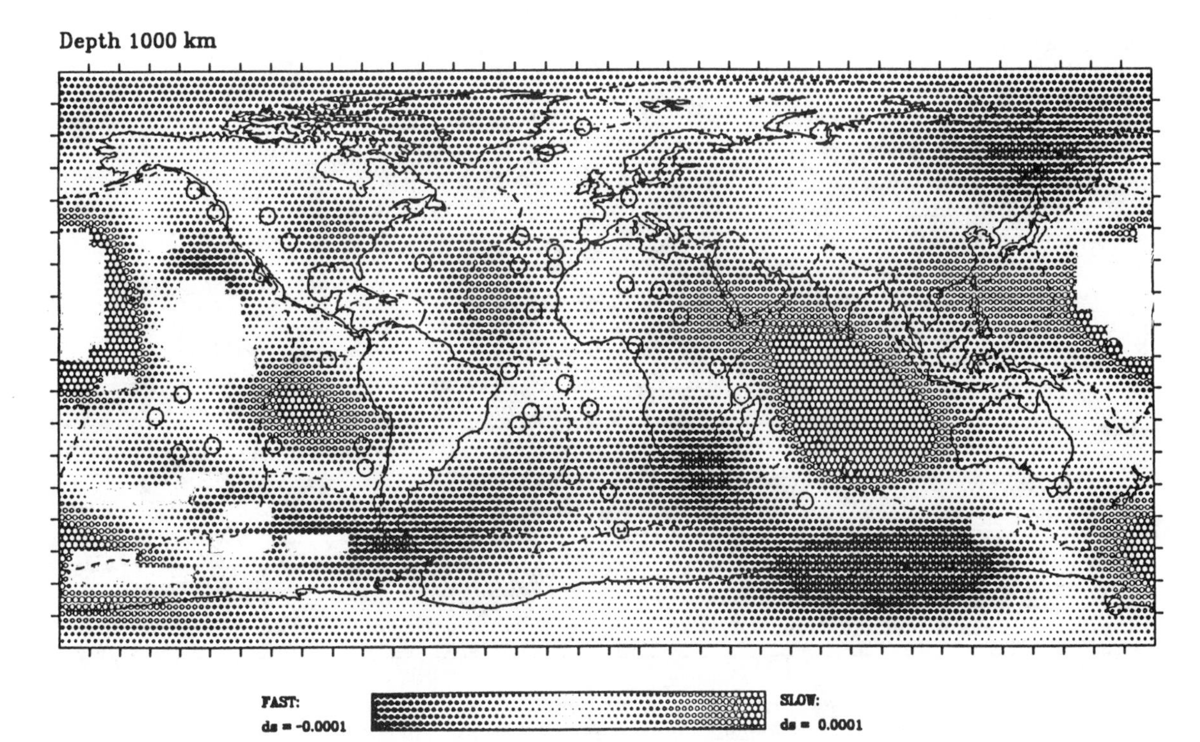

FIGURE 13-16
Smoothed (l = 1–6) compressional velocities at a depth of 1000 km (after Hager and Clayton, 1986).

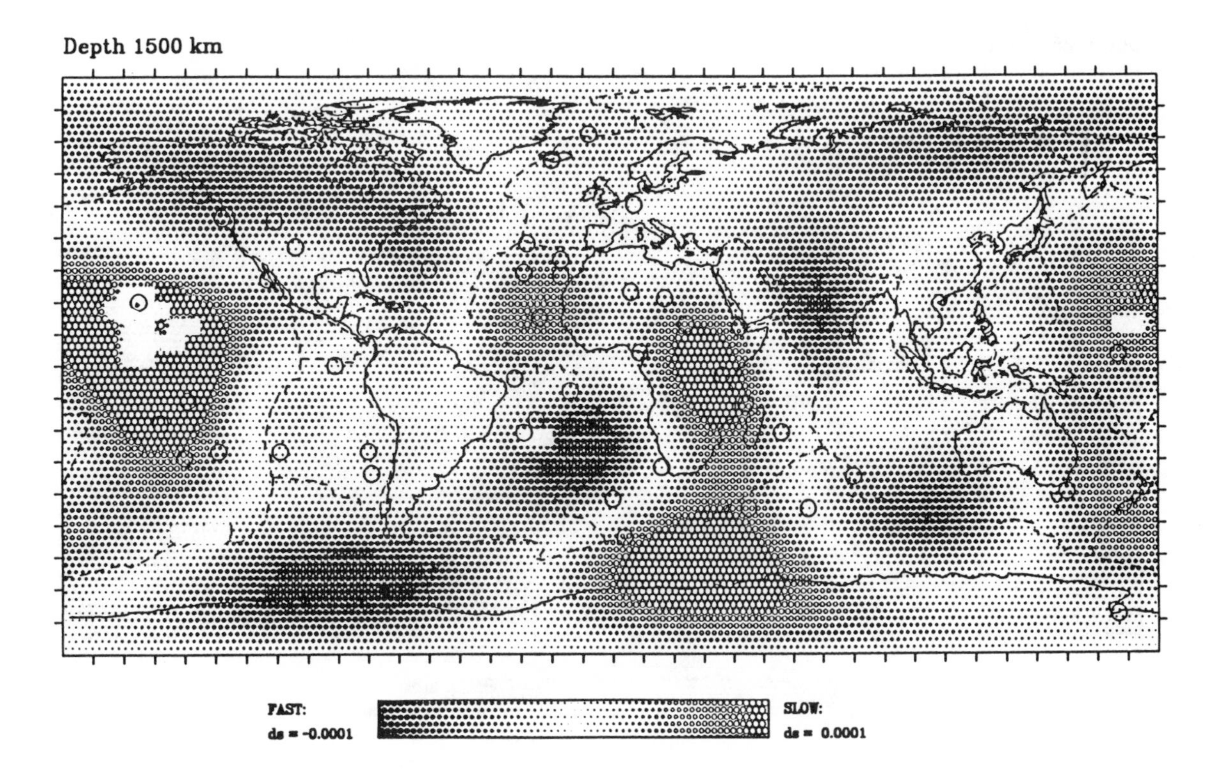

FIGURE 13-17
Smoothed compressional velocities at a depth of 1500 km.

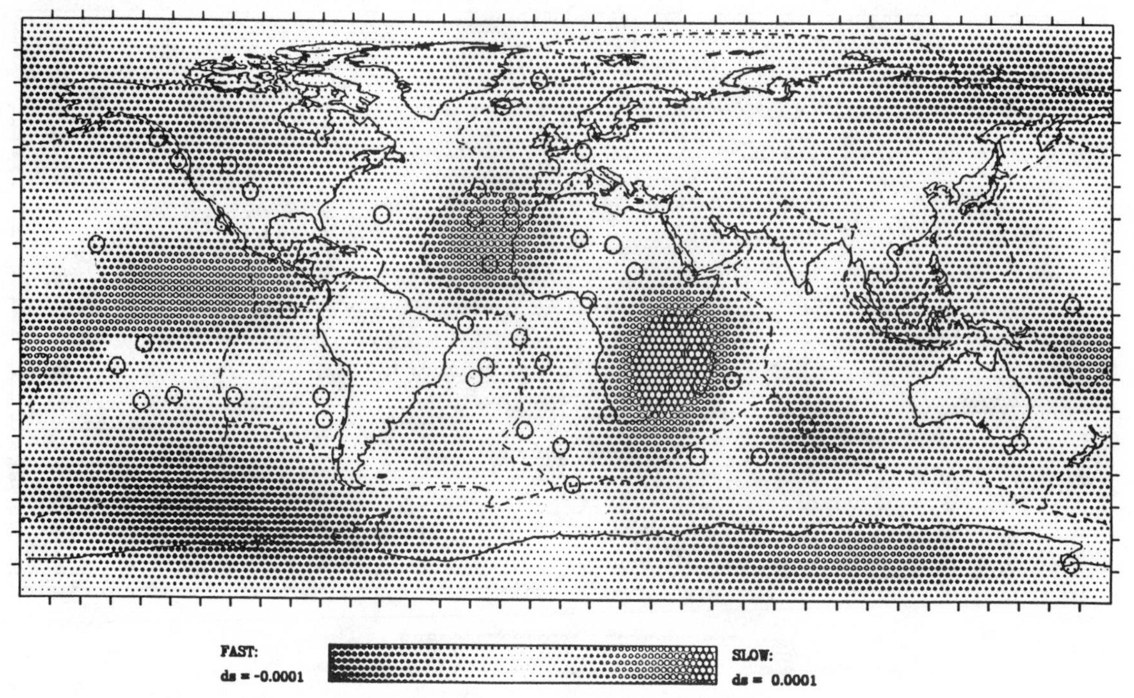

FIGURE 13-18
Smoothed compressional velocity at a depth of 2000 km.

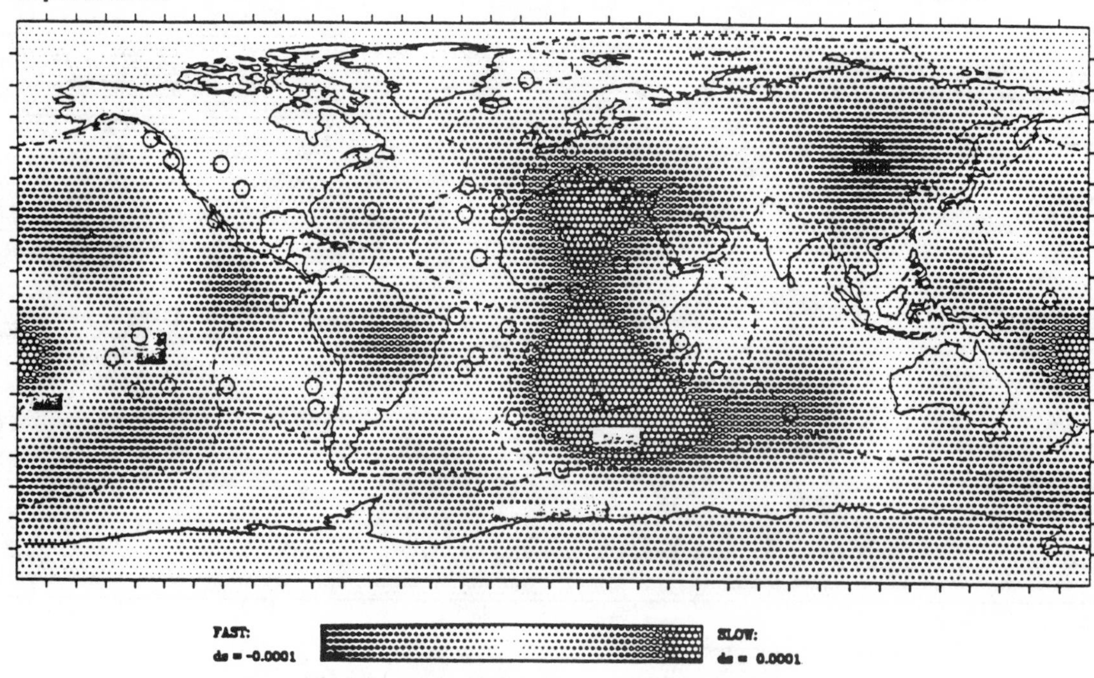

FIGURE 13-19
Smoothed compressional velocity at a depth of 2500 km.

Anderson, D. L. (1979) The deep structure of continents, *J. Geophys. Res., 84,* 7555–7560.

Forsyth, D. W. (1975) The early structural evolution and anisotropy of the oceanic upper mantle, *Geophys. J. R. Astron. Soc., 43,* 103–162.

Grand, S. P. and D. V. Helmberger (1984a) Upper mantle shear structure of North America, *Geophys. J. Roy. Astron. Soc., 76,* 399–438.

Grand, S. P. and D. V. Helmberger (1984b) Upper mantle shear structure beneath the Northwest Atlantic Ocean, *J. Geophys. Res., 89,* 11,465–11,475.

Graves, R. and D. V. Helmberger (1987) *J. Geophys. Res., 93,* 4701–4711.

Hager, B. H. and R. Clayton, Constraints on the structure of mantle convection using seismic observations, flow models and the geoid (in press), 1987.

Jordan, T. H. (1975) The continental tectosphere, *Rev. Geophys. Space Phys., 13,* 1–12.

Morris, E. M., R. W. Raitt and G. G. Shor (1969), *J. Geophys. Res., 74,* 4300–4316.

Nakanishi, I. and D. L. Anderson (1983) Measurements of mantle wave velocities and inversion for lateral heterogeneity and anisotropy: Part I, Analysis of great circle phase velocities, *J. Geophys. Res., 88,* 10,267–10,283.

Nakanishi, I. and D. L. Anderson (1984a) Aspherical heterogeneity of the mantle from phase velocities of mantle waves, *Nature, 307,* 117–121.

Nakanishi, I. and D. L. Anderson (1984b) Measurements of mantle wave velocities and inversion for lateral heterogeneity and anisotropy: Part II, Analysis by single-station method, *Geophys. J. Roy. Astron. Soc., 78,* 573–617.

Nataf, H.-C., I. Nakanishi and D. L. Anderson (1984) Anisotropy and shear velocity heterogeneities in the upper mantle, *Geophys. Res. Lett., 11,* 109–112.

Nataf, H.-C., I. Nakanishi and D. L. Anderson (1986) Measurements of mantle wave velocities and inversion for lateral heterogeneities and anisotropy: III, Inversion, *J. Geophys. Res., 91,* 7261–7308.

Okal, E. A. and D. L. Anderson (1975) A study of lateral inhomogeneities in the upper mantle by multiple ScS travel-time residuals, *Geophys. Res. Lett., 2,* 313–316.

Sipkin, S. A. and T. H. Jordan (1976) Lateral heterogeneity of the upper mantle determined from the travel times of multiple ScS, *J. Geophys. Res., 81,* 6307–6320.

Tanimoto, T. and D. L. Anderson (1984) Mapping convection in the mantle, *Geophys. Res. Lett., 11,* 287–290.

Tanimoto, T. and D. L. Anderson (1985) Lateral heterogeneity and azimuthal anisotropy of the upper mantle: Love and Rayleigh waves 100–250 sec., *J. Geophys. Res., 90,* 1842–1858.

Toksöz, M. N. and D. L. Anderson (1966) Phase velocities of long-period surface waves and structure of the upper mantle, 1. Great circle Love and Rayleigh wave data, *J. Geophys. Res., 71,* 1649–1658.

Walck, M. C. (1984) The P-wave upper mantle structure beneath an active spreading center; the Gulf of California, *Geophys. J. Roy. Astron. Soc., 76,* 697–723.

Woodhouse, J. H. and A. M. Dziewonski (1984) Mapping the upper mantle: Three-dimensional modeling of earth structure by inversion of seismic waveforms, *J. Geophys. Res., 89,* 5953–5986.

14

Anelasticity

"As when the massy substance of the Earth quivers."

—CHRISTOPHER MARLOWE, *TAMBURLAINE THE GREAT*

Real materials are not perfectly elastic. Stress and strain are not in phase, and strain is not a single-valued function of stress. Solids creep when a sufficiently high stress is applied, and the strain is a function of time. These phenomena are manifestations of anelasticity. The attenuation of seismic waves with distance and of normal modes with time are examples of anelastic behavior, as is postglacial rebound. Generally, the response of a solid to a stress can be split into an elastic or instantaneous part and an anelastic or time-dependent part. The anelastic part contains information about temperature, stress and the defect nature of the solid. In principle, the attenuation of seismic waves can tell us about such things as dislocation density and defect mobility. These, in turn, are controlled by temperature, pressure, stress and the nature of the lattice defects. If these parameters can be estimated from seismology, they in turn can be used to estimate other anelastic properties such as viscosity. For example, the dislocation density of a crystalline solid is a function of the nonhydrostatic stress. These dislocations respond to an applied oscillatory stress, such as a seismic wave, but they are out of phase because of the finite diffusion time of the atoms around the dislocation. The dependence of attenuation on frequency can yield information about the dislocations. The longer-term motions of these same dislocations in response to a higher tectonic stress gives rise to a solid-state viscosity.

SEISMIC WAVE ATTENUATION

Seismic waves attenuate or decay as they propagate. The rate of attenuation contains information about the anelastic properties of the propagation medium.

A propagating wave can be written

$$A = A_o \exp i(\omega t - kx)$$

where A is the amplitude, ω the frequency, k the wave number, t is travel time, x is distance and $c = \omega/k$ the phase velocity. If spatial attenuation occurs, then k is complex and

$$A = A_o \exp i(\omega t - kx) \cdot \exp -k^*x$$

where k and k^* are now the real and imaginary parts of the wave number. k^* is called the spatial attenuation coefficient.

The elastic moduli, say $\hat{M}$, are now also complex:

$$\hat{M} = M + iM^*$$

The specific quality factor, a convenient dimensionless measure of dissipation, is

$$Q_M^{-1} = M^*/M$$

This is related to the energy dissipated per cycle $\Delta\varepsilon$:

$$Q = 4\pi\langle\varepsilon\rangle/\Delta\varepsilon$$

where $\langle\varepsilon\rangle$ is the average stored energy (O'Connell and Budiansky, 1978). This is commonly approximated as

$$Q = 2\pi\varepsilon_{max}/\Delta\varepsilon$$

where ε_{max} is the maximum stored energy. Since phase velocity, c, is

$$c = \frac{\omega}{k} = \sqrt{M/\rho}$$

it follows that

$$Q^{-1} = 2\frac{k^*}{k} = \frac{M^*}{M} \quad \text{for } Q >> 1$$

In general $c(\omega)$, $M(\omega)$, $k(\omega)$ and $Q(\omega)$ are all functions of frequency.

For standing waves, or free oscillations, we write a complex frequency, $\omega + i\omega^*$, where ω^* is the temporal attenuation coefficient and

$$Q^{-1} = 2\frac{\omega^*}{\omega}$$

In general, all the elastic moduli are complex, and each wave type has its own Q and velocity. For an isotropic solid the imaginary parts of the bulk modulus and rigidity are denoted as K^* and G^*. Most mechanisms of seismic-wave absorption affect G, the rigidity, more than K, and usually

$$Q_K >> Q_G$$

Important exceptions over certain frequency ranges have to do with thermoelastic mechanisms and composite systems such as fluid-filled rocks.

Frequency Dependence of Attenuation

In a perfectly elastic homogeneous body, the elastic-wave velocities are independent of frequency. In an imperfectly elastic, or anelastic, body the velocities are dispersive, that is, they depend on frequency. This is important when comparing seismic data taken at different frequencies or when comparing seismic and laboratory data.

A variety of physical processes contribute to attenuation in a crystalline material: motions of point defects, dislocations, grain boundaries and so on. These processes all involve a high-frequency, or unrelaxed, modulus and a low-frequency, or relaxed, modulus. At sufficiently high frequencies the defects, which are characterized by a time constant, do not have time to contribute, and the body behaves as a perfectly elastic body. Attenuation is low and Q is high in the high-frequency limit. At very low frequencies the defects have plenty of time to respond to the applied force and they contribute an additional strain. Because the stress cycle time is long compared to the response time of the defect, stress and strain are in phase and again Q is high. Because of the additional relaxed strain, however, the modulus is low and the relaxed velocity is low. When the frequency is comparable to the characteristic time of the defect, attenuation reaches a maximum, and the wave velocity changes rapidly with frequency.

These characteristics are embodied in the *standard linear solid,* which is composed of a spring and a dashpot (or viscous element) arranged in a parallel circuit, which is then attached to another spring. At high frequencies the second, or series, spring responds to the load, and this spring constant is the effective modulus and controls the total extension. At low frequencies the other spring and dashpot both extend, with a time constant τ characteristic of the dashpot, the total extension is greater, and the effective modulus is therefore lower. This system is sometimes described as a viscoelastic solid.

The Q^{-1} of such a system is

$$Q^{-1}(\omega) = \frac{k_2}{k_1}\frac{\omega\tau}{1 + (\omega\tau)^2}$$

where k_2 and k_1 are, respectively, the spring constants (or moduli) of the series spring and the parallel spring and τ is the relaxation time, $\tau = \eta / k_2$ where η is the viscosity. Clearly, $Q^{-1}(\omega)$ is a maximum, $Q^{-1}_{\max}$, at $\omega\tau = 1$, and

$$Q^{-1}(\omega) = 2Q^{-1}_{\max}\frac{\omega\tau}{1 + (\omega\tau)^2} \tag{1}$$

and $Q^{-1}(\omega) \rightarrow 0$ as $\omega\tau$ when $\omega \rightarrow 0$ and was $(\omega\tau)^{-1}$ as $\omega \rightarrow \infty$. The resulting absorption peak is shown in Figure 14-1.

The phase velocity is approximately given by

$$c(\omega) = c_0\left(1 + Q^{-1}_{\max}\frac{(\omega\tau)^2}{1 + (\omega\tau)^2}\right) \tag{2}$$

where c_0 is the zero-frequency velocity. The high-frequency or elastic velocity is

$$c_\infty = c_0\,(1 + Q^{-1}_{\max}) \tag{3}$$

Far away from the absorption peak, the velocity can be written

$$c(\omega) \approx c_o\left(1 + \frac{1}{2}\frac{k_1}{k_2}Q^{-2}\right) \quad \text{for } \omega\tau << 1$$

$$\approx c_\infty\left(1 - \frac{k_1^2}{(2k_1 + k_2)k_2}Q^{-2}\right) \quad \textit{for } \omega\tau >> 1$$

and the Q effect is only second order. In these limits, velocity is nearly independent of frequency, but Q varies as ω or ω^{-1}. In a dissipative system, Q and c cannot both be independent of frequency, and the velocity depends on the attenuation. When Q is constant, or nearly so, the fractional change in phase velocity is proportional to Q^{-1} rather than Q^{-2} and becomes a first-order effect.

By measuring the variation of velocity or Q in the laboratory as a function of frequency or temperature, the nature of the attenuation and its characteristic or relaxation time, τ, can often be elucidated. For activated processes,

$$\tau = \tau_o \exp E^*/RT \tag{4}$$

where E^* is an activation energy. Most defect mechanisms at seismic frequencies and mantle temperatures can be described as activated relaxation effects. These include stress-induced diffusion of point and line (dislocation) defects. At very high frequencies and low temperatures, other mechanisms come into play, such as resonances of defects, and these cannot be so described. For activated processes, then,

$$Q^{-1}(\omega) = 2Q^{-1}_{\max}\frac{\omega\tau_o \exp E^*/RT}{(1 + (\omega\tau_o)^2 \exp 2E^*/RT)} \tag{5}$$

and the relaxation peak can be defined either by changing ω

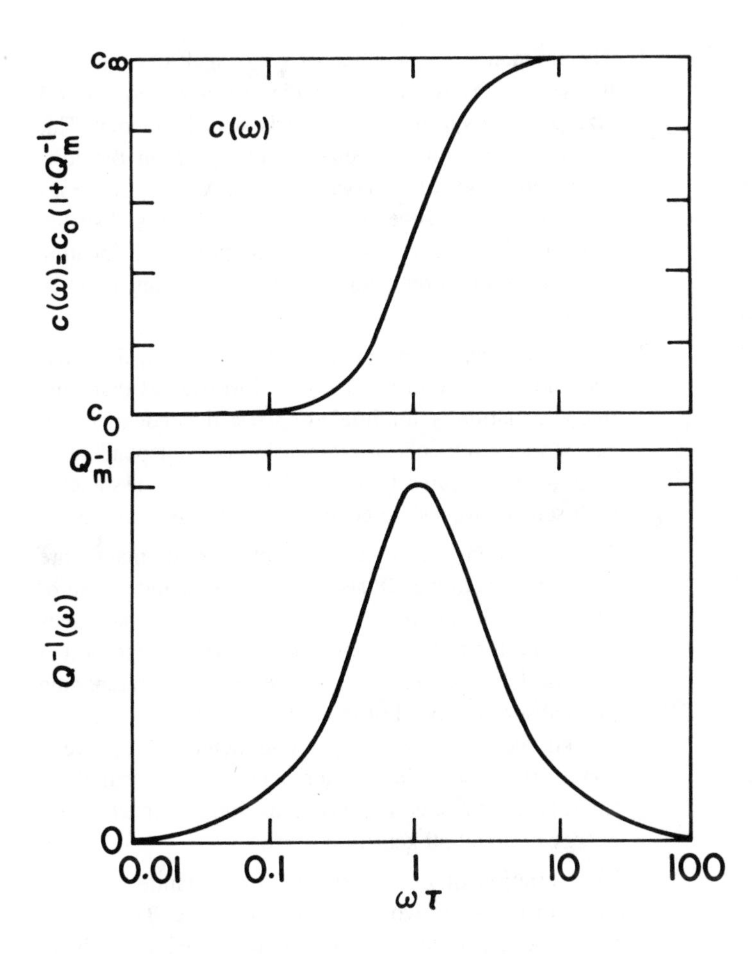

FIGURE 14-1
Specific attenuation function $Q^{-1}(\omega)$ and phase velocity $c(\omega)$ as a function of frequency for a linear viscoelastic solid with a single relaxation time, τ (after Kanamori and Anderson, 1977).

or changing T. Activation energies typically lie in the range of 10 to 100 kcal/mole.

At high temperatures, or low frequencies,

$$Q^{-1}(\omega) = 2Q_{\max}^{-1}\omega\tau_o \exp E^*/RT \qquad (6)$$

This is contrary to the general intuition in the seismological literature that attenuation increases with temperature. However, if τ differs greatly from seismic periods, it is possible that we may be on the low-temperature or high-frequency portion of the absorption peak, and

$$Q^{-1}(\omega) = 2Q_{\max}^{-1} / (\omega\tau_0 \exp E^*/RT), \quad \omega\tau >> 1 \qquad (7)$$

In that case Q does decrease with an increase in T, and in that regime Q increases with frequency. This appears to be the case for short-period waves in the mantle. It is also generally observed that low-Q and low-velocity regions of the upper mantle are in tectonically active and high heat-flow areas. Thus, seismic frequencies appear to be near the high-frequency, low-temperature side of the absorption peak in the Earth's upper mantle. Melting, in fact, may occur before the condition

$$\omega\tau = \omega\tau_o \exp E^*/RT = 1$$

is satisfied. More generally,

$$\tau = \tau_o \exp(E^* + PV^*)/RT \qquad (8)$$

where V^*, the activation volume, controls the effect of pressure on τ and Q. Because of the pressure and temperature gradient in the mantle, we can expect τ and therefore $Q(\omega)$ to vary with depth. Increasing temperature drives the absorption peak to higher frequencies (characteristic frequencies increase with temperature). Increasing pressure drives the peak to lower frequencies.

Absorption in a medium with a single characteristic frequency gives rise to a bell-shaped Debye peak centered at a frequency $\omega = \tau^{-1}$, as shown in Figure 14-1. The specific dissipation function, Q^{-1}, and phase velocity satisfy the differential equation for the standard linear solid and can be written

$$Q^{-1}(\omega) = 2Q_{\max}^{-1}\omega\tau/(1 + \omega^2\tau^2)$$

$$c^2(\omega) = c_o^2(1 + \omega^2\tau^2c_\infty^2/c_o^2)/[(1 + \omega^2\tau^2)^2 + 2\omega^2\tau^2Q_{\max}^{-1}]^{1/2}$$

The high-frequency (c_∞) and low-frequency velocities are related by

$$\frac{c_\infty^2 - c_o^2}{c_oc_\infty} = 2Q_{\max}^{-1}$$

so that the total dispersion depends on the magnitude of the peak dissipation. For a Q of 200, a typical value for the upper mantle, the total velocity dispersion is 2 percent.

Solids in general, and mantle silicates in particular, are not characterized by a single relaxation time and a single Debye peak. A distribution of relaxation times broadens the peak and gives rise to an absorption band (Figure 14-2). Q can be weakly dependent on frequency in such a band. Seismic Q values are nearly constant with frequency over much of the seismic band. A nearly constant Q can be explained by involving a spectrum of relaxation times and a superposition of elementary relaxation peaks. If τ is distributed continuously between τ_1 and τ_2 (see Anderson and others, 1977), then

$$Q^{-1}(\omega) = (2Q_{\max}^{-1}/\pi)\tan^{-1}[\omega(\tau_1 - \tau_2)/(1 + \omega^2\tau_1\tau_2)]$$

and

$$c(\omega) = c_o(1 + (Q_{\max}^{-1}/2\pi)\ln[(1 + \omega^2\tau_1^2) \div (1 + \omega^2\tau_2^2)])$$

For $\tau_1 << \omega^{-1} << \tau_2$ the value of Q^{-1} is constant and equal to $Q_{\max}^{-1}$. The total dispersion in this case is

$$\frac{c_\infty - c_o}{c_o} = (Q_{\max}^{-1}/\pi)\ln(\tau_1/\tau_2)$$

which depends on the ratio τ_1/τ_2, which is the width of the absorption band. The spread in τ can be due to a distribution of τ_o or of E^*.

Attenuation Mechanisms

The actual physical mechanism of attenuation in the mantle is uncertain, but it is likely to be a relaxation process involving a distribution of relaxation times. Many of the attenuation mechanisms that have been identified in solids occur at relatively low temperatures and high frequencies and can therefore be eliminated from consideration. These include point-defect and dislocation resonance mechanisms, which typically give absorption peaks at kilohertz and megahertz frequencies at temperatures below about half the melting point. The so-called grain-boundary and cold-work peak and the "high-temperature background" occur at lower frequencies and higher temperatures. These mechanisms involve the stress-induced diffusion of dislocations. The Bordoni peak occurs at relatively low temperature in cold-worked metals but may be a higher-temperature peak in silicates. It is apparently due to the motion of dislocations since it disappears upon annealing.

Even in the laboratory it is often difficult to identify the mechanism of a given absorption peak. The effects of amplitude, frequency, temperature, irradiation, annealing, deformation and impurity content must be studied before the mechanism can be identified with certainty. This information is not available for the mantle or even for the silicates that may be components of the mantle. Nevertheless, there is some information which helps constrain the possible mechanism of attenuation in the mantle.

1. The frequency dependence of Q is weak over most of the seismic band. At frequencies greater than about 1 Hz, Q appears to increase linearly with frequency. This is consistent with the behavior expected on the low-temperature side of a relaxation band. A weak frequency dependence is best explained by invoking a distribution of relaxation times. A distribution of dislocation lengths, grain sizes and activation energies may be involved.
2. Although it has not been specifically studied, there has been no evidence brought forward to suggest that seismic attenuation is amplitude or stress dependent. Laboratory measurements of attenuation are independent of amplitude at strains less than 10^{-6}. Strains associated with seismic waves are generally much less than this.
3. The radial and lateral variations of Q in the mantle are our best clues to the effects of temperature and pressure. The lower-Q regions of the mantle are in those areas where the temperatures are highest. This suggests that most of the upper mantle is on the low-temperature side of an absorption band or in the band itself. At a depth of 100 km the temperature of the continental lithosphere is about 200 K less than under oceans. Q is roughly 7 times larger under continents. This implies an activation energy of about 50 kcal/mole.
4. The variation of Q with depth in the mantle covers a range of less than two orders of magnitude. This means that the effects of temperature and pressure are relatively modest or that they tend to compensate each other.
5. Losses in shear are more important than losses in compression. This is consistent with stress-induced motion of defects rather than a thermoelastic mechanism or other mechanisms involving bulk dissipation.

Spectrum of Relaxation Times

Relaxation mechanisms lead to an internal friction peak of the form

$$Q^{-1}(\omega) = \Delta\int_{-\infty}^{\infty} D(\tau)[\omega\tau/(1 + \omega^2\tau^2)]\,d\tau \tag{9}$$

where ω is the applied frequency, τ is a characteristic time, $D(\tau)$ is called the retardation spectrum and Δ is the modulus defect $(G_u - G_r)/G_r$, the relative difference between the high-frequency, unrelaxed shear modulus G_u and the low-frequency, relaxed modulus G_r. The modulus defect is also a measure of the total reduction in modulus that is obtained in going from low temperature to high temperature. For a dislocation model the modulus defect is due to the strain contributed by dislocation bowing. The dislocations bow to an equilibrium radius of curvature that is dictated by the applied stress. The rate at which they do so is controlled by the propagation of kinks or diffusion of point defects near

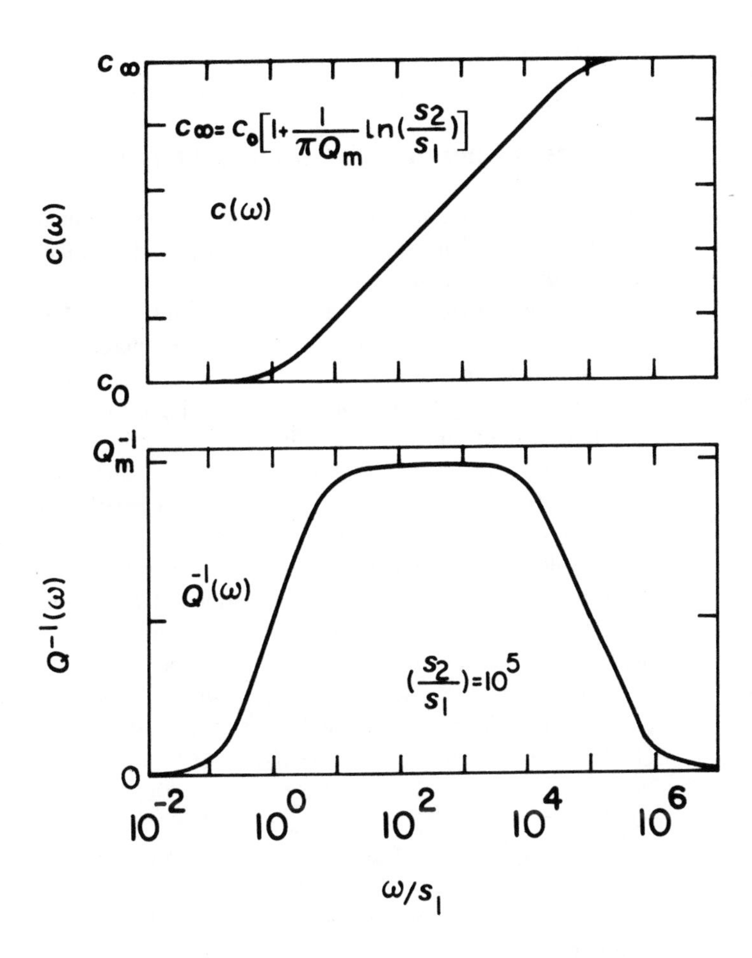

FIGURE 14-2
Band-limited constant-Q model with cutoff frequencies s_1 and s_2 ($s = 1/\tau$) (after Kanamori and Anderson, 1977).

the dislocation and therefore is an exponential function of temperature.

A convenient form of the retardation spectrum is

$$D(\tau) = [\alpha/(\tau_2^\alpha - \tau_1^\alpha)]\tau^{\alpha-1} \tag{10}$$

for $\tau_1 < \tau < \tau_2$ and zero elsewhere (Minster and Anderson, 1981). τ_1 and τ_2 are the shortest and longest relaxation times, respectively, of the mechanism being considered.

When $\tau_1 \approx \tau_2$, equation 9 reduces to the well-known Debye peak. When $\alpha = 0$ and $\tau_2 >> \tau >> \tau_1$, a frequency-independent Q results. The more general case of $0 < \alpha < 1$ has been discussed by Anderson and Minster (1979, 1980). This leads to a weak dependence, $Q(\omega) \approx \omega^\alpha$, for $\tau_1 < \tau < \tau_2$ and has characteristics similar to Jeffreys' (1970) modification of Lomnitz's law. Outside the absorption band, $Q(\omega) \approx \omega^{\pm 1}$ for $\omega\tau \gtrless 1$; this result holds for all relaxation mechanisms.

The Q corresponding to frequencies ω such that

$$\omega\tau_1 << 1 << \omega\tau_2$$

is given by

$$Q(\omega) = \cot\frac{\alpha\pi}{2} + \frac{2J_u}{\pi\alpha\delta J}$$

$$\times\ [(\omega\tau_2)^\alpha - (\omega\tau_1)^\alpha]\cos\frac{\alpha\pi}{2}$$

$$\approx \frac{2J_u}{\pi\alpha\delta J}(\omega\tau_2)^\alpha\cos\frac{\alpha\pi}{2} \tag{11}$$

since $\tau_2 >> \tau_1$ and $\cot \alpha\pi/2$ is small. J_u, J_r, and $\delta J = J_r - J_u$ are the unrelaxed, relaxed, and defect compliances, respectively. The compliance is strain divided by stress.

For high frequencies, $1 << \omega\tau_1 << \omega\tau_2$,

$$Q(\omega) = (1 - \alpha)(J_u/\alpha\delta J)(\tau_2/\tau_1)\alpha(\omega\tau_1)$$

At low frequencies,

$$Q(\omega) = (1 + \alpha)(J_r/\alpha\delta J)(\omega\tau_2)^{-1}$$

For a thermally activated process the characteristic times depend on temperature. It is clear that at very low temperature the characteristic times are long and $Q \propto \omega$. At very high temperature, Q is predicted to increase as ω^{-1}. Equation 11 holds for intermediate temperatures.

The dispersion appropriate for the frequency-dependent Q model considered above gives, in the case $Q >> 1$,

$$G_1/G_2 = 1 - \cot(\alpha\pi/2)(\omega_2/\omega_1)Q_2^{-1}$$

where Q_2 is the Q at ω_2. The rigidities at the two frequencies are $G_1(\omega_1)$ and $G_2(\omega_2)$. This applies for $\omega\tau < \omega\tau_2$. For low frequencies, $G = G_r$. For high frequencies, $G = G_u$. For simple dislocation networks the difference between the relaxed and unrelaxed moduli is about 8 percent (Minster and Anderson, 1981).

ACTIVATED PROCESSES

For simple point-defect mechanisms τ_o^{-1} is close to the Debye frequency, about 10^{13} Hz. For mechanisms involving dislocations or dislocation-point defect interactions, τ_o depends on other parameters such as dislocation lengths, Burger's vectors, kink and jog separations, Peierls stress, and interstitial concentrations. The activation energy, depending on the mechanism, is a composite of activation energies relating to self-diffusion, kink or jog formation, Peierls energy, bonding energy of point defects to the dislocation, core-diffusion and so on. E^* for creep of metals is often just the self-diffusion activation energy, E_{SD}.

Nonelastic processes in geophysics are commonly expressed in terms of viscosity, for long-term phenomena, and Q, for oscillatory and short-term phenomena. It is convenient to express these in terms of the characteristic time. For creep,

$$\tau = \sigma/G\dot{\varepsilon} = \eta/G$$

where σ, G, $\dot{\varepsilon}$ and η are the deviatoric stress, rigidity, strain rate and viscosity, respectively. Thus, the characteristic time can be simply computed from creep theories and experiments.

Relaxation theories of attenuation can be described by $Q(\omega\tau)$. In the mantle, where Q is low, we infer that the characteristic relaxation time is close to $1/\omega$, the reciprocal of the measurement frequency. High values of Q indicate that $\omega >> 1/\tau$ or $\omega << 1/\tau$. In these cases $Q \approx \omega$ or ω^{-1}. When Q is slowly varying we are probably in the midst of an absorption band that is conveniently explained as a result of a distribution of relaxation time.

Since Q is small in the higher-temperature regions of the mantle, it is likely that at seismic periods the mantle is on the low-temperature side of an absorption band. Thus

$$Q^{-1} = 2Q_{\max}^{-1}/\omega\tau = (2Q_{\max}^{-1}/\omega\tau_o)\exp(-E^*/RT) \quad (12)$$

in regions of rapidly varying attenuation. The seismic data can be used to estimate the relaxation time in the upper mantle and its variation with depth. The high-temperature background (HTB) is a dominant mechanism of attenuation in crystalline solids at high temperature and low frequency. This usually satisfies a $Q \approx \omega$ or $Q \approx \omega^\alpha$ relation. This can be interpreted in terms of a distribution of relaxation times. The former is valid for all relaxation mechanisms when the measurement frequency is sufficiently high. In the following $\hat{\tau}_o$ will be used for the pre-exponential creep, or Maxwell, relaxation time and τ_o will be used for the attenuation, or Q, characteristic time.

Activation Energies for Climb and Glide

The creep of ductile materials such as metals is rate-limited by self-diffusion. This is the case for either pure diffusional creep or dislocation climb. In these circumstances the activation energy for creep is the same as that for self-diffusion. In olivine the activation energy for creep is appreciably higher than that for self-diffusion of O^{2-} or Si^{4+}. This suggests that creep may be controlled by kink or jog formation.

Kinks are offsets in a dislocation line in the glide plane of a crystal. Kinks separate portions of a dislocation that are in adjacent potential valleys, separated by a Peierls energy hill. Segments of a dislocation line that are normal to the glide plane are called jogs. At finite temperature, dislocations contain equilibrium concentrations of kinks and jogs. The kinks and jogs undergo a diffusive drift under an applied stress, producing glide and climb, respectively.

The activation energy for creep when dislocation climb is due to nucleation and lateral drift of jogs is $(1/2)(E_{SD}^* + 2E_j^*)$ or $(1/2)(E_{SD}^* + E_{CD}^* + 4E_j^*)$, depending on whether the dislocations are longer or shorter than the equilibrium jog spacing. Here E_{SD}^*, E_{CD}^* and E_j^* are the activation energies for self-diffusion, core diffusion and jog formation, respectively (Hirth and Lothe, 1968).

E_{SD}^* is 90 kcal/mole for both oxygen and silicon diffusion in olivine (Jaoul and others, 1979). The jog formation energy is somewhat greater than kink formation energy (Hirth and Lothe, 1968). The activation energy controlling high-temperature deformation of olivine has been estimated to be 135 kcal/mole from the climb of dislocation loops in olivine (Kohlstedt and Goetze, 1974; Kohlstedt, 1979) and 122–128 kcal/mole from creep data on olivine (Goetze and Brace, 1972; Goetze, 1978).

In the absence of interstitial defects near the dislocation, the activation energy for glide is E_k^* if the kink density is high and $2E_k^*$ when the dislocation length is shorter than the equilibrium distance between kinks. E_k^* is the kink formation energy, which is about 26 kcal/mole (Ashby and Verall, 1978). The activation energy for glide is therefore 52 kcal/mole if the kink spacing is large or if double kink formation is required for glide. If point defects must diffuse with the dislocation line, then their diffusivity may be rate limiting. The activation energy for Mg-Fe interdiffusion in olivine is about 50–58 kcal/mole (Misener, 1974). Dislocation glide in olivine will therefore probably have an activation energy of the order of 50–60 kcal/mole, unless there are slower-moving species in the vicinity of the dislocation. Olivine in contact with pyroxene is likely to have excess silicon interstitials as a dominant point defect. The activation energy in this case may be due to drag of silicon interstitials, about 90 kcal/mole. There is also a small term

TABLE 14-1
Material Properties for Olivine

Property	Symbol	Value	Units
Burger's vector	b	6×10^{-8}	cm
Oxygen ion volume	Ω	1×10^{-23}	cm^3
Shear modulus	G	8×10^{11}	dy/cm^2
Silicon diffusivity			
Pre-exponential	D_{Si}	1.5×10^{-6}	cm^2/s
Activation energy	E_{Si}	90	kcal/mole
Oxygen diffusivity			
Pre-exponential	D_{ox}	3.5×10^{-3}	cm^2/s
Activation energy	E_{ox}	89	kcal/mole
Mg-Fe diffusivity			
Pre-exponential	D_{Mg}	3.4×10^{-3}	cm^2/s
Activation energy	E_{Mg}	47	kcal/mole
Subgrain size ÷ Dislocation length	K'	15	—
Kink energy	E_k	26	kcal/mole
Jog energy	E_j	35	kcal/mole

Ashby and Verall (1978).

representing the binding energy of interstitials to the dislocation. This is ordinarily a few kilocalories per mole, but data are lacking for silicates. The kink and jog formation energies in silicates are also highly uncertain, and the above estimates for olivine are approximate.

Point defects can also diffuse along the core of a dislocation. The activation energy for dislocation core diffusion should be similar to that for surface diffusion. This, for metals, is usually about ½ to ⅔ of the lattice diffusion value. If this rule applies to olivine, the core diffusion activation energy would be about 50 to 60 kcal/mole. High values for the core and surface activation energies are expected from the heat of vaporization, which is much higher for silicates than for metals. If we adopt $E^*_{SD} = E^*_{CD} = 90$ kcal/mole and $E^*_{creep} = 125$ kcal/mole, then the implied jog formation energy is 35 kcal/mole. This is much greater than is typical of metals. It is also expected, since the kink and jog formation energies are proportional to Gb^3, which is much greater for silicates than for metals.

Because of the high values for core diffusion and jog formation, we expect relatively high values for the creep activation energy for silicates and for a non-correspondence with the self-diffusional activation energy, except for very small grain sizes and low stresses, where diffusional creep may be important.

A summary of the physical properties of olivine is assembled in Table 14-1. These will be used in subsequent calculations.

Geophysical Constraints on the Activation Parameters

Some of the activation parameters for creep and attenuation can be estimated directly from geophysical data. The characteristic time scales of relaxation processes in the mantle depend on temperature, through the activation energy, and on the pressure, through the activation volume. Stress has an indirect effect on relaxation times since it controls such parameters as dislocation density and subgrain size. The characteristic frequencies of atomic processes and processes involving dislocation motions are typically 10^{10} to 10^{13} Hz. The characteristic relaxation time for creep in the upper mantle, the ratio of viscosity to rigidity, is about 10^{10} s. For temperatures of 1500–1600 K, appropriate for the upper mantle, the required activation energy is 145–170 kcal/mole. This is similar to the activation energy for creep of olivine and values inferred for the climb of dislocations in olivine (Goetz and Kohlstedt, 1973).

Judging from the attenuation of surface waves, the characteristic time controlling the attenuation of seismic waves in the upper mantle is of the order of 100 s. This implies an activation energy in the range 90–100 kcal/mole for the above τ_o. This is close to the activation energy found for diffusion of O^{2-} and Si^{4+} in olivine (Table 14-1). A higher value for τ_o gives a smaller value for E^*.

The trade-off between τ_o and E^* is shown in Figure 14-3. The curves to the right cover a range of estimated

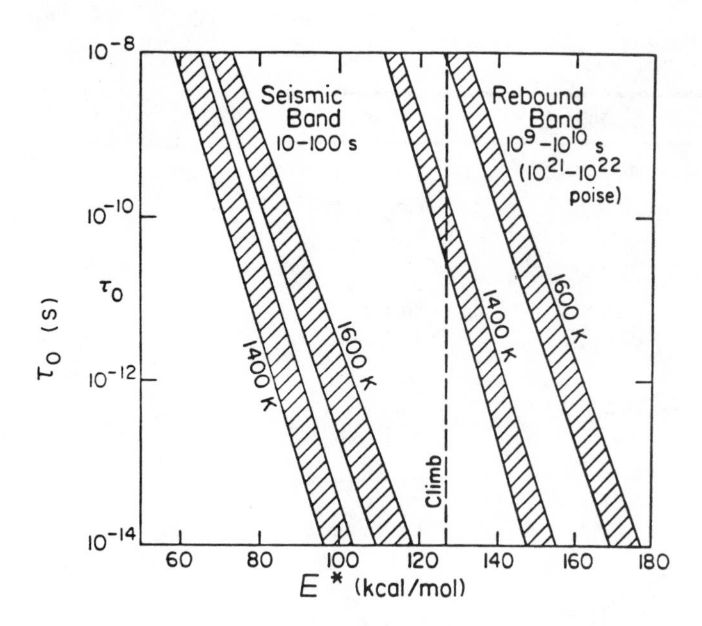

FIGURE 14-3
τ_o versus activation energy combinations that satisfy the seismic data (left) and rebound data (right) for a range of upper-mantle temperature. The dashed line is the expected activation energy for dislocation climb in olivine (Anderson and Minster, 1980).

upper-mantle temperatures and viscosities. The curves to the left represent the seismic band. Theories of creep and attenuation predict τ_o-E^* pairs. Those that fall in the areas shown are geophysically plausible mechanisms. Low values of τ_o require low values for E^*. The characteristic pre-exponential times are not the same for creep and attenuation.

It is not simple to estimate the pre-exponential characteristic time appropriate for attenuation in the mantle. High-temperature background damping processes in metals give τ_o in the range of 10^{-12}–10^{-14} s (Woirgard, 1976). Lower temperature peaks give characteristic times one or two orders of magnitude larger. Theoretically, the times scale as the diffusivity, which is typically three orders of magnitude lower in silicates than in metals. On the other hand, subgrain sizes and dislocation lengths in the mantle can be expected to be one or two orders of magnitude greater than typical laboratory grain sizes in metals. Since $\tau_o \approx l^2/D$, the characteristic time for the mantle may therefore be of the order of 10^{-6} to 10^{-8} s. There are very few data on silicates. The relaxation times have been estimated from forsterite data as about 10^{-8} s (Jackson, 1969). This implies an activation energy of 58–74 kcal/mole if the upper-mantle seismic absorption band is centered at 100 s.

Although τ is proportional to l^2 for most attenuation and creep mechanisms, there are other factors that affect the characteristic times. Steady-state creep is controlled by self-diffusion of the slowest moving species. This affects both D_o and E^*. E^* for creep, which also includes the energy of jog formation, is therefore greater than E^* for attenuation, which need not involve self-diffusion. For a polygonized network the climb of dislocations in cell walls is rate-limiting, but the characteristic time for creep involves both the motion of the long mobile dislocations in the cells and the shorter dislocations in the walls. For attenuation, the separate population of dislocations would give rise to widely separated internal friction peaks. A wide separation of characteristic times can therefore be obtained from a single dislocated solid.

Experiments on dislocated single crystals of silicates are needed to isolate the attenuation mechanisms in the mantle. Some of the theoretical predictions of this chapter have been confirmed by experiments on rocks (Berckhemer and others, 1979, 1982; Kampfmann and Berckhemer, 1985). The effects of partial melting are evident in these data.

Estimates of τ_o and E^* for Olivine

The characteristic relaxation time for creep of olivine can be obtained from high-temperature laboratory experiments:

$$\dot{\varepsilon} = A\sigma^n \exp(-E^*/RT)$$

Using the Kohlstedt-Goetze (1974) data with $n = 3$ and $E^* = 125$ kcal/mole and the relation

$$\hat{\tau}_o = \sigma G/\dot{\varepsilon}_o$$

we obtain $\hat{\tau}_o = 3 \times 10^{-9}$ s for a stress of 10 bars. The low-stress data have a σ^2 trend. In this case $\hat{\tau}_o$ is 10^{-9} s.

From the seismic rigidity and the postglacial rebound viscosity, the relaxation time of the upper mantle is $10^{10\pm1}$ s. Using an activation energy of 125 kcal/mole and a range of upper-mantle temperatures from 1100 to 1300°C, the inferred value for $\hat{\tau}_o$ is 10^{-6} to 10^{-10} s. This is consistent with the laboratory creep data for differential stresses in the range of a fraction of a bar to about 5 bars. The mantle $\hat{\tau}_o$ values also overlap the lower values for $\hat{\tau}_o$ determined from dislocation mechanisms rate-limited by silicon diffusion.

The effect of pressure can be estimated from the observation that both viscosity and Q do not vary by more than two orders of magnitude in the mantle. This constrains the activation volume to be between 4 and 9 cm³/mole. Since the activation volume of oxygen, the largest major ion in the mantle, is about 11 cm³/mole and decreases with pressure, and since the effective activation volume depends on all diffusing ions for coupled diffusion, the result from the mantle is consistent with diffusional control for both dislocation creep and attenuation.

A crude estimate of the activation energy controlling attenuation can be obtained from the observation that Q can vary by about an order of magnitude over relatively short distances both laterally and vertically in the upper mantle. Using 200°C as a reasonable difference in temperature, the inferred activation energy is about 52 kcal/mole. This, in turn, requires a τ_o of about 10^{-5}–10^{-6} s for the attenuation mechanism.

These results support previous conclusions that both creep and attenuation in the mantle are activated processes that are controlled by diffusion. They are also consistent with numerous studies on a variety of materials finding that high-temperature creep and internal friction are controlled by the diffusive motion of dislocations. The large modulus defect implied by the upper-mantle low-velocity, low-Q zone is also consistent with a dislocation relaxation mechanism, perhaps aided by partial melting. Point-defect internal friction peaks are generally relatively sharp, occur at low temperatures and do not exhibit the low Q's and large velocity dispersion that characterize the mantle and the high-temperature internal friction peaks.

Although attenuation and creep, or viscosity, are different mechanisms, they are both activated processes that depend strongly on temperature, pressure and defect concentrations. We therefore expect the low-Q region of the upper mantle to be a low-viscosity region (Anderson, 1966). By the same reasoning we expect the lower mantle to exhibit high viscosity. High pressure decreases the mobility of point defects, leading to high viscosity and longer relaxation times. Parts of D″ may have high thermal gradients, and viscosity may decrease in these regions.

The high viscosity of the lower mantle, expected from both Q and ΔV^* considerations, probably means that lower mantle convection is more sluggish than that associated with plate tectonics.

DISLOCATION DAMPING

The passage of a stress wave through a crystalline solid can cause a stress-induced diffusive motion and reordering of such point defects as vacancies, interstitial or solute impurity atoms, and substitutional atoms. This reordering can take place in a single crystal if the point defect produces distortions that have a lower symmetry than that of the lattice. In polycrystals defects undergo a mass transport between regions of different dilatational strain. A characteristic feature of atomic diffusion is the relatively long time of relaxation and its strong dependence on temperature. The relaxation time for the process is of the order of the inverse jumping frequency of the defect. This frequency, ω_R, depends on the temperature as

$$\omega_R = \omega_o e^{E^*/RT}$$

where E^* is an energy of activation and R is the gas constant. The activation energies for elements in metals are of the order of tens of kilocalories per mole. For very rapid varying stresses or for very low temperatures, the defects do not have a chance to move, and stress and strain are in phase and no losses occur. For very slowly varying stresses or for very high temperatures, the diffusion frequency can be greater than the applied frequency, again giving no delay and no loss. For intermediate frequencies and temperatures the motion of defects is out of phase with the applied stress, and a typical relaxation peak occurs. Point-defect attenuation mechanisms occur at much higher frequencies than seismic-wave frequencies and, acting alone, are negligible in seismic problems. However, point defects interacting with dislocations may be important in seismic-wave attenuation.

Dislocations play a significant role in both the plastic behavior and damping properties of single crystals. They are also involved in the properties of polycrystalline substances. There are basically three kinds of mechanical losses that are associated with dislocations: hysteresis, resonance, and relaxation losses. Some dislocation mechanisms give rise to damping that is independent of amplitude and varies with frequency; other mechanisms depend on the amplitude of the stress wave and vary only slowly with frequency. The decrement and modulus changes due to dislocation motions depend on the orientation of the crystal since only the shear component of an applied stress in the slip direction contributes to dislocation motion.

The motion of dislocations that are pinned down by impurities or other dislocations is a common loss mechanism in metal single crystals. The damping depends sensitively on the amount of previous cold work and on the purity of the metal. Increasing the number of impurity atoms or increasing the number of dislocations by cold work provides more anchoring points, and the internal friction decreases. Decreases in internal friction are accompanied by increases in the elastic moduli. The pinning of dislocations by point defects is especially effective at low temperatures and high frequencies since the ability of an impurity atom to follow an alternating stress is controlled by the slow process of diffusion. Amplitude dependence is attributed to the breakaway of the dislocation line from the impurity atoms at large strain amplitudes.

The name hysteresis is given to losses associated with the occurrence of irreversible changes of state. Typical of hysteresis are those in which dislocations are irreversibly torn away from pinning points, such as impurity atoms or other point defects. This mechanism is frequency independent and depends on the amplitude of applied stress. It is suppressed by annealing, by increasing the impurity concentration or by irradiation with neutrons or fast electrons, all of which serve to more effectively pin the dislocation and prevent breakaway. The binding energy of an impurity atom to a dislocation is rather small, and, therefore, the dependence of hysteresis damping on temperature is rather small.

Resonance motion of dislocations within their equilibrium potential valleys can contribute to the damping of waves with frequencies comparable to the resonant frequency of the pinned dislocation segment. Because of the presumed shortness of dislocation loops, this will be important only at high frequencies, say megahertz. Therefore, we concentrate on relaxation, or diffusional, processes.

The general expression for the characteristic time of stress-induced diffusion of a dislocation line is

$$\tau = (kTL^2/D_oGb^3)\exp(E^*/RT)$$

(Hirth and Lothe, 1968). In the bowed-string approximation for climb, L is the dislocation length, D_o is the pre-exponential self-diffusion coefficient, E^* is the activation energy for self-diffusion (SD) and b is the Burger's vector. If the climb is rate-limited by core diffusion (CD), then

$$L^2 = l^4/\lambda b$$

where l is the dislocation length, $\lambda = b\exp(E_j/RT)$ is the jog separation and D_o and E^* refer to core diffusion. The effective activation energy is therefore the sum of the core diffusion and jog formation energies. If $D_o(\mathrm{CD}) = 10^3 D_o(\mathrm{SD})$ and $l = 10^3\ b$, the τ_o for core diffusion is 10^3 times that for self-diffusion. At modest temperature, however, $\tau(\mathrm{CD}) < \tau(\mathrm{SD})$ if $E_{\mathrm{CD}} < E_{\mathrm{SD}}$. In general, relaxation peaks with high activation energies occur at higher temperature and lower frequency than those with lower activation energies. The presence of point defects, kinks or jogs along the dislocation line changes the characteristic time.

An estimate of the relaxation time for climb can be obtained by assuming that the dislocation length equals the subgrain size, which in turn is related to the tectonic stress

$$L = 15Gb/\sigma$$

Using D_{Si} and σ between 1 and 10 bars we obtain τ_o of $1-10^{-2}$ s. Neither the self-diffusion nor the core-diffusion characteristic times can be brought into the upper-mantle seismic band with the appropriate activation energy. The actual mobile dislocation lengths may be an order of magnitude smaller than the subgrain size. With an activation energy of 90 kcal/mole the relaxation time at upper mantle temperatures is 10^8 to 10^6 s, still well outside the seismic band but straddling the Chandler period. Dislocation climb in subgrains is therefore unlikely to contribute to seismic attenuation but may contribute to damping of the Chandler wobble.

The dislocation lengths in cell walls are an order of magnitude smaller than in the cells. This makes a further reduction in τ_o to about 10^{-5} to 10^{-7} s, which gives relaxation times just outside the seismic band at mantle temperatures and $E^* = 90$ kcal/mole, but the relaxation strength is very small because of the limited bow-out possible for small dislocations. Therefore, it appears that dislocation climb, an important mechanism of steady-state creep, can be ruled out as a mechanism for attenuation in the mantle. We therefore turn our attention to dislocation glide.

In general, dislocations glide much faster than they climb, and the relaxation time is therefore much reduced. Glide is rate-limited by lattice (Peierls) stresses or by point defects. For a gliding dislocation, rate-limited by the diffusion of interstitial defects,

$$L^2 = C_i l^2$$

where C_i is the concentration of interstitials along the dislocation line at sufficiently high temperature,

$$C_i = C_D \exp(-E_B/RT)$$

and C_D is the bulk concentration of interstitial point defects or impurity atoms and E_B is the binding energy of the point defects to the dislocation line. The diffusivity is that appropriate for the diffusion of the impurity or interstitial ions.

Using $D_o = 10^{-3}$ cm²/s and $C_D \approx C_i = 10^{-3}$, we obtain τ_o of 10^{-6} to 10^{-8} s for typical dislocation lengths appropriate for mantle stresses of 1 to 10 bars. These, combined with activation energies appropriate for diffusion of cations in silicates, give relaxation times in the seismic band at upper-mantle temperatures. The impurity content refers to that in the subgrains. Most impurities in the mantle are probably at grain boundaries, and therefore the grain interiors are relatively pure. The above estimate for C_D is therefore not unreasonable.

The relaxation strength is given by

$$2Q^{-1}_{\max} = (1/6\sqrt{5})\,\rho_m l^2$$

for a collection of randomly oriented gliding dislocations (Minster and Anderson, 1981). In the subgrains $\rho_m = 1/l^2$ for a rectangular grid of dislocations, and the relaxation strength is about 7.5 percent for a Q of about 26. The same relationship holds for a dislocation network in the cell walls, but because of the small volume in the walls the average relaxation strength is very small for the wall contribution to the attenuation. The relaxation strength is proportional to the area swept out by the gliding dislocations and gives the difference between the high-frequency or low-temperature modulus and the relaxed modulus.

The Köster peak occurs in cold-worked metals containing interstitial impurities, some of which have concentrated along the dislocations. It occurs at high temperature and low frequency and usually gives a τ_o of 10^{-13} to 10^{-14} s for metals with oxygen, nitrogen or hydrogen interstitials. It is believed to be caused by an interaction between dislocations and interstitial point defects in the neighborhood of the dislocation. With reasonable choices of parameters, this type of mechanism can explain attenuation in the mantle.

The grain-boundary peak (GBP) occurs in metals at about one-half the melting temperature at 1 Hz; actually, several peaks are often observed between about 0.3 and 0.6 of melting temperature. The activation energies range from about 0.5 to 1.0 times the self-diffusion activation energy and are higher for the high-temperature peaks. An increase in solute concentration increases the magnitude of the higher temperature peak (Woirgard, 1976). These peaks are all superimposed on an even stronger high-temperature background (HTB) which dominates at very high temperature. In synthetic forsterite a strong peak occurs at about 0.25–0.5 Hz at 1000°C (Berckhemer and others, 1982) with an activation energy of 57 kcal/mole. At upper-mantle temperatures this absorption would shift to about 10^{-2} s and therefore would not be important at seismic frequencies. In addition the peak is much less pronounced in natural olivine, and the HTB gives greater absorption than the GBP at temperatures greater than about 1500°C. The GBP may be responsible for attenuation in the lithosphere.

The Bordoni peak was first identified in deformed (cold-worked) metals at low temperature. It is unlikely to be important at mantle temperatures and seismic frequencies, but it would be useful to know its properties for mantle minerals since it contains information about the Peierls energy. The Hirth and Lothe (1968) theory for the characteristic relaxation time gives

$$\tau = \frac{b^2kT}{2E_kD_k}\exp(2E_k/RT)$$

With nominal values for olivine the relaxation time at mantle temperatures is 10^{-7} s, thus this is a high-frequency, low-temperature mechanism. The controlling activation energy is $2E_k$ or about 52 kcal/mole for olivine. This combination of τ_o and E^* make it unlikely that the Peierls energy alone controls attenuation in the mantle. It may be responsible for the "grain boundary" peak in olivine, for which Jackson (1969) obtained τ_o of 10^{-13} s and E^* of 57 kcal/mole. The value for E_k, however, is uncertain.

Values of $\hat{\tau}_o$ of order 10^{-7} to 10^{-12} s for the upper mantle are implied by relaxation times of 10^9–10^{10} s at temperatures of 1400–1600 K and an activation energy of 125 ± 5 kcal/mole. These values are consistent with both laboratory creep data and creep in a polygonized network model of olivine if the stresses are less than about 10 bars. If kilobar-level stresses existed in the upper mantle, then the inferred viscosity and relaxation times would be at least six orders of magnitude lower than observed. On the other hand, there is no contradiction with kilobar-level stresses being maintained in the lithosphere for 10^6 years if temperatures are less than about 900 K. Kilobar-level stresses also shift the absorption band to much higher frequencies because of the high dislocation density and short dislocations, or small grains, at high stress.

The same dislocations contribute to phenomena with quite different time scales. The climb of jogged dislocations is rate-limited by self-diffusion and jog nucleation. In silicates, and other materials with high Peierls energy, the activation energy for creep can be appreciably greater than for self-diffusion. At finite temperature the creep rate will therefore be slower, and the characteristic time longer, than for materials rate-limited by self-diffusion alone. The effective activation energy for creep is either E_j or $2E_j$ times greater than for self-diffusion, depending on whether the dislocation length is smaller or greater than the equilibrium spacing of thermal jogs. These two situations lead to a σ^3 or σ^2 creep law (Hirth and Lothe, 1968). If the activation energy for core diffusion is less than for self-diffusion, the effective activation energy is reduced by one-half the difference of the two. This is a relatively small effect and appears to be negligible for olivine.

On the other hand, the glide of dislocations is controlled by a much smaller activation energy and is therefore much faster than climb at finite temperature. This is the case whether glide is controlled by kink nucleation or diffusion of interstitials. Both creep and attenuation depend on dislocation length and therefore tectonic stress. They are both also exponentially dependent on temperature although the controlling activation energies are different. In principle, the viscosity and stress state of the mantle can be estimated from the seismic Q.

Grain-boundary Losses

In addition to thermoelastic and scattering losses associated with grain boundaries in polycrystals, there are relaxation effects related to the dislocation mechanism at grain boundaries. Boundaries between grains of different orientation or composition are regions of vacancies, impurities, high strain, general disorder and, at high temperatures, partial melting. Losses associated with grain boundaries are high compared to other processes. The simplest kind of grain boundary is simply a plane separating slightly misoriented crystals. Such a boundary can be considered to be a series of edge dislocations that can be made to glide or to migrate upon the application of a shearing stress.

More complicated boundaries can migrate by diffusion of atoms if the temperature is high enough to overcome the activation energy of diffusion. As in other relaxation mechanisms, the temperature of the maximum attenuation shifts as the driving frequency is changed. For high-frequency vibrations, when the relaxation of stress by diffusion cannot keep up with the applied stress, the internal friction increases with temperature and with the grain-boundary area per unit volume, that is, inversely with grain size.

The energy dissipated across a slip boundary is proportional to the product of relative displacement and shear stress. At low temperatures relative displacement between grains is small, and at high temperatures the shear stress across boundaries is essentially relaxed at all times. Appreciable energy dissipation by grain-boundary slip occurs only at intermediate temperatures where there is both displacement and shear stress between grains.

Grain-boundary damping was initially attributed to sliding at grain boundaries, which relaxed the shear stress across the boundary. The boundary was interpreted as a region of disorder that would behave as a film of viscous fluid. The activation energy measured for grain-boundary damping, however, shows good correlation with values measured for grain-boundary diffusion and creep. This suggests that the grain-boundary "sliding" is actually due to ordering and reordering of atoms under a stress bias in the boundary by diffusion. Representative values for the activation energies associated with grain-boundary damping are 46 kcal/mole for iron and 22 kcal/mole for silver; these are close to measured values for grain-boundary diffusion. At low stress levels a stress-induced migration of arrays of dislocations by dislocation climb is probably a more appropriate description of grain-boundary losses than is grain-boundary sliding. This climb would be controlled by self-diffusion rates in the boundary region.

Although grain-boundary damping is clearly an activated process, the laboratory peaks are too broad to be fit by a unique value for the relaxation time, even in pure substances. A range of activation energies is indicated. The differences in orientation and composition between grains lead to boundaries with different structure and mechanical relaxation times. A spectrum of activation energies can be expected for polycrystalline substances leading to broad internal-friction peaks. Few data are available for rocks, but a broad spread of activation energies is probable. This could lead to a Q roughly independent of frequency over a wide frequency band.

The effect on the grain-boundary peak caused by substitutional and impurity defects suggests that it is due to grain-boundary migration rather than sliding. Substitutional impurities increase the activation energy for the grain-boundary peak and reduce its height. Migration of atoms from one grain to another driven by the applied stress is identical to grain-boundary diffusion.

Grain-boundary damping is not really a separate phenomenon from those previously discussed but represents the increased effectiveness of these mechanisms in the boundary regions. The most plausible explanation of grain-boundary damping is essentially the same as stress-induced migration of atoms or reordering of the atoms and atomic bonds in and across the boundary. Partial melting is a variant of grain-boundary attenuation (Nur, 1971; Mavko and Nur, 1975).

High-temperature Thermally Activated Internal Friction and Transient Creep

At high temperatures, attenuation increases rapidly with temperature:

$$Q^{-1} = Q_o^{-1} e^{-E^*/RT}$$

This high-temperature background generally dominates at one-half the melting temperature and above. It increases exponentially with temperature and is a function of frequency and grain size but is independent of amplitude, at least to strains less than about 10^{-5}. The general functional form is

$$\begin{aligned} Q^{-1}(\omega) &= Q_o^{-1}(\omega\tau)^{-\alpha} \\ &= Q_o^{-1}(\omega\tau_o)^{-\alpha} \exp(-\alpha E^*/RT) \qquad (13) \end{aligned}$$

where Q_o^{-1} and α are constants, E^* is an apparent activation energy and τ is a characteristic time. α is generally between ¼ and ½. When E^* is known from diffusion or creep experiments, the parameter α can be determined from the temperature dependence of Q^{-1} as well as from its frequency dependence.

Attenuation is closely related to transient creep; specifically, equation 13 implies a transient creep response of the form

$$\begin{aligned} \varepsilon(t) &= \varepsilon_o(t/\tau)^{\alpha} \\ &= \varepsilon_o(t/\tau_o)^{\alpha} \exp(-\alpha E^*/RT) \qquad (14) \end{aligned}$$

where t is the time. This is in addition to the instantaneous elastic response. The transient response is appropriate for short-term and small-strain situations such as seismic-wave attenuation, tidal friction, damping of the Chandler wobble and postearthquake rebound. Note that the apparent activation energy, αE^*, is less than the true activation energy.

Many materials satisfy the transient creep law (equation 14) with α between ⅓ and ½. Such values are obtained for example from transient creep experiments on rocks at high temperature. Equations 13 and 14 are essentially equivalent: If attenuation is known as a function of frequency, transient creep can be determined and vice versa. Such equivalence can be applied fruitfully to the study and interpretation of seismic-wave attenuation and creep in the Earth.

When $\alpha = ⅓$, equation 14 is the well-known Andrade creep equation. This equation cannot hold for all time since it predicts infinite creep rate at $t = 0$ and a continuously decreasing strain rate as time increases. Real materials at high temperature with a constantly applied stress reach a steady-state creep rate at long time. As Jeffreys (1958) has pointed out, thermal convection is precluded for a material having the rheology of equation 14, and he has used this to argue against continental drift. This prohibition, however, only applies to relatively short times before steady state is established. For very small stresses the strain rate does approach zero at long times for some transient creep mechanisms, such as dislocation bowing under stresses less than those required for dislocation multiplication. The critical stress is $\sigma = Gb/l$ where G is the shear modulus, b is the Burger's vector and l is the dislocation length. For dislocation lengths greater than 10^{-3} cm, multiplication, and therefore steady-state creep, will occur for stresses as low as 10 bar.

Equations 13 and 14 are usually considered as merely phenomenological descriptions of the behavior of solids at high temperatures. However, they can be constructed from a linear superposition of elastic and viscous elements and are thus a generalization of the standard linear solid or the standard anelastic solid.

The creep response of the standard linear solid with a single characteristic relaxation time (actually the strain retardation time), τ, is

$$\varepsilon(t) = \sigma[J_u + \delta J(1 - \exp(-t/\tau))]$$

where σ is the stress, J_u the unrelaxed compliance (the reciprocal of the conventional high-frequency elastic modulus), and δJ is the difference between the relaxed (long-time or low-frequency) and unrelaxed (short-time or high-frequency) compliances.

A more general model involves a spectrum of retardation times. For a distribution or spectrum of characteristic times, $D(\tau)$, the creep is

$$\varepsilon(t) = \sigma\left[J_u + \delta J \int_0^\infty D(\tau)[1 - \exp(-t/\tau)]d\tau\right] \quad (15)$$

(Minster and Anderson, 1981). For a single characteristic time $D(\tau)$ is a delta function. The retardation spectrum should yield physically realizable creep behavior, particularly at short and long times. $D(\tau)$ will be assumed to be limited to a finite band (τ_1,τ_2)—the cutoffs are determined by the physics of the problem and depend on physical variables (such as diffusivity), geometric variables (such as diffusion paths), as well as the thermodynamic conditions.

The distribution function

$$D(\tau) = \begin{cases} \dfrac{\alpha}{\tau_2^\alpha - \tau_1^\alpha}\dfrac{1}{\tau^{1-\alpha}} & \text{for } \tau_1 \leq \tau \leq \tau_2 \\ 0 & \text{otherwise} \end{cases} \quad (16)$$

yields a creep response which can be approximated by

$$\varepsilon(t) \approx \sigma J_u\left[1 + \frac{\delta J}{J_u}\Gamma(1 - \alpha)\left(\frac{t}{\tau_2}\right)^\alpha\right],$$

$$\tau_1 << \tau << \tau_2 \quad (17)$$

(Anderson and Minster, 1979). Comparison with equation 14 yields

$$\varepsilon_o = \frac{\delta J}{J_u}\Gamma(1 - \alpha) = \frac{\delta M}{M_r}\Gamma(1 - \alpha) \quad (18)$$

where M_r refers to the relaxed modulus. Note that the characteristic time appearing in equation 14 is actually the upper cutoff τ_2.

Furthermore, this model yields a finite initial creep rate

$$\varepsilon = \sigma\delta J \frac{\alpha}{1 - \alpha}\frac{\tau_1^{\alpha-1} - \tau_2^{\alpha-1}}{\tau_2^\alpha - \tau_1^\alpha} \quad (19)$$

A finite creep rate at the origin was achieved differently by Jeffreys (1965), who proposed what he called the modified Lomnitz law:

$$\varepsilon(t) = \frac{\sigma}{M_u}\left[1 + \frac{q}{\alpha}\left\{\left(1 + \frac{t}{\tau}\right)^\alpha - 1\right\}\right] \quad (20)$$

so that

$$\varepsilon(t) = \frac{\sigma}{M_u}\left[1 + \frac{q}{\alpha}\left(\frac{t}{\tau}\right)^\alpha\right] \quad \text{for } t >> \tau \quad (21)$$

By identification with equation 17 and 19 we obtain the correspondence

$$\tau = [\Gamma(2 - \alpha)]^{1/(1-\alpha)}\tau_1$$

$$q = [\Gamma(2 - \alpha)]^{1/(1-\alpha)}\frac{\delta J}{J_u}\left(\frac{\alpha}{1 - \alpha}\right)\left(\frac{\tau_1}{\tau_2}\right)^\alpha \quad (22)$$

Note that the time constant in Jeffreys' equation is related to the shortest relaxation time in the spectrum, τ_1, which is less than 1 s for many materials at high temperature. The parameter q is related to the ratio τ_1/τ_2, which can be very small. For a dislocation bowing mechanism τ is proportional to the dislocation length squared. A spread of two orders of magnitude in dislocation lengths gives τ_1/τ_2 equal to 10^{-4}.

The Q corresponding to equation 17 for frequencies, ω, such that

$$\omega\tau_1 << 1 << \omega\tau_2$$

is given by

$$Q(\omega) = \cot\frac{\alpha\pi}{2} + \frac{2J_u}{\pi\alpha\delta J}[(\omega\tau_2)^\alpha - (\omega\tau_1)^\alpha]\cos\frac{\alpha\pi}{2}$$

$$\approx \frac{2J_u}{\pi\alpha\delta J}(\omega\tau_2)^\alpha\cos\frac{\alpha\pi}{2} \quad (23)$$

since $\tau_2 >> \tau_1$ and $\cot \alpha\pi/2$ is small. Thus, by comparison with equation 13,

$$Q_o = \frac{2J_u}{\pi\alpha\delta J}\cos\frac{\alpha\pi}{2} \quad (24)$$

Therefore Q is predicted to increase with frequency as ω^α. Geophysical data give α in the range of 0.15 to 0.5 for periods greater than about 1 s.

For high frequencies $1 \ll \omega\tau_1 \ll \omega\tau_2$,

$$Q(\omega) = (1 - \alpha)(J_u/\alpha\delta J)(\tau_2/\tau_1)^\alpha(\omega\tau_1) \quad (25a)$$

At low frequencies,

$$Q(\omega) = (1 + \alpha)(J_r/\alpha\delta J)(\omega\tau_2)^{-1} \quad (25b)$$

For thermally activated processes the characteristic times depend on temperature:

$$\tau = \tau_o\exp(E^*/RT)$$

Inserting this into equations 17 and 23 yields the experimental equations. It is clear that at very low temperature the characteristic times are long and $Q \approx \omega$. At very high temperatures Q is predicted to increase as ω^{-1}. Equation 23 holds for intermediate temperatures and intermediate frequencies.

Regardless of its origin the high-temperature background damping seems to dominate other mechanisms when measurements are performed at sufficiently high temperature. Therefore, it is particularly important in geophysical discussions.

PARTIAL MELTING

Several seismic studies have indicated that increased absorption, particularly of S-waves, occurs below volcanic zones and is therefore presumably related to partial melting. Regional variations in seismic absorption are a powerful tool in mapping the thermal state of the crust and upper mantle. Preliminary results indicate that, where the absorption is anomalously high, V_s is more affected than V_p. It has also been suggested that partial melting is the most probable cause of the low-velocity layer in the upper mantle of the Earth (Anderson and Sammis, 1970). Thus the role of partial melting in the attenuation of seismic waves may be a critical one, at least in certain regions of the Earth.

Studies of the melting of polycrystalline solids have shown that melting begins at grain boundaries, often at temperatures far below the melting point of the main constituents of the grains. This effect is caused by impurities that have collected at the grain boundaries during the initial solidification. Walsh (1969) computed the internal friction for phenomenological models of a partially melted solid. His calculations included mechanical effects only, ignoring thermoelastic and thermochemical attenuation in two-phase systems.

Attenuation according to Walsh's model should exhibit the following properties:

1. Q_K^{-1}, the attenuation of the purely compressional component of seismic waves due to this mechanism, will be negligible at seismic frequencies.
2. Under shear, this material exhibits the properties of the standard linear solid with relaxation time roughly given by

$$1/\tau = a\mu/\eta$$

   where μ is grain rigidity and η is melt viscosity. The ratio μ/η typically ranges from 10^4 to 10^8 Hz, so that this mechanism would be important for inclusion aspect ratios a, of 10^{-8} to 10^{-4}. These aspect ratios would be reasonable for a melt phase that wets the grain boundary, thus forming a thin film. Furthermore, the relaxation strength is strongly dependent on the number of sites of melting, and could exceed 1 even for a very small volume of melt.
3. This shear attenuation mechanism will behave as a thermally activated relaxation.
4. Shear attenuation will increase very sharply with the onset of melting.

The concentration of melt as a function of temperature and pressure is perhaps the most important unknown quantity, followed closely by the viscosity of the melt. As some partial melting is likely to occur in the Earth's mantle, this mechanism is a possible cause of seismic attenuation, particularly at very low frequencies. Melt concentrations of as little as 10^{-2} may cause significant attenuation at seismic frequencies.

Spetzler and Anderson (1968) studied the effect of partial melting in the system NaCl–H_2O. At the eutectic temperature the system is partially molten, having a melt content that is proportional to the original salinity. The internal friction for longitudinal waves increased abruptly by 48 percent at the eutectic point for 1 percent partial melting and 71 percent for 2 percent partial melting. The corresponding increases in internal friction for shear were 37 percent and 73 percent. Further melting caused a gradual further increase in Q^{-1}. The fractional increases in Q^{-1} across the eutectic point were much greater than the fractional drops in velocity. Figure 14-4 shows the Q in the ice-brine-NaCl system for a concentration of 2 percent NaCl for longitudinal vibrations of a rod. Plotted are data for the fundamental and first two overtones. Note the abrupt drop in Q as partial melting is initiated at the eutectic temperature. There is a corresponding, but much less pronounced, drop in velocity at the same temperature.

BULK ATTENUATION

Most of the mechanisms of seismic-wave attenuation operate in shear. Shear losses, generally, are much larger than compressional losses, and therefore shear waves attenuate more rapidly than compressional waves. Bulk or volume attenuation can become important in certain circumstances. One class of such mechanisms is due to thermoelastic relaxation. An applied stress changes the temperature of a sample relative to its surrounding, or of one part of a specimen relative to another. The heat flow that then occurs in order to equalize the temperature difference gives rise to energy dissipation and to anelastic behavior. The change in temperature is simply

$$(\partial T/\partial P)_S = -\alpha T/C_P$$

and the difference between the unrelaxed and relaxed moduli is the difference between the adiabatic and isothermal moduli. Under laboratory conditions this is typically 1 percent for oxides and silicates. α is the coefficient of thermal expansion.

In a propagating longitudinal wave the regions of compression (hot) and dilatation (cold) are separated by a half-wavelength, $\lambda/2$, where $\lambda = c/f$. The relaxation time is

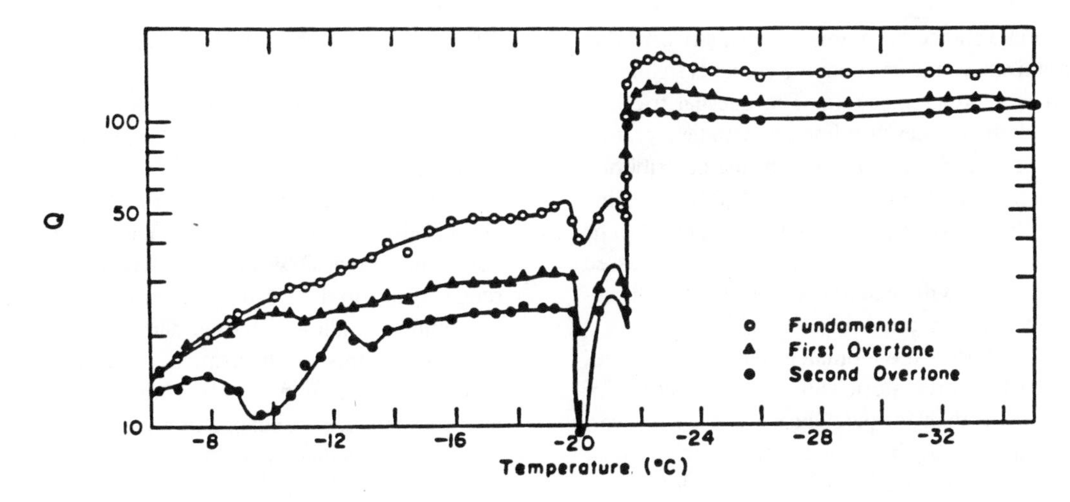

FIGURE 14-4
Q of ice containing 2 percent NaCl. At low temperatures this is a solid solution. At temperatures higher than the eutectic the system is an ice-brine mixture (Spetzler and Anderson, 1968).

$$\tau = \lambda^2/(2\pi)^2\kappa$$

where κ is the thermal diffusivity. Attenuation is maximum at

$$\omega\tau = c\lambda/2\pi\kappa$$

Peak damping occurs at very high frequencies ($>10^{10}$ Hz), and this mechanism is therefore unimportant at seismic frequencies and most ultrasonic experiments. The hot and cold parts of the wave are too far apart for appreciable heat to flow during a cycle. On the other hand experiments in the gigahertz range, as in Brillouin scattering, may actually be measuring the isothermal rather than the adiabatic moduli. This may be counterintuitive since static, or long-term, deformation experiments also give the isothermal moduli. The explanation is that heat flow depends both on the time scale and the length scale. Although only a short time is available in the high-frequency experiments, the length scale, or wavelength, is also very short.

A variant of this mechanism has to do with intergranular thermal currents. In a polycrystalline material another length scale is determined by the grain size. When a stress is applied to an inhomogeneous or polycrystalline material, different parts are compressed differently and intergranular thermal currents are set up. The magnitude of the effect depends on the anisotropy and heterogeneity of the grains. The characteristic relaxation time is now in proportion to d^2/κ where d is the grain size, and Q is a unique function of fd^2/κ. The theory for this mechanism (Zener, 1948) shows that the absorption peak is broader than a Debye peak on the high-frequency side, and Q increases as $\omega^{1/2}$, rather than as ω for a Debye peak appropriate for the standard linear solid. At low frequencies $Q \approx \omega^{-1}$, as for a Debye peak. The relaxation time is

$$\tau = d^2/3\pi^2\kappa$$

In polycrustal materials the flow of heat may take place between adjacent grains because of differential compression. This can occur even in solids where the grains are of identical composition if they are anisotropic and unoriented. This latter situation gives

$$Q^{-1} = \left(\frac{C_P - C_V}{C_V}\right) R \frac{\omega\omega_o}{\omega^2 + \omega_o^2}$$

where the first term on the right is the relative difference between the specific heats, R is a factor that increases with the anisotropy of the grains and ω_o is the intergranular relaxation frequency. Again the damping is a maximum when $\omega_o\tau \approx 1$ or approximately

$$\omega^{-1} = \tau \approx 10^3 d^2$$

For very anisotropic crystals this mechanism gives $Q \approx 10^3$ at maximum damping. The grain size in the upper mantle is uncertain, but for d = 1 mm to 1 m the critical wave period for maximum attenuation is 10^{-1} to 10^7 s. The range of grain sizes and thermal properties in a common rock would tend to broaden the relaxation peak. In the nearly adiabatic high-frequency case, the thermal currents are confined to the immediate vicinity of the grain boundaries and the internal friction varies as $\omega^{-1/2}$; this for wavelengths short compared to the grain dimensions. An approximate calculation gives

$$Q^{-1} \approx \Delta\Omega(\mathcal{K}/C_v)(2\omega)^{-1/2}$$

where Ω is the ratio of total surface area to volume. In the other extreme, isothermal vibrations (small grain sizes and low frequency), the internal friction is proportional to the first power of frequency. $\mathcal{K}$ is thermal conductivity.

By appropriate choice of parameters, primarily the distribution of grain sizes, one can obtain a Q that is frequency independent over a broad spectrum. The internal friction, Q^{-1}, increases slightly faster than linearly with temperature for thermoelastic mechanisms. Except for the contribution of open cracks, which presumably dominates the damping at low pressure, the effect of pressure can be expected to be slight. Intergranular thermoelastic stresses are suppressed by pressure primarily to the extent that anisotropy and mechanical heterogeneity are reduced.

Thermoelastic losses are intimately related to changes in volume and, therefore, are usually more effective for compressional waves than for pure shear waves. However, when inhomogeneities are imbedded in a material, shear waves as well as compressional waves generate local dilatational stress inhomogeneities and, therefore, can have thermoelastic components in their damping. In fact, it is possible to contrive geometries that lead to greater shear wave attenuation than longitudinal wave attenuation.

Thermoelastic losses undoubtedly exist in materials as inhomogeneous as those thought to make up the mantle. Anisotropic grains with random orientation, the close association of grains with different elastic and thermal properties and the possible existence of molten regions all would tend to introduce local fluctuations in stress with an attendant thermal diffusion. The wide spectrum of grain sizes, anisotropies, orientations and physical properties of rock-forming minerals tend to smear out the thermoelastic peak and broaden the Q spectrum over a wide frequency band.

Table 14-2 shows the peak frequency expected for a range of possible grain sizes, using the properties of forsterite. The maximum internal friction will occur in the range of seismic frequencies if the grain size in the mantle is in the range of 1 to 100 cm. The magnitude of the peak internal friction is harder to estimate. Most experiments have been performed on cubic metals with isotropic coefficients of thermal expansion, so that the anisotropy factor R has reflected only the elastic anisotropy. In rocks, particularly multicomponent rocks, anisotropy of thermal expansion will be significant, and the elastic anisotropy will probably exceed that of metals. The effect of variations in grain size in rocks would be to weaken the frequency dependence of the internal friction. Taking these effects into account, it is estimated that the internal friction from this mechanism, could reach 10^{-3} to 10^{-2} in the mantle. For shear waves, the attenuation would probably be somewhat less but could still be significant. Substantial shear strains can be introduced in a heterogeneous solid by a purely longitudinal applied stress.

The peak attenuation depends on the contrast, amongst the grains, in bulk modulus and thermal expansion. If a composite contains grains that are very anisotropic or differ greatly in their properties, the bulk attenuation can be substantial. For a factor of 2 contrast in the coefficient of thermal expansion or bulk modulus, Q_K can be as low as 300 and 3000, respectively (Budiansky and O'Connell, 1980). Bulk attenuation of this type depends on the differences in the properties of the constituent grains and is therefore a possibly useful constraint on the petrology of the mantle.

In an application of the theory, Budiansky and O'Connell (1980) showed that a lower mantle composed of stishovite and magnesiowüstite and having grain sizes in the range 3 mm to 1 m can be ruled out on the basis of the maximum Q_K allowed by the damping of the radial mode ${}_0S_0$. Q_K, for this mechanism, is not a thermally activated process and therefore is only a weak function of temperature and pressure. A change in mineralogy, however, associated with a change in pressure or temperature, or a change in grain size and orientation, due to a change in stress, can affect Q_K. The peak frequency is related to the grain size.

There is another effect associated with polycrystalline material. A volumetric strain produces local shear deformations within the individual grains (Heinz and others, 1982). This leads to Q_K^{-1} of the order of 1 percent of Q_G^{-1}, and the variation of Q_K with temperature and frequency will follow the variation of Q_G. Q_K is difficult to measure in the Earth since it is generally higher than the Q_G, which therefore dominates the attenuation of seismic energy over most frequency bands.

The radial modes (${}_0S_0$, ${}_1S_0$, ${}_2S_0$, etc.) are affected by Q_K and are less sensitive to Q_G. In fact, these modes are observed to have very high Q. Unfortunately, they have little resolving power, and it is not even clear if it is the mantle or the core that is responsible. In regions of the mantle or the spectrum where Q_G is large, it may be possible to study Q_K from the attenuation of compressional waves.

RELAXATIONS IN SYSTEMS UNDERGOING PHASE TRANSFORMATIONS

In systems subject to solid-liquid or solid-solid phase transitions, small stresses can induce a reaction from one phase to another, causing stress relaxation and bulk attenuation. For two phases in equilibrium, the attenuation is given by the standard relaxation formula

$$Q^{-1} = \Delta \frac{\omega\tau}{1 + (\omega\tau)^2}$$

Δ is evaluated from the thermodynamic properties of the system by using

$$K_i = (\partial P/\partial \rho)_i = \left(\frac{-\gamma V}{\rho(\partial V/\partial P)_T}\right)_i$$

where V is the molar volume, K_i is the bulk modulus and γ is the ratio of specific heats:

TABLE 14-2
Frequency Dependence of Thermoelastic Attenuation as a Function of Grain Size in Forsterite

Grain Size (cm)	Peak Frequency (Hz)
10^{-2}	10^{3}
10^{-1}	10
1	10^{-1}
10	10^{-2}
10^{2}	10^{-5}

$$\gamma_i = (C_P/C_V) = \left[\frac{C_P}{C_P + T(\partial V/\partial T)_P^2/(\partial V/\partial P)_T}\right]_i$$

This theory holds when both phases are present and in equilibrium. At the temperature of onset of a phase change, discontinuous volume changes occur that require a modified analysis. For most solid-liquid transitions, and for solid-solid transitions of geophysical interest, the relaxation strength will be large, of the order of magnitude 10^{-1} to 1. The peak frequency is determined by the kinetics of the appropriate reaction.

The subscript i refers to u (unrelaxed or infinite frequency) or r (relaxed or zero frequency). At infinite frequency K_u is simply a volume average of the bulk moduli of the components, at constant extent of reaction. That is, the amount of each phase does not change with temperature or pressure. K_r includes the normal elastic deformation plus the volume change due to reaction. This gives a smaller effective modulus.

The Vaisnys (1968) theory treats the attenuation of acoustic waves propagating through a system undergoing phase transition where a finite reaction rate introduces a lag between the applied pressure change and the response. Stevenson (1983) treated the case of infinite reaction rate but a diffusive lag of heat and matter across the phase boundary. Irreversible entropy production accompanies the imposition of a pressure or temperature perturbation on a two-phase system consisting of a dilute suspension of a solid phase in a liquid. The minimum Q in this case occurs at a frequency

$$\omega_o = 4\pi Dl\phi$$

where D is the solute diffusivity, l is the dimension of the solid particles and ϕ is volume concentration of particles. $Q \approx \omega^{1/2}$ in the high frequency limit, as appropriate for a diffusive mechanism, and $Q \approx \omega$ in the low-frequency limit, in common with other mechanisms. For well separated particles there is also a $Q \approx \omega^{-1}$ regime. This mechanism might be important in magma chambers and in subliquidus parts of the core. In principle, it also applies for solid-solid phase changes but requires small grain sizes, high diffusivities and low-frequency waves.

PHYSICS OF ATTENUATION IN FLUIDS

The absorption of sound in a liquid at low frequencies is given by

$$k^*/f^2 = (2\pi^2/\rho c^3)\left\{\frac{4}{3}\eta_s + (c^2\alpha^2\mathcal{K}T/C_P^2) + \eta_v\right\}$$

where k^* is the spatial attenuation coefficient, α is the coefficient of thermal expansion, $\mathcal{K}$ the thermal conductivity, η_s the shear viscosity, η_v the volume viscosity and f is frequency. The first two terms on the right-hand side represent the classical Stokes-Kirchoff absorption due to shear stresses and thermal conductivity. The first term is important in the attenuation of longitudinal waves in liquid metals but is absent for strains involving pure compression. The second term is important in liquid metals at laboratory frequencies, but is negligible at core conditions and seismic frequencies. The remaining term is called the excess absorption and is due to bulk or volume viscosity. In liquid metals at low pressure, η_v is similar in magnitude to η_s.

Several mechanisms contribute to bulk viscosity in simple fluids. One is structural relaxation due to perturbations in the short-range order caused by the sound wave. Another is due to concentration fluctuations in alloy systems. The characteristic times for shear and structural relaxation are likely to be similar as they are both controlled by short-range diffusion. The relaxation time due to concentration fluctuations or chemical reactions may be quite different.

Longitudinal waves in liquid metals typically are within an order of magnitude of

$$Q = 10^{11}/f$$

which, under normal conditions is composed of roughly equal parts contributed by shear viscosity, thermal conductivity and bulk viscosity. The attenuation in liquid metals at seismic frequencies would be quite negligible. However, the effects of pressure and alloying lengthen the characteristic relaxation times.

The bulk viscosity of a pure liquid metal is due to structural rearrangement as there are no internal degrees of freedom. Two-state theories have been developed to explain structural relaxation associated with pressure changes. The assumption is that a liquid is a mixture of two structural states with differing mole volumes and energies.

The two-state theory (Flinn and others, 1974) can be used to estimate the structural volume viscosity of iron just above the melting point, T_m, and its relation to the shear viscosity:

$$\eta_v = \frac{x_1x_2V}{RT_m}\left(\frac{\Delta V}{V} - \frac{\alpha\Delta H}{C_P}\right)^2 \tau K^2$$

and

$$\eta_v/\eta_s = \frac{(2n - 12)(12 - n)}{n^2}\left(\frac{VK}{RT_m}\right)^2\left(\frac{\Delta V}{V} - \frac{\alpha\Delta H}{C_P}\right)^2$$

$$x_2 = (12 - n)/n = 1 - x_1$$

where x_1 and x_2 are the mole fractions of the two states, which differ by ΔV and ΔH in molar volume and enthalpy, and n is the coordination number. For liquid iron at low pressure we take $n = 9$; $V = 8$ cm^3/mol; $\Delta V/V = 0.1$; $\Delta H/RT_m = 3$; the bulk modulus $K = 1.3$ Mbar (10 percent less than solid iron); $\alpha = 70 \times 10^{-6}$/K, $C_P = 10.2$ cal/mol K; and the relaxation time, $\tau = 5 \times 10^{-12}$ s. This relaxation time is typical of liquid metals. These parameters give

$$\eta_v = 3 \times 10^{-2} \text{ poise} = 3 \text{ cP}$$

and

$$\eta_s = (5/3)\eta_v = 5 \text{ cP}$$

for molten iron at low pressure. The measured value for η_s is 4.8–7.0 cP, and the near equality of η_s and η_v is consistent with measurements on other liquid metals. Thus, the two-state theory gives satisfactory results at low pressure, which leads us to investigate the properties at core conditions.

The main effect of pressure and temperature is on the bulk modulus K and the relaxation time τ. The bulk modulus of the core is 4–10 times larger than iron at normal conditions, and relaxation times have been estimated to be of the order of 10^{-8}–10^{-9} s. Therefore, bulk viscosities four to six orders of magnitude larger than at low pressures, or η_v of 10^2–10^4 P, are expected. The associated shear viscosities would be in the range of 10–100 P. The neglect of the effect of temperature and pressure on the other parameters would decrease these estimates slightly but probably by less than an order of magnitude. An upper limit for η_s at the top of the outer core is 10^6 P based on the lack of damping of the 18.6-year principal nutation (Toomre, 1974). A strong magnetic field in the core can result in a magnetic viscosity that is in addition to these structural relaxations. The magnetic viscosity is anisotropic, controlled by the orientation of the magnetic field. Seismic waves traveling in a core with a magnetic field of the order of 10^5 gauss would have a measurable anisotropy. Slow shear waves would also propagate in the "liquid" core.

Effect of Alloying

Based on density and cosmic abundances, the core is likely to contain up to 10 percent nickel and other siderophiles and a similar amount of a light element such as oxygen or sulfur. There are additional volume relaxation mechanisms associated with alloys and solutions. In particular the relaxation times can be much longer than those due to shear and structural relaxations. This might be significant in explaining the apparent frequency dependence of Q of P-waves in the outer core. There is a suggestion of a linear increase of Q_p with frequency, which, if verified, indicates that the seismic band is on the high-frequency side of the absorption band in the outer core (Anderson and Given, 1982). As this cannot be true for shear waves unless the unrelaxed shear modulus of the core is much lower than the rigidity of the inner core, we may need a low-frequency volume relaxation mechanism. This contribution to the bulk attenuation and volume viscosity depends on concentration and may eventually shed light on the nature of alloying elements in the core.

Flinn and others (1974) found a highly temperature-dependent absorption mechanism operating in binary melts that is not present in the pure metals. The extra absorption in liquid metal alloys is probably due to concentration fluctuations. Theoretically, it is inversely proportional to the diffusivity or directly proportional to the shear viscosity. As with the structural viscosity, it is also proportional to the square of the bulk modulus and can therefore be expected to increase with pressure. More importantly, it is a function of the curvature of the Gibbs potential versus concentration relation and, therefore, becomes very large in the vicinity of a critical point. Unfortunately, it is not possible to estimate the relaxation time of this mechanism in the core.

We can conclude that structural and concentration fluctuations may be responsible for bulk viscosity in the core and contribute to the excess absorption of the radial modes.

Relaxation Theory of Bulk Attenuation in the Core

The Q due to bulk losses in a relaxing solid or viscoelastic fluid is

$$Q_K^{-1}(\omega) = (\Delta K/K)\omega\tau_v/(1 + \omega^2\tau_v^2) \tag{26}$$

where $\Delta K/K = \Delta$ is the modulus defect and τ_v is the volume relation time. The modulus defect is the fractional difference between the high-frequency and low-frequency moduli. This has not been measured for liquid metals but is typically about 0.3 for other liquids. The decrease in bulk modulus on melting of metals ranges from about 5 to 30 percent. This may be of the same order as the modulus defect for molten metals. The bulk moduli of glasses are often taken as approximately the same as the high-frequency moduli for fluids. This gives a modulus defect of 0.2–0.7 for most liquid-glass systems. These estimates give Q_{min} between 10 and ~3 at $\omega\tau = 1$. The increase in bulk modulus at the inner core–outer core boundary is about 5 percent. If this is taken as $\Delta K/K$, we obtain an estimate of 40 for the minimum Q_K of the core at $\omega\tau = 1$. A Q_K of 2×10^3 at a period of 20 min (the period of $_0S_0$) implies, equation (26), that $\tau \approx 4$ s or 10^4 s. The smaller value would be consistent with the solid-like behavior of the inner core for short-period body waves. The location of the inner core–outer core boundary would then be frequency dependent, and the inner core would apparently be smaller at free-oscillation periods. Such a frequency dependence has, in fact, been suggested by Gutenberg (1957). A frequency-dependent inner-core boundary could explain the discrepancy between the free-oscillation and body-wave determinations of the properties of the inner core. For example, the shear velocity obtained for the inner core from short-period waves is about 20 percent smaller than that obtained from the free-oscillation data for a fixed inner core radius. These results can equally well be interpreted in terms of a 20 percent decrease in inner-core radius in going from body-wave to free-oscillation periods. Glass-like transitions, or transitions from fluid-like to solid-like behavior, typically take place over a small range of temperature. The sharpness of the inner-core boundary could be explained by such a transition occurring over a small range of pressure.

Interpreting the data in this way requires that the seismic data for the outer core be on the low-frequency side of the relaxation peak (Anderson, 1980). For the outer core,

$$Q_K^{-1} = \frac{\Delta K}{K}(\omega\tau_v)$$

and for the inner core,

$$Q_K^{-1} = (\Delta K/K)/\omega\tau_v$$

The boundary of the inner core, then, corresponds to $\omega\tau = 1$, and this is where attenuation would be a maximum. For a 1-s body wave the inferred relaxation time at the inner core–outer core boundary is $(2\pi f)^{-1}$ or about 0.2 s. This is close to the value deduced from Q_K in the outer core.

ABSORPTION-BAND Q MODEL FOR THE EARTH

Attenuation in solids and liquids, as measured by the quality factor Q, is typically frequency dependent. In seismology, however, Q is usually assumed to be independent of frequency. The success of this assumption is a reflection of the limited precision, resolving power and bandwidth of seismic data and the trade-off between frequency and depth effects rather than a statement about the physics of the Earth's interior.

Frequency-independent Q models provide an adequate fit to most seismic data including the normal-mode data (Anderson and Archambeau, 1964; Anderson and others, 1965; Anderson and Hart, 1979). There is increasing evidence, however, that short-period body waves may require higher Q values. Some geophysical applications require estimates of the elastic properties of the Earth outside the seismic band. These include calculations of tidal Love numbers, Chandler periods and high-frequency moduli for comparison with ultrasonic data. The constant-Q models cannot be used for these purposes. For these reasons it is important to have a physically sound attenuation model for the Earth.

The theory of seismic attenuation has now been worked out in some detail (Anderson and Minster, 1979; Minster and Anderson, 1981). Although a mild frequency dependence of Q can be expected over a limited frequency band, Q or Q^{-1} should be a linear function of frequency at higher and lower frequencies.

For a solid characterized by a single relaxation time τ, Q^{-1} is a Debye function with maximum absorption at $\omega\tau = 1$. For a solid with a spectrum of relaxation times, the band is broadened and the maximum attenuation is reduced. For a polycrystalline solid with a variety of grain sizes, orientations and activation energies, the absorption band can be appreciably more than several decades wide. If, as seems likely, the attenuation mechanism in the mantle is an activated process, the relaxation times should be a strong function of temperature and pressure. The location of the absorption band, therefore, changes with depth. The theory of attenuation in fluids indicates that Q in the outer core should also depend on frequency.

The theoretical considerations in this chapter indicate that Q can be a weak function of frequency only over a limited bandwidth. If the material has a finite elastic modulus at high frequency and a nonzero modulus at low frequency, there must be high- and low-frequency cutoffs in the relaxation spectrum. Physically this means that relaxation times cannot take on arbitrarily high and low values. The relationship between Q and bandwidth indicates that a finite Q requires a finite bandwidth of relaxation times and therefore an absorption band of finite width. Q can be a weak function of frequency only in this band.

We can approximate the absorption band in the following way:

$$Q = Q_m(f\tau_2)^{-1}, \quad f < 1/\tau_2$$

$$Q = Q_m(f\tau_2)^{\alpha}, \quad 1/\tau_2 < f < 1/\tau_1$$

$$Q = Q_m(\tau_2/\tau_1)^{\alpha}(f\tau_1), \quad f > 1/\tau_1$$

where $f(= \omega/2\pi)$ is the frequency, τ_1 is the short-period cutoff, τ_2 is the long-period cutoff and Q_m is the minimum Q, which occurs at $f = 1/\tau_2$ if $\alpha > 0$. These parameters are shown in Figure 14-5. This approximation of an absorption band was used by Anderson and Given (1982) to model the attenuation in the mantle and core.

The relaxation time, τ, for an activated process depends exponentially on temperature and pressure. The characteristic time τ_0 depends on l^2 and D for diffusion-controlled mechanisms where l and D are characteristic length and a diffusivity, respectively. Characteristic lengths, such as dislocation or grain size, are a function of tectonic stress, which is a function of depth. The location of the band, therefore, depends on tectonic stress, temperature and pressure. The width of the band is controlled by the

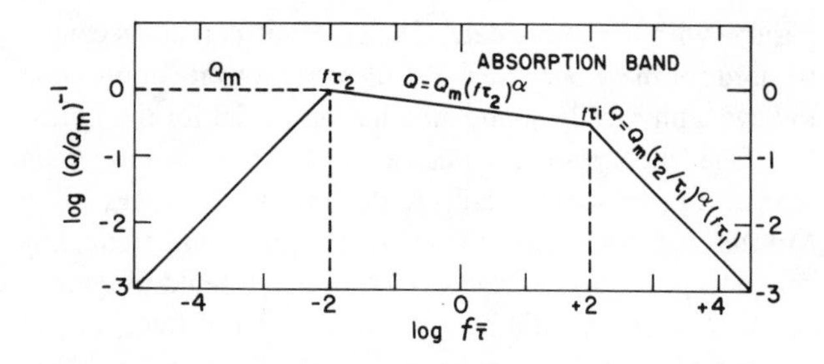

FIGURE 14-5
Schematic illustration of an absorption band (Anderson and Given, 1982).

distribution of relaxation times, which in turn depends on the distribution of grain sizes or dislocation lengths.

With an activation energy of 60 kcal/mole, τ decreases by about an order of magnitude for a temperature rise of 200°C. An activation volume of 10 cm^3/mole causes τ to increase by an order of magnitude for a 30-kbar increase in pressure. In regions of low temperature gradient, the maximum absorption shifts to longer periods with increasing depth. This condition is probably satisfied throughout most of the mantle except in thermal boundary layers. Characteristic lengths such as grain size and dislocation length decrease with increasing stress. A decrease in tectonic stress by a factor of 3 increases the relaxation time by about an order of magnitude. A decrease of stress with depth moves the absorption band to longer periods.

The effect of pressure dominates over temperature for most of the upper mantle, and tectonic stress probably decreases with depth. Therefore the absorption band is expected to move to longer periods with increasing depth. A reversal of this trend may be caused by steep stress or temperature gradients across boundary layers, or by enhanced diffusion due to changes in crystal structure or in the nature of the point defects. For a given physical mechanism there are relationships between the width of the band (τ_2/τ_1), Q_m and α; that is, these three parameters are not independent.

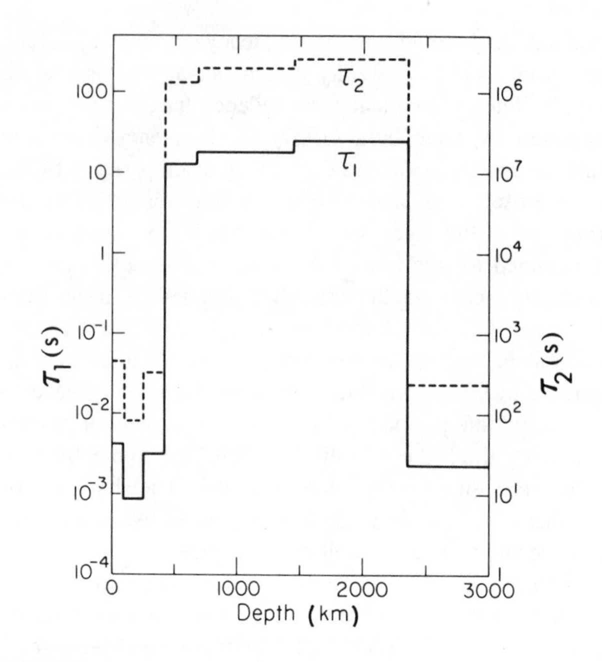

FIGURE 14-6
τ_1 and τ_2 as a function of depth in the mantle for the absorption-band model (ABM) for Q_s (Anderson and Given, 1982).

For fixed α, τ_2/τ_1 increases as Q_m^{-1} decreases. If we assume that the parameters of the absorption band, Q_m^{-1}, τ_1/τ_2 and α, are constant throughout the mantle, we can use the seismic data to determine the location of the band as a function of depth, or $\tau_1(z)$. This assumption is equivalent to assuming that the activation energy, E^*, and activation volume, V^*, are fixed. By assuming that the characteristics of the absorption band are invariant with depth, we are assuming that the width of the band is controlled by a distribution of characteristic relaxation times rather than a distribution of activation energies. Although this assumption can be defended, to some extent it has been introduced to reduce the number of model parameters. If a range of activation energies is assumed, the shape of the band (its width and height)

TABLE 14-3
Absorption Band Parameters for Q_s

Radius (km)	Depth (km)	τ_1 (s)	τ_2/τ_1	Q_s(min) (τ_1)	Q_s (100 s)	α
1230	5141	0.14	2.43	35	1000	0.15
3484	2887	—	—	—	—	—
4049	2322	0.0025	10^5	80	92	0.15
4832	1539	25.2	10^5	80	366	0.15
5700	671	12.6	10^5	80	353	0.15
5950	421	0.0031	10^5	80	330	0.15
6121	250	0.0009	10^5	80	95	0.15
6360	11	0.0044	10^5	80	90	0.15
6371	0	0	∞	500	500	0

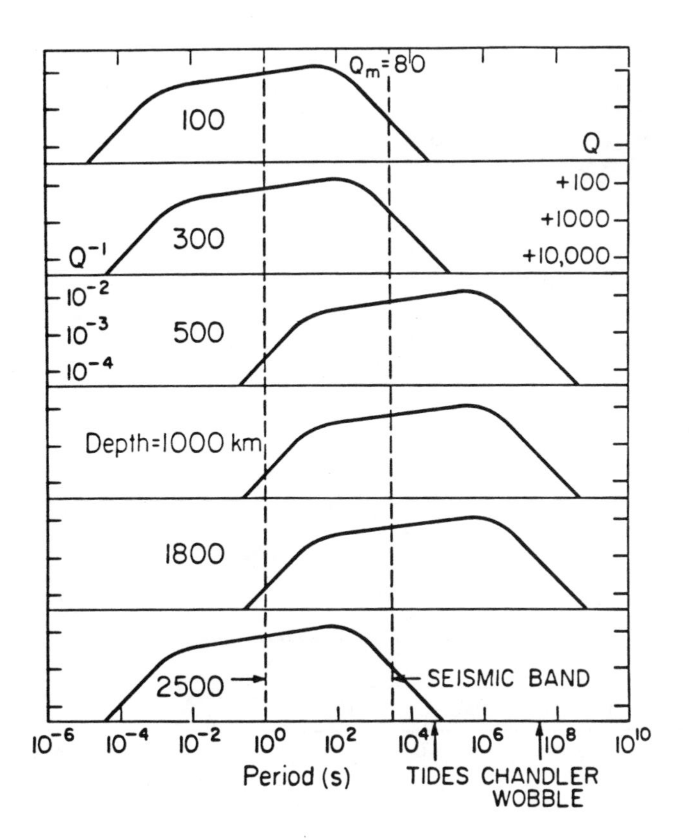

FIGURE 14-7
The location of the absorption band for Q_s as a function of depth in the mantle (Anderson and Given, 1982).

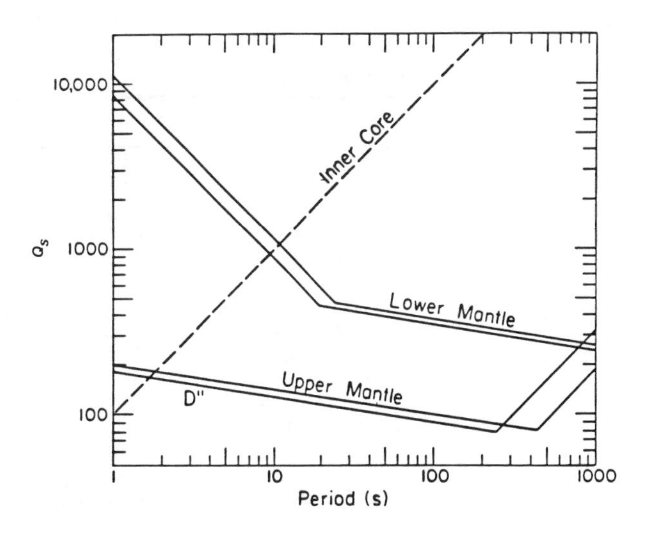

FIGURE 14-8
Q_s as a function of period for the mantle and inner core for ABM. Note the similarity of the upper mantle and the lowermost mantle (D''). These may be thermal (high-temperature gradient) and mechanical (high-stress) boundary layers associated with mantle convection (Anderson and Given, 1982).

varies with temperature and pressure. The parameters of the absorption bands are given in Table 14-3. The locations of the bands as a function of depth are shown in Figures 14-6, 14-7 and 14-8. I will refer to the absorption-band model as ABM. The parameter α can be adjusted by trial and error: $\alpha = 0.3$ is too high and $\alpha = 0.15$ is about right.

The variation of τ_1 and τ_2 with depth in the mantle is shown in Figure 14-6. Note that both decrease with depth in the uppermost mantle. This is expected in regions of steep thermal gradient. They increase slightly below 250 km and abruptly at 400 km. No abrupt change in τ_1 or τ_2 occurs at 670 km. Apparently, a steep temperature gradient and high tectonic stresses can keep the absorption band at high frequencies throughout most of the upper mantle, but these effects are overridden below 400 km where most mantle minerals are in the cubic structure. A phase change, along with high pressure and low stress, may contribute to the lengthening of the relaxation times. Relaxation times change only slightly through most of the lower mantle.

TABLE 14-4
Q_s and Q_p as Function of Depth and Period for Model ABM

Depth	Q_s				Q_p			
(km)	1s	10s	100s	1000s	1s	10s	100s	1000s
5142	100	1000	10,000	10^5	511	454	3322	3.3×10^4
4044	—	—	—	—	4518	493	1000	10^4
2887	—	—	—	—	7530	753	600	6000
2843	184	130	92	315	427	345	247	846
2400	184	130	92	315	427	345	247	846
2200	11,350	1135	366	259	2938	2687	942	668
671	8919	892	353	250	2921	2060	866	615
421	5691	569	330	234	741	840	819	603
421	190	134	95	254	302	296	244	659
200	157	111	90	900	270	256	237	2365
11	200	141	100	181	287	262	207	377
11	500	500	500	500	487	767	1168	1232

TABLE 14-5
Average Mantle Q Values as Function of Period, Model ABM

Region	0.1s	1s	4s	10s	100s	1000s
			Q_s			
Upper mantle	379	267	210	173	127	295
Lower mantle	1068	721	520	382	211	266
Whole mantle	691	477	360	280	176	274
			Q_p			
Upper mantle	513	362	315	354	311	727
Lower mantle	586	1228	1262	979	550	671
Whole mantle	562	713	662	639	446	687
			$Q({}_0S_2)$			
	3232 s			12 h		14 mos
	596			514		463

TABLE 14-6
Relaxation Times (τ_1, τ_2) and Q_K for Bulk Attenuation at Various Periods

Region	τ_1 (s)	τ_2 (s)	Q_K 1s	10s	100s	1000s
Upper mantle	0	3.33	479	1200	1.2×10^4	1.2×10^5
Lower mantle	0	0.20	2000	2×10^4	2×10^5	2×10^6
Outer core	15.1	66.7	7530	753	600	6000
	9.04	40.0	4518	493	1000	10,000
Inner core	3.01	13.3	1506	418	3000	3×10^4

Q_K (min) = 400, $\alpha = 0.15$.

The location of the absorption band is almost constant from a depth of about 400 to 2000 km (Figure 14-6). Most of the lower mantle therefore has high Q_s for body waves and low Q_s for free oscillations (Tables 14-4 and 14-5). The low-order spheroidal modes require a distinctly different location for the absorption band in the lower 500 km of the mantle. These data can be satisfied by moving the band to the location shown in the lower part of Figure 14-7. This gives a low-Q region at the base of the mantle at body-wave periods.

Under normal conditions, fluids in general and molten metals in particular satisfy the relaxation equations, with $\omega\tau \ll 1$ giving $Q \propto \omega^{-1}$. Pressure serves to increase mean relaxation time $\bar{\tau}$, and the high absorption of short-period P-waves in the inner core suggests that $\bar{\tau}$ may be of the order of 1 s in this region (Anderson, 1980). The high Q of the outer core at body-wave frequencies suggests that the absorption band is at longer or shorter periods.

If Q in the core is assumed to be a Debye peak, with fixed Q_m and width, we find that $Q_m = 400$ and $\bar{\tau}$ decreases with depth from 32 to 19 s in the outer core to 6 s in the inner core.

The net result is a slowly varying Q_p, 406 to 454, for the inner core between 0.1 and 10 s, increasing to 3320 at 100 s and 3.3×10^4 at 1000 s. Most studies of Q_p of the inner core at body-wave frequencies give values between 200 and 600 (Anderson and Archambeau, 1964). In the outer part of the outer core Q_p for ABM decreases from 7530 at 1 s to 753 at 10 s and then increases to 6000 at 1000 s. The relative location of the band is fixed by the observation that short-period P-waves see a high-Q outer core and a low-Q inner core. The absorption band model predicts relatively low Q for 10–50-s P-waves in the outer core. Data are sparse, but long-period P-waves in the outer core may be attenuated more than short-period waves (Anderson and others, 1965).

The attenuation of shear waves, Love and Rayleigh waves, toroidal oscillation and most of the spheroidal modes are controlled almost entirely by Q_s in the mantle. Once the mantle Q_s is determined, the Q of P-waves and the high-Q spheroidal and radial modes can be used to estimate Q_K, as shown in Figure 14-9 and listed in Table 14-6. The relationship between Q_p, Q_s and Q_K is

$$Q_p^{-1} = LQ_s^{-1} + (1 - L)Q_K^{-1}$$

where $L = (4/3)(\beta/\alpha)^2$ and β and α are the shear and compressional velocities. For a Poisson solid with $Q_K^{-1} = 0$,

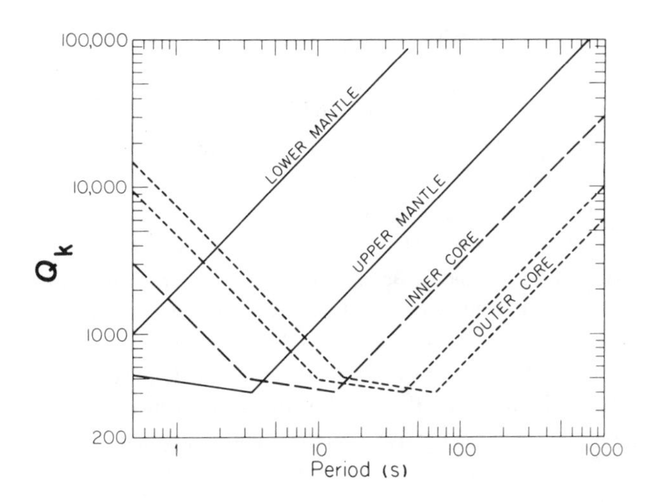

FIGURE 14-9
Q_K as a function of frequency for various regions in the Earth (Anderson and Given, 1982).

$$Q_P = (9/4)Q_s$$

The ABM average Earth model was used as a starting model to satisfy a variety of seismic data for the Eurasian shield by Der and others (1986). Their model is given in Table 14-7. The differences in the upper mantle, where both studies have their highest resolution, are probably due to regional differences. The differences in the lower mantle probably reflect the intrinsic uncertainty in Q with available data.

The study of Q is in its infancy. The results given in this section are very preliminary. The variation of Q with frequency and the lateral variations are very uncertain. Even the average radial variation is uncertain because of the trade-offs between frequency and depth when inverting surface waves and free oscillations.

TABLE 14-7
Q_s Model for the Eurasian Shield

Depth	Q_s			
(km)	0.01 Hz	0.1 Hz	1 Hz	10 Hz
50–100	216	246	365	665
100	216	246	365	665
150	123	138	200	380
200	123	138	200	380
250	123	138	530	950
350	123	356	530	950
425	307	356	530	950
659	307	356	530	950
660	565	860	2145	5825
1000	565	860	2145	5825
1571	580	1055	3860	16,155
2891	580	1055	3860	16,155

Der and others (1986).

General References

Anderson, D. L. (1967) The anelasticity of the mantle, *Geophys. J. R. Astr. Soc.*, *14*, 135–164.

Anderson, D. L. and J. W. Given (1982) Absorption band Q model for the Earth, *J. Geophys. Res.*, *87*, 3893–3904.

Anderson, D. L., H. Kanamori, R. Hart and H. Liu (1977) The Earth as a seismic absorption band, *Science*, *196*, 1104–1106.

Anderson, D. L. and J. B. Minster (1979) The frequency dependence of Q in the Earth and implications for mantle rheology and Chandler wobble, *Geophys. J. R. Astron. Soc.*, *58*, 431–440.

Friedel, J. (1964) *Dislocations*, Addison-Wesley.

Goetze, C. (1977) A brief summary of our present day understanding of the effect of volatiles and partial melt on the mechanical properties of the upper mantle. In *High-Pressure Research* (M. Manghnani and S. Akimoto, eds.), 3–23, Academic Press, New York.

Hirth, J. P. and J. Lothe (1968) *Theory of dislocations*, McGraw-Hill, New York, 780 pp.

Jackson, D. D. and D. L. Anderson (1970) Physical mechanisms of seismic wave attenuation. *Rev. Geophys.*, *8*, 1–63.

Kanamori, H. and D. L. Anderson (1977) Importance of physical dispersion in surface-wave and free-oscillation problems; a review, *Rev. Geophys. Space Phys.*, *15*, 105–112.

Minster, B. and D. L. Anderson (1981) A model of dislocation-controlled rheology for the mantle, *Phil. Trans. Roy. Soc. Lond. A*, *299*, 319–356.

Nowick, A. S. and B. S. Berry (1972) *Anelastic Relaxation in Crystalline Solids*, Academic Press, New York, 677 pp.

Stevenson, D. (1983) Anomalous bulk viscosity of two-phase liquids and implications for planetary interiors, *J. Geophys. Res.*, *88*, 2445–2455.

Stocker, R. and M. Ashby (1973) On the rheology of the upper mantle, *Rev. Geophys. Space Phys.*, *11*, 391–426.

Van Bueren, H. (1960) *Imperfections in crystals*, North-Holland, Amsterdam, 676 pp.

Weertman, J. (1968) Dislocation climb theory of steady-state creep, *Trans. Amer. Soc. Metals*, *61*, 681–694.

References

Anderson, D. L. (1966) Earth's viscosity, *Science*, *151*, 321–322.

Anderson, D. L. (1980) Bulk attenuation in the Earth and viscosity of the core, *Nature*, *285*, 204–207.

Anderson, D. L. and C. Archambeau (1964) The anelasticity of the Earth, *J. Geophys. Res.*, *69*, 2071.

Anderson, D. L., A. Ben-Menahem and C. B. Archambeau (1965) Attenuation of seismic energy in the upper mantle, *J. Geophys. Res.*, *70*, 1441–1448.

Anderson, D. L. and J. Given (1982) Absorption band Q model for the Earth, *J. Geophys. Res.*, *87*, 3893–3904.

Anderson, D. L. and R. S. Hart (1979) Q of the Earth, *J. Geophys. Res.*, *83*, 5869–5882.

Anderson, D. L. and J. B. Minster (1979) The frequency dependence of Q in the Earth and implications for mantle rheology and Chandler wobble, *Geophys. J. R. Astr. Soc.*, *58*, 431–440.

Anderson, D. L. and J. B. Minster (1980) Seismic velocity, attenuation and rheology of the upper mantle. In *Source Mechanism and Earthquake Prediction* (C. J. Allègre, ed.), 13–22, Centre national de la recherche scientifique, Paris.

Anderson, D. L. and C. G. Sammis (1970) Partial melting in the upper mantle, *Phys. Earth Planet. Inter.*, *3*, 41–50.

Ashby, M. F. and R. A. Verall (1978) Micromechanisms in flow and fracture, and their relevance to the rheology of the upper mantle, *Phil. Trans. R. Soc. Lond. A 288*, 59–95.

Berckhemer, H., F. Auer and J. Drisler (1979) High-temperature anelasticity and elasticity of mantle peridotite, *Phys. Earth Planet. Inter.*, *20*, 48–59.

Berckhemer, H., W. Kampfmann, E. Aulbach and H. Schmeling (1982) Shear modulus and Q of forsterite and dunite near partial melting from forced oscillation experiments *Phys. Earth Planet. Inter.*, *29*, 30–41.

Budiansky, B. and R. O'Connell (1980) *Solid Earth Geophys. Geotechn.*, *AMD 42*, Amer. Soc. Mech. Eng. 1980.

Der, Z. A., A. C. Lees and V. F. Cormier (1986) *Geophys. J. Roy. Astron. Soc.*, *87*, 1103–1112.

Flinn, J., P. Gupta and T. Litovitz (1974) *J. Chem. Phys. 60*, 4390.

Goetze, C. (1978) The mechanisms of creep in olivine, *Phil. Trans. R. Soc. Lond. A*, *288*, 99–119.

Goetze, C. and W. F. Brace (1972) Laboratory observations of high temperature rheology of rocks, *Tectonophysics*, *12*, 583–600.

Goetze, C. and D. L. Kohlstedt (1973) Laboratory study of dislocation climb and diffusion in olivine, *J. Geophys. Res.*, *78*, 5961–5971.

Gutenberg, B. (1957) The "boundary" of the Earth's inner core, *Trans. Am. Geophys. Un.*, *38*, 750–753.

Heinz, D. L., R. Jeanloz and R. J. O'Connell (1982) Bulk attenuation in a polycrystalline Earth, *J. Geophys. Res.*, *87*, 7772–7778.

Hirth, J. and J. Lothe (1968) *Theory of Dislocations*, McGraw-Hill, New York, 780 pp.

Jackson, D. (1969) Grain boundary relaxation and the attenuation of seismic waves, Thesis, M.I.T., Cambridge, Mass.

Jaoul, O., C. Froideveaux and M. Poumellec (1979) Atomic diffusion of ^{18}O and ^{30}Si in forsterite: Implication for the high temperature creep mechanism, Abstract, I.U.G.G. XVII General Assembly, Canberra, Australia.

Jeffreys, H. (1958) Rock creep, *Mon. Nat. R. Astr. Soc.*, *118*, 14–17.

Jeffreys, H. (1965) *Nature*, *208*, 675.

Jeffreys, H. (1970) *The Earth*, 5th ed., Cambridge University Press, London.

Jeffreys, H. and S. Crampin (1970) On the modified Lommitz law of damping, *Mon. Not. R. Astr. Soc.*, *147*, 295–301.

Kampfmann, W. and H. Berckhemer (1985) High temperature experiments on the elastic and anelastic behaviour of magmatic rocks, *Phys. Earth Planet. Inter.*, *40*, 223–247.

Kohlstedt, D. L. (1979) Creep behavior of mantle minerals, Abstract, I.U.G.G. XVII General Assembly, Canberra, Australia.

Kohlstedt, D. L. and C. Goetze (1974) Low-stress high-temperature creep in olivine single crystals, *J. Geophys. Res.*, *79*, 2045–2051.

Mavko, G. and A. Nur (1975) Melt squirt in the asthenosphere, *J. Geophys. Res.*, *80*, 1444–1448.

Minster, J. B. and D. L. Anderson (1981) A model of dislocation-controlled rheology for the mantle, *Phil. Trans. Roy. Soc. London*, *299*, 319–356.

Misener, D. V. (1974) Cationic diffusion in olivine to 1400°C and 35 kbar, *Carnegie Instn. Wash. Publs.*, *634*, 117–129.

Nur, A. (1971) Viscous phases in rocks and the low velocity zone, *J. Geophys. Res.*, *76*, 1270–1278.

O'Connell, R. J. and B. Budiansky (1978) Measures of dissipation in viscoelastic media, *Geophys. Res. Lett.*, *5*, 5–8.

Spetzler, H. and D. L. Anderson (1968) The effect of temperature and partial melting on velocity and attenuation in a simple binary system, *J. Geophys. Res.*, *73*, 6051–6060.

Toomre, A. (1974) On the "nearly diurnal wobble" of the Earth, *Geophys. J. R. Astron. Soc.*, *38*, 335–348.

Vaisnys, R. (1968) Propagation of acoustic waves through a system undergoing phase transformations, *J. Geophys. Res.*, *73*, 7675–7683.

Walsh, J. B. (1969) New analysis of attenuation in partially melted rock, *J. Geophys. Res.*, *74*, 4333–4337.

Woirgard, J. (1976) Modèle pour les pics de frottement interne observés à haute température sur les monocristaux. *Phil. Mag.*, *33*, 623–637.

Zener, C. (1948) *Elasticity and Anelasticity of Metals*, University of Chicago Press, Chicago.

15

Anisotropy

And perpendicular now and now transverse,
Pierce the dark soil and as they pierce and pass
Make bare the secrets of the Earth's deep heart.

—SHELLEY, "PROMETHEUS UNBOUND"

Anisotropy is responsible for the largest variations in seismic velocities; changes in the preferred orientation of mantle minerals, or in the direction of seismic waves, cause larger changes in velocity than can be accounted for by changes in temperature, composition or mineralogy. Therefore, discussions of velocity gradients, both radial and lateral, and chemistry and mineralogy of the mantle must allow for the presence of anisotropy. Anisotropy is not a second-order effect. Seismic data that are interpreted in terms of isotropic theory can lead to models that are not even approximately correct. On the other hand a wealth of new information regarding mantle mineralogy and flow will be available as the anisotropy of the mantle becomes better understood.

INTRODUCTION

The Earth is usually assumed to be isotropic to the propagation of seismic waves. It should be stressed that this assumption is made for mathematical convenience. The fact that a large body of seismic data can be satisfactorily modeled with this assumption does not prove that the Earth is isotropic. There is often a direct trade-off between anisotropy and heterogeneity. An anisotropic structure can have characteristics, such as travel times and dispersion curves, that are identical, or similar, to a different isotropic structure. A layered solid, for example, composed of isotropic layers that are thin compared to a seismic wavelength will behave as an anisotropic solid. The velocity of propagation depends on direction. The effective long-wavelength elastic constants depend on the thicknesses and elastic properties of the individual layers. The reverse is also true: An anisotropic solid with these same elastic constants can be modeled exactly as a stack of isotropic layers. The same holds true for an isotropic solid permeated by oriented cracks or aligned inclusions. This serves to illustrate the trade-off between heterogeneity and anisotropy. Not all anisotropic structures, however, can be modeled by laminated solids.

The crystals of the mantle are generally anisotropic, and rocks from the mantle show that these crystals exhibit a high degree of alignment. There is also evidence that crystal alignment is uniform over large areas of the upper mantle. At mantle temperatures, crystals tend to be easily recrystallized and aligned by the prevailing stress and flow fields.

The effects of anisotropy are often subtle and, if unrecognized, are usually modeled as inhomogeneities, for example, as layering or gradients. The most obvious manifestations of anisotropy are:

1. Shear-wave birefringence—the two polarizations of S-waves arrive at different times;
2. Azimuthal anisotropy—the arrival times, or apparent velocities of seismic waves at a given distance from an event, depend on azimuth;
3. An apparent discrepancy between Love waves and Rayleigh waves.

Even these are not completely unambiguous indicators of anisotropy. Effects such as P–S conversion, dipping interfaces, attenuation, and density variations must be properly taken into account.

There is now a growing body of evidence that much of the upper mantle may be anisotropic to the propagation of seismic waves. The early evidence was the discrepancy between dispersion of Rayleigh waves and Love waves (Anderson, 1961, 1967; Harkrider and Anderson, 1962) and the azimuthal dependence of oceanic Pn velocities (Hess, 1964; Raitt and others, 1969, 1971). Azimuthal variations have now been documented for many areas of the world, and the Rayleigh-Love discrepancy is also widespread. Shear-wave birefringence, a manifestation of anisotropy, has also been reported. The degree of anisotropy varies but is typically about 5 percent. It is not known to what depth the anisotropy extends, but there is abundant evidence for it at depths shallower than 200 km.

It has been known for some time that the discrepancy between mantle Rayleigh and Love waves could be explained if the vertical P and S velocities in the upper mantle were 7–8 percent less than the horizontal velocities (Anderson, 1967). Models without an upper-mantle low-velocity zone, such as the Jeffreys model, could satisfy the dispersion data if the upper mantle was anisotropic. The surface-wave data, which are sensitive to the properties of the upper mantle, imply a low-velocity zone if the mantle is assumed to be isotropic. The Love-Rayleigh discrepancy has survived to the present, and average Earth models have been proposed that have *SV* in the upper mantle less than *SH* by about 3 percent. Some early models, however, were based on separate isotropic inversions of Love and Rayleigh waves (pseudo-isotropic inversions) and therefore did not indicate the true anisotropy. There is a trade-off between anisotropy and structure. In particular, the very low upper-mantle average shear velocities, 4.0–4.2 km/s, found by many isotropic and pseudo-isotropic inversions, are not a characteristic of models resulting from full anisotropic inversion. The *P*-wave anisotropy also makes a significant contribution to Rayleigh wave dispersion. This has been ignored in many inversion attempts.

Since intrinsic anisotropy requires both anisotropic crystals and preferred orientation, the anisotropy of the mantle contains information about the mineralogy and the flow. For example, olivine, the most abundant upper-mantle mineral, is extremely anisotropic for both *P*-wave and *S*-wave propagation. It apparently is easily oriented by the ambient stress or flow field. Olivine-rich outcrops show a consistent preferred orientation over large areas. In general, the seismically fast axes of olivine are in the plane of the flow with the *a* axis, the fastest direction, pointing in the direction of flow. The *b* axis, the minimum velocity direction, is generally normal to the flow plane. Pyroxenes are also very anisotropic. The petrological data are summarized in Peselnick and others (1974) and Christensen and Salisbury (1979).

The magnitude of the anisotropy in the mantle is comparable to that found in ultramafic rocks (Figure 15-1). Soft layers or oriented fluid-filled cracks also give an apparent anisotropy, but these need to be invoked only for very low velocities. Much seismic data that are used in upper-mantle modeling are averages over several tectonic provinces or

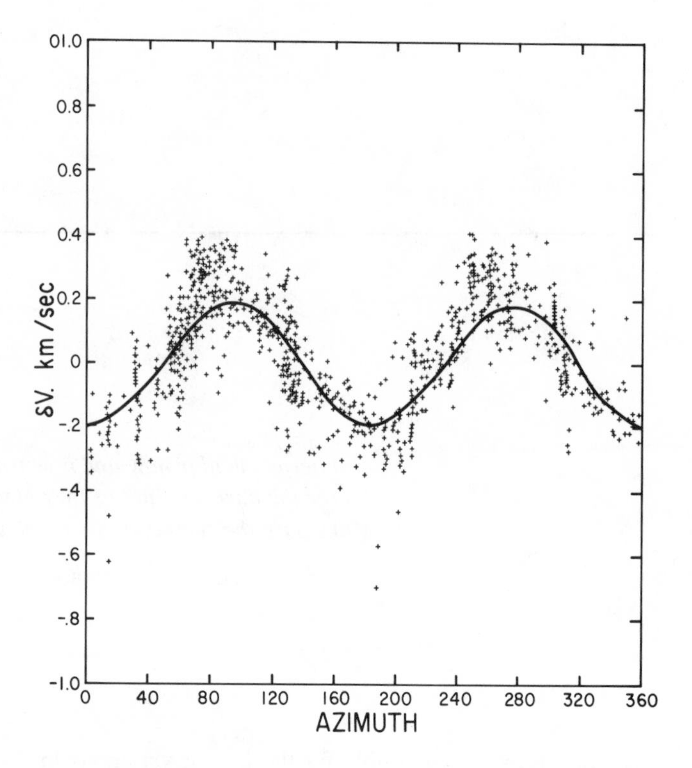

FIGURE 15-1
Azimuthal anisotropy of Pn waves in the Pacific upper mantle. Deviations are from the mean velocity of 8.159 km/s. Data points from seismic-refraction results of Morris and others (1969). The curve is the velocity measured in the laboratory for samples from the Bay of Islands ophiolite (Christensen and Salisbury, 1979).

over many azimuths. Azimuthal anisotropy may therefore be averaged out, but differences between vertical and horizontal velocities are not.

The presence of anisotropy is not only of theoretical interest. If the upper mantle is, in fact, anisotropic, then isotropic inversion of seismic data will result in erroneous structures because of improper parameterization. Such important seismological problems as the thickness of the oceanic lithosphere, the presence and nature of a low-velocity zone, the depth extent of differences between oceans and continents and the deep penetration of slabs depend critically on the validity of the assumption of isotropy. Finding the effect of anisotropy on surface waves and free oscillations and the trade-offs between structure and anisotropy is conveniently accomplished by the use of partial derivatives (Anderson, 1964), which have played an important role in Earth structure modeling and inversion of seismic data. Partial-derivative plots succinctly summarize the effect of the various parameters on normal-mode periods or surface-wave dispersion, and these will be discussed later in the chapter.

Seismic velocity variations due to anisotropy are potentially much greater than those due to other effects such as composition and temperature. It is therefore important to understand anisotropy well before one attempts to infer chemical, mineralogical and temperature variations from seismic data. A change in preferred orientation with depth or from one tectonic region to another can be easily misinterpreted.

The study of mantle anisotropy to infer mineralogy and flow is an example of how many disciplines are involved in the recognition and solution of a major problem in geophysics. The mineral physicist measures the single-crystal elastic constants of mantle candidate minerals. The field geologist maps rock fabrics and notices the orientation of minerals relative to bedding planes, dikes and sills. The experimental tectonophysicist deforms rocks and single crystals in the laboratory in order to understand the processes of creep, slip and recrystallization. The thermodynamicist develops theories of crystal behavior in non-hydrostatic stress fields. Mathematicians derive the theory for wave propagation in anisotropic media. Seismologists develop a variety of methods for mapping anisotropy in the mantle. Convection modelers calculate flow and stress fields for a range of assumptions about flow in the mantle. Only when all of these elements are in place can one completely interpret seismic data in terms of mantle convection. The study of seismic anisotropy is a rich and vigorous field for a variety of subdisciplines in geology and geophysics.

ORIGIN OF MANTLE ANISOTROPY

Nicholas and Christensen (1987) elucidated the reason for strong preferred crystal orientation in deformed rocks. First, they noted that in homogeneous deformation of a specimen composed of minerals with a dominant slip system, the preferred orientations of slip planes and slip directions coincide respectively with the orientations of the flow plane and the flow line. Simple shear in a crystal rotates all the lines attached to the crystal except those in the slip plane. This results in a bulk rotation of crystals so that the slip planes are aligned, as required to maintain contact between crystals. The crystal reorientations are not a direct result of the applied stress but are a geometrical requirement. Bulk anisotropy due to crystal orientation is therefore induced by plastic strain and is only indirectly related to stress. The result, of course, is also a strong anisotropy of the viscosity of the rock, and presumably attenuation, as well as elastic properties. This means that seismic techniques can be used to infer flow in the mantle. It also means that mantle viscosity inferred from postglacial rebound is not necessarily the same as that involved in plate tectonics and mantle convection.

Peridotites from the upper mantle display a strong preferred orientation of the dominant minerals, olivine and orthopyroxene. They exhibit a pronounced acoustic-wave anisotropy that is consistent with the anisotropy of the constituent minerals and their orientation. In igneous rocks preferred orientation can be caused by grain rotation, recrystallization in a nonhydrostatic stress field or in the presence of a thermal gradient, crystal setting in magma chambers, flow orientation and dislocation-controlled slip. Macroscopic fabrics caused by banding, cracking, sill and dike injection can also cause anisotropy.

Plastic flow induces preferred orientations in rock-forming minerals. The relative roles of deviatoric stresses and plastic strain have been long debated. In order to assure continuity of a deforming crystal with its neighbors, five independent degrees of motion are required (the Von Mises criterion). This can be achieved in a crystal with the activation of five independent slip systems or with a combination of fewer slip systems and other modes of deformation. In silicates only one or two slip systems are activated under a given set of conditions involving a given temperature, pressure and deviatoric stresses. The homogeneous deformation of a dominant slip system and the orientation of slip planes and slip directions tend to coincide with the flow plane and the flow direction (Nicholas and Christensen, 1987).

Mantle peridotites typically contain more than 65 percent olivine and 20 precent orthopyroxene. The high-V_p direction in olivine (Figure 15-2) is along the *a* axis [100], which is also the dominant slip direction at high temperature. The lowest V_p crystallographic direction is [010], the *b* direction, which is normal to a common slip plane. Thus, the V_p pattern in olivine aggregates is related to slip orientations. There is no such simple relationship with the V_s anisotropy and, in fact, the *S*-wave anisotropy of peridotites is small.

Orthopyroxenes also have large *P*-wave anisotropies and relatively small *S*-wave anisotropies, and have the principal V_p directions related to the slip system. The high-V_p direction coincides with the [100] pole of the unique slip plane and the intermediate V_p crystallographic direction coincides with the unique [001] slip line (Figure 15-2). In natural peridotites the preferred orientation of olivine is more pronounced than the other minerals. Olivine is apparently the most ductile and easily oriented upper-mantle mineral, and therefore controls the seismic anisotropy of the upper mantle. The anisotropy of β-spinel, a high-pressure form of olivine that is expected to be a major mantle component below 400 km, is also high. The γ-spinel form of olivine, stable below about 500 km, is much less anisotropic. Recrystallization of olivine to spinel forms can be expected to yield aggregates with preferred orientation but with perhaps less pronounced *P*-wave anisotropy. β-spinel has a strong *S*-wave anisotropy (24 percent variation with direction, 16 percent maximum difference between polarizations). The fast shear directions are parallel to the slow *P*-wave directions, whereas in olivine the fast *S*-directions correspond to intermediate *P*-wave velocity directions.

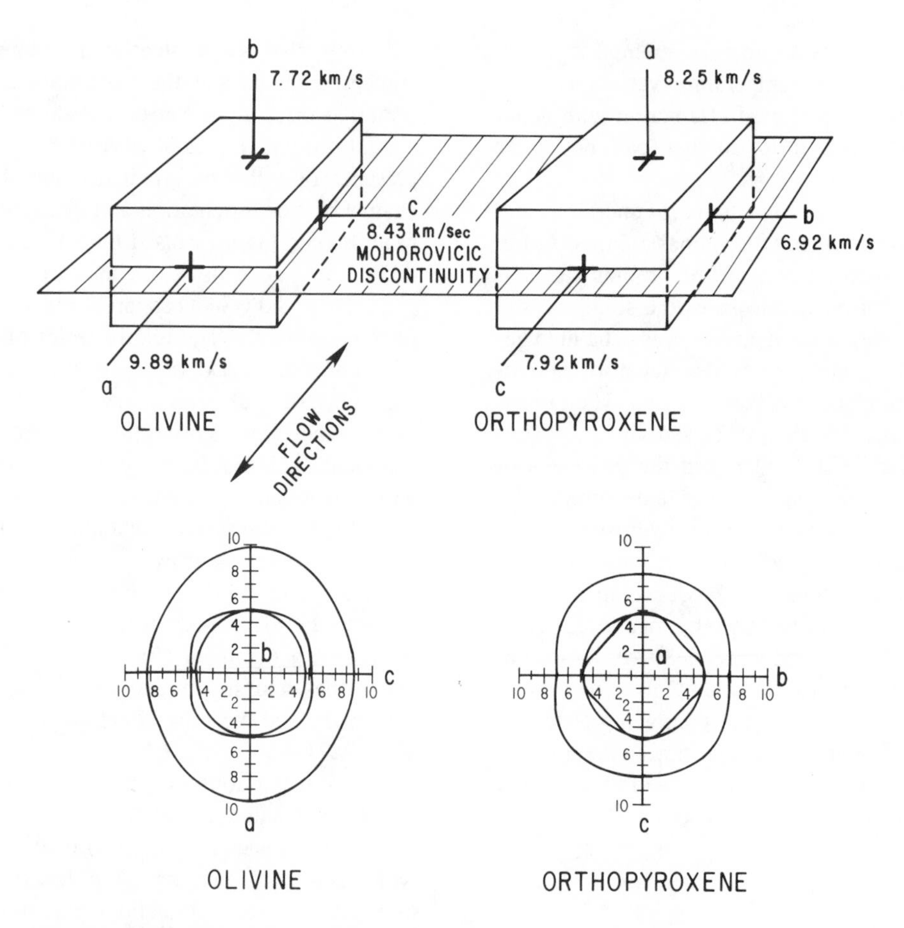

FIGURE 15-2
Olivine and orthopyroxene orientations within the upper mantle showing compressional velocities for the three crystallographic axes, and compressional and shear velocities in the olivine *a*-*c* plane and orthopyroxene *b*-*c* plane (after Christensen and Lundquist, 1982).

Orthopyroxene transforms to a cubic garnet-like structure that is stable over much of the transition region part of the upper mantle. This mineral, majorite, is expected to be relatively isotropic. Therefore, most of the mantle between 400 and 650 km depth is expected to have relatively low anisotropy, with the anisotropy decreasing as olivine transforms successively to β-spinel and γ-spinel. At low temperatures, as in subduction zones, the stable form of pyroxene is an ilmenite-type structure that is extremely anisotropic. Thus, the deep part of slabs may exhibit pronounced anisotropy, a property that could be mistaken for deep slab penetration in certain seismic experiments.

Different slip systems in olivine are activated at different temperatures (Avé Lallement and Carter, 1970; Carter, 1976; Nicholas and Poirier, 1976). At very high temperature olivine slips essentially with a single slip system and peridotites develop very strong fabrics. Although it is the anisotropy of individual crystals and their degree of orientation that controls the seismic anisotropy, it is dislocation physics and geometric constraints, combined with external variables such as the stress, flow and temperature, that ultimately control the degree of orientation. Thus, seismic data have the potential to infer not only mineralogy but also present and paleostress fields.

Petrofabric studies combined with field studies on ophiolite harzburgites give the following relationships:

1. Olivine *c* axes and orthopyroxene *b* axes lie approximately parallel to the inferred ridge axis in a plane parallel to the Moho discontinuity.
2. The olivine *a* axes and the orthopyroxene *c* axes align subparallel to the inferred speading direction.
3. The olivine *b* axes and the orthopyroxene *a* axes are approximately perpendicular to the Moho.

These results indicate that the compressional velocity in the vertical direction increases with the orthopyroxene content, whereas horizontal velocities and anisotropy decrease with increasing orthopyroxene content.

The maximum compressional wave velocity in orthopyroxene (along the *a* axis) parallels the minimum (*b* axis) velocity of olivine. For olivine *b* axis vertical regions of the mantle, as in ophiolite peridotites, the vertical *P*-velocity increases with orthopyroxene content. The reverse is true for other directions and for average properties. Appreciable shear-wave birefringence is expected in all directions even if the individual shear velocities do not depend much on azimuth. The total *P*-wave variation with azimuth in olivine- and orthopyroxene-rich aggregates is about 4 to 6 percent, while the *S*-waves only vary by 1 to 2 percent (Figure 15-3). The difference between the two shear-wave polarizations, however, is 4 to 6 percent. The azimuthal variation of *S*-waves can be expected to be hard to measure because the maximum velocity difference occurs over a small angular difference and because of the long-wavelength nature of shear waves. However, Tanimoto and Anderson (1984) measured azimuthal variations of surface-wave velocities that are comparable to the above predictions. The azimuthal variation of Rayleigh waves involves the azimuthal variation of both the *P*-waves and the *SV*-waves. The above relations between *P*-wave and *S*-wave anisotropy are not general and should not be applied to deeper parts of the mantle. In particular, the shear-wave anisotropy in the ilmenite structure of pyroxene, expected to be important in the deeper parts of subducted slabs, is quite pronounced and bears a different relationship to the *P*-wave anisotropy than that in peridotites. One possible manifestation of slab anisotropy is the variation of travel times with take-off angle from intermediate- and deep-focus earthquakes. Fast in-plane velocities, as expected for oriented olivine and probably spinel and ilmenite, may easily be misinterpreted as evidence for deep slab penetration. The mineral assemblages in cold slabs are also different from the stable phases in normal and hot mantle. The colder phases are generally denser and seismically fast. Anisotropy and isobaric phase changes have been ignored in most studies purporting to show deep slab penetration into the lower mantle. There is a complete trade-off, however, between the length of a high-velocity slab and its velocity contrast and anisotropy.

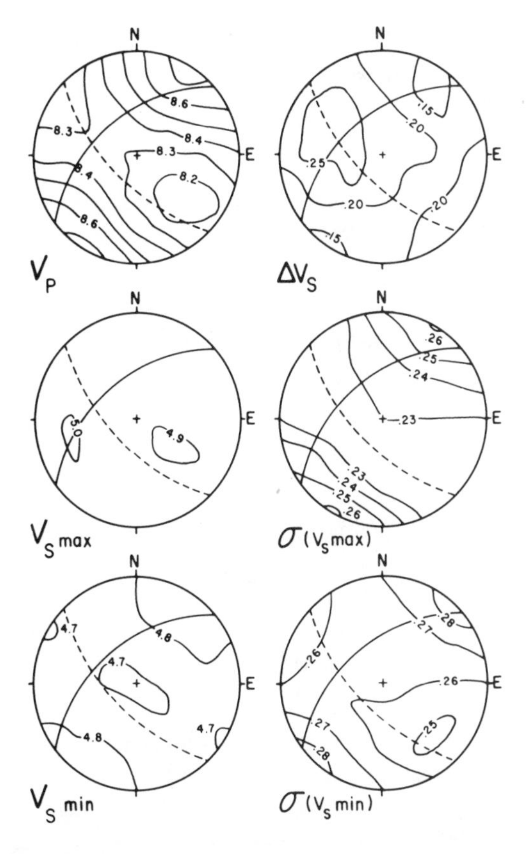

FIGURE 15-3
Equal area projection of V_p and two shear velocities measured on samples of peridotite. Dashed line is vertical direction, solid great circle is the horizontal (after Christensen and Salisbury, 1979).

ANISOTROPY OF CRYSTALS

Because of the simplicity and availability of the microscope, the optical properties of minerals receive more attention than the acoustic properties. It is the acoustic or ultrasonic properties, however, that are most relevant to the interpretation of seismic data. Being crystals, minerals exhibit both optical and acoustic anisotropy. Aggregates of crystals, rocks, are also anisotropic and display fabrics that can be analyzed in the same terms used to describe crystal symmetry. Tables 15-1, 15-2 and 15-3 summarize the acoustic anisotropy of some important rock-forming minerals. Pyroxenes and olivine are unique in having a greater *P*-wave anisotropy than *S*-wave anisotropy. Spinel and garnet, cubic crystals, have low *P*-wave anisotropy. Hexagonal crystals, and the closely related class of trigonal crystals, have high shear-wave anisotropies. This is pertinent to the deeper part of cold subducted slabs in which the trigonal ilmenite form of pyroxene may be stable. Deep-focus earthquakes exhibit a pronounced angular variation in both *S*- and *P*-wave velocities and strong shear-wave birefringence. Cubic crystals do not necessarily have low shear-wave anisotropy. The major minerals of the shallow mantle are all extremely anisotropic. β-spinel and clinopyroxene are stable below 400 km, and these are also fairly anisotropic. Below 400 km the major mantle minerals at high temperature, γ-spinel and, probably, garnet-majorite are less anisotropic. At the temperatures prevailing in subduction zones, the cold high-pressure forms of orthopyroxene, clinopyroxene and garnet are expected to give high velocities and anisotropies. If these are lined up, by stress or flow or recrystallization, then the slab itself will be anisotropic. This effect will be hard to distinguish from a long isotropic slab, if only sources in the slab are used.

TABLE 15-1
Anisotropic Properties of Rock-Forming Minerals

Mineral	Symmetry	*P* direction Max.	*P* direction Min.	Max *S* Direction/ Polarization	Anisotropy[†] (percent) *P*	Anisotropy[†] (percent) *S*
Olivine	Orthorhombic	[100]	[010]	45°/135°*	25	22
Garnet	Cubic	[001]	~50°*	[110] / [110]	0.6	1
Orthopyroxene	Orthorhombic	[100]	[010]	[010] / [001]	16	16
Clinopyroxene	Monoclinic	[001]	[101]	[011] / [$0\bar{1}1$]	21	20
Muscovite	Monoclinic	[110]	[001]	[010] / [100]	58	85
Orthoclase	Monoclinic	[010]	[101]	[011] / [$01\bar{1}$]	46	63
Anorthite	Triclinic	[010]	[101]	[011] / [$01\bar{1}$]	36	52
Rutile	Tetragonal	[001]	[100]	[100] / [010]	28	68
Nepheline	Hexagonal	[001]	[100]	[011] / [$0\bar{1}1$]	24	32
Spinel	Cubic	[101]	[001]	[100] / [010]	12	68
β-Mg_2SiO_4	Orthorhombic	[010]	[001]	—	16	14

Babuska (1981), Sawamoto and others (1984).
*Relative to [001].
[†] $[(v_{max} - v_{min})/v_{mean}] \times 100$.

The elastic properties of simple oxides and silicates are predominantly controlled by the oxygen anion framework, especially for hexagonally close-packed and cubic close-packed structures, but also by the nature of the cations occurring within the oxygen interstices. Corundum (Al_2O_3) consists of a hexagonal close-packed array of oxygen ions (radius 1.4Å) into which the smaller aluminum ions (0.54Å) are inserted in interstitial positions. Forsterite (Mg_2SiO_4) consists of a framework of approximately hexagonal close-packed oxygen ions with the Mg^{2+} cations (0.72Å) occupying one-half of the available octahedral sites (sites surrounded by six oxygen ions) and the Si^{4+} cations (0.26Å) occupying one-eighth of the available tetrahedral sites (sites surrounded by four oxygens). The packing of the oxygens depends on the nature of the cations.

Leibfried (1955) calculated the elastic constants for hexagonal close-packed (h.c.p.) and face-centered cubic (f.c.c.) structures from a central force model in which only the nearest neighbor interactions are considered. The elastic constants *C* are expressed in terms of the bulk modulus *K*:

$$c_{ij} = a_{ij}K$$

The coefficients a_{ij} are numbers that depend on the crystal symmetry only. The bulk modulus is

$$K = [4f/3(2)^{1/2}d]$$

where *f* is the force constant and *d* is the nearest neighbor distance. The coefficients a_{ij} are

$$\frac{1}{48}\begin{bmatrix} 72 & 36 & 36 & 0 & 0 & 0 \\ 36 & 72 & 36 & 0 & 0 & 0 \\ 36 & 36 & 72 & 0 & 0 & 0 \\ 0 & 0 & 0 & 36 & 0 & 0 \\ 0 & 0 & 0 & 0 & 36 & 0 \\ 0 & 0 & 0 & 0 & 0 & 36 \end{bmatrix} \quad \text{(f.c.c.)}$$

and

$$\frac{1}{48}\begin{bmatrix} 87 & 33 & 24 & 0 & 0 & 0 \\ 33 & 87 & 24 & 0 & 0 & 0 \\ 24 & 24 & 96 & 0 & 0 & 0 \\ 0 & 0 & 0 & 24 & 0 & 0 \\ 0 & 0 & 0 & 0 & 24 & 0 \\ 0 & 0 & 0 & 0 & 0 & 27 \end{bmatrix} \quad \text{(h.c.p.)}$$

This theoretical model refers to the static lattice and ignores thermal effects. This model was tested against experimental data by Gieske and Barsch (1968). Table 15-4 tabulates results for the theoretical model and several crystals that exhibit hexagonal or cubic packing of the oxygen anions. The moduli are normalized to the bulk modulus; the pressure derivatives are normalized to the pressure derivative of the bulk modulus, *K'*. The simple theoretical model, ignoring cations, does a fairly good job of predicting the relative

TABLE 15-2
Elastic Constants of Cubic Crystals

Mineral	c_{11}	c_{12} (Mbar)	c_{44}	ρ (g/cm³)
Garnet	2.966	1.085	0.916	3.705
γ-Mg_2SiO_4	3.27	1.12	1.26	3.559
MgO	2.97	0.95	1.56	3.580
$MgAl_2O_4$	2.82	1.55	1.54	3.578

Chang and Barsch (1969), Liu and others (1975), Suzuki and Anderson (1983), Weidner and others (1984).

TABLE 15-3
Elastic Constants of Orthorhombic, Trigonal, Tetragonal and Monoclinic Crystals

Mineral	c_{11}	c_{22}	c_{33}	c_{44}	c_{55}	c_{66}	c_{12}	c_{13}	c_{23}	c_{14}	c_{25}
					Orthorhombic						
α-Mg_2Si_4	3.28	1.91	2.30	0.66	0.81	0.81	0.64	0.68	0.72	—	—
β-Mg_2SiO_4	3.60	3.83	2.73	1.12	1.18	0.98	0.75	1.10	1.05	—	—
γ-Mg_2SiO_4*	3.46	3.46	3.27	1.26	1.26	1.08	0.94	1.12	1.12	—	—
Bronzite†	2.30	1.65	2.06	0.83	0.76	0.79	0.70	0.57	0.50	—	—
					Trigonal						
$MgSiO_3$-ilmenite	4.72	(4.72)	3.82	1.06	(1.06)	(1.52**)	1.68	0.70	—	−0.27	−0.24
Al_2O_3	4.97	(4.97)	4.98	1.47	(1.47)	(1.67**)	1.63	1.11	—	−0.23	0
					Tetragonal						
SiO_2-stishovite	4.53	(4.53)	7.76	2.52	(2.52)	3.02	2.11	2.03	—	—	—
					Monoclinic						
Diopside	2.23	1.71	2.35	0.74	0.67	0.66	0.77	0.81	0.57	—	0.07
		$c_{15} = 0.17$			$c_{35} = 0.43$			$c_{46} = 0.073$			

Kumazawa (1969), Levien and others (1979), Weidner and others (1984).
*Rotated 45° about c axis (three independent constants).
†84.5 percent $MgSiO_3$.
**$c_{66} = (c_{11} - c_{12})/2$.
( ) Not independent of other entries.

sizes of the elastic constants and even, in some cases, the pressure derivatives. The ratio K/G for the theoretical models is about 1.7, which corresponds to a Poisson's ratio of 0.25. In real crystals the K/G ratio depends on the nature of the cations, coordination, the radius ratios (r_{cation}/r_{anion}), the valencies and the covalencies. Transition ions cause a large increase in K/G.

THEORY OF ANISOTROPY

In an isotropic elastic solid there are two types of elastic waves. One type is variously called compressional, longitudinal or dilational and is characterized by particle motion in the direction of propagation. The second type of wave—the transverse, shear or distortional—has particle motion in a plane normal to the direction of propagation. For anisotropic media waves are neither purely longitudinal nor transverse, except in certain directions. The particle displacement has components both along and transverse to the direction of propagation. In an anisotropic medium the velocity of shear waves depends not only on direction but also on the polarization. In general, there are two shear waves that propagate with different velocities. This is known as shear-wave birefringence. Since particle motions are no longer simply related to ray directions or wave fronts, the waves are called quasi-*P*, quasi-longitudinal or quasi-transverse waves. I will continue, however, to refer to these as *P*-waves and *S*-waves since the faster, or primary, waves are *P*-like in their particle motions; the secondary or slower waves are still *S*-like.

Stress and strain in a solid body can each be resolved into six components: three extensional and three shearing. According to Hooke's law, each stress component can be expressed as a linear combination of the strain components and vice versa. For the most anisotropic material there are 21 constants of proportionality, the elastic constants or moduli or stiffnesses. As the symmetry increases, the number of elastic constants decreases. For an isotropic body, only two moduli are independent. For small displacements the stresses, T_i, are proportional to the strains, S_j:

$$T_i = c_{ij}S_j$$

where c_{11}, for example, is an elastic constant expressing the proportionality between the T_1 stress and the S_1 strain. Since the internal energy U is a perfect differential,

$$c_{ij} = c_{ji}$$

There are at most 21 elastic constants for the most unsymmetrical crystal, a triclinic crystal. For an isotropic solid the constants are independent of the choice of axes and reduce to two: These are the Lamé constants λ and μ,

$$\lambda + 2\mu = c_{22} = c_{33} = K + (4/3)\mu$$

$$\lambda = c_{12} = c_{13} = c_{23} = c_{21}$$

$$= c_{31} = c_{32} = c_{33} - 2c_{44}$$

$$\mu = c_{44} = c_{55} = c_{66}$$

TABLE 15-4
Elastic Constants ($a_{ij} = c_{ij}/K$) of Hexagonal Close Packed (h.c.p.) and Face Centered Cubic (f.c.c.) Arrays of Oxygen Ions Compared with Elastic Constants and Pressure Derivatives (c'_{ij}/K') of Oxides and Silicates

Hexagonal		Al_2O_3		$MgSiO_3$(il)	Mg_2SiO_4(ol)	
ij	a_{ij}	c_{ij}/K	c'_{ij}/K'	c_{ij}/K	c_{ij}/K	c'_{ij}/K'
11	1.81	1.97	1.79	2.23	1.86	1.59
22	1.81	1.97	1.79	2.23	1.58	1.64
33	2.00	1.98	1.58	1.80	2.59	1.84
44	0.50	0.58	0.80	0.50	0.64	0.72
55	0.50	0.58	0.80	0.50	0.64	0.55
66	0.56	0.66	0.61	0.72	0.53	0.69
12	0.69	0.65	0.57	0.79	0.57	0.58
13	0.50	0.44	0.65	0.33	0.54	0.67
23	0.50	0.44	0.65	0.33	0.52	0.70
14	0.00	−0.09	0.02	−0.13		
25	0.00	0.00				

Cubic		MgO(rs)		Al_2MgO_4(sp)		$SmAlO_3$(pv)
ij	a_{ij}	c_{ij}/K	c'_{ij}/K'	c_{ij}/K	c'_{ij}/K'	c'_{ij}/K'
11	1.50	1.85	2.1	1.44	1.13	1.69
33	1.50	1.85	2.1	1.44	1.13	1.69
12	0.75	0.58	0.44	0.73	0.92	0.74
13	0.75	0.58	0.44	0.73	0.92	0.63
44	0.75	0.97	0.26	0.76	0.21	0.75
66	0.75	0.97	0.26	0.76	0.21	0.63

Cubic		Garnet		$SrTiO_3$(pv)	
ij	a_{ij}	c_{ij}/K	c'_{ij}/K'	c_{ij}/K	c'_{ij}/K'
11	1.50	1.74	1.43	1.85	1.78
33	1.50	1.74	1.43	1.85	1.78
12	0.75	0.63	0.78	0.60	0.61
13	0.75	0.63	0.78	0.60	0.61
44	0.75	0.54	0.30	0.72	0.22
66	0.75	0.54	0.30	0.72	0.22

and all other constants are zero. The Lamé constant μ is the same as the rigidity or shear modulus, G.

The bulk modulus, K, is the ratio between an applied hydrostatic pressure P and the fractional change in volume $\Delta\ (= S_1 + S_2 + S_3)$:

$$K = P/\Delta = \lambda + (2/3)\mu$$

$$P = T_1 = T_2 = T_3$$

$$T_4 = T_5 = T_6 = 0$$

$$S_1 = S_2 = S_3 = \Delta/3 = -P/(3\lambda + 2\mu)$$

Young's modulus, E, is the ratio between an applied longitudinal stress and the longitudinal extension when the lateral surfaces are stress-free as in a bar:

$$E = \mu(3\lambda + 2\mu)/(\lambda + \mu)$$

Poisson's ratio, σ, is the negative of the ratio between the lateral contraction and the longitudinal extension:

$$\sigma = \lambda/[2(\lambda + \mu)]$$

When the strains are expressed in terms of the stresses,

$$S_i = s_{ij}T_j$$

$$s_{ij} = s_{ji}$$

For the isotropic case,

$$s_{11} = s_{22} = s_{33} = 1/E$$

$$s_{12} = s_{13} = s_{23} = -\sigma s_{11} = -\sigma/E$$

$$s_{44} = s_{55} = s_{66} = 1/\mu$$

The S_{ij} are called compliances.

The wave equation is usually written in terms of strains and elastic constants. Christoffel showed that the equations of motion could be written in terms of the displacements (u,v,w) and a series of moduli, λ_{ij}, which are functions of the elastic constants or stiffnesses c_{ij} and direction cosines (l,m,n) of the normal to the plane wave:

$$\rho \frac{\partial^2 u}{\partial t^2} = \lambda_{11}\frac{\partial^2 u}{\partial s^2} + \lambda_{12}\frac{\partial^2 v}{\partial s^2} + \lambda_{13}\frac{\partial^2 w}{\partial s^2}$$

$$\rho \frac{\partial^2 v}{\partial t^2} = \lambda_{12}\frac{\partial^2 u}{\partial s^2} + \lambda_{22}\frac{\partial^2 v}{\partial s^2} + \lambda_{23}\frac{\partial^2 w}{\partial s^2}$$

$$\rho \frac{\partial^2 w}{\partial t^2} = \lambda_{13}\frac{\partial^2 u}{\partial s^2} + \lambda_{23}\frac{\partial^2 v}{\partial s^2} + \lambda_{33}\frac{\partial^2 w}{\partial s^2}$$

where ρ is density, t is time, and

$$\lambda_{11} = l^2c_{11} + m^2c_{66} + n^2c_{55} + 2mnc_{56} + 2nlc_{15} + 2lmc_{16}$$

$$\lambda_{12} = \lambda_{21} = l^2c_{16} + m^2c_{26} + n^2c_{45} + mn(c_{46} + c_{25}) + nl(c_{14} + c_{56}) + nlm(c_{12} + c_{66})$$

$$\lambda_{13} = \lambda_{31} = lc_{15}^2 + m^2c_{46} + n^2c_{35} + mn(c_{45} + c_{36}) + nl(c_{13} + c_{55}) + lm(c_{14} + c_{56})$$

$$\lambda_{23} + \lambda_{32} = l^2c_{56} + m^2c_{24} + n^2c_{34} + \mathrm{mn}(c_{44} + c_{23}) + nl(c_{36} + c_{45}) + lm(c_{25} + c_{46})$$

$$\lambda_{22} = l^2c_{66} + m^2c_{22} + n^2c_{44} + 2mnc_{24} + 2nlc_{46} + 2lmc_{26}$$

$$\lambda_{33} = l^2c_{55} + m^2c_{44} + n^2c_{33} + 2mnc_{34} + 2nlc_{35} + 2lmc_{45}$$

The solution for an isotropic medium indicates three waves propagated, but it is only in special cases that the particle motions will be perpendicular to the direction of propagation. The three velocities satisfy the determinant

$$\begin{vmatrix} \lambda_{11} - \rho V^2 & \lambda_{12} & \lambda_{13} \\ \lambda_{12} & \lambda_{22} - \rho V^2 & \lambda_{23} \\ \lambda_{13} & \lambda_{23} & \lambda_{33} - \rho V^2 \end{vmatrix} = 0$$

(Evaluating such an expression is described briefly in the Appendix.)

A distance s along the normal to a plane wave is

$$s = lx + my + nz$$

The particle velocity ξ of a point on the surface has direction cosines α, β, γ with respect to the x, y and z axes:

$$\xi = \alpha u + \beta v + \gamma w$$

The direction cosines are related to the λ_{ij} constants and a solution V_i by the equations

$$\left.\begin{aligned} \alpha\lambda_{11} + \beta\lambda_{12} + \gamma\lambda_{13} &= \alpha\rho V_i^2 \\ \alpha\lambda_{12} + \beta\lambda_{22} + \gamma\lambda_{23} &= \beta\rho V_i^2 \\ \alpha\lambda_{13} + \beta\lambda_{23} + \gamma\lambda_{33} &= \gamma\rho V_i^2 \end{aligned}\right\} \quad i = 1, 2, 3$$

For a cubic crystal, there are three elastic constants c_{11}, c_{12}, and c_{44}, and the expressions for the λ_{ij} are

$$\lambda_{11} = l^2c_{11} + (m^2 + n^2)c_{44}$$

$$\lambda_{12} = lm(c_{12} + c_{44})$$

$$\lambda_{13} = nl(c_{12} + c_{44})$$

$$\lambda_{23} = mn(c_{12} + c_{44})$$

$$\lambda_{22} = (l^2 + n^2)c_{44} + m^2c_{11}$$

$$\lambda_{33} = (l^2 + m^2)c_{44} + n^2c_{11}$$

The determinant for the three velocities is

$$\begin{vmatrix} l^2c_{11} + (m^2 + n^2)c_{44} - \rho V^2 & lm(c_{12} + c_{44}) & nl(c_{12} + c_{44}) \\ lm(c_{12} + c_{44}) & (l^2 + n^2)c_{44} + m^2c_{11} - \rho V^2 & mn(c_{12} + c_{44}) \\ nl(c_{12} + c_{44}) & mn(c_{12} + c_{44}) & (l^2 + m^2)c_{44} + n^2c_{11} - \rho V^2 \end{vmatrix} = 0$$

There are three orientations for a cubic crystal for which a longitudinal and two shear waves can be transmitted. These are

[100] $l = 1;\ m = n = 0$

[110] $l = 1/\sqrt{2};\ n = 0$

[111] $l = m = n \pm 1/\sqrt{3}$

For the first orientation,

$$(c_{11} - \rho V^2)(c_{44} - \rho V^2)(c_{44} - \rho V^2) = 0$$

and the velocities and associated particle velocities in the [100] direction are

$$V_1 = \sqrt{\frac{c_{11}}{\rho}};\ \alpha = 1,\ \xi \text{ along } [100]$$

$$V_2 = \sqrt{\frac{c_{44}}{\rho}}; \beta = 1, \xi \text{ along } [010]$$

$$V_3 = \sqrt{\frac{c_{44}}{\rho}}; \gamma = 1, \xi \text{ along } [001]$$

For the shear waves, the particle velocity ξ can be in any direction in the (100) plane. V_1 is a P-type wave; V_2 and V_3 are shear-type waves. For the [110] direction,

$$V_1 = \sqrt{\frac{c_{11} + c_{12} + 2c_{44}}{2\rho}}; \alpha = \beta = \frac{1}{\sqrt{2}}; \xi \text{ along } [110]$$

$$V_2 = \sqrt{\frac{c_{44}}{\rho}}; \gamma = 1; \xi \text{ along } [001]$$

$$V_3 = \sqrt{\frac{c_{11} - c_{12}}{2\rho}}; \alpha = \frac{1}{\sqrt{2}} = -\beta; \xi \text{ along } [1\bar{1}0]$$

All three elastic constants can be measured from the longitudinal and two shear velocities of the [110] direction. For the [111] direction,

$$V_1 = V_{\text{long}} = \sqrt{\frac{c_{11} + 2c_{12} + 4c_{44}}{3\rho}};$$

$$\alpha = \beta = \gamma = \frac{1}{\sqrt{3}};$$

$$\xi \text{ along } [111]$$

$$V_2 = V_3 = V_{\text{shear}} = \sqrt{\frac{c_{11} - c_{12} + c_{44}}{3\rho}};$$

$$\alpha + \beta + \gamma = 0;$$

$$\xi \text{ in the } (111) \text{ plane}$$

Important cubic minerals in the mantle are garnet, majorite (a high-pressure form of pyroxene), and (Mg,Fe)O, a possibly important phase in the lower mantle.

For a hexagonal crystal, or a material exhibiting transverse isotropy, waves transmitted along the unique axis and any axis perpendicular to it are separated into a longitudinal and two shear waves. The elastic constants for a hexagonal crystal are

$$c_{11} = c_{22}, c_{12}, c_{13} = c_{23}, c_{33}, c_{44} = c_{55},$$

$$c_{66} = \left(\frac{c_{11} - c_{12}}{2}\right)$$

Hence there are five independent constants. The elements of the velocity equation take the form

$$l^2 c_{11} + m^2 \left(\frac{c_{11} - c_{12}}{2}\right) + n^2 c_{44} - \rho V^2;$$

$$lm \left(\frac{c_{11} + c_{23}}{2}\right); nl(c_{13} + c_{44})$$

$$lm \left(\frac{c_{11} + c_{12}}{2}\right); l^2 \left(\frac{c_{11} - c_{12}}{2}\right) + m^2 c_{11}$$

$$+ n^2 c_{44} - \rho V^2; mn(c_{13} + c_{44})$$

$$nl(c_{13} + c_{44}); mn(c_{13} + c_{44});$$

$$(l^2 + m^2) c_{44} + n^2 c_{33} - \rho V^2$$

For transmission along the unique axis ($n = 1$), the waves transmitted have the velocities and particle directions

$$V_1 = \sqrt{\frac{c_{33}}{\rho}}; \xi \text{ along } [001]$$

$$V_2 = V_3 = \sqrt{\frac{c_{44}}{\rho}}; \xi \text{ in the } (001) \text{ plane}$$

For the [100] direction or any other direction perpendicular to the [001] axis,

$$V_1 = \sqrt{\frac{c_{11}}{\rho}}; \xi \text{ along } [100]$$

$$V_2 = \sqrt{\frac{c_{44}}{\rho}}; \xi \text{ along } [001]$$

$$V_3 = \sqrt{\frac{c_{11} - c_{12}}{2\rho}}; \xi \text{ along } [010]$$

Measurements along these two directions will determine four of the five elastic constants. To determine the fifth one, a wave must be propagated in an intermediate direction.

The most general elastic constant matrix is

$$c_{ijkl} \begin{bmatrix} c_{11} & c_{12} & c_{13} & c_{14} & c_{15} & c_{16} \\ c_{12} & c_{22} & c_{23} & c_{24} & c_{25} & c_{26} \\ c_{13} & c_{23} & c_{33} & c_{34} & c_{35} & c_{36} \\ c_{14} & c_{24} & c_{34} & c_{44} & c_{45} & c_{46} \\ c_{15} & c_{25} & c_{35} & c_{45} & c_{55} & c_{56} \\ c_{16} & c_{26} & c_{36} & c_{46} & c_{56} & c_{66} \end{bmatrix}$$

This is the elastic constant matrix for a triclinic crystal. Because of symmetry conditions, and relationships between some of the elastic constants, there are only nine constants for an orthorhombic crystal, five for a hexagonal or transversely isotropic solid, three for a cubic crystal and two for isotropic media (see Table 15-5). In a cubic crystal, for example,

$$c_{11} = c_{22} = c_{33}$$

$$c_{12} = c_{21} = c_{13} = c_{31} = c_{23} = c_{32}$$

$$c_{44} = c_{55} = c_{66}$$

and the other c_{ij} are zero.

In a transversely isotropic solid with a vertical axis of symmetry, we can define four elastic constants in terms of *P*- and *S*-waves propagating perpendicular and parallel to the axis of symmetry:

$$A = c_{11} = \rho V_{PH}^2;\ C = c_{33} = \rho V_{PV}^2;$$

$$N = (c_{11} - c_{12})/2 = \rho V_{SH}^2;\ L = c_{44} = \rho V_{SV}^2$$

The fifth elastic constant, *F*, requires information from another direction of propagation. *PH, SH* are waves propagating and polarized in the horizontal direction and *PV, SV* are waves propagating in the vertical direction. In the vertical direction $V_{SH} = V_{SV}$; the two shear waves travel with the same velocity, and this velocity is the same as *SV* waves traveling in the horizontal direction. There is no azimuthal variation of velocity in the horizontal, or symmetry, plane.

Love waves are composed of *SH* motions, and Rayleigh waves are a combination of *P* and *SV* motions. In isotropic material $N = L = \mu$, $\rho V_s^2 = \mu$, $\rho V_p^2 = K + (4/3)\mu$, and Love waves and Rayleigh waves require only two elastic constants to describe their velocity. In general, more than two elastic constants at each depth are required to satisfy seismic surface-wave data, even when the azimuthal variation is averaged out, and complex vertical variations are allowed.

The upper mantle exhibits what is known as "polarization anisotropy," a phenomenon related to shear-wave birefringence. In general, four elastic constants are required to describe Rayleigh-wave propagation in a homogeneous transversely or equivalent transversely isotropic mantle.

For *SH* waves,

$$\rho V_{SH}^2 = l^2 N + n^2 L$$

where *l* and *n* are the direction cosines from the horizontal and vertical directions. For a transversely isotropic layer (layer 1) over a transversely isotropic half-space (layer 2), the velocity of Love waves can be derived from

$$\tan k\delta_1 d = -\mathrm{i}\,\frac{L_2\delta_2}{L_1\delta_1}$$

where *k* is wave number, *d* is layer thickness, and

$$\delta = (N/L)^{1/2}\left(\frac{V^2}{V_{SH}^2} - 1\right)^{1/2}$$

Thus Love waves, although composed of *SH*-type motion, require both of the two shear-type moduli for their description. From the ray point of view, Love waves can be viewed

TABLE 15-5
Schematic Elastic Constant Matrices

Monoclinic						Orthorhombic						Trigonal (1)					
a	b	c	·	·	d	a	b	c	·	·	·	a	b	c	d	·	·
b	e	f	·	·	g	b	d	e	·	·	·	b	a	c	−d	·	·
c	f	h	·	·	i	c	e	f	·	·	·	c	c	e	·	·	·
·	·	·	j	k	·	·	·	·	g	·	·	d	−d	·	f	·	·
·	·	·	k	m	·	·	·	·	·	h	·	·	·	·	·	f	d
d	g	i	·	·	n	·	·	·	·	·	i	·	·	·	·	d	x
Trigonal (2)						**Tetragonal (1)**						**Tetragonal (2)**					
a	b	c	d	g	·	a	b	c	·	·	·	a	b	c	·	·	g
b	a	c	−d	−g	·	b	a	c	·	·	·	b	a	c	·	·	−g
c	c	e	·	·	·	c	c	d	·	·	·	c	c	d	·	·	·
d	−d	·	f	·	−g	·	·	·	e	·	·	·	·	·	e	·	·
g	−g	·	·	f	d	·	·	·	·	e	·	·	·	·	·	e	·
·	·	·	−g	d	x	·	·	·	·	·	f	g	−g	·	·	·	f
Hexagonal						**Cubic**						**Isotropic**					
a	b	c	·	·	·	a	b	b	·	·	·	a	b	b	·	·	·
b	a	c	·	·	·	b	a	b	·	·	·	b	a	b	·	·	·
c	c	d	·	·	·	b	b	a	·	·	·	b	b	a	·	·	·
·	·	·	e	·	·	·	·	·	c	·	·	·	·	·	x	·	·
·	·	·	·	e	·	·	·	·	·	c	·	·	·	·	·	x	·
·	·	·	·	·	x	·	·	·	·	·	c	·	·	·	·	·	x

$x = (a - b)/2$.

as constructive interference of *SH*-polarized waves with both horizontal and vertical components of propagation, that is, upgoing and downgoing waves.

Rayleigh waves involve the coefficients c_{11}, c_{33}, c_{44} and $c_{13} + c_{44}$, or A, C, L and F. It is convenient to introduce the nondimensional parameter

$$\eta = F/(A - 2L)$$

a parameter that controls the variation of velocity away from the symmetry axis and that is important in Rayleigh-wave dispersion. Figure 15-4 shows how velocities vary with direction as a function of η. The other shear wave, with particle motion parallel to the symmetry plane, has a simple ellipsoidal phase velocity surface.

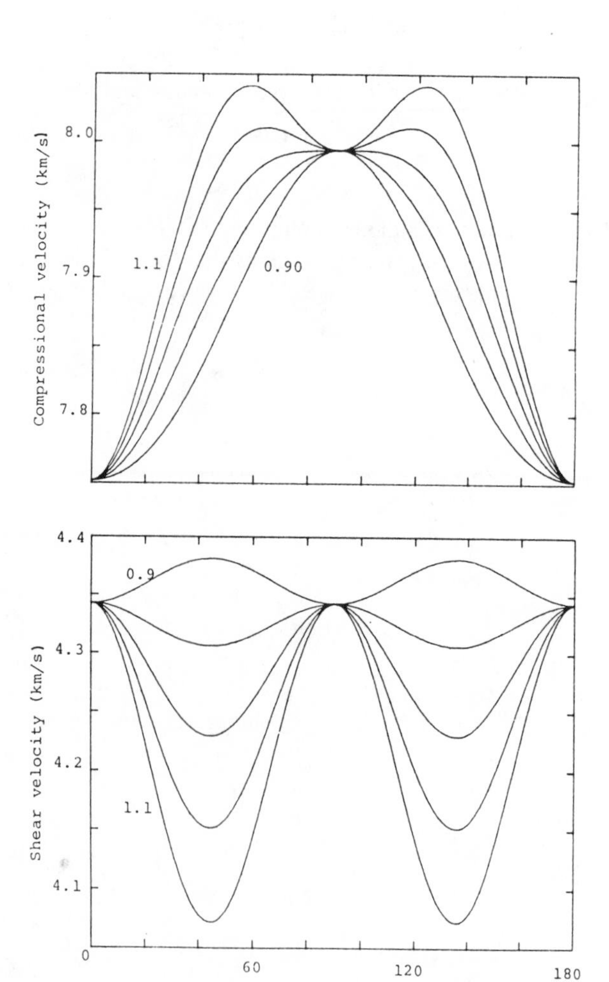

FIGURE 15-4
P and *S* velocities as a function of angle of incidence relative to the symmetry plane and the anisotropic parameter, which varies from 0.9 to 1.1 at intervals of 0.05. Parameters are $V_{PV} = 7.752$, $V_{PH} = 7.994$, $V_{SV} = 4.343$, all in km/s (after Dziewonski and Anderson, 1981).

In a weakly anisotropic medium the azimuthal dependence of the velocities of body waves (Smith and Dahlen, 1973) can be written

$$\rho V_{PH}^2 = A + B_c \cos 2\Theta + B_s \sin 2\Theta + C_c \cos 4\Theta + C_s \sin 4\Theta$$

$$\rho V_{SH}^2 = D - C_c \cos 4\Theta - C_s \sin 4\Theta$$

$$\rho V_{SV}^2 = F + G_c \cos 2\Theta + G_s \sin 2\Theta$$

where

$$B_c = (c_{11} - c_{22})/2; \; B_s = c_{16} + c_{26}$$

$$G_c = (c_{55} - c_{44})/2; \; G_s = c_{54}$$

$$C_c = (c_{11} + c_{22})/8 - c_{12}/4 - c_{66}/2$$

$$C_s = (c_{16} - c_{26})/2$$

The plane wave surfaces for *P*- and *SV*-waves are:

$$2\rho V_{P,SV}^2 = 2L + (A - L)l^2 + (C - L)n^2 \pm \{[(A - L)l^2 + (C - L)n^2]^2 + [(F + L)^2 - (A - L)(C - L)] \cdot \sin^2 2\Theta\}^{1/2}$$

where Θ is measured from the symmetry axis, and $l = \sin\Theta$, $n = \cos\Theta$ (Thomsen, 1986). These represent one quasi-longitudinal and two quasi-transverse waves for each direction of propagation. The three are polarized in mutually orthogonal directions. Of the two quasi-transverse waves, one has a polarization vector with no component in the symmetry axis direction. It is denoted by *SH*, the other by *SV*. The directional dependence of the three phase velocities can be written

$$\rho V_p^2(\Theta) = 1/2\,[c_{33} + c_{44} + (c_{11} - c_{33}) \times \sin^2\Theta + D(\Theta)]$$

$$\rho V_{SV}^2(\Theta) = 1/2\,[c_{33} + c_{44} + (c_{11} - c_{33}) \times \sin^2\Theta - D(\Theta)]$$

$$\rho V_{SH}^2(\Theta) = c_{66} \sin^2\Theta + c_{44} \cos^2\Theta$$

where ρ is density, and phase angle Θ is the angle between the wavefront normal and the unique (vertical) axis. $D(\Theta)$ is the quadratic combination

$$D(\Theta) \equiv [(c_{33} - c_{44})^2 + 2\,[2(c_{13} + c_{44})^2 - (c_{33} - c_{44})(c_{11} + c_{33} - 2c_{44})]\sin^2\Theta + [(c_{11} + c_{33} - 2c_{44})^2 - 4(c_{13} + c_{44})^2]\sin^4\Theta]^{1/2}$$

Thomsen (1986) introduced the "anisotropy factors"

$$\varepsilon \equiv \frac{c_{11} - c_{33}}{2c_{33}}$$

$$\gamma \equiv \frac{c_{66} - c_{44}}{2c_{44}}$$

$$\delta^* \equiv \frac{1}{2c_{33}^2} [2(c_{13} + c_{44})^2 - (c_{11} - c_{44}) \times (c_{11} + c_{33} - 2c_{44})]$$

giving

$$V_p^2(\Theta) = \alpha_o^2 [1 + \varepsilon \sin^2\Theta + D^*(\Theta)]$$

$$V_{SV}^2(\Theta) = \beta_o^2 [1 + \frac{\alpha_o^2}{\beta_o^2} \varepsilon \sin^2\Theta - \frac{\alpha_o^2}{\beta_o^2} D^*(\Theta)]$$

$$V_{SH}^2(\Theta) = \beta_o^2 [1 + 2\gamma \sin^2\Theta]$$

with

$$D^*(\Theta) \equiv \frac{1}{2}\left(1 - \frac{\beta_o^2}{\alpha_o^2}\right) \times \left\{\left[1 + \frac{4\delta^*}{(1 - \beta_o^2/\alpha_o^2)^2} \sin^2\Theta \cos^2\Theta + \frac{4(1 - \beta_o^2/\alpha_o^2 + \varepsilon)\varepsilon}{(1 - \beta_o^2/\alpha_o^2)^2} \sin^4\Theta\right]^{1/2} - 1\right\} \Big/ 2$$

In the approximation of weak anisotropy, the quadratic D^* is approximately

$$D^* \approx \frac{\delta^*}{(1 - \beta_o^2/\alpha_o^2)} \sin^2\Theta \cos^2\Theta + \varepsilon \sin^4\Theta$$

where Θ, the phase angle, is the normal to the wavefront. This gives, valid for weak anisotropy:

$$V_p(\Theta) = V_{PV} [1 + \delta \sin^2\Theta \cos^2\Theta + \varepsilon \sin^4\Theta]$$

$$V_{SV}(\Theta) = V_{SV} \left[1 + \frac{\alpha_o^2}{\beta_o^2} (\varepsilon - \delta) \sin^2\Theta \cos^2\Theta\right]$$

$$V_{SH}(\Theta) = V_{SV} [1 + \gamma \sin^2\Theta]$$

where

$$\delta \equiv \frac{1}{2}\left(\varepsilon + \frac{\delta^*}{1 - \beta_o^2/\alpha_o^2}\right) = \frac{(c_{13} + c_{44})^2 - (c_{33} - c_{44})^2}{2c_{33}(c_{33} - c_{44})}$$

In the linear approximation the group velocity, U, for the ray at angle ϕ to the symmetry axis (vertical in this case)

$$U_p(\phi) = V_p(\Theta)$$

$$U_{SV}(\phi) = V_{SV}(\Theta)$$

$$U_{SH}(\phi) = V_{SH}(\Theta)$$

Therefore at a given ray (group) angle ϕ, if one calculates the corresponding wavefront normal (phase) angle Θ, then one may find the ray (group) velocity. The relationship between group angle ϕ and phase angle Θ is, in the linear approximation,

$$\tan \phi_p = \tan \Theta_p[1 + 2\delta + 4(\varepsilon - \delta) \sin^2\Theta_p]$$

$$\tan \phi_{SV} = \tan \Theta_{SV}[1 + 2 \frac{\alpha_o^2}{\beta_o^2} (\varepsilon - \delta) (1 - 2 \sin^2\Theta_{SV})]$$

$$\tan \phi_{SH} = \tan \Theta_{SH}[1 + 2\gamma]$$

For a transversely isotropic medium, with vertical symmetry axis, the different sheets of the slowness surface n, can be described in terms of the parameters $w = \tan \beta \tan \alpha_p$, where β and α_p are the angles between the axis of symmetry and the wave-normal and (quasi-) longitudinal displacement, respectively. One obtains as polar expressions

$$n_p^2(w)/n_p^2(0) = N/\{1 + [G + H + (1 - H^2)A/c_{33}]w + Qw^2\}, \quad n_p^2(0) = \rho/c_{33}$$

$$n_{SV}^2(w)/n_{SV}^2(0) = N/\{1 + [G + H + (GH - 1)A/c_{44}]w + w^2\}$$

$$n_{SH}^2(w)/n_{SH}^2(0) = N/\{1 + (G + H/\lambda)w + w^2/\lambda\},$$

$$n_{SV}^2(0) = n_{SH}^2(0) = \rho/c_{44}$$

$$\tan^2\beta = (HW + w^2)/(1 + Gw)$$

where $N = 1 + (G + H)w + w^2$, $A = c_{13} + c_{44}$, $G = (c_{11} - c_{44})/A$, $H = (c_{33} - c_{44})/A$, $Q = c_{11}/c_{33}$, and $\lambda = c_{44}/c_{66}$ (Backus, 1965; Helbig, 1972). The *SH*-sheet is always an ellipsoid of rotation with axes $\sqrt{\rho/c_{44}}$ and $\sqrt{\rho/c_{66}}$. The *P*-wave sheet is generally not an ellipsoid.

The energy transport of seismic waves in anisotropic media is not in general normal to the plane of constant phase. The energy of a plane wave travels at the phase velocity perpendicular to the plane but also has a component of motion parallel to the plane. It is the energy, or group velocity that controls the travel time of a pulse of seismic energy from source to receiver. In anisotropic media, wave surfaces and group velocity surfaces are characterized by the presence of cusps, regions of rapidly varying body-wave amplitudes and directions, and multiple arrivals of a single wave type. Only in the case of small anisotropy do the familiar relations between ray directions, group directions, wave fronts, polarization and co-planarity hold. The direction of propagation is no longer the unique direction to which all of the other directions can be simply related.

TRANSVERSE ISOTROPY OF THE UPPER MANTLE

A solid characterized by an axis of symmetry is termed transversely isotropic and exhibits the same symmetry as a hexagonal crystal. It is described by five elastic constants.

Pure longitudinal and shear waves propagate in the symmetry plane and along the symmetry axis, and measurements of velocities in these two orthogonal directions determine four of the five elastic constants. At intermediate directions there are three coupled elastic wave modes, and the velocities of these involve the fifth constant. The five elastic constants can also be determined by measuring the toroidal and spheroidal normal-mode spectra. Toroidal modes are sensitive to the two shear-type moduli, and spheroidal modes are sensitive to four of the five moduli.

Transverse isotropy, although a special case of anisotropy, has quite general applicability in geophysical problems. This kind of anisotropy is exhibited by laminated or finely layered solids, solids containing oriented cracks or melt zones, peridotite massifs, harzburgite bodies, the oceanic upper mantle and floating ice sheets. A mantle containing small-scale layering, sills or randomly oriented dikes will also appear to be macroscopically transversely isotropic. If flow in the upper mantle is mainly horizontal, then the evidence from fabrics of peridotite nodules and massifs suggests that the average vertical velocity is less than the average horizontal velocity, and horizontally propagating *SH*-waves will travel faster than *SV*-waves. In regions of upwelling and subduction, the slow direction may not be vertical, but if these regions are randomly oriented, the average Earth will still display the spherical equivalent of transverse isotropy. Since the upper mantle is composed primarily of the very anisotropic crystals olivine and pyroxene, and since these crystals tend to align themselves in response to flow and nonhydrostatic stresses, it is likely that the upper mantle is anisotropic to the propagation of elastic waves. Although the preferred orientation in the horizontal plane can be averaged out by determining the velocity in many directions or over many plates with different motion vectors, the vertical still remains a unique direction. It can be shown that if the azimuthally varying elastic velocities are replaced by the horizontal averages, then many problems in seismic wave propagation in more general anisotropic media can be reduced to the problem of transverse isotropy.

If anisotropy persists to moderate depth, then it must be allowed for in gross Earth and regional inversions as well as in more local studies. The large-scale mantle motions responsible for plate tectonics, combined with the ease of dislocation creep at the high temperatures in the upper mantle, can be expected to orient the crystals in the mantle. In a crystalline solid the crystals must be oriented at random in order to be isotropic. There is no particular reason for believing that this is true in the mantle. Since isotropy is a degenerate case, it cannot be assumed that models resulting from isotropic inversion are even approximately correct.

The inconsistency between Love- and Rayleigh-wave data, first noted for global data, has now been found in regional data sets. It appears that lateral heterogeneity is not responsible for the Love-Rayleigh wave discrepancy and that anisotropy is an intrinsic and widespread property of the uppermost mantle. The crust and exposed sections of the upper mantle exhibit layering on scales ranging from meters to kilometers. Such layering in the deeper mantle would be beyond the resolution of seismic waves and would show up as an apparent anisotropy. This, plus the preponderance of aligned olivine in mantle samples, means that at least five elastic constants are probably required to properly describe the elastic response of the upper mantle. It is clear that inversion of *P*-wave data, for example, or even of *P* and *SV* data cannot provide all of these constants. Even more serious, inversion of a limited data set, with the assumption of isotropy, does not necessarily yield the proper structure. The variation of velocities with angle of incidence, or ray parameter, will be interpreted as a variation of velocity with depth. In principle, simultaneous inversion of Love-wave and Rayleigh-wave data can help resolve the ambiguity.

The theory of surface-wave propagation in a layered transversely isotropic solid was developed in the early 1960s (Anderson, 1961, 1962, 1967; Harkrider and Anderson, 1962). The effect of sphericity was treated by Takeuchi and Saito (1972). Propagation in the axial directions of a medium displaying orthorhombic symmetry was treated by Anderson (1966) and Toksöz and Anderson (1963). For Love waves, isotropic theory can be generalized easily to the anisotropic case. The shear moduli determined from isotropic inversion of Love waves is a simple function of the two anisotropic shear moduli; therefore, an isotropic model can always be found that will satisfy Love-wave and toroidal-mode data for an anisotropic structure. No such simple transformation is possible for Rayleigh waves. Models found from isotropic inversion of Rayleigh-wave data are not necessarily even approximately similar to the real anisotropic Earth. If four or five elastic constants plus density are necessary to describe the Earth at a given depth and only two or three parameters are allowed to vary, it is obvious that the problem is underparameterized. An isotropic inversion scheme will result in perturbations of the available parameters and may result in a model exhibiting oscillatory or rough structure that is not a characteristic of the real Earth. If accurate spheroidal and toroidal data are available, systematic deviations from predicted periods for the best-fitting isotropic model may be symptomatic of anisotropy. Other symptoms may be unreasonable Pn and Sn velocities, velocity ratios, or velocity and density reversals. Some of the above are characteristics of most gross Earth inversion attempts, using isotropic theory.

The discrepancy between Earth models resulting from separate isotropic inversion of fundamental mode Love-wave data (controlled by the horizontally propagating *SH*-wave velocity or V_{SH}) and Rayleigh-wave data (controlled by V_{SV} and the *P*-velocity) is well known. When the same data are simultaneously inverted using anisotropic theory, the resulting model is quite different. In particular, oceanic data yield a much thinner seismic lithosphere or LID. The Love-Rayleigh wave discrepancy is a fairly direct indication of anisotropy since the two wave types are generally measured over the same path. More subtle is the large difference between near-vertical ScS travel times (controlled by vertically traveling *SH* waves, equivalent to horizontally traveling *SV* waves, V_{SV}, in a transversely isotropic solid with a vertical symmetry axis) and the times predicted from Love wave models (V_{SH}) for the oceanic upper mantle. Using isotropic theory one would conclude that large ocean-continent differences extend to much greater depths than the 200–400 km or so indicated by other techniques. This, however, is just the Love-Rayleigh discrepancy in disguise.

The azimuthal variation of Pn velocity is one of the most direct indications of anisotropy. Such data show that both the oceanic and continental lithospheres are markedly anisotropic. Upon subduction the oceanic plate likely retains its anisotropy, with fast velocities in the plane of the slab. This anisotropy is harder to detect because the problem is now three-dimensional, and rays at different azimuths and take-off angles traverse different parts of the mantle after they leave the slab. If deep-focus earthquakes are studied with upgoing rays, the earthquakes will be located too shallow, and the velocity contrast between slab and normal mantle will appear to be large. If only horizontal and downgoing rays are considered, and isotropic theory is applied, the earthquakes will appear to be deeper than they are, and it may appear that a high-velocity slab must extend to great depth below the earthquake. This is another example of the subtle effects of anisotropy that can result in erroneous conclusions.

There will probably never be enough seismic data to completely characterize the anisotropy of a given region of the Earth. Velocities of waves of all polarizations in many directions are required. Many natural rocks, and layered media, closely approximate a transversely isotropic solid, although the symmetry axis is not necessarily vertical. Quite often there is one preferred direction controlled by gravity, stress or thermal gradient that tends to be a unique direction, affecting the orientation of one of the crystallographic axes or the bedding plane in layered formations. Such a solid can be characterized by seven parameters: five elastic constants and two orientation angles, say strike and dip of the unique axis.

The theory of wave propagation in material having tilted hexagonal symmetry is therefore of interest (Christensen and Crosson, 1968).

If Θ' is the angle of the symmetry axis from vertical and ϕ is the azimuth in the horizontal plane from the plane containing the symmetry axis,

$$V_p^2(\phi) = C_p^2 + D_o + D_2 \cos 2\phi + D_4 \cos 4\phi$$

$$D_o = c_{11} + (c_{13} + 2c_{44} - c_{11})\sin^2\Theta' + (3/8)(c_{11} + c_{33} - 2(c_{13} + 2c_{44}))\sin^4\Theta'$$

$$D_2 = (c_{13} + 2c_{44} - c_{11})\sin^2\Theta' + (1/2)(c_{11} - 2(c_{13} + 2c_{44}))\sin^4\Theta'$$

$$D_4 = (1/8)(c_{11} + c_{33} - 2(c_{13} + 2c_{44}))\sin^4\Theta'$$

$$C_p^2 = \text{Square of isotropic velocity}$$

This is a first-order perturbation theory giving the deviation of the squared phase velocity, V_p^2, from the square of an assumed isotropic velocity, C_p^2, and applies when the group and phase velocities are approximately equal; that is, the direction of energy propagation is essentially normal to the wave fronts. Christensen and Crosson (1968) showed that the azimuthal variation of Pn velocity in the Pacific could be adequately explained by the assumption of transverse isotropy with a nearly horizontal symmetry axis. The azimuthal variation of Pn also depends on the dip of the Moho or the thickness of the crust.

Smith and Dahlen (1973) gave expressions for the azimuthal dependence of surface waves for the most general case of an anisotropic medium with 21 independent c_{ij} elastic coefficients for the case of weak anisotropy. Montagner and Nataf (1986) showed that the average over all azimuths reduces to a term that involves five independent combinations of the elastic coefficients. The general case therefore reduces to an equivalent transversely isotropic solid when the appropriate azimuthal average is taken. The elastic coefficients of the equivalent medium are

$$A = 3(c_{11} + c_{22})/8 + c_{12}/8 + c_{66}/2$$

$$C = c_{33}$$

$$F = (c_{13} + c_{23})/2$$

$$L = (c_{44} + c_{55})/2$$

$$N = (c_{11} + c_{22})/8 - c_{12}/4 + c_{66}/2$$

Table 15-6 gives these elastic coefficients for the transversely isotropic equivalent of several models for the upper mantle. Also given are the anisotropic parameters (Anderson, 1961): $\phi = C/A$, $\xi = N/L$, $\eta = F/(A - 2L)$. η is the anisotropy of *S*-waves, ϕ is the anisotropy of *P*-waves, and η is the fifth parameter required to fully describe transverse isotropy and ρ is the density in g/cm^3.

Table 15-7 gives some results for the azimuthal variation of seismic velocity in the uppermost mantle.

TRANSVERSE ISOTROPY OF LAYERED MEDIA

A material composed of isotropic layers appears to be transversely isotropic for waves that are long compared to the layer thicknesses. The symmetry axis is obviously perpendicular to the layers. All transversely isotropic material,

TABLE 15-6
Elastic Coefficients of Equivalent Transversely Isotropic Models of the Upper Mantle

	Olivine Model		Petrofabric		PREM	
	(1)	(2)	(3)	(4)	(5)	(6)
			Mbar			
A	2.416	2.052	2.290	2.208	2.251	2.176
C	2.265	3.141	2.202	2.365	2.151	2.044
F	0.752	0.696	0.721	0.724	0.860	0.831
L	0.659	0.723	0.770	0.790	0.655	0.663
N	0.824	0.623	0.784	0.746	0.708	0.661
			km/s			
V_{PH}	8.559	7.888	8.324	8.174	8.165	8.049
V_{PV}	8.285	9.757	8.166	8.460	7.982	7.800
V_{SH}	5.000	4.347	4.871	4.752	4.580	4.436
V_{SV}	4.470	4.684	4.828	4.889	4.404	4.441
			Mbar/Mbar			
ξ	1.250	0.861	1.018	0.944	1.081	0.997
ϕ	0.937	1.531	0.961	1.071	0.956	0.939
η	0.686	1.151	0.963	1.153	0.914	0.977
			g/cm³			
ρ	3.298	3.298	3.305	3.305	3.377	3.360

(1) Olivine based model; *a*-horizontal, *b*-horizontal, *c*-vertical (Nataf and others, 1986).
(2) *a*-vertical, *b*-horizontal, *c*-horizontal.
(3) Petrofabric model; horizontal flow (Montagner and Nataf, 1986).
(4) Petrofabric model; vertical flow.
(5) PREM, 60 km depth (Dziewonski and Anderson, 1981).
(6) PREM, 220 km depth.

however, cannot be approximated by a laminated solid. For example, in layered media, the velocities parallel to the layers are greater than in the perpendicular direction. This is not generally true for all materials exhibiting transverse or hexagonal symmetry. Backus (1962) derived other inequalities which must be satisfied among the five elastic constants characterizing long-wave anisotropy of layered media.

For a layered medium composed of two kinds of isotropic material, the equivalent transversely isotropic solid has

$$N = \mu_1 d_1 + \mu_2 d_2$$

$$L = \frac{\mu_1 \mu_2}{\mu_1 d_2 + \mu_2 d_1}$$

where μ_i are the layered rigidities and d_i are their thicknesses, normalized to the total doublet thickness (Anderson, 1967). For a material composed of N' laminations, each of different rigidity and thickness, the stack can be replaced, in the long-wavelength limit for horizontal propagation, by a layer having an equivalent rigidity μ' and thickness d':

$$\mu' = \left\{ \frac{\left(\prod^{N'} \mu_i\right)\left(\sum^{N'} d_i \mu_i\right)}{\sum^{N'} \left(d_i \prod \mu_j\right)} \right\}^{1/2}$$

$$d' = \left\{ \frac{\left(\sum^{N'} d_i \mu_i\right) \sum^{N'} d_i \left(\prod^{N'} \mu_j\right)}{\left(\prod^{N'} \mu_i\right)} \right\}^{1/2}$$

for $k \sum^{N'} d_i \ll 1$, where $j \neq i$. Π and Σ are, respectively, the product and summation operators.

Thus, Love-wave propagation in transversely isotopic material can be computed with isotropic programs simply by scaling the parameters. It also follows that Love waves alone cannot be used to detect transverse isotropy. A similar nonuniqueness, although usually not exact, occurs for many problems involving anisotropy. For example, the thickness of the oceanic lithosphere, the deep structure of continents and the possibility of deep slab penetration all involve data that can have a dual interpretation, one involving the presence of anisotropy.

Although a finely layered solid acts as a transversely isotropic medium for long waves, the five independent elastic constants cannot take on arbitrary values. Backus (1962) proved that, for a layered solid,

$$c_{11}, c_{44}, c_{66} > 0$$

$$c_{12} + c_{66} > 0$$

$$(c_{12} + c_{66})c_{33} \geq c_{13}^2$$

$$c_{66} > c_{44}$$

TABLE 15-7
Anisotropic Parameters for the Uppermost Mantle: Variation with Azimuth

	$V_p^2 = A + B\cos 2\Theta + C\cos 4\Theta$ $V_{SV}^2 = D + E\cos 2\Theta$		
	(1)	(2)	(3)
Moduli (km²/s²)			
A	67.84	66.14	66.83
B	5.91	3.66	3.68
C	0.19	0.71	0
D	21.82	21.62	21.62
E	0.91	0.37	0
Velocities (km/s)			
$V_p(0)$	8.60	8.40	8.40
$V_p(90)$	7.88	7.95	7.95
$V_{SV}(0)$	4.77	4.69	4.65
$V_{SV}(90)$	4.57	4.61	4.65

(1) Pacific Ocean uppermost mantle (Kawasaki and Konno, 1984).
(2) 22 percent aligned olivine in isotropic matrix (Shearer and Orcutt, 1986).
(3) South Pacific upper mantle (Shearer and Orcutt, 1986).

Berryman (1979) derived the additional inequalities

$$c_{11} \geq c_{44}$$

$$(c_{11} - c_{44})(c_{33} - c_{44}) \geq (c_{13} + c_{44})^2$$

$$c_{33} > c_{44}$$

For two layers one can show

$$c_{11} \geq c_{33}/2$$

(Postma, 1955), and for most cases of physical interest,

$$c_{11} \geq c_{33}$$

Isotropy in the symmetry plane yields

$$c_{11} = c_{12} + 2c_{66}$$

There is always the question in seismic interpretations whether a measured anisotropy is due to intrinsic anisotropy or to heterogeneity, such as layers, sills or dikes. The above relations can be used to test these alternatives, or at least, to possibly rule out the laminated-solid interpretation. The magnitude of the anisotropy often can be used to rule out an apparent anisotropy due to layers if the required velocity contrast between layers is unrealistically large.

Some of the above inequalities simply state that velocities along the layers are faster than velocities perpendicular to the layers. No such restrictions apply to the general case of crystals exhibiting hexagonal symmetry or to aggregates composed of crystals having preferred orientations. In a laminated medium, with a vertical axis of symmetry, the *P* and *SH* velocities decrease monotonically from the horizontal to the vertical, and the *SV* velocity is minimum in the vertical and horizontal directions. These also are not general characteristics of transversely isotropic media.

THE EFFECT OF ORIENTED CRACKS ON SEISMIC VELOCITIES

The velocities in a solid containing flat oriented cracks depend on the elastic properties of the matrix, porosity, aspect ratio of the cracks, the bulk modulus of the pore fluid, and the direction of propagation (Anderson and others, 1974). Substantial velocity reductions, compared with those of the uncracked solid, occur in the direction normal to the plane of the cracks. Shear-wave birefringence also occurs in rocks with oriented cracks.

Figure 15-5 gives the intersection of the velocity surface with a plane containing the unique axis for a rock with ellipsoidal cracks. The short-dashed curves are the velocity surfaces, spheres, in the crack-free matrix. The long-dashed curves are for a solid containing 1 percent by volume of aligned spheroids with $\alpha = 0.5$ and a pore-fluid bulk modulus of 100 kbar. The solid curves are for the same parameters as above but for a relatively compressible fluid in the pores with modulus of 0.1 kabar. The shear-velocity surfaces do not depend on the pore-fluid bulk modulus. Note the large compressional-wave anisotropy for the solid containing the more compressible fluid.

The ratio of compressional velocity to shear velocity is strongly dependent on direction and the nature of the fluid phase. In general, the V_p/V_s ratio is normal (1.73) or greater

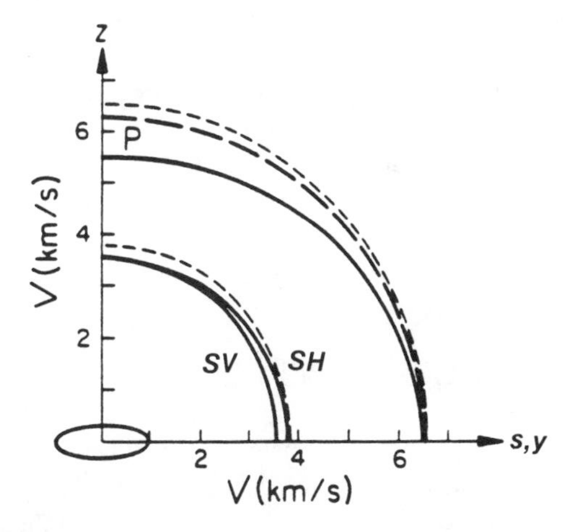

FIGURE 15-5
Velocities as a function of angle and fluid properties in granite containing aligned ellipsoidal cracks (orientation shown at origin) with porosity = 0.01 and aspect ratio = 0.05. The short dashed curves are for the isotropic uncracked solid, the long dashes for liquid-filled cracks (K_L = 100 kbar) and the solid curves for gas-filled cracks (K_L = 0.1 kbar) (after Anderson and others, 1974).

than normal when it is measured along the plane of the cracks. In the direction perpendicular to the cracks, the V_p/V_s ratio is nearly normal for liquid-filled cracks but decreases rapidly as the bulk modulus of the fluid phase approaches that of a gas.

A cracked solid with flat aligned cracks behaves as a transversely isotropic solid with velocities

$$V_p^2(\Theta) = \alpha^2[1 + 2\delta \sin^2\Theta \cos^2\Theta + 2\varepsilon \sin^4\Theta]$$

$$V_{s1}^2(\Theta) = \beta^2[1 + 2(\alpha/\beta)^2(\varepsilon - \delta)\sin^2\Theta \cos^2\Theta]$$

$$V_{s2}^2(\Theta) = \beta^2[1 + 2\gamma \sin^2\Theta]$$

(Thomsen, 1988). This is for small crack density, $e = 3\phi/(4\pi c/a)$ where ϕ is the porosity and c/a is the crack aspect ratio. The anisotropy parameters, in terms of Poisson's ratio of the uncracked solid, σ, or the bulk moduli of the fluid and solid, K_F and K, respectively, are

$$\delta = (16/3)[(1 - K_F/K)(1 - \sigma)D - (1 - 2\sigma)/(2 - \sigma)]e$$

$$\varepsilon = (8/3)(1 - K_F/K)De$$

$$\gamma = (8/3)[(1 - \sigma)/(2 - \sigma)]e$$

$$D^{-1} = 1 - (K_F/K) + (K_F/K)[16(1 - \sigma^2)/9(1 - 2\sigma)](e/\phi)$$

Oriented cracks are important in crustal seismic studies and indicate the direction of the prevailing stress or a paleostress field. In general, a reflection experiment will generate two sets of shear waves that considerably complicate *S*-wave seismograms and record sections. The orientation of the shear waves and their velocities will be controlled by the orientation of the cracks. The magnitude of the velocity difference will be controlled by the crack density and the nature of the pore fluid. Cracks may form and open up as a result of tectonic stresses, and seismic velocity variations and anisotropy may be a tool for earthquake prediction. The dilatancy-diffusion model of earthquake prediction (Anderson and Whitcomb, 1973; Whitcomb and others, 1973) is based on the pressure changes and fluid flow in crustal cracks.

INVERSION RESULTS FOR THE UPPER MANTLE

Normal-mode periods, teleseismic travel times and great-circle surface-wave dispersion data are known as the gross Earth data set. By combining data from many earthquakes and stations, it is hoped that lateral variations and azimuthal effects can be averaged out. Such problems as regional variations, asimuthal anisotropy and their depth extent can then be discussed in terms of variations from the average Earth. It has been surprisingly difficult to find a spherically symmetric Earth model that satisfies the entire gross Earth data set. The normal-mode models did not satisfy body-wave data until it was recognized that absorption made the "elastic" constants frequency dependent (Jeffreys, 1965; Liu and others, 1976; Randall, 1976; Kanamori and Anderson, 1977) as originally proposed by Jeffreys (1965, 1968). Even when absorption was allowed for, gross Earth models did not satisfy the complete data set. The most obvious problem is the well-known Rayleigh wave–Love wave discrepancy. The Earth models of Jordan and Anderson (1974), Gilbert and Dziewonski (1975), and Anderson and Hart (1976) were the result of isotropic inversion of large normal-mode data sets. These models did not satisfy shear-wave travel-time data or short-period (< 200) Love- and Rayleigh-wave data. The inclusion of attenuation made it possible to reconcile some of the free-oscillation and body-wave data (Anderson and Hart, 1976; Anderson and others, 1977, Hart and others, 1977). The Earth models derived in these studies satisfied a large variety of data, but they still disagreed with the mantle Love- and Rayleigh-wave observations. This suggests that the assumption of isotropy in the upper mantle may be in error.

In a 1981 report Adam Dziewonski and I inverted a large data set consisting of about 1000 normal-mode periods, 500 summary travel-time observations, 100 normal-mode Q values, mass and moment of inertia to obtain the radial distribution of elastic properties, Q values and density in the Earth's interior. By allowing for transverse isotropy in the upper 200 km of the mantle, we were able to satisfy, to high precision, teleseismic travel times and normal-mode periods and, at the same time, Love- and Rayleigh-wave dispersion to periods as short as 70 s. The parameters of the upper mantle of this model, PREM, are given in the Appendix. The model is isotropic below a depth of 220 km. The upper mantle is characterized by a 2–4 percent anisotropy in velocity and a slight variation of the five elastic constants with depth. A similar structure satisfies dispersion data for Pacific Ocean paths.

In the PREM inversion, a satisfactory fit to the gross Earth data set, including mantle Love and Rayleigh waves, was achieved with a linear gradient in all five elastic constants between Moho and 220 km. *PH, PV* and *SH* decrease slightly, and *SV* increases slightly with depth. The overall anisotropy decreases with depth. This is in marked contrast to isotropic inversions, which invariably give pronounced shear-wave low-velocity zones. The anisotropic models have average anisotropies in the upper 200 km of the mantle of about 3 percent.

The introduction of anisotropy into the upper mantle introduces more degrees of freedom into the inversion problem. We were able to fit the gross Earth data set with an Earth model that had 13 radial subdivisions. The density and elastic-wave velocities in each region were described by low-order polynomials. A total of 92 parameters were sufficient to satisfy the data. The locations of the boundaries are additional parameters, making a total of 105 parameters. Some of the parameters such as mass and radius of the Earth, radius of the inner core and average depth to Moho were determined from other data. We also attempted to fit the same dataset with isotropic inversion but were unsuccessful.

In the anisotropic modeling, the upper mantle, to a depth of 220 km, required 12 parameters for its description. These are the density, the five elastic constants and a linear gradient of each. In the isotropic modeling this region had to be split into two, giving also 12 parameters, which involves a two-parameter description of density and the two elastic constants in each region. The isotropic inversion also resulted in a large and unreasonable mean crustal thickness. Even the best-fitting isotropic models, however, were unable to fit the short-period (< 200 s) Love- and Rayleigh-wave data. The overall fit to the normal-mode data set was also inferior to the anisotropic model. It appears, therefore, that the superior fit achieved by anisotropic modeling is not due to an increase in the number of parameters. It appears rather to be the result of a more appropriate parameterization. The anisotropic parameters are only a small fraction of the total number of parameters in the model.

The presence of even a small amount of anisotropy completely changes the nature of the surface-wave and normal-mode problem. In particular, the apparent lack of sensitivity of many of the spheroidal modes to the compressional velocity structure is due to the degeneracy in the isotropic case. The normal-mode dataset appears to be adequate to resolve the five elastic constants of an equivalent transversely isotropic upper mantle. The anisotropic models fit the data better, they removed the Rayleigh-Love discrepancy, and the resulting models for the upper mantle were substantially different from the isotropic models. If Love-wave and Rayleigh-wave data cannot be satisfied by an isotropic model, there is no recourse but to assume that at least five elastic constants control the dispersion. One cannot assume that toroidal and spheroidal modes are controlled by only one of the shear moduli or that Rayleigh waves are not sensitive to the compressional-wave velocity.

Because of the apparent pervasiveness of anisotropy, it cannot be assumed that isotropic inversion of limited data sets, such as Rayleigh waves or *P*-waves, yield even approximately correct models for the upper mantle. Isotropy must be demonstrated by, for example, combined inversion of Love and Rayleigh waves, or *P*, *SH* and *SV* data. Lacking this, models that exhibit shear velocities less than about 4.3 km/s in the upper mantle must be viewed with suspicion since data leading to such models can be explained by a small degree of anisotropy, anisotropy that is generally required by the broader data set.

In general, the spheroidal modes are more sensitive to V_{SV} than to V_{SH}, but this does not mean that an anisotropic structure can be approximated by an isotropic structure using the V_{SV} velocity for the shear structure (Anderson, 1961, 1967). The three compressional parameters η, V_{PV} and V_{PH} are also required. As a rule of thumb, the compressional velocity is important at depths shallower than one-sixth the wave length in an isotropic structure. In an anisotropic structure, the individual contributions of η, V_{PV} and V_{PH} persist to depths comparable to the depths influenced by the shear structure.

For the fundamental toroidal modes (Love waves) the main controlling parameter is the horizontal *SH* velocity. The vertical shear-wave velocity, V_{SV}, however, is important for the overtones. The partial derivatives as a function of depth oscillate, and V_{SV} and V_{SH} are alternately important. The toroidal modes involve *SH* particle motion. The velocity, however, varies from V_{SH} in the horizontal direction to V_{SV} in the vertical, or radial direction. The toroidal overtones can be viewed as constructively interfering body waves; since the condition for constructive interference involves the wavelength and the angle of emergence, it is clear that both components of velocity are important.

GLOBAL MAPS OF TRANSVERSE ISOTROPY AS A FUNCTION OF DEPTH

Azimuthal anisotropy can reach 10 percent in the shallowest mantle, as it is measured from Pn waves and from *P*-delays. Polarization anisotropy up to 5 percent is inferred from surface-wave studies in order to fit Love waves and Rayleigh waves. Azimuthal anisotropy can be averaged out. We are then left with only polarization anisotropy and can use a transversely isotropic parameterization. This involves six inversion parameters: ρ, V_{PH}, V_{SV}, ξ, ϕ and η, where ρ is the density, V_{PH} is the horizontal *P*-wave velocity, V_{SV} the vertically polarized horizontal *S*-wave velocity, ξ the anisotropy of *S*-waves, ϕ the anisotropy of *P*-waves, and η is the fifth elastic parameter.

Resolution kernels show that only V_{SV} and ξ can be resolved from the fundamental-mode Love and Rayleigh waves. However, changes in ρ, V_{PH}, ϕ, and η affect these modes substantially. For example, a 5 percent *P*-anisotropy (ϕ) has the same effect on Rayleigh-wave phase velocity as a 0.1-km/s change in *SV* velocity. We must thus bring in further a priori information. If lateral variations in velocity are due to temperature variations, we can relate them to density changes using laboratory data. Similarly, if anisotropy is caused by the preferred orientation of olivine crystals, we can relate *P*-anisotropy to *S*-anisotropy. Nataf and others (1986) inverted a large data set of fundamental-mode Love and Rayleigh velocities to obtain the global distribution of heterogeneity and anisotropy. Shear-velocity (V_{SV}) heterogeneities are shown in Figures 15-6 to 15-8. With a few exceptions, they exhibit a strong correlation with surface tectonics down to about 200 km. Deeper in the mantle the correlation vanishes, and some long-wavelength anomalies appear. At 50 km-100 km depth heterogeneities are closely related to surface tectonics. All major shields show up as fast regions (Canada, Africa, Antarctica, West Australia, South America). All major ridges show up as slow regions (East Pacific, triple junctions in the Indian Ocean and in the Atlantic, East African rift). The effect of the fast mantle beneath the shields is partially offset by the thick crust for waves that sample both. Old oceans also appear to be fast, but not as fast as shields. A few regions seem to be anomalous, considering their tectonic setting: a slow region around French Polynesia in average age ocean, a fast region centered southeast of South America. At 100 km depth the overall variations are smaller than at shallower depth. Ve-

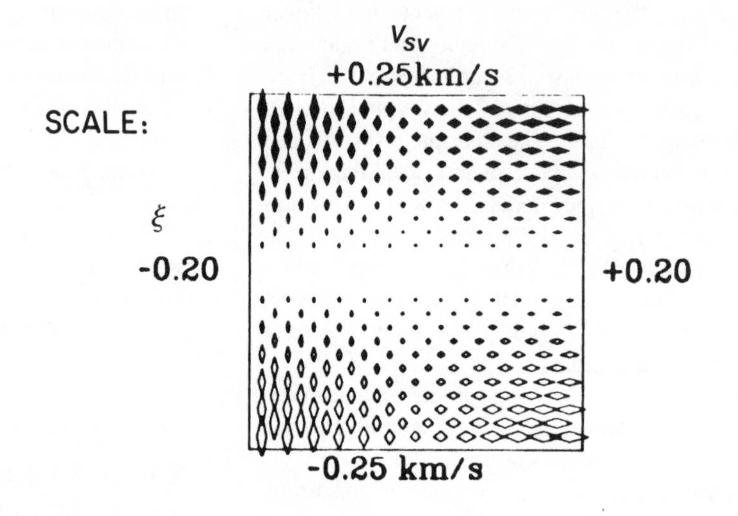

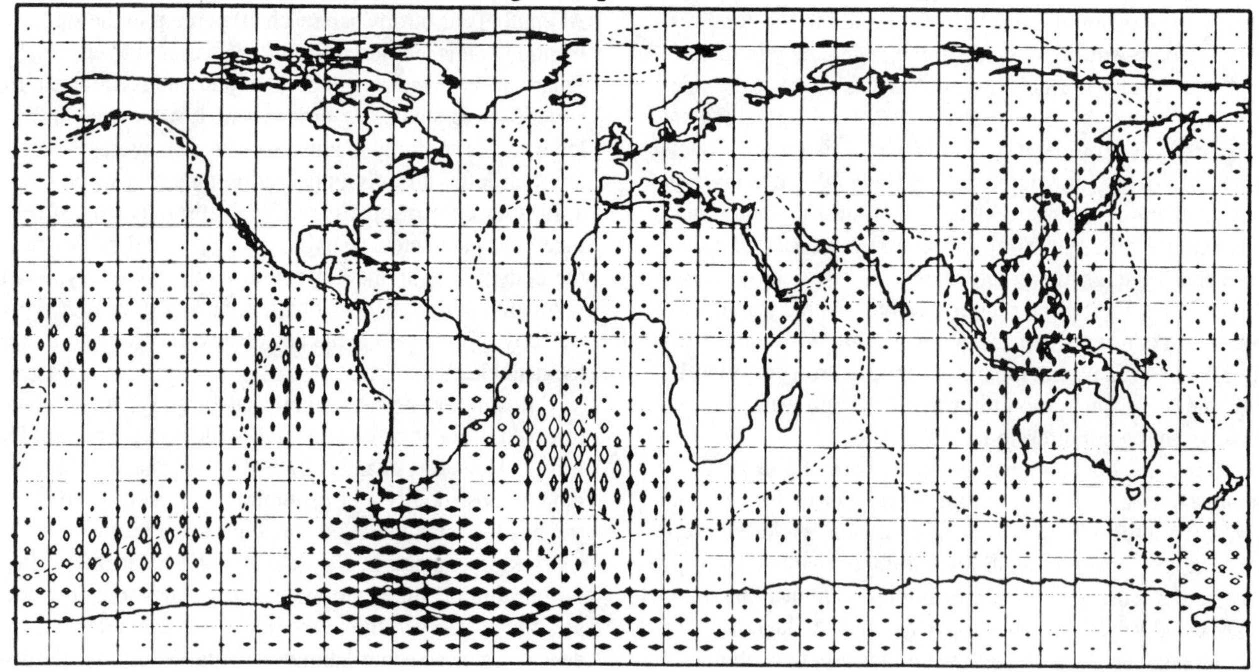

FIGURE 15-6
Seismic flow map at 280 km depth. This combines information about shear velocity and polarization anisotropy ξ. Open symbols are slow, solid symbols are fast. Vertical diamonds are $SV > SH$, presumably due to vertical flow. Horizontal diamonds are $SH > SV$. Slow velocities are at least partially due to high temperatures and, possibly, partial melting. Regions of fast velocity are probably high density as well. The south-central Atlantic and the East Pacific Rise appear to be upwelling buoyant regions. Similar features occur in the central Pacific and the Afar region. The western Pacific and northeastern Indian Ocean appear to be regions of downwelling.

locities under the ocean are close to the average, except under the youngest ocean. Triple junctions are slow. Below 200 km depth, the correlation with surface tectonics starts to break down. Shields are fast, in general, but ridges do not show up systematically. The East African region, centered on the Afar, is slow. The south-central Pacific is faster than most shields. An interesting feature is the belt of slow mantle at the Pacific subduction zones. This may be a manifestation of the volcanism and marginal sea formation induced by the sinking ocean slab. Below 340 km, the same belt shows up as fast mantle; the effect of cold subducted material that was formerly part of the surface thermal boundary layers. Many ridge segments are now fast. At larger depths the resolution becomes poor, but these trends seem to persist.

At intermediate depths, regions of uprising (ridges) or downwelling (subduction zones) have an $SV > SH$ anisotropy, in agreement with olivine crystals aligned in a vertical flow. Shallow depths (50 km) show very large anisotropy variations ($\pm$ 10 percent). From observed Pn anisotropy and measured anisotropy of olivine, such values are not unreasonable. At 100 km the amplitude of the variations is much smaller ($\pm$ 5 percent), but the pattern is similar. The Mid-Atlantic Ridge has $SV > SH$, whereas the other

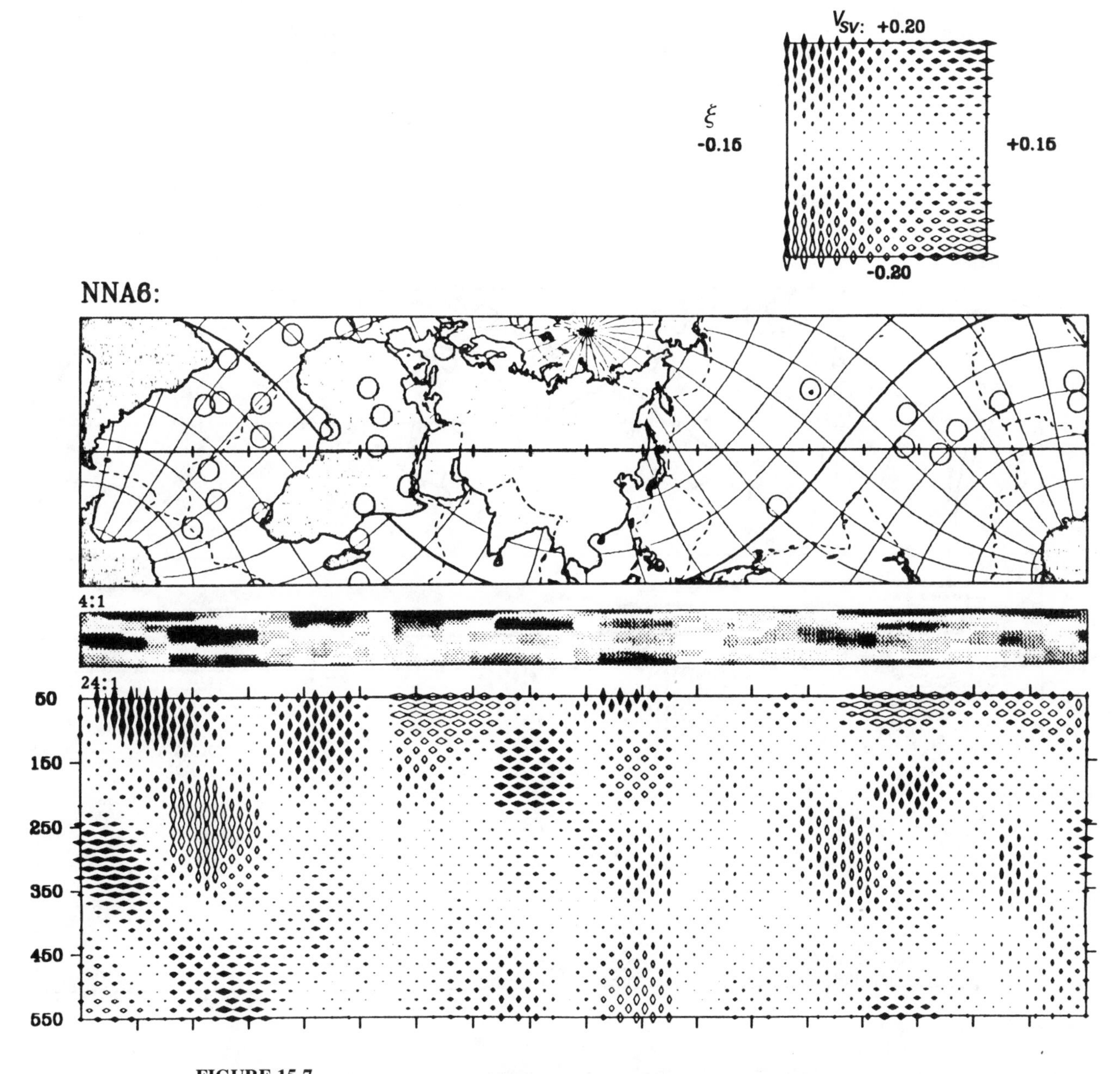

FIGURE 15-7
S velocity from 50 to 550 km along the great-circle path shown. Cross-sections are shown with two vertical exaggerations. Velocity variations are much more extreme at depths less than 250 km than at greater depths. The circles on the map represent hotspots.

ridges show no clearcut trend. Under the Pacific there appear to be some parallel bands trending northwest-southeast with a dominant $SH > SV$ anomaly. This is the expected anisotropy for horizontal flow of olivine-rich aggregates. At 340 km, most ridges have $SH < SV$ (vertical flow). Antarctica and South America have a strong $SV < SH$ anomaly (horizontal flow). North America and Siberia are almost isotropic at this depth. They exhibit, however, azimuthal anisotropy as discussed below. The central Pacific and the eastern Indian Ocean have the characteristics of vertical flow. These regions have faster than average velocities at shallow depths and may represent sinkers.

AGE-DEPENDENT TRANSVERSE ISOTROPY

There have been many surface-wave studies of the structure of the oceanic upper mantle. The general agreement be-

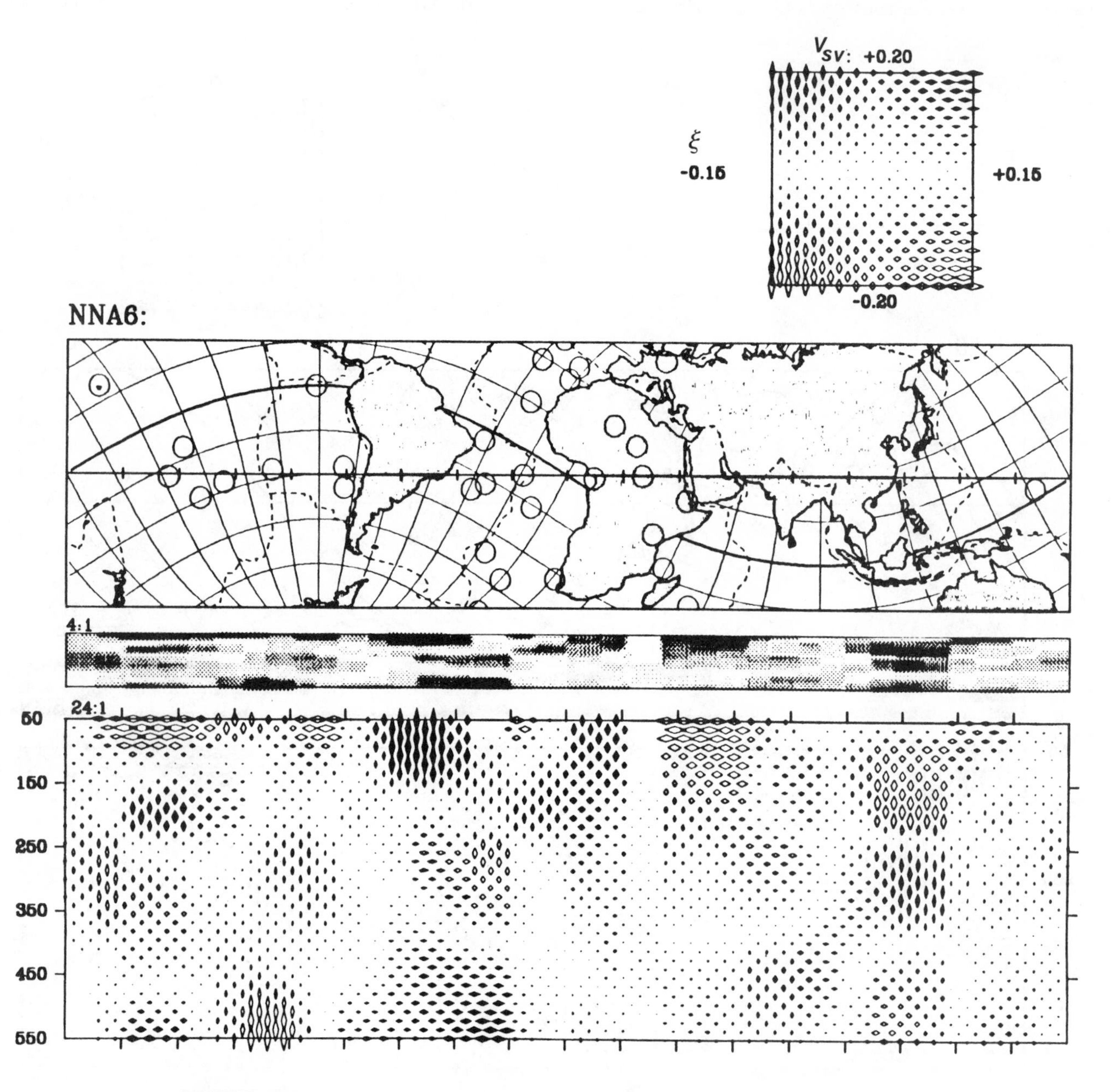

FIGURE 15-8
S velocity in the upper mantle along the cross-section shown. Note low velocities at shallow depth under the western Pacific, replaced by high velocities at greater depth. The eastern Pacific is slow at all depths. The Atlantic is fast below 400 km.

tween the various studies and the calculation of resolving kernels suggested that we were in the model refinement stage and that no major surprises were in store. There was general agreement, for example, that the seismic lithosphere, or LID, is about 100 km thick in old ocean basins, and that, at all ages, it is much thicker than the flexural lithosphere. Although there are formalisms for estimating uniqueness and resolving power of a given set of geophysical data, these are applied only after decisions and assumptions have already been made about model parameterization and what class of models is considered appropriate. Regan and Anderson (1984) showed that the self-consistent inversion of oceanic surface-wave data gives results that are drastically different from previous results. The neglect of anelastic dispersion and anisotropy results in erroneous structures even though the structure appears to be well resolved using elastic, isotropic resolution kernels. In a particularly dramatic example Anderson and Dziewonski (1982) showed that anisotropic models that satisfy both Rayleigh- and Love-wave data bear little resemblance to models based on Rayleigh-wave data alone or on separate isotropic inversion of Love- and Rayleigh-wave data.

If the mantle is anisotropic, the use of Rayleigh waves alone is of limited usefulness in the determination of mantle structure because of the trade-off between anisotropy and structure. Love waves provide an additional constraint. If it is assumed that available surface-wave data are an azimuthal average, we can treat the upper mantle as a transversely isotropic solid with five elastic constants. The azimuthal variation of long-period surface waves is small.

The combined inversion of Rayleigh waves and Love waves across the Pacific has led to models that have age-dependent LID thicknesses, seismic velocities and anisotropies. In general, the seismic lithosphere increases in the thickness with age and $V_{SH} > V_{SV}$ for most of the Pacific. However, $V_{SH} > V_{SV}$ for the younger and older parts of the Pacific, suggesting a change in the flow regime.

The variation of velocities and anisotropy with age suggests that stress- or flow-aligned olivine may be present. The velocities depend on temperature, pressure and crystal orientation. An interpretation based on flow gives the velocity depth relations illustrated in Figure 15-9. The upper left diagram illustrates a convection cell with material rising at the midocean range (R) and flowing down at the trench (T).

In the lower left of Figure 15-9 is the schematic temperature profile for such a cell. The seismic velocities decrease with temperature and increase with pressure. Combining the effects of temperature and pressure, one obtains a relation between velocity and depth. At the ridge the temperature increases very rapidly with depth near the surface; thus, the effects of temperature dominate over those of pressure, and velocities decrease. Deeper levels under the ridge are almost isothermal; thus, the effect of pressure dominates, and the velocities increase. At the trench the temperature gradient is large near the base of the cell and nearly isothermal at shallower depths. Therefore, the velocity response is a mirror image of that at the ridge. Midway between the ridge and the trench (M), the temperature increases rapidly near the top and bottom of the cell. Thus, the velocities decrease rapidly in these regions.

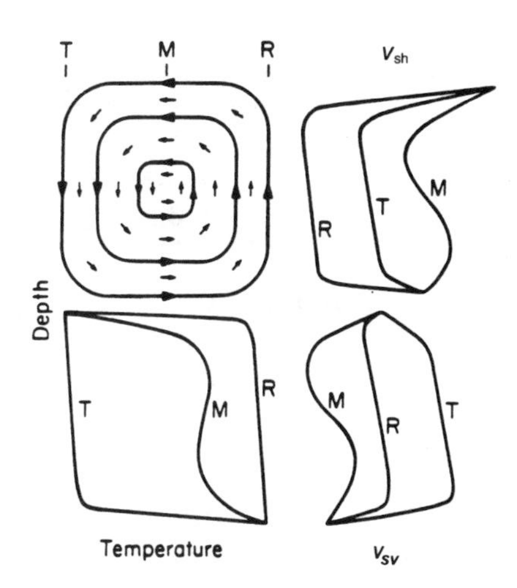

FIGURE 15-9
Schematic representation of seismic velocities due to temperature, pressure, and crystal orientation assuming a flow-aligned olivine model. The upper left diagram shows a convection cell with arrows indicating flow direction. The trench is indicated by T, the ridge by R, and the midpoint by M. The lower left diagram shows temperature depth profiles for the trench, ridge and midpoint. The upper and lower right diagrams show the nature of the velocity depth structure of V_{SH} and V_{SV} respectively due to pressure, temperature, and crystal orientation.

The crystal orientation, if alignment with flow is assumed, is with the shortest and slowest axis (*b* axis) perpendicular to the flow. Thus, at the ridge and the trench where the flow is near vertical, the *b* axis is horizontal, and midway between where the flow is horizontal, the *b* axis is vertical. V_{PH} and V_{SH} between the upward and downward flowing edges of the convection cell are controlled by the velocities along the *a* axis and *c* axis; V_{PV} and V_{SV} depend on the velocity along the *b* axis. Thus, at midpoint, and wherever flow is horizontal, $SH > SV$ and $PH > PV$. At the ridge and at the trench the flow is vertical, rapidly changing to horizontal at the top and bottom of the cell. For vertical flow the horizontal velocity is controlled by the *b*-axis and *c*-axis velocities, so $SH < SV$ and $PH < PV$. The values at the top and bottom of the cell rapidly change to the horizontal flow values. Between the midpoint and the trench or ridge, the transition from horizontal to vertical flow velocities becomes sharper, and the depth extent of constant vertical velocities increases.

The velocity depth structures derived for the oceanic provinces (Table 15-8, Figures 15-10 and 15-11) show that this could be a viable interpretation. For the average ocean model and for the upper 100 km of the mantle in the young-

TABLE 15-8
Upper-Mantle Velocities (km/s) for Oceanic Age Provinces; Reference Period is 1 s

	Water and Sediments		Crust		LID		LVZ		220 km	400 km
			1	2	1	2	Top	Bottom		
					0–20 Ma					
H (km)	3.45	0.02	1.51	4.64	6	14				
V_{PV}	1.52	1.65	5.21	6.80	8.02	8.21	7.67	7.57	8.47	8.82
V_{PH}	1.52	1.65	5.21	6.80	8.19	8.21	7.90	7.77	8.47	8.82
V_{SV}	0.0	1.00	3.03	3.90	4.40	4.60	4.20	4.31	4.60	4.72
V_{SH}	0.0	1.00	3.03	3.90	4.61	4.60	4.45	4.28	4.60	4.72
η	1.00	1.00	1.00	1.00	0.90	1.00	0.92	1.00	1.00	1.00
					20–50 Ma					
H (km)	4.67	0.13	1.58	5.15	6	24				
V_{PV}	1.52	1.65	5.21	6.80	8.02	8.42	7.77	7.59	8.47	8.82
V_{PH}	1.52	1.65	5.21	6.80	8.19	8.42	7.88	7.77	8.47	8.82
V_{SV}	0.00	1.00	3.03	3.90	4.40	4.72	4.28	4.32	4.60	4.72
V_{SH}	0.00	1.00	3.03	3.90	4.61	4.72	4.39	4.29	4.60	4.72
η	1.00	1.00	1.00	1.00	0.90	1.00	0.93	1.00	1.00	1.00
					50–100 Ma					
H (km)	5.40	.23	1.60	5.19	6	34				
V_{PV}	1.52	1.65	5.07	6.70	8.02	8.39	8.04	7.46	8.56	8.91
V_{PH}	1.52	1.65	5.07	6.70	8.19	8.48	8.15	8.03	8.56	8.91
V_{SV}	0.0	1.00	2.96	3.84	4.40	4.70	4.43	4.25	4.64	4.77
V_{SH}	0.0	1.00	2.96	3.84	4.61	4.75	4.58	4.43	4.64	4.77
η	1.00	1.00	1.00	1.00	0.90	1.00	0.83	0.96	1.00	1.00
					>100 Ma					
H (km)	5.75	.30	1.6	5.19	6	44				
V_{PV}	1.52	1.65	5.01	6.63	8.02	8.27	8.12	7.66	8.56	8.91
V_{PH}	1.52	1.65	5.01	6.63	8.10	8.31	8.12	8.03	8.56	8.91
V_{SV}	0.0	1.00	2.93	3.80	4.40	4.63	4.48	4.36	4.64	4.77
V_{SH}	0.0	1.00	2.93	3.80	4.61	4.66	4.56	4.31	4.64	4.77
η	1.00	1.00	1.00	1.00	0.90	1.00	0.87	0.97	1.00	1.00
Q_{μ}	∞	600	600	600	600	600	80	80	143	143

Regan and Anderson (1984).

est regions, $PH > PV$ and $SH > SV$. This is consistent with horizontal flow. For the 100–200 km depth range for the youngest regions (0–5, 5–10 Ma), $SH > SV$. The vertical flow expected in the ridge-crest environment would exhibit this behavior. The temperature gradients implied are 5–8°C per kilometer for older ocean. The young ocean results are consistent with reorientation of olivine along with a small temperature gradient. With these temperature and flow models, the velocity of Love waves along ridges is expected to be extremely slow. The velocity of Rayleigh waves is predicted to be high along subduction zones. For midplate locations, Love-wave velocities are higher and Rayleigh-wave velocities are lower than at plate boundaries.

The average Earth model (Table 15-9) takes into account much shorter period data than used in the construction of PREM. Note that a high-velocity LID is required by this shorter wavelength information. This is the seismic lithosphere. It is highly variable in thickness, and an average Earth value has little meaning.

AZIMUTHAL ANISOTROPY

Anisotropy of the upper mantle most likely originates from preferred orientation of olivine crystals. Studies to date indicate that the *a* axes of olivine-rich aggregates cluster around the flow direction, the *a* and *c* axes concentrate in the flow plane, and the *b* axes align perpendicular to the flow plane (Nicolas and others, 1923; Nicolas and Poirier, 1976). For *P*-waves the *a*, *b*, and *c* axes are, respectively, the fast, slow, and intermediate velocity directions. If the flow plane is horizontal, the azimuthal *P*-wave velocity

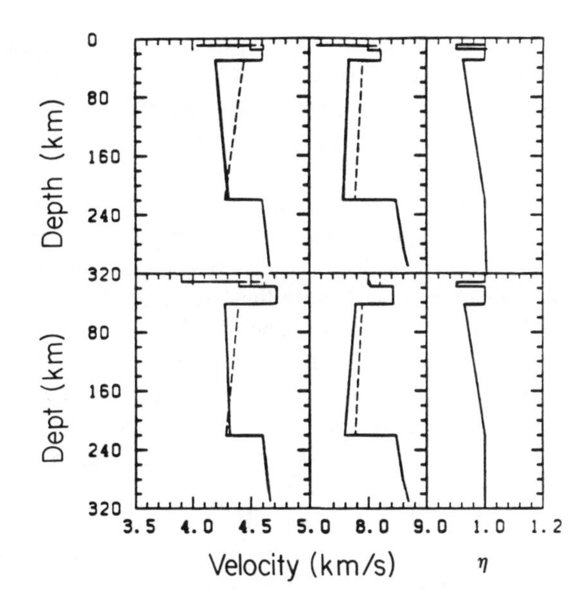

FIGURE 15-10
Velocity depth profiles for the 0–20 Ma (upper set) and the 20–50 Ma (lower set) old oceanic age provinces. From left: *PV*, *SV*, and ETA; dashes are *PH* and *SH*.

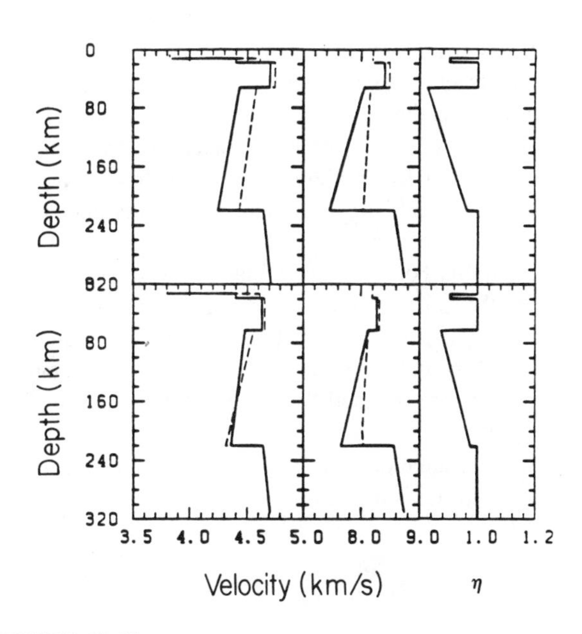

FIGURE 15-11
Velocity depth profiles for the 50–100 Ma (upper set) and the > 100 Ma (lower set) old oceanic regions.

variation in the horizontal plane is 17 percent for a single olivine crystal or an olivine aggregate with 100 percent preferred orientation. For natural olivine-rich aggregates there is some dispersion in the orientation of the crystallographic axes, and the *P*-wave anisotropy is reduced to 4 percent. The corresponding azimuthal *SV* variations are 10 percent for single-crystal olivine and 2 percent for natural aggregates. Rayleigh waves are most sensitive to *SV* velocities and have less sensitivity to *P*-waves. The strong anisotropy of *P*-waves, however, means that the azimuthal anisotropy of Rayleigh waves will be affected by the anisotropies of both *P*- and *S*-waves. The azimuthal variation of both *P*- and *SV*-waves can be described well by a cos 2Θ variation.

The variation of *SH* velocity in the flow plane is four-lobed, that is, a cos 4Θ type variation. The total variation is about 14 percent in single-crystal olivine and about 1 percent for aggregates. The fast directions are at 45° to the *a* and *c* axes; that is, the fast direction for Love waves is not in the flow direction and is not parallel to the fast direction for Rayleigh waves. The presence of other minerals and the azimuthal averaging of long-period waves will dilute the anisotropy. The *SH* (Love) anisotropy can be expected to be hard to detect, since the total variation occurs in only 45°, as distinct from the 90° over which the *P* and *SV* variations occur. Good azimuthal coverage is therefore required.

For vertical flow (vertical *a* axis orientation), the *SV* and *P* anisotropies in the horizontal plane are weak for olivine-rich aggregates, the average *P*-wave velocity in the horizontal plane is relatively low, and *SH* < *SV*. The fast direction for *P* is in the flow plane, and *SV* is nearly isotro-

TABLE 15-9
Upper-Mantle Velocities for the Average Earth Model

	Thickness (km)	V_{PV} (km/s)	V_{PH} (km/s)	V_{SV} (km/s)	V_{SH} (km/s)	η	Q_μ
Water	3.00	1.45	1.45	0.00	0.00	1.00	∞
Crust1	12.00	5.80	5.80	3.20	3.20	1.00	600
Crust2	3.40	6.80	6.80	3.90	3.90	1.00	600
LID	28.42	8.02	8.19	4.40	4.61	.90	600
LVZ top		7.90	8.00	4.36	4.58	.80	80
LVZ bottom		7.95	8.05	4.43	4.44	.98	80
220 km		8.56	8.56	4.64	4.64	1.00	1.43
400 km		8.91	8.91	4.77	4.77	1.00	143

Regan and Anderson (1984).

pic. *SH* has two lobes in the horizontal plane, oriented 45° to the intersection of the flow plane with the horizontal. Regions of ascending and descending flow, such as at ridges and trenches, will therefore be characterized by small azimuthal variation for Rayleigh waves and low Love-wave velocities. In regions of descending flow, Rayleigh waves are fast because of low temperatures and the effects of crystal orientation. For both horizontal and vertical flow, the nature of the anisotropy is approximately cos 2Θ for Rayleigh waves, and the fast direction is in the flow plane. For Love waves the nature of the anisotropy is cos 4Θ, and the fast directions are oriented 45° to the fast directions for Rayleigh waves.

The azimuthal dependence of velocity in a weakly anisotropic material is of the form

$$V_p^2(\Theta) = C_p^2 + D_1 + D_2 \sin 2\Theta + D_3 \cos 2\Theta + D_4 \sin 4\Theta + D_5 \cos 4\Theta \quad (1)$$

Here C_p is the isotropic wave speed and D_1 through D_5 are combinations of the elastic constants. Christensen and Crosson (1968) specialized these results to the case of transverse isotropy with arbitrary tilt of the symmetry axis, and showed that velocity data collected near the Mendocino and Molokai fracture zones can be explained under this assumption, with a nearly horizontal axis. Transverse isotropy implies that a direction of sagittal symmetry should be present in the observations, and in that case equation (1) is actually of the form

$$V_p^2(\Theta) = C_p^2 + D_1^* + D_3^* \cos 2(\Theta - \beta) + D_5^* \cos 4(\Theta - \beta) \quad (2)$$

where β is the azimuth of sagittal symmetry. A detailed discussion of the use of these relations is provided by Crampin and Bamford (1977).

Christensen and Crosson (1968) found that in many cases (1) the olivine *b* axes (low velocity) in ultramafic rocks tend to be concentrated normal to the schistosity or banding with the *a* and *c* axes forming girdles normal to the *b* axis concentration or (2) the *a* axes are bundled, with the *b* and *c* axes arranged in girdles in the orthogonal plane. Either arrangement results in transverse isotropy for compressional waves.

Early measurements of anisotropy were practically confined to the Pacific Ocean. In these cases, the high-velocity direction coincides well with the spreading direction. Bamford (1977) and Fuchs (1977, 1983) found a fairly large degree of anisotropy (7–8 percent) in southern Germany and noted that the fast axis may be correlated with the direction of absolute motion of the European plate.

Long-period (100–250 s) Love and Rayleigh waves were used by Tanimoto and Anderson (1984, 1985) to map heterogeneity and azimuthal anisotropy in the upper mantle. Spherical harmonic descriptions of heterogeneity up to $l = m = 6$ and azimuthal anisotropy up to $l = m = 3$ and cos 2Θ terms were derived. Azimuthal anisotropy obtains values as high as 1½ percent. This is actually fairly high considering the wavelengths and the averaging that is involved.

There is good correlation of fast Rayleigh wave directions with the upper-mantle return-flow models derived from kinematic considerations by Hager and O'Connell (1979). This is consistent with the fast (*a* axis) of olivine being aligned in the flow direction. The main differences between the kinematic return-flow models and the Rayleigh-wave azimuthal variation maps occur in the vicinity of hotspots. Hawaii, for example, appears to deflect the return flow. A large part of the return flow associated with plate tectonics appears to occur in the upper mantle, and this in turn requires a low-viscosity channel in the upper mantle. Figure 15-12 is a map of the azimuthal results for 200-s Rayleigh waves expanded up to $l = m = 3$. The lines are oriented in the maximum velocity direction, and the length of the lines is proportional to the anisotropy. The azimuthal variation is low under North America and the central Atlantic, between Borneo and Japan, and in East Antarctica. Maximum velocities are oriented northeast-southwest under Australia, the eastern Indian Ocean, and northern South America and east-west under the central Indian Ocean; they vary under the Pacific Ocean from north-south in the southern central region to more northwest-southeast in the northwest part. The fast direction is generally perpendicular to plate boundaries. There is little correlation with plate motion directions, and little is expected since 200-s Rayleigh waves are sampling the mantle beneath the lithosphere. The lack of correlation suggests that the flow in the asthenosphere is not strongly coupled to the overlying plate, and this in turn suggests the presence of a low-viscosity asthenosphere.

Hager and O'Connell (1979) calculated flow in the upper mantle by taking into account the drag of the plates and the return flow from subduction zones to spreading centers. Flow lines for a model that includes a low-viscosity channel in the upper mantle are shown in Figure 15-12b. Flow under the large fast-moving plates is roughly antiparallel to the plate motions. Thermal buoyancy is ignored in these calculations, and there is no lateral variation in viscosity.

Hager and O'Connell also calculated shallow mantle flow for a model with a higher upper-mantle viscosity, 4×10^{20} poises. In this model there is some coupling between the plate and the underlying mantle; consequently, material is dragged along with the plate. There are major differences in the orientation of flow lines relative to a model with a low-viscosity shallow mantle, which effectively decouples mantle flow from plate drag. The Rayleigh-wave fast directions are similar to those of the low-viscosity model but diverge from those of the high-viscosity model, particularly under Africa, Australia, India, Greenland, the South Atlantic, and the Tasman Sea.

In the kinematic flow model the flow is nearly due south under Australia, shifting to southwest under the eastern Indian Ocean, or directly from the subduction zones to the nearest ridge. In the anisotropic map the inferred flow is more southwestward under Australia, nearly parallel to the plate motion, shifting to east-west in the eastern Indian Ocean. The southeastern Indian Ridge is fast at depth, suggesting that this ridge segment is shallow. The Mid-Indian Ridge, the Indian Ocean triple junction, and the Tasman Sea regions are slow, suggesting deep hot anomalies in these regions. These deep anomalies are offset from those implicit in the kinematic model and apparently are affecting the direction of the return flow.

Similarly, the flow under the northern part of the Nazca plate is diverted to the southwest relative to that predicted, consistent with the velocity anomaly observed near the southern part of the Nazca-Pacific ridge. The anisotropy due north of India indicates north-south flow, perpendicular to the plate motion of Eurasia and the theoretical return-flow direction. One interpretation is that the Indian plate has subducted beneath the Tibetan plateau and extends far into the continental interior.

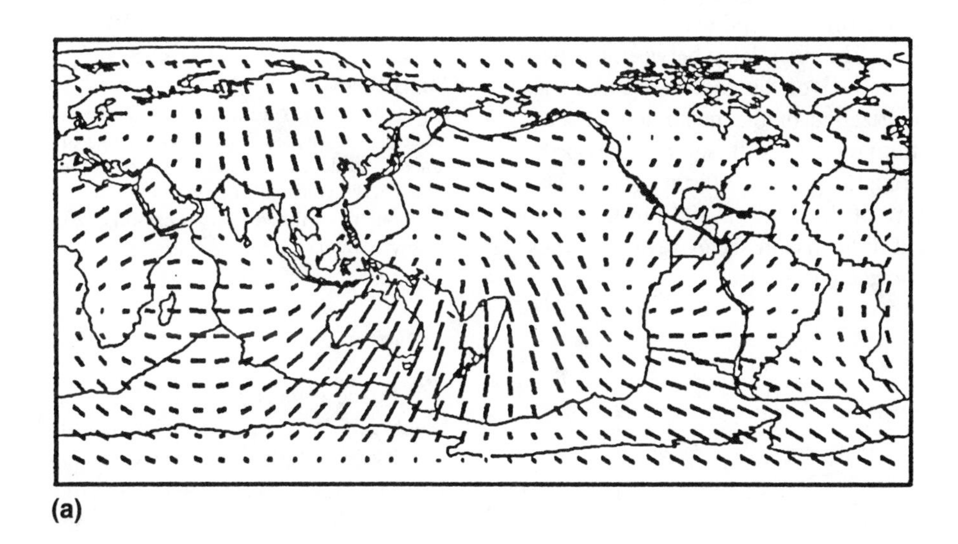

(a)

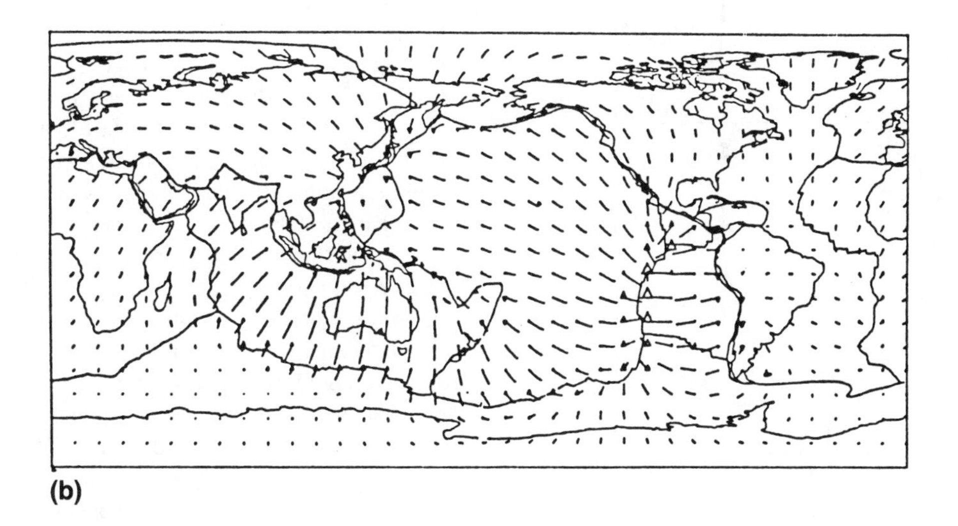

(b)

FIGURE 15-12
(a) Azimuthal anisotropy of 200-s Rayleigh waves. The map includes cos 2Θ, sin 2Θ and l = 1, 2 and 3 terms. The lines indicate the fast phase velocity direction. The length of the lines is proportional to the anisotropy (Tanimoto and Anderson, 1984). **(b)** Flow lines at 260 km depth for the upper-mantle kinematic flow model of Hager and O'Connell (1979). The model includes a low-viscosity channel (10^{19} poises) in the upper mantle.

SHEAR-WAVE SPLITTING AND SLAB ANISOTROPY

In an anisotropic solid there are two shear waves, having mutually orthogonal polarizations, and they travel with different velocities. This is known as shear-wave splitting or birefringence. Since shear waves are secondary arrivals and generally of long period, it requires special studies to separate the two polarizations from each other and from other later arrivals. Deep-focus events are the most suitable for this purpose, and several studies have clearly demonstrated the existence of splitting.

Ando and others (1983) studied nearly vertically incident shear waves from intermediate and deep-focus events beneath the Japanese arc. The time delay between the two nearly horizontal polarizations of the shear waves was as much as 1 s. The polarization of the maximum-velocity shear waves changed from roughly north-south in the northern part of the arc to roughly east-west further south. The anisotropic regions were of the order of 100 km in extent and implied a 4 percent difference in shear-wave velocities in the mantle wedge above the slab.

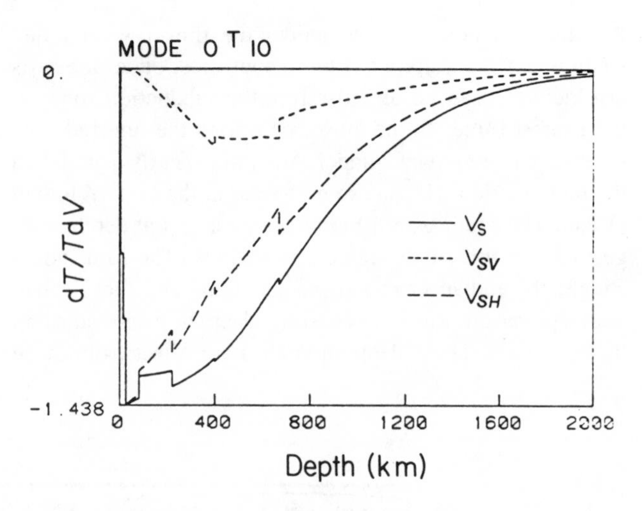

FIGURE 15-14
Partial derivatives for a relative change in period of toroidal mode (Love wave) ${}_0T_{10}$ due to a change in shear velocity as a function of depth. The solid line gives the isotropic partial derivative. The dashed lines give the effect of perturbations in two components of the velocity. Period is 720 s (after Anderson and Dziewonski, 1982).

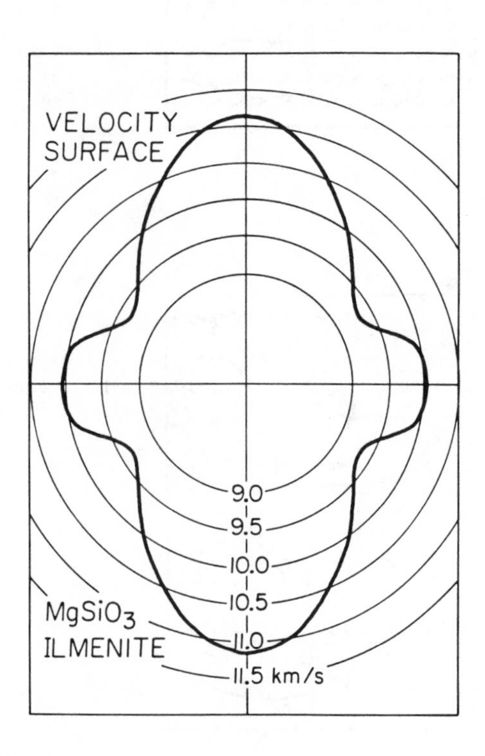

FIGURE 15-13
Variation of compressional velocity with direction in the ilmenite form of $MgSiO_3$. This is a stable mineral below about 500 km in cold slabs. $MgSiO_3$-ilmenite is a platy mineral and may be oriented by stress, flow and recrystallization in the slab. Ice (in glaciers) and calcite (in marble) have similar crystal structures and are easily oriented by flow, giving anisotropic properties to ice and marble masses. The deep slab may also be anisotropic.

Fukao (1984) studied ScS splitting from a deep-focus event (535 km) in the Kuriles recorded in Japan. The uniformity in polarization across the Japanese arc is remarkable. The faster ScS phase had a consistent polarization of north-northwest–south-southeast and an average time advance of 0.8 ± 0.4 s over the slower ScS wave. The splitting could occur anywhere along the wave path, but the consistency of the results over the arc and the difference from the direct S results, from events beneath Japan, suggests that the splitting occurs in the vicinity of the source. The fast polarization direction is nearly parallel to the dip direction of the Kurile slab and the fast *P*-wave direction of the Pacific plate in the vicinity of the Kurile Trench. The stations are approximately along the strike direction of the deep Kurile slab. All of this suggests that the splitting occurs in the slab beneath the earthquake. This earthquake has been given various depths ranging from 515 to 544 km, the uncertainty possibly resulting from deep-slab anisotropy. If the slab extends to 100 km beneath the event, the observed splitting could be explained by 5 percent anisotropy. This event shows a strong *S*-wave residual pattern with the fast directions along the strike direction. The residuals vary by about 6 s. The waves showing the earliest arrival times spend more time in the slab than the nearly vertical ScS waves. They also travel in different azimuths. If the fast shear-velocity directions are in the plane of the slab, this will add to the effect caused by low temperatures in the slab. Thus, a large azimuthal effect can accumulate along a relatively short travel distance in the slab. If the slab is 5 percent faster due to temperature and 5 percent anisotropic,

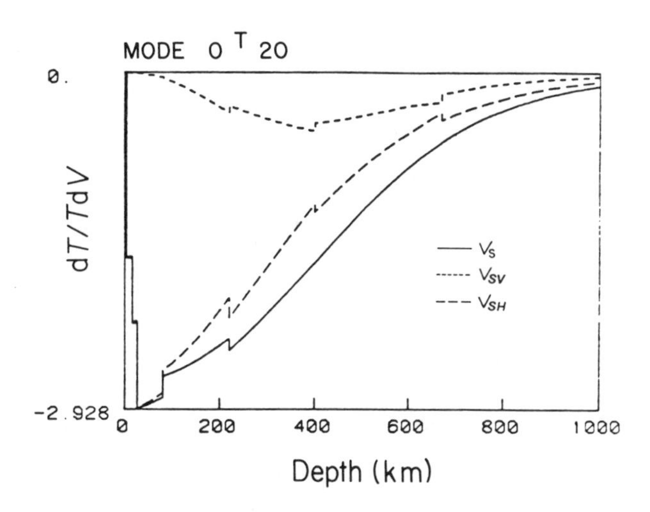

FIGURE 15-15
Same for $_0T_{20}$ (period 360 s).

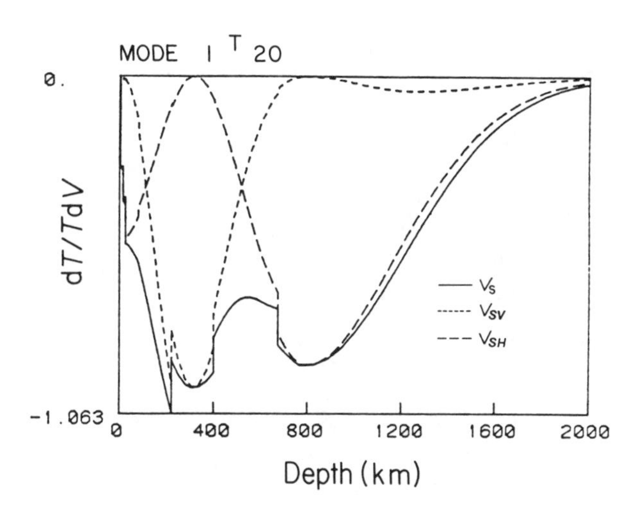

FIGURE 15-16
Same for $_1T_{20}$, first toroidal overtone (period 241 s).

then rays travelling 300 km along the strike direction will arrive 6 s earlier than waves that exit the slab earlier. Actually, the anisotropy implied by the vertical ScS waves just gives the difference in shear-wave velocities in that (arbitrary) direction and is not a measure of the total azimuthal *S*-wave velocity variation, which can be much larger. In any event the presence of near-source anisotropy can cause a residual sphere pattern similar to that caused by a long cold isotropic slab.

In a later study Ando (1984) measured ScS polarization anisotropy for intermediate and deep-focus events around the margin of the Pacific. Shear-wave splitting was commonly observed, and the patterns were relatively consistent for a given earthquake. The arrival-time differences were typically 1.0 ± 0.4 s, corresponding to 4 percent anisotropy for 100-km path lengths. For events in the Tonga-Fiji region, the direction of polarization of the fast ScS wave is roughly parallel to the trend of the deep seismic zone and quite different from the trend of the anisotropy for direct *S*-waves, which sample the mantle above the slab. This again suggests an anisotropic slab and again would complicate the interpretation of the azimuthal pattern of arrival times in terms of deep slab penetration.

The strong anisotropy of the oceanic lithosphere is expected to be maintained after subduction, and this is verified by seismic studies (Hitahara and Ishikawa, 1985). Up to 4 percent anisotropy has been inferred in both the vertical and horizontal planes under Japan down to depths of 200 km. The complex configuration of seismic zones and inferred flow patterns in the mantle wedge above the slab are reflected in a variable magnitude and direction of the anisotropy. This anisotropy is superimposed on the generally fast velocities found in the slab.

The mineral assemblage in the deeper parts of the slab are, of course, different from those responsible for the anisotropy in the plate and in the shallower parts of the slab. Laboratory experiments, however, show that recrystallization of olivine at high pressure results in oriented spinel. The orientation of high-pressure phases is possibly controlled both by the ambient stress field and the orientation of the "seed" low-pressure phases. Results to date, although fragmentary, are consistent with the fast crystallographic axes being in the plane of the slab. The most anisotropic minerals at various depths are olivine (< 400 km), β-spinel (400–500 km) and $MgSiO_3$-ilmenite (> 500 km). The last phase is not expected to be stable at the higher temperatures in normal mantle, being replaced by the more isotropic garnet-like phase majorite. Thus, the deep slab cannot be modeled as simply a colder version of normal mantle. It differs in mineralogy and therefore in intrinsic velocity and anisotropy. Ilmenite is one of the most anisotropic of mantle minerals (Figure 15-13), especially for shear waves. If it behaves in aggregate as do ice and calcite, which are similar structures, then a cold slab can be expected to be extremely anisotropic.

PARTIAL DERIVATIVES

The sensitivity of various period surface waves to parameters of the Earth are conveniently summarized in partial derivative diagrams. Some of these are shown in Figures 15-14 to 15-19.

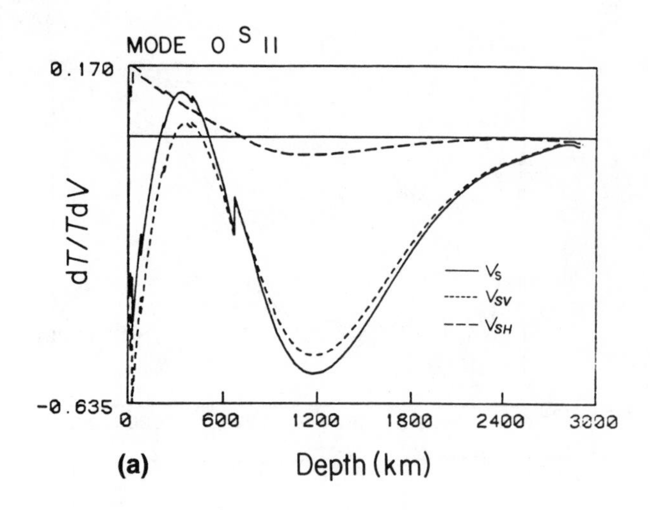

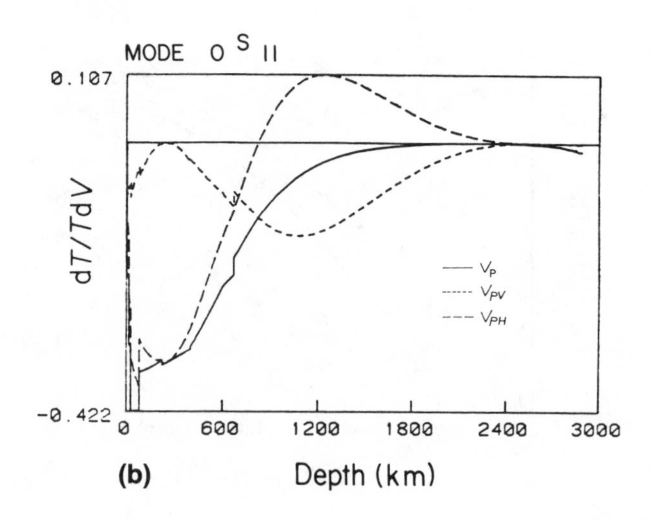

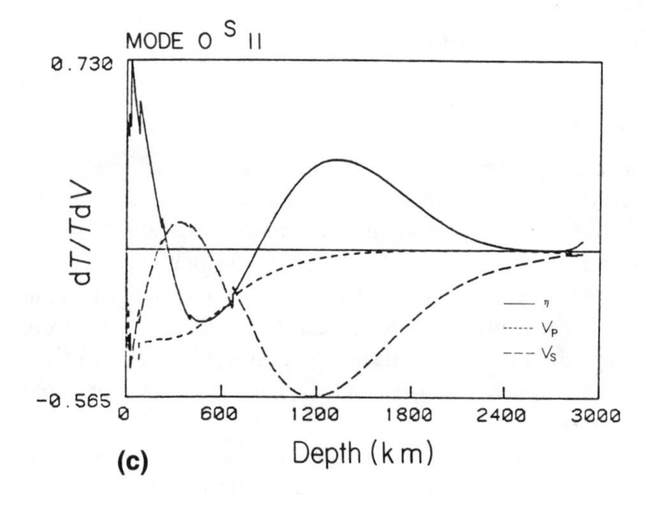

FIGURE 15-17
Partial derivatives for a relative change in period of spheroidal mode ${}_0S_{11}$, period about 537 s. **(a)** shear-wave partial derivatives. **(b)** compressional-wave partial derivatives. Solid lines in A and B are for isotropic perturbations. **(c)** isotropic compressional-wave and shear-wave partials (dashed lines) and the η partial (solid line). Note that Rayleigh waves are sensitive to V_{SV}, V_{PV}, V_{PH} and η. The isotropic partial derivative (solid line) shows that compressional-wave velocities are only important near the top of the structure. At depth the *PV* and *PH* partials are nearly equal and opposite; individually they are significant but in the isotropic case they nearly cancel. Changes of opposite sign of the component velocities cause an additive effect, and the net partial is nearly as significant as the *SV* partial.

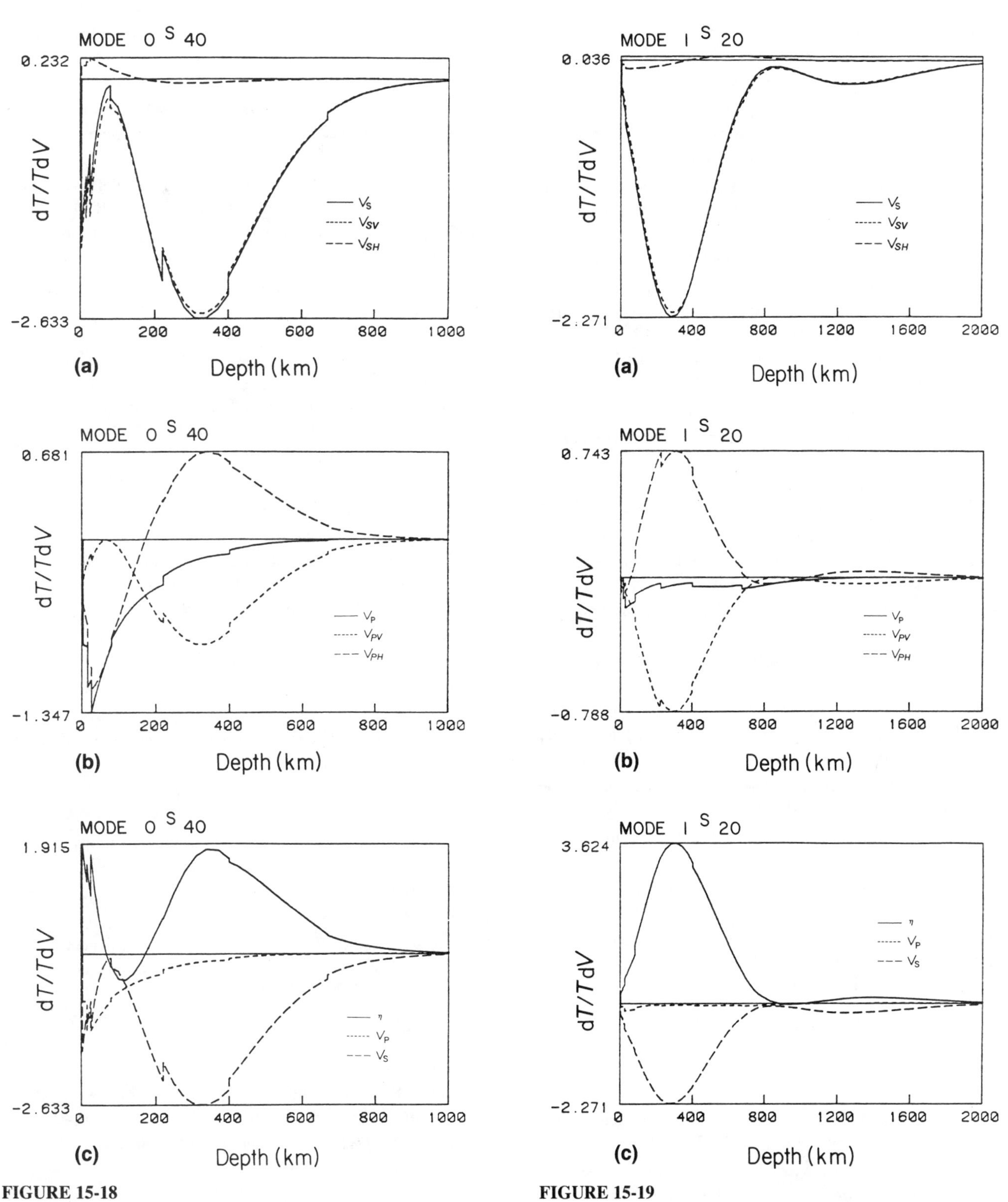

FIGURE 15-18
(a–c) Partial derivatives for a relative change in period of spheroidal mode ${}_0S_{40}$. See Figure 15-17 for explanation.

FIGURE 15-19
(a–c) Partial derivatives for a relative change in period of spheroidal mode ${}_1S_{20}$. See Figure 15-17 for explanation.

General References

Anderson, D. L. (1961) *J. Geophys. Res., 66,* 2953.

Anderson, D. L. (1966) Recent evidence concerning the structure and composition of the Earth's mantle. In *Physics and Chemistry of the Earth, 6,* 1–131, Pergamon, Oxford.

Anderson, D. L. and A. M. Dziewonski (1982) Upper mantle anisotropy: Evidence from free oscillations, *Geophys. J. R. Astron. Soc., 69,* 383–404.

Christensen, N. I. and R. S. Crosson (1968) Seismic anisotropy in the upper mantle, *Tectonophysics, 6,* 93–107.

Christensen, N. I. and S. Lundquist (1982) Pyroxene orientation within the upper mantle, *Bull. Geol. Soc. Am., 93,* 279–288.

Dziewonski, A. M. and D. L. Anderson (1981) Preliminary Reference Earth Model, *Phys. Earth Planet. Inter., 25,* 297–356.

Forsyth, D. W. (1975) The early structural evolution and anisotropy of the oceanic upper mantle, *Geophys. J. R. Astron. Soc., 43,* 103–162.

Mason, P. (1958) *Physical Acoustics and the Properties of Solids,* Van Nostrand, New York.

Mitchell, B. J. and G. Yu (1980) Surface wave dispersion, regionalized velocity models, and anisotropy of the Pacific crust and upper mantle, *Geophys. J. R. Astron. Soc., 63,* 497–514.

Nataf, H.-C., I. Nakanishi and D. L. Anderson (1986) Measurements of mantle wave velocities and inversion for lateral heterogeneities and anisotropy, Part III. Inversion, *J. Geophys. Res., 91,* 7261–7307.

Nicolas, A. and J. P. Poirier (1976) *Crystalline Plasticity and Solid State Flow in Metamorphic Rocks,* Wiley-Interscience, London, 437 pp.

Regan, J. and D. L. Anderson (1984) Anisotropic models of the upper mantle, *Phys. Earth Planet. Int., 35,* 227–263.

Tanimoto, T. and D. L. Anderson (1984) Mapping convection in the mantle, *Geophys. Res. Lett., 11,* 287–290.

Vetter, E. and J. Minster (1981) Pn velocity anisotropy in southern California, *Bull. Seismol. Soc. Am., 71,* 1511–1530.

Yu, G.-K. and B. J. Mitchell (1979) Regional shear velocity models of the Pacific upper mantle from observed Love and Rayleigh wave dispersion, *Geophys. J. R. Astron. Soc., 57,* 311–341.

References

Anderson, D. L. (1961) Elastic wave propagation in layered anisotropic media, *J. Geophys. Res., 66,* 2953–2963.

Anderson, D. L. (1962) Love wave dispersion in heterogeneous anisotropic media, *Geophysics, 27,* 445–454.

Anderson, D. L. (1964) Universal dispersion tables, 1. Love waves across oceans and continents on a spherical Earth, *Seismol. Soc. America Bull., 54,* 681–726.

Anderson, D. L. (1966) Recent evidence concerning the structure and composition of the Earth's mantle. In *Physics and Chemistry of the Earth, 6,* 1–131, Pergamon, Oxford.

Anderson, D. L. (1967) Latest information from seismic observations. In *The Earth's Mantle* (T. F. Gaskell, ed.), 355–420, Academic Press, New York.

Anderson, D. L. and A. M. Dziewonski (1982) Upper mantle anisotropy: Evidence from free oscillations, *Geophys. J. R. Astron. Soc., 69,* 383–404.

Anderson, D. L. and R. Hart (1976) An Earth model based on free oscillations and body waves, *J. Geophys. Res., 81,* 1461–1475.

Anderson, D. L., H. Kanamori, R. Hart and H.-P. Liu (1977) The Earth as a seismic absorption band, *Science, 196,* 1104.

Anderson, D. L., B. J. Minster and D. Cole (1974) The effect of oriented cracks on seismic velocities, *J. Geophys. Res., 79,* 4011–4015.

Anderson, D. L. and J. Whitcomb (1973) The dilatancy-diffusion model of earthquake prediction. In *Proceedings of the Conference on Tectonic Problems of the San Andreas Fault System* (A. M. Nur, ed.), 417–426, Stanford U. Publ. Geol. Sci., 13.

Ando, M., Y. Ishikawa and F. Yamazaki (1983) Shear wave polarization anisotropy in the upper mantle beneath Honshu, Japan, *J. Geophys. Res., 10,* 5850–5864.

Ando, M. J. (1984) *Phys. Earth, 32,* 179.

Avé Lallemant, H. G. and N. L. Carter (1970) Syntectonic recrystallization of olivine and modes of flow in the upper mantle, *Geol. Soc. Am. Bull., 81,* 2203–2220.

Babuska, V. (1981) *J. Geophys., 50,* 1–6.

Backus, G. E. (1962) Long-wave elastic anisotropy produced by horizontal layering, *J. Geophys. Res., 67,* 4427–4440.

Backus, G. E. (1965) Possible forms of seismic anisotropy of the upper-most mantle under oceans, *J. Geophys. Res., 70,* 3429–3439.

Bamford, D. (1977) Pn velocity anisotropy in a continental upper mantle, *Geophys. J. R. Astron. Soc., 49,* 29–48.

Berryman, J. G. (1979) Long-wave elastic anisotropy in transversely isotropic media, *Geophysics, 44,* 896–917.

Carter, N. L. (1976) Steady state flow of rocks, *Rev. Geophys. Space Phys., 14,* 301–359.

Chang, Z. P. and G. R. Barsch (1969) Pressure dependence of the elastic constants of single-crystalline magnesium oxide, *J. Geophys. Res., 74,* 3291–3294.

Christensen, N. I. and M. H. Salisbury (1979) Seismic anisotropy in the upper mantle: Evidence from the Bay of Islands ophiolite complex, *J. Geophys. Res., 84,* B9, 4601–4610.

Crampin, S. and D. Bamford (1977) Inversion of P-wave velocity anisotropy, *Geophys. J. R. Astron. Soc., 49,* 123–132.

Fuchs, K. (1977) Seismic anisotropy of the subcrustal lithosphere as evidence for dynamic processes in the upper mantle, *Geophys. J. R. Astron. Soc., 49,* 167–179.

Fuchs, K. (1983) Recently formed elastic anisotropic and petrological models for the continental subcrustal lithosphere in southern Germany, *Phys. Earth Planet. Inter., 31,* 93–118.

Fukao, Y. (1984) ScS evidence for anisotropy in the earth's mantle, *Nature, 309,* 695–698.

Gieske, J. H. and G. R. Barsch (1968) *Phys. Stat. Sol. 29,* 121.

Gilbert, F. and A. M. Dziewonski (1975) *Philos. Trans. R. Soc. London, Ser. A, 278,* 187.

Hager, B., and R. O'Connell (1979) Kinematic models of large-scale flow in the Earth's mantle, *J. Geophys. Res., 84,* 1031–1048.

Harkrider, D. G. and D. L. Anderson (1962) Computation of surface wave dispersion for multilayered anisotropic media, *Seismol. Soc. America Bull., 52,* 321–332.

Hart, R., D. L. Anderson and H. Kanamori (1977) The effect of attenuation on gross Earth models, *J. Geophys. Res., 82,* 1647–1654.

Helbig, K. (1984) Transverse anisotropy in exploration seismics, *Geophys. J. R. Astron. Soc., 76,* 79–88.

Hess, H. (1964) Seismic anisotropy of the uppermost mantle under oceans, *Nature, 203,* 629–631.

Hirahara, K. and Y. Ishikawa (1984) Travel time inversion for three-dimensional P-wave velocity anisotropy, *J. Earth Phys., 32,* 197–218.

Jeffreys, H. (1958) *Mon. Not. R. Astr. Soc., 118,* 14–17.

Jeffreys, H. (1965) *Nature, 208,* 675.

Jeffreys, H. (1968). *Mon. Not. R. Astr. Soc., 141,* 255.

Jordan, T. and D. L. Anderson (1974) Earth structure from free oscillations and travel times, *Geophys. J. R. Astron. Soc., 36,* 411–459.

Kanamori, H. and D. L. Anderson (1977) Importance of physical dispersion in surface-wave and free-oscillation problems, a review, *Rev. Geophys. Space Phys., 15,* 105–112.

Kawasaki, I. and F. Konno (1984) Azimuthal anisotropy of surface waves and the possible type of seismic anisotropy due to preferred orientation of olivine in the uppermost mantle beneath the Pacific Ocean, *J. Phys. Earth, 32,* 229–244.

Kumazawa, M. (1969) The elastic constants of single-crystal orthopyroxene, *J. Geophys. Res., 74,* 5973–5980.

Levien et al. (1979) *Phys. Chem. Minerals, 4,* 105–113.

Liebfried, G. (1955) *Encyclopedia of Physics* VII, Pt. 2, 104, Springer-Verlag, Berlin.

Liu, H.-P., D. L. Anderson and H. Kanamori (1976) Velocity dispersion due to anelasticity; implications for seismology and mantle composition, *Geophys. J. R. Astr. Soc., 47,* 41–58.

Liu, H.-P., R. N. Schock and D. L. Anderson (1975) Temperature dependence of single-crystal spinel ($MgAlO_4$) elastic constants from 293 to 423°K measured by light-sound scattering in the Raman-Nath region, *Geophys. J. Roy. Astron. Soc., 42,* 217–250.

Montagner, J.-P. and H.-C. Nataf (1986) A simple method for inverting the azimuthal anisotropy of surface waves, *J. Geophys. Res., 91,* 511–520.

Morris, E. M., R. W. Raitt and G. G. Shor (1969) *J. Geophys. Res., 74,* 4300–4316.

Nataf, H.-C., I. Nakanishi and D. L. Anderson (1986) Measurements of mantle wave velocities and inversion for lateral heterogeneities and anisotropy; Part III, Inversion, *J. Geophys. Res., 91,* 7261–7307.

Nicolas, A. and N. I. Christensen (1987) Formation of anisotropy in upper mantle peridotite, *Rev. Geophys.*

Nicolas, A., F. Boudier, and A. M. Boullier (1973) Mechanisms of flow in naturally and experimentally deformed peridotites, *Am. J. Sci., 273,* 853–876.

Nicolas, A. and J. P. Poirier (1976) *Crystalline Plasticity and Solid State Flow in Metamorphic Rocks,* Wiley, London, 437 pp.

Peselnick, L., A. Nicolas and P. R. Stevenson (1974) Velocity anisotropy in a mantle, peridotite from Ivrea zone: Application to upper mantle anisotropy, *J. Geophys. Res., 79,* 1175–1182.

Postma, G. W. (1955) *Geophys., 20,* 780.

Raitt, R. W., G. G. Shor, T. J. G. Francis and H. K. Kirk (1971) Mantle anisotropy in the Pacific Ocean, *Tectonophysics, 12,* 173–186.

Raitt, R. W., G. G. Shor, T. J. G. Francis and G. B. Morris (1969) Anisotropy of the Pacific upper mantle, *J. Geophys. Res., 74,* 3095–3109.

Randall, M. J. (1976) Attenuative dispersion and frequency shifts of the Earth's free oscillations, *Phys. Earth Planet. Inter., 12,* P1–P4.

Sawamoto, H., D. J. Weidner, S. Sasaki and M. Kumazawa (1984) Single-crystal elastic properties of the modified spinel phase of magnesium orthosilicate, *Science, 224,* 749–751.

Shearer, P. M. and J. Orcutt (1986) Compressional and shear wave anisotropy in the oceanic lithosphere, *Geophys. J. Roy. Astron. Soc., 87,* 967–1003.

Smith, M. L. and F. A. Dahlen (1973) The azimuthal dependence of Love and Rayleigh wave propagation in a slightly anisotropic medium, *J. Geophys. Res., 78,* 3321–3333.

Suzuki, I. and O. L. Anderson (1983) Elasticity and thermal expansion of a natural garnet up to 1,000 K, *J. Phys. Earth, 31,* 125–138.

Takeuchi, H. and M. Saito (1972) Seismic surface waves. In *Seismology: Surface Waves and Earth Oscillations, Methods in Computational Physics,* 11 (B. A. Bolt, ed.), 217–295, Academic Press, New York.

Tanimoto, T. and D. L. Anderson (1984) Mapping convection in the mantle, *Geophys. Res. Lett., 11,* 287–290.

Thomsen, L. (1986) Weak elastic anisotropy, *Geophys., 51,* 1954.

Thomsen, L. A. (1988) Elastic anisotropy due to aligned cracks, *Geophysics.*

Toksöz, M. N. and D. L. Anderson (1963) Generalized two-dimensional model seismology with application to anisotropic earth models, *Jour. Geophys. Res., 68,* 1121–1130.

Weidner, D. J., H. Sawamoto, S. Sasaki and M. Kumazawa (1984) *J. Geophys. Res., 87,* 4740–4746.

Whitcomb, J. H., J. D. Garmany and D. L. Anderson (1973) Earthquake prediction; variation of seismic velocities before the San Francisco earthquake, *Science, 180,* 632–635.

16

Phase Changes and Mantle Mineralogy

It is my opinion that the Earth is very noble and admirable . . . and if it had contained an immense globe of crystal, wherein nothing had ever changed, I should have esteemed it a wretched lump of no benefit to the Universe.

—GALILEO

The densities and seismic velocities of rocks are relatively weak functions of temperature, pressure and composition unless these are accompanied by a drastic change in mineralogy. The physical properties of a rock depend on the proportions and compositions of the various phases or minerals—the mineralogy. These, in turn, depend on temperature, pressure and composition. In general, one cannot assume that the mineralogy is constant as one varies temperature and pressure. Lateral and radial variations of physical properties in the Earth are primarily due to changes in mineralogy. To interpret seismic velocities and density variations requires information about both the stable phase assemblages and the physical properties of minerals.

SPHERICAL IONS AND CRYSTAL STRUCTURE

It is often useful to think of a crystal as a packing of different size spheres, the small spheres occupying interstices in a framework of larger ones. In ionic crystals each ion can be treated as a ball with certain radius and charge (Table 16-1). The arrangement of these balls, the crystal structure, follows certain simple rules. The crystal must contain ions in such a ratio so that the crystal is electrically neutral. Maximum stability is associated with regular arrangements that place as many cations around anions as possible, and vice versa, without putting ions with similar charge closer together than their radii allow while bringing cations and anions as close together as possible. In other words, we pack the balls together as closely as possible considering their size and charge. Many crystals are based on cubic close packing or hexagonal close packing of the larger ions.

It is a simple matter of geometry to calculate the ratio of the radii of two types of spheres, A and B, that permits a certain number of B to fit around A, and vice versa. If A is very small compared to B, taken as the anion, only two B ions can be arranged to touch A, and the coordination number is 2. When A reaches a critical size, three touching B ions can surround it in a trigonal-planar group, the only regular threefold coordinated structure. The limiting value of the radius ratio for this packing is

$$R_A/R_B = 0.155$$

As A grows still further we reach the point where A can be surrounded by four B as in the SiO_4^{4-} tetrahedron, the basis for many silicates. The R_A/R_B range for this arrangement is 0.225–0.414. For R_A/R_B between 0.414 and 0.732, A can be surrounded by four B in a square planar arrangement or six B in an octahedron arrangement. Rocksalt structures such as NaCl and MgO exhibit octahedral coordination, a common substructural element in silicates. Low-pressure minerals commonly have silicon in tetrahedral coordination and the metals (such as Mg, Fe, Ca, Al) in octahedral coordination. High-pressure phases, such as SiO_2-stishovite, $MgSiO_3$-ilmenite and perovskite, have silicon in octahedral coordination. The mineral majorite,

$${}^{VI}Mg_3\ {}^{VI}[MgSi]\ {}^{IV}Si_3O_{12}$$

has the silicons split between octahedral and tetrahedral sites.

When R_A is almost the size of R_B ($R_A/R_B = 0.732$–1.0), the cation (A) can be surrounded by eight anions (B) as in a square bipyramid. The anions are at the corners of a

TABLE 16-1
Ionic Radii for Major Mineral-Forming Elements (Å)

Ion	Coordination Number	Ionic Radius	Ion	Coordination Number	Ionic Radius
Al^{3+}	IV	0.39	Fe^{3+}	IV	0.49(HS)*
	V	0.48		VI	0.55(LS)
	VI	0.53		VI	0.65(HS)
Ca^{2+}	VI	1.00	Mg^{2+}	IV	0.49
	VII	1.07		VI	0.72
	VIII	1.12		VIII	0.89
	IX	1.18	Fe^{2+}	IV	0.63(HS)
	X	1.28		VI	0.61(LS)
	XII	1.35		VI	0.77(HS)
Si^{4+}	IV	0.26	Ti^{4+}	V	0.53
	VI	0.40		VI	0.61
Na^{+}	VI	1.02	K^{+}	VI	1.38
	VIII	1.16		VIII	1.51
O^{2-}	II	1.35	F^{-}	II	1.29
	III	1.36		III	1.30
	IV	1.38		IV	1.31
	VI	1.40		VI	1.33
	VIII	1.42	Cl^{-}	VI	1.81

*HS, high spin; LS, low spin.

cube, and the cation is in the center of the cube. CsCl structures fall in this category. The garnet structure has eight M^{2+} ions about each oxygen.

If $R_A = R_B$, twelve B ions can surround each A ion in close packing. The ions can be arranged in two ways, hexagonal close packing or cubic close packing, but the coordination number is 12 in either case. The ideal perovskite structure exhibits twelvefold coordination of the M^{2+} ions around the oxygen ions.

When A and B have the same charge, the anion and cation coordinations are the same because of the requirement of charge neutrality. For AB_2 compounds neutrality necessitates that the coordination numbers not be the same.

Higher coordination makes for denser packing, and when ionic crystals are compressed, structures with greater coordination numbers tend to form. Common crustal and upper-mantle minerals, however, have such an open packing structure that rearrangements (phase changes) not involving coordination changes usually occur, leading to a more efficient packing of ions, before the coordination-changing transformations can take place. High temperature tends to decrease the coordination.

In the context of rigid spherical ions, the increase of coordination with pressure means that R_A/R_B must increase, necessitating a decrease of R_B, say the oxygen ion, or an increase in R_A. The A–B distance increases as the coordination increases, since more A ions must fit around the B ion, and this is usually assigned to an increase in the cationic radius. The increase in density is due to the decrease in B–B distances from closer packing of the oxygen ions. The increase of A–B and the decrease in B–B means that the attractive potential is decreased and the repulsive, or overlap, potential is increased. This leads to an increase in the bulk modulus or incompressibility.

To a first approximation, then, ionic crystal structures, such as oxides and silicates, consist of relatively large ions, usually the oxygens, in a closest-pack arrangement with the smaller ions filling some of the interstices. The large ions arrange themselves so that the cations do not "rattle" in the interstices. The "nonrattle" requirement of tangency between ions is another way of saying that ions pack so as to minimize the potential energy of the crystal.

In addition to geometric rules of sphere packing and overall charge neutrality, there are additional rules governing ionic crystals that have been codified by Linus Pauling. The considerations discussed so far are equivalent to Pauling's first rule. The second rule states that an ionic structure will be stable to the extent that the sum of the strengths of the electrostatic bonds that reach an anion from adjacent cations equals the charge on the anion. This is the *electrostatic valence principle* or condition of *local* charge neutrality. In general, in a stable ionic crystal the charge on any cation is neutralized by adjacent anions. Cations with large charges must therefore have high coordination numbers and tend to occur in the large interstices or holes in the structure. On the other hand highly charged ions are usual small in radius and thus, on the basis of radius ratio, seek to occupy the small holes. Which tendency wins depends on other rules.

Pauling's third rule states that the sharing of edges and particularly of faces by two anion polyhedra decreases the stability of the crystal structure. By this rule, highly charged cations prefer to maintain as large a separation as possible and to have anions intervening between them so as to screen them from each other. This deceases a crystal's potential energy by minimizing the replusive forces existing between nearby cations. Multivalent cations tend to avoid the face-sharing anion cubes of the CsCl structure and prefer the edge-sharing NaCl structure.

The fourth rule, an extension of the third, states that in crystal structures containing different cations, those of high valency and small coordination number tend not to share polyedron elements with each other.

The fifth rule states that the number of essentially different kinds of constituents in a crystal tends to be small; that is, the number of types of interstitial sites in a periodically regular packing of anions tends to be small.

These considerations can be used to understand the stability of crystal lattices. For example, magnesiowüstite, (Mg,Fe)O, is a 6-coordinated phase, making this a low-pressure structure. Packing is relatively inefficient, having a very large volume per oxygen ion relative to other mantle minerals. However (Mg,Fe)O is stable to extremely high pressure, probably through most of the lower mantle. The radius ratio of MgO is 0.51, putting it well within the range (0.41–0.73) of expected octahedral coordination. The CsCl structure, with 8-coordination, is displayed by many alkali halides and is the high-pressure form of others that normally display the rocksalt structure. The packing of eight cations around an anion occurs for R_A/R_B greater than 0.732. CsCl itself, for example, has a ratio of 0.93, although at high temperature it adopts the NaCl structure. Likewise RbCl, radius ratio of 0.81, crystallizes in the NaCl structure at low pressure and CsCl at high pressure, being close to the boundary of the radius ratio for these structures. If MgO were to adopt the CsCl structure, the O^{2-} ions would be in contact at the cube faces and the Mg^{2+} ions would be unshielded across cube faces. Each cube would share a face with six others. This makes the CsCl structure unattractive to multicharged ions, such as Mg^{2+}, and pressure apparently is unable to force MgO to bring its ions into closer proximity to achieve a closer packing. The radius ratio $^{VIII}Mg/O$ is 0.63, still outside the range for a CsCl structure. In the NiAs structure, another alternative AB structure, the octahedra share faces, whereas in the NaCl structure they share only edges. Consequently, the NiAs structure is not favored by ionic crystals. Thus, MgO has little option but to remain in the rocksalt structure, in spite of the relatively open structure.

Garnet is another mantle mineral that is stable over a large pressure range. The garnet structure consists of independent SiO_4 and AlO_6 polyhedra, which share corners to form a framework within which each M^{2+} ion is surrounded by an irregular polyhedron (a distorted cube) of eight oxygen atoms. In $Mg_3Al_2Si_3O_{12}$ (pyrope) two edges of the silicon tetrahedron and six edges of the aluminum octahedron are shared with the magnesium cube, leaving four unshared edges in the tetrahedron, six in the octahedron and six in the cube. The high percentage of shared edges leads to a tightly packed arrangement, a high density and an apparently stable lattice. In deference to Pauling, most edges are unshared. The packing of oxygen atoms is so efficient that the volume per oxygen atom (15.7 Å^3) is less than in most other high-pressure silicates except ilmenite (14.6), perovskite (13.5) and stishovite (11.6). The M^{2+}-coordination in garnet is 8, so garnet can be considered a high-pressure phase. Because of its relatively low density, it probably remains in the upper mantle. This is an argument against eclogite subduction into the lower mantle.

The polyhedra in garnet are considerably distorted, giving a wide range of Mg–O (in pyrope) and O–O distances and requiring a large unit cell (eight units of $M_3^{2+}M_2^{3+}(SiO_4)^3$ in a cubic unit cell). This distortion reflects Pauling's admonishment against edge sharing involving highly charged ions. Local charge balance is a factor in the structure of pyrope, $Mg_3Al_2Si_3O_{12}$, and other garnets. Each O^{2-} ion bonds to one Si^{4+}, an Al^{3+} and two Mg^{2+} ions. The total of the electrostatic bonds leading to an O^{2-} equals +2. If Mg^{2+} were to occupy the octahedral sites, such local charge balance would be impossible. The elastic properties of silicate garnets are relatively insensitive to the nature of the $^{VIII}M^{2+}$ ion.

The garnet structure is particularly important since [MgSi] and [FeSi] can substitute for $[Al]_2$ at high pressure, giving a majorite-garnet solution that may be a dominant phase in the transition region. The large, high-coordination site in garnet allows the garnet structure to accommodate a wide variety of cations including minor elements and elements that are usually termed incompatible, particularly the heavy rare-earth elements (HREE). The crystallization of garnet from a magma can remove these elements, giving a diagnostic HREE depletion signature to such magmas.

Atoms are particularly close packed in body-centered (bcc) and face-centered cubic (fcc) structures. In a body-centered cubic crystal each atom has eight neighbors, and in a face-centered cubic crystal each has twelve neighbors. The atoms can be more closely packed in the face-centered structure. The distance between neighboring atoms in the body-centered case is $a\sqrt{3}/2$, and in the face-centered case it is $a/\sqrt{2}$ where the volume of the unit cell is $a^3/2$ and $a^3/4$, respectively. Assuming spherical ions with radii equal to half these interatomic distances, the volumes of the spheres are $\sqrt{3}\pi a^3/16$ and $\pi a^3/(12\sqrt{2})$ for bcc and fcc. The fraction of the unit cell occupied by spherical ions is 0.68 (bcc) and 0.742 (fcc). The fcc structure has the spheres packed as closely as possible. A rigid sphere can be surrounded by twelve equally spaced neighbors since in this case each sphere touches all of its neighbors.

There are two ways in which one plane of close-packed spheres can fit snugly on top of a similar plane. One gives the fcc structure, and the other gives hexagonal close pack (hcp) structure. If a is the distance between atoms arranged in a hexagon on a plane and c is the distance to the next plane above or below, the distance to the nearest neighbor out of plane is $(a^2/3 + c^2/4)^{1/2}$. For closest packing this equals a, giving $c/a = 1.633$. The volume of the unit cell is $a^2c\sqrt{3}/2$ so that the volume per atom is $a^2c\sqrt{3}/4$. Departures from the ratio $c/a = 1.633$ represent departures from the closest pack.

In simple cubic packing (scp) an ion sits at each corner of a cube. This is an open structure, and unusual properties might be anticipated compared to close-packed structures. The fluorite structures (CaF_2) and CsCl-structures (CsCl, CsI, TlCl) are based on scp of the anions. Cations are at the center of every cube for CsCl and every other cube for CaF_2, the cations being surrounded in each case by eight anions. When the cations approach the size of the anions, the structure resembles bcc in its overall packing for the CsCl structures.

In the rocksalt or NaCl structure the cations by themselves, or the anions, lie on a face-centered cubic lattice. In NaCl, for example, the sodium ions lie in an fcc lattice and chlorine ions are half-way between the Na ions at the centers of the cube edges and at the center of the cube. The alkali halides and oxides of magnesium, calcium, strontium and barium have the NaCl structure. The CsCl structure is not particularly important in mantle mineralogy.

In the rutile (TiO_2) structure each cation has six anion neighbors, two in the plane above, two in the same plane, and two in the plane below. Each anion has three cation neighbors. Stishovite is a high-pressure form of SiO_2 having the rutile structure. Stishovite also forms by the disproportionation of $2MgSiO_3$ to Mg_2SiO_4(β or γ) plus SiO_2(st).

As we go to more complex compounds, we have a large variety of possible crystal structures, but many of the more important ones are based on relatively simple packing of the oxygen ions with the generally smaller cations fitting into the interstices. Some cubic minerals mimic the structures of perovskites ($CaTiO_3$), spinels (Al_2MgO_4) or garnets (such as $Mg_3Al_2Si_3O_{12}$-pyrope).

The perovskite structure, $M^{2+}N^{3+}O_3$, has all the atoms arranged in a cubic lattice with the M^{2+} at the corners, the N^{3+} at the center and the O's at the face centers. Generally, the M^{2+} ions (say Mg^{2+}) and the oxygen ions together constitute a cubic close pack structure. All the interatomic distances are determined in terms of one parameter, the side a of the unit cell. The M^{2+}–O, N^{3+}–O and M^{2+}–M^{3+} distances are approximately $a/\sqrt{2}$, $a/2$ and $a\sqrt{3}/2$, respectively. These are also approximately the sum of the appropriate ionic radii. In $MgSiO_3$-perovskite there is a considerable range in the individual distances. Although each magnesium is surrounded by twelve oxygens, the Mg–O distances are not all the same. There is therefore a tendency of the structure to distort and not be exactly cubic. Some perovskites are ferroelectric: The displacements of the ions from the positions that they would have in the cubic structure results in a permanent electric dipole for the crystal. If the M^{2+} ion were the same size as the oxygen ions and precisely fit its twelvefold site, then the line joining the centers of the oxygens would equal twice the sum of the ionic radii or 1.414 times the cube's edge. The cube's edge in turn equals twice the sum of the oxygen and M^{4+} radii. The ideal relationship between radii for ions in the perovskite structure is $R(O) + (M^{2+}) = 1.414\,(R(O) + R(M^{4+})t)$ with $t = 1$ (the tolerance factor). In perovskites t generally lies between 0.8 and 1.0. "High-temperature" superconductors have the perovskite structure, with conducting layers alternating with resistive layers of atoms.

Spinel, Al_2MgO_4, is an example of a large class of important compounds, including ferrites, that have important magnetic properties. There are eight magnesium ions per cube of side a. They occupy the centers of four out of the eight small cubes of side $a/2$ into which the larger cube can be divided. Each of the other four small cubes contains four aluminum ions. There are then 16 aluminum ions in the cube of side a. Each aluminum is surrounded by six oxygens, and each magnesium is attached to only four oxygens, an unusual coordination for magnesium. Not all spinels have these site assignments for atoms of different valencies. In some, half of the trivalent atoms are located in tetrahedral sites, and the other half of the trivalent atoms and the divalent atoms are distributed in the octahedral sites. These are called inverse spinels. Examples of inverse spinels include the ferrites Fe_2MgO_4 and Fe_3O_4. The spinel structure is essentially a cubic close pack (fcc) of oxygen ions with metal cations occupying one-eighth of the tetrahedral sites and one-half of the octahedral sites. γ-Mg_2SiO_4 can be viewed approximately as the substitution of Mg^{2+} and Si^{4+} for Al^{3+} and Mg^{2+}.

Garnets are also cubic minerals, and some have important magnetic and optical properties. For silicate garnets (M^{2+} = Mg, Ca, Fe . . .) the unit cube contains 24 M^{2+} ions, 16 aluminums, 24 silicons and 96 oxygens. The coordinations are (for M^{2+} = Mg)

$$^{VIII}Mg_3\,^{VI}Al_2\,^{IV}Si_3O_{12}$$

Many substitutions are possible for all the cations, and garnets in the crust and mantle are important repositories for trace elements, particularly those having ionic radii similar to magnesium, calcium and aluminum. The tetrahedral silicate groups, four oxygens tetrahedrally arranged around a central silicon atom, are independent of each other. Garnet is therefore called an island silicate. The elastic properties of garnets are almost independent of the nature of the M^{2+} ion, in contrast to other silicates.

Hexagonal and trigonal crystals are closely related. Calcite ($CaCO_3$) and corundum (Al_2O_3) are trigonal crystals, and one high-pressure form of $MgSiO_3$ is similar to ilmenite ($FeTiO_3$), another trigonal crystal. In the calcite structure each M^{2+} atom is bonded to six oxygens, and each oxygen is bonded to one M^{4+} and two M^{2+} atoms. We may think of the calcite structure as a distorted NaCl structure. The oblate CO_3 group replaces the spherical chloride ion. Calcite is the main constituent of the metamorphic rock marble, a rock with strongly anisotropic properties because of the alignment of the individual calcite crystals. The strong alignment of calcite and of ice, a hexagonal crystal, in natural masses suggests that $MgSiO_3$-ilmenite will also be strongly aligned in the mantle.

In corundum (Al_2O_3) the oxygens occur on equilateral triangles, similar to calcite, but there is no atom at the center of the triangle, in the same plane. The aluminum atoms are not all in a plane. Each aluminum is surrounded by six oxygens. Cr_2O_3 mixes in all proportions with Al_2O_3, Cr^{3+} and Al^{3+} having similar radii, and the mixture yields the gem ruby with its characteristic red color. It has important optical properties. Al_2O_3 may be important in the lower mantle. It is found in some kimberlites.

In ilmenite ($FeTiO_3$) half the aluminums are replaced by iron, half by titanium. This breaks the symmetry, and ilmenites are expected to be more anisotropic than corundums. The oxygens in corundum and ilmenite structures are in approximate hcp.

Some of the more common low-pressure silicates are also based on simple packing of the oxygen ions. In olivine, for example, the oxygens are in approximate hcp. The transformation to the spinel form results in a slight decrease in the Mg–O distance, a slight increase in the Si–O distance and a decrease in the larger of the O–O near-neighbor distances, resulting in an 8 percent decrease in the volume per oxygen and a change in oxygen packing from approximately hcp to approximately bcc. The coordinations of the ions remain the same. Note that

$$^{VI}Mg_2^{2+} \; ^{IV}Si^{4+} \; O_4\text{-spinel}$$

is not analogous to true spinel

$$^{VI}Al_2^{3+} \; ^{IV}Mg^{2+} \; O_4$$

either in the coordination of magnesium or the valency of the ions in the tetrahedral and octahedral sites. Aluminate spinels have some anomalous elastic properties, presumably related to the IV coordination of the M^{2+} ions, which cannot be assumed to carry over to the silicate spinels. In fact, because of the very small size of the ^{IV}Mg ion it must be treated as a different element than ^{VI}Mg.

In this chapter we use mineral names such as spinel, ilmenite and perovskite to refer to structural analogs in silicates rather than to the minerals themselves. This has become conventional in high-pressure petrology and mineral physics, but it can be confusing to those trained in conventional mineralogy with no exposure to the high-pressure world.

Interatomic Distances in Dense Silicates

The elastic properties of minerals depend on interatomic forces and hence on bond type, bond length and packing. As minerals undergo phase changes, the ions are rearranged, increasing the length of some bonds and decreasing others. The interatomic distances and the average volume per oxygen atom are given in Table 16-2 for many of the crystal structures that occur in the mantle. For a given coordination the cation-anion distances are relatively constant. This, in fact, is the basis for ionic radius estimates. Cation-anion distances increase with coordination, as required by packing considerations. It is clear that the increases of density and bulk modulus, K_S, are controlled by the increase in packing efficiency of the oxygen ions.

TABLE 16-2
Average Interatomic Distances (Angstroms), Volume Per Oxygen Ion (Å^3) and Bulk Modulus (GPa) in Mantle Minerals

Mineral or Structure	Mg–O	Si–O	O–O	V/O^{2-}	K_S
MgO	2.11	—	2.98	18.7	163
Olivine	2.11	1.63	2.66–2.99	18.1	129
Pyroxene	2.09	1.63		17.3	108
β-spinel	2.08	1.65	2.69–2.95	16.8	174
γ-spinel	2.06	1.67	2.73–2.91	16.5	184
Ilmenite	2.08	1.80	2.54–2.89	14.6	212
Perovskite	2.06–2.20**	1.75–1.82	2.53–2.71	13.5	262
Stishovite	—	1.78	2.16	11.6	316
Garnet	2.27*	1.64	2.5–2.78	15.7	174
Al_2O_3	†	—	2.52	17.0	254

*Al–O distance in garnet is 1.89.
**4 shortest distances.
†Al–O distance is 1.91.

Crystals and Magmas

Up to this point I have treated crystals as isolated entities. Most mantle crystals are formed from a melt, and it is instructive to consider them from this point of view. The distribution of ions and the nature of the crystals formed depend on properties of both the melt and the solids. The first silicate crystals to form from a cooling magma usually have the Si^{4+} ions as widely separated as possible, in accordance with Pauling's third rule. Oxygen ions touch at most one Si^{4+} ion. No two tetrahedra drawn about each Si^{4+} share a corner, that is, an oxygen. Crystal structures in which each $[SiO_4]^{4-}$ tetrahedron is isolated from all others are called *island silicates* or *nesosilicates* (from the Greek word for "island") or *orthosilicates*. Olivines and garnets are such structures. Bloss (1971) gives a good summary.

As island silicates crystallize from the melt, the remaining liquid is enriched in Si^{4+}, permitting silicates with higher Si^{4+} to O^{2-} ratios to form. The Si^{4+} ions cannot be so widely spaced, and some of the oxygens touch two Si^{4+} ions. β-spinel has linked SiO_4 tetrahedra and is therefore a double-island or *sorosilicate*. With progressive crystallization the Si/O ratio increases further, and Si^{4+} occurs in so many interstices in the crystals that do form that each tetrahedron shares a corner with two others to form $[SiO_3]^{2-}$ chains as in the pyroxenes, which are single-chain or *metasilicates*. With increased cooling, *double-chain silicates* form with $[Si_4O_{11}]^{6-}$ units and a large number of shared corners. Amphiboles are double-chain silicates. At even lower temperatures the Si^{4+} to O^{2-} ratio is higher still, and the $(OH)^-$ to O^{2-} ratio is also high. *Layer silicates* such as micas and talc form under these conditions.

In some structures, the *framework silicates,* the proportion of Si^{4+} to O^{2-} is so high that all tetrahedra share all their corners. The various crystalline forms of SiO_2—quartz, tridymite, cristobalite—are framework structures that crystallize at low temperature. The feldspars are also framework silicates, but Al^{3+} tetrahedra as well as Si^{4+} tetrahedra are involved, so they can also crystallize early.

Most of the structures discussed above have relatively open structures and are unstable at moderate pressures. They also tend to have low seismic velocities and to be anisotropic. The increased packing efficiency of high-pressure mineral phases makes their description in terms of arrangements of tetrahedra and octahedra less useful than a description in terms of dominant sublattices of close-packed ions with the remaining ions occupying available interstices. Nevertheless, it is likely that most minerals in the mantle have crystallized from magmas at low pressure and have subsequently converted to high-pressure phases. Therefore, it is of interest to know the conditions that initially determine the relative proportions of the various constituents of minerals at low pressure. The trace and minor elements follow similar rules. They tend to replace major ions of similar size and valency, or occupy interstices of appropriate size or, for the very incompatible elements, remain in the melt to the end, coating the major crystals with exotic phases.

MINERALS AND PHASES OF THE MANTLE

As far as physical properties and major elements are concerned, the most important upper mantle minerals are olivine, orthopyroxene, clinopyroxene and aluminum-rich phases such as plagioclase, spinel and garnet. Olivine and orthopyroxene are the most refractory phases and tend to occur together, with only minor amounts of other phases, in peridotites. Clinopyroxene and garnet are the most fusible components and also tend to occur together as major phases in rocks such as eclogites.

All of the above minerals are unstable at high pressure and therefore only occur in the upper part of the mantle. Clinopyroxene, diopside plus jadeite, may be stable to depths as great as 500 km. Olivine transforms successively to β-spinel, a distorted spinel-like structure, near 400 km and to γ-spinel, a true cubic spinel, near 500 km. At high pressure it disproportionates to $(Mg,Fe)SiO_3$ in the perovskite structure plus (Mg,Fe)O, magnesiowüstite, which has the rocksalt structure. The FeO component is strongly partitioned into the (Mg,Fe)O phase.

Orthopyroxene, $(Mg,Fe)SiO_3$, transforms to a distorted garnet-like phase, majorite, with an increase in coordination of some of the magnesium and silicon:

$$^{VIII}(Mg,Fe)_3{}^{VI}Mg^{VI}Si^{VI}Si_3O_{12}$$

where the Roman numerals signify the coordination. This can be viewed as a garnet with MgSi replacing the Al_2. This is a high-temperature transformation. At low temperature the following transformations occur with pressure:

$$2MgSiO_3(opx) \rightarrow Mg_2SiO_4\ (\beta\text{-sp}) + SiO_2(st)$$

$$\rightarrow Mg_2SiO_4(\gamma\text{-sp}) + SiO_2(st) \rightarrow 2\ MgSiO_3\ (ilmenite)$$

$$\rightarrow 2\ MgSiO_3\ (perovskite)$$

A different sequence occurs for $CaMgSi_2O_6$ (diopside clinopyroxene). The ionic radius of calcium is much greater than aluminum, and this is expected to make the transition pressure to the garnet structure much higher than for orthopyroxene.

In the presence of Al_2O_3, or garnet, the pyroxene garnets form solid solutions with ordinary aluminous garnets, the transition pressure decreasing with Al_3O_3 content.

The mineralogy in the transition region, at normal mantle temperatures, is expected to be β- or γ-spinel plus garnet solid solutions. At colder temperatures, as in subduction zones, the mineralogy at the base of the transition region is probably γ-spinel plus ilmenite solid solution.

The garnet component of the mantle is stable to very

high pressure, becoming, however, less aluminous and more siliceous as it dissolves the pyroxenes. At low temperature and at pressures equivalent to those in the lower part of the transition region, the garnet as well as the pyroxenes are probably in ilmenite solid solutions.

The ilmenite structure of orthopyroxene can be regarded as a substitution of ^{VI}Mg ^{VI}Si for $^{VI}Al_2$ in the corundum structure. The transformation of $CaMgSi_2O_6$ clinopyroxene to ilmenite, if it occurs, is probably a higher pressure transition. $CaMgSi_2O_6$ may transform to the perovskite structure without an intervening field of ilmenite:

$$^{VIII}(Ca_{0.5}Mg_{0.5})_3\ ^{VI}(Ca_{0.5}Mg_{0.5})\ ^{VI}Si\ ^{IV}Si_3O_{12}$$

$$(\text{"majorite"}) \rightarrow\ ^{VIII-XII}(Ca_{0.5}Mg_{0.5})_4\ ^{VI}Si_4O_{12}\ (\text{perovskite})$$

The ionic radii (in angstroms) of some of the ions involved in the above reactions are

^{VI}Al, 0.53	^{VIII}Mg, 0.89	^{VI}Ca, 1.00
^{XII}Ca, 1.35	^{VI}Si, 0.40	$^{VI}[CaSi]$, 0.70
^{VI}Mg, 0.72	^{XII}Mg, 1.07	^{VIII}Ca, 1.12
^{IV}Si, 0.26	$^{VI}[MgSi]$, 0.56	

The ionic radius of $^{VI}[MgSi]$ is similar to ^{VI}Al, and therefore a solid-solution series between garnet and majorite at relatively low pressure is expected. The ionic radius of $^{VI}[CaSi]$ is much greater, and therefore diopside is expected to require higher pressures for its garnet transformation unless only the Mg^{2+} enters the octahedral site. The replacement of Al_2 for $Ca_{0.5}$ $Mg_{0.5}Si$ in the perovskite structure is also expected to be difficult, so it is possible that clinopyroxene-garnet disproportionates to calcium-rich perovskite plus Al_2O_3, the excess Al_2O_3 probably combining with $MgSiO_3 \cdot x\ Al_2O_3$ in the garnet, ilmenite or perovskite structure, depending on pressure. The disparity in ionic radii between $^{VIII-XI}Ca$, $^{VIII-XII}Mg$, ^{IV}Si and ^{VI}Al probably means that there will be three separate perovskite phases in the lower mantle: Ca-rich, (Mg, Fe)-rich and Al-rich. The presence of Al_2O_3 decreases the density of perovskite but increases the density of ilmenite and garnet.

Ferrous iron (Fe^{2+}) is expected to readily substitute for Mg^{2+} in all the phases discussed so far. The relative partitioning of Fe^{2+} amongst phases, however, is expected to vary with pressure. Garnet is by the far the most Fe^{2+}-rich phase in the upper mantle, followed by olivine, clinopyroxene and orthopyroxene. As the pyroxenes dissolve in the garnet they dilute the Fe/Mg ratio, and γ-spinel may be the most iron-rich phase in the lower part of the transition region. In the lower mantle Fe^{2+} favors (Mg,Fe)O over perovskite. When the Fe^{2+} high spin–low spin transition occurs, somewhere deep in the lower mantle, solid solution between Fe^{2+} and Mg^{2+} is probably no longer possible because of the disparity in ionic radii, and a separate FeO-bearing phase, such as Fe(L.S.)O, is likely. At high pressure this is expected to dissolve extensively in any molten iron that traverses this region on the way to the core, or to be stripped out of any mantle that comes into contact with the core in the course of mantle convection. An Fe-O-poor lower mantle is therefore a distinct possibility. The corollary is an iron-FeO core (see Chapter 4).

As far as we know perovskite and MgO are stable throughout the lower mantle, which is consistent with seismic radial homogeneity of most of the lower mantle. $MgSiO_3$-perovskite is the most abundant mineral in the mantle. The seismic properties of the lower mantle are broadly consistent with (Mg,Fe) SiO_3-perovskite, although other phases may be present, such as (Mg,Fe) O.

In general FeO decreases the pressure of phase transitions in the mantle, including olivine–β-spinel, β–γ-spinel and pyroxene-garnet. Al_2O_3 widens the stability fields of garnet and ilmenite. Ca^{2+} is a large ion and in simple compounds is expected to cause phase transitions at lower pressures than the equivalent Mg^{2+} compound. However, the large size of Ca^{2+} makes it difficult to substitute for Al^{3+} (in the coupled CaSi substitution for Al_2). At high temperature clinopyroxene contains excess Mg^{2+} compared to pure diopside,

$$CaMgSi_2O_6 + xMg_2Si_2O_6 = CaMg_{2x+1}\ Si_2O_6$$

Therefore, orthopyroxene probably reacts out of the mantle at shallower depths than clinopyroxene.

The difficulty of substituting $^{VI}[CaSi]$ for $^{VI}[Al_2]$ suggests the following structural formula for diopside-garnet:

$$^{VIII}Ca_2\ ^{VIII}Mg\ ^{VI}[MgSi]\ ^{IV}Si_3O_{12}$$

This has the virtue of requiring no coordination change for Ca^{2+} in going from the diopside to garnet structure.

PHASE EQUILIBRIA IN MANTLE SYSTEMS

The lateral and radial variations of seismic velocity and density depend, to first order, on the stable mineral assemblages and, to second order, on the variation of the velocities with temperature, pressure and composition. Temperature, pressure and composition dictate the compositions and proportions of the various phases. In order to interpret observed seismic velocity profiles, or to predict the velocities for starting composition, one must know both the expected equilibrium assemblage and the properties of the phases.

To a first approximation, olivine, orthopyroxene, clinopyroxene and an aluminous phase (feldspar, spinel, garnet) are stable in the shallow mantle. β-spinel, majorite, garnet and clinopyroxene are stable in the vicinity of 400 km, near the top of the transition region. γ-spinel, majorite or γ-spinel plus stishovite, Ca-perovskite, garnet and ilmenite are stable between about 500 and 650 km. Garnet, ilmenite, Mg-perovskite, Ca-perovskite and magnesiowüs-

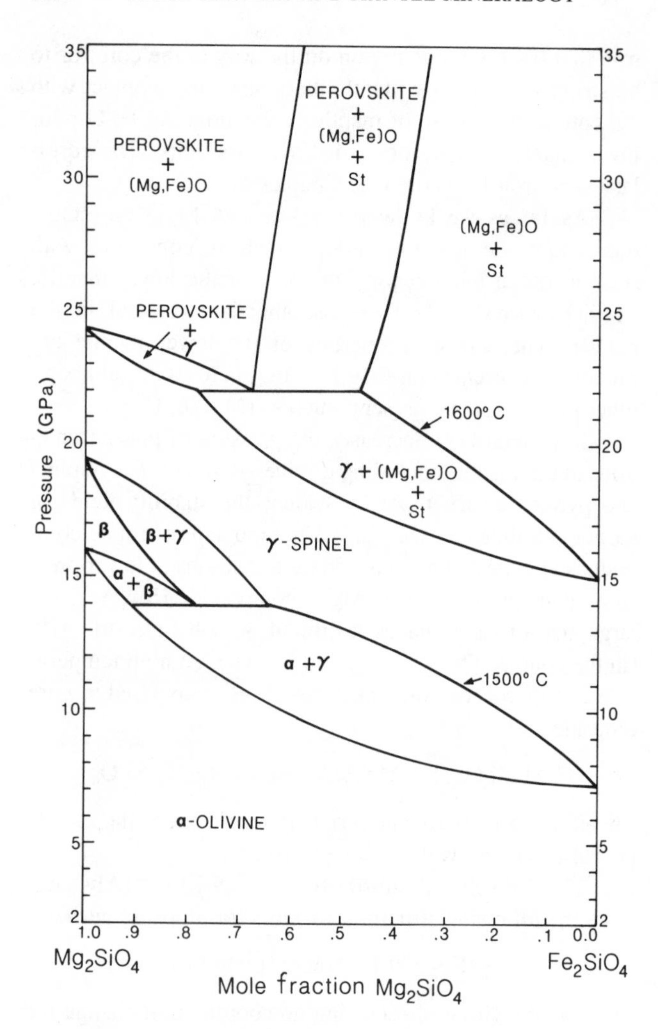

FIGURE 16-1
Phase relations in Mg_2SiO_4–Fe_2SiO_4 system (J. Bass and D. L. Anderson, unpublished).

tite are stable near the top of the upper mantle, and perovskites and magnesiowüstite ± Al_2O_3 are stable throughout most of the lower mantle. The details of the stable assemblages depend on composition and temperature. Lateral variations in velocity due to temperature-induced phase changes can be as important as pressure-induced phase changes are in the radial direction.

Phase equilibria can be discussed in terms of the olivine system, the pyroxene system and the pyroxene-garnet system. Pyroxenes can tolerate a certain amount of Al_2O_3 and garnets, at high pressure, dissolve pyroxene, so pyroxenes and garnets must be treated together. On the other hand the olivine system is almost pure Mg_2SiO_4–Fe_2SiO_4, although FeO can be exchanged with pyroxene-garnet.

Phase equilibria in mantle systems are summarized below in a series of figures based on available experiments and calculations. A useful summary of early work is contained in Ringwood (1975). The olivine system is shown in Figure 16-1.

Pyroxene System

Enstatite (en) and diopside (di) do not form a complete solid-solution series, but en dissolves a certain amount of di, and di contains an appreciable amount of en at moderate temperature and pressure. The amount of mutual solubility increases with temperature and decreases with pressure, and this provides a method for estimating the temperature of equilibration of mantle-derived xenoliths. Pyroxenes also react with garnet:

$$3MgSiO_3 \cdot xAl_2O_3 \rightleftharpoons xMg_3Al_2Si_3O_{12} + 3(1 - x)MgSiO_3$$

$$\text{aluminous enstatite} \rightleftharpoons \text{garnet} + \text{enstatite}$$

which proceeds to the right with increasing pressure (Figure 16-2). In principle, the measurement of the compositions

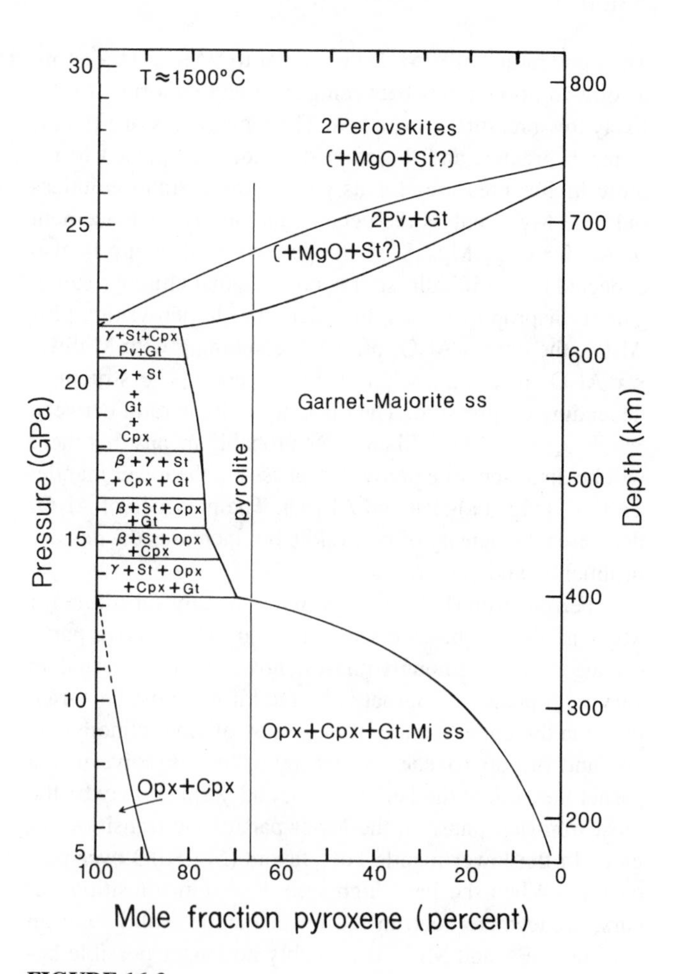

FIGURE 16-2
Phase relations in orthopyroxene-garnet system (J. Bass and D. L. Anderson, unpublished).

of coexisting pyroxenes and garnets provides information about pressures and temperatures in the mantle. At higher pressure garnet dissolves the enstatite, and this requires a change in coordination of one-fourth of the Mg and Si. Pressures in excess of about 100 kbar are required for this change in coordination.

Natural clinopyroxenes, particularly in eclogites, are solid solutions between diopside and jadeite called omphacite:

$$^{VIII}Ca^{VI}Mg^{IV}Si_2O_6 + {}^{VIII}Na^{VI}Al^{IV}Si_2O_6$$

At modest pressure, about 20 kbar, the solid solution series is complete (Figure 16-3). Natural clinopyroxenes from kimberlite eclogites contain up to 8 weight percent Na_2O. Clinopyroxenes from peridotites typically have much less Na_2O and jadeite.

Most garnets contain very little sodium; however, at high pressure Na_2O can enter the garnet lattice (Ringwood, 1975). In particular, the following garnets have been synthesized (Ringwood and Major, 1971):

$$(NaCa_2)\,(AlSi)Si_3O_{12}$$

$$(Na_2Ca)Si_2Si_3O_{12}$$

Natural garnets associated with diamonds in kimberlite pipes contain up to 0.26 percent Na_2O. In the transition region, the sodium is probably contained in a complex garnet solid solution.

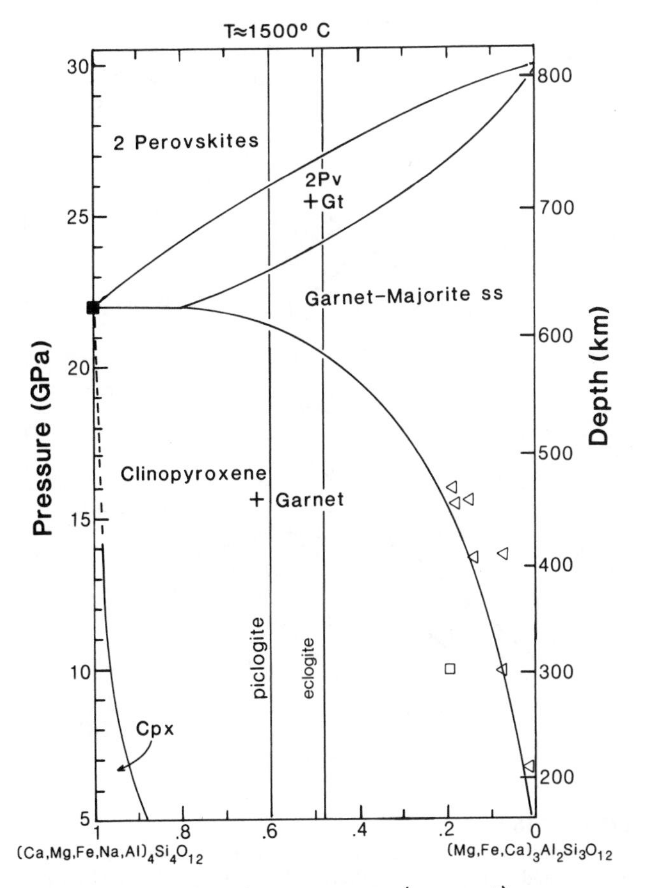

FIGURE 16-3
Phase relations in clinopyroxene-garnet system (J. Bass and D. L. Anderson, unpublished).

The other high-pressure forms of pyroxene include ilmenite, β plus stishovite, γ plus stishovite, ilmenite and perovskite, depending on pressure, temperature and content of calcium, aluminum and iron. The pressures at which clinopyroxene and orthopyroxene disappear are strong functions of the other variables.

The phase behavior of garnet + clinopyroxene + orthopyroxene, the peridotite assemblage, is substantially different from the behavior of garnet + clinopyroxene, the basalt-eclogite assemblage. When only clinopyroxene + garnet are present, the clinopyroxene dissolves in the garnet with increasing pressure and eventually a homogeneous garnet solid solution is formed. When orthopyroxene is also present, the garnet and clinopyroxene compositions move toward orthopyroxene with increasing pressure; that is, the orthopyroxene component dissolves in both the garnet and the clinopyroxene and eventually disappears. At this point an MgSi-rich, aluminum-deficient garnet coexists with a magnesium-rich diopside. Garnet then moves toward the clinopyroxene composition as diopside dissolves in the garnet. Therefore, in contrast to the bimineralic eclogite system, the garnet takes a detour toward $MgSiO_3$ before it heads toward $CaMgSi_2O_6$. In either case, orthopyroxene disappears at a relatively low pressure, being either completely absorbed in the garnet structure or partially converting to β + stishovite or γ + stishovite, depending on the garnet, or Al_2O_3, content of the initial mixture. Recent high-pressure, high-temperature experiments have helped elucidate the phase relations in $MgSiO_3$, the main component of the orthopyroxene system. A preliminary synthesis of available results is given in Figure 16-4. Note that the high-temperature sequence of transitions is different from the low-temperature sequence. The resulting densities and seismic velocities are also quite different. The garnet form of $MgSiO_3$, majorite, is dominant at high temperature. Majorite is similar to pyrope garnet with ^{VI}Mg ^{VI}Si replacing $^{VI}Al_2$ yielding

$$^{VIII}Mg_3\ {}^{VI}[MgSi]^{IV}Si_3O_{12}$$

Since $^{VI}[MgSi]O_3$ (periclase plus stishovite) has elastic properties similar to $^{VI}Al_2O_3$, we expect majorite to have elastic properties similar to pyrope.

$^{VI}Mg^{VI}SiO_3$-ilmenite, a high-pressure, low-temperature form of enstatite, has considerable higher elastic moduli since all of the silicon is in sixfold coordination as in stishovite. Ilmenite is extremely anisotropic. The low-

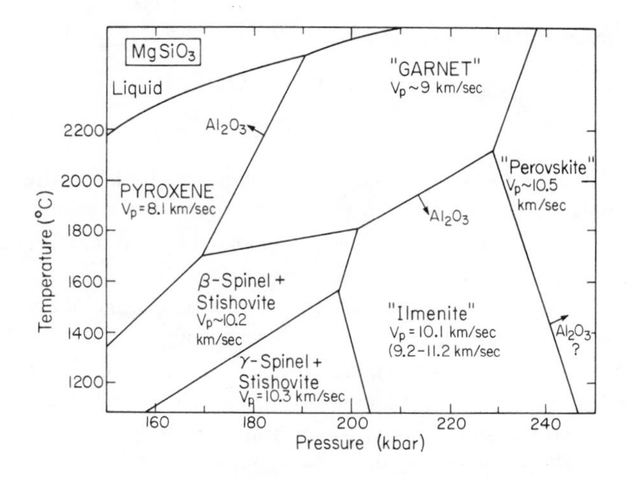

FIGURE 16-4
Provisional phase relations in $MgSiO_3$ (modified from Kato and Kumazawa, 1985, Sawamoto, 1986, and Akaogi and others, 1986). The arrows show the direction that the phase boundaries are expected to move when Al_2O_3, or garnet, is added. The approximate compressional velocities are shown for each phase.

pyroxene with increasing pressure. At very high pressure, garnet transforms to an Al_2O_3-rich perovskite (Al-pv) (Liu, 1974). The lower mantle below about 750 km probably consists of three perovskites plus magnesiowüstite (3 pv + mw).

Note that the phase assemblages along the "cold slab" adiabat are different from those along the "normal mantle" adiabat at almost all pressures, and this is true also for temperature contrasts much smaller than the 800°C chosen for purposes of illustration. The $\alpha-\beta$ transition is elevated by 30 kbar and the $\beta-\gamma$ transition is elevated by 40 kbar in the cold slab. In the case of enstatite, not only is the upper stability limit decreased at low temperature, but the low-temperature phase assemblages are different. The dense and fast phase γ-spinel has a much broader stability field in cold mantle.

Different phase assemblages are also encountered as one increases the temperature from the "normal mantle" adiabat. The partial melt field (not shown) is encountered at temperature minerals (spinel + stishovite, ilmenite) are 10 percent to 20 percent higher in velocity than the high-temperature minerals (pyroxene, majorite).

The low-temperature assemblages, β + stishovite, γ + stishovite and ilmenite are of the order of 10 percent faster than majorite. The locations of the phase boundaries depend on Al_2O_3 content, and the expected migration direction is given in the figure by the arrows. Note that the garnet form of pyroxene is stabilized by high temperature and high Al_2O_3 content. Also, ordinary pyroxene is stable to relatively high pressures in hot mantle. This is partially responsible for the low velocities in the upper mantle under oceanic and tectonic regions.

The CMAS System

The system CaO-MgO-Al_2O_3-SiO_2 (CMAS) is shown in Figure 16-5 in simplified form. The phases of Mg_2SiO_4 are olivine (α), β-spinel (β), γ-spinel (γ) and perovskite plus MgO (using structural names throughout). The major phases of $MgSiO_3$ are enstatite, majorite, ilmenite and perovskite (Mg-pv). At low temperatures and when little or no Al_2O_3 is present, there are also extensive fields of β + stishovite (st) and γ + stishovite. High temperatures and the presence of Al_2O_3 stabilize the majorite (mj) structure, and a broad majorite-garnet solid-solution field occurs between enstatite, ilmenite and perovskite. Diopside and jadeite, components of clinopyroxene, are stable to higher pressures than enstatite. Diopside collapses to a dense calcium-rich phase, probably perovskite (Ca-pv), at pressures less than required to transform enstatite to perovskite (Ringwood, 1975, 1982). Garnet (gt) itself is stable throughout most of the upper mantle, although it dissolves

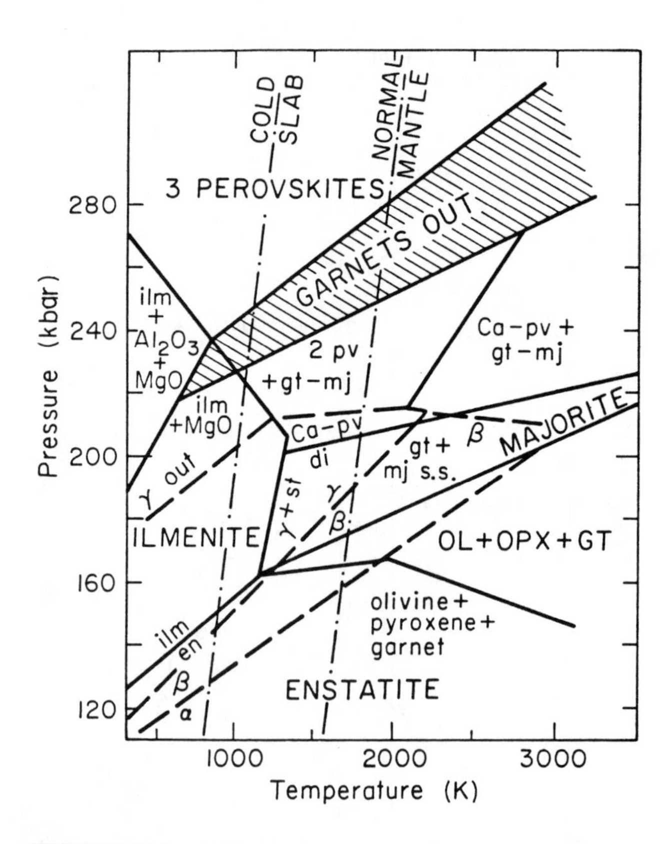

FIGURE 16-5
Equilibrium phase boundaries for mantle minerals modified from Kuskov and Galimzyanov (1985) and Ito and Takahashi (1986). Dashed lines are olivine system boundaries; solid lines are pyroxene-garnet system boundaries. Approximate adiabats are shown for "normal" mantle and "cold slab" mantle. Note the different phase assemblages, at constant pressure, for the two adiabats. Most boundaries are for pure end-members in the system CaO-MgO-Al_2O_3-SiO_2.

low pressure. The fields of the low-pressure, low-density assemblages are expanded at high temperature, but the change in density is not symmetric about the average, or normal, temperature. The thermal expansion coefficient increases with temperature, so there is a larger decrease of density for a given increase in temperature than for the corresponding decrease. For an internally heated mantle, upwellings are broader than downwellings, so the lateral changes in physical properties are expected to be more diffuse than for the slab.

The peridotite mineral assemblage, olivine + pyroxenes + garnet, stable in the shallow mantle, has transformed by 190 kbar to β- or γ-spinel + majorite + garnet ± clinopyroxene ± stishovite with a substantial increase in density and seismic velocity. This is for a normal mantle adiabat. At this pressure, but at temperatures 400°C colder, this assemblage is replaced by ilmenite, γ-spinel and magnesiowüstite ± calcium-rich perovskite. At 230–260 kbar the normal mineral assemblage is two perovskites + garnet, while the cold assemblage is three perovskites or calcium-perovskite + garnet + ilmenite ± oxides. By 280 kbar the assemblages should be independent of temperature over the temperature range expected in the lower mantle. Temperature-induced variations in density and velocity should, therefore, be relatively small below 770 km. Note the very broad stability fields of garnet-majorite solid solution in high-temperature mantle and γ-spinel and ilmenite in cold mantle. Also note the negative Clapeyron slope for the ilmenite-perovskite transition.

CALCULATION OF PHASE RELATIONS

The standard free energy of a reaction is given by

$$\Delta G^\circ(P,T) = \Delta H^\circ_T - T\Delta S^\circ_T + \int_1^P \Delta V(P,T)\mathrm{d}P$$

where ΔH°_T *and* ΔS°_T are the enthalpy and entropy of reaction, respectively, at temperature T and are given by

$$\Delta H^\circ_T = \Delta H^\circ_{T_o} + \int_{T_o}^T \Delta C_p \,\mathrm{d}T$$

$$\Delta S^\circ_T = \Delta S^\circ_{T_o} + \int_{T_o}^T (\Delta C_p/T)\mathrm{d}T$$

where T_o is a standard temperature and ΔC_p is the heat capacity difference between products and reactants. At equilibrium

$$\Delta G^\circ\,(P,T) = 0$$

The slope of the reaction, in P,T space, is given by

$$\mathrm{d}P/\mathrm{d}T = \Delta S/\Delta V$$

Values of ΔS and ΔV are given in Table 16-3.

TABLE 16-3
Thermochemical Data for Phase Transitions

Transition	ΔV°_{298} (cm^3/mol)	ΔS°_{1000} (cal/mol K)	ΔS°_{298} (cal/mol K)
Mg_2SiO_4			
$\alpha \to \beta$	−3.13	−2.5	−3.1
$\beta \to \gamma$	−0.89	−1.5	−0.9
$\alpha \to \gamma$	−4.02	−4.0	−4.0
$\beta \to$ ox	−4.03	+1.92	—
$\gamma \to$ ox	−3.14	+3.36	−0.35
$\gamma \to$ pv + mw	−3.84	—	−3.2
γ+st → 2ilm	−0.79	+2.71	−1.9
β+st → 2ilm	−1.89	+0.48	—
$MgSiO_3$			
2px → β+st	−7.99	−5.63	−4.95
2px → γ+st	−9.09	−7.86	−5.85
px → ilm	−4.94	−2.58	−3.6
ilm → pv	−1.91	—	−2.0
px → pv	−6.83	—	−5.9
px → gt	−2.74	−3.5	—
SiO_2(q) → st	−9.70	−3.24	—

Navrotsky and others (1979), Watanabe (1982), Akaogi and others (1984), Ito and Navrotsky (1985).

ISOBARIC PHASE CHANGES AND LATERAL VARIATIONS OF PHYSICAL PROPERTIES

As the ability to map the three-dimensional structure of the Earth improves, it becomes important to understand the factors that influence the lateral heterogeneity in density and seismic velocities. Much of the radial structure of the Earth is due to changes in mineralogy resulting from pressure-induced equilibrium phase changes or changes in composition. The phase fields depend on temperature as well as pressure so that, for example, a given mineral assemblage will occur at a different depth in colder parts of the mantle. The elevation of the olivine-spinel phase boundary in cold slabs is probably the best known example of this effect. The other important minerals of the mantle, orthopyroxene, clinopyroxene and garnet, also undergo temperature-dependent phase changes to denser phases with higher elastic moduli. These phases include majorite, ilmenite, spinel plus stishovite, and perovskite. The pronounced low-velocity zone under oceans and tectonic regions and its suppression under shields is another example of phase differences (partial melting) associated with lateral temperature gradients.

Temperature provides more than just a perturbation to the depths of phase boundaries. Variations in temperature, at constant pressure, also cause changes in the stable mineral assemblages, and these isobaric phase changes result in larger changes in the physical properties than are caused by the effect of temperature alone. In general, the sequence of phase changes that occurs with increasing pressure also occurs with decreasing temperature. There are also some mineral assemblages that do not exist under normal conditions of pressure and temperature but occur only under the extremes of temperature found in cold slabs or near the solidus in hot upwellings. Generally, the cold assemblages are characterized by high density and high elastic moduli.

The magnitude of the horizontal temperature gradients in the mantle are unknown, but slab modeling suggests about 800°C over about 50 km. In an internally heated material the upwellings are much broader than slabs or downwellings. Tomographic results show extensive low-velocity regions associated with ridges and tectonic regions, consistent with broad high-temperature regions. The cores of convection cells have relatively low thermal gradients. We therefore expect the role of isobaric phase changes to be most important and most concentrated in regions of subducting slabs. The temperature drop across a downwelling is roughly equivalent to a pressure increase of 50 kbar, using typical Clapeyron slopes of upper-mantle phase transitions.

In the recent geophysics literature it is often assumed that lateral variations in density and seismic velocity are due to temperature alone. By contrast, it is well known that radial variations are controlled not only by temperature and pressure but also by pressure-induced phase changes. Phase changes such as partial melting, basalt-eclogite, olivine-spinel-postspinel, and pyroxene-majorite-perovskite dominate the radial variations in density and seismic velocity. It would be futile to attempt to explain the radial variations in the upper mantle, particularly across the 400- and 650-km discontinuities, in terms of temperature and pressure and a constant mineralogy. All of the above phase changes, plus others, also occur as the temperature is changed at constant pressure or depth. Yet it is common practice to ignore these temperature-induced phase changes in attempting to explain geophysical anomalies associated with the geoid and slabs, and the ocean-continent contrast. This has led to models requiring deep slab penetration, since temperature alone is not sufficient to explain the magnitude of the anomalies if slabs are confined to the upper mantle. Likewise, the integrated travel-time contrast between shields and oceans implies a thick continental root unless lateral phase changes, such as partial melting, are allowed for. In any case the large lateral variations in velocity above 300 km, and particularly above 200 km, make it difficult to resolve variations below 400 km, and those that are resolved are small and uncorrelated with shields. Lateral density and velocity variations associated with phase changes are not confined to narrow depth intervals, nor are they all associated with simple elevation of phase boundaries in cold mantle. The combination of cold temperature and high pressure can stabilize assemblages that are not present in warmer mantle and can broaden, in depth, the stability fields of high-density minerals.

There are large geoid and seismic velocity anomalies associated with subducting slabs. Ordinary temperature effects are so small that slab penetration deep into the lower mantle has been invoked in order to explain the size of the anomalies. This, in turn, has been used by some to support models of whole-mantle convection and to reject chemically stratified models with slabs confined to the upper mantle.

Because of lateral temperature gradients we have to be concerned with lateral, or isobaric, phase changes as well as radial phase changes. Some of these phase changes, their approximate depth extent in "normal" mantle and the density contrasts include the following:

50–60 km	basalt → eclogite (15 percent)
50–60 km	spinel peridotite → garnet peridotite (3 percent)
50–200 km	partial melting (10 percent)
400–420 km	olivine → β-spinel (7 percent)
300–400 km	orthopyroxene → majorite (10 percent)
500–580 km	→ β + st(4.5 percent) → γ + st(1.6 percent)
500–580 km	β-spinel → γ-spinel (3 percent)

400–500 km	clinopyroxene → garnet (10 percent)
500 km	garnet-majorite s.s. → ilmenite s.s. (5 percent)
700 km	ilmenite → perovskite (5 percent)

For a multicomponent mantle the above percentages must be multiplied by the fraction of mantle involved.

Most of these reactions have positive Clapeyron slopes and are therefore elevated in cold slabs and depressed in hot upwellings. The density contrasts are much larger than those associated with thermal expansion and, in any case, add to the thermal expansion effect. This is important in geoid modeling. The associated velocity contrasts are important in modeling lateral changes in seismic velocity and slab anomalies. For a coefficient of thermal expansion of $3 \times 10^{5}/°C$, it requires a temperature change of $10^{3}°C$ to change the density by 3 percent. Since the thermal coefficients of density and elastic moduli decrease with compression, it is even more difficult to obtain large lateral variations at depth with temperature alone.

SLABS

One result of the neglect of isobaric phase changes is the conclusion that density and velocity anomalies in the upper mantle are not sufficient to explain the magnitude of slab-related geoid and seismic anomalies. Hager (1984) calculated the geoid signal by associating slabs with an average density contrast of 0.1 g/cm^3. This can explain the geoid signal for whole-mantle convection models. When phase changes are included, the average density contrast of slabs is about 3 times greater, being about 0.4 g/cm^3 midway through the upper mantle. In the whole-mantle flow models, this agreement is a result of the large density-geoid kernels associated with whole-mantle convection. In chemically layered models the density-geoid response goes to zero at chemical boundaries, and the average value of the upper-mantle kernel is much less than for the whole-mantle case. Therefore, a large density contrast is consistent with stratified convection.

There are several phase changes in cold subducting material that contribute to the increase in the relative density of the slab. The basalt-eclogite transition is elevated, contributing a 15 percent density increase for the basaltic portion of the slab in the upper 60 km or so. The absence of melt in the slab relative to the surrounding asthenosphere increases the density and velocity anomaly of the slab in the upper 300 km or so of the mantle. The olivine-spinel and pyroxene-majorite phase changes are elevated by some 100 km above the 400-km discontinuity, contributing about 10 percent to the density contrast in the upper mantle. The $\beta-\gamma$ transition is also elevated, adding several percent to the density of the slab between 400 and 500 km. The ilmenite form of pyroxene is 5 percent denser than garnetite, increasing the density contrast of the cold slab between 500 and 670 km, relative to hot mantle, by about a factor of 2 or 3 over that computed from thermal expansion. In addition to these effects, the formation of a detached thermal boundary layer below the 650-km discontinuity can also increase the total mass anomaly associated with subduction. The latter can increase or decrease the geoid anomaly associated with subduction, depending on the nature of the chemical layering (Richards and Hager, 1984).

The seismic anomaly associated with slabs is also much greater than can be accounted for by the effect of temperature on velocity. Phase changes, including partial melting outside the slab, may be the cause. The associated density contrast between slab and normal mantle is greater than between the plate and the underlying mantle as estimated from thermal expansion alone.

The ilmenite form of $MgSiO_3$ is a stable phase, at low temperature, in the lower part of the transition region (Figure 16-4). Ilmenite is about 8 percent denser than garnet-majorite and has a V_p about 10 percent greater. Although ilmenite is only 4 percent slower than perovskite, it is 7 percent less dense. Ilmenite becomes stable at slab temperatures somewhere between 450 and 600 km and is predicted to remain stable to depths greater than the perovskite phase boundary in higher-temperature mantle. Although the slab is predicted to be locally less dense than the lower mantle at about 650 km, it will probably depress the boundary between the upper and lower mantles because of the accumulated density excess and the negative Clapeyron slope between ilmenite-spinel and perovskite-MgO. The $CaMgSi_2O_6$-component of the mantle transforms to perovskite at a slightly lower pressure than the comparable transformation in $MgSiO_3$, and the transformation pressure is lower at low temperature. These all contribute to the velocity and density anomaly of the slab at the base of the transition region.

In the discussion so far we have assumed that the slab is identical in composition to the adjacent mantle. Yet the slab is probably laminated: The upper layer is basalt/eclogite, and the second layer is probably olivine-orthopyroxene harzburgite. These undergo their own series of phase changes and, when cold, remain denser than garnet peridotite to at least 500 km.

The composition of the lower oceanic lithosphere is unknown. The alternatives are basalt-depleted peridotite, or harzburgite, undepleted peridotite and basalt/eclogite. The last could be the result of underplating of the oceanic lithosphere by melts from the mantle. Cold harzburgite averages about 0.1 g/cm^3 denser than warm pyrolite between 400 and 600 km (Ringwood, 1982). At 600 km it becomes less dense. Eclogite is denser than peridotite at the same temperature to depths of 500–560 km (Anderson, 1982; Anderson and Bass, 1986) and, when cold, to 680 km (Irifune and Ringwood, 1986). Cold eclogite is about 4–5 percent

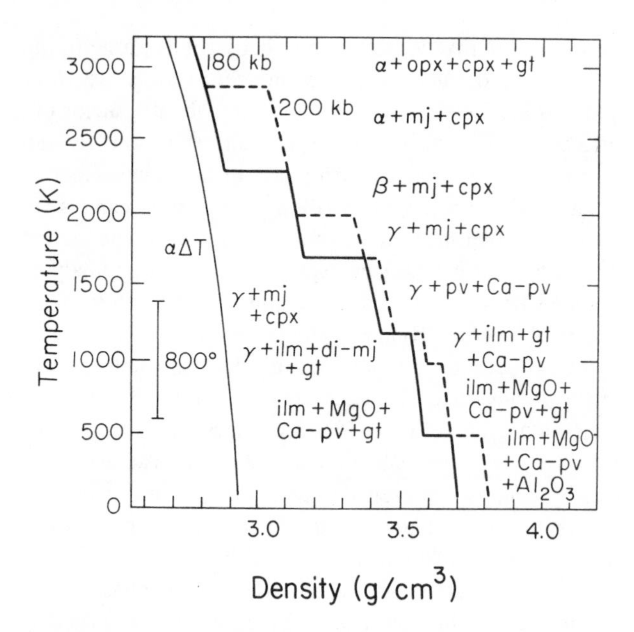

FIGURE 16-6
Approximate variation of zero-pressure density with temperature taking into account thermal expansion (curve to left) and phase changes at two pressures. The effect of pressure on density is not included. $\alpha\Delta T$ is not linear since α increases with T and generally decreases from low-density to high-density phases. The inclusion of density jumps associated with phase changes increases the average effect of temperature by a factor of 3 to 4. At phase boundaries the effect is much larger. The phases stable over each temperature interval are also shown. The abbreviations are α(olivine), β(β-spinel), γ(γ-spinel), opx (orthopyroxene), cpx (clinopyroxene), gt (garnet), mj ($MgSiO_3$-majorite), pv ($MgSiO_3$-perovskite), Ca-pv ($CaSiO_3$-perovskite), ilm ($MgSiO_3$-ilmenite) (after Anderson, 1987b).

denser than warm peridotite above 550 km depth. If the 650-km discontinuity is a chemical boundary, this boundary will be depressed by the integrated density excess in overlying cold mantle even if the deeper part of the slab is buoyant. The "650-km discontinuity" is expected to be an irregular boundary in a chemically stratified mantle and to be much deeper under slabs. In any event it appears that the different phase assemblages in the slab relative to warm mantle will contribute to the density contrast. An increase of intrinsic density between upper and lower mantle and a negative Clapeyron slope will inhibit slab penetration into the lower mantle. Hager (1984) invoked an increase in viscosity at 650 km to partially support the slab in order to explain the geoid highs associated with subduction zones. A chemical change has a similar effect.

The approximate zero-pressure density as a function of temperature, at two pressures, is shown in Figure 16-6. Temperature is plotted increasing upward to emphasize the fact that decreasing temperature has effects similar to increasing pressure—lateral temperature changes are similar to vertical pressure changes. The curve labeled $\alpha\Delta T$ is the approximate effect of thermal expansion alone on density. The bar labeled 800°C shows the expected change in temperature across a subducted slab and is approximately half the maximum expected lateral temperature changes in the mantle. Note that a temperature change of 800°C placed anywhere in the field of temperatures expected in the mantle will cross one, two or even three phase boundaries, each of which contributes a density change in addition to the $\alpha\Delta T$ term from thermal expansion. Changes in elastic properties are associated with these phase changes.

Figure 16-7 shows the approximate zero-pressure room-temperature density for the CMAS system with a low-pressure mineralogy appropriate for garnet peridotite with olivine > orthopyroxene > clinopyroxene ≈ garnet. Note that the low-temperature phase assemblages are denser than high-temperature assemblages (normal mantle) until about 210 kbar and that the differences are particularly pronounced between about 300 and 550 km. The density change associated with a temperature change of 800 K ranges from 7 to 17 percent in the temperature interval 1000 to 2300 K at pressures of 180 and 200 kbar. This includes thermal expansion and isobaric phase changes. Thermal expansion alone gives 2 to 3 percent. Note that the density anomaly associated with the slab is far from constant with depth. Furthermore, the density anomaly of a slab with respect to the adjacent mantle is quite different from the den-

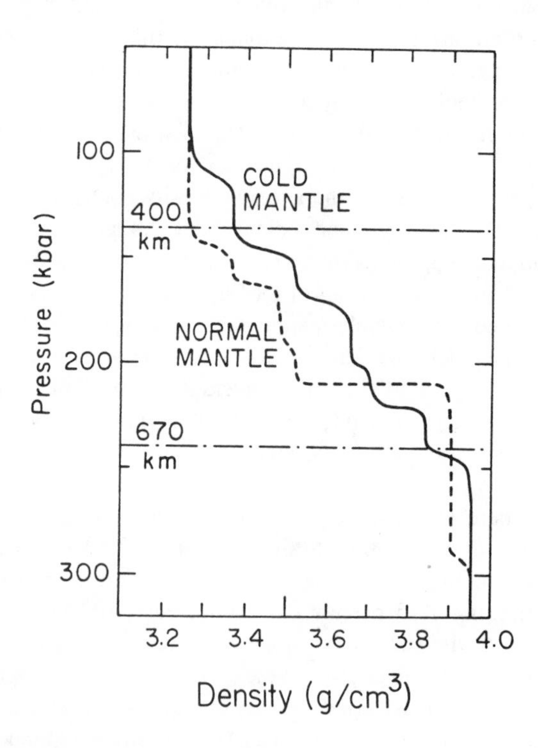

FIGURE 16-7
Zero-pressure room-temperature density of FeO-free peridotite (olivine > orthopyroxene > clinopyroxene ≈ garnet) using mineral assemblages appropriate for the temperatures in warm ("normal") mantle and slab ("cold") mantle and the phase relations of Figure 16-5 (after Anderson, 1987b).

sity contrast between the surface plate and the underlying mantle, as estimated from the bathymetry-age relation for oceanic plates.

The important phase changes in the mantle mostly have Clapeyron slopes that correspond to depth variations of 30–100 km per 1000°C. The widths of the phase changes, at constant temperature when solid-solid effects are taken into account, are, for example, 40 kbar for complete transformation of α-olivine to γ-spinel, about 10 kbar for α–β and 40 kbar for β–γ. The comparable changes, at constant pressure, occur over a temperature interval of 300°C, which is less than half the temperature contrast across a subducting slab.

By use of Figure 16-8 or 16-5 we see that several phase boundaries are crossed in going, at constant pressure, from normal mantle temperatures to colder temperatures. For example, near 150 kbar, olivine and pyroxenes are stable at high temperature, β-spinel, majorite and clinopyroxene are stable at lower temperatures and β ± stishovite ± clinopyroxene is the stable assemblage at cold temperature. At 230 kbar calcium-rich perovskite, $MgSiO_3$-perovskite and magnesiowüstite is the normal assemblage, and ilmenite replaces $MgSiO_3$-perovskite at cold temperature. Garnet is stable over a very large pressure and temperature range. This is important since large amounts of garnet will decrease both the radial and lateral variations in physical properties.

The question of whether eclogite becomes less dense than peridotite at pressures near the base of the transition region and top of the lower mantle has been controversial. One side has maintained that eclogite becomes less dense than peridotite at the base of the transition region and is less dense than the top of the lower mantle (Anderson, 1979a,b; Anderson and Bass, 1984). The primary reason is the large stability interval of garnet. Ringwood and coworkers have maintained until recently that eclogite is denser than peridotite at all pressures. Very recently they were able to extend the pressure range of their experiments (Irifune and others, 1986), and they show that the densities of eclogite of MORB composition and peridotite do indeed intersect at high pressure. Quartz-free eclogites will actually be less dense than they calculate since MORB-eclogite has a large SiO_2-stishovite component. The crustal component of the slab, even if it survives subduction to 650 km, does not control its own destiny. It is the integrated density of the slab, intrinsic density contrast across the 650-km discontinuity and pressure-temperature locations of phase boundaries in the slab and lower mantle that determine whether slabs will be able to sink into the lower mantle. In a chemically layered mantle the "650-km" discontinuity will be depressed by the subducted slab and therefore, in this sense, the upper mantle protrudes into the lower mantle or, more precisely, protrudes below the depth usually assigned to the upper mantle–lower mantle boundary. Whether it can protrude deeply enough to be entrained in lower-mantle flow

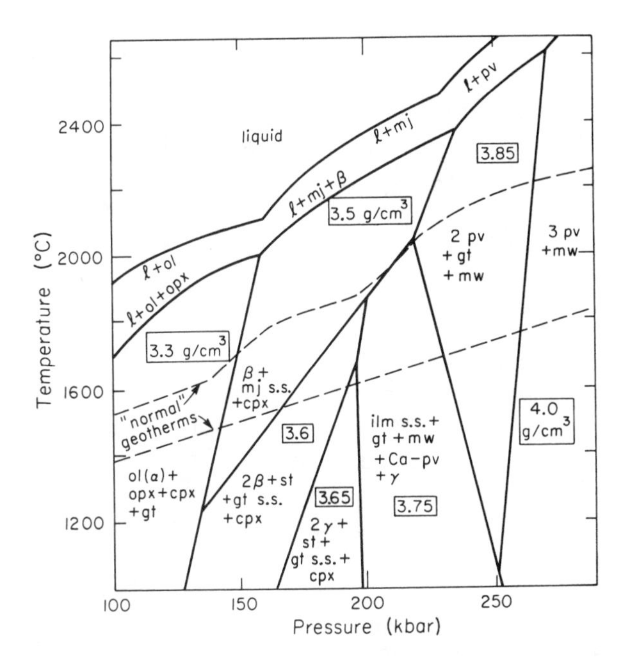

FIGURE 16-8
Tentative phase relations in the system MgO-SiO_2-CaO-Al_2O_3 with an ol > opx > cpx ≈ gt mineralogy at low pressure, based on a synthesis of a variety of subsolidus and melting experiments on peridotites. The geotherms bracket most estimates of temperatures in "normal" or average mantle. Warmer parts of the mantle may be near the solidus; the interiors of slabs may be 800°C colder. The phase diagram is based on incomplete and sometimes inconsistent reconnaissance experiments and must be taken as provisional until systematic and reversed experiments are performed. Based on experiments and interpretations by Kato and Kumazawa (1985, 1986), Akaogi and Akimoto (1977), Ohtani and others (1986a,b), Kanzaki (1986), Ito and Takahashi (1986), Yamada and Takahashi (1984), Irifune and others (1986), and Sawamoto (1986). "2β" and "2γ" mean that both Mg_2SiO_4 and $MgSiO_3$ have transformed to a spinel assemblage (Anderson, 1987; after Anderson, 1987b).

or to overcome the negative Clapeyron slope or whether, even in this case, it can become denser than the lower mantle, are questions that have yet to be completely addressed. The seismic problem of determining whether slabs are continuous across the upper-lower mantle boundary (the depressed "650-km discontinuity") is complicated by the likely presence of a detached thermal boundary layer at the top of the lower mantle under such conditions. This would be dense and fast and would be hard to distinguish from a penetrating slab. In this regard, the complete lack of seismicity below about 690 km and the apparent crumpling of the slab near this depth are significant (Giardini and Woodhouse, 1984).

In peridotite the density in the lower part of the transition region (Figure 16-8 and 16-9) is dominated by β-spinel (ρ_o = 3.47 g/cm³) or γ-spinel (3.56 g/cm³) and majorite (3.52 g/cm³) or ilmenite (3.82 g/cm³). In eclogite the mineralogy is garnet (3.56 g/cm³), calcium-perovskite

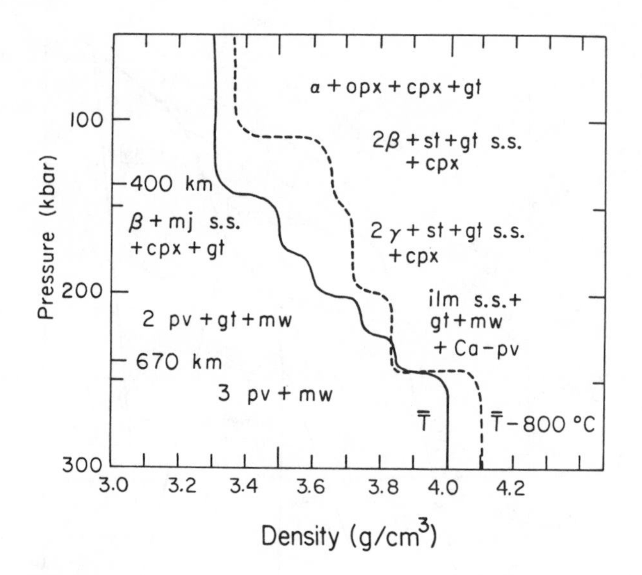

FIGURE 16-9
Zero-pressure density for peridotite at two temperatures (the mean of the "normal" geotherms and a temperature 800°C colder). Phase changes and thermal expansion are included. The density differences would be different for an eclogitic mantle or for a slab that differs in chemistry from the surrounding mantle. Abbreviations are α(olivine), opx (orthopyroxene), cpx (clinopyroxene), gt (garnet), gt s.s. (garnet plus majorite, mj), st (stishovite), ilm s.s. ($MgSiO_3$ ilmenite + gt-ilmenite), mw (magnesiowüstite) (Anderson, 1987b).

(~4.1 g/cm³) and possibly Al_2O_3 (3.99 g/cm³) and MgO (3.58 g/cm³). An eclogitic slab will therefore have a density between about 4 g/cm³ (calcium-perovskite + Al_2O_3) and 3.56 g/cm³ (garnet), which is less than the uncompressed density of the lower mantle (Butler and Anderson, 1978). Therefore, neither harzburgite nor eclogite have densities high enough to sink into the lower mantle, and we expect a barrier to slab penetration, and thickening and lateral flow of subducted material, at the base of the transition region. A large jump in viscosity near 650 km (Hager, 1984) also provides a barrier to slab penetration, but in a chemically homogeneous mantle, upper mantle material eventually circulates into the lower mantle. In either case, the slab does not slide easily into the lower mantle as implied in the thermal models of Creager and Jordan (1984, 1986). The high density of the slab, relative to adjacent mantle, contributes to the geoid and seismic-velocity anomalies associated with subduction zones and, in general, will increase the dips of Wadati-Benioff zones compared to strictly thermal models of the slab.

There is little information on the variation in depth of the 650-km discontinuity, but it is unlikely to vary by more than 100 km from its mean depth. A determination of the actual depth of the discontinuity under slabs will constrain the integrated density contrast of the slab and the nature of the boundary. An interesting question is whether the average composition of the slab is the same as the surrounding mantle, as assumed in most discussions. An eclogitic slab, for example, sinking through a peridotitic mantle would cause a smaller depression of a chemical interface at 650 km than a peridotitic slab.

Velocity anomalies associated with slabs are about +4 percent at 50 km, +10 percent at 100 km, +3 to 10 percent from 200 to 300 km, +4 to 7 percent at 400 km, +2 to 5 percent at 450 km, +4 to 5 percent at 500 km, +12 ± 4 percent at 620 km, +9 ± 3 percent at 650 km and +8 ± 2 percent between 580 and 660 km (Fitch, 1975; Hirahara, 1977; Huppert and Frohlich, 1981; Hirahara and Ishikawa, 1984; Engdahl and Gubbins, 1988). A ΔT of 800°C and a velocity contrast of 6 percent gives $\partial V_p/\partial T = -6.4 \times 10^{-4}$km/s °C. A 12 percent velocity contrast would double this figure. Allowing for the averaging of seismic waves across the temperature gradient would also raise the implied temperature derivatives. The values of $\partial V_p/\partial T$ for MgO, Al_2O_3, olivine, spinel and garnets fall in the range -3 to -5.2×10^{-4}km/s °C, which are less than the values implied by the seismic data for a purely thermal effect. Furthermore, temperature derivatives are expected to decrease rapidly with pressure. Changes in phase or composition from "normal" mantle are therefore implied since purely thermal effects are small. Thus, both the phase equilibria calculations and the seismic data support the presence of isobaric phase changes across the slab.

Seismic velocity differences between low-pressure assemblages involving olivine and orthopyroxene and high-pressure assemblages involving β- and γ-spinel and majorite or stishovite are of the order of 10 percent, and these are expected to be elevated in the cold slab by 100 to 150 km. Therefore, the slab velocity anomaly should be particularly large between about 300 and 400 km. The observed velocity contrast is smaller than the temperature-plus-phase change effect in an olivine-orthopyroxene mantle, suggesting that there is about 50 percent of "inert" component in the slab, material that does not transform over this pressure interval. Garnet is stable to about 600 km, and clinopyroxene (diopside ± jadeite) is stable to about 500 km. The implication is that the slab is garnet-clinopyroxene-rich relative to most garnet peridotites, which are samples from the shallow mantle. The possibility that the average composition of the slab is not the same as the average composition of the upper mantle complicates the interpretation of slab anomalies and the calculation of the depression of chemical discontinuities by subducted slabs.

The observed slab anomalies are comparable to and greater than the velocity jumps associated with mantle discontinuities. This strongly suggests that lateral velocity changes in the mantle, particularly near subducted slabs, involve isobaric phase changes as well as temperature variations. The velocity jump at the 400-km discontinuity is about 4–5 percent. This also suggests that garnet and clinopyroxene are important components of the mantle near

400 km; otherwise the velocity jump would be much greater.

The high gradient in seismic velocity between about 400 and 600 km depth implies a gradual phase change or series of phase changes occurring over this depth interval. The candidate transformations are majorite to β + stishovite, β to γ, γ + stishovite to ilmenite, and clinopyroxene to garnet or perovskite. The increase of the slab seismic anomaly below about 500 km is consistent with major isobaric phase changes occurring below this depth.

The measured or estimated compressional velocities of the important phases of upper-mantle minerals are shown in Figure 16-10. Also shown are the estimated velocities for peridotite at two temperatures, taking into account the different stable phase assemblages. The major differences occur between about 130 and 225 kbar. The heavy lines show the approximate stability pressure range for the various phases. Garnet and clinopyroxene represent less than 20 percent of the chosen peridotite composition. Since these minerals are stable to about 260 and 200 kbar, respectively, the effect of density and velocity changes associated with phase changes as both a function of temperature and pressure will be lower for less olivine and orthopyroxene-rich mantle. The deformation of the 650-km discontinuity, if this is a chemical boundary, will also be less.

Creager and Jordan (1986) used intermediate-depth earthquakes (149 to 256 km) to calibrate the effect of the underlying slab on travel times of deeper focus earthquakes. It is clear from Figure 16-10 that the velocity anomaly below intermediate-depth events is not similar to that expected below the deeper events (585 and 624 km) used in their analysis. (These depths correspond to pressures of 200–220 kbar.) Not only are the average velocity contrasts different, but the anisotropies of the stable phase assemblages are also quite different. Any degree of preferred orientation in the slab and the adjacent mantle will affect the differential travel times of rays leaving the source in different directions. A more direct determination of velocity anomalies in the vicinity of intermediate-depth earthquakes is about 5–6 percent (Engdahl and Gubbins, 1988). The velocity anomaly in the vicinity of deep-focus earthquakes appears to be much greater, in agreement with the above expectations. There is a direct trade-off between the magnitude of the velocity contrast in the vicinity of deep-focus earthquakes and normal mantle, and the depth extent of the slab beneath these earthquakes in the interpretation of residual sphere anomalies. A 10 percent velocity contrast and a slab extending only 100 km beneath the source gives a travel-time anomaly of 1.5 s for rays sweeping out of a cone of 45° to the plane of the slab. This is about the range of travel-time anomalies observed. The smaller velocity anomalies assumed by Creager and Jordan require much deeper slab penetration. Creager and Jordan assumed a contrast of 10 percent in their 1984 paper, but their calculations are obviously in error since their deep slab model predicts approximately 8-s anomalies, much greater than observed. Their results, when corrected, are consistent with a short slab.

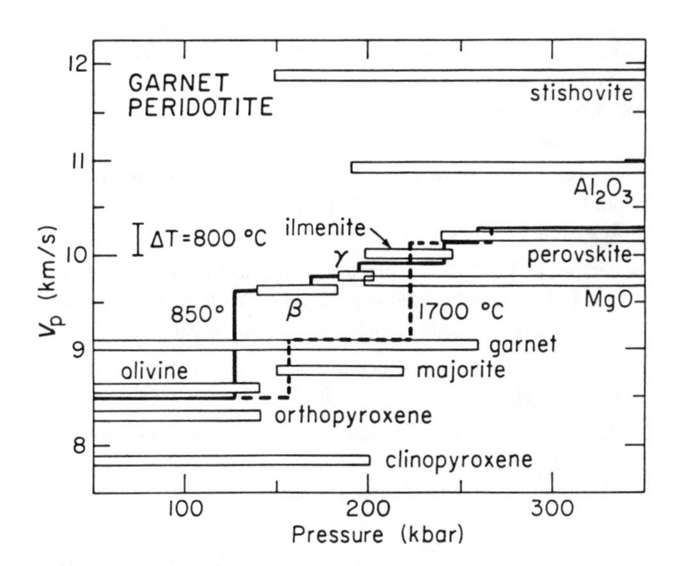

FIGURE 16-10
Compressional velocities, at standard conditions, and stability fields of mantle minerals. The approximate velocity of garnet peridotite, at standard conditions ($P = 0$, $T \approx 20°C$) for phases stable at two temperatures (850°C, 1700°C) is also shown. The small bar gives the approximate change in V_p for a 800°C change in temperature ($\partial V_p/\partial T = -4.1 \times 10^{-4}$ km/s°C).

The small bar in Figure 16-10 shows a typical change of V_p for an 800°C temperature change, assuming no phase changes. Note that the effect of the more important phase changes is to double or triple the effect of temperature. The largest effects occur between about 130 and 230 kilobars, the pressure range of deep-focus earthquakes. Note that clinopyroxene and garnet, minor constituents of peridotite and pyrolite but major components of eclogite and piclogite (Bass and Anderson, 1984), have the largest stability fields of low-pressure minerals. The presence of clinopyroxene and garnet reduces the size of the phase-change effects in the upper part of the transition region, particularly near 400 km. The smallness of the velocity jump near 400 km, both radially and laterally, indicates the presence of substantial amounts of a "neutral" component—garnet and clinopyroxene—near this depth. These are the dominant minerals in eclogite. On the other hand, eclogite experiences major transformations and velocity increases between 200 and 250 kbar, a pressure range pertinent to the velocity anomaly near deep-focus earthquakes.

General References

Akaogi, M. and S. Akimoto (1977) Pyroxene-garnet solid solution equilibrium, *Phys. Earth Planet. Int.*, *15*, 90–106.

Akaogi, M., A. Navrotsky, T. Yagi and S. Akimoto (1986) Pyroxene-garnet transformation, U.S.-Japan Seminar, Jan. 13–16, 1986, Program with Abstracts, 46.

Akimoto, S. (1972) The system $MgO-FeO-SiO_2$ at high pressure and temperature, *Tectonophysics, 13,* 161–187.

Anderson, D. L. (1967) Latest information from seismic observations. In *The Earth's Mantle* (T. F. Gaskell, ed.), 355–420, Academic Press, New York.

Anderson, D. L. (1981) A global geochemical model for the evolution of the mantle. In *Evolution of the Earth* (R. J. O'Connell, ed.), 6–18, American Geophysical Union, Washington, D.C.

Anderson, D. L. (1982) Chemical composition of the mantle, *Jour. Geophys. Res., 88,* B41–B52.

Anderson, D. L. (1982) The chemical composition and evolution of the mantle. In *High-Pressure Research in Geophysics* (S. Akimoto and M. H. Manghnani, eds.), 301–318, D. Reidel, Dordrecht.

Anderson, D. L. (1987a) A seismic equation of state II, *Phys. Earth Planet. Int.*

Anderson, D. L. and J. Bass (1986) Transition region of the Earth's upper mantle, *Nature, 320,* 321–328.

Bloss, F. D. (1971) *Crystallography and Crystal Chemistry,* Holt Rinehart and Winston, New York, 545 pp.

Frohlich, C. and M. Barazangi (1980) A regional study of mantle velocity variations beneath eastern Australia and the southwestern Pacific using short-period recordings of P, S, PcP, ScP and ScS waves produced by Tonga deep earthquakes, *Phys. Earth Planet. Inter., 21,* 1–14.

Fyfe, W. S. (1964) *Geochemistry of Solids,* McGraw-Hill, New York, 199 pp.

Hager, B. H. and R. Clayton (1987) Constraints on the structure of mantle convection using seismic observations, flow models, and the geoid (in press).

Ito, E. and H. Yamada (1982) Stability relations of silicate spinels, ilmenite and perovskites. In *High-Pressure Research in Geophysics* (S. Akimoto and M. Manghnani, eds.), 405–419, D. Reidel, Dordrecht.

Ito, E. and E. Takahashi (1986) Ultra high-pressure phase transformations and the constitution of the deep mantle, in *High-Pressure Research in Mineral Physics* (ed. M. Manghnani and Y. Syono), *Geophys.* Monograph 34, Amerian Geophys. Union, Washington, D.C., 486 pp.

Jeanloz, R. and E. Knittle (1986) Reduction of mantle and core properties to a standard state by adiabatic decompression. In *Chemistry and Physics of the Terrestrial Planets* (S. K. Saxena, ed.), 275–305, Springer-Verlag, New York.

Kanzaki, M. (1986) Ultrahigh-pressure phase relations in the system $MgSiO_3$-$Mg_3Al_2Si_3O_{12}$, *Phys. Earth Planet. Int.*

Kato, T. and M. Kumazawa (1985) Garnet phase of $MgSiO_3$ filling the pyroxene-ilmenite gap at very high temperature, *Nature, 316,* 803–805.

Kato, T. and M. Kumazawa (1986) Melting and phase relations in the Mg_2SiO_4–$MgSiO_3$ system at 20 GPa under hydrous conditions, *J. Geophys. Res., 91,* 9351–9355.

Kuskov, O. L. and R. Galimzyanov (1986) Thermodynamics of stable mineral assemblages of the mantle transition zone. In *Chemistry and Physics of the Terrestrial Planets* (S. K. Saxena, ed.) 310–361, Springer-Verlag, New York.

Liu, L. (1974) Silicate perovskite from phase transformation of pyrope garnet, *Geophys. Res. Lett., 1,* 277–280.

Ohtani, E. (1983) Melting temperature distribution and fractionation in the lower mantle, *Phys. Earth Planet. Inter., 33,* 12–25.

Ohtani, E. (1985) The primordial terrestrial magma ocean and its implications for stratification of the mantle, *Phys. Earth Planet. Inter., 38,* 70–80.

Ohtani, E., T. Kato and H. Sawamoto (1986a) Melting of a model chondritic mantle to 20 GPa, *Nature, 322,* 352–354.

Ohtani, E., C. T. Herzberg and T. Kato (1986b) Majorite stability, *Earth Planet. Sci. Lett.*

Pauling, L. C. (1979) *J. Am. Chem. Soc., 51,* 1010.

Ringwood, A. and A. Major (1970) The system Mg_2SiO_4–Fe_2SiO_4 at high pressures and temperatures, *Phys. Earth Planet. Int., 3,* 89–108.

Ringwood, A. E. (1975) *Composition and Petrology of the Earth's Mantle,* McGraw-Hill, New York, 618 pp.

Ringwood, A. E. (1982) Phase transformations and differentiation in subducting lithosphere, *Jour. Geol., 90,* 611–643.

Ringwood, A. E. and A. Major (1971) Synthesis of majorite and other high pressure garnets and perovskites, *Earth Planet. Sci. Lett., 12,* 411–418.

Sawamoto, H. (1986) Phase equilibrium of $MgSiO_3$ under high pressure and high temperature, U.S.–Japan Seminar, Jan. 13–16, 1986, Program with Abstracts, 50–51.

Yamada, H. and E. Takahashi (1984) Subsolidus phase relations between coexisting garnet and two pyroxenes at 50 to 100 kb in the system CaO-MgO-Al_2O_3-SiO_2. In *Kimberlites II: The Mantle and Crust Relationships* (J. Kornprobst, ed.), 247–255, Elsevier, Amsterdam.

References

Akaogi, M., N. Ross, P. McMillen and A. Navrotsky (1984) The Mg_2SiO_4 polymorphs; thermodynamic properties from oxide melt solution calorimetry, phase relations, and models of lattice vibrations, *Am. Mineral., 69,* 499–512.

Anderson, D. L. (1979a) The upper mantle transition region: Eclogite? *Geophys. Res. Lett., 6,* 433–436.

Anderson, D. L. (1979b) Chemical stratification of the mantle, *Jour. Geophys. Res., 84,* 6297–6298.

Anderson, D. L. (1987b) Thermally induced phase changes, lateral heterogeneity of the mantle, continental roots, and deep slab anomalies, *J. Geophys. Res., 92,* 13,968–13,980.

Anderson, D. L. and J. D. Bass (1984) Mineralogy and composition of the upper mantle, *Geophys. Res. Lett., 11,* 637–640.

Butler, R. and D. L. Anderson (1978) Equation of state fits to the lower mantle and outer core, *Phys. Earth Planet. Int., 17,* 147–162.

Creager, K. C. and T. H. Jordan (1984) Slab penetration into the lower mantle, *J. Geophys. Res., 89,* 3031–3049.

Creager, K. C. and T. H. Jordan (1986) Slab penetration into the lower mantle beneath the Mariana and other island arcs of the northwest Pacific, *J. Geophys. Res., 3,* 3573–3580.

Engdahl, E. R. and D. Gubbins (1988) Simultaneous travel-time inversion for earthquake location and subduction zone structure in the central Aleutian Islands, *J. Geophys. Res., 92,* 13,855.

Fitch, T. J. (1975) Compressional velocity in source regions of deep earthquakes, *Earth Planet. Sci. Lett., 26,* 156–166.

Giardini, D. and J. H. Woodhouse (1984) Deep seismicity and modes of deformation in Tonga subduction zone, *Nature, 307,* 505–509.

Hager, B. H. (1984) Subducted slabs and the geoid; constraints on mantle rheology and flow, *J. Geophys. Res., 89,* 6003–6015.

Hirahara, K. (1977) A large-scale three-dimensional seismic structure under the Japan islands and the Sea of Japan, *J. Phys. Earth, 25,* 393–417.

Hirahara, K. and Y. Ishikawa (1984) Travel time inversion for three-dimensional P-wave velocity anisotropy, *J. Phys. Earth, 32,* 197–218.

Huppert, L. and C. Frohlich (1981) The P velocity within the Tonga Benioff zone, *J. Geophys. Res., 86,* 3771–3782.

Irifune, T. and A. E. Ringwood (1986) Phase transformations in primitive MORB and pyrolite compositions to 25 GPa and some geophysical implications, High-Pressure Research in Mineral Physics, Proceedings of the U.S.-Japan Seminar (ed. M. Manghnani and Y. Syono), *Geophys.* Monograph 39, American Geophysical Union, Washington, D.C., 486 pp.

Irifune, T., T. Sekine, A. E. Ringwood and W. O. Hibberson (1986) The eclogite-garnet transformation at high pressure and some geophysical implications, *Earth Planet. Sci. Lett., 77,* 245–256.

Ito, E. and A. Navrotsky (1985) $MgSiO_3$ ilmenite; calorimetry, phase equilibria, and decomposition at atmospheric pressure, *Am. Mineral., 70,* 1020–1026.

Navrotsky, A., F. S. Pintchovski and S. Akimoto (1979) Calorimetric study of the stability of high pressure phases in the systems $CoO\text{-}SiO_2$ and "FeO"-SiO_2, and calculation of phase diagrams in $MgO\text{-}SiO_2$ systems, *Phys. Earth Planet. Int., 19,* 275–292.

Richards, M. A. and B. H. Hager (1984) Geoid anomalies in a dynamic Earth, *J. Geophys. Res., 89,* 5987–6002.

Watanabe, H. (1982) Thermochemical properties of synthetic high-pressure compounds relevant to the Earth's mantle. In *High-Pressure Research in Geophysics* (S. Akimoto and M. Manghnani, eds.), 441–464, Elsevier, Amsterdam.

Appendix

TABLE A-1
Earth Model PREM and its Functionals Evaluated at a Reference Period of 1 s.
Above 220 km the Mantle is Transversely Isotropic (see Table A-2);
the Parameters Given are "Equivalent" Isotropic Moduli and Velocities.

Level	Radius (km)	Depth (km)	Density (g/cm^3)	V_p (km/s)	V_s (km/s)	Q_μ	Q_K	Q_P	Φ (km^2/s^2)	K_s (kbar)	μ (kbar)	σ	Pressure (kbar)	dK/dP	B.P.	Gravity (cm/s^2)
1	0.	6371.0	13.08	11.26	3.66	85	1328	431	108.90	14253	1761	0.440	3638.5	2.33	0.99	0.
2	100.0	6271.0	13.08	11.26	3.66	85	1328	431	108.88	14248	1759	0.440	3636.1	2.33	0.99	36.5
3	200.0	6171.0	13.07	11.25	3.66	85	1328	431	108.80	14231	1755	0.440	3628.9	2.33	0.99	73.1
4	300.0	6071.0	13.06	11.24	3.65	85	1328	432	108.68	14203	1749	0.440	3617.0	2.33	0.99	109.6
5	400.0	5971.0	13.05	11.23	3.65	85	1328	432	108.51	14164	1739	0.441	3600.3	2.33	0.99	146.0
6	500.0	5871.0	13.03	11.22	3.64	85	1328	433	108.29	14114	1727	0.441	3578.8	2.33	0.99	182.3
7	600.0	5771.0	13.01	11.20	3.62	85	1328	434	108.02	14053	1713	0.441	3552.7	2.33	0.99	218.6
8	700.0	5671.0	12.98	11.18	3.61	85	1328	436	107.70	13981	1696	0.441	3522.0	2.34	0.99	254.7
9	800.0	5571.0	12.94	11.16	3.59	85	1328	437	107.33	13898	1676	0.442	3486.6	2.34	0.99	290.6
10	900.0	5471.0	12.91	11.13	3.57	85	1328	439	106.91	13805	1654	0.442	3446.7	2.34	0.99	326.4
11	1000.0	5371.0	12.87	11.10	3.55	85	1328	440	106.45	13701	1630	0.442	3402.3	2.34	0.99	362.0
12	1100.0	5271.0	12.82	11.07	3.53	85	1328	443	105.94	13586	1603	0.443	3353.5	2.34	1.00	397.3
13	1200.0	5171.0	12.77	11.03	3.51	85	1328	445	105.38	13462	1574	0.443	3300.4	2.34	1.00	432.5
14	1221.5	5149.5	12.76	11.02	3.50	85	1328	445	105.25	13434	1567	0.443	3288.5	2.34	1.00	440.0
15	1221.5	5149.5	12.16	10.35	0.	0	57822	57822	107.24	13047	0	0.500	3288.5	3.75	1.03	440.0
16	1300.0	5071.0	12.12	10.30	0.	0	57822	57822	106.29	12888	0	0.500	3245.4	3.65	1.02	463.6
17	1400.0	4971.0	12.06	10.24	0.	0	57822	57822	105.05	12679	0	0.500	3187.4	3.54	1.01	494.1
18	1500.0	4871.0	12.00	10.18	0.	0	57822	57822	103.78	12464	0	0.500	3126.1	3.46	1.01	524.7
19	1600.0	4771.0	11.94	10.12	0.	0	57822	57822	102.47	12242	0	0.500	3061.4	3.40	1.00	555.4
20	1700.0	4671.0	11.87	10.05	0.	0	57822	57822	101.12	12013	0	0.500	2993.4	3.35	1.00	586.1
21	1800.0	4571.0	11.80	9.98	0.	0	57822	57822	99.71	11775	0	0.500	2922.2	3.32	1.00	616.6
22	1900.0	4471.0	11.73	9.91	0.	0	57822	57822	98.25	11529	0	0.500	2847.8	3.30	1.00	647.0
23	2000.0	4371.0	11.65	9.83	0.	0	57822	57822	96.73	11273	0	0.500	2770.4	3.29	1.00	677.1
24	2100.0	4271.0	11.57	9.75	0.	0	57822	57822	95.14	11009	0	0.500	2690.0	3.29	1.00	706.9
25	2200.0	4171.0	11.48	9.66	0.	0	57822	57822	93.48	10735	0	0.500	2606.8	3.29	1.00	736.4
26	2300.0	4071.0	11.39	9.57	0.	0	57822	57822	91.75	10451	0	0.500	2520.9	3.30	1.00	765.5
27	2400.0	3971.0	11.29	9.48	0.	0	57822	57822	89.95	10158	0	0.500	2432.4	3.32	1.00	794.2
28	2500.0	3871.0	11.19	9.38	0.	0	57822	57822	88.06	9855	0	0.500	2341.6	3.34	1.00	822.4
29	2600.0	3771.0	11.08	9.27	0.	0	57822	57822	86.10	9542	0	0.500	2248.4	3.36	1.00	850.2
30	2700.0	3671.0	10.97	9.16	0.	0	57822	57822	84.04	9220	0	0.500	2153.1	3.39	1.00	877.4

TABLE A-1 *(continued)*

Level	Radius (km)	Depth (km)	Density (g/cm^3)	V_p (km/s)	V_s (km/s)	Q_μ	Q_K	Q_p	Φ (km^2/s^2)	K_s (kbar)	μ (kbar)	σ	Pressure (kbar)	dK/dP	B.P.	Gravity (cm/s^2)
31	2800.0	3571.0	10.85	9.05	0.	0	57822	57822	81.91	8889	0	0.500	2055.9	3.41	1.00	904.1
32	2900.0	3471.0	10.73	8.92	0.	0	57822	57822	79.68	8550	0	0.500	1956.9	3.44	1.00	930.2
33	3000.0	3371.0	10.60	8.79	0.	0	57822	57822	77.36	8202	0	0.500	1856.4	3.47	1.00	955.7
34	3100.0	3271.0	10.46	8.65	0.	0	57822	57822	74.96	7846	0	0.500	1754.4	3.49	1.00	980.5
35	3200.0	3171.0	10.32	8.51	0.	0	57822	57822	72.47	7484	0	0.500	1651.2	3.52	1.00	1004.6
36	3300.0	3071.0	10.18	8.36	0.	0	57822	57822	69.89	7116	0	0.500	1546.9	3.54	0.99	1028.0
37	3400.0	2971.0	10.02	8.19	0.	0	57822	57822	67.23	6743	0	0.500	1441.9	3.56	0.99	1050.6
38	3480.0	2891.0	9.90	8.06	0.	0	57822	57822	65.04	6441	0	0.500	1357.5	3.57	0.98	1068.2
39	3480.0	2891.0	5.56	13.71	7.26	312	57822	826	117.78	6556	2938	0.305	1357.5	1.64	0.99	1068.2
40	3500.0	2871.0	5.55	13.71	7.26	312	57822	826	117.64	6537	2933	0.304	1345.6	1.64	1.00	1065.3
41	3600.0	2771.0	5.50	13.68	7.26	312	57822	823	116.96	6440	2907	0.303	1287.0	1.64	1.01	1052.0
42	3630.0	2741.0	5.49	13.68	7.26	312	57822	822	116.76	6412	2899	0.303	1269.7	1.64	1.01	1048.4
43	3630.0	2741.0	5.49	13.68	7.26	312	57822	822	116.76	6412	2899	0.303	1269.7	3.33	1.01	1048.4
44	3700.0	2671.0	5.45	13.59	7.23	312	57822	819	115.08	6279	2855	0.302	1229.7	3.29	1.01	1040.6
45	3800.0	2571.0	5.40	13.47	7.18	312	57822	815	112.73	6095	2794	0.301	1173.4	3.24	1.01	1030.9
46	3900.0	2471.0	5.35	13.36	7.14	312	57822	811	110.46	5917	2734	0.299	1118.2	3.20	1.00	1022.7
47	4000.0	2371.0	5.30	13.24	7.09	312	57822	807	108.23	5744	2675	0.298	1063.8	3.17	1.00	1015.8
48	4100.0	2271.0	5.25	13.13	7.05	312	57822	803	106.04	5575	2617	0.297	1010.3	3.15	1.00	1010.0
49	4200.0	2171.0	5.20	13.01	7.01	312	57822	799	103.88	5409	2559	0.295	957.6	3.13	1.00	1005.3
50	4300.0	2071.0	5.15	12.90	6.96	312	57822	795	101.73	5246	2502	0.294	905.6	3.13	1.00	1001.5
51	4400.0	1971.0	5.10	12.78	6.91	312	57822	792	99.59	5085	2445	0.292	854.3	3.14	1.00	998.5
52	4500.0	1871.0	5.05	12.66	6.87	312	57822	788	97.43	4925	2388	0.291	803.6	3.16	1.00	996.3
53	4600.0	1771.0	5.00	12.54	6.82	312	57822	784	95.26	4766	2331	0.289	753.5	3.19	0.99	994.7
54	4700.0	1671.0	4.95	12.42	6.77	312	57822	779	93.06	4607	2273	0.288	704.1	3.23	0.99	993.6
55	4800.0	1571.0	4.89	12.29	6.72	312	57822	775	90.81	4448	2215	0.286	655.2	3.27	0.99	993.1
56	4900.0	1471.0	4.84	12.16	6.67	312	57822	770	88.52	4288	2157	0.284	606.8	3.32	0.99	993.0
57	5000.0	1371.0	4.78	12.02	6.61	312	57822	766	86.17	4128	2098	0.282	558.9	3.38	0.99	993.2
58	5100.0	1271.0	4.73	11.88	6.56	312	57822	761	83.76	3966	2039	0.280	511.6	3.45	0.99	993.8
59	5200.0	1171.0	4.67	11.73	6.50	312	57822	755	81.28	3803	1979	0.278	464.8	3.52	0.99	994.6
60	5300.0	1071.0	4.62	11.57	6.44	312	57822	750	78.72	3638	1918	0.275	418.6	3.59	0.99	995.7

TABLE A-1 *(continued)*

Level	Radius (km)	Depth (km)	Density (g/cm^3)	V_p (km/s)	V_s (km/s)	Q_μ	Q_K	Q_p	Φ (km^2/s^2)	K_s (kbar)	μ (kbar)	σ	Pressure (kbar)	dK/dP	B.P.	Gravity (cm/s^2)
61	5400.0	971.0	4.56	11.41	6.37	312	57822	743	76.08	3471	1856	0.273	372.8	3.67	0.98	996.9
62	5500.0	871.0	4.50	11.24	6.31	312	57822	737	73.34	3303	1794	0.270	327.6	3.75	0.98	998.3
63	5600.0	771.0	4.44	11.06	6.24	312	57822	730	70.52	3133	1730	0.266	282.9	3.84	0.97	999.8
64	5600.0	771.0	4.44	11.06	6.24	312	57822	730	70.52	3133	1730	0.266	282.9	2.98	0.97	999.8
65	5650.0	721.0	4.41	10.91	6.09	312	57822	744	69.51	3067	1639	0.273	260.7	3.00	0.97	1000.6
66	5701.0	670.0	4.38	10.75	5.94	312	57822	759	68.47	2999	1548	0.279	238.3	3.03	0.98	1001.4
67	5701.0	670.0	3.99	10.26	5.57	143	57822	362	64.03	2556	1239	0.291	238.3	2.40	0.37	1001.4
68	5736.0	635.0	3.98	10.21	5.54	143	57822	362	63.32	2523	1224	0.291	224.3	2.38	0.37	1000.8
69	5771.0	600.0	3.97	10.15	5.51	143	57822	362	62.61	2489	1210	0.290	210.4	2.37	0.37	1000.3
70	5771.0	600.0	3.97	10.15	5.51	143	57822	362	62.61	2489	1210	0.290	210.4	8.09	1.98	1000.3
71	5821.0	550.0	3.91	9.90	5.37	143	57822	363	59.60	2332	1128	0.291	190.7	7.88	1.92	999.6
72	5871.0	500.0	3.84	9.64	5.22	143	57822	364	56.65	2181	1051	0.292	171.3	7.67	1.86	998.8
73	5921.0	450.0	3.78	9.38	5.07	143	57822	365	53.78	2037	977	0.293	152.2	7.46	1.79	997.9
74	5971.0	400.0	3.72	9.13	4.93	143	57822	366	50.99	1899	906	0.294	133.5	7.26	1.73	996.8
75	5971.0	400.0	3.54	8.90	4.76	143	57822	372	48.97	1735	806	0.298	133.5	3.37	0.83	996.8
76	6016.0	355.0	3.51	8.81	4.73	143	57822	370	47.83	1682	790	0.297	117.7	3.33	0.82	995.2
77	6061.0	310.0	3.48	8.73	4.70	143	57822	367	46.71	1630	773	0.295	102.0	3.30	0.80	993.6
78	6106.0	265.0	3.46	8.64	4.67	143	57822	365	45.60	1579	757	0.293	86.4	3.26	0.79	992.0
79	6151.0	220.0	3.43	8.55	4.64	143	57822	362	44.50	1529	741	0.291	71.1	3.23	0.78	990.4
80	6151.0	220.0	3.35	7.98	4.41	80	57822	195	37.80	1270	656	0.279	71.1	−0.73	−0.12	990.4
81	6186.0	185.0	3.36	8.01	4.43	80	57822	195	38.01	1278	660	0.279	59.4	−0.72	−0.12	989.1
82	6221.0	150.0	3.36	8.03	4.44	80	57822	195	38.21	1287	665	0.279	47.8	−0.70	−0.12	987.8
83	6256.0	115.0	3.37	8.05	4.45	80	57822	195	38.41	1295	669	0.279	36.1	−0.68	−0.13	986.6
84	6291.0	80.0	3.37	8.07	4.46	80	57822	195	38.60	1303	674	0.279	24.5	−0.67	−0.13	985.5
85	6291.0	80.0	3.37	8.07	4.46	600	57822	1447	38.60	1303	674	0.279	24.5	−0.67	−0.13	985.5
86	6311.0	60.0	3.37	8.08	4.47	600	57822	1447	38.71	1307	677	0.279	17.8	−0.66	−0.13	984.9
87	6331.0	40.0	3.37	8.10	4.48	600	57822	1446	38.81	1311	680	0.279	11.2	−0.65	−0.13	984.3
88	6346.6	24.4	3.38	8.11	4.49	600	57822	1446	38.89	1315	682	0.278	6.0	−0.64	−0.13	983.9
89	6346.6	24.4	2.90	6.80	3.90	600	57822	1350	25.96	753	441	0.254	6.0	−0.00	−0.00	983.9
90	6356.0	15.0	2.90	6.80	3.90	600	57822	1350	25.96	753	441	0.254	3.3	0.00	0.00	983.3
91	6356.0	15.0	2.60	5.80	3.20	600	57822	1456	19.99	520	266	0.281	3.3	0.00	0.00	983.3
92	6368.0	3.0	2.60	5.80	3.20	600	57822	1456	19.99	520	266	0.281	0.3	−0.00	−0.00	982.2
93	6368.0	3.0	1.02	1.45	0.	0	57822	57822	2.10	21	0	0.500	0.2	−0.00	−0.00	982.2
94	6371.0	0.	1.02	1.45	0.	0	57822	57822	2.10	21	0	0.500	0.0	0.00	0.00	981.5

Dziewonski, A. M. and D. L. Anderson (1981) Preliminary Reference Earth Model, *Phys. Earth Planet. Inter.*, *25*, 297–356

TABLE A-2

Crust and Upper Mantle of PREM Including Directional Velocities, Anisotropic Elastic Constants and "Equivalent" Isotropic Velocities. Evaluated at Reference Periods of 1 s (top) and 200 s (bottom).

Radius (km)	Depth (km)	Density (g/cm^3)	V_{PV} (km/s)	V_{PH} (km/s)	V_{SV} (km/s)	V_{SH} (km/s)	η	Q_μ	Q_K	A (kbar)	C (kbar)	L (kbar)	N (kbar)	F (kbar)	V_p (km/s)	V_s (km/s)
6151.0	220.0	3.35950	7.80050	8.04862	4.44110	4.43629	0.97654	80	57822	2176	2044	663	661	831	7.98970	4.41885
6171.0	200.0	3.36167	7.82315	8.06310	4.43649	4.45423	0.96877	80	57822	2186	2057	662	667	835	8.00235	4.42580
6191.0	180.0	3.36384	7.84581	8.07760	4.43189	4.47218	0.96099	80	57822	2195	2071	661	673	839	8.01494	4.43285
6211.0	160.0	3.36602	7.86847	8.09209	4.42728	4.49013	0.95321	80	57822	2204	2084	660	679	843	8.02747	4.44000
6231.0	140.0	3.36819	7.89113	8.10659	4.42267	4.50807	0.94543	80	57822	2213	2097	659	685	847	8.03992	4.44724
6251.0	120.0	3.37036	7.91378	8.12108	4.41806	4.52602	0.93765	80	57822	2223	2111	658	690	851	8.05231	4.45458
6271.0	100.0	3.37254	7.93644	8.13558	4.41345	4.54397	0.92987	80	57822	2232	2124	657	696	854	8.06463	4.46201
6291.0	80.0	3.37471	7.95909	8.15006	4.40885	4.56191	0.92210	80	57822	2242	2138	656	702	857	8.07688	4.46953
6291.0	80.0	3.37471	7.95911	8.15008	4.40884	4.56193	0.92209	600	57822	2242	2138	656	702	857	8.07689	4.46954
6311.0	60.0	3.37688	7.98176	8.16457	4.40424	4.57987	0.91432	600	57822	2251	2151	655	708	860	8.08907	4.47715
6331.0	40.0	3.37906	8.00442	8.17906	4.39963	4.59782	0.90654	600	57822	2260	2165	654	714	863	8.10119	4.48486
6346.6	24.4	3.38076	8.02212	8.19038	4.39603	4.61184	0.90047	600	57822	2268	2176	653	719	866	8.11061	4.49094
6346.6	24.4	2.90000	6.80000	6.80000	3.90000	3.90000	1.00000	600	57822	1341	1341	441	441	459	6.80000	3.90000
6356.0	15.0	2.90000	6.80000	6.80000	3.90000	3.90000	1.00000	600	57822	1341	1341	441	441	459	6.80000	3.90000
6356.0	15.0	2.60000	5.80000	5.80000	3.20000	3.20000	1.00000	600	57822	875	875	266	266	342	5.80000	3.20000
6368.0	3.0	2.60000	5.80000	5.80000	3.20000	3.20000	1.00000	600	57822	875	875	266	266	342	5.80000	3.20000
6368.0	3.0	1.02000	1.45000	1.45000	0.	0.	1.00000	0	57822	21	21	0	0	21	1.45000	0.
6371.0	0.	1.02000	1.45000	1.45000	0.	0.	1.00000	0	57822	21	21	0	0	21	1.45000	0.
6151.0	220.0	3.35950	7.72930	7.97975	4.34748	4.34277	0.97654	80	57822	2139	2007	635	634	849	7.92008	4.32495
6171.0	200.0	3.36167	7.75230	7.99380	4.34297	4.36033	0.96877	80	57822	2148	2020	634	639	853	7.93236	4.33212
6191.0	180.0	3.36384	7.77531	8.00786	4.33846	4.37790	0.96099	80	57822	2157	2034	633	645	856	7.94457	4.33934
6211.0	160.0	3.36602	7.79831	8.02192	4.33395	4.39547	0.95321	80	57822	2166	2047	632	650	859	7.95673	4.34665
6231.0	140.0	3.36819	7.82132	8.03598	4.32943	4.41304	0.94543	80	57822	2175	2060	631	656	863	7.96882	4.35406
6251.0	120.0	3.37036	7.84432	8.05004	4.32492	4.43061	0.93765	80	57822	2184	2074	630	662	866	7.98084	4.36155
6271.0	100.0	3.37254	7.86732	8.06410	4.32041	4.44818	0.92987	80	57822	2193	2087	630	667	869	7.99279	4.36914
6291.0	80.0	3.37471	7.89031	8.07815	4.31590	4.46574	0.92210	80	57822	2202	2101	629	673	871	8.00468	4.37682
6291.0	80.0	3.37471	7.94982	8.14037	4.39645	4.54911	0.92209	600	57822	2236	2133	652	698	859	8.06715	4.45717
6311.0	60.0	3.37688	7.97251	8.15480	4.39186	4.56700	0.91432	600	57822	2246	2146	651	704	862	8.07928	4.46480
6331.0	40.0	3.37906	7.99522	8.16924	4.38726	4.58490	0.90654	600	57822	2255	2160	650	710	865	8.09135	4.47253
6346.6	24.4	3.38076	8.01295	8.18051	4.38368	4.59887	0.90047	600	57822	2262	2171	650	715	867	8.10074	4.47863
6346.6	24.4	2.90000	6.79151	6.79151	3.88904	3.88904	1.00000	600	57822	1338	1338	439	439	460	6.79151	3.88904
6356.0	15.0	2.90000	6.79151	6.79151	3.88904	3.88904	1.00000	600	57822	1338	1338	439	439	460	6.79151	3.88904
6356.0	15.0	2.60000	5.79328	5.79328	3.19101	3.19101	1.00000	600	57822	873	873	265	265	343	5.79328	3.19101
6368.0	3.0	2.60000	5.79328	5.79328	3.19101	3.19101	1.00000	600	57822	873	873	265	265	343	5.79328	3.19101
6368.0	3.0	1.02000	1.44996	1.44996	0.	0.	1.00000	0	57822	21	21	0	0	21	1.44996	0.
6371.0	0.	1.02000	1.44996	1.44996	0.	0.	1.00000	0	57822	21	21	0	0	21	1.44996	0.

Dziewonski, A. M. and D. L. Anderson (1981) Preliminary Reference Earth Model, *Phys. Earth Planet. Inter.*, *25*, 297–356

TABLE A-3
Conversion Factors

To Convert	To	Multiply by
angstrom, Å	cm	10^{-8}
	nm	10
bar	atm	0.987
	dyne/cm^2	10^6
	MPa	10^{-1}
calorie (g), cal	joule	4.184
dyne	g cm/s^{-2}	1
	newton	10^{-5}
erg	cal	2.39×10^{-8}
	watt-second	1
	joule	10^{-7}
gamma	gauss	10^{-5}
	tesla	10^{-9}
gauss	tesla	10^{-4}
heat-flow unit (H.F.U.)	μcal/cm^2/s	1
	m W/m^2	41.84
micrometer, μm	cm	10^{-4}
poise	g/cm/s	1
	Pa-s	0.1
stoke	cm^2/s	1
watt	J/s	1
year	s	3.156×10^7

TABLE A-4
Physical Constants

Speed of light	c	2.998×10^8 m/s
Electronic charge	e	-1.602×10^{19} C
Permeability of vacuum	μ_0	$4\pi \times 10^{-7}$ N/A^2
Permittivity of vacuum	ε_0	8.854×10^{-12} F/m
Planck constant	h	6.626×10^{-34} J
Boltzmann constant	k	1.381×10^{-23} J/K
Stefan-Boltzmann constant	σ	5.67×10^{-8} W/m^2/K^4
Gravitational constant	G	6.673×10^{-11} m^3/kg/s^2
Electron rest mass	m_e	0.911×10^{-10} kg
Avogadro's number	N_A	6.022×10^{23}/mol
Gas constant	R	8.314 J/mol/K

TABLE A-5
Earth Parameters

Equatorial radius, a	6378.137 km
Polar radius, c	6356.752 km
Equivolume sphere radius	6371.000 km
Surface area	5.1×10^8 km^2
Geometric flattening, $(a-c)/a$	1/298.257
Ellipticity, $(a^2-c^2)/(a^2+c^2)$	1/297.75
GM	3.986005×10^{14} m^3 s^{-2}
Mass, M	5.97369×10^{24} kg
Mean density	5.5148×10^3 kg m^{-3}
Moments of inertia	
Polar, C	8.0378×10^{37} kg m^2
Equatorial, A	8.0115×10^{37} kg m^2
C/Ma^2	0.33068
Dynamic ellipticity (precession constant), $H = \dfrac{C-A}{C}$	= 1/305.51
Ellipticity coefficient, $J_2 = \dfrac{C-A}{Ma^2}$	10826.3×10^{-7}
Angular velocity, Ω	7.2921×10^{-5} rad s^{-1}
Angular momentum, $C\Omega$	5.8604×10^{33} kg m^2 s^{-1}
Normal gravity at equator	9.7803267 m s^{-2}
Normal gravity at poles	9.832186 m s^{-2}

Determinants

The determinant of the second order

$$\begin{vmatrix} a & b \\ c & d \end{vmatrix}$$

stands for the expression $(ad-bc)$.

The determinant of a 3×3 matrix, or array of numbers, may be defined in terms of determinants of the second order. The determinant

$$\begin{vmatrix} a & b & c \\ d & e & f \\ g & h & k \end{vmatrix}$$

stands for

$$a \begin{vmatrix} e & f \\ h & k \end{vmatrix} - b \begin{vmatrix} d & f \\ g & k \end{vmatrix} + c \begin{vmatrix} d & e \\ g & h \end{vmatrix}$$

That is, a product of each element in a row or column, and a second-order determinant formed by elements from the other rows and columns. Definitions of determinants of higher order may be given in a similar way.

Index

Italics indicate an extended discussion. Items in parentheses are abbreviations or alternate entries to be checked.